Lecture Notes in Electrical Engineering

Volume 827

The book series *Lecture Notes in Electrical Engineering* (LNEE) publishes the latest developments in Electrical Engineering - quickly, informally and in high quality. While original research reported in proceedings and monographs has traditionally formed the core of LNEE, we also encourage authors to submit books devoted to supporting student education and professional training in the various fields and applications areas of electrical engineering. The series cover classical and emerging topics concerning:

- Communication Engineering, Information Theory and Networks
- Electronics Engineering and Microelectronics
- Signal, Image and Speech Processing
- Wireless and Mobile Communication
- Circuits and Systems
- Energy Systems, Power Electronics and Electrical Machines
- Electro-optical Engineering
- Instrumentation Engineering
- Avionics Engineering
- Control Systems
- Internet-of-Things and Cybersecurity
- Biomedical Devices, MEMS and NEMS

For general information about this book series, comments or suggestions, please contact leontina. dicecco@springer.com.

To submit a proposal or request further information, please contact the Publishing Editor in your country:

China

Jasmine Dou, Editor (jasmine.dou@springer.com)

India, Japan, Rest of Asia

Swati Meherishi, Editorial Director (Swati.Meherishi@springer.com)

Southeast Asia, Australia, New Zealand

Ramesh Nath Premnath, Editor (ramesh.premnath@springernature.com)

USA, Canada:

Michael Luby, Senior Editor (michael.luby@springer.com)

All other Countries:

Leontina Di Cecco, Senior Editor (leontina.dicecco@springer.com)

** This series is indexed by EI Compendex and Scopus databases. **

More information about this series at https://link.springer.com/bookseries/7818

Jason C. Hung · Neil Y. Yen ·
Jia-Wei Chang
Editors

Frontier Computing

Proceedings of FC 2021

Set 1

 Springer

Editors
Jason C. Hung
Department of Computer Science
and Information Engineering
National Taichung University of Science
and Technology
Taichung, Taiwan

Neil Y. Yen
School of Computer Science
and Engineering
The University of Aizu
Aizuwakamatsu, Japan

Jia-Wei Chang
Department of Computer Science
and Information Engineering
National Taichung University of Science
and Technology
Taichung, Taiwan

ISSN 1876-1100   ISSN 1876-1119 (electronic)
Lecture Notes in Electrical Engineering
ISBN 978-981-16-8051-9   ISBN 978-981-16-8052-6 (eBook)
https://doi.org/10.1007/978-981-16-8052-6

This Springer imprint is published by the registered company Springer Nature Singapore Pte Ltd.
The registered company address is: 152 Beach Road, #21-01/04 Gateway East, Singapore 189721,
Singapore

Preface

This LNEE volume contains the papers presented at the International Conference on Frontier Computing (FC 2021) virtually held online on July 14, 2021. This event is the 11th event of the series, in which fruitful results can be found in the digital library or conference proceedings of FC 2010 (Taichung, Taiwan), FC 2012 (Xining, China), FC 2013 (Gwangju, Korea), FC 2015 (Bangkok, Thailand), FC 2016 (Tokyo, Japan), FC 2017 (Osaka, Japan), FC 2018 (Kuala Lumpur, Malaysia), FCABH 2019 (Taichung, Taiwan), FC 2019 (Kitakyushu, Japan), FC 2020 (virtual event), and FC 2021 (virtual event). This conference is expected to bring together researchers and practitioners from both academia and industry to meet and share cutting-edge development in the field. One colocated event, namely International Conference on Machine Learning on FinTech, Security and Privacy (MLFSP 2021), is jointly held with high appreciations from the participants.

The papers accepted for inclusion in the conference proceedings primarily cover the topics: database and data mining, networking and communications, Web and Internet of things, embedded system, soft computing, social network analysis, security and privacy, optics communication, and ubiquitous and pervasive computing. Many papers have shown their academic potential and value and indicate promising directions of research in the focused realm of this conference. We believe that the presentations of these accepted papers will be more exciting than the papers themselves and lead to creative and innovative applications. We hope that the attendees and readers will find these results useful and inspiring to your field of specialization and future research.

On behalf of the organizing committee, we would like to thank the members of the organizing and the program committees, the authors, and the speakers for their dedication and contributions that make this conference possible especially to FC conference group and Korean Institute of Information Technology, Korea Institute of Information Technology and Innovation (KIITI), and SIEC Korea Chapter. We appreciate the contributions from these experts and scholars to enrich our event. We would take the chance to thank and welcome all participants, and hope that all of them enjoy the technical discussions within the conference period, build a strong friendship, and establish ties for future collaborations.

Jason C. Hung
Neil Y. Yen
Jia-Wei Chang

Contents

An Analysis of the Research Situation of China's National Security—Based on the Perspective of Social Network Analysis

ZhengJi Wu[(⊠)] [iD] and LiangBin Yang [iD]

University of International Relations, Beijing, China

Abstract. [Objective/Meaning] As a new discipline, national security has developed rapidly in China in recent years. It is more intuitive and clearer to use social networks to conduct research situation analysis. [Methods] The main methods of this study include: keyword analysis, bi-cluster thematic analysis, citation curve analysis, author role research, research institution role research. [Procedures] This article is based on the keywords of "National Security|National Security Concept" in the CNKI database from 1997 to 2016, using SCI, CSSCI, Peking University core journals and other databases as the journal source, using the network analysis software Ucinet, Gephi professional functions and the drawing software Origin8 conducts research and analysis on keywords, authors, research institutions, and the number of citation documents, draws network maps, citation curves, and explores the development trend of China's national security studies. [Results/Conclusions] Although China's national security studies are an emerging discipline, preliminary construction has been completed. From the traditional level of national security studies to "network security", "data security", and "ideology" Security "and other soft power security transitions. It is precisely because of the numerous branches and rapid updates of the discipline of international security that the relationships between scholars involved in research are not as close as in other disciplines. However, the number of institutions participating in the research is very large. Among them, Peking University, the Chinese Academy of Sciences, and the Institute of International Relations are the main institutions.

Keywords: Social network · National security · Citation curve · Visualization · Origin8 · Ucinet

Funded projects: National Security Research Project of High-precision and Discipline Construction (School Fund): Research on Entity Recognition and Influence Mechanism of Social Networks from the Perspective of National Security (NO: 2019GA37); The Beijing Municipal Education Commission's 2019 Talent Cultivation Co-construction Project-College Student Research and Training Project: Data Science and Big Data Technology Research Project of School of Information Technology.

1   Introduction

National security is the basic interest of a country. It is an objective state in which a country is free from danger, that is, an objective state in which the country has no external threats or violations and no internal chaos or illness. Contemporary national security includes basic content in 12 areas: political security, homeland security, military security, economic security, cultural security, social security, technological security, cyber security, ecological security, resource security, nuclear security, overseas interest security [1].

In recent years, China's national security situation has become increasingly complex. Whether in the virtual world or in the real world, there are uncertainties and challenges everywhere from the personal safety of the people to the national sovereignty. Therefore, it becomes more and more important to scientifically understand national security issues and study national security as a theory which has practical significance. On April 9, 2018, the Ministry of Education issued the "Implementation Opinions on Strengthening the National Safety Education in Large, Middle and Primary Schools", requiring that national safety education be generally carried out from elementary schools, middle schools to universities, and at the same time set up national security first-level disciplines in universities to conduct national safe education [2]. Meanwhile social network analysis theory has made great progress in community structure investigation, statistical significance test, visualization, and exploration of cultural background in social networks [3].

Based on the importance of China's national security science and its adaptability to social network analysis, this article attempts to analyze the development of China's national security science from a macro perspective and analyze its situation with the help of social network analysis theory in order to provide useful reference for the development of China's national security science.

2   Current Situation

1) National Security

China first proposed the concept of "national security" in the government work report of the First Conference of the Sixth People's Congress in June 1983 and treated it as an emerging discipline in 1990s. However, due to the hard work of scholars, our country's national security science has developed rapidly, and it has completed changed from "traditional binary" to "non-traditional endless", from Marxism's "One-Dominant" to "Multiple", from "Ideological Conflict" to "Multiple Drivers Inside and Outside", and from "flat" to "three-dimensional" [4]. In 2002, our country proposed a new security concept with a systematic core of "mutual trust, mutual benefit, equality, and collaboration", and the main changes compared with the tradition are: 1) Considering both internal and external aspects, emphasizing that the country's core values, political order, and survival methods are not infringed. 2) Emphasizing the diversity of security levels which global security, regional security and people's security are all considered. 3) Emphasizing the expansion of the security field. Economic security, information security, environmental security and social security are included in the security category [5–7].

The "Excerpts from Xi Jinping's Discussion on Overall Security Concept" not only made a profound analysis of China's national security science, but also made a very clear interpretation of the focus of future research. The overall national security of our country is a huge complex, and the insecurity factors it faces are comprehensive, including many factors at home and abroad. Our country's overall security should have Chinese characteristics, with cultural security as a breakthrough, build an overall security system with top-level design [8, 9].

2.1 Social Network Analysis

With the advent of the information age, social network analysis technology has received much attention from academic circles at home and abroad. At present, the hot spots in the international academic community can be roughly divided into three parts [10]: 1) Research on the structural characteristics of social networks 2) Research on group interaction in social networks 3) Research on information dissemination in social networks.

China also attaches great importance to the research of social networks. In terms of structural analysis, in 2008, Cheng Xueqi and other researchers from the Institute of Computing Technology of the Chinese Academy of Sciences analyzed the characteristics of the community structure and improved the discovery method of network hierarchical overlapping communities [11]. In terms of group characteristics, Yang Shanlin and others from Hefei University of Technology divided decision-making behaviors into three categories based on the herd and proposed a cellular automaton-based evolutionary model for group decision-making herd behaviors in 2009 [12]. In terms of information dissemination, in 2000 years, Xing Xiusan from Beijing Institute of Technology proposed the theory of non-equilibrium statistical information with the information (entropy) evolution equation that expresses the law of information evolution as the core, deduced the nonlinear evolution equation of Shannon information (entropy), introduced Statistical physical information and derived its nonlinear evolution equation [13]. At the same time, in 2014, Fang Binxing and others published the book "Analysis of Online Social Networks", combining the above three aspects to systematically elaborate the basic theory, key methods and technologies in the analysis of online social networks [14].

3 Data Sources and Research Methods

3.1 Data Sources

The literature comes from articles in the CNKI database. The search method used in this study was based on the keyword "national security" or "national security" and selected the 250 most cited papers in Chinese literature, and after screening, a total of 247 documents were used as experimental materials. These documents include keywords, author names and institutions, citation author names and institutions, and the number of citations for each document from the date of publication to 2019 and other attributes for visual analysis.

3.2   Research Methods

Social networks, that is, the structure of social relations, can reflect the social relations between actors. In recent years, they have been widely used to study citation relations. In this paper, we used the related social networks analysis indicators such as centrality, cohesion subgroups, and related concepts of citation curves. Meanwhile, we used gCluto to perform bi-clustering and visualization of high-frequency keywords, and professional function drawing software Origin8 to fit the number of cited documents. Citation curve analysis, statistical product and service solution software SPSS performs corresponding analysis on highly cited literature topics and citation curve types, and network analysis tool Ucinet and visualization tool NetDraw was used to analyze the acquired literature.

gCluto is the graphical clustering toolkit, which is the front end of Cluto data clustering. The advantages are: 1. Make Cluto cluster in a user-friendly graphical way. 2. Provides several methods for the visualization of interactive clustering results such as: visualization matrix, visualization hills, etc.

Origin8 is a popular professional function drawing software produced by OriginLab. It is recognized as a fast, flexible and easy-to-learn engineering drawing software. Origin8 has two main functions: data analysis and drawing.

Gephi is an open source free cross-platform complex network analysis software based on JVM, which is mainly used for various networks and complex systems, interactive visualization and detection of dynamic and hierarchical open-source tools. Its advantages are: 1. Simple operation and easy to use. 2. Diversified functions and beautiful visual graphics.

The Ucinet software is written by a group of network analysts at the University of California, Irvine. The network analysis integration software includes NetDraw for 1D and 2D analysis data analysis and the 3D display analysis software Mage that is under development and application.

4   Keyword Analysis

As the author's refining of an entire article, keywords can often clearly and intuitively express the theme that the entire article wants to elaborate. Therefore, keyword extraction and analysis play a great role in understanding the theme of the article. At the same time, keyword analysis of a certain number of documents also helps us to grasp the discipline development trends.

4.1   High Frequency Keyword Network

In order to study the development trends of China's national security and the internal links between the keywords involved, this study divided 247 articles into two parts according to publication time: 1995–2007 and 2007–2019 to perform the bi-cluster keyword analysis Keyword analysis. For these two parts, define keywords that appear more than or equal to 2 as high-frequency keywords and pair the high-frequency keywords in pairs to obtain a co-occurrence matrix. Use Netdraw to draw a keyword co-occurrence network map (Figs. 1 and 2).

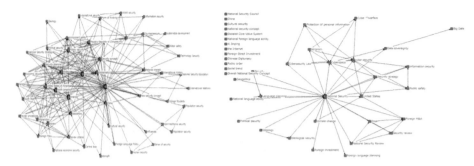

Fig. 1. 1995–2007 high frequency keyword co-occurrence network map

Fig. 2. 2008–2019 high frequency keyword co-occurrence network map

It can be seen from the graph that the focus of China's national security development has shifted in these two time periods. Excepted the keywords "national security" and "national security concept" used in the search in this article, the keywords at the core of 1995–2007 are "politics", "military security", "environmental security", "globalization", etc. This shows that the focus of China's national security studies at this time is more inclined to some national basic security aspects, which is also consistent with the achievements made by our country at that time.

The key words in the network center from 2008 to 2019 are "cyber security", "United States", "climate change" and so on. This fully shows that with the rapid development of the times, in a modern, networked, and diversified world, a country needs to focus on the issues that related to the national security.

4.2 Bi-cluster Research Topics

Bi-cluster analysis is a method to discover the potential local patterns by clustering the rows and columns of the matrix at the same time. This method was created due to the need for large amounts of biological gene data for analysis. Compared with traditional clustering methods, bi-clustering can extract the information of rows and columns and discover their potential connections, which is especially effective for sparse and high-dimensional matrices. In this study, after deleting the keyword "national security" used in the search, the co-occurrence matrix was reconstructed and imported into gCLUTO software for bi clustering. The clustering method is repeated dichotomy, the number of clusters is 10, the optimization function is I2, and the similarity coefficient is the cosine function. After clustering and drawing a visual mountain chart, the number of clusters with obvious differences was found to be 5 types, so relocate the number of clusters to 5 and reclassify, combined with the similarity index of results given by gCLUTO software, it was found that most of the ISim values Greater than 0.5, indicating that the clustering effect is well-behaved, and the similarity is high in various groups (Table 1).

Table 1. Bi-cluster result similarity index

Clusters	Size	ISim	ISdev	ESim	ESdev
From 1995 to 2007					
0	4	0.943	0.008	0.354	0.050
1	6	0.713	0.087	0.369	0.016
2	10	0.696	0.092	0.387	0.052
3	10	0.650	0.048	0.421	0.043
4	23	0.492	0.055	0.387	0.068
From 2008 to 2019					
0	4	0.959	0.011	0.006	0.004
1	3	0.758	0.095	–0.000	0.000
2	9	0.617	0.078	0.018	0.010
3	4	0.447	0.042	0.041	0.049
4	19	0.022	0.036	–0.000	0.000

The clustering results in the visualized mountain peak diagram in Fig. 3 are represented by mountain peaks. The steeper the mountain peaks, the stronger the similarity of the data within the class. At the same time, the peak is composed of 5 colors of red, yellow, green, blue, and light blue. The closer to red, the lower the standard deviation of data similarity within the class, and the closer to blue, the closer the standard deviation of data similarity to class High, when the color tends to be more monotonous, it means that the similarity between the data within the class is greater. The color blocks in the visualization matrix in Fig. 3 represent the frequency of the original keywords. The larger the value, the darker the color. At the same time, the matrix also rearranges the rows of the matrix, arranging the keywords belonging to the same cluster together. Figure 3 lists the high-frequency keyword bi-cluster visualization matrix and mountain charts for the period 1995–2007.

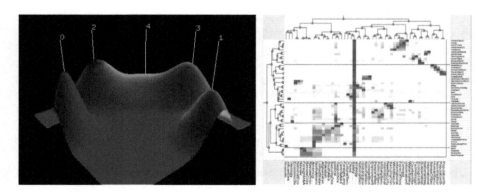

Fig. 3. High-frequency keyword bi-clustering visualization matrix and mountain peak plots from 1995 to 2007

Combining the above figures, the highly cited literature related to national security can be summarized into the following topics:

1) From 1995 to 2007

Theme 0: Foreign Mergers and Acquisitions Safety and Review Standards. The words "foreign capital mergers and acquisitions" and "review standards and procedures" have appeared in the keywords many times. This shows that China attached great importance to the security of finance and foreign investment.

Theme 1: The Concept of China's National Security and the Thinking Form of International Security. The emergence of these words shows that at the time, China not only focused on the development of basic theoretical knowledge of its national security science, but also remember to look abroad, always paying attention to the trend of international security. This also reflects the characteristics of the development of national security as an emerging discipline in my country.

Theme 2: Security Issues Between China and the United States. As the two largest economies in the world, they never forget the relationship with each other.

Theme 3: Political Security. Reflects the importance of political security for the initial construction of China's national security studies.

Theme 4: Ecological Security and Sustainable Development. The emergence of this theme shows that at that time, China recognized the importance of environmental and ecological security issues. A good ecological environment has a very important influence on national security.

2) From 2008 to 2019

Theme 0: Foreign Mergers and Acquisitions and Finsa. In the period of 2008–2019, national security issues such as foreign mergers and acquisitions appeared again, indicating that our national security science attaches great importance to the direction of international financial security.

Theme 1: Language Planning and National Language Skills. The link between language and national security has gradually attracted the attention of scholars in recent years. National language ability is the language ability that a country has to deal with domestic and foreign affairs. Language education is an important factor for ensuring safety and plays a very important role in promoting mutual understanding. It is precisely because of its importance and China's current lack of national language proficiency (such as the small number of foreign languages and the lack of high-level foreign language talents) that this branch of national security has received widespread attention in recent years.

Theme 2: Cyber Security. The rapid development of the Internet in recent years has made the Internet the main battlefield for people to socialize. It is precisely because the role of the network in people's lives is very important that the branch of network security has received extensive attention from scholars in the field of national security in the past 10 years.

Theme 3: Data Sovereignty. Now that we are in a highly informationize world, data becomes more and more important. The maintenance of data sovereignty can not only affect the maintenance of China's domestic security, but also help China gain a foothold in the international arena.

Theme 4: Ideological Security. The traditional concept of national security mainly refers to the military and political security of the country, while ideological security includes many fields such as economy, information, ecology, culture, science and technology. The proposal of ideological security means that China has completed the construction of the initial traditional national security concept, and pay more attention to the impact of "ideology" and "cultural soft power" on national security.

4.3   Citation Curve Analysis of Highly Cited Literature

Each document has its own life curve. Since the date of publication, its content has been sought after in line with the academic upsurge of the time to its gradual aging and lost its use value. It is precisely because the literature has such characteristics that scholars have proposed measurement indicators such as half-life and Price index to study the process of document aging from different perspectives in order to hope to dig out its underlying laws and explore document topics and documents relationship between cited quantity and citation life.

1) Citation curve distribution

This study refers to Qu Wenjian's summary of the modern classification of citation curves [15], divided the curves as 6 categories: classical citation curve, multi-peak citation curve, exponential growth citation curve, exponential decline citation curve, waveform citation curve and sleeping beauty citation curve.

According to the classification characteristics of the citation curve and with the help of manual classification of topics and the fit function of Origin8 software for the number of literature citations per year, the top 50 citations are classified and summarized in Table 2.

Table 2. Cross-tabulation of citation curves and research topics for highly cited documents

	Classical	Multi-peak	curve	Sleeping beauty	Exponetial growth	Exponetial decline	Sum
1995–2007							
Theme 0	4	1	0	0	0	0	5
Theme 1	2	9	2	2	0	0	15
Theme 2	0	1	0	0	0	0	1
Theme 3	2	9	0	0	0	0	11
Theme 4	0	0	0	0	0	0	0
Sum	8	20	2	2	0	0	32
Average Annual citation	75.94	97.96	29.97	34.5	0	0	—

(continued)

Table 2. (*continued*)

	Classical	Multi-peak	curve	Sleeping beauty	Exponetial growth	Exponetial decline	Sum
2008–2019							
Theme 0	0	6	0	0	0	0	6
Theme 1	2	2	0	0	0	0	4
Theme 2	2	1	0	0	0	0	3
Theme 3	0	0	0	0	1	0	1
Theme 4	4	0	0	0	0	0	4
Sum	8	9	0	0	1	0	18
Average Annual citation	100.8	55.62	0	0	46.25	0	—

Looking at the distribution of citation curves in these two periods, the following points can be summarized:

1) Most of the highly cited documents of national security science have a life cycle
No matter which period you are in, you can find that the articles with life characteristics such as the classic citation curves and the multi-peak citation curves have a great high proportion among the whole articles, for example, in the 1995–2007 period, there are 32 articles in total and 28 articles have a life cycle.

2) 1995–2007 period
The documents that match the multi-peak citation curve account for the most, up to 62.5%, indicating that during 1995–2007, highly cited documents of national security sciences often have multiple peaks, which delayed the decline cycle and extended the document life cycle. At the same time, the frequency of citations for the 1995–2007 period is very high. Among them, the citations that match the multi-peak citation curve type have reached an average of 97.96 citations per year, which means these articles would be cited almost every three days. Thus, these literatures are huge for the construction of China's national security studies and the help for future generations.

3) 2008–2019 period
The number of documents that match the classic citation curve type and the compound multi-citation citation curve type in the 2008–2019 period are close, but the average annual citation curve of the compound classic citation curve type is up to 100.8 times, indicating that the past 10 years China's national security research shows a prosperous scene, and the subject of the research is also very popular. At the same time, drawing on the average annual citation frequency of the documents with multi-peak citation curve types from 1995 to 2007, I think we can look forward to the citation frequency of the documents with multi-peak citation curve types from 2008 to 2019 in the next few years.

5 Author Knowledge Role Recognition Research

By conducting research on author knowledge role recognition, it can help us understand the development of national security in China, the academic exchanges between authors, and the role of authors in their disciplines.

The research object of this study is the citation relationship between the authors of the top 50 cited articles. Since there are more than one author in some literatures, there are a total of 75 authors. Table 3 is the first 5th author's ranking.

Table 3. The top 5 ranking of the authors from top 50 articles

Ranking	Author
1	Liu Jiayi
2	Chen Guojie
3	Zeng Changxin, Shen Weishou
4	Guo Zhongwei
5	Zheng Hangsheng, Hong Dayong

5.1 Analysis of Community Graph

After determining the list of authors, then retrieve all the author's literature resources and citation documents in the field of national security from the CNKI database, so as to obtain the amount of documents cited by the author in the field and the amount of cited documents, thus we can use Gephi to establish the author's citation network diagram (Fig. 4).

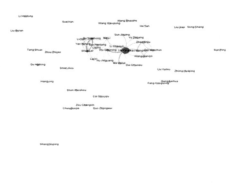

Fig. 4. Author citation network diagram

In this network, nodes represent each author, the darker the color, the more active the author. It can be seen from the figure that the relationship between these 75 authors is not very tight (density is only 0.0137). By investigating the areas that the authors are involved in and good at, they can be roughly divided into national legal and political security fields with Liu Yuejin and Zhao Ronghui as the core, and national economic security with Gu Haibing, Liu Jiayi, Wang Shaozhe, Shen Lei, Gu Shuzhong as the core Field, as well as the national ecological environment security field with Geng Leihua and Chen Guojie as the core. This phenomenon not only reflects the large number of branches in the field of national security science in China, but also indicating the differences between the branches are large, showing a phenomenon of diversification. On the other hand, it shows that the mature branches in the field of national security science in China today include political security, economic security and ecological security.

5.2   Centrality Analysis

1) Degree centrality

In a network, if a point is directly connected to many other points, then we think that this point is in a more central position in the network where it is located. In this study, due to the use of directed graphs, it is necessary to distinguish between the concept of point-in and point-out. Point-in-degree represents the situation where the author's literature is cited by others, that is, the frequency of the author's citation, the greater the value, the more the author plays a role of "knowledge source" in the network. Conversely, the point-out degree represents the frequency with which authors cite other people's documents. The larger the value, the more the author is in the role of a "knowledge sink" in the network.

This study used Ucinet to analyze the author's degree centrality and summarized the top 5 authors with the in-degree and out-degree (Table 4). It can be seen that Zhao Ronghui, Ma Weiye, Li Shaojun and Wang Shaozhe have a higher in-degree but lower out-degree, which belong to the role of high-quality article creators in this network. Wu Qingrong, Zhao Shiju and Dai Manchun are highly out-degree, and they play a more role in absorbing the high-quality opinions of others in the network. And Shen Lei, Gu Shuzhong and Liu Yuejin have a both high in-degree and out-degree, which means that not only can they publish excellent articles, but also be good at absorbing the views of others, which is the core of the entire network.

Table 4. Ranking of top 5 authors in in-degree and out-degree

Author	In-Degree	Author	Out-Degree
Liu Yuejin	7	Shen Lei	11
Zhao Ronghui	6	Gu Shuzhong	11
Gu Shuzhong	4	Liu Yuejin	10
Ma Weiye	4	Wu Qingrong	10
Li Shaojun	4	Zhao Shiju	6

2) Betweenness centrality

Betweenness centrality is another major indicator for centrality analysis. If the number of shortest paths at any two points is greater, the centrality will be greater. In actual analysis, betweenness centrality often represents the degree to which a person controls resource. The larger the value, the more the person is in a core position and controls the spread of knowledge.

With the help of Ucinet's analysis function, we analyze the centrality of each author and list the top 5 authors in Table 5. As mentioned earlier, due to the low network density of this reference relationship, the value of betweenness centrality is not very large. However, we found that Liu Yuejin and Zhao Ronghui ranked high, and judged that they should belong to the core position of the branch they are good at.

Table 5. Ranking of top 5 authors in betweenness centrality

Author	Betweenness centrality
Liu Yuejin	27.500
Zhao Ronghui	7.500
Wang jianqin	3.000
Gu Haibing	3.000
Shen Lei	2.500

3) Closeness centrality

Closeness centrality is opposite to the first two indicators above. It measures that a node is not limited to the capabilities of other nodes in the network. In actual analysis, the higher the ranking, the more the node has the ability to acquire knowledge, that is, it is in a core position.

The author's mutual citation graph is directed graph, so it is necessary to distinguish between its internal closeness centrality and external closeness centrality. Table 6 lists the ranking of its top 5 authors.

Table 6. Ranking of top 5 authors in closeness centrality

Author	Internal closeness centrality	Author	External closeness centrality
Ma Weiye	1.469	Wu Qingrong	1.586
Li Shaojun	1.469	Zhao Shiju	1.491
Dai Chaowu	1.468	Dai Manchun	1.490
Sun Jinping	1.468	Gu Shuzhong	1.471
Liu Yuejin	1.449	Shen Lei	1.471

Since the density of the whole graph is low, only 0.0137, the overall values of the centrality present a lower level. However, it can also be seen that "Li Shaojun", "Dai Chaowu" and some others ranked high, indicating that they are relatively independent and not controlled by other authors. This is also consistent with the in-degree and out-degree tables analyzed above.

5.3   Cohesive Subgroup Analysis

When the relationship between certain actors in the network is relatively close, forming a sub-structure, this small group is called a cohesion sub-group. Here we conduct an analysis of the cohesive subgroup, thereby trying to have a preliminary understanding of the state of the substructure within the citation network.

With the help of Ucinet's CONCOR iterative algorithm function, the network nodes are partitioned to obtain the number of subgroups that appear in the overall citation network, and the subgroup visualization graph (Fig. 5) is obtained. Depend on the density matrix (Table 8) and the scale density matrix (Table 9), I draw a simplified diagram (Table 7).

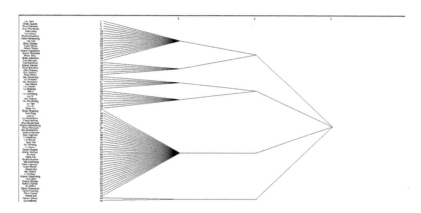

Fig. 5. Subgroup visualization

Express the above chart as a table.

Table 7. Author classification chart

Subgroup A: Liu Jiayi、Chen Guojie、Zou Changxin、Shen Weishou、Guo Zhongwei…
Subgroup B: Wang Jianqin、Wu Qingrong、Dai Manchun、Zhao Ronghui…
Subgroup C: Hu Jingguang、Ma li、Gu Shuzhong、Liu Yi、Liu Weidong…
Subgroup D: Dai Chaowu、Liu Yuejin、Sun Jinping、Ma Weiye…
Subgroup E: Sun Ping、Liu Dongzhou、Luo Li、Zhu Chuanrong、Zheng Hangsheng…

It can be seen from the classification that the entire group is divided into 5 sub-groups. Combined with the division of knowledge roles and the fields and point analysis of each author's expertise, the following subgroups can be defined as follows.

Although the members of subgroup A have a certain amount of out and in degrees, but the other members that each member contacts are very fixed. Combining the areas, they are good at, we can find that the members of subgroup A are more active in the fields they involve while low activity in other branches of national security.

The out degrees of the members of subgroup B are often greater than their in degree, that is, they often cited other schoolers' articles rather than be cited, which means that they are good at absorbing multi-faceted and multi-domain knowledge, and plays the role of "knowledge sink" in the network.

Although the members of subgroups C and D are involved in different branch areas, their in-degree is often greater than their point-out degree, that is, their articles always be cited by other person and they belong to the role of "knowledge source". Meanwhile, these members also communicate very closely with members in fields they do not involve, indicating that they are in a relatively important position in the citation network.

The members of subgroup E have both low in-degree and low out-degrees, indicating they are in the boundary of the network.

While using the CONCOR program to group, we get the density matrix between each subgroup. In order to grasp the relationship between each subgroup, I normalized the density matrix by getting the value of the subgroup density less than the overall density to 0, and the value greater than 1 to 1. According to the above rules, get the following matrix.

Table 8. Normalized density matrix between subgroups

	A	B	C	D	E
A	1.000	0.000	0.000	0.000	0.000
B	0.000	1.000	0.000	1.000	0.000
C	0.000	1.000	1.000	0.000	0.000
D	0.000	1.000	0.000	1.000	0.000
E	0.000	0.000	0.000	0.000	0.000

With the help of Gephi's drawing function, the simplified image matrix is as follows (Graph 6).

Graph 6. Simplified image matrix between subgroups

It can be seen from the chart that subgroup B not only plays the role of "knowledge sink" among these 75 authors, but also acts as a bridge, connecting subgroup D and subgroup C, making these two parts' authors can exchange their knowledge to each other. The authors of subgroup A only maintain resource exchanges under the same discipline branch, while the authors of subgroup E neither communicate with other subgroups nor communicate within subgroups, so they are not shown in the figure.

6 Analysis of National Security Research Institutions

Through perform the social network analysis on research institutions, it can help us to understand what the main institutions of national security research in China are, the role they played and the relationship between them. This can help us to be targeted in the construction of the institutions which involved national security discipline in the future.

6.1 Community Graph Analysis

Similar to the network diagram of the author's cross-citation relationship described above, this study selected author institutions, citation institutions and cited institutions involved in the top 50 cited documents. Figure 7 is the network diagram of the relationship between institutions.

Fig. 7. Network diagram of relationship between institutions (The two pictures below are enlarged versions of the dark parts in the picture above)

For the research of institutions, we still use the drawing function of Gephi, where nodes represent the names of institutions participating in national security research, and edges represent the citation relations between institutions. The darker the node, the greater the degree of node, and the more active it is. With reference to Fig. 7, we can see that there are many research institutes related to national security in China. Among them, Peking University, Renmin University of China, The University of International Relations, China University of Political Science and Law, etc. These institutions are the backbone of China's national security research and the leader in this discipline.

6.2   Centrality Analysis

1) Degree centrality

Put the acquired data into Ucinet for degree centrality analysis and integrate the top 5 institutions with in-degree and out-degrees into Table 9. It can be seen from the table that institutions such as "Renmin University of China", "China University of Political Science and Law" and etc. have higher in-degree and lower out-degree, indicating that they have played a role of knowledge source in the construction and provided many high-quality resources for China's national security studies. The out-degree of "Jilin University", "Nanjing Audit Institute", and "Zhongnan University of Economics and Law" are significantly greater than their in-degree, indicating that they are more active in the flow of knowledge and are actively learning the innovative ideas of other institutions. The "Peking University" and "University of International Relations" are not only having a high in-degree, but also have a high out-degree, indicating that they are not only made outstanding contributions to the construction of China's national security, but also maintained the habit of absorbing high-quality views from the outside world which are the core force for the development of this discipline.

Table 9. Ranking of Top 5 Institutions in in-degrees and out-degrees

Institution	In-Degree	Institution	Out-Degree
Renmin University of China	45	Peking University	30
Peking University	39	Jilin University	20
University of International Relations	31	Nanjing Audit University	16
Institute of Geographical Sciences and Natural Resources Research, Chinese Academy of Sciences	28	State Environmental Protection Administration	14
China University of Political Science and Law	26	University of International Relations	12

2) Betweenness centrality

Put the obtained data into Ucinet for intermediate centrality analysis and integrate the top 5 institutions into Table 10. It can be seen from the figure that "Peking University", "The University of International Relations", "Renmin University of China", "China

University of Political Science and Law" are ranked high, indicating that they have played a very important role in knowledge dissemination and exchange, and these institutions' in-degree is often greater than theirs out-degree, indicating that they play a role in controlling other nodes in the relationship network.

The betweenness centrality of the network has a potential of 16.02%, and the value is low, indicating that the betweenness centrality of the institutions is relatively low in terms of the entire network. That is, in national security research, most institutions do not need other institutions as intermediaries to obtain information.

Table 10. Ranking of top 5 institutions in betweenness centrality

Author	Betweenness centrality
Peking University	5896.308
University of International Relations	2794.206
Renmin University of China	2765.697
China University of Political Science and Law	2130.063
Suzhou University	2001.177

3) Closeness centrality
Closeness centrality represents the ability of a node to be free from control by others. Import the acquired data into Ucinet for closeness centrality analysis and put the top 5 institutions in the Table 11.

Table 11. Ranking of top 5 institutions in closeness centrality

Author	In-closeness centrality	Author	Out-closeness centrality
Institute of Zoology, Chinese Academy of Sciences	2.055	Jilin University	1.003
Anhui University	2.055	Northwest University	0.973
Nanjing Institute of International Relations	1.992	Dalian University of Technology	0.973
National Audit Office	1.977	South China University of Technology	0.972
China Institute of Modern International Relations	1.947	Harbin Institute of Technology	0.972

Due to the low density of this network relationship graph, the overall centrality value is not very high, but it can also be seen that the top rankings are "Institute of Zoology", "Chinese Academy of Sciences", "Anhui University", "Nanjing Institute of International Relations", etc., which also corresponds to the characteristic that the point out degree is basically 0 and the point out degree value is higher.

7   Conclusion and Prospect

Although national security science has only been established for less than 30 years as an emerging discipline, due to the strong support of the country, the active participation of many scientific research institutions and the importance and complexity of its discipline, this discipline has achieved great success in China. Since 1999 the University of International Relations first pioneered the National Security Science which has achieved a series of achievements in the construction of disciplines and the "National Security Science Foundation" has been listed as a ministerial-level textbook for funding. After that, many scientific research institutions have blossomed and formed a national research group with "Peking University", "The University of International Relations" and "Chinese Academy of Sciences" as the core. During this period, a large number of scholars and professors who have made unremitting efforts for the vigorous development of this discipline have emerged, such as "Liu Yuejin" professor from "The University of International Relations" and "Zhao Ronghui" professor from "Shanghai International Studies University". It is precisely because of the hard work of these institutions and scholars that China's national security research and construction can have a rapid development in less than 20 years and complete the transformation from such as "resource security", "military security", "political security" and other traditional security to such as "language ability", "network security", "data security", "ideological security" and other national soft power security transition. However, from the beginning of the establishment of national security to the present 30 years, the theme of "America" has been enthusiastically discussed by scholars. At the same time, combined with the phenomenon that the literature type of multi-peak citation curve in the period of 2008–2019 has been cited less frequently, which because the literature has a long life cycle and has not reached its peak, I think that in the next 10 years, the research theme of China's national security would be concentrated in "security issues between China and the United States" and China's soft power security construction such as "Network security". Here, I sincerely hope that our national security studies can continue to flourish in the future construction and continue to prosper!

References

1. Liu, Y.: Studying for National Security-The Exploring Course of National Security Discipline and Several Problems. Jilin University Press, Changchun (2014)
2. Opinions of the Ministry of Education on Strengthening the Implementation of National Safety Education in Colleges and Universities. Ministry of Education of the of China 360A12-04-2018-0004-1
3. Scott, J.: Social Network Analysis (2010)
4. Hu, H.: Research on China's national security issues: history, evolution and trends. J. Renmin Univ. China **4**, 150–152 (2014)
5. China's position paper on the new security concept. People's Republic of China Ministry of Foreign Affairs, 31 July 2002 (2002)
6. Report of the 16th CPC National Congress. Chinese Government (2002)
7. Gao, F.: Analysis of China's overall security concept. Foreign Affairs Coll. 11–12 (2015)

8. An excerpt from Xi Jinping's discussion on the overall national security concept. CPC Central Party School and Document Research Institute (2018)
9. Ye, Z.: The Chinese implication of Xi Jinping's overall security concept. Peking Univ. Sch. Int. Relat. **2014**(6), 17–20 (2014)
10. Fang, B., Jia, Y., Han, Y.: Social network analysis of core scientific issues, research status and future prospects. Beijing Univ. Posts Telecommun. Natl. Univ. Defense Technol. (2), 188–194 (2015)
11. Shen, H.W., Cheng, X.Q., Cai, K., et al.: Detect overlapping and hierarchical community structure in networks. Phys. A-Stat. Mech. Appl. **29**(9), 115–124 (2009)
12. Yang, S., Zhu, K., Fu, C., Lu, G.: Cellular automaton-based simulation of group decision crowd behavior. Hefei Univ. Technol. **29**(9), 115–124 (2009)
13. Xing, X.: Non-equilibrium statistical information theory. Beijing Inst. Technol. **53**(9), 2852–2863 (2004)
14. Fang, B., Xu, J., Li, J., et al.: Online social network analysis. Beijing Electronic Industry Press (2014)
15. Qu, W., Zhu, L., Yu, Y.: Aggregation Analysis of Topics of Highly Cited Documents Based on Citation Curve Features
16. Li, J., Jiang, M., Li, T.: Research on analysis framework of citation curve—Taking Nobel prize winner's citation curve as an example. Chin. Libr. J. **40**(210), 43–45 (2014)
17. Liu, W.: Analysis of internet citations based on the literature of CNKI competitive intelligence. Online J. Libr. Inf. Work **2012**(6), 2–3 (2012)
18. Peng, H., Zeng, X., Han, K.: Analysis of research situation of logistics network based on CNKI. Lanzhou Univ. Finance Econ. **2019**(3), 10–11 (2019)
19. Wang, J.: Research on knowledge role recognition of authors in accounting based on social network analysis. Chong Qing Univ. 22–34 (2014)
20. Yang, L., Zhou, X., Liu, Y., Hu, L., Zeng, J.: Visual analysis of research status and trends in the field of international cybersecurity in the past 10 years. Univ. Int. Relat. 2–4 (2016)
21. Tsvetovat, M., Kouzsov, A., et al.: Social network analysis methods and practices (2013)

A Study of Online Electronic Voting System Based Blockchain

Byeongtae Ahn[(✉)]

Liberal and Arts College, Anyang University, 22, 37-Beongil, Samdeok-Ro, Manan-Gu, Anyang 430-714, South Korea
ahnbt@anyang.ac.kr

Abstract. As the offline paper voting, which has been conducted until recently, continues to be a problem in terms of security and reliability, the interest in online electronic voting with increased safety and convenience is increasing. Several countries around the world are now adopting and activating online electronic voting. However, existing electronic voting has not been officially introduced in most countries due to interdependence and procedural security flaws. If 100% of these interdependencies and security can be trusted, the high-cost paper voting method will gradually decrease. Therefore, in this paper, blockchain technology is applied to compensate for the problem of electronic voting. This system can verify and hold blocks independently using a P2P method without a central authority to prevent falsification with reliability. In addition, by removing temporal and spatial constraints, the voter turnout can be increased and voting is possible directly on smartphones and PCs.

Keywords: Electronic voting · BlockChain · Ethereum · Smart contract · Paper voting

1 Introduction

Reliability and security issues for voting around the world are increasing every time. Two months prior to the 2016 US presidential election, the fact that voting machines in Arizona and Illinois were hacked from foreign hackers was also publicized. These issues raised doubts about the integrity of the voting results and raised the need to improve the electronic voting system [1].

Compared to existing paper voting, e-voting is not only capable of real-time counting, but also has fewer errors, so Estonia introduced and implemented e-voting since 2005. However, due to the nature of interdependence in electronic voting, even if an error occurs at one point in the system, it is difficult to accurately identify the point of occurrence of the error because they are mutually affected because different parts of the system depend on each other [2]. In addition, the problems of ensuring reliability and preventing forgery, which are the limitations of the existing online and offline voting systems, are constantly increasing. Therefore, in this paper, we designed a blockchain-based electronic voting system that can reduce the cost of offline voting and prevent fraudulent elections. This system was made to run in an Ethereum-based distributed computing network environment and enabled on the web or d-App.

J. C. Hung et al. (Eds.): FC 2021, LNEE 827, pp. 20–25, 2022.
https://doi.org/10.1007/978-981-16-8052-6_2

In Sect. 2 of this paper, related research is introduced, and in Sect. 3, a blockchain-based electronic voting system is designed. And Sect. 4 presents conclusions and future tasks.

2 Related Studies

Electronic voting is currently being implemented in several countries, and in particular, Estonia is the first country to apply the voting system on a national scale using the Internet. The structure of the existing Internet-based electronic voting I-voting system uses E2E encryption to vote, but Estonia's electronic voting system utilizes blind signature technology [3, 4]. Estonia's electronic voting system uses nationally issued ID cards and the keys they hold to vote. For voting, Estonian voters can use a card reader and client software to access the website for electronic voting and create a legally valid signature. During the voting process, two RSA key pairs were generated, one for authentication and one for digital signing.

FollowMyVote is a blockchain-based electronic voting system that is implemented online and uses blockchain to prove to voters and watchers that the vote has not disappeared. The FollowMyVote system adopts a way to obtain voting rights by downloading applications and authenticating their identities. The system uses two key pairs using elliptic-curve encryption to maintain voter anonymity. One is for identification, and the other is used for voting [5, 6].

Zhao and Chan proposed a bitcoin lottery-based system [7]. This system is a new approach that eliminates the need for a central authority to decrypt the vote after the election period and does not need to encrypt the vote.It uses a random number value to hide the relationship between the voting action and the voter. The authenticity of the vote was determined by proving that it was 0 [8]. Later, Takabatake proposed a voting system using similar proof of knowledge [9].

Most recently, Bistarelli et al. proposed an electronic voting protocol using Bitcoin, and this system divided the election organization into two subjects, one performing the authentication function and the other the token distribution function to grant voting rights. Through this, the privacy of voters was protected in the voting process, but it was difficult to monitor the behaviors of two separate entities organized in the protocol and had limitations in expanding the voting scale [10].

Blockchain eliminates the central bank system that is applied to existing banks, and decentralizes the data stored in the central ledger to store data on the users' computers in the chain. We have built a peer-to-peer (P2P) system that can trade with. The reason why experts call the blockchain a very secure technology that cannot be forged or altered is the hash value. Before transmission occurs, a unique hash value is generated in the block, which has the previous hash value and the current hash value in the block head, so if someone tries to forge/modulate the data value of a specific block for malicious purposes The hash value of the block is changed. This is a very safe technique because it is determined that the hash values of all blocks in the blockchain, which are generally composed of hundreds of thousands or more, cannot be changed because the previous hash values of the blocks connected to the front and back must also be changed.

Blockchain is a structure that allows all users who participate in the blockchain to access not only my data, but all users' data by storing all data on the user's computer, rather than storing user data on a central server. Blockchain uses a hash function to build a very secure data storage method where data cannot be forged or tampered with. The inside of the block of the blockchain used by Bitcoin is shown in Fig. 1.

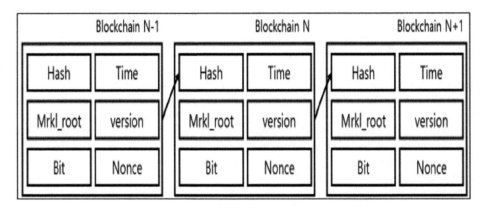

Fig. 1. Blockchain connected by a hash function

Inside the block, the previous hash value is stored, and you can see it as if it were chained. If a malicious user attempts to forge/modify the data of a specific block, the currently stored hash value is changed, and accordingly, hash values connected to front and back must also be changed. It is impossible to change the hash value of all blocks in the existing blockchain. Although it is a decentralized system, users can use it without worrying about data forgery or forgery.

3   Design of Electronic Voting System

In this paper, we designed an electronic voting system based on Ethereum using blockchain technology. This system is an electronic voting method that prevents illegal elections and allows safe and convenient voting online. This method solved the high-cost problem of paper voting, which is an existing offline method, and solved the security problem for electronic voting in an online method. In addition, it increased the convenience of voters who are inconvenient or difficult to vote, and is an electronic voting method specialized in Korea. Fig. 2 shows the flow chart of this system.

The user can vote on a smartphone or PC, and the results of the vote are stored on a server and transmitted to a smart contract. These smart contracts are generated as byte codes through the Ethereum virtual machine and stored as blocks in the Ethereum main net. Voting results are provided in smart contracts through a solidity language that supports Turing completeness. This method improves reliability by ensuring the anonymity and convenience of voting as well as providing complete defense against malicious attacks. This system is designed as a web-based interface to increase

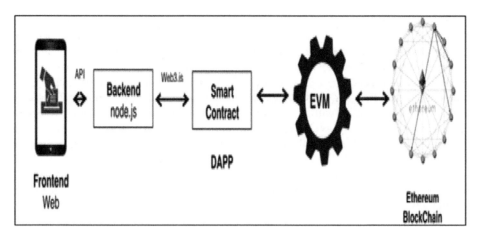

Fig. 2. Data flow architecture of system

accessibility, and because it is modular, it is easy to integrate with other tools including authentication tools.

Figure 3 shows the overall configuration of the system.

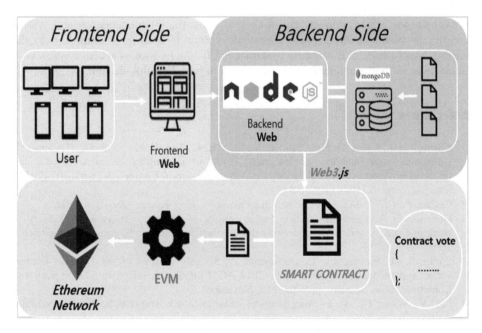

Fig. 3. Structure of electronic vote system

The user accesses a website using a smart phone and a computer and uses an electronic voting system. At this time, the identity may be verified through the personal information stored in the block for identification, and the electronic voting may be accessed. When using the electronic voting system, a node.Js server capable of asynchronous processing was used to eliminate the overload that could cause many problems. In addition, for fast data update and storage, memory DB mongoDB is used to improve compatibility with the server. The web and the server were made to access the smart contract using Web3.js, and the normally signed voting information was distributed and stored in the Ethereum blockchain. At this time, the smart contract is generated as a byte code through the Ethereum virtual machine and stored in a block.

The design of this system increases the efficiency of voting management tasks such as shortening voting and counting time, preventing invalid votes, etc. In particular, it has the advantage of improving turnout and low budget.

4 Conclusion and Future Works

In this paper, we designed an electronic voting system based on Ethereum using blockchain technology. The system can be used not only for large-scale public elections, but also for voting in small groups, and it has improved the reliability of electronic voting. In addition, the cost of voting was reduced and the turnout was improved.

In this paper, we proposed an Ethereum-based electronic voting system using blockchain and smart contracts and tokens to solve the interdependence and system integrity problems of the existing electronic voting system. And, compared to the existing bitcoin system, it was easy to implement functions such as access control. As a future task, we will implement a practical system based on this design.

References

1. Cheng, C.-H., Chen, C.-H., Chen, Y.-S., Guo, H.-L., Lin, C.-K.: Exploring Taiwanese's smartphone user intention: an integrated model of technology acceptance model and information system successful model. Int. J. Soc. Humanist. Comput. 3(2), 97–107 (2019). https://doi.org/10.1504/IJSHC.101591
2. Ciazzo, F., Chow, M.: A blockchain implemented voting system, December 2016
3. Government Accountability Office: Federal efforts to improve security and reliability of electronic voting systems are under way, but key activities need to be completed, September 2005
4. Springall, D., Finkenauer, T., Durumeric, Z.: Security analysis of the Estonian internet voting system. In: Proceedings of the 2014 ACM SIGSAC Conference on Computer and Communications Security, pp. 703–715, November 2014
5. Liu, Y., Wang, Q.: An e-voting protocol based on blockchain. IACR Cryptology ePrint Archive, 1043 (2017)
6. Krimmer, R.: Electronic voting 2006. GI Lecture Notes in Informatics, P-86, Bonn (2006)
7. Goodman, N.J.: Internet voting in a local election in Canada. In: Grofman, B., Trechsel, A., Franklin, M. (eds.) The Internet and Democracy in Global Perspective. SIPC, vol. 31, pp. 7–24. Springer, Cham (2014). https://doi.org/10.1007/978-3-319-04352-4_2

8. Brightwell, I., Cucurull, J., Galindo, D., Guashch, S.: An overview of the iVote 2015 voting system (2015). https://www.elections.nsw.gov.au

9. Jeong, C.R., et al.: Analysis of requirements for construction of electronic voting system based on blockchain. In: CISC-W 2017, pp. 31–34, December 2017

10. Takabatake, Y., Kotani, D., Okabe, Y.: An anonymous distributed electronic voting system using Zerocoin. Institute of Electronics, Information and Communication Engineers (IEICE), Technical Report IA2016-54, pp. 127–131, November 2016

11. Nakamoto, S.: Bitcoin: a peer-to-peer electronic cash system (2008)

Study of Effective Storage System for MPEG-7 Document Clustering

Byeongtae Ahn[(⊠)] [iD]

Liberal and Arts College, Anyang University, 22, 37-Beongil, Samdeok-Ro,
Manan-Gu, Anyang 430-714, South Korea
ahnbt@anyang.ac.kr

Abstract. To use multimedia data in restricted resources of mobile environment, any management method of MPEG-7 documents is needed. At this time, some XML clustering methods can be used. But, to improve the performance efficiency better, a new clustering method which uses the characteristics of MPEG-7 documents is needed. A new clustering improved query processing speed at multimedia search and it possible document storage about various application suitably. In this paper, we suggest a new clustering method of MPEG-7 documents for effective management in multimedia data of large capacity, which uses some semantic relationships among elements of MPEG-7 documents. And also we compared it to the existed clustering methods.

Keywords: MPEG-7 · XML · Clustering · Semantic block · Multimedia

1 Introduction

Recently, as various multimedia applications such as MP3, video mail, and digital multimedia broadcasting (DMB) have appeared in mobile devices, the management of multimedia data under limited resources has become a very important research topic. However, in order to effectively manage a large amount of multimedia data under limited resources, it is essential to manage metadata for multimedia data.

Recently, MPEG-7 has been adopted as an international standard for multimedia data technology methods in order to be able to handle multimedia effectively [1]. MPEG-7 is a standard defining multimedia data description method, that is, multimedia metadata in XML format [2]. Therefore, using MPEG-7 can effectively handle multimedia data [3, 4].

However, in order to handle MPEG-7 data more efficiently under limited resources, an appropriate clustering method is required, and MPEG-7 documents can be expressed in XML format. Therefore, existing XML clustering methods can be used for MPEG-7 document management. Fortunately, various clustering methods for XML documents have recently been proposed [5–8]. However, MPEG-7 documents have different characteristics from general XML documents. In other words, there are various semantic relations between the elements of the MPEG-7 document. Therefore, in order to manage MPEG-7 documents more efficiently, a new clustering method reflecting this is required.

© The Author(s), under exclusive license to Springer Nature Singapore Pte Ltd. 2022
J. C. Hung et al. (Eds.): FC 2021, LNEE 827, pp. 26–33, 2022.
https://doi.org/10.1007/978-981-16-8052-6_3

Existing studies on XML clustering can be largely divided into a method targeting documents and a method targeting schema. Here, as in this paper, we will briefly examine the characteristics and problems of clustering methods targeting XML documents as well as targeting XML schemas. Document-based clustering is a method of determining an effective clustering policy by receiving only an XML document as an input without any information about the XML schema. This method mainly handles XML documents by recognizing them as a tree.

Guillaume and Murtagh [5] proposed an XML document clustering method. In this paper, the problem of clustering XML documents is regarded as a problem of finding the optimal partition using graph theory. In other words, documents in the database can be divided into multiple clusters by modeling each document in a database as a node and finding the optimal partition in a graph modeling links between documents as weighted edges between nodes. In addition, the accuracy of clustering is improved by adding keyword links between documents that share the same keyword. However, other semantic relationships between elements are not used.

Francesca et al. [6] proposed a method of extracting a matching common structure between elements of two trees representing XML documents. Through this, they extract the representative structure of the cluster through the process of merging and simplifying the trees. Then, a hierarchical clustering algorithm is applied based on the weight value defined as the size of the common structure. However, in the case of a complex tree, it takes a lot of time to merge.

Schema-based clustering is a method used when the schema representing the frame of an XML document can be known. In general, many well-defined applications have schemas for most XML documents they handle. MPEG-7 also has a well-defined schema [2]. This schema determines the structure of XML documents in advance. Therefore, by using this, a more effective clustering policy can be devised.

The XClust proposed by Lee et al. [7] measures the similarity between two DTDs representing a schema, and clusters DTDs in the same domain using a hierarchical technique based on the similarity. In this method, after modeling the DTD in the form of a tree, the linguistic and structural similarity is calculated by comparing the names, properties, and paths to the root node of each node constituting the DTD. In addition, contextual similarity is determined through the similarity between direct descendants and terminal nodes. The similarity between DTDs is calculated as a matrix through the obtained similarity between nodes, and the DTD instance documents are gradually clustered from DTD pairs with high similarity using this. This method has high accuracy in consideration of contextual factors between nodes of each DTD, but it takes a long time and has a problem that accurate similarity cannot be obtained between DTDs of different sizes.

OrientX proposed by Xiaofeng Meng et al. [8] applies various clustering using schema. This is an improved method of dividing the XML document into subtrees and storing them in one record by grouping them into logical page units in consideration of association. First, the schema is analyzed to form a semantic block. This semantic block is a logical unit for storage as a group of related elements. OrientX uses an empirical method to obtain these semantic blocks. However, since only the grammatical elements of the schema (e.g. cardinality) are considered, it takes a lot of time to construct a semantic block in a complex schema structure, and contextual similarity between each element is not supported. In this paper, we propose a new clustering method that can efficiently manage MPEG-7 documents by improving this method. The paper is organized as follows. In Sect. 2, the proposed MPEG-7 clustering method is analyzed, and in Sect. 2.2, the proposed clustering method is compared with the existing methods. Finally, Sect. 3 proposes conclusions and future tasks.

2   MPEG-7 Document Clustering

This chapter proposes a new clustering method applicable to the management of MPEG-7 documents based on the analysis of the previous studies.

2.1   Motivation of the Proposal

In OrientX introduced in the previous chapter, schema analysis is performed for clustering. A schema graph is drawn through the schema, where a node in the schema graph becomes the root of the semantic block. At this time, there are two cases where the root of the semantic block becomes the following. First, it is the case of the root of the schema graph. Second, it is the case of having child nodes while having * or + as cardinality. In this method, an instance of such a semantic block is viewed as one logical record, and if it belongs to the same semantic block, it is stored in close proximity (clustered). Figure 1 shows an example of schema and XML document [9].

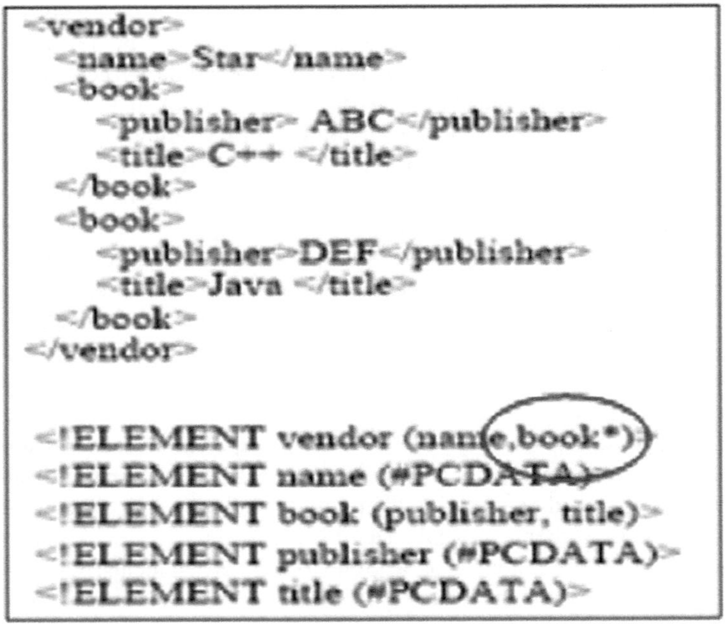

```
<vendor>
   <name>Star</name>
   <book>
      <publisher> ABC</publisher>
      <title>C++ </title>
   </book>
   <book>
      <publisher>DEF</publisher>
      <title>Java </title>
   </book>
</vendor>


<!ELEMENT vendor (name,book*)>
<!ELEMENT name (#PCDATA)>
<!ELEMENT book (publisher, title)>
<!ELEMENT publisher (#PCDATA)>
<!ELEMENT title (#PCDATA)>
```

Fig. 1. Schema & Xml document

Figure 2 shows the result of applying OrientX's clustering storage policy.

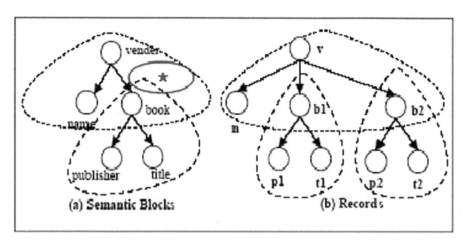

Fig. 2. OrientX clustering storage

We intend to apply OrientX's clustering method applied to XML document management to MPEG-7 document management.

This is because MPEG-7 documents are also XML documents and can be saved by applying the OrientX method. However, when MPEG-7 documents are saved only with the OrientX method, two main problems arise. First, since MPEG-7 schema has a very complex structure, it is difficult to construct a semantic block simply by grammatical meaning. Second, it is a storage ignoring the association between MPEG-7 elements [10].

Therefore, this paper presents two solutions. First, we introduce the R-CT attribute that represents the organic association between each element. Second, various semantic block generation rules are used in consideration of the complex MPEG-7 schema. Learn more about how to do it in the next section.

2.2   Clustering Application Procedure

We propose a three-step clustering application procedure for MPEG-7 document clustering. The application of the clustering method we propose consists of three steps. First, step 1 redefines the schema by adding an R-CT attribute value to indicate the elements related to the existing MPEG-7 schema. Step 2 constructs a semantic block based on the redefined schema. In the last 3 steps, clustering is performed using a clustering algorithm through the configured semantic blocks. Figure 3 shows the procedure for applying this three-step clustering. A detailed description of this is made in the following sections.

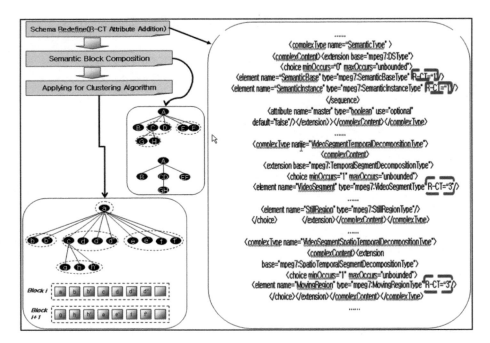

Fig. 3. Clustering step of Mpeg-7 document

Figure 4 is the MPEG-7 schema for this. It consists of various descriptors (D: Descriptor), DS: Description Scheme, and data types defined in MPEG-7. In other words, movingRegionType and StillRegionType to describe each key image, MediaInformationType, CreationInformationType, and TextAnnotationType to describe information about other videos were used, centering on the VideoSegment type to describe metadata about the video [11].

```
        - <choice minOccurs="0" maxOccurs="unbounded">
            <element name="VisualDescriptor" type="mpeg7:VisualDType" />
          </choice>
        - <choice minOccurs="0" maxOccurs="unbounded">
            <element name="TemporalDecomposition" type="mpeg7:VideoSegmentTemporalDecompositionType" />
            <element name="SpatioTemporalDecomposition" type="mpeg7:VideoSegmentSpatioTemporalDecompositionType" />
          </choice>
        </sequence>
      </extension>
    </complexContent>
  </complexType>
+ <complexType name="DSType" abstract="true">
- <complexType name="SegmentType" abstract="true">
  - <complexContent>
    - <extension base="mpeg7:DSType">
      - <sequence>
        - <choice minOccurs="0">
            <element name="MediaInformation" type="mpeg7:MediaInformationType" />
            <element name="MediaLocator" type="mpeg7:MediaLocatorType" />
          </choice>
          <element name="StructuralUnit" type="mpeg7:ControlledTermUseType" minOccurs="0" />
        + <choice minOccurs="0">
        + <element name="TextAnnotation" minOccurs="0" maxOccurs="unbounded">
        </sequence>
      </extension>
    </complexContent>
  </complexType>
+ <complexType name="MediaTimeType">
+ <complexType name="VisualDType" abstract="true">
+ <complexType name="VisualDSType" abstract="true">
- <complexType name="VideoSegmentTemporalDecompositionType">
  - <complexContent>
    - <extension base="mpeg7:TemporalSegmentDecompositionType">
      - <choice minOccurs="1" maxOccurs="unbounded">
          <element name="VideoSegment" type="mpeg7:VideoSegmentType" />
        </choice>
      </extension>
    </complexContent>
  </complexType>
- <complexType name="VideoSegmentSpatioTemporalDecompositionType">
  - <complexContent>
    - <extension base="mpeg7:SpatioTemporalSegmentDecompositionType">
      - <choice minOccurs="1" maxOccurs="unbounded">
          <element name="MovingRegion" type="mpeg7:MovingRegionType" />
        </choice>
      </extension>
    </complexContent>
  </complexType>
```

Fig. 4. Mpeg-7 schema

Figure 5 is an example of an instance document generated according to this MPEG-7 schema. It is largely divided into 3 video segments and each has a key image.

```
- <Video id="TRECVID2003_253">
  + <MediaInformation id="news1_media">
  + <CreationInformation>
  + <TextAnnotation>
  - <MediaTime>
      <MediaTimePoint>T00:00:00:0F30000</MediaTimePoint>
      <MediaDuration>PT6M10S16105N30000F</MediaDuration>
    </MediaTime>
  + <VisualDescriptor xsi:type="GoFGoPColorType" aggregation="Average">
  - <SpatioTemporalDecomposition>
    - <MovingRegion id="ManMR1">
      - <TextAnnotation>
          <FreeTextAnnotation>Man1 (MovingRegion1)</FreeTextAnnotation>
        </TextAnnotation>
      - <TemporalDecomposition gap="true" overlap="false">
        - <StillRegion id="ManKeySR1">
          - <MediaLocator>
              <MediaUri>image.jpg</MediaUri>
            </MediaLocator>
          - <TextAnnotation>
              <FreeTextAnnotation>Man (still region)</FreeTextAnnotation>
            </TextAnnotation>
            <MediaTimePoint>T00:00:13:15405F30000</MediaTimePoint>
          - <VisualDescriptor xsi:type="ScalableColorType" numOfCoeff="16" numOfBitplanesDiscarded="0">
              <Coeff>1 2 3 4 5 6 7 8 9 0 1 2 3 4 5 6</Coeff>
            </VisualDescriptor>
          </StillRegion>
        </TemporalDecomposition>
      </MovingRegion>
    + <MovingRegion id="ManMR2">
    + <MovingRegion id="ManMR3">
    </SpatioTemporalDecomposition>
  - <TemporalDecomposition gap="false" overlap="false">
    - <VideoSegment id="shot253_1">
      - <MediaTime>
          <MediaTimePoint>T00:00:00:0F30000</MediaTimePoint>
          <MediaDuration>PT27S20830N30000F</MediaDuration>
        </MediaTime>
      - <TemporalDecomposition>
        - <VideoSegment id="shot253_1_RKF">
          - <MediaTime>
              <MediaTimePoint>T00:00:13:15405F30000</MediaTimePoint>
            </MediaTime>
          </VideoSegment>
        </TemporalDecomposition>
      </VideoSegment>
```

Fig. 5. Mpeg-7 instance document

3   Conclusion and Future Works

In this paper, a new clustering method based on XML schema is proposed for efficient storage of MPEG-7 documents.

The advantages of the clustering method proposed in this paper are as follows. First, it improves query processing speed by supporting clustering based on MPEG-7 schema. Second, by generating the most suitable semantic block for MPEG-7 applications, it can be used to implement an MPEG-7 document storage system suitable for various applications. Based on these, unlike existing systems, it is possible to search and manage multimedia effectively in a mobile terminal. In addition, consistent processing is possible for a large amount of MPEG-7 document retrieval.

In this paper, we proposed two semantic block generation rules generated based on various MPEG-7 applications. However, in this part, we need more creation rules that are more suitable for the actual meaning. Therefore, in order to support the creation of a complete semantic block, additional research is needed to enable the setting of various R-CT attributes. At this time, it is necessary to adapt to the increasingly complex and diverse MPEG-7 applications.

References

1. Abdullah, A., Veltkamp, R.C., Wiering, M.A.: Fixed partitioning and salient points with MPEG-7 cluster correlograms for image categorization, Technical Report UU-CS-2009-008, Department of Information and Computing Sciences, Utrecht University, The Netherlands (2019)
2. Abdullah, A., Wiering, M.A.: CIREC: cluster correlogram image retrieval and categorization using MPEG-7 descriptors. In: IEEE Symposium on Computational Intelligence in Image and Signal Processing, pp. 431–437 (2019)
3. Abrishami, H., Roohi, A.H., Taghizadeh, T.: Wavelet correlogram: a new approach for image indexing and retrieval. J. Pattern Recogn. Soc. 38(12), 2506–2518 (2019)
4. Smith, J.R., Chang, S.F.: VisualSEEk: a fully automated content-based image query system. In: Proceedings of ACM Multimedia, pp. 87–98 (2020)
5. Bay, H., Tuytelaars, T., van Gool, L.J.: SURF: speeded up robust features. In: Proceedings of the Ninth European Conference on Computer Vision (ECCV), vol. III, pp. 404–417 (2019)
6. Carson, C., Thomas, M., Belongie, S., Hellerstein, J.M., Malik, J.: Blobworld: a system for region-based image indexing and retrieval. In: VISUAL, pp. 509–516 (2020)
7. Chang, C., Lin, C.: LIBSVM: a library for support vector machines (2021). http://www.csie. ntu.edu.tw/cjlin/libsvmS
8. Chen, Y., Wang, J.Z.: Image categorization by learning and reasoning with regions. J. Mach. Learn. Res. 5, 913–939 (2020)
9. Csurka, G., Dance, C.R., Fan, L., Willamowski, J., Bray, C.: Visual categorization with bags of keypoints. In: Workshop on Statistical Learning in Computer Vision ECCV, pp. 1–22 (2020)
10. Deselaers, T., Keysers, D., Ney, H.: Features for image retrieval: a quantitative comparison. Lect. Notes Comput. Sci. 2021, 40–45 (2021)
11. Deselaers, T., Keysers, D., Ney, H.: Classification error rate for quantitative evaluation of content-based image retrieval systems. In: Proceedings of the 17th International Conference on Pattern Recognition, ICPR 2004, vol. 2, pp. 505–508. IEEE Computer Society (2021)

An Automatic Pollen Grain Detector Using Deep Learning

Chengyao Xiong[1], Jianqiang Li[1], Yan Pei[2(✉)], Jingyao Kang[1], Yanhe Jia[3],
and Caihua Ye[4]

[1] Faculty of Information, Beijing University of Technology, Beijing 100124, China
{chengyaoxiong,kangjingyao}@emails.bjut.edu.cn, lijianqiang@bjut.edu.cn
[2] Computer Science Division, University of Aizu, Aizu-wakamatsu 965-8580, Japan
peiyan@u-aizu.ac.jp
[3] School of Economics and Management,
Beijing Information Science and Technology University, Beijing 100192, China
yhejia@bistu.edu.cn
[4] Beijing Meteorological Service Center, Beijing, China

Abstract. In this paper, we propose a deep learning framework to automatically detect pollen grains instead of the manual counting of pollen numbers under an optical microscope. Specifically, we first establish a large-scale dataset of pollen grains, which contains 3000 images of five subcategories. All the images in our dataset are scanned by an optical microscope. Then, a pollen grain detector (PGD) based on deep learning is designed to eliminate the effects of noise and capture subtle features of pollen grains. Finally, extensive experiments are conducted and show that the proposed PGD method achieves the best performance (84.52% mAP).

Keywords: Automatic pollen identification · Pollen grain detector · Deep learning · Object detection · Pollen allergy

1 Introduction

As a common and frequently occurring disease in clinical practice, allergy has become one of the three diseases for key prevention and treatment in the 21-st century by the World Health Organization [13]. Pollen is the most important one of numerous allergens. A large number of plant pollen allergies floating in the air can induce a series of allergic diseases such as allergic rhinitis, bronchial asthma, dermatitis, and so on. The above-mentioned allergic diseases caused by pollen allergies are also called hay fever [11]. With the intensification of urbanization in human society and the expansion of plant cultivation areas, hay fever has become a seasonal epidemic disease with a very high incidence. In the United States, the population incidence rate is about 5%, and even as high as 15% in some areas. The incidence rate in Europe has reached 20%, and will up to nearly 35% in the next 20 years [5]. And in China, with the continuous increase in recent years, the

J. C. Hung et al. (Eds.): FC 2021, LNEE 827, pp. 34–44, 2022.
https://doi.org/10.1007/978-981-16-8052-6_4

incidence rate is around 0.5% to 1% and reaches 5% in the high incidence area [4]. Therefore, accurate identification of pollen grains is of great significance to patients with hay fever. With the continuous development of deep learning and its application in various fields [8,9,14,21,23], it is undoubtedly a good choice to apply deep learning to automatic pollen detection.

Traditional pollen identification generally includes three steps: pollen collection, sample dyeing, recognition, and microscope counting. In the recognition and counting, experienced biologists are required to manually locate the pollen grains in the microscope image, which is a time-consuming and labor-intensive process. Therefore, many studies have been proposed automatic pollen identification methods with computer vision technology [2,3,15,19,20,22]. These methods mainly rely on image processing to detect and extract pollen objects. However, there are many impurities in the microscope image (such as air bubbles, dust, etc.), which could be misclassified as pollen under simple image processing.

To address this problem, we firstly built a pollen dataset with bounding box annotation for automatic detection of the pollen grain. This dataset contains 5 kinds of common airborne allergen pollens collected from March to October (i.e., the main flowering period of the whole year) in Beijing, thus has rather good representativeness. Based on this dataset, we propose a CNN-based pollen grain detector (PGD), applying matching strategy and hard negative mining, to eliminate the effects of pollen clumping, cracking deformities, numerous impurities in pollen images. Meanwhile, the focal loss is adopted to solve the imbalance problem to improve the prediction accuracy of hard samples. The experiment results show the mAP of our PGD reached 84.52%, which has achieved the best detection performance.

Following this introduction section, previous related works that have already employed for automatic pollen grain classification are presented and reviewed in Sect. 2. In Sect. 3 we then provide our dataset, image processing method and pollen grain detector model. Section 4 shows the experimental results and related parameters of our model. Finally, some conclusions are presented in Sect. 5.

2   Related Works

At present, existing methods mainly designed multiple descriptors for various characteristics of pollen grains to segment pollen grains and background in the image. These studies usually only use predefined features, such as contour, shape, texture and brightness [2,15–17,19,20,22]. For example, Travieso et al. proposed a contour feature descriptor based on a hidden Markov model [22], which can greatly reduce the dimensionality of image feature vectors. However, the noise interference of the secondary image will cause the change of the original image information, making the later-learned descriptor become unstable, which affects the accuracy of similarity matching. A descriptor based on brightness and shape [19] was proposed by Rodriguez Damian et al. This method uses the Hough transform to roughly estimate the contour of the pollen grains and then combines the active contour model to obtain an accurate boundary. However, these features are only applicable to specific data types of specific tasks with poor scalability and limited application scope.

Since manual features require a lot of experienced knowledge, automatic feature extraction based on the convolutional neural network has become the mainstream for automatic pollen identification tasks. Amar et al. [6] utilized a convolutional neural network with 6 convolutional layers to further improve the performance of pollen detection. Battiato et al. [1] proposed a pollen detection method, which uses an image processing method to extract pollen image features according to the color of stained pollen grains and then sends them to a CNN classifier (such as AlexNet, VGG) for detection and classification. The processing of this method is still complex and its accuracy is low. Besides that, some CNN-based pollen detection methods literally achieved good experimental results, however, the data they adopted and detected are only the pollen samples that have been carefully processed in the laboratory environment. In fact, pollen samples in reality usually contain a lot of impurities, such as plant debris, other spores, bubbles, and so on, which will have a great impact on pollen detection. Besides, pollen data has its particularity. Different types of pollen grains can look similar in a specific perspective, and that will cause a lot of trouble in determining label information of pollen grains.

In this paper, we establish a real-world pollen dataset with a bounding box labeled by experts. Base on this, we propose an object detection network (named PGD), to extract subtle features of different species of the pollen grain. Besides, the focal loss is also adopted in our PGD, which is expected to improve the prediction accuracy of hard samples.

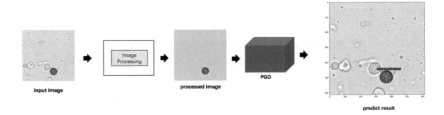

Fig. 1. The overall prediction pipeline. The original image is processed to remove air bubbles and other impurities, the processed image enters the PGD model, and the pollen is detected

3   Material and Our Method

Our pollen detection process is shown in Fig. 1. An image needs to be processed before prediction. And then the prediction result is obtained through PGD. We introduce the data set and image processing method in Sect. 3.1. In Sect. 3.2, we will introduce our PGD in detail.

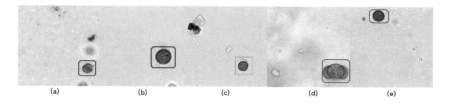

Fig. 2. Images of various pollen samples in PDD. Pollen grains locate in the red box. (a) Artemisia, (b) Chenopodiaceae, (c) Moraceae, (d) Gramineae, (e) Pinaceae

3.1 Dataset and Image Processing

Dataset. In this work, we have built a dataset that contains 3000 pollen grain objects. The glass slides with pollen grains are provided by the Beijing Meteorological Service of China. Pollen collection is done by a botanist, who applies a thin layer of adhesive to the glass slide, next place it in a sampler, then expose the glass slide to the air, and take it back 24 h later to complete the collection. The next step is to use an appropriate amount of dye, put it on the part of the glass slide coated with adhesive, melt it slowly with low heat, and then add a cover glass.

After collecting and staining the pollen, a pathological section scanner is used to scan the glass slide. The pathological section scanner is an automated optical microscope that can connect to a PC device and the scanned image will be presented on the PC. We put the glass slide into the scanner, set the scanning area, the number of focal planes, the distance between the focal planes, the magnification, and other parameters to complete the focusing operation, and finally get the scanned pathological section image. We then divide a whole slide image into 512×512 pixels images by using OpenSlide[1] and NDPITools[2]. Finally, the images that contain pollen grains were selected and sent to botanical experts to mark the pollen labels and locations. After the above steps, the Pollen Detection Dataset (PDD) was built. Five common types of allergenic pollen in Beijing have been included in PDD, which are: (1) Artemisia, (2) Chenopodiaceae, (3) Moraceae, (4) Gramineae, (5) Pinaceae. (see Fig. 2).

Image Processing. Some transparent bubbles in the pollen image have similar size, texture, and shape to pollen grains, thus may make the PGD model confused. To eliminate the influence of bubbles, the image needs to be processed before using PGD for pollen detection. Since the dyed pollen in the image appears magenta (see Fig.), we extract pollen color to remove bubbles. The operation steps are as follows. First, we use the mean shift algorithm[3] to smooth the image. Then, we use the Gaussian blur to reduce noise and convert the RGB

---

[1] https://openslide.org/api/python/.

[2] https://www.imnc.in2p3.fr/pagesperso/deroulers/software/ndpitools/.

[3] https://docs.opencv.org/2.4/modules/imgproc/doc/filtering.html?highlight=means hiftfiltering.

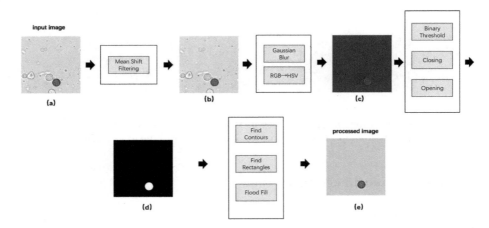

Fig. 3. The details of image processing. (a) Image sample under optical microscope, (b) smooth image processed by mean shift algorithm, (c) HSV image after Gaussian blur and color space conversion, (d) the binary image after the opening and closing operations, (e) the image after contour extraction, calculating the smallest rectangle containing the contour and flood fill algorithm

image to an HSV image. After that, we binarize the image according to Eq. 1, where $(HSV)_{min}$ and $(HSV)_{max}$ are the thresholds we set. Next, a closing operator and an opening operator are employed using a 30×30 kernel and a 10×10 kernel, respectively, for the binary image.

$$dst = H_{min} \leq src(H) \leq H_{max} \cap S_{min} \leq src(S) \leq V_{max} \cap V_{min} \leq src(V) \leq V_{max}$$
$$(1)$$

After this, we get a binary image named mask. After extracting the contour of the mask image and calculating the minimum vertical boundary rectangle of the contour, the pollen color is finally extracted. The whole image processing process is demonstrated in Fig. 3. And from Fig. 3 (e), the image obtained after processing, we can observe that the bubbles in Fig. 3 (a) are removed.

3.2 Our Automatic Pollen Grain Detector

As shown in Fig. 4, more details of our PGD model will be shown in this section. Specifically, there are three main parts:

Default Boxes: For each feature map cell, we use two corresponding default bounding boxes, each is used to predict the offsets and category scores of this box. For an m × n feature map, each default box is used to calculate c category scores and 4 location information. Therefore, we need to use $2 \times (c+4)$ convolution kernels to filter the feature map, and the output size of the m × n feature map is $2 \times m \times (c+4)$. In our data set, since the pollen is mostly round, so the two

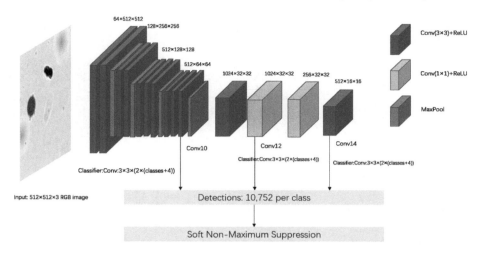

Fig. 4. The architecture of our PGD. PGD receives a 512×512 size picture. Three branches are selected for detection to reduce parameters, because they have receptive field similar to the size of pollens.

default boxes are all set to squares rather than rectangles, which can also reduce the model parameters.

Matching Strategy and Hard Negative Mining: If the IoU of the default box and a ground truth box is greater than 0.5, the default box is regarded as a positive example, otherwise it is a negative example. This will make a significant imbalance between the negative and positive cases because the number of negative cases is much larger than the number of positive cases. Therefore, we follow the work [18] to sort negative cases according to the confidence loss, and the ratio of positive and negative examples is controlled as 1:3.

Loss Function: Let $x_{ij}^p = \{1, 0\}$ be an indicator for matching the i-th default box to the j-th ground truth box of category p. The overall loss function is a weighted sum of the localization loss (loc) and the confidence loss (conf):

$$L_{(x,c,l,g)} = \frac{1}{N}(L_{conf}(x,c) + \alpha L_{loc}(x,l,g)) \qquad (2)$$

where N is the number of matched default boxes. The weight term α is set to 1 by cross-validation. Our classification loss function is Focal Loss [10]. This function can reduce the weight of samples that are easy to classify so that the model can focus more on samples that are difficult to classify during the training process.

$$L_{conf}(x,c) = - \sum_{i \in \{Pos, Neg\}} \beta(1 - \hat{c}_i^p)^\gamma log(\hat{c}_i^p) \qquad (3)$$

The localization loss is a Smooth L1 [7] loss between the predicted box (l) and the ground truth box (g) parameters. The offsets for the default bounding box (d) center(cx, cy) and its width (w) and height (h) can be regressed by optimizing the following loss:

$$L_{loc}(x, l, g) = \sum_{i \in Pos}^{N} \sum_{m \in \{cx, cy, w, h\}} x_{ij}^k smooth_{L1}(l_i^m - \hat{g}_j^m)$$

$$\hat{g}_j^{cx} = (g_j^{cx} - d_i^{cx})/d_i^w, \hat{g}_j^{cy} = (g_j^{cy} - d_i^{cy})/d_i^w$$

$$\hat{g}_j^w = log(g_j^w/d_i^w), \hat{g}_j^h = log(g_j^h/d_i^h)$$

(4)

4   Experiments

In this section, we will show the training details and the experimental results.

4.1   Training Detail

To solve the problem of data imbalance, we used the following data Augmentation strategies:

– Random horizontal flip
– Randomly selects the size of the crop box, analyzes the IoU value of the crop box and the GT box in the picture, and updates the GT box as a rectangular box where the crop box and the GT box overlap.

For the parameter setting in training, the Xavier method is employed to initialize all the parameters, and the SGD optimizer is adopted with an initial learning rate set as 10^{-4}, momentum as 0.9, weight decay as 0.0005, and batch size as 32. 30k training iterations are carried out in total, among which, 10^{-4} learning rate is used for the first 10K training iterations, then 10^{-5} for the next 10K, finally 10^{-6} and 10^{-7} for the last two 5K iterations, respectively. In the Focal Loss function, balanced variant β is set to 0.25, and focusing parameter γ is set to 2. We use conv10, conv12, conv14 to predict both location and confidences. Two default bounding boxes are associated with each feature map cell. The scale of these default boxes is (35,52), (76,107), (153, 188), respectively.

Table 1. Object detection comparison on our pollen detection dataset. This table shows that the PGD has the highest detection accuracy of Moraceae, Artemisia, Pinaceae and Chenopodiaceae, and it has the highest mAP value.

Methods	AP-Artemisia	AP-Chenopodiaceae	AP-Moraceae	AP-Gramineae	AP-Pinaceae	mAP
Faster R-CNN [18]	90.84%	91.67%	70.27%	73.06%	88.95%	82.96%
SSD [12]	98.11%	91.26%	58.46%	*75.10%*	88.32%	82.25%
PGD	*99.11%*	*91.94%*	*71.15%*	66.94%	*93.49%*	*84.52%*

4.2   Results

We adopt the same evaluation metrics that are used by PASCOL VOC, report the mean average precision (mAP) at IoU = 0.5, and average precision of every pollen category at IoU = 0.5. Table 1 shows our experimental results using four models. Faster R-CNN [18] obtains an mAP score of 82.96 %. Similar to Faster R-CNN, SSD300 [12] obtains an mAP score of 82.25%. While our PGD outperforms three other models with an mAP score of 84.52%. For Artemisia, Chenopodiaceae, Moraceae, and Pinaceae, PGD has the best AP scores as 99.11%, 91.94%, 71.15%, and 93.49% respectively. As shown in Fig. 5, the Precision & Recall Curves also indicates PGD outperforms two other state-of-the-art methods. The above experimental results all verify the effectiveness of our PGD.

In addition, we use the k-fold cross-validation method to perform 20 trials on the pollen dataset, and use mAP as a parameter to perform the Wilcoxon Signed Rank Test. We perform bilateral tests on PGD and Faster R-CNN, PGD and SSD, and the p-values obtained are 1.11×10^{-4} and 9.57×10^{-5}. It proves that there is a significant difference between the results of PGD and these two models.

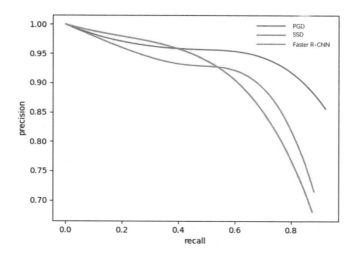

Fig. 5. Precision & Recall Curves. It shows that the PGD model has the highest mAP value and its performance is better than the other two models.

Furthermore, we perform some positive prediction examples by PGD in Fig. 6. Figure 6(a) shows that PGD can still detect pollen well even though pollen grains are covered by impurities. Figure 6(b), Fig. 6(c), and Fig. 6(d) show that PGD has good detection performance for densely distributed pollen grains. Figure 6(e) and Fig. 6(f) show that PGD can perform well when the pollen grain is at the edge of the image and some part of it is missing. From these positive examples, we believe that our PGD has an excellent generalization ability to detect pollen grains in different conditions.

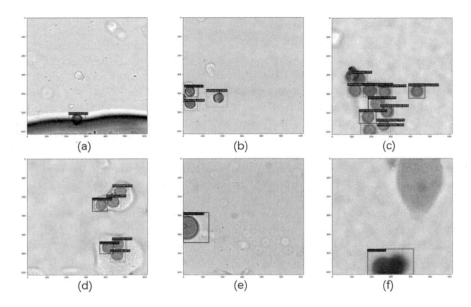

Fig. 6. Positive examples predicted by PGD. PGD still performs well when pollen is densely distributed, covered by debris or blurred in focus.

5   Conclusions

In this paper, we propose an automatic pollen grain detector (PGD) based on deep learning, which aims to automatically detect and classify pollen grains. Specifically, we first establish a pollen dataset. Then, a new structure of PGD, including matching strategy and hard negative mining, is designed to extract subtle pollen features. Besides, the focal loss is applied in our PGD to solve the imbalance problems. Extensive experiments on real-world datasets show that our PGD has obtained state-of-the-art detection results. This interdisciplinary research developed an automatic pollen detection and identification method by integrating the knowledge of botany, medicine, and computer graphics, therefore, enables the timely forecast of pollen grains concentration in the air and guides the prevention of pollen allergy.

Acknowledgement. This study is supported by Beijing Municipal Science and Technology Project with no. Z191100009119013.

References

1. Battiato, S., Ortis, A., Trenta, F., Ascari, L., Politi, M., Siniscalco, C.: Detection and classification of pollen grain microscope images. In: Proceedings of the IEEE/CVF Conference on Computer Vision and Pattern Recognition Workshops, pp. 980–981 (2020)

2. Carrión, P., Cernadas, E., Gálvez, J.F., Damián, M., de Sá-Otero, P.: Classification of honeybee pollen using a multiscale texture filtering scheme. Mach. Vis. Appl. **15**(4), 186–193 (2004)
3. Chica, M.: Authentication of bee pollen grains in bright-field microscopy by combining one-class classification techniques and image processing. Microsc. Res. Tech. **75**(11), 1475–1485 (2012)
4. Dai, L.P., Lu, C.: The pollen and its measurement technique in spring. Meteorol. Mon. **26**(12), 49–52 (2000)
5. D'amato, G., et al.: Pollen-related allergy in Europe. Allergy **53**(6), 567–578 (1998)
6. Daood, A., Ribeiro, E., Bush, M.: Pollen grain recognition using deep learning. In: Bebis, G., et al. (eds.) ISVC 2016. LNCS, vol. 10072, pp. 321–330. Springer, Cham (2016) . https://doi.org/10.1007/978-3-319-50835-1_30
7. Girshick, R.: Fast R-CNN. In: Proceedings of the IEEE International Conference on Computer Vision, pp. 1440–1448 (2015)
8. Imran, A., Li, J., Pei, Y., Akhtar, F., Wang, Q.: Cataract detection and grading with retinal images using SOM-RBF neural network. In: 2019 IEEE Symposium Series on Computational Intelligence (SSCI), pp. 2626–2632 (2019)
9. Imran, A., Li, J., Pei, Y., Yang, J.J., Wang, Q.: Comparative analysis of vessel segmentation techniques in retinal images. IEEE Access **7**, 114862–114887 (2019)
10. Lin, T.-Y., Goyal, P., Girshick, R., He, K., Dollár, P.: Focal loss for dense object detection. In: Proceedings of the IEEE International Conference on Computer Vision, pp. 2980–2988 (2017)
11. Liu, H.: Protection guidance for patients with pollen allergy. Lishizhen Med. Materia Medica Res. **17**(006), 1091–1091 (2006)
12. Liu, W., et al.: SSD: single shot multibox detector. In: Leibe, B., Matas, J., Sebe, N., Welling, M. (eds.) ECCV 2016. LNCS, vol. 9905, pp. 21–37. Springer, Cham (2016). https://doi.org/10.1007/978-3-319-46448-0_2
13. Liu, Z., et al.: Detection of Dermatophagoides farinae in the dust of air conditioning filters. Int. Archiv. Allergy Immunol. **144**(1), 85–90 (2007)
14. Mahmood, T., Li, J., Pei, Y., Akhtar, F., Rehman, K.U.: A brief survey on breast cancer diagnostic with deep learning schemes using multi-image modalities. IEEE Access **8**, 165779–165809 (2020)
15. Marcos, J.V., et al.: Automated pollen identification using microscopic imaging and texture analysis. Micron **68**, 36–46 (2015)
16. Nguyen, N.R., Donalson-Matasci, M., Shin, M.C.: Improving pollen classification with less training effort. In: 2013 IEEE Workshop on Applications of Computer Vision (WACV), pp. 421–426. IEEE (2013)
17. Nikolov, D.N., Tsankova, D.D.: Features extraction for pollen recognition using Gabor filters. Food Sci. Appl. Biotechnol. **1**(2), 86–95 (2018)
18. Ren, S., He, K., Girshick, R., Sun, J.: Faster R-CNN: towards real-time object detection with region proposal networks. arXiv preprint arXiv:1506.01497 (2015)
19. Rodriguez-Damian, M., Cernadas, E., Formella, A., Sa-Otero, R.: Pollen classification using brightness-based and shape-based descriptors. In: Proceedings of the 17th International Conference on Pattern Recognition, ICPR 2004, vol. 2, pp. 212–215. IEEE (2004)
20. Rodriguez-Damian, M., Cernadas, E., Formella, A., Fernández-Delgado, M., De Sa-Otero, P.: Automatic detection and classification of grains of pollen based on shape and texture. IEEE Trans. Syst. Man Cybern. Part C (Appl. Rev.) **36**(4), 531–542 (2006)

21. Sun, J., et al.: A deep learning method for MRI brain tumor segmentation. In: Hung, J., Yen, N., Chang, J.W. (eds.) FC 2019. LNCS, vol. 551, pp. 161–169. Springer, Singapore (2020). https://doi.org/10.1007/978-981-15-3250-4_19
22. Travieso, C.M., Briceño, J.C., Ticay-Rivas, J.R., Alonso, J.B.: Pollen classification based on contour features. In: 2011 15th IEEE International Conference on Intelligent Engineering Systems, pp. 17–21. IEEE (2011)
23. Zhang, L., et al.: Automatic cataract detection and grading using deep convolutional neural network. In: 2017 IEEE 14th International Conference on Networking, Sensing and Control (ICNSC), pp. 60–65 (2017)

Hiding of Personal Information Areas Through a Dynamic Selection Strategy

Sang-Hong Lee⬥ and Seok-Woo Jang$^{(\boxtimes)}$ ⬥

Anyang University, Anyang 14028, Republic of Korea
shleedosa@anyang.ac.kr

Abstract. Video content with exposed personal information is freely distributed to the general public through the Internet without user approval, which is a problem. In this paper, we propose a method of detecting areas representing personal information from continuous color images through an artificial neural network and hiding the detected target area appropriately for the surrounding environment. The proposed method first applies a color model and deep learning to robustly detect a target area representing personal information exposed from an image. The detected area is then tracked quickly based on position prediction and simultaneously blocked. In this paper, the hiding of the target object is performed by adaptively selecting one of the mosaic processing, image blurring, and virtual object insertion techniques in consideration of surrounding conditions. Experimental results show that the proposed method accurately extracts the personal information area, efficiently tracks the extracted area, and dynamically selects the hiding technique. The method proposed in this paper is expected to be useful in related fields such as object detection, big data analysis, and biometric recognition.

Keywords: Image content · Adaptive selection · Object hiding

1 Introduction

Due to the advent of digital cameras with excellent performance and high quality, the development of large-capacity storage devices that are inexpensive and very fast, and information delivery trends from text data-oriented to video data-oriented, a large amount of high-quality color video content is rapidly spreading [1]. These video contents are usefully used for big data analysis in a variety of practical applications, such as IoT-based building monitoring, artificial intelligence-based traffic control, human computer integration-based video security, and computer vision-based object tracking [2].

However, video content including personal information such as a person's face, a specific part of an exposed body, and a resident registration number is also a problem because it is freely distributed through the Internet without any restrictions. In particular, the psychological pain of those who learn that their personal information has been exposed to a large number of unspecified people will be very significant, and the exposed personal information may be abused for purposes such as sending spam text messages and spam e-mails.

© The Author(s), under exclusive license to Springer Nature Singapore Pte Ltd. 2022
J. C. Hung et al. (Eds.): FC 2021, LNEE 827, pp. 45–51, 2022.
https://doi.org/10.1007/978-981-16-8052-6_5

Therefore, research is needed to effectively protect the exposure of unwanted personal information to the outside world by automatically hiding only the target areas where personal information is exposed from the images being entered into the system through image blurring effects, block-based mosaic processing, and virtual object insertion [3].

Existing studies of detecting target areas from received video content and hiding detected target areas can be found in relevant literature. However, the existing hiding methods are still low in completeness, and contain various restrictions on the surrounding environment and circumstances. In particular, a technique for presenting multiple hiding methods and adaptively selecting the surrounding contextual hiding method among these methods, has not yet been proposed in the field of image security.

Therefore, in this paper, a target area containing personal information, such as a human face, in the input image is robustly detected using a deep learning technique, and mosaic processing, image blurring, and virtual object insertion are performed while efficiently tracking the detected target area. We present a new approach to hiding by dynamically selecting the most suitable method among the techniques. Figure 1 shows the overall flow chart of the personal information hiding algorithm through the dynamic selection strategy proposed in this paper.

Therefore, in this paper, we present a novel approach to robustly detect target areas containing personal information, such as human faces, in received images, and dynamically select the most suitable method among mosaic processing, image blurring, and virtual object insertion techniques while efficiently tracking the detected target areas. Figure 1 shows the overall flowchart of the privacy domain-hiding algorithm through the dynamic selection strategy proposed in this paper.

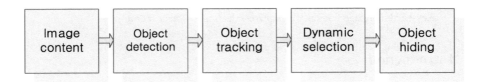

Fig. 1. Overall flowchart of the suggested method.

As shown in Fig. 1, the method proposed in this paper first accurately detects a target area containing personal information from color images continuously input using a deep learning technique. Then, the detected area is quickly tracked by applying a position prediction algorithm. Finally, the target area being tracked is dynamically blocked by adaptively selecting a suitable method from among mosaic processing, image blurring, and virtual object insertion techniques in consideration of the surrounding environment.

2 Object Detection and Tracking

In this paper, the human face area is set as the target area representing the exposed personal information. In the proposed method, the RGB color space of the input image is first converted into the YC_bC_r color space [4]. An elliptical skin color distribution model adaptive to the input image, defined as in Eq. (1), is then generated through learning, and only skin color pixels are extracted by applying the generated model to the image.

$$\frac{(x - ec_x)^2}{a^2} + \frac{(y - ec_y)^2}{b^2} = 1 \tag{1}$$

$$\begin{bmatrix} x \\ y \end{bmatrix} = \begin{bmatrix} \cos \theta(C_b' - C_x) + \sin \theta(C_r' = -c_y) \\ -\sin \theta(C_b' - C_x) + \cos \theta(C_r' - c_y) \end{bmatrix}$$

Subsequently, in the algorithm proposed in this paper, a deep learning technique [5] is applied to accurately detect only the face area from the skin color area acquired in the previous step. In the deep learning structure used in the proposed system, the image pyramid is created in six steps. When finally detecting the target area, the final result is derived by summing the detection results of each step of the pyramid.

In the deep learning model, max-margin object detection (MMOD) optimizes all sub-windows without performing sub-sampling. In other words, the target region is detected by applying the window scoring function F(x, y) to all sub-windows [6]. In Eq. (2), ϕ denotes a feature vector obtained from the moving window position r of the image x. w is a weight vector that performs learning to reduce false positives. By applying Eq. (2), a result corresponding to the detection score can be obtained. Finally, the location of the target area is finally extracted by summing the areas with the highest score in the hierarchical pyramid image.

$$y^* \arg \max_{y \in Y} F(x, y) = \arg \max_{y \in y} \sum_{r \in y} <w, \phi(x, r)> \tag{2}$$

In this paper, the target region detected using deep learning is quickly tracked by applying the Kernelized Correlation Filters (KCF) technique [7]. In general, KCF is based on the idea of a typical correlation filter, and it is known that it produces good performance by significantly improving the computational speed using a kernel technique and a circulant matrix.

$$\alpha = F^{-1}\left(\frac{F(y)}{F(k(x_1, x_2, \sigma) + \lambda)}\right) \tag{3}$$

$$k(x_1, x_2, \sigma) = \exp\left(-\frac{1}{\sigma^2}\left(\|x_1\|^2 + \|x_2\|^2\right) - 2F^{-1}(F(x_1) \oplus F^*(x_2))\right)$$

In this study, when a face area containing personal information is detected in a color image that is continuously input, the target area is learned for stable tracking by using the corresponding image and the location of the detected personal information area. Learning in this study is carried out as Eq. (3). In Eq. (3), x denotes the current frame, which is a training image, and y denotes the target area to be tracked. λ represents a parameter to prevent overfitting. α means the model of the target area to be tracked. k denotes the kernel matrix. In the approach presented in this study, the target area is continuously tracked from the next input color image using the model α of the target area obtained by learning.

3   Dynamic Hiding Strategy

In this paper, using a deep learning algorithm, the target area including the initially detected personal information is continuously tracked through KCL while adaptively hiding at the same time. In other words, the best method suitable for target area hiding is selected from among the three hiding methods, namely mosaic processing, image blurring, and virtual object insertion, in consideration of user preference and surrounding environment. Then, the exposed personal information area can be effectively protected by blocking the target area with the selected hiding technique.

In the mosaic-based hiding technique, a grid-shaped mosaic [8] is created, and then the generated mosaic is naturally overlapped on the target area detected in the previous step, so that the area including personal information is not exposed to the outside. The proposed method first obtains the minimum enclosing rectangle (MER) of the extracted target area to cover the target area with a mosaic. Then, the corresponding region of the image contained within the obtained MER is divided into blocks with the same horizontal and vertical sizes in equal units, and then each block is filled with color values generated by the mosaic processing technique.

The image blur-based hiding technique covers the exposed personal information area by applying a two dimensional Gaussian function [9] to an image. Image blurring is performed through convolution. Blurring using a Gaussian function used in the proposed system is one of the representative hiding techniques. In general, if blurring using a Gaussian function is applied to the target area where personal information is exposed, it is possible to obtain a hiding result with a relatively small sense of heterogeneity with the area located around the target area.

The virtual object-based hiding blocks the exposed personal information area more intimately by overlapping the character object on the detected target area. However, most of the existing methods of inserting virtual objects are very inefficient because they have to manually position the virtual object at a desired point in the image and manually adjust the virtual object several times according to the size and shape of the area to cover the virtual object. Therefore, in this paper, the location and size of the designated virtual object can be automatically adjusted according to the detected target area.

In the proposed method, image blurring, mosaic processing, and virtual object insertion techniques are dynamically selected in consideration of user preferences and surrounding conditions. In this paper, first, when a user selects one of the three hiding techniques, the target area is hided with the selected technique. If the user's preference values for the three hiding techniques are similar, a hiding technique is adaptively selected by reflecting the surrounding environment in which the image was captured. To this end, in this study, the complexity metric of the surrounding environment as shown in Eq. (4) is calculated using the weighted sum of distance features and lighting features.

$$\Omega(\alpha, \beta; t) = \alpha \times \left(1 - \left|\frac{F_{illum}(t)}{\|F_{illum}(t)\|} - \frac{1}{2}\right| \times 2\right) + \beta \times \left(1 - \left|F_{dis\tan ce}(t) - \frac{1}{2}\right| \times 2\right)$$

(4)

$$F_{illum}(t) = \frac{1}{256 \times P \times Q} \times \sum_{i=0}^{P-1}\sum_{j=0}^{Q-1} Y(i,j)$$

$$F_{distance}(t) = \frac{1}{P \times Q} \times \{MER_W \times MER_H\}$$

4 Experimental Results

The computer used for the experiment consists of an Intel Core(TM) i7-6700 3.4 Ghz CPU, 16 GB main memory, 256 GB SSD, and a Galaxy Geforce GTX 1080 Ti graphics card equipped with NVIDIA's GPU GP104. On the personal computer used, Microsoft's Windows 10 operating system was installed. In addition, Microsoft's Visual Studio version 2017 was used as the integrated development environment of the introduced method, and the proposed algorithm was developed using the OpenCV computer vision library and Dlib C++ library. In this paper, various types of images including the exposed personal information area were collected and used to evaluate the performance of the proposed hiding algorithm. Most of these image data were captured in a natural environment with no specific constraints.

$$M_{accuracy} = \frac{TARGET_{hiding}}{TARGET_{total}} \times 100(\%)$$

(5)

In this paper, the performance of the proposed target area hiding method was quantitatively evaluated in terms of accuracy. In this study, the same metric as Eq. (5), defined as the ratio of the number of target regions accurately detected and blocked in the input color image and the number of target regions contained in the entire image data, was used. $TARGET_{hiding}$ in Eq. (5) represents the number of target regions that are correctly hided using the proposed algorithm. In addition, $TARGET_{total}$ means the total number of target areas representing personal information belonging to the image data to be tested. The quantitative measure used in this study is defined as a percentage.

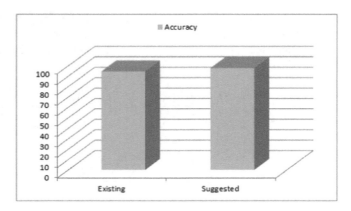

Fig. 2. Performance comparison.

Figure 2 presents a graph of the performance measurement results of the target area hiding method in terms of accuracy. As can be seen in Fig. 2, the algorithm using the dynamic selection strategy proposed in this paper blocks the target area including personal information more effectively than the existing method.

5   Conclusion

In this paper, we propose an algorithm that robustly detects the target area containing personal information excluding the background part from various color images, and then effectively protects the detected target object area by a blocking method suitable for the surrounding situation. In the method, the target object area containing personal information is robustly segmented based on the color of the human skin. Then, a morphology operation is applied to the selected skin pixel image to remove noise, and labeling is performed to extract individual skin regions. Then, only the human face region is robustly detected using a deep learning-based object detection algorithm from the extracted skin region. Then, by effectively covering the detected face area by selecting a blocking method suitable for the surrounding situation, it is possible to protect the personal information from being exposed to the outside.

In the future, we plan to verify the robustness of the system by applying the deep learning-based multi-level object domain blocking method proposed in this paper to many images captured in more diverse environments. In addition, the proposed blocking algorithm will be more stabilized by tuning various parameters used in the proposed system.

Acknowledgement. This work was supported by the National Research Foundation of Korea (NRF) grant funded by the Korea government (MSIT) (2019R1F1A1056475).

References

1. AlSkaif, T., Bellalta, B., Zapata, M.G., Barcelo Ordinas, J.M.: Energy efficiency of MAC protocols in low data rate wireless multimedia sensor networks: a comparative study. Ad Hoc Netw. **56**, 141–157 (2017)
2. Preishuber, M., Hutter, T., Katzenbeisser, S., Uhl, A.: Depreciating motivation and empirical security analysis of chaos-based image and video encryption. IEEE Trans. Inf. Forensics Secur. **13**(9), 2137–2150 (2018)
3. Liu, Y., Chen, J.: Unsupervised face frontalization for pose-invariant face recognition. Image Vis. Comput. **106**, 1–10 (2020)
4. Zhu, S.Y., et al.: High-quality color image compression by quantization crossing color spaces. IEEE Trans. Circuits Syst. Video Technol. **29**(5), 1474–1487 (2019)
5. Amin, S.U., Alsulaiman, M., Muhammad, G., Mekhtiche, M.A., Hossain, M.S.: Deep learning for EEG motor imagery classification based on multi-layer CNNs feature fusion. Future Gener. Comput. Syst. **101**, 542–554 (2019)
6. Chrysos, G.G., Antonakos, E., Snape, P., Asthana, A., Zafeiriou, S.: A comprehensive performance evaluation of deformable face tracking in-the-wild. Int. J. Comput. Vis. **126**(2), 198–232 (2018)
7. Wang, J., Liu, W., Xing, W., Zhang, S.: Visual object tracking with multi-scale superpixels and color-feature guided kernelized correlation filters. Sig. Process.: Image Commun. **63**, 44–62 (2018)
8. Guo, D., Tang, J., Cui, Y., Ding, J., Zhao, C.: Saliency-based content-aware lifestyle image mosaics. J. Vis. Commun. Image Represent. **26**, 192–199 (2015)
9. Wang, R., Li, W., Zhang, L.: Blur image identification with ensemble convolution neural networks. Sig. Process. **155**, 73–82 (2019)

An Enhanced Adaptive Neighbourhood Adjustment Strategy on MOEA/D for EEG Signal Decomposition-Based Big Data Optimization

Meng Xu[1], Yuanfang Chen[2], Dan Wang[1(✉)], and Jiaming Chen[1]

[1] Faculty of Information Technology, Beijing University of Technology, Beijing, China
{xumeng,billchen}@emails.bjut.edu.cn,wangdan@bjut.edu.cn
[2] Beijing Institute of Mechanical Equipment, Beijing, China

Abstract. Multi-objective evolutionary algorithms (MOEAs) have shown good performance in many complex mathematics benchmarks and real applications problems. However, MOEAs still have some challenges in the big data optimization problem with thousands of variables. In previous studies, an adaptive neighbourhood adjustment strategy on MOEA/D had a good optimization performance in CEC2009 competition test instances, in which variables had high dimensions. In this paper, an enhanced adaptive neighbourhood adjustment strategy on MOEA/D, called MOEA/D-EANA, has been proposed for 2015 EEG signal decomposition-based big data optimization. It combines fitness-rate-rank based multi-armed bandit operators (FRRMAB) selection strategies with the advantages of adaptive neighbourhood adjustment strategies. These operators enhance the diversity of solutions in our previous algorithm. A set of big data optimization problems, including six single objective problems and six multi-objective problems, are tested in the experiments. Computational results show that our proposed algorithm achieves promising performance on all test problems.

Keywords: Multi-objective evolutionary algorithms · Big data optimization · Adaptive neighbourhood adjustment

1 Introduction

Big data optimization problems are a large challenge in the real world, particularly in the massive and complex data set [1,2]. This problem consists of hundreds of thousands of decision variables, various mathematical properties of the objective functions and require considerable computational time. Big data optimization problems are a hot topic for multiple domains such as health care, financial, logistics, and many others.

The characteristics of big data bring challenges to some computational problems. Abbas et al. have recently published one of the big data optimization

J. C. Hung et al. (Eds.): FC 2021, LNEE 827, pp. 52–62, 2022.
https://doi.org/10.1007/978-981-16-8052-6_6

problems in artifacts removal from electroencephalographic (EEG), introduced in Big Data Com-petition 2015 [3,4]. The natural EEG signal is desired to be decomposed into two different parts using algorithms. The first part of the source signal and the latter part corresponds to the artifacts or noise. So researchers proposed many multi-objective optimization approaches to model this decomposition and presented two metrics to quantify the amount of EEG information lost during the cleaning process [5–8].

The Multi-objective based on decomposition (MOEA/D) transforms a MOP into a set of single-objective optimization subproblems and optimizes them simultaneously. Using each subproblem neighbourhood to the subproblem to optimize itself makes it possible to solve MOPs faster than other multi-objective algorithms [9]. An adaptive neighbourhood adjustment strategy of MOEA/D (MOEA/D-ANA) is proposed [10]. It can be considered a local search approach, which had better performance on convergence while not considerate enough about the diversity of solutions. Our proposed algorithm combined composite operators selection with adaptive neighbourhood adjustment strategy and balanced the performance on convergence and diversity. This algorithm we called MOEA/D-EANA. It will help to improve the performance of big data optimization problems.

This paper is organized as follows. In Sect. 2, the related work of our previous work is shown. In Sect. 3, the MOEA/D-EANA is described in detail in Sect. 3. In Sect. 4, the problem of big-data optimization is introduced. In Sect. 5, the results on six datasets are reported. In Sect. 6 summarizes the work in this paper.

2   Related Work

MOEA/D is a classical algorithm adapted to solve application problems such as medical care, logic, task scheduling, etc. This algorithm uses the neighbourhood of each subproblem, helps each subproblem faster find the solutions of MOP. In the previous work, the neighbourhood size had proved to affect the performance of MOEA/D, and its variants [9,10]. The large neighbourhood size helps improve the performance of exploration. On the contrary, the small neighbourhood size improves the exploitation [11]. In original MOEA/D, each weight vector corresponds to each subproblem(individual). Individual density could be obtained through the European distance of weight vector. For a subproblem, the neighbourhood size is fixed. In the evolution process, the number of neighbours around each individual will not be fixed. When a subproblem obtained good solutions, it means the more individual evolve in their directions, the density of this individual is dense, and vice versa. This fixed neighbourhood size did not fit the evolution process. So we proposed the adaptive neighbourhood adjustment strategy. So the adaptive neighbourhood size could help enhance the performance of MOEA/D, the details of this algorithm could be found in [10].

This strategy is introduced in Algorithm 1, where id is the diversity of individual, and T is neighbourhood size. If $id = 1$, this means that a subproblem corresponds to an individual, the neighbourhood size is not changed. If id is small, we think that the density of individual is sparse, so these individual needs enlarge the neighbourhood size. If id is too large, the density of individuals is dense, so it reduces the neighbourhood size.

Algorithm 1. The strategy of adaptive neighbourhood size adjustment

---

Input:
    the current point of each subproblem x^i, i=1,2,...N;
    the weight vectors λ;
    initialize $id^i = 0, i = 1, 2, ..., N$;
Output:
    the neighbourhood size T;
1: Compute the max value of objectives by the $z^{nad} = \max\limits_{1\leq j\leq m, 1\leq i\leq N}\{f_j(x^i)\}$;
2: Compute the min value of objectives by the $z^{min} = \min\limits_{1\leq j\leq m, 1\leq i\leq N}\{f_j(x^i)\}$;
3: **for** each $i \in [1, N]$ **do**
4:     **for** each $j \in [1, N]$ **do**
5:         Compute the vertical distance from the solution and weight vector by equation (1). Before obtaining the value of vertical distance, the objectives need normalization.;

$$D(x, \lambda_i) = \left\| \overline{f}(x) - \frac{\lambda^T f(x)\lambda}{\lambda^T \lambda}\lambda \right\|, where \overline{f}(x) = \frac{f(x) - z^{min}}{z^{min} - z^{nad}}; \quad (1)$$

6:     **end for**
7:     Get the number of individuals in a subproblem around:
        $k = min \quad D(x, \lambda_i),$
        $id^k + +$
8:     **if** $id^i < 1$ **then**
9:         select a large neighbourhood size, $T^{i+1} = T^i + N^*$;
10:     **else** $\{id^i > 1\}$
11:         select a small neighbourhood size, $T^{i+1} = T^i - N^*$;
12:     **else**
13:         $T^{i+1} = T^i$
14:     **end if**
15: **end for**
16: **return** T;

---

3   The Description of MOEA/D-EANA

The advantage of the adaptive neighbourhood adjustment strategy is that it is a problem specific by changing the neighbourhood size for each subproblem's performance at different times. However, in big data optimization, too many variables can lead to aggregation of solutions, so the composite selection operators is used to improve the diversity of the previous algorithm.

The composite operator selection strategy could lead to more diverse solutions compared to the single genetic strategy. MOEA/D-EANA selected three candidate operators: DE/rand/1, DE/rand/2 [13,14,19] and CMX (Center of Mass Crossover) [15]. DE/rand/1 is the most traditional operator, DE/rand/2, increasing population diversity. CMX operator can search the places where the DE difference operator cannot explore and increase population diversity. In the evolutionary process, each subproblem is selected based on the probability of successful selection in the previous times to choose the operator operation to be used this time. This strategy selected different operators for different subproblems is beneficial to the evolution of the whole population. The details of the three operators expression in Table 1.

Table 1. Computational results achieved by MOEA/D-EANA, MOEA/D-ANA, and the baseline algorithm for single objective optimization problems.

Operator	Formula expression
DE/rand/1	$v_{i,g} = x_{r_1,g} + F \times (x_{r_2,g} - x_{r_3,g})$
DE/rand/2	$v_{i,g} = x_{r_1,g} + F \times (x_{r_2,g} - x_{r_3,g}) + F \times (x_{r_4,g} - x_{r_5,g})$
CMX	$o = \frac{1}{3} \sum_{i=1}^{3} x_{i,g},\ v_{i,g} = 2o - x_{r_1,g}$

where $F = 0.5$, $X_{r_1,g}$, $X_{r_2,g}$, $X_{r_3,g}$, $X_{r_4,g}$, and $X_{r_5,g}$ are different solutions randomly selected from neighbourhood, $V_{i,g}$ is the solution of offspring.

In this paper, the fitness-rate-rank based multi-armed bandit operators selection strategy (FRRMAB) is used to select a good operator in each iterations [12]. In the FRRAMB, the fitness improvement rate saved in a sliding window similar to the queue implementation. It designs a matrix where the first row is used to store the operator numbers used in the previous generations, such as $op_1, op_2 and op_3$, and the second row is used to record the fitness improvement rate (FIR) of the offspring generated by each operator, and the FIR can be used with the following mathematics:

$$FIR = \frac{pf_i - cf_i}{pf_i} \tag{2}$$

where pf_i is the fitness of the parent and cf_i is the fitness of the offspring. The number of each operator and FIR entered into the matrix in pairs.

In the initialization of the algorithm, this matrix is set to be empty. In the evolution process, the number operators and fitness improvement rate (FIR) are entered in this matrix, as shown in the Fig. 1. When the matrix is filled, the first column of data will be lost, and the data of the whole matrix will move forward one unit, and the last column will be added with new data. The reason for discarding the first column is that it has the least impact on the operator number in the recent generation. Then the operator number corresponding to FIR is

Fig. 1. The matrix of operators and FIR

counted and evaluated by the fitness reputation ratio (FRR), the mathematical expression of FRR is as follows:

$$reward_i = \sum_{i=1}^{W} FIR_i \quad i = 1, 2, 3 \tag{3}$$

$$FRR_i = reward_i / \sum_{j=1}^{n} reward_j \tag{4}$$

Therefore, the operation selection to be selected next time can be expressed by the following formula:

$$op_i = \arg\max_{i=\{1,2,3\}} \left[FRR_i + C \times \sqrt{\frac{2 \times \ln \sum_{j=1}^{3} n_j}{n_i}} \right] \tag{5}$$

where the C is the expansion factor. The basic scheme of MOEA/D-EANA is described as Algorithm 2.

Algorithm 2. The scheme of MOEA/D-EANA

Initialize:
    the weight vector λ^i;
    the solution x^i;
    the ideal point $z_j{}^*, z_i = \min\{f_i(x^1), \ldots, f_i(x^N)\}$;
    initialize neighborhood size: $T = (T^1, T^2, \cdots, T^n)$;
Evolution:
1: **for** each $i \in [1, N]$ **do**
2:    To produce new offspring y from one of three operators by the Eq. 5
3:    Update the reference point z_i
4:    Evaluation the fitness and update FIR_i in the sliding windows.
5:    Using Algorithm 1 to select neighbourhood size from the neighbourhood mating pool for each subproblems i.
6: **end for**
7: Stopping criterion: if the stopping criteria is satis-fied, the stop and output the Pareto set and Pareto fitness, otherwise, go to step 1
8: **return** PF and PS;

---

4   Big Optimization Problems

The background of CEC 2015 Competition is about the processing of electroen-cephalographic (EEG) signals [16,17]. There are three different datasets (D4, D12, and D19) in the competition, which are based on different time series. The length of each time series is 256. If the dataset has 2 time series, the corre-sponding problem will have 512 variables to be optimized. The number of time series used in the competition are 4, 12, and 19, and the corresponding opti-mization problems have 1024, 3072, and 4864 variables, respectively. For the three datasets, a noise component is added to obtain three new datasets called D4N, D12N, and D19N, respectively. Thus, there are six single objective opti-mization problems and six MOPs. In this paper, we try to use the proposed MOEA/D-EANA to solve these problems.

The ICA [4,18] model of this problem can be shown below. It assumed that the X is a large matrix, its dimension is $N * M$, N presents the number of time series, and M is the length of each time series. In this competition, each time series M is of length 256, and the number of times series N are 4,12 and 19. S has the same dimension with N. A is a small dimension $N * N$ as a transformation matrix.

$$X = A \times S \tag{6}$$

This problem aims to decompose S into two matrixes with the same dimension: S_1 and S_2, which have the same dimensionality. Then we can get $S = S_1 + S_2$ and $X = A \times S_1 + A \times S_2$, It assumes that C is the Person correlation coefficient between X and

$$C_{i,j} = \frac{\text{covar}(X, A \times S_1)}{\sigma(X) \times \sigma(A \times S_1)} \tag{7}$$

where $cov(.)$ is the covariance matrix, and σ is the standard deviation. so this model abstracted as bi-objective function.

$$\min f_1 = \frac{1}{N^2 - N} \sum_{i,j \neq i} C_{i,j}^2 + \frac{1}{N} \sum_i (1 - C_{i,i})^2 \tag{8}$$

$$\min f_2 = \frac{1}{M \times N} \sum_{i,j} (S_{i,j} - S_{1i,j}) \tag{9}$$

The first objective function is to generate C to maximize the diagonal elements of C, while minimizing non-diagonal elements to zeros. The second objective function is used to minimize the distance of S and S_1. Each element of S_1 varies from -8 to 8. In addition, the number of variables is 1024, 3072 and 4864.

5   Experiment and Results

In this section, we experimented with verifying our proposed algorithm. These experiments are executed on a PC desktop with a 2.3 GHz CPU and I5 processor. No server platform has been performed.

5.1   The Parameters of Big Data Optimization Problems

For big data optimization problems, the datasets had six different EEG records with several records contain noise, the white noise level of 0.1, which namely D4, D4N, D12, D12N, D19, D19N. We compared MOEA/D-EANA with MOEA/D-ANA and NSGA-II(baseline). The parameter setting as follows:

The population size N is 50;
Each algorithm will run 10 times independently;
Stopping condition: all algorithms function evaluation is 100000;
The ideal points are $z^* = [1.01, 22.34]$;
F and CR are 0.5;
$T = 0.1 \times N$ and $N^* = 0.02 \times N$;
The window sliding $W = 20$.

5.2   The Performance Metrics

The score function is the metric of big optimization problems, five solutions be sampled uniformly from the obtained Pareto set, including the two extreme solutions. The score function calculated based on the 2015 Big Data Optimization Competition, the score metric value is larger, the better performance of the S_1 close to S, the score of can be expressed by the following formula:

$$score = \begin{cases} bfv_{bl} - bfv & if\ bfv < bfv_{bl} \\ -1000(bfv_{bl} - bfv) & otherwise \end{cases} \tag{10}$$

where bfv and bfv_{bl} are the best fitness value obtained by MOEA/D-EANA with the baseline algorithm.

5.3   The Results of Single Objective and Multi-objective Objective

Table 2 lists the computational results achieved by MOEA/D-EANA, MOEA/D-ANA, and the baseline algorithm (NSGA-II) for single optimization problems, and mean value is the average best function value. The proposed MOEA/D-EANA achieves much better performance from the results than MOEA/D-ANA and NSGA-II on all test sets. MOEA/D-ANA performs better than NSGA-II.

For multi-objective optimization, five points are selected from the obtained Pareto front, the proposed algorithm compared to the baseline algorithm (NSGA-II), if the solution achieved by our method has better performance than the baseline algorithm, the score is the mean distance. If the compared solution

Table 2. Computational results achieved by MOEA/D-EANA, MOEA/D-ANA, and the baseline algorithm for single objective optimization problems.

Problem	MOEA/D-EANA mean	MOEA/D-ANA mean	Baseline(NSGA-II) mean
D4	**6.16E−02**	9.88E−01	1.87
D4N	**5.92E−02**	8.71E−01	1.74
D12	**1.93E−03**	8.46E−02	2.93
D12N	**1.82E−03**	7.23E−02	2.82
D19	**1.45E−03**	2.54E−03	3.19
D19N	**1.24E−03**	2.62E−03	3.17

dominates our solution, a large negative constant value (−1000) is multiplied by the distance. The score value is equal to the value of distance. Therefore, a higher score means that the algorithm is better than the baseline algorithm.

The scores of MOEA/D-EANA and MOEA/D-ANA based on the baseline algorithm(NSGA-II) is shown in Table 3, the MOEA/D-EANA achieve better performance than MOEA/D-ANA and NSGA-II. The $''N''$ which containing noise is lower than not containing noise EEG signal, it means that as the noise can affect the algorithm's performance, the score gets lower. As the number of variables increases, the lower the score, which means that the algorithm becomes increasingly difficult to solve the problem [20]. Figure 2 compares five solutions selected from the Pareto set of MOEA/D-EANA and the baseline algorithm. It can be seen that MOEA/D-EANA achieves a better score value than the baseline.

Table 3. Scores obtained by MOEA/D-EANA and MOEA/D-ANA for multi-objective optimization problems.

Problem	MOEA/D-EANA	MOEA/D-ANA
D4	**2.26E+01**	2.25E+01
D4N	**2.22E+01**	2.18E+01
D12	**2.10E+01**	2.07E+01
D12N	**1.89E+01**	1.75E+01
D19	**1.57E+01**	1.50E+01
D19N	**1.49E+01**	1.20E+01

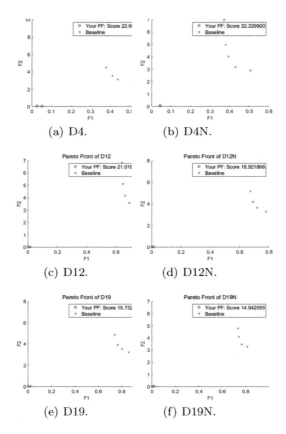

Fig. 2. The comparison score of five point from the Pareto set of MOEA/D-EANA and the baseline algorithm (NSGA-II).

6   Conclusion

In this paper, the MOEA/D-EANA of the framework was proposed for optimizing the big data 2015 benchmark problems with both single and multi-objective problems, which ensembles the adaptive neighbourhood size adjustment and composite operator selection. In the experiments, six EEG datasets from the big data optimization problems are used to validate the performance of MOEA/D-EANA. The experimental results show that MOEA/D-EANA had better performance than the baseline algorithm and MOEA/D-ANA. For future work, the technology of dimension reduction are used in MOEA/D-EANA. It will could reduce the number of variables and computational time in big data optimization problem.

References

1. Hassanien, A.E., Azar, A.T., Snasael, V., Kacprzyk, J., Abawajy, J.H. (eds.): Big Data in Complex Systems. SBD, vol. 9. Springer, Cham (2015). https://doi.org/10.1007/978-3-319-11056-1

2. Abdi, Y., Feizi-Derakhshi, M.R.: Hybrid multi-objective evolutionary algorithm based on search manager framework for big data optimization problems. Appl. Soft Comput. **87**, 105991 (2020)

3. Abbass, H.A. : Calibrating independent component analysis with Laplacian reference for real-time EEG artifact removal. In: Loo, C.K., Yap, K.S., Wong, K.W., Beng Jin, A.T., Huang, K. (eds.) ICONIP 2014. LNCS, vol. 8836, pp. 68–75. Springer, Cham (2014). https://doi.org/10.1007/978-3-319-12643-2_9

4. Goh, S.K., Abbass, H.A., Tan, K.C., Mamun, A.A.: Artifact removal from EEG using a multi-objective independent component analysis model. In: Loo, C.K., Yap, K.S., Wong, K.W., Teoh, A., Huang, K. (eds.) ICONIP 2014. LNCS, vol. 8834, pp. 570–577. Springer, Cham (2014). https://doi.org/10.1007/978-3-319-12637-1_71

5. Zhang, Y., Liu, J., Zhou, M., et al.: A multi-objective memetic algorithm based on decomposition for big optimization problems. Memetic Comput. **8**(1), 1–17 (2016)

6. Majdouli, M.A.E., Bougrine, S., Rbouh, I., et al.: A comparative study of the EEG signals big optimization problem using evolutionary, swarm and memetic computation algorithms. In: Proceedings of the Genetic and Evolutionary Computation Conference Companion, pp. 1357–1364. ACM, Berlin (2017)

7. Bejinariu, S.I., Costin, H., Rotaru, F., et al.: Fireworks algorithm based single and multi-objective optimization. Bull. Polytech. Inst. Iasi Autom. Control Comput. Sci. Sect. **62**(66), 19–34 (2016)

8. Wang, H., et al.: A hybrid multi-objective firefly algorithm for big data optimization. Appl. Soft Comput. **69**, 806–815 (2018)

9. Zhang, Q., Li, H.: MOEA/D: a multiobjective evolutionary algorithm based on decomposition. IEEE Trans. Evol. Comput. **11**(6), 712–731 (2007)

10. Xu, M., Zhang, M., Cai, X., Zhang, G.: Adaptive neighbourhood size adjustment in MOEA/D-DRA. Int. J. Bio-Inspir. Comput. **17**(1), 14–23 (2021)

11. Zhao, S.Z., Suganthan, P.N., Zhang, Q.: Decomposition-based multiobjective evolutionary algorithm with an ensemble of neighborhood sizes. IEEE Trans. Evol. Comput. **16**(3), 442–446 (2012)

12. Li, K., Fialho, A., Kwong, S., et al.: Adaptive operator selection with bandits for a multiobjective evolutionary algorithm based on decomposition. IEEE Trans. Evol. Comput. **18**(1), 114–130 (2014)

13. Das, S., Suganthan, P.N., et al.: Differential evolution: a survey of the state-of-the-art. IEEE Trans. Evol. Comput. **15**(1), 4–31 (2011)

14. Lin, Q., Liu, Z., Yan, Q., et al.: Adaptive composite operator selection and parameter control for multiobjective evolutionary algorithm. Inf. Sci. **339**(C), 332–352 (2016)

15. Khan, W., Zhang, Q.: MOEA/D-DRA with two crossover operators. In: 10th UK Workshop on Computational Intelligence (UKCI), pp. 1–6. IEEE, UK (2010)

16. Goh, S.K., Tan, K.C., Al-Mamun, A., Abbass, H.A.: Evolutionary big optimization (BigOpt) of signals. In: 2015 IEEE Congress on Evolutionary Computation (CEC), pp. 3332–3339. IEEE, May 2015

17. Aslan, S.: A comparative study between artificial bee colony (ABC) algorithm and its variants on big data optimization. Memetic Comput. **12**(2), 129–150 (2020)

18. Goh, S.K., Abbass, H.A., Tan, K.C., Al-Mamun, A.: Decompositional independent component analysis using multi-objective optimization. Soft Comput. **20**(4), 1289–1304 (2015). https://doi.org/10.1007/s00500-015-1587-7
19. Xu, M., Cui, Z., Zhang, M., Zhang, G.: Experimental comparison of different differential evolution strategies in MOEA/D. In: 2017 13th International Conference on Natural Computation, Fuzzy Systems and Knowledge Discovery (ICNC-FSKD), pp. 201–207. IEEE, July 2017
20. Vlahogianni, E.I.: Computational intelligence and optimization for transportation big data: challenges and opportunities. In: Lagaros, N., Papadrakakis, M. (eds.) Engineering and Applied Sciences Optimization. COMPUTMETHODS, vol. 38, pp. 107–128. Springer, Cham (2015). https://doi.org/10.1007/978-3-319-18320-6_7

Detection of HVDC Interference on Pipeline Based on Convolution Neural Network

Zerui Ma[1], Jianqiang Li[2], Jing Li[1(✉)], Xi Xu[2], and Yanan Wang[2]

[1] Faculty of Science, Beijing University of Technology, Beijing, China
leejing@bjut.edu.cn
[2] Faculty of Information, Beijing University of Technology, Beijing, China
lijianqiang@bjut.edu.cn

Abstract. High voltage direct current (HVDC) interference is the stray current discharged from HVDC transmission line to the ground during operation, which will cause great damage to the pipeline. Pipeline corrosion protection personnel use pipeline potential as monitoring data to monitor interference from HVDC in real time and determine the health of the pipeline. At present, the corrosion protection industry relies on human experience to judge HVDC interference, but no machine learning methods have been used to detect HVDC interference. In this study, one-dimensional convolutional neural network (1-D CNN) was used to analyze the time-series data of pipeline potential, and a classifier of HVDC interference was constructed to realize the automatic detection of HVDC interference. The experimental results show that the classification accuracy of 1-D CNN in the pipeline spontaneous potential time series data reaches 91.6%, which is better than the general feature extraction method, and can effectively detect HVCD interference.

Keywords: Convolution Neural Network · HVDC · Pipeline protection

1 Introduction

1.1 Research Background

Oil and gas pipelines are the medium used to transport oil and gas over long distances. Oil and gas resources are generally stored in oil and gas fields or reservoirs, and transported to cities and regions in need through pipelines. Oil and gas pipeline is the blood vessel in the national energy security system, which bears a very important task. Therefore, the safety protection of pipeline is very important. Oil and gas pipelines are buried deep in the ground and are often affected by two factors, one is from man-made mechanical damage, the other is from chemical corrosion [1]. Mechanical damage refers to the damage to the outer wall or internal structure of the underground pipeline caused by construction workers and facilities during the construction process due to the unclear location of the underground pipeline, resulting in pipeline damage and oil and gas leakage. The damage is often accidental, but the damage is enormous. Chemical corrosion is due to the influence of pH value and stray current in the soil, the pipeline and the soil electrochemical reaction, resulting in the oxidation reaction of its metal

J. C. Hung et al. (Eds.): FC 2021, LNEE 827, pp. 63–70, 2022.
https://doi.org/10.1007/978-981-16-8052-6_7

structure caused by corrosion. This damage can occur from the time the pipe is buried in the ground, and is not obvious and usually takes a long time to show up.

In order to be able to detect more secret chemical corrosion, intelligent continuous monitoring devices are generally set along the pipeline to detect potential changes in different positions of the pipeline, so that the potential changes of the pipeline can be detected in real time and the health status of the pipeline can be analyzed in time. There are three types of factors affecting pipeline potential: high voltage direct current transmission line interference (HVDC), direct current track interference and alternating current track interference [2]. Among them, the interference of high-voltage transmission line refers to the interference caused by the current released into the soil during the operation of the high-voltage transmission system to the pipeline. The duration of such interference is short, but the soil potential change is very obvious, which will cause great damage to the pipeline. In this paper, based on the time series data generated by the intelligent detection device, the interference detection of high-voltage transmission lines is realized by using one-dimensional convolutional neural network.

The potential change caused by the interference of HVDC transmission is shown in the Fig. 1. Its characteristic is that in a period of time, the power off potential and the power on potential will produce a relatively large offset to the positive or negative direction, and in this process, the potential remains relatively stable. The fluctuation of HVDC is random, so it is impossible to predict the disturbance through the change of waveform. It can only be monitored in real time through technical means. When the potential waveform changes occur, it can be monitored and recorded, so as to evaluate the corrosion situation of the pipeline.

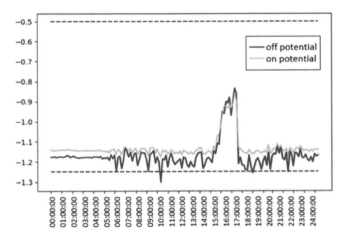

Fig. 1. Potential curve under HVDC interference

1.2 Related Work

At present, the research progress of stray current in oil and gas pipeline is divided into several aspects. In order to evaluate the corrosion of pipelines, it is necessary to analyze the interference sources of real-time detection data and record the relevant electro-chemical parameters of pipelines when the stray current is generated. When interference occurs, parameter changes such as potential are analyzed manually, and researchers rely on professional knowledge and experience to determine the specific cause of stray current generation [3]. Some studies [4, 5] also use signal analysis for data processing and research. For example, denoising and signal separation methods are used to separate potential data generated by AC interference, and Fourier transform method is used to extract the symbolic features of potential changes. In addition, some studies [6, 7] focus on the correlation between stray current and corrosion degree of metal pipes. There is a correlation between stray current data and load during operation of grounding pole, and factors such as chemical properties of soil also have an impact on this. It can be seen that the current research on the stray current of pipelines still remains on the traditional method, and there are few automatic detection of stray current on pipelines, especially the detection of stray current generated by HVDC.

For time series data similar to pipeline potential, there are many methods at present. With the proposal of deep learning, one-dimensional convolutional neural network has become one of the most common methods for feature extraction of time series data. One-dimensional convolutional neural network is widely used in the classification and detection of time series data. For example, in literature [8], one-dimensional convolutional neural network is used to extract the features of seismic waves, so as to extract the special waveforms corresponding to earthquakes from the interference waveforms. In the literature [9], the timing data of electrocardiogram can detect the corresponding sleep apnea events, and the study on electrocardiogram can also judge the health status of the heart, and the study on intelligent diagnosis methods of atrial fibrillation. It is used in the industrial field to analyze the fault state of the machine [10]. It is also applied to power system transient stability assessment [11].

In this study, for the stray current interference signals generated by HVDC, the one-dimensional convolutional neural network is used to extract the features and detect the characteristic waveform of HVDC.

2 Materials and Method

2.1 Dataset

The data in this study were taken from the real-time cathodic protection monitors of an oil and gas pipeline system. There were 240 monitoring points in total for nearly 6 months, and the data were obtained by the intelligent test pile for cathodic protection. In order to detect the health condition of the pipeline in real time, a large number of cathodic protection intelligent test piles are buried along the oil and gas long-distance pipeline, which are used to detect the potential, current and insulation performance of the pipeline. Each test pile is equipped with a wireless communication chip, which can transmit data to the server in real time through the Internet of Things, and then the server will analyze the transmitted data using machine and manual methods to analyze the possible interference problems in the pipeline.

The collected data include pipeline potential, soil environment, pipeline current, etc. Among them, the power-on potential and power-off potential of the pipeline are sensitive to HVDC interference, so we use these two values as the main data sources for detection. The typical curve of the pipe potential after the interference of HVDC is shown in the Fig. 2(a). Point A represents the beginning of interference, C represents the peak, and B represents the end of interference. Figure 2(b) is the potential curve without HVDC interference.

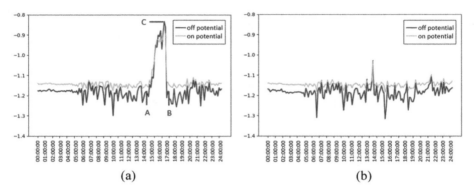

(a)   (b)

Fig. 2. HVDC interference and normal potential curve

2.2   Data Processing

In order to realize the effective analysis of the data, we first clean the data and manually delete the obviously wrong data. These data are usually distributed discretely within the time range and have significant differences with the values of the surrounding moments. In addition, the test pile may send data repeatedly to the remote server, and we deleted the completely duplicated data. The data were marked by professionals in such a way that the potential curve presented by the device was significantly offset if HVDC interference occurred during a certain period of time, and the data marked for that time range were characteristic data.

In order to use the convolutional neural network to extract the features of the data, the data is firstly segmented. Usually, the discharge time of HVDC transmission line is about 30 min-4 h, and a data is collected every 10 min. In order to improve the classification accuracy of the model, we used a random way to divide the time segments. Starting from the first data after 0 o'clock every day, every L data is divided into a time segment until the data of the day is clipped. The criteria for determination are as follows: if the segment contains HVDC data, the segment shall be marked as a positive sample; otherwise, it shall be marked as a negative sample. Finally, the data were artificially screened to eliminate the segments marked with obvious errors in the positive and negative samples. According to different L, we can get different sample numbers. Sample number statistics are shown in the Table 1.

Table 1. Sample size of dataset

L	Positive	Negative	Total
15	825	823	829
20	828	814	837
25	836	823	845
30	834	828	843
40	844	837	850
50	839	832	841
100	825	821	831

2.3   1-D CNN

Convolutional neural network is usually used in image correlation analysis, and convolutional neural network is gradually applied to feature analysis of one-dimensional time series data. In this study, one-dimensional convolutional neural network is used to detect the time sequence segments of HVDC interference. When HVDC interference occurs, the potential data will have obvious characteristic changes.

Simple 1-D CNN principle as shown in Fig. 3. The input feature vectors is the data to be analysed, convolution layers is used to extract the local regional characteristics of input vector. after multiple convolution operations and pooling operations, the feature maps is transformed into a feature vector by flattening or global pooling. Multiple fully connected layers are used to further extract feature. The final output vector represents the result of the classification.

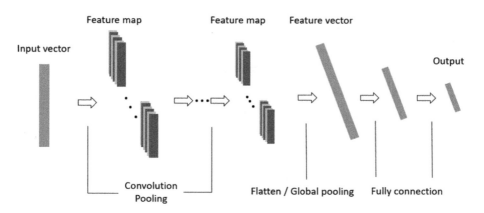

Fig. 3. The structure of 1-D CNN

Due to the small dimension of data to be classified, appropriate convolutional kernel and layer number should be taken into consideration when designing convolutional neural network. According to the corresponding literature, any larger convolution kernel can be replaced by several smaller convolution kernels. A one-dimensional convolution kernel of length 3 can be used as the basic structure of the network. After every two convolution kernels, the maximum pooling layer is added. The number of layers of the network is appropriately increased according to the length of the input sequence. The design networks of different length sequences are shown in the table. In order to utilize the extracted features as much as possible, the feature map is transformed into a one-dimensional feature vector by using flattening operation. In order to improve the speed of convergence, ReLU is used as the activation function to conduct nonlinear processing on the features of each layer (Table 2).

Table 2. Parameter setting of 1-D CNN

L	Layer	Size	Stride
15, 20, 25	CONV1	3	1
	CONV2	3	1
	POOL1	2	2
	FC1	–	–
	FC2	–	–
30, 40, 50, 100	CONV1	3	1
	CONV2	3	1
	POOL1	2	2
	CONV3	3	1
	CONV4	3	1
	POOL2	2	2
	FC1	–	–
	FC2	–	–

Taking the data with L of 30 as an example, the parameters, input and output of the model are shown in Fig. 4.

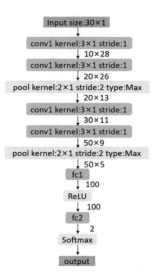

Fig. 4. The inputs and outputs of each layer in 1-D CNN

3   Experimental Result and Analysis

There are two groups of potential data: power-on potential and power-off potential. According to experience, we chose the power-off potential as input. At the same time, we also do a comparative experiment, which proves that the characteristics of the power-off potential are more obvious than the power-on potential. The results of the experiment are shown in Table 3.

Table 3. Performance comparison between on-potential data and off-potential data

L	On-potential data			Off-potential data		
	Acc	Rec	Pre	Acc	Rec	Pre
15	0.825	0.823	0.829	0.882	0.875	0.887
20	0.828	0.814	0.837	0.893	0.883	0.899
25	0.836	0.823	0.845	0.905	0.891	0.914
30	0.834	0.828	0.843	0.913	0.904	0.919
40	0.844	0.837	0.850	0.916	0.922	0.912
50	0.839	0.832	0.841	0.894	0.901	0.886
100	0.825	0.821	0.831	0.885	0.862	0.893

It can be found from the experimental results that the performance of the model on the power-off potential is better than that on the power-on potential, because the power-on potential is affected by the protection current, and the characteristics of HVDC are relatively weakened. When L is considered as the power-off potential, the best result of the experiment is when L is selected at 40. This is in line with our expectation, because as the length of the segment L increases, the more data L contains, the more complete the segment containing HVDC data will be. When L exceeds 40, the accuracy rate drops somewhat, which may be due to the decrease of data volume and the performance decline caused by model overfitting. In addition, the feature distribution is relatively scattered due to the large L, and the network can not capture HVDC features well.

Secondly, the comparison between the one dimensional convolutional neural network classifier and other classifier models is presented. DTW [12] and Wavelet transform [13] are common feature extraction methods in time series data classification. Segments of fixed length L = 40 were selected, and different classification methods were used to extract and classify the features of the data. The results of comparison are shown in Table 4. The experimental results show that 1-D CNN has better performance than DTW and Wavelet transform. This indicates that convolutional neural network can be selected as the HVDC interference detection method.

Table 4. Comparison with other time series data classification methods

Methods	Acc	Rec	Pre
DTW+KNN	0.846	0.839	0.850
Wavelet transform + SVM	0.855	0.861	0.849
1-D CNN	0.916	0.922	0.912

References

1. Pan, J.: Risk analysis of oil and gas pipelines. Oil Gas Storage Transp. **14**(005), 3–10 (1995)
2. Ji, X, Qin, C.: Stray current corrosion of underground gas pipeline and its monitoring. Shanghai Gas (04), 12–15(2007)
3. Li, X., Teng, W., Xiao, J., Lao, D.: Monitoring and analysis of electrical interference of east china UHVDC grounding electrodes to gas pipeline. Corros. Prot. **41**(04), 38–42 (2020)
4. Dong, L., Yao, Z., Ge, C., Shi, C., Chen, J.: Fourier analysis of the fluctuation characteristics of pipe-to-soil potential under metro stray. Surf. Technol. **50**(02), 294–303 (2021)
5. Qu, Z., Zhou, Y., Zeng, Z., et al.: Detection of the abnormal events along the oil and gas pipeline and multi-scale chaotic character analysis of the detected signals. Meas. Sci. Technol. **19**(2), 025301 (2008)
6. Yu, G., Xue, C., Yuan, Z., et al.: Advanced analysis of HVDC electrodes interference on neighboring pipelines. J. Power Energy Eng. **03**(04), 332–341 (2015)
7. Yu, Z., Liu, L.: Analysis of stray current monitoring data of Donghuang oil pipeline due to Gaoqing HVDC ground electrode interference. In: International Conference on Applied Superconductivity and Electromagnetic Devices, Tianjin (2018)
8. Zhao, M., Chen, S., Dave, Y.: Automatic classification and recognition of seismic waveform based on deep learning convolutional neural network. Chin. J. Geophys. **62**(01), 374–382 (2019)
9. Urtnasan, E., Park, J.-U., Joo, E.-Y., Lee, K.-J.: Automated detection of obstructive sleep Apnea events from a single-lead electrocardiogram using a convolutional neural network. J. Med. Syst. **42**(6), 1–8 (2018). https://doi.org/10.1007/s10916-018-0963-0
10. Ye, Z., Yu, J.: Gearbox fault diagnosis method based on multi-channel one-dimensional convolutional neural network feature learning. J. Vibr. Shock **39**(20), 55–66 (2020)
11. Liu, B., Huang, X., Fang, G.: High impedance fault identification in distribution network based on one-dimensional convolution neural network. Electr. Energy Manag. Technol. (09), 99–103 (2020)
12. Liu, H.: The classification of MEA signal spike by wavelet transform. Comput. Digit. Eng. (04), 35–38 (2006)
13. Chen, Q., Hu, G.: A new DTW optimal bending window learning method. Comput. Sci. **39** (08), 191–195 (2012)

Detection of Human Relaxation Level Based on Deep Learning

Zhouzheng Wang[1,2(✉)] and ChenYang Hu[1,2]

[1] Faculty of Information Technology, Beijing University of Technology,
Beijing, China
wangzhuozheng@bjut.edu.cn
[2] Intelligent Signal Processing Laboratory, Beijing University of Technology,
Beijing, China

Abstract. As the pace of life accelerates, the psychological pressure on people is increasing day by day. It is a very meaningful measure to use wearable devices to monitor heart activity in real time, and to monitor the degree of physical and mental relaxation by acquiring heart rate variability (HRV). Based on the physiological similarity, the pulse data which is collected by the wearable smart bracelet is processed as the heart rate signal. The paper analyzes the HRV signal to obtain the three-dimensional input characteristics of the CNN network: time domain features, frequency domain features, and nonlinear features. The designed CNN network in this paper consists of two layers of convolutional layers, maximum pooling layer, average pooling layer, and fully connected layer. According to the three classifications, the CNN network outputs the level of the relaxation index H. The prediction results are embodied in the confusion matrix. The experimental results show that compared with the traditional human body relaxation state evaluation model, the results obtained by using CNN network prediction have higher accuracy. The research results of this subject are expected to initially realize the daily monitoring applications of wearable devices in homes and hospitals.

Keywords: Wearable smart bracelet · Heart rate variability · CNN network · Relaxation index

1 Introduction

With the acceleration of the pace of life, more and more people are plagued by psychological pressure, and the number of patients with mental and mental illnesses is increasing. How to effectively assess the individual's physical and mental relaxation state has become a hot research topic in recent years. By accurately assessing the relaxed state of the human body, individuals can learn about their own vital sign changes and psychological stress changes in time, so as to adopt effective intervention and adjustment methods to avoid the trouble caused by stress. Therefore, based on this purpose, this study aims to monitor human physiological signals at any time to predict human relaxation levels. At present, it is a more objective and convincing method to assess the degree of physical and mental relaxation of the human body through

J. C. Hung et al. (Eds.): FC 2021, LNEE 827, pp. 71–80, 2022.
https://doi.org/10.1007/978-981-16-8052-6_8

physiological parameters. When the heart is doing cyclical diastolic and contraction movements, the vibration generated by it will form a pulse wave with certain characteristics, which contains a wealth of human pathological and physiological information [1–3]. Based on the physiological similarity, this article processes the collected pulse signal as the heart rate signal. The original pulse data were processed to obtain HRV signals with multiple features.

HRV refers to the small variation characteristics between heartbeat intervals. It is a method to check the heartbeat interval variation. It is a sensitive and non-invasive index that can quantitatively evaluate the function of the heart's autonomic nerves. The HRV signal contains a lot of information about cardiovascular regulation. The extraction and analysis of this information can qualitatively assess the tension and balance of cardiac sympathetic nerve and vague nerve activity and its influence on cardiovascular system activity. Therefore, HRV is an effective feature for monitoring individual relaxation levels.

In the field of detection and evaluation of the degree of psychological relaxation, a lot of methods have emerged at home and abroad. Picard et al. [4] studied the stress level of 9 call center employees. These employees wore a skin conductance sensor on the wrist for a week and reported the stress level of each call. The improved support vector machine (SVM) was used for training and testing. The recognition accuracy rate of stress level reached 78.03%. Hongyang Sun of Shanghai Jiao tong University et al. [5] used IAPS image library and mental arithmetic tasks to induce three types of emotions and three levels of psychological stress in the laboratory, and established a database of related physiological signals. The particle swarm optimization (PSO) algorithm and the K nearest neighbor (KNN) algorithm were combined to classify and recognize three kinds of emotions and three kinds of pressures, and the recognition accuracy of the three kinds of emotions reaches 75%.

In this paper, the RR interval was calculated based on pulse data. Heart rate variability (HRV) was analyzed using RR interval data in time domain, frequency domain and nonlinear domain to obtain a suitable feature set. The feature set was sent to the CNN network to assess the psychological relaxation of individuals.

2　Related Work

2.1　Hardware Construction

This article uses Zigbee technology to network the coordinator and watch. The overall hardware network is shown in Fig. 1. The overall hardware network consists of two steps: the first step is the initialization of the coordinator. At first, the JLink_V490 software needs to be installed on the server side. The jlink debug line connects the coordinator with the server, and then J-Flash software is used to download the SB.jflash file and the SB_Code.bin file to the coordinator. The second step is to burn the program of the terminal device. At first, use the J-Flash software to open the MB.jflash project, and then load the watch motherboard program MB_Code.bin into the J-Flash project in the new project interface, and next connect the jlink debug line to the watch board in the corresponding hole positions, finally complete the burning of the program through the Auto button in the J-Flash software interface.

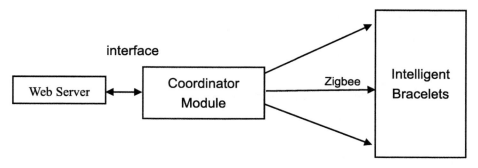

Fig. 1. Schematic diagram of hardware networking.

2.2 Software Debugging

According to the communication protocol of the coordinator, this article adopts post-man software or group decompression and relaxation training system for online debugging. The original data collected by the smart bracelet is transmitted to the server through the Zigbee network and serial port, and the 28-dimensional original characteristic data is calculated by the algorithm program.

3 Data Analysis

The normal heart rate signal is shown in Fig. 2. The duration between two R waves is represented by the RR interval, and the normal RR interval data is between 600 and 1200 ms. For the abnormal value not in this range, the value less than 600 ms is calculated as 600, and the value greater than 1200 ms is calculated as 1200. Based on the RR interval, the heart rate variability analysis includes time domain, frequency domain and nonlinear analysis methods.

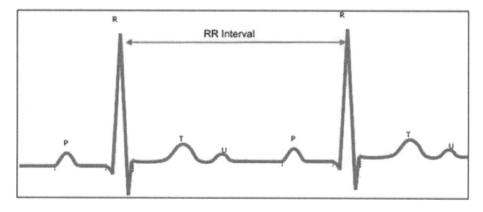

Fig. 2. Heart rate signal graph.

3.1   HRV Time Domain Analysis

Among the 28-dimensional features obtained, there are 18-dimensional time-domain indicators, and several typical features are shown in Table 1.

Table 1. Typical time domain characteristic parameters.

Feature name	Description
mean_nni	Average value of RR interval
SDNN	The standard deviation of the time interval between consecutive normal heartbeats
SDSD	The standard deviation of the difference between adjacent RR intervals
rmssd	The square root of the mean of the sum of squared differences between adjacent RR intervals. Reflect the impact of high frequency (rapid or parasympathetic) on HRV
pnni_50	The ratio obtained by dividing nni_50 (the interval difference between consecutive RR intervals greater than 50 ms) by the total number of RR intervals

Among them, SDNN is the standard deviation of all RR intervals, the unit is ms, and the normal reference value is 141 ± 39 ms. SDNN reflects the total evaluation of heart rate regulation by the autonomic nervous system, and the specific calculation formula is as follows:

$$SDNN(ms) = \sqrt{\frac{\sum_{i=1}^{N}\left(NN_i - \overline{NN}\right)^2}{N-1}} \qquad (3.1)$$

3.2   HRV Frequency Domain Analysis

Perform a fast Fourier transform on a relatively stable RR interval or instantaneous heart rate variability signal (usually greater than 256 heartbeat points) in the time domain to obtain a power spectrum with frequency (Hz) as the abscissa and power spectral density as the ordinate. The frequency domain characteristic parameter has 7 dimensions, among which there are several typical indexes. Some are shown in Table 2 below. The heart rate spectrum curve of normal people in the basic state is between 0–0.4 Hz, 0.003–0.04 Hz is the very low frequency (VLF), 0.04–0.15 Hz is the low frequency (LF), 0.15–0.4 Hz is the high frequency (HF), 0–0.40 Hz is the total power (TP). The recording time for short-range recording is 5 min.

Table 2. Division of frequency bands for short-range recording and spectrum analysis.

Index	Unit	Description	Frequency band
5 min total power	ms × ms	Change of NN interval in 5 min	≤ 0.40
VLF	ms × ms	Very low frequency power	≤ 0.04
LF	ms × ms	Low frequency power	0.04–0.15
HF	ms × ms	High frequency power	0.15–0.40
LF/HF		LF to HF ratio	

The physical and mental control index is determined by the balance between the sympathetic nervous system and the parasympathetic nervous system. It is specifically expressed as the value of LF/HF. Normally, the ratio is about 6:4.

The original pulse data is collected through a wearable smart bracelet, and appropriate features are selected through feature extraction and sent to the neural network for training. There are three different states of relaxation to divide of human body: fatigue, calm, and excitement, corresponding to different ranges of the relaxation index H. This paper extracts the time domain feature SDNN and the frequency domain feature LF/HF and mean_nni to analyze the relaxation index.

4   CNN Neural Network

Convolutional Neural Networks (CNN) is a type of Feedforward Neural Networks which includes convolution calculations and deep structure. It is one of the main representative algorithms for deep learning. The CNN network has the characteristics of strong self-learning ability, good self-adaptation ability and the ability to optimize the network structure through the back propagation algorithm. It is a good pattern classifier.

4.1   CNN Network Structure

The network structure with 5 hidden layers designed in this paper is shown in Fig. 3.

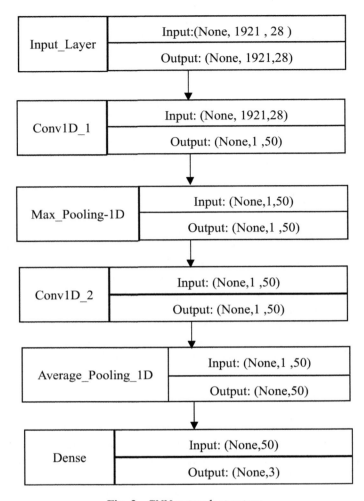

Fig. 3. CNN network structure.

The first is the input layer. The CSV file composed of selected features is used as the initial input. It passes through the one-dimensional convolutional layer, the maximum pooling layer, the one-dimensional convolutional layer, the average pooling layer, and the fully connected layer. Finally, the three classification results are obtained, and then print out the accuracy, loss function and confusion matrix of the training set and test set.

4.2 The Design of CNN Network

There are 1921 experimental data in this paper, and two convolutional layers are designed. The size of the convolution kernel of the first convolution layer is 1, that is, the step size is also 1, and the ReLU (The Rectified Linear Unit) activation function is used. It is characterized by fast convergence and simple gradient calculation. The expression of the ReLU function is as follows:

$$f(x) = max(0, x) \tag{4.1}$$

In the above formula, $x = w^T x + b$, where w is the weight matrix and b is the bias matrix. That is, for the input vector x from the upper layer of the neural network that enters the neuron, the neuron using the linear rectification activation function produces an output.

The pooling layer is sandwiched between successive convolutional layers to compress the amount of data and parameters and reduce overfitting. The first pooling layer uses maximum pooling, that is, each window selects the largest number as the value of the corresponding element of the output matrix. The convolution kernel and activation function of the second convolution layer are the same as those of the first convolution layer. The following second pooling layer uses average pooling, that is, the average value of the elements in each window is used as the value of the corresponding element of the output matrix. The last is the full link layer, where Softmax is used as the activation function, and its expression is as follows:

$$p(x_i) = \frac{exp(x_i)}{\sum_{i=1}^{n} exp(x_i)} i = 1, 2, 3\ldots\ldots n \tag{4.2}$$

Where x_i is the input and $p(x_i)$ is the output. Numerator: Through the exponential function, the real number output is mapped from zero to positive infinity. Denominator: Add all the results together and normalize them. The optimizer chooses adam. The purpose of the Softmax function is to convert multi-class output into probability. The classification evaluation indicators are precision, recall, and F1 score.

5 Results and Analysis

Based on the CNN network to predict the degree of physical and mental relaxation, this paper has done two sets of comparative experiments. The first is to change the epoch of the model and the number of convolutional layers of the CNN network for horizontal comparison; then the CNN model is compared with the LSTM (Long Short Term Memory Network) and SVM (Support Vector Machine) models longitudinally.

5.1 Horizontal Comparative Experiment

5.1.1 Influence of Convolution Layer Number on Results
When the data volume was 1921, the epoch was 300, and the batch-size was 200, the number of convolutional layers was changed from 2 to 4. Comparison results of accuracy and loss function are shown in Table 3.

Table 3. Influence of the number of convolution layers on the results.

The layer number of convolution	Accuracy	Loss
2	**0.811**	**0.257**
3	0.768	0.384
4	0.757	0.421

As can be seen from the table, when the number of convolution layers is 2, the accuracy is the highest, which is 81.1%. Increasing the number of convolution layers will decrease the accuracy of prediction, which may be caused by the limited amount of data. It can be seen that the more layers of convolution is not always better. The number of convolution layers should be reasonably set according to the specific experimental situation.

5.1.2   Influence of Iteration Number on Results

The number of iterations is the quotient of the amount of data and the size of the batch. When the data volume is 1921, the epoch is 300, the convolutional layer is two layers, the batch size is 100,200,300 respectively, i.e. the number of iterations is 20, 10, and 7 respectively. The results of accuracy and loss function comparison are shown in Table 4.

Table 4. Influence of the number of iterations on the results.

The number of iterations	Accuracy	Loss
20	0.712	0.459
10	**0.811**	**0.384**
7	0.783	0.347

As can be seen from the table, when the number of iterations is 10, the effect is the best. The number of iterations should be reasonably optimized according to the size and composition of the data set.

5.2   Longitudinal Comparative Experiment

In the experiment, LSTM network and SVM (support vector machine) [6] were selected as the reference of CNN network to carry out comparative experiments. Precision, recall and F1-score were selected as evaluation indexes. The experimental results are shown in Table 5.

Table 5. Comparison of different results of CNN, LSTM and SVM.

Model	Precision	Recall	F1-score
CNN	**0.735**	**0.910**	**0.802**
LSTM	0.651	0.814	0.763
SVM	0.602	0.798	0.639

As can be seen from the table, CNN network has the highest accuracy, recall rate, F1 score, and the best classification effect and stability. When observing different indicators, the value of recall rate is the highest, indicating the highest recognition rate for positive samples.

5.3 Experimental Results of CNN Network

When the CNN network is used for experiments, 80% of the data set is taken as the training set and 20% as the test set. The accuracy and loss rates on the training set and test set are shown in Fig. 4, and the confusion matrix is shown in Fig. 5.

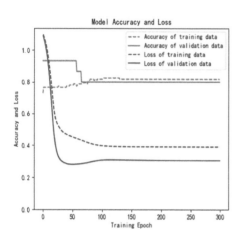

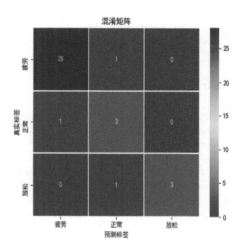

Fig. 4. Accuracy and loss degree under CNN.   **Fig. 5.** Confusion matrix under CNN.

Figure 4 shows that the accuracy of CNN network on the training set reaches more than 80%, and the test set curve and the training set curve are basically consistent when the epoch is greater than 50, indicating that the prediction effect is very good. For the confusion matrix, when there are 36 prediction samples, the prediction classification of 34 samples is correct, indicating that the classification effect of the model is very nice.

6 Results and Prospects

In this experiment, the relaxation index was predicted through CNN network, and the prediction of H relaxation index was basically realized, with an accuracy of about 80%. What is more, compared with LSTM and SVM, it is found that CNN network has higher accuracy. However, there are still some shortcomings in the work that need to be further improved in the future researches. Therefore, this paper put forward the following prospects for future researches:

1) Further expansion of training samples. Increase the number of subjects on the basis of different states of subjects. Increasing the number of subjects can not only

improve the accuracy of classification, but also help to reduce errors and overfitting in the classification model.

2) Excavate more characteristics that can be used to identify psychological stress. In addition to the HRV pressure features extracted in this paper, more hidden features can be mined, and then these features can be fused and sent into the classifier to improve the recognition rate of the index.

3) Study more objective labeling methods. In this paper, the correlation of H relaxation index is labeled subjectively according to different characteristics, and we need to find a more optimized labeling method to improve the scientificity of results.

Acknowledgement. This research was funded by the Beijing Municipal Natural Science Foundation, grant number 4192005.

References

1. Song, X., Wang, Y.: Pulse wave: a bridge between traditional Chinese medicine and western medicine. J. Integr. Tradit. West. Med. (09), 891–896 (2008)
2. Xie, G.: Pressure Identification and Characteristic Contrast Analysis Based on HRV Signal. Southwest University (2017)
3. Yang, N.: Research and Implementation of Psychological Stress Recognition Algorithm Based on HRV. Xi'an Polytechnic University (2019)
4. Hernanderz, J., Morris, R., Picard, R.W.: Call center stress recognition with person-specific modes. Affect. Comput. Intell. Interact. **6974**(1), 125–134 (2011)
5. Sun, H., Xu, Z., Wang, J., Lei, P., Wu, K., Chai, X.: Emotion recognition based on PSO-KNN algorithm and multiple physiological parameters under stress. Chin. J. Med. Devices **37**(02), 79–83 (2013)
6. Cortes, C., Vapnik, V.: Support-vector networks. Mach. Learn. **20**(3), 273–297 (1995)

Human Physiological Signal Detection Based on LSTM

ZhuoZheng Wang[1,2(✉)], YuYang Wang[1,2], and ChenYang Hu[1,2]

[1] Faculty of Information Technology, Beijing University of Technology,
Beijing, China
wangzhuozheng@bjut.edu.cn
[2] Intelligent Signal Processing Laboratory, Beijing University of Technology,
Beijing, China

Abstract. In this paper, a wearable device is used to obtain the human pulse signal and analyze the heart rate variability. Heart rate variability is affected by respiration or breathing process, so it can estimate the human body's breathing rate and then get the body's relaxation index through the neural network. The Zigbee network and serial port transmit the collected raw data to the server, perform HRV analysis at a fixed sampling frequency, and calculate various characteristic data. A Long Short-Term Memory Network algorithm is proposed to classify the physical and mental relaxation index into three categories when performing deep learning to analyze feature data. This experiment shows that compared with SVM, LSTM is faster, and the optimal parameters of the model are found by the method of controlling variables. The accuracy reaches more than 90%, which aligns with expectations, verifying the algorithm's effectiveness. Wearable watches can be used for daily monitoring in hospitals or families and have good application value.

Keywords: Zigbee · HRV · LSTM

1 Introduction

In the context of the rapid development of the times, people in all fields of life are concerned about convenient life, physical health, and life safety. In terms of the convenience of life, wearable devices allow people to make calls and take photos anytime, anywhere. In terms of life safety, it can monitor people's fatigue and emotions while driving and can issue warnings in time to reduce traffic accidents. In terms of physical health, with the aging of the population, busy young people cannot always pay attention to the physical condition of the elderly at home. Wearable devices provide safety guarantees for the elderly. The smartwatch detects any physical condition of the elderly and feeds it back to the family members' mobile phones. Due to the impact of the COVID-19 epidemic, many young people also attach great importance to their health problems. It is practical to analyze their H relaxation index by wearing smartwatches and adjusting their status in time.

Many previous studies on wearable devices, such as G. Paolini [1], presented research about a 5.8 GHz system for vital signs monitoring, specifically human breath. The system was designed to be fully wearable; it could be mounted inside a plastic case

© The Author(s), under exclusive license to Springer Nature Singapore Pte Ltd. 2022
J. C. Hung et al. (Eds.): FC 2021, LNEE 827, pp. 81–90, 2022.
https://doi.org/10.1007/978-981-16-8052-6_9

and worn by the user under test at chest–level position. B. Xiang [2] presented iMask, a noninvasive and cost-effective wireless wearable respirator that measured breathing parameters in real-time. W.S. Johnston [3] demonstrated the feasibility of extracting accurate breathing rate information from a photoplethysmographic signal recorded by a reflectance pulse oximeter sensor mounted on the forehead subsequently processed by a simple time-domain filtering and frequency domain Fourier analysis.

In recent years, many articles have had their methods of analyzing heart rate variabilities, such as In 2017, H. Dubey [4] designed a wearable PPG system and proposed a technique that uses spectral kurtosis along with the state-of-the-art respiratory-induced frequency, intensity, and amplitude features. In 2018, Y. Shang [5] mentioned the error backpropagation network and its improved algorithm are used to realize the initial recognition of HRV signals. In 2020, A. C. Podaru [6] designed a device capable of simultaneously recording the ECG, PPG, and a pulse signal acquired from the earlobe biological signals to determine the heart rate variability (HRV) parameters. In 2021, M. Hussain [7] proposed in-vehicle breathing rate monitoring by exploiting channel state information (CSI) available in Wi-Fi signals.

In the analysis of heart rate variability, time domain and frequency domain are usually used to analyze the impact of HRV on the user's stress level. In the frequency domain analysis, the user's psychological stress state is analyzed by observing the increase and decrease of the low-frequency and high-frequency parts of HRV. In the time domain analysis, the user's psychological stress state is analyzed by observing the changes in SDNN, heart rate, and RMSSD.

In previous studies, there are very few articles that analyze the human mental state through neural networks, and many of them use machine learning to analyze the degree of relaxation of the human body. This article uses the LSTM network in deep learning to analyze the degree of relaxation of the human body. Compared with the previous SVM network, this network has higher accuracy and the model can converge faster. First, the pulse signal is measured by a wearable watch, and the required feature data is calculated through time-domain analysis and frequency domain analysis. According to a certain standard, the tester's physical and mental relaxation index is scored from 0–100, and the scores are divided into three categories, which are sent to the LSTM neural network for learning. Among them, the score is positively correlated with the degree of physical and mental relaxation.

2   Data Acquisition and Analysis

2.1   Data Acquisition

During the experiment, psychosocial stress was induced with the three states of sitting, walking, and jogging. The wearable smartwatch can obtain pulse data. The watch and the coordinator were networked through Zigbee technology, and the data was sent to the server. The sampling frequency is 200 Hz.

The watch is shown in Fig. 1(a). The outer end has a power interface and a handset interface. The time can be displayed in the upper left corner of the page, and the battery capacity can be displayed in the upper right corner of the page. In Fig. 1(b), the watch has four options. The system settings can query the device number, device name,

Bluetooth MAC address, and Zigbee address, as shown in Fig. 1(c). On the physiological signal page, the wearer's heart rate curve can be observed in real-time. The music library provides main melody music and background music for users to choose from. Training topics are divided into three types, namely Meditation breathing, Active breathing, and Sleeping breathing. The coordinator is the manager of the network organization [8], as shown in Fig. 2.

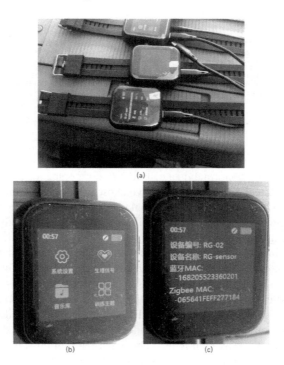

Fig. 1. Watch style display, (a) wearable watch (b) watch page function (c) system setting page in watch page.

Fig. 2. Coordinator

The RR interval is obtained by analyzing the pulse data. The RR interval reflects the distance and time of the R wave peak in two adjacent heartbeats. The original pulse data is shown in Fig. 3. The unit is ms.

0,0,3,12,22,31,39,43,48,52,58,65,73,83,92,103,111,120,125,126,124,116,103,84,59,31,18,0,0,0,0,0,0,0,0,0,0,0,0,0,0,0,
0,0,0,0,0,0,0,0,0,0,0,0,0,1,23,57,126,214,326,464,626,810,914,1016,1023,1023,1023,1023,1023,1023,1023,1023,1023,1
023,1023,1023,1023,1023,1023,1023,1023,1023,1006,984,947,900,859,822,788,757,726,696,665,632,597,560,52
0,479,436,391,346,298,249,199,149,99,51,25,0,0,0,0,0,0,0,0,4,18,36,68,98,127,153,178,202,224,243,259,272,281,286,28
8,286,284,278,273,265,258,249,239,229,218,206,194,182,170,157,143,125,108,88,69,51,33,18,9,0,0,0,0,0,0,0,0,0,0,0,0
,0,0,0,0,0,0,0,0,0,0,0,0,54,120,257,424,615,815,924,1023,1023,1023,1023,1023,1023,1023,1023,1023,1023,1023,1023,1
023,1023,1023,1023,1023,1023,1023,1023,1003,982,935,889,842,798,754,713,674,637,601,564,530,494,458,419,377,3
32,287,240,193,149,105,65,32,10,0,0,0,0,0,0,0,0,10,20,39,64,90,116,143,168,193,216,237,255,271,287,299,311,318,323,3
24,321,317,308,301,290,280,270,259,248,236,225,213,200,186,171,155,139,122,105,91,73,59,41,23,14,0,0,0,0,0,0,0,0,0,
0,0,0,0,0,0,0,0,0,0,0,0,0,0,0,0,36,100,213,363,540,739,874,987,1023,1023,1023,1023,1023,1023,1023,1023,1023,1023,1023
,1023,1023,1023,1023,1023,1023,1023,1023,1023,1016,995,970,924,880,836,792,750,709,670,632,597,560,525,488,45
0,411,369,326,281,235,190,147,107,72,39,16,3,0,0,0,0,0,6,23,44,73,105,138,171,203,233,259,283,302,320,334,347,355,

Fig. 3. Raw pulse data

2.2 Data Analysis

During the experiment, abnormal data caused by poor contact of the watch needs to be processed for abnormal values. Studies have shown that the RR interval data outside the range of 600–1200 are caused by external factors, such as severe shaking, abnormal contact, and sudden instability of network signals, rather than the body's breathing rate. Therefore, the range of filtering outliers is selected between 600 and 1200.

After the RR interval is processed, time domain and frequency domain analysis are performed to obtain 27-dimensional data and 1822 rows of data. Through reading the literature, we can see that in the 24-h measurement data, the SDNN value in the time domain has a more significant impact. In the 5-min measurement data, SDNN in the time domain, HF, LF, LF/HF, LFNU, and HFNU in the frequency domain has a more significant impact on the H relaxation index. In contrast, other data have a more negligible effect on the relaxation index. The experiment in this article uses short-term measurement data and scores the H relaxation index according to the reference value. The reference value is shown in Table 1 below. Since HRV is affected by gender, age, and environment, this expected value is measured in a sitting environment with male experimenters aged 20–40. This paper classifies the H relaxation index into three categories to explore the impact of characteristic data on the H relaxation index. The classification rules are shown in Table 2 below.

Table 1. Typical characteristic reference value

Feature name	Reference threshold (unit)
SDNN	20–70 (ms)
LF	754–1586 (ms^2)
HF	772–1178 (ms^2)
LFNU	30–78 (nu)
HFNU	26–32 (nu)

Table 2. Three classifications of H relaxation index

H relaxation index	Label
0–40	0
41–70	1
71–80	2

The H relaxation index is used as an index to measure the body's degree of mind and mind relaxation. The H relaxation index score is calculated according to the threshold under certain conditions. The physical and mental control index depends on the balance between the sympathetic nervous system and the parasympathetic nervous system, specifically, the ratio of LF/HF. So it can be scored manually through some typical feature data.

3   Heart Rate Variability Analysis

HRV signal contains much information about cardiovascular regulation. The extraction and analysis of this information can quantitatively evaluate the tension and balance of cardiac sympathetic nerve and vagus nerve activity and its influence on cardiovascular system activity. The detection of HRV has been widely used in clinical practice at home and abroad. When the sympathetic nerve activity increases, the HRV decreases, and when the parasympathetic activity increases, the HRV increases.

After HRV refers to the slight variation characteristics between heartbeat intervals, it is a method to check the heartbeat interval variation. It is a sensitive and noninvasive index that can quantitatively evaluate the cardiac autonomic nerve function [9]. HRV analysis methods generally include time-domain analysis, frequency domain analysis, linear analysis, and nonlinear analysis.

3.1   Time-Domain Feature Analysis

The RR interval values arranged in time sequence or heartbeat sequence are directly analyzed by statistics or geometry for the collected time-series signal of RR interval.

The time-domain characteristic parameters have 18 dimensions, of which there are several typical indicators, as shown in Table 3 below.

Table 3. Time-domain characteristic parameters

Feature name	Description
MEAN_NNI	The mean of RR-intervals
SDNN	The standard deviation of the time interval between successive normal heart beats
SDSD	The standard deviation of differences between adjacent RR-intervals
RMSSD	The square root of the mean of the sum of the squares of differences between adjacent NN-intervals. Reflects high frequency (fast or parasympathetic) influences on HRV
PNNI_50	The proportion derived by dividing nni_50 (The number of interval differences of successive RR-intervals greater than 50 MS) by the total number of RR-intervals

SDNN refers to the average of all normal sinus heartbeat intervals (NN), the unit is ms, and the calculation formula is as follows:

$$SDNN(ms) = \sqrt{\frac{\sum_{i=1}^{N}\left(NN_i - \overline{NN}\right)^2}{N-1}} \qquad (1)$$

RMSSD refers to the root mean square of the difference between adjacent NN intervals throughout the entire process, and the unit is MS. The calculation formula is as follows:

$$RMSSD(ms) = \sqrt{\frac{\sum_{i=1}^{N-1}\left(\Delta NN_i\right)^2}{N-1}} \qquad (2)$$

SDSD refers to the standard deviation of the difference between the lengths of adjacent NN intervals in the whole process, the unit is MS, and the calculation formula is as follows:

$$SDSD(ms) = \sqrt{\frac{\sum_{i=1}^{N}\left(\Delta NN_i - \overline{\Delta NN}\right)^2}{N-1}} \qquad (3)$$

PNN50 refers to the ratio of the number of adjacent NN intervals greater than 50 ms to the total number of NN intervals in the record of all NN intervals, expressed as a percentage. The SDNN reflects the total evaluation of the autonomic nervous system on the regulation of heart rate, while RMSSD and PNN50 reflect the Vagus nerve tension [10].

3.2 Frequency-Domain Feature Analysis

The principle of frequency domain analysis decomposes the randomized chance of RR interval or instant heart rate signal into different frequency components of energy, which converts the heart rate curve to the frequency domain to analyze in other words. It provides the basic information of energy varying with frequency. The data in each group are processed by FTT operation to get the power spectrum [5]. From the spectrum curve, observe the sympathetic nerve's regulation and the Vagus nerve on the heart rate.

The frequency-domain characteristic parameter has seven dimensions, among which there are several stock indexes, as shown in Table 4 below:

Table 4. Frequency-domain characteristic parameters

Feature name	Description
LF	variance (= power) in HRV in the low Frequency (.04 to .15 Hz)
HF	variance (= power) in HRV in the High Frequency (.15 to .40 Hz by default)
LF/HF	lf/hf ratio is sometimes used by some investigators as a quantitative mirror of the sympathy/vagal balance

4   Experiment Results and Conclusion

This article mainly explores the Long short-term memory (LSTM) network results on the H relaxation index multi-classification problem. Compared with SVM, LSTM has a better classification effect and higher accuracy. LSTM is a special RNN, mainly to solve gradient disappearance and gradient explosion in the training process of long sequences. Compared with ordinary RNN, LSTM can perform better in longer sequences.

4.1   Design of LSTM Network

In network design, many experiments have shown that setting the number of LSTM layer units to 256 has the best effect. The optimizer Adam has a more noticeable effect than RMSprop instead of stochastic gradient descent. The activation function uses Softmax for multi-classification. The layered design of the model and the parameters of each layer are shown in Table 5 below:

Table 5. Model architecture

Layer (type)	Output Shape	Param
lstm_1 (LSTM)	(None, 5, 256)	264192
lstm_2 (LSTM)	(None, 256)	525312
dense_1 (Dense)	(None, 3)	771

4.2 The Effect of Modification of Different Parameters on Experimental Results

In this experiment, 80% of the data set is used as the training set. The running time and the accuracy of the test set are observed by modifying the iterations and batch size, as shown in Table 6 below. The model loss curve for 300 iterations is shown in Fig. 4 below. Nevertheless, when the number of iterations exceeds 200, the loss of the verification set is overfitted.

Table 6. Experimental results of different parameters

epoch	batch_size	time (s)	Test accuracy	Test loss	batch_size	time (s)	Test accuracy	Test loss
50		26.9	0.9	0.27		23.2	0.86	0.34
100		43.6	0.92	0.23		42.2	0.9	0.25
150	**500**	61.6	0.91	0.21	1000	61.4	0.91	0.24
200		79.2	0.92	0.2		80.1	0.92	0.22
250		97.2	0.93	0.19		102.76	0.93	0.2
300		118	**0.94**	**0.19**		121.2	0.92	0.2

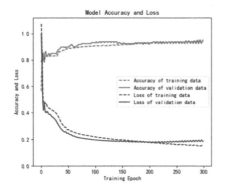

Fig. 4. (Left)The accuracy and loss rate of the LSTM model

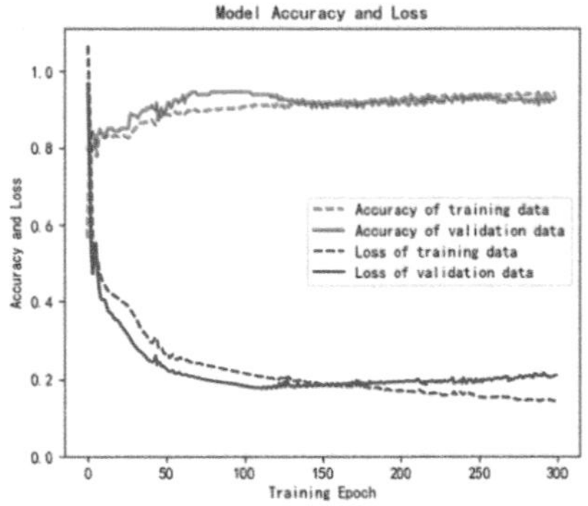

Fig. 5. (Right)The accuracy and loss rate of the Bidirectional LSTM model

4.3 Bidirectional LSTM Networks

This section adds a two-way LSTM network layer to the network design, keeps the batch size unchanged at 500, and modifies the number of iterations to get the accuracy and loss of the test set, as shown in the following Table 7. The model loss curve for 300 iterations is shown in Fig. 5. Similarly, when the number of iterations exceeds 100, the loss of the validation set is overfitted.

Table 7. Bidirectional LSTM experiment results

epoch	batch_size	Running time (s)	Test accuracy	Test loss
50	500	36.6	0.9	0.25
100	500	69	0.92	0.23
150	500	102	0.91	0.21
200	500	129	**0.92**	**0.2**
250	500	163.8	0.92	0.2
300	500	201	0.92	0.2

4.4   Comparison of LSTM Network and SVM Network

Comparing the LSTM network with SVM, the Table 8 lists the Running time, Recall and F1-score value of the three.

Table 8. LSTM network and SVM network experiment results

	Running time (s)	Test accuracy	recall	f1-score
LSTM	79.2	**0.92**	**0.84**	**0.84**
Bidirectional LSTM	129	0.92	0.82	0.82
SVM	0.3	0.84	0.6	0.57

4.5   Conclusion

Experiments show that when the number of iterations is 300, and the batch size is 500, the accuracy of LSTM is 0.94, and the running time is 118 s, but there will be overfitting. Therefore, when the number of iterations is 200, and the batch size is 500, the LSTM accuracy rate is 0.92, and the running time is 79.2 s. Therefore, the LSTM network proposed in this paper is better than the previous SVM method to classify the body relaxation index. In general, we have the following conclusions:

(1) It is feasible to conduct data analysis on wearable watches through neural networks.
(2) Through the improvement of the LSTM network structure, the addition of a Bidirectional LSTM network did not make the accuracy rate better than the previous results and more time requirements.
(3) The application of wearable watches in hospitals and homes has practical significance.

5   Future Work

This experiment realized the multi-classification problem of the H relaxation index through the LSTM network, the accuracy rate is about 90%, and it can be used in clinical or daily life. However, there are still some shortcomings, and further research is needed:

1. The sample data set is too small. The 1989 original data collected this time are far from enough, and a more extensive data set is needed to improve the model's accuracy.
2. Find other feature inputs that affect the H relaxation index and add them to the original data to make the original data more complex to improve the model's accuracy.
3. Explore the possibility of the Bidirectional LSTM network structure to improve the experimental results.

Acknowledgement. Funding: This research was funded by the Beijing Municipal Natural Science Foundation, grant number 4192005.

References

1. Paolini, G., Feliciani, M., Masotti, D., Costanzo, A.: Toward an energy-autonomous wearable system for human breath detection. In: 2020 IEEE MTT-S International Microwave Biomedical Conference (IMBioC), pp. 1–3 (2020)
2. Xiang, B., et al.: Wireless wearable respirator for accurate measurement of breathing parameters. In: 2019 IEEE 2nd International Conference on Electronic Information and Communication Technology (ICEICT), pp. 106–112 (2019)
3. Johnston, W.S., Mendelson, Y.: Extracting breathing rate information from a wearable reflectance pulse oximeter sensor. In: The 26th Annual International Conference of the IEEE Engineering in Medicine and Biology Society, pp. 5388–5391 (2004)
4. Dubey, H., Constant, N., Mankodiya, K.: RESPIRE: a spectral kurtosis-based method to extract respiration rate from wearable PPG signals. In: 2017 IEEE/ACM International Conference on Connected Health: Applications, Systems and Engineering Technologies (CHASE), pp. 84–89 (2017)
5. Shang, Y., Yang, N., Zhu, Y., Song, X.: Application of artificial neural network in HRV analysis. In: 2018 IEEE 4th International Conference on Computer and Communications (ICCC), pp. 2138–2141 (2018)
6. Podaru, A.C., David, V.: A simple method for determining the HRV parameters. In: 2020 International Conference and Exposition on Electrical And Power Engineering (EPE), pp. 151–155 (2020)
7. Hussain, M., Akbilek, A., Pfeiffer, F., Napholz, B.: In-vehicle breathing rate monitoring based on WiFi Signals. In: 2020 50th European Microwave Conference (EuMC), pp. 292–295 (2021)
8. Ameen, S.Y., Nourildean, S.W.: Coordinator and router investigation in IEEE802.15.14 ZigBee wireless sensor network. In: 2013 International Conference on Electrical Communication, Computer, Power, and Control Engineering (ICECCPCE), pp. 130–134 (2013)
9. Catai, A.M., Pastre, C.M., Godoy, M.F., et al.: Heart rate variability: are you using it properly? Standardisation checklist of procedures. Braz. J. Phys. Ther. **24**(2), 91–102 (2020)
10. Xian, J., Chen, X.: Design of fatigue and excitation test scheme based on heart rate variability. Electron. Technol. Softw. Eng. **21**, 80–82 (2018)

A Novel Diagnosis Method of Depression Based on EEG and Convolutional Neural Network

Zhuozheng Wang[1]([✉]), Zhuo Ma[1], Zhefeng An[2], and Fubiao Huang[3]

[1] Faculty of Information Technology, Beijing University of Technology,
Beijing, China
wangzhuozheng@bjut.edu.cn
[2] Advising Center for Student Development, Beijing University of Technology,
Beijing, China
[3] Department of Occupational Therapy,
China Rehabilitation Research Center, Beijing, China

Abstract. Depression as a common mental illness, has become the second largest killer of human beings. However a significant number of patients are not even aware they have depression. Therefore, combined with deep learning method, a novel diagnosis method of depression based on Electroencephalography (EEG) signals is proposed in this paper, which adopts two-dimensional Convolutional Neural Network (2D-CNN) to build a binary classification model. Firstly, the EEG signals are converted into RGB three-channel color brain maps as the input of 2D-CNN. Secondly, 2D-CNN is applied to automatically extract EEG features and classify them. Moreover, the effectiveness and reliability of the proposed algorithm are assessed on the depression dataset. In addition, the proposed method is compared with Support Vector Machine (SVM) classifier and Long and Short Term Memory (LSTM) network. The experimental results show that the proposed 2D-CNN algorithm has the best performance, and the accuracy can reach up to 92%. This method provide a novel approach for the diagnosis of depression.

Keywords: Electroencephalography (EEG) signals · Depression · Two-Dimensional Convolutional Neural Network (2D-CNN)

1 Introduction

1.1 Background

Depression is a common mental illness with high recurrence rate, high disability rate and high suicide rate [1]. According to the statistics of the World Health Organization (WHO) [2], nearly 350 million people suffer from depression around the world, and it is estimated that more than 95 million people in China are suffering from depression. However, the public awareness of depression is far lower than that of other mental illnesses, and many people with depression don't

J. C. Hung et al. (Eds.): FC 2021, LNEE 827, pp. 91–102, 2022.
https://doi.org/10.1007/978-981-16-8052-6_10

even know they are sick. Experts say the diagnosis rate of depression is less than 20% at present, which often leads to patients being undiagnosed. If the patients with depression are not treated in time, they are likely to develop into refractory diseases, which can lead to self-harm or even suicide in severe cases. It is reported that the incidence of depression (and suicide) has begun to appear at younger ages (college students, even primary and middle school students) [3]. Therefore, the science popularization, prevention and treatment of depression need to be paid more attention. Currently, the most widely used method for diagnosing depression is based on the Baker Depression Scale [4] or patient self-reported information and the physician's clinical experience. However, the accuracy of diagnosis may be affected by many factors, such as the proficiency of the physician, the degree of cooperation of the patient and so on. So, there is an urgent need to find an objective and effective method to identify depression.

Electroencephalography (EEG) is the overall response of the electrophysiological activities of human brain nerve cells in the cerebral cortex or scalp surface [5], which comprehensively reflects the functional state of the brain and contains a large number of physiology and disease information. Therefore, EEG can be used as a potential biomarker for the diagnosis and treatment of some diseases, such as depression. In addition, EEG data has the characteristics of high temporal resolution, low cost, easy to access and use, so EEG has become the first choice in the field of brain research [6]. Reference [7] have shown that different frequency ranges and spatial distributions of EEG are related to different functional states of the brain. So it is possible to use EEG signals to diagnose and identify patients with depression. EEG signals are usually divided into 5 bands, as shown in the Table 1.

Table 1. Different frequency bands of EEG signals and corresponding body states

Brain wave type	Frequency range	State of the human body
Delta (δ)	0.5–4 Hz	Deep sleep with no dreams
Theta (θ)	4–8 Hz	Emotional stress, especially disappointment or frustration
Alpha (α)	8–13 Hz	Relaxed, calm, eyes closed but awake
Beta (β)	13–30 Hz	Concentration, excitement, anxiety
Gamma (γ)	30–100 Hz (usually 40 Hz)	Raise awareness and meditate

1.2   Related Works

In recent years, with the development of computer technology, the processing technology of complex EEG data has achieved rapid development. A number of researchers have combined EEG signals with traditional feature extraction algorithms and machine learning [8] to distinguish between depressed patients and normal subjects. Reference [9] proposed a spectral spatial EEG feature extractor named kernel feature filter and library co-space model, which was used to extract

EEG features from patients with depression and normal controls. Support Vector Machine (SVM) was used for classification, and the accuracy is about 80%. Reference [10] selected Alpha, Alpha1, Alpha2, Beta, Delta and Theta asymmetry as features, adopted multi-cluster feature selection (MCFS) for feature selection and used machine learning method to classify the features. The result showed that the combination of Alpha2 and Theta asymmetry had the highest classification accuracy of 88.33% in SVM. Reference [11] in order to solve the problem of misdiagnosis between patients with depression and schizophrenia, the features of resting state EEG signal of patients with depression and schizophrenia were extracted respectively, including: (1) information entropy, sample entropy and approximate entropy (2) statistical attributes (3) relative power spectral density of each rhythm. Then, using these features to form eigenvectors, combined with SVM and Naive Bayes classifier, the classification of schizophrenia and depression patients was studied. The results showed that classification model composed of SVM and eigenvectors of each rhythm rPSD had a good effect on the classification of patients with schizophrenia and depression. Reference [12] proposed a multimodal model, which used the feature level fusion technology to integrate the EEG data of different EEG patterns to establish a depression recognition model. The accuracy of three machine learning classifiers was compared, among which the KNN classification accuracy was the highest, reaching 86.98%. In addition to machine learning, deep learning [13] has also been used in the classification of EEG signals in recent years. In order to improve the accuracy of emotion recognition of EEG signals, reference [14] applied ConvNet to emotion recognition, The classification accuracy of valence and arousal were 81.406% and 73.36%, respectively. Reference [15] put forward an emotion recognition method based on Convolutional Neural Network (CNN) and Long and Short Term Memory(LSTM) network, and conduct emotion recognition experiments on emotion dataset of EEG signals. The average classification accuracy reached 88.15%.

1.3 Our Works

Due to poor performance and generalization ability, these traditional machine learning methods have difficulties in expressing complex functions, and it takes a lot of time to extract effective features. Compared with traditional machine learning, Convolutional Neural Network (CNN) doesn't need to carry out feature extraction before training and testing data and can still guarantee classification accuracy with simplified steps. In recent years, CNN, which is widely used in the field of image and video, has been gradually applied to the processing of EEG signals. EEG signals contain time, frequency and space information, while traditional machine learning only considers time or frequency information. This paper takes the spatial location information of EEG signals into account, so 2D-CNN which can extract spatial features is chosen as the classification model. Then compared with SVM, LSTM and 1D-CNN, the proposed 2D-CNN algorithm has the highest accuracy, reaching 92%.

The structure of the rest of the paper can be summarized as follows: In the Sect. 2, the data used in this study, several commonly used EEG features for classification and how to extract features from 2D-CNN are briefly introduced.

The Sect. 3 proposes a depression recognition algorithm based on CNN. The network structure of 2D-CNN is briefly introduced, and a 2D-CNN model for diagnosis of depression is established. In the Sect. 4, how to transform EEG data into a color brain map and how to partition the dataset are introduced. In addition, the proposed method is verified by the depression dataset, and compared with other methods. Finally, the Sect. 5 provides the conclusions and prospects for the future.

2   Data Analysis

2.1   Data Source

The data source of this paper are provided by the psychiatric department in 3A-grade hospital in China. The experimental data were collected from 16 electrodes (FP1, FP2, F3, F4, F7, F8, T3, T4, C3, C4, T5, T6, P3, P4, O1, O2) at the frequency 100 Hz. The subjects are divided into 16 patients with depression and 16 healthy people, and there is no significant difference in gender or age between the two groups. Then they are asked to sit in a quiet room at resting state with their eyes closed and awake. This process took 2 to 4 min. In order to meet the sample size requirements of deep learning, the signal is clipped into a segment of 100 sample points (1 s). In addition, there is 50% overlap rate between adjacent segments, as shown in Fig. 1. So there are 15,053 segments of 32 subjects in total.

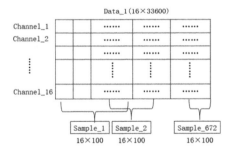

Fig. 1. 16 channels EEG data of one subject.

2.2   Data Preprocessing

The EEG signals should be preprocessed before feature extraction. Firstly, the EEG signals are filtered by bandpass, and the bandpass of the filter is set to 0.5–80 Hz to remove interference such as high-frequency noise, low-frequency drift and EMG. Then the notch filter is carried out to filter out 50 Hz power-line interference. Finally, the data is divided into 1 s time segments to improve the speed of feature extraction and increase the amount of data.

2.3   Feature Extraction

At present, there are many methods for feature extraction and analysis of EEG signals, including time domain analysis, frequency domain analysis and time-frequency analysis [16]. Pearson Correlation Coefficient (PCC) [17] is widely used to measure the degree of correlation between two variables. It has no requirement on the range of values between different variables, and the correlation measures the trend. The PCC between two variables is calculated as follows:

$$\rho_X, Y = \frac{cov(X,Y)}{\sigma_X \sigma_Y} = \frac{E[(X - \mu_X)(Y - \mu_Y)]}{\sigma_X \sigma_Y} \tag{1}$$

Power Spectral Density (PSD) [18], which defines how the power of a signal or time series is distributed with frequency, is a measure of the mean square value of a random variable. The PSD of the EEG signals is the Fourier Transform of its autocorrelation function y(i). The autocorrelation function is as follows:

$$y(i) = \frac{1}{N} \sum_{t=0}^{N-1-i} x(t)x(t+i) \tag{2}$$

where i = 0, 1, $\cdots$, N − 1. So the PSD is as follows:

$$P(\omega_k) = \sum_{t=-(N-1)}^{N-1} y(t)e^{i\omega_k t} \tag{3}$$

where k = −(N − 1), −(N − 2), $\cdots$, 0, 1, $\cdots$, N − 1.

When using SVM or LSTM to classify the above features, manual feature extraction is required in advance, while Convolutional Neural Network (CNN) can automatically extract features. The convolutional layer in the CNN is responsible for feature extraction. The eigenvalue can be obtained through the convolutional kernel and image convolution. In addition to the eigenvalue itself, the feature map of the convolution output also contains relative position information. The convolution process is shown in the Fig. 2. Each convolution kernel can extract specific features, and different convolution kernels can extract different features. This is all done automatically by the CNN.

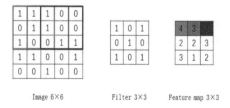

Image 6×6   Filter 3×3   Feature map 3×3

Fig. 2. The convolution process.

3   Model Building

3.1   Traditional Method

Support Vector Machine (SVM) [19], which is essentially a binary classifier, has been widely used in the field of feature classification. It maps the samples from the input space to the higher dimensional eigenspace by selecting the appropriate inner product function, and then solves a quadratic programming problem with linear constraints to obtain the global optimal solution. This method effectively avoids the dimensional disaster, ensures the convergence speed, and doesn't have the local minimum problem. In this paper, the polynomial kernel function is selected as the kernel function of SVM, and the parameter gamma is set to 'auto'. Long and Short Term Memory (LSTM) [20] network is a special RNN proposed by Hockritel and Schmidhuber, which can learn long-term dependence and solve the problems of gradient disappear and gradient explosion in long-sequence training. There are three gates of input, forget, and output in the memory cells that are designed to preserve the previous information, update the cell state, and control the flow information. Although SVM and LSTM can classify EEG signals, they both need to extract features from the data first. They only consider the temporal and spectral feature of EEG signals, but do not consider the spatial feature and location information.

3.2   Convolutional Neural Network

Convolutional Neural Network (CNN) was originally proposed by LeCun [21] as a typical feedforward neural network for image processing. CNN as a deep learning model, has the following advantages: (1) Features can automatically be extracted from the input data directly without constructing a set of features for classification, (2) The weights can be shared to avoid the overfitting, (3) The sparse connection of CNN can reduce the number of training parameters. The whole CNN contains three different types of layers, which are Convolution Layer, Pooling Layer and Fully Connected Layer.

- Convolution Layer consists of a number of hidden units. Each hidden unit is a matrix, which can convolve the input data. Each hidden unit convolves to produce a feature map.
- Pooling Layer is used for the subsampling of data to reduce the data dimension after the output of the convolutional layer, which can effectively reduce the amount of calculation and prevent overfitting.
- Full Connection Layer can correlate all the previously extracted features and map them to the output space.

In this study, 2D-CNN is selected, in which the Max Pooling is used for the Pooling Layer. The structure of each layer and the size and number of filters are shown in the Fig. 3. Rectified Linear Unit (ReLU) [22] is selected as the activation function of the hidden layer. Compared with other activation functions such as

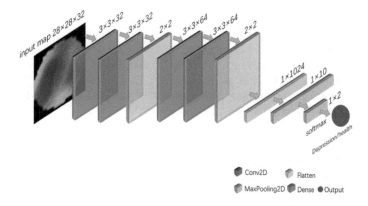

Fig. 3. The proposed 2D-CNN structure.

Sigmoid and *tanh*, ReLU has the following advantages: (1) For linear functions, ReLU has a stronger ability of expression, (2) For the nonlinear function, the gradient of ReLU in the non-negative range is constant, so there is no problem of gradient disappear, so that the convergence rate of the model is kept in a stable state. The ReLU expression is shown as below:

$$f(x) = \begin{cases} x, x > 0 \\ 0, x \leq 0 \end{cases} \tag{4}$$

Moreover, in order to solve the overfitting, this paper introduced the Dropout layer [23]. During the training of the actual network model, Dropout sets some neurons of hidden layer to 0, and then these neurons become ineffective during the forward propagation. In the Dense layer, softmax is selected as the classifier to classify the output of CNN. The softmax expression is as follows:

$$S(x) = \frac{1}{1 + e^{-x}} \tag{5}$$

4   Experimental Process and Results

4.1   Converting EEG Signals into Images

In previous studies, EEG classification was mainly based on extracting some features of EEG frequency domain and then classification with traditional classifiers. Because the input of 2D-CNN is image, this paper transforms EEG data into image. On the basis of frequency domain, spatial information is also considered in this study and the EEG signals are finally input to 2D-CNN in the form of images. The position of the electrode on the electrode cap reflects the spatial information of the EEG signals. In this paper, Azimuth Equidistant Projection (AEP) [24] is used to project the position of the electrodes in three-dimensional space onto a two-dimensional plane. The width and height of the image represent the spatial distribution of cortical activity. Research [25] shows that there are significant differences between patients with depression and normal people in the $\theta(4\text{–}8\,\text{Hz})$, $\alpha(8\text{–}13\,\text{Hz})$ and $\beta(13\text{–}30\,\text{Hz})$ bands. Therefore, we extract the θ, α and β bands of EEG signals respectively. Repeat the process for θ, α and β bands to generate three topographic maps, which are combined to form an image with three (color) channels, as shown in the Fig. 4.

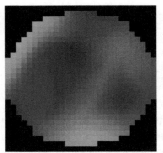   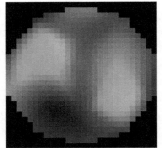

(a) Brain Map of Depressed People   (b) Brain Map of Healthy People

Fig. 4. RGB three-channel color brain map.

4.2   Training

Firstly, all samples and labels are randomly shuffled to ensure uniform distribution of samples. Then, the first 80% is selected as the training set, the last 20% as the test set, and 10% of the training set as the validation set. After several tests, the parameters used in depression diagnosis is shown in Table 2. In addition, Adam is selected as the optimizer, and Categorical Crossentropy is chosen as the loss function.

Table 2. Parameters used in the diagnosis model of depression

Description	Value
The number of filters in convolution layer	32/64
The convolution kernel size in convolution layer	3*3
The pooling size in maxpooling layer	2*2
Dropout	0.25
Batch size	100
Epoch	100

4.3   Evaluation Criteria

The index to evaluate classification effect is generally accuracy, which is defined as the proportion of the correct classified samples of a given sample to the total sample. However, this indicator performed poorly when the positive and negative samples are unbalanced. Therefore, in this study, multiple evaluation indicators are used to comprehensively reflect the performance of the model, including accuracy, precision, recall and F1-score, which are calculated using the following equations:

$$Accuracy = \frac{|TP| + |TN|}{|TP| + |FP| + |TN| + |FN|} \tag{6}$$

$$Precision = \frac{|TP|}{|TP| + |FP|} \tag{7}$$

$$Recall = \frac{|TP|}{|TP| + |FN|} \tag{8}$$

$$F1\text{-}score = 2 * \frac{Precision * Recall}{Precision + Recall} \tag{9}$$

Where TP, FP, TN and FN represent true positive, false positive, true negative and false negative respectively. True positives represent the number of people who predicted depression and actually are depressed, false positives represent the number of healthy samples predicted to be depressed, true negatives represent the number of samples predicted to be healthy and actually healthy, and false negatives represent the number of samples predicted to be healthy but actually depressed. Meanwhile, the loss function Categorical Crossentropy is used in 2D-CNN to measure the performance of the model. The equation of the loss function is as follows:

$$Loss = -\sum_{i=1}^{N} Y_i \cdot log\hat{y}_i \tag{10}$$

The accuracy and loss curves of the 2D-CNN model on training set and validation set are shown in the Fig. 5.

Confusion matrix is an effective visualization tool for the performance of classification methods. Each row in the confusion matrix represents the truth, while each column represents the predicted label [26]. The confusion matrix records the test classification results, and this study conducts tests on 1,300 samples, as shown in the Fig. 6.

In order to evaluate the effectiveness of the proposed method, comparative experiments were conducted on the same dataset, and the results are shown in Table 3 and Table 4, including the classification results of Pearson Correlation Coefficient, and Power Spectral Density on SVM, LSTM and 1D-CNN. The experimental results show that the accuracy of the proposed 2D-CNN is significantly higher than other methods, and the loss is the lowest.

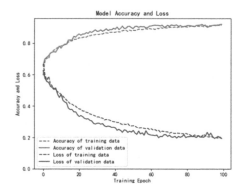

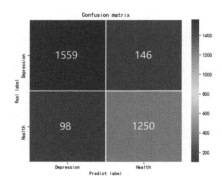

Fig. 5. Accuracy and loss of 2D-CNN.   **Fig. 6.** Confusion matrix of 2D-CNN.

Table 3. The evaluation criteria of the proposed method and other methods

Method	Feature	Accuracy %	Precision	Recall	F1-score	Support
SVM	PCC	0.8871	0.88	0.88	0.88	3053
	PSD	0.9057	0.91	0.91	0.91	3053
LSTM	PCC	0.8970	0.90	0.90	0.90	3053
	PSD	0.8786	0.88	0.88	0.88	3053
1D-CNN	PCC	0.87	0.87	0.87	0.87	3053
	PSD	0.88	0.88	0.88	0.88	3053
2D-CNN	Feature map	**0.92**	**0.92**	**0.92**	**0.92**	3053

Table 4. The accuracy and loss of LSTM, 1D-CNN and 2D-CNN

Neural network	Accuracy	Loss
LSTM	0.8970	0.259
1D-CNN	0.88	0.31
2D-CNN	**0.92**	**0.18**

5   Conclusion and Future Work

In order to improve the performance of depression diagnosis, a approach based on 2D-CNN is proposed in this paper. Experimental results show that the performance of this method is better than SVM, LSTM and 1D-CNN and the accuracy can reach 92% and the loss is only 0.18. Therefore, this paper provides an effective method for the diagnosis and recognition of depression. The results show that the combination of frequency domain information and spatial information can improve the classification effect. The depression diagnosis approach based on deep learning also has many difficulties to overcome. Firstly, there is no systematic deep learning tuning theory knowledge. The tuning of model parameters needs to be based on experience. Secondly, depression diagnosis requires the system to be able to identify the depression in a timely and fast manner, but the training of deep learning model is time consuming. Therefore, in the future work, we will optimize the deep neural network to improve the classification accuracy and reduce the training time. Moreover, the degree of depression can also be classified in more detail, such as normal, mild, moderate, severe, etc. that will be adopted to the multiple classifications scenarios.

Funding: This research was funded by the Beijing Municipal Natural Science Foundation, grant number 4192005.

References

1. Cuevas, G.: Functional brain networks of trait and state anxiety in late-life depression. Am. J. Geriatr. Psychiatry **29**(4S), S52–S53 (2021)
2. Brundtland, H.: Mental health: new understanding. New Hope **286**(19), 2391–2391 (2001)
3. Wang, L.: Distinguish between depression and depression - protect your child's physical and mental health. Report. Observed **000**(009), 103–103 (2018)
4. Meesters, Y.: Sensitivity to change of the beck depression inventory versus the inventory of depressive symptoms. J. Affect. Disord. **281**, 338–341 (2021)
5. Gannouni, S.: EEG-based BCI system to detect fingers movements **10**(12) (2020)
6. Fernando, S.: Depression biomarkers using non-invasive EEG: a review **105**, 83–93 (2019)
7. Wei, Y.: Comparative analysis of electroencephalogram in patients with neurological disorders and depression. J. Shanxi Med. Univ. **36**(1), 96–97 (2005)
8. Linardatos, P.: Explainable AI: a review of machine learning interpretability methods. Entropy (Basel, Switzerland) **23**(1), 18 (2020)
9. Liao, S.: Major depression detection from EEG signals using kernel eigen-filterbank common spatial patterns. Sensors **17**, 1385
10. Schirrmeister, R.: Deep learning with convolutional neural networks for brain mapping and decoding of movement-related information from the human EEG. arXiv preprint arXiv:170305051
11. Lai, H.: Classification of resting state EEG signals in patients with depression and schizophrenia. J. Biomed. Eng. **36**(06), 916–923 (2019)
12. Hanshu, C.: Feature-level fusion approaches based on multimodal EEG data for depression recognition. Inf. Fusion **59**, 127–138 (2020)
13. Tripathi, S., Acharya, S.: Using deep and convolutional neural networks for accurate emotion classification on DEAP dataset. In: AAAI, pp. 4746–4752
14. Wang, J., Zhu, H., Wang, S.-H., Zhang, Y.-D.: A review of deep learning on medical image analysis. Mob. Netw. Appl. **26**(1), 351–380 (2020). https://doi.org/10.1007/s11036-020-01672-7
15. Lu, G.: Emotion recognition of EEG signals based on CNN and LSTM. J. Nanjing Univ. Posts Telecommun. (Nat. Sci. Ed.) (01), 1–7 (2021)
16. Frassineti, L.: Multiparametric EEG analysis of brain network dynamics during neonatal seizures. J. Neurosci. Methods **348**, 355–358 (2021)

17. Edelmann, D.: On relationships between the Pearson and the distance correlation coefficients **169**, 108960 (2021)
18. Alam, R.: Differences in power spectral densities and phase quantities due to processing of EEG signals. Sensors (Basel, Switzerland) **20**(21) (2020)
19. Guler, I.: Multiclass support vector machines for EEG-signals classification. IEEE Trans. Inf. Technol. Biomed. **11**(2), 117–126 (2007)
20. Chen, J.: Gated recurrent unit based recurrent neural network for remaining useful life prediction of nonlinear deterioration process. Reliab. Eng. Syst. Saf. **185**, 372–382 (2019)
21. LeCun, Y.: Handwritten digit recognition with a back-propagation network. Adv. Neural Inf. Process. Syst. **2**, 396–404 (1997)
22. Glorot, X., Bordes, A.: Deep sparse rectifier neural networks. In: Proceedings of the Fourteenth International Conference on Artificial Intelligence and Statistics (AISTATS), Fort Lauderdale, FL, USA, 11–13 April 2011 (2011)
23. Ioffe, S.: Batch Normalization: Accelerating Deep Network Training by Reducing Internal Covariate Shift. arXiv arXiv:1502.03167 (2015)
24. Ahmad, A.: 3D to 2D bijection for spherical objects under equidistant fisheye projection. Comput. Vis. Image Underst. **125**, 172–183 (2014)
25. Hosseinifard, B., Moradi, M.: Classifying depression patients and normal subjects using machine learning techniques. In: 2011 19th Iranian Conference on Proceedings of the Electrical Engineering (ICEE). IEEE (2011)
26. Chen, Z.: Gearbox fault identification and classification with convolutional neural networks. Shock Vib. **2015**, 1–10 (2015)

Applying Decision Tree to Detect Credit Card Fraud

Wen-Chih Chang[✉], Yi-Hong Guo, Ya-Ling Yang,
Ming-Chien Hsu, Yi-Hsuan Chu, Ting-Yi Chu,
and Long-Cheng Meng

Department of Information Management, Chung Hua University,
Hsinchu, Taiwan
earnest@g.chu.edu.tw

Abstract. The topic we are participating in is the credit card fraud detection
Artificial Intelligence Open Challenge 2019 Autumn Competition. The number of
times the prediction results can be uploaded in this competition is five times a day.
The competition period is from September 6, 2019 to 2019. On November 22, the
end time is November 22, 2019. Our goal is to accurately predict whether the
transaction is a fraudulent transaction. In this topic, we learn how to process data,
analyze and predict, explore the analytical performance of different machine
learning models, identify potential fraudulent transactions, and respond early to
find key factors that can be judged as fraudulent, such as: When the transaction is
located in a foreign country, the chance of being stolen is higher, the mobile phone
is often used by others, the chance of being stolen is higher, the chance of using the
Internet transaction is higher, the use of Fallback is easier to be stolen, and the
chance of being stolen The key factors such as the high probability of being stolen
and brushing. We applied the XGBoost model. This model has higher accuracy
and faster training speed, which can greatly reduce the training time. Let us find
that the XGBoost model is more suitable for predicting the data set of credit card
fraud among other models. Effectively increase the prediction accuracy to 51.97%

Keywords: Credit card fraud · Artificial intelligence · XGBoost model ·
Big data · Decision tree

1 Introduction

1.1 Background and Motivation

In order to improve the ability of data analysis and prediction and cultivate the acuity of
data observation, we decided to participate in the competition organized by T-Brain AI
actual combat bar. The competition is divided into two major goals: predictive analysis
and data classification. We chose Artificial Intelligence Open Challenge- "Credit Card
Counterfeit Detection" competition. This competition is provided by the bank to help
banks find counterfeit counterfeit transactions. Regardless of the size of the transaction
amount, it can determine the characteristics of counterfeit counterfeit transactions, and
can predict it Whether the transaction is fraudulent, prevent the occurrence of fraudulent brush as soon as possible.

© The Author(s), under exclusive license to Springer Nature Singapore Pte Ltd. 2022
J. C. Hung et al. (Eds.): FC 2021, LNEE 827, pp. 103–113, 2022.
https://doi.org/10.1007/978-981-16-8052-6_11

The data type of the data set for this competition is relatively complex. The training set has a huge amount of data, diverse attributes, and data overfitting and serious problems. The goal of this prediction is to analyze whether the credit card will be stolen. It is suitable for us to in-depth study of data processing and analysis., And use the past experience to compete with other participants to improve the ranking and improve our ability to process data.

1.2   Research Purpose

This topic mainly predicts whether the credit card used by customers of Yushan Bank is stolen. First, you must understand the key factors of the credit card transaction type, transaction amount, installment number... etc., and finally analyze whether the customer will be To steal, the data provided for this competition can be used to solve related problems. The purpose of this research is as follows:

1. Learn how to conduct in-depth analysis and prediction of data.
2. Cultivate the acuity of data observation.
3. Explore the analytical performance of different machine learning models.
4. Assist the bank to determine the key factors of potential fraud.

1.3   Research Process

After clarifying the background, research motivation and purpose of the problem, you can begin to collect and analyze the relevant literature to establish the scope of the research, based on the reference and arrangement of the literature, as the theoretical basis of the research framework of this topic.

The research process is mainly divided into seven stages. The first stage needs to confirm the purpose, the second stage collects and organizes data, and sorts out the data that are helpful to the forecast. The third stage is data pre-processing, and the collected data is subjected to feature engineering. And formalization. The fourth stage is for visual design. The important attributes in the data set are presented and observed using visual methods such as drawing charts to help us clearly understand the characteristics and correlations between attributes. The fifth stage is to establish a model and use different models to carry out Compare and use Gridsearch cross-comparison to find the most suitable machine learning model for this data set. The sixth stage evaluation model. After the model is established, use the decision tree model, confusion matrix, and classification report as the evaluation index of the machine learning model. Finally, the seventh stage presents the prediction results, which are explained in two parts below.

2   Related Works

2.1   Decision Tree

Decision tree was first proposed by Hunt in 1996. It is a model that uses the concept of tree branch as a decision mode. Most of it can be used in classification prediction. It is divided into the following three types: classification tree, regression tree, and classification regression tree.

The classification tree is used to predict the strain number of the category type, the regression tree (Regression Tree) uses regression analysis to predict the value, and the classification and regression tree (Classification and Regression Tree, CART) was developed by American statistician Leo Breiman in 1984. It is proposed that it combines the characteristics of classification tree and regression tree, which can predict the type of classification and also predict the numerical data. The feature is that only two branches are generated for each classification, which is used to plan and analyze the data set, and it is not limited. Variable types have greater flexibility in analysis.

The establishment of the decision tree is divided into three steps:

1. Take the data matrix as the root node
2. Use different decision tree algorithms to find the largest attribute as the criterion for branching. Common methods include GINI (Gini coefficient) and Entropy.
3. If the accuracy of the judgment result of the algorithm rule used in the second step does not meet the condition, then calculate the next branch attribute.

GINI is an attribute selection index that uses binary division of attributes to build a decision tree, so each node has only 2 branches connected to child nodes. For example: CART decision tree uses this coefficient to select better data attributes to construct a complete decision tree.

2.2 Random Forest

Random Forest was proposed by American statistician Leo Breiman in 2001. The principle is to combine multiple CART decision trees, add a preset number of data attributes, and randomly select split points to build decision trees for training. The trees look similar to increase the final calculation result.

It is known from the principle that the random forest has multiple decision tree combinations. This machine learning is called Ensemble Method, which can also be called ensemble or ensemble. It must meet the condition that each classifier is different, and the accuracy is greater than 0.5. The overall machine learning algorithm (Ensemble learning) integrates the prediction results of multiple weak learners to enhance accuracy. The integration method creates a set of learners for training data, and votes based on the prediction results of each learner, so it is more stable and more accurate. There will be no deviation or overtraining. The operation steps of integrated learning are as follows:

1. Create multiple data sets
2. Construct multiple learners
3. Integrated learner (voting)

2.3 Support Vector Machine

Support Vector Machine (SVM), proposed by Corinna Cortes and Vapnik in 1993, is mainly for the classification of binary data, and the principle of minimum classification error rate is used for two-category segmentation. SVM uses the current data as training, selects several support vectors from the analyzed training data to represent all the data, removes some extreme values, and packs the selected support vectors into a model. SVM is a supervised learning classifier, it can be used to analyze the classification pattern in the data, the main purpose is to do classification and regression analysis. Take the following figure as an example. Given a training data set, each data (black dots, white dots) is given a classification answer 0 and 1, each data is a vector, and the length of the vector is 2(ex:x1,x2), SVM can construct a classification model to classify all data into 2 categories. If new test data is input into the model later, it can also be automatically classified into 0 or 1 category by the model.

2.4   Artificial Neural Network

Artificial Neural Network (ANN) is a mathematical model or calculation model inspired by biology that simulates the structure and function of the human brain. Each neuron is connected with other neurons and will be impacted by the state of other neurons, thereby determining whether it will be stimulated or not. ANN is a black box model that can find the relationship between input data and output data and form a pattern to make predictions for different input data.

2.5   eXtreme Gradient Boosting

Extreme gradient boosting (eXtreme Gradient Boosting, referred to as XGBoost), proposed by the initiator Chen Tianqi in 2014, Gradient means that the XGBoost module is developed in C++ of the Gradient Boosting Machine. Boosting belongs to the overall learning classifier, which will reduce the accuracy of many classifications. The model trees are combined to become a highly accurate model, which will be continuously revised, and a new tree will be generated every time it is revised. XGBoost proposes the concept of block, which sorts the data before training. Then save it as a block structure, and reuse the data in the structure in the subsequent process, so that the module can be parallelized and reduce the training time.

3   Analysis Model

3.1   Data Set Description

The files are provided for this competition data, namely training set, test set, and prediction result data set. The training set data can be used for data pre-processing and visualization design and model building. After the model is established, use the processed training set to predict the data in the test set, and finally generate a data set of prediction results, we will also introduce the fields in the data set. Figures 1 and 2 shows the training set and test set. Table 1 is the data description of each column.

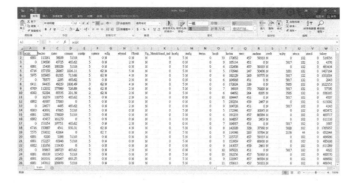

Fig. 1. Training data set

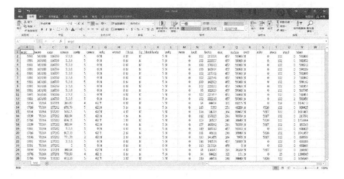

Fig. 2. Test data set

3.2 Models and Kits

We used some models and kits in this research.

The Model:

1. XGBoost model: used to predict the accuracy rate, the accuracy rate is high and the training speed is very fast, and the XGBClassifier type is selected.
2. Random Forest model: used to predict the accuracy rate, combine multiple CART trees, and add randomly allocated training data to increase the final calculation results.

The Module:

1. CSV module: used to read and write CSV format. CSV file format is a common format for importing and exporting electronic forms and databases.
2. Numpy module: mainly used for data processing, which can quickly manipulate huge data in multi-dimensional arrays.
3. Sklearn module: used for machine learning. Used to divide training set and test set. Make a classification report to show the accuracy and recall rate of each category

Table 1. Data information

Id	Column description
acqic	The bank code of the store's credit card POS machine
bacno	User Id
cano	The credit card number and sponsor code of the transaction
conam	The organizer has uniformly converted the data unit to NewTaiwan dollars
contp	Sponsor code, 0–6
csmcu	The currency used for the transaction
ecfg	Whether the transaction is an online transaction, yes = Y, no = N
etymd	The transaction type of the transaction, the organizer code, 0–10
flbmk	Whether the transaction is completed by swiping the magnetic stripe, yes = Y, no = N
flg_3dsmk	Whether the transaction uses 3DS, yes = Y, no = N, (3DS is a safe design for online card or cardless payment)
fraud_ind	Whether the transaction has been stolen, yes = 1, no = 0 (only the train file has a note)
hcefg	The payment type of the transaction, the sponsor code, 0–9 (excluding 4)
insfg	Whether the transaction has installment, yes = Y, no = N
iterm	The transaction is divided into several periods, 0–8 periods
locdt	The number of days after the authorization base date, train (1–90), test (91–120)
loctm	In seconds, the time format is 24-h system, (235610 means 23:56:10)
mcc	Merchant category code, organizer code
mchno	The transaction store code, organizer code
ovrlt	The credit card amount exceeds the credit limit, excess = Y, none = N
scity	According to the city location of the transaction store, the organizer code
stocn	According to the country location of the transaction store, the organizer code
stscd	Sponsor code, 0–4
txkey	Refer to (locdt and loctm two fields) coding

The Package:

1. Pandas package: supports a variety of text and data loading, and can quickly understand the data structure
2. Matplotlib suite: used to build charts, can be used with numpy.
3. Seaborn suite: used to show the relationship between data.

3.3 Data Pre-processing

1. We applied heat map (Fig. 3) to discover every data column relation
2. Loctm authorization time: We found that the time field is composed of hour, minute, and second in order, so we cut the field and created an Ampm field to store it in hours.

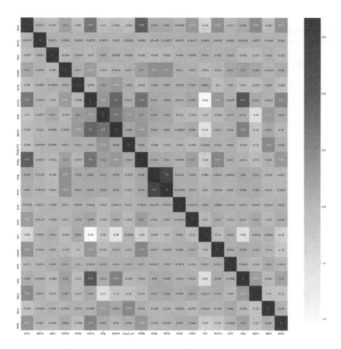

Fig. 3. The attribute relations with heat map

3. Conam consumption amount: The forum administrator has stated that the unit is Taiwan dollar. We take foreign consumption as the starting point that it is easier to be stolen, and hope to find out the decimal point of the transaction amount for Taiwan dollar consumption and foreign consumption.
4. Bacno return account: Accounts that have a precedent for fraudulent brushing are more likely to be stolen again.

Figure 4 shows the distribution of fraud id abd Fig. 5 shows the distribution of etymd attribute stealing.

Figure 6 shows the first time drawing of the machine learning model decision tree with some blank data. We Used XGBoost model to fill in empty values and redraw machine learning model decision tree in Fig. 7.

```
In [14]: ampm_fraud_ind = pd.crosstab(train['ampm'], train['fraud_ind'])
         ampm_fraud_ind.plot(kind ='bar')
```

Out[14]: <matplotlib.axes._subplots.AxesSubplot at 0x1fe0006b860>

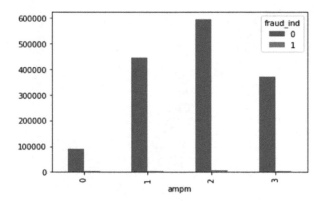

Fig. 4. View the distribution of attributes

```
In [28]: sns.countplot(train["etymd"],hue=train["fraud_ind"])
         #發現2跟8比較多盜刷
```

Out[28]: <matplotlib.axes._subplots.AxesSubplot at 0x1fe14a51860>

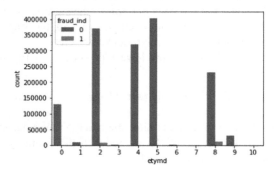

Fig. 5. View the distribution of etymd attribute stealing

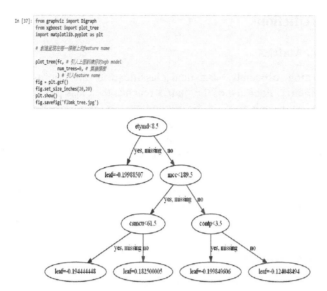

Fig. 6. Drawing machine learning model decision tree

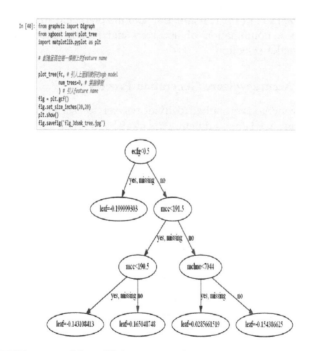

Fig. 7. Using XGBoost model to fill in empty values and redraw machine learning model decision tree

4   Model Prediction

4.1   Confusion Matrix

In machine learning, especially statistical classification, the confusion matrix is also called the error matrix. Each list of the matrix reaches the class prediction of the sample by the classifier, and each row of the second matrix expresses the true category to which the version belongs. The reason why it is called the "confusion matrix" is because it is easy to see whether the machine learning is confused or not.

The confusion matrix is divided into four categories.

TP: The real situation is "Yes", and the module predicts the number of "Yes".
TN: The real situation is "No", and the number of modules that predicts "No".
FP: The real situation is "No", and the number of modules predicting "Yes".
FN: The real situation is "Yes", the module predicts the number of "No".

4.2   Classification Report

The classification report has three elements. They are the precision, recall and F1_score.

Precision: is the prediction accuracy.
Recall: the same ratio of the real situation and the predicted situation of the module.
F1_score: It is a combination of accuracy and recall, and is often used as an indicator for model selection.

4.3   Prediction Accuracy/Score Generation Process

Figure 8 shows the method we applied to divide two set (test set and training set). With the chosen data columns, we kept modulating the model and predicted the credit card fraud.

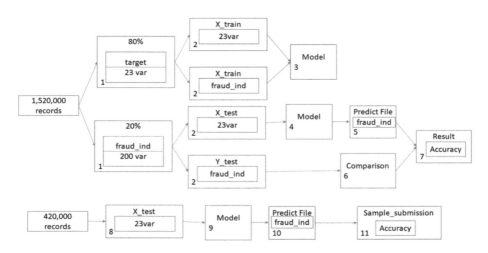

Fig. 8. Forecast accuracy rate generation flowchart

5 Conclusion

We use the complete 1.52 million data for analysis, but due to the imbalance of the data, the training module will be misjudged, and the accuracy rate is only about 10%. Later, we observed and processed the field information. Delete some fields that are not very helpful for data prediction, and then merge some fields with high relevance to effectively reduce the misjudgment rate, and use this result for training, and the accuracy rate begins to improve significantly.

At the beginning, the random forest model was used to predict, but the accuracy of the data trained by the random forest was limited. After the accuracy rate reached 37%, it could not be effectively improved. Therefore, we used the calculation model XGBoost learned in the previous competition. Further predictions finally made a breakthrough and effectively improved to 51% prediction accuracy.

References

1. Cortes, C., Vapnik, V.: Suppport-vector netowrks. Mach. Learn. **20**, 273–297 (1995)
2. https://www.Kaggle.com/c/santander-customer-transaction-prediction/kernels
3. Hunt, E.B., Marin, J., Stone, P.J.: Experiments in Induction. Academic Press, New York (1966)
4. Breiman, L., Friedman, J., Stone, C.J., Olshen, R.A.: Classification and Regression Trees. Taylor & Francis (1984)
5. McCulloch, W.S., Pitts, W.: A logical calculus of the ideas immanent in nervous activity. Bull. Math. Biophys. **5**(4), 115–133 (1943). https://doi.org/10.1007/BF02478259, ISSN 0007-4985
6. Story and Lessons behind the evolution of XGBoost. Accessed 01 Aug 2016
7. XGBoost on FPGAs. Accessed 01 Aug 2019
8. XGBoost4J. Accessed 01 Aug 2019

Implementation of Traffic Sign Recognition System on Raspberry Pi

Chuan-Feng Chiu[✉], Zheng-Qing Liu, Xian-Yi Wu, Hong-Jun Yan,
Qi-Xuan Xu, and Yun-Kai Lin

Department of Information Management,
Minghsin University of Science and Technology, Xinfeng, Taiwan
cfchiu@must.edu.tw

Abstract. In this paper we intent to design and develop an intelligent traffic sign recognition system on Raspberry Pi. In the real situation, we could find the fact that drivers are often drive the car tiredly and cause dangerous situation. In order to avoid the dangerous accidents, automatic drive applications are appeared to assist drivers. However, the automatic driving car does not recognize the traffic sign correctly in real-time and depend on non-real time traffic sign information so that the dangerous accidents are appeared. So, in this paper we will implement a system that have recognition mechanism based on deep learning to distinguish the traffic signs. With the developing application, we would not depend on the map updating frequently to know the traffic sign situation. This mechanism also helps drivers to have secure driving behavior. On the other hand, we also design a prototype of automatic car based on Raspberry pi development board to evaluate our proposed method.

Keywords: Self-drive car · Traffic sign detection · Traffic sign recognize · Deep learning

1 Introduction

Because of the rapid development of hardware and computation speed, the action or the behavior of the daily life which need more human power had changed. For example, in order to reduce the cost of human power in the traditional manufacturing industries, the automatic manufacturing machine would be used to lower manufacturing cost. On the other hand, such kind of changing could reduce the error by human beings also. In recent year, AI is gained more attention. In the application of AI, autonomous driving is one of popular application. Autonomous driving is an enhancing driving behavior, the autonomous car could drive on the road without human control. Autonomous car uses a lot of sensors including GPS, camera LiDAR etc. to detect environment situation to have safety driving experiences. The idea of autonomous car is inspired from 1920. Until 1980, the prototyping of autonomous car is appeared. CMU propose the Navlab [1] and ALV [2] project in 1984 and proposed the first prototyping of autonomous car NavLab1 which had 20 mile/hour maximum speed. In 2014, Telsa proposed the Model S electric car with semi-automatic driving functionality. Many car manufactures also start to develop the autonomous car including Ford, Waymo, Intel, Volkswagen etc..

J. C. Hung et al. (Eds.): FC 2021, LNEE 827, pp. 114–120, 2022.
https://doi.org/10.1007/978-981-16-8052-6_12

Although many companies start to develop autonomous car and have demonstration, the current status focus on the safety and correctly driving alone the road lane. However, the actual situation on the road is very different so that current development of autonomous still have a lot of challenges. Therefore, in our opinion an autonomous driving assistance is more practical for drivers. So, in this paper we focus on the traffic sign recognition application. Now most traffic sign symbol is embedded in the map and the map should be updated frequently. The real-time traffic sign would be reflected on time. Therefore, we intent to implement a real-time traffic recognition system that could monitoring the real-time traffic sign on the road and reducing the wrong appearance of traffic sign on the map. In our implementation we would design a car prototype to validate the traffic sign recognition system also.

The paper is organized as following. In Sect. 2, we would have brief review of related works. In Sect. 3 we would reveal the detail of our implementation. Finally, we have conclusion in Sect. 4.

2   Related Works

Autonomous car could use different sensors and technology to detect the situation of the road that including GPS, camera, inertial sensors, LiDAR, distance measurement etc.. Using these technology or combined technology, autonomous car could reduce the traffic accidents and have better driving performance including path planning etc.. On the other hand, vehicles could communicate with each other by Vehicle-to-Vehicle net to have additional information for driving. These technologies could improve the development of the autonomous car. In the following we focus on the computer vision or image processing technology which is related with our implementation directly. Road lane detection is the first noticed issue. By the computer vision technology, autonomous car could detect the edge and situation of the road and could drive on the road safety and steady to avoid the tiredly driving causing dangerous situation [3, 4]. But the real situation of the road is un-predicable, such kind of autonomous driving does not use in real environment and still in experimental situation. In the rapid development of deep learning technology on computer vision, many researchers start to use the deep learning technology to detect the road lane [5–8]. In these approach, CNN [8] is the most popular model. In CNN model, sliding window is applied to scan the whole image and feed the scanned result to convolution layer, pooling layer and fully connected layer to have correct object detection result. However, CNN have high computation cost and would not process in real-time easily so that it could not apply to autonomous car development directly. In order to resolve the computation cost problem, R-CNN [9] is proposed to reduce the computation cost to have better performance by selecting the most important area in the image to avoid scanning whole image comparing with traditional CNN model. Fast R-CNN [10] is proposed to have more improvement on performance comparing with R-CNN and CNN model by pass scanning the overlapping area in the image. Fast R-CNN use nearest neighbour interpolation method to compute the object position in the image. However, this method could have position offset to get error object position which result in lower accuracy. So Mask R-CNN [11] is proposed to resolve the problem of Fast R-CNN in advance by using bilinear interpolation method. Fast R-CNN and Mask R-CNN are the two-stage object detection model. In order to have improved performance in advance, the one-stage model is proposed and YOLO [12–15] is the famous technology. YOLO could reduce the computation cost by reduce the count of image scanning. So, the better performance is achieved comparing with the above approach. [16, 17] use the YOLO to detect the objects on the road like human and cars etc.. In our implementation, we would use the YOLO as the traffic sign detection mechanism to have better performance on the Raspberry Pi platform.

3 The Proposed System

In this section we describe the detail of the implementation of traffic sign recognition system which is implementing on Raspberry Pi platform. The overall design has two major parts including prototype of hardware and the corresponding software module. We describe the design of the proposed prototype of hardware in the following first. We use the Raspberry Pi 4 as the implementation platform. The implementing hardware include L298N motor driver, cameras and ultrasonic sensor. The Raspberry Pi 4 is the main computation and control kernel to coordinate the sensors and component action. On the other hand, the traffic sign detection and recognition are also processing by Raspberry Pi 4. The motor driver would control the actions of go forward, backward, stop and turn left or right. The camera would capture the traffic sign to deliver to the Raspberry Pi computation kernel to perform traffic sign detection and recognition. The above components are integrating as a testing prototyping car showed in Fig. 1. Figure 2 shows the layout design of the proposed implementation.

Fig. 1. The Back-End and Front-End of the implementation prototyping

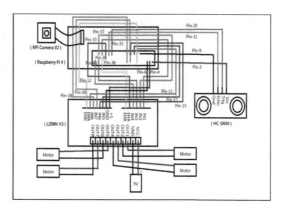

Fig. 2. The implementing hardware layout of traffic sign enabled Car

In order to achieve the implementation of the traffic sign detection and recognition on Raspberry Pi, we proposed the design of the software architecture first. Figure 3 shows the proposed design. We implement the software architecture based on the Raspbian operating system which is based on linux kernel. In order to control and monitor the behavior of camera, motor driver and ultrasonic sensors, we design a common sensor abstract processing layer to unify the message passing between the proposed application and sensors. In the Sensor Abstract Processing Layer, we would encode each different sensor with different channel. Based on such kind of design, we could unify the processing for developing Control and Monitoring modules. In the future, we could use the unify design to integrate with more different equipments related with cars. In our implementation, Control and Monitoring would capture the images from camera sensor and feed to Traffic Sign Detection and Recognition module. On the other hand, the Control and Monitoring would communicate with motor drivers to trigger the motion of the car and ultrasonic to detect obstacle.

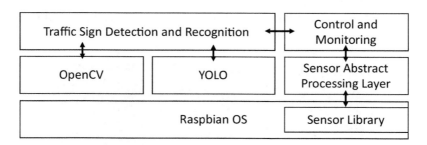

Fig. 3. The implementation software architecture

The Traffic Sign Detection and Recognition module is implemented based on OpenCV and YOLO libraries. We use YOLO which is an object detection with high performance to detect and extract traffic sign from the images captured by camera. After extracting the traffic sign in a bounding box, we use Haar Feature-based Cascade

Classifier [18, 19] to recognize the traffic sign. Figure 4 shows our traffic sign detection and recognition implementation. In order to make the detection and recognition correctly, we collect positive and negative samples to training the traffic sign detection and recognition. In order to have higher accuracy, we also extended the sample by different capturing image angles and different environment situation like light factors. Figure 5 shows the part of our training positive and negative samples.

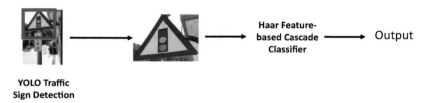

Fig. 4. The implementation flowchart of traffic sign detection and recognition

Fig. 5. The part of training positive and negative samples

In the following some experimental results are showed. Figure 6(a) and (b) shows the experimental recognition result of different traffic sign. The implementation car could recognize correctly. Figure 6(c) shows the detection of the traffic light with red light and the implementation car could stop in front of the motorcycle waiting zone and stop line. Figure 6(d) shows the experimental result of the weak light situation and the implementation could recognize successfully.

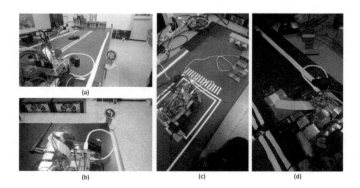

Fig. 6. Implementation experimental result

4   Conclusion

In this paper we implement a real-time traffic sign detection and recognition system on the Raspberry Pi platform. The implementation could avoid the error or non-state-of-art appearance of traffic sign on the map. We use Raspberry Pi as the platform and embed sensors to develop a prototype of car which has the traffic sign recognition assistance functional. For the implementation, we use data argument technology to have more training dataset to have better recognition performance. For the final result of the implementation, we validate the proposed implementation also and the implemented prototype of car could recognize the traffic sign correctly. In the future, we would apply the super resolution technology to traffic sign detection which achieve more long traffic sign detection distance and we would put the functionality to HUD system to act as a real driving assistant system.

References

1. Carnegie Mellon: Navlab: The Carnegie Mellon University Navigation Laboratory. The Robotics Institute. http://www.cs.cmu.edu/afs/cs/project/alv/www/index.html. Accessed 1 June 2019
2. Kanade, T., Thorpe, C., Whittaker, W.: Autonomous land vehicle project at CMU. In: Proceedings of the 1986 ACM Fourteenth Annual Conference on Computer Science (CSC 1986), pp. 71–80. ACM, USA (1986)
3. Szelisk, R.: Computer Vision Algorithms and Applications. Springer, Heidelberg (2011). https://doi.org/10.1007/978-1-84882-935-0
4. Jo, J., Tsunoda, Y., Stantic, B., Liew, A.-C.: A likelihood-based data fusion model for the integration of multiple sensor data: a case study with vision and Lidar sensors. In: Kim, J.-H., Karray, F., Jo, J., Sincak, P., Myung, H. (eds.) Robot Intelligence Technology and Applications 4. AISC, vol. 447, pp. 489–500. Springer, Cham (2017). https://doi.org/10.1007/978-3-319-31293-4_39
5. Huval, B., et al.: An empirical evaluation of deep learning on highway driving. Comput. Vis. Pattern Recogn. 17 (2015)
6. Sermanet, P., Eigen, D., Zhang, X., Mathieu, M., Fergus, R., LeCun, Y.: Overfeat: integrated recognition, localization and detection using convolutional networks. In: International Conference on Learning Representations (2013)
7. Szegedy, C., Toshev, A., Erhan, D.: Deep neural networks for object detection. In: Proceedings of the 26th International Conference on Neural Information Processing Systems (NIPS 2013), vol. 2, pp. 2553–2561 (2013)
8. Krizhevsky, A., Sutskever, I., Hinton, G.E.: ImageNet classification with deep convolutional neural networks. In: Proceedings of the 25th International Conference on Neural Information Processing Systems (NIPS 2012), vol. 1, pp. 1097–1105 (2012)
9. Ren, S., He, K., Girshick, R., Sun, J.: Faster R-CNN: towards real-time object detection with region proposal networks. In: Proceedings of the 28th International Conference on Neural Information Processing Systems (NIPS 2015), vol. 1, pp. 91–99 (2015)
10. Girshick, R.: Fast R-CNN. In: IEEE International Conference on Computer Vision (ICCV), pp. 1440–1448 (2015)
11. He, K., Gkioxari, G., Dollár, P., Girshick, R.: Mask R-CNN. In: IEEE International Conference on Computer Vision (ICCV), pp. 2980–2988 (2017)

12. Redmon, J., Divvala, S., Girshick, R., Farhadi, A.: You only look once: unified, real-time object detection. In: IEEE Conference on Computer Vision and Pattern Recognition (CVPR), pp. 779–788 (2016)
13. Redmon, J., Farhadi, A.: YOLO9000: better, faster, stronger. In: IEEE Conference on Computer Vision and Pattern Recognition (CVPR), pp. 6517–6525 (2017)
14. Redmon, J., Farhadi, A.: YOLOv3: an incremental improvement. Computer Vision and Pattern Recognition (2018)
15. Bochkovskiy, A., Wang, C.Y., Liao, H.Y.M.: YOLOv4: optimal speed and accuracy of object detection. Computer Vision and Pattern Recognition (2020)
16. Kulkarni, R., Dhavalikar, S., Bangar, S.: Traffic light detection and recognition for self driving cars using deep learning. In: Fourth International Conference on Computing Communication Control and Automation (ICCUBEA), pp. 1–4 (2018)
17. Corovic, A., Ilic, V., Duric, S., Marijan, M., Pavkovic, B.: The real-time detection of traffic participants using YOLO algorithm. In: 26th Telecommunications Forum (TELFOR). pp. 1–4 (2018)
18. Viola, P., Jones, M.J.: Rapid object detection using a boosted cascade of simple features. In: The 2001 IEEE Conference on Computer Vision and Pattern Recognition, vol. 1. p. 1 (2001)
19. Lienhart, R., Maydt, J.: An extended set of Haar-like features for rapid object detection. In: International Conference on Image Processing, vol. 1, p. 1 (2002)

Integrating Object Detection and Semantic Segmentation into Automated Pallet Forking and Picking System in AGV

Ruei-Jhih Hong[1], Yong-Ren Li[1], Min-Hsien Hung[2],
Jia-Wei Chang[2(✉)], and Jason C. Hung[2]

[1] Industrial Technology Research Institute, Hsinchu County, Taiwan
{rayhong,itriRen}@itri.org.tw
[2] Department of Computer Science and Information Engineering,
National Taichung University of Science and Technology,
Taichung City, Taiwan
{jhung,jwchang}@nutc.edu.tw

Abstract. The integration of the Internet of Things and artificial intelligence technologies in picking and distributing automated guided vehicles (AGVs) will reduce the ineffective walking time and reduce the workload of logistics operators during the picking process. It is expected to improve their work efficiency and speed. In the proposed system, calculating the travel path and rotation diameter of the self-propelled forklift with algorithms can make AGVs moving to the correct position for pallet forking and picking. In addition, AGVs achieved good object recognition and automatic pallet forking and picking by object detection and semantic segmentation. We believe the AGVs can reduce the workload of logistics operators and save the working time to improve their efficiency.

Keywords: Object detection · Semantic segmentation · Pallet forking and picking · Automated guided vehicles

1 Introduction

With the development of automation, Internet of Things and artificial intelligence technology, intelligent logistics has been the current trend. As the labor force decreases and the number of logistics operators decreases, it is necessary to develop picking and sorting trucks with automatic forklift pallets to help reduce the workload of logistics operators and improve the overall efficiency and speed. In the past, the traditional logistics picking and sorting mainly used manpower with trolleys to sort incoming or outgoing goods. The intelligent picking and sorting self-propelled vehicles are different from the previous sorting mode, as the self-propelled vehicles help the staff to carry and move to the sorting location independently, which will effectively reduce the staff walking and transporting time.

Currently, there are many technologies on the market that support forklift autonomous movement, such as magnetic rail, SLAM, laser positioning, WIFI positioning and other movement technologies, which can effectively assist forklift autonomous

© The Author(s), under exclusive license to Springer Nature Singapore Pte Ltd. 2022
J. C. Hung et al. (Eds.): FC 2021, LNEE 827, pp. 121–129, 2022.
https://doi.org/10.1007/978-981-16-8052-6_13

movement. In the part of automatic fork picking, such as fork picking of common pallets, the technology that needs to be considered is not only the accuracy of the fork carriage, but also the pallet position, the angle of the hole, the depth of fork picking and the way of moving, and other factors need to be calculated.

Therefore, the focus of this study is on how the forklift can accurately fork different types of pallets autonomously. In the pallet placement of Taiwan's logistics environment, there are usually different types of items around, so it is more difficult to identify the pallets in this kind of field than general items.

2   Purpose

To make the forklift to automatically fork the pallet accurately, this study will use the machine learning model training and depth camera to identify the hole position of the pallet to achieve the accurate automatic forking. With the shortage of manpower in logistics centers and the increasing workload, it is necessary to use the assistance of automated forklifts to help logistics personnel to complete their work more efficiently. Therefore, with the existing forklifts running automatically, the most important thing is to develop functions that can automatically fork different types of pallets and reduce the cost of introducing them into the logistics system. The function of automatic fork picking is very important to the logistics operation. At present, the domestic self-propelled forklifts are mainly made by SICK 3D camera from Germany, which is not only too costly, but also cannot match with different types of pallets. However, it is the main challenge of this study to combine the existing self-propelled function with the automatic forklift function, and to recognize different types of pallets effectively to complete the automatic forklift.

3   Related Work

3.1   YOLOv4

YOLOv4 [1] is a fast and accurate object detection system that can use a small number of GPUs, which is extremely suitable for combining with this research. YOLOv4 is composed of Input, Backbone, Neck, and Dense Prediction. The content of Input is image. Backbone chose to use CSPDarknet53 [2] because its receptive field is larger than the input network resolution. In the COCO data set test, the number of acceptable parameters is the largest, and the FPS value is the highest. Neck chose to use the SPP block [3] because it can not only greatly increase the receptive field but hardly reduce the network speed. Instead of choosing to use FPN [4] in YOLOv3 [5], Neck uses PANet [6] as a parameter aggregation method for different detectors and backbone levels. The head part chose to use YOLOv3.

YOLOv3 uses Darknet53 as Backbone, while YOLOv4 base on Cross Stage Partial Network (CSPNet), combined with ResNext50 and Darknet53 to form CSPResNext50 and CSPDarknet53. The experiment in Table 1 proves that the CSPResNext50 model is suitable for target classification. CSPDarknet53 model has higher accuracy in target detection, and can use Mish [7] and other techniques to improve the classification accuracy. Therefore, YOLOv4 uses CSPDarknet53 as the Backbone.

Table 1. Parameters of neural networks for image classification.

Backbone model	Input network resolution	Receptive field size	Parameters	Average size of layer output (WxHxC)	BFLOPS	FPS
CSPResNext50	512×512	425×425	20.6 M	**1058 K**	31 (15.5 FMA)	62
CSPDarknet53	512×512	725×725	**27.6 M**	950 K	52 (26.0 FMA)	**66**
EfficientNet-B3	512×512	**1311×1311**	12.0 M	668 K	11 (5.5 FMA)	26

The purpose of CSPNet is to enable the network structure to obtain more gradient fusion information and reduce the amount of calculation. The author of CSPNet believes that the repetition of gradient information in network optimization leads to a high amount of calculation for prediction. Therefore, Cross Stage Partial connections (CSP) is used to divide the feature map of the base layer into two parts, and then merges them through a cross-stage hierarchical structure. This method improves the learning ability of CNN, reduces the weight of the model and maintains accuracy, reduces the amount of computation, and increases the computation speed.

YOLOv4 adjusts for Spatial Pyramid Pooling (SPP) original architecture, to maintain the output of the spatial dimension. The method is to use multiple kernels of different sizes for pooling and concatenate the generated feature maps to output. Compared to a single-size single pooling kernel, using SPP can effectively increase the backbone of the receptive field, and separate salient features.

Path Aggregation Network (PANet) is an improvement based on the FPN used by YOLOv3. PANet adds an extra layer of upsampling to the number of layers connected in series and merges the originally added part to strengthen the ability of feature extraction and integration.

3.2   EfficientNetV2

EfficientNetV2 [8] is a new convolutional neural network model in the EfficientNets series. Compared with other models in the same series, EfficientNetV2 has faster training speed and better parameter efficiency. EfficientNetV2 combines Training-Aware NAS and Scaling to jointly optimize training speed and parameter efficiency. EfficientNetV2 improves the progressive learning method, dynamically adjusting the regularization (such as dropout) according to the size of the training image, which not only improves the training speed but improves the accuracy.

EfficientNetV2 studied the bottlenecks of EfficientNet [9] and pointed out that when the size of the training image is large, the training speed will be very slow. Therefore, EfficientNetV2 applies FixRes [10] to train with smaller image sizes but does not fine-tune anything after training. This approach can reduce the amount of calculation, have a larger batch size, and speed up training.

The depth-wise convolution used by EfficientNet in the shallow layer of the network causes the training speed to be very slow. Table 2 is an experiment on EfficientNet-B4. It is found that gradually replacing the shallow MBConv with Fused-MBConv can improve the training speed. However, replacing all stages with Fused-MBConv will significantly increase the number of parameters and increase the amount of calculation, and the training speed will also be reduced. So EfficientNetV2 uses NAS search technology to automatically search for the best combination of MBConv and Fused-MB Conv.

Table 2. Replacing MBConv with Fused-MBConv

	Params (M)	FLOPs (B)	Top-1 Acc	TPU imgs/sec/core	V100 Imgs/sec/gpu
No fused	19.3	4.5	82.8%	262	155
Fused stage 1–3	20.0	7.5	83.1%	362	216
Fused stage 1–5	43.4	21.3	83.1%	327	223
Fused stage 1–7	132.0	34.4	81.7%	254	206

EfficientNetV2 uses EfficientNet as the backbone. Compared with EffifientNet, EfficientNetV2 widely uses MBConv, uses a smaller expansion ratio in MBConv to reduce memory usage, and adds Fused-MBConv in the shallow layer. Table 3 is the structure of EfficientNetV2, which uses a 3x3 kernel and increases the number of layers in each stage to compensate for the reduced receptive field caused by the smaller kernel size. Considering the problem of parameters and memory usage, EfficientNetV2 completely deletes the last stride-1 stage of EfficientNet.

3.3   EfficientFCN

EfficientFCN [11] can realize efficient and high-accuracy semantic segmentation while considering the overall context of the input image. The backbone of EfficientFCN is an ImageNet pre-training network without any dilated convolution. It has introduced a holistically-guided decoder (HGD) and can obtain high-resolution feature maps through multi-scale features. Table 4 compared with the most advanced method based on dilatedFCN, EfficientFCN uses only 1/3 of the calculation amount and fewer parameters on the PASCAL Context [12], PASCAL VOC 2012 [13], ADE20K [14] data sets result in equal or even better.

Table 3. EfficientNetV2-S architecture

Stage	Operator	Stride	Channels	Layers
0	Conv3x3	2	24	1
1	Fused-MBConv 1, k3x3	1	24	2
2	Fused-MBConv 4, k3x3	2	38	4
3	Fused-MBConv 4, k3x3	2	64	4
4	MBCon4, k3x3, SE0.25	2	128	6
5	MBCon6, k3x3, SE0.25	1	160	9
6	MBCon6, k3x3, SE0.25	2	272	15
7	Conv 1x1 & pooling & FC	-	1792	1

Table 4. Comparisons with classical encoder-decoder methods.

Method	Backbone	OS	mIoU%	Parameters (MB)	GFlops (G)
FCN-32s	ResNet101	32	43.4	54.0	44.6
dilatedFCN-8s	Dilated-ResNet101	8	47.2	54.0	223.6
UNet-Bilinear	ResNet101	8	49.3	60.7	87.9
UNet-Deconv	ResNet101	8	49.1	62.8	93.2
EfficientFCN	ResNet101	8	55.3	55.8	69.6

HGD decomposes the feature upsampling task into a series of holistic codewords from the high-level feature map to capture the global context. And linearly combine codewords at each spatial position to perform feature upsampling with rich semantics. The purpose of HGD is to restore the three feature maps of the last three blocks of the ResNet encoder backbone to high-resolution (OS = 8, Output Stride) feature maps, which are mainly divided into three parts: multi-scale feature fusion, holistic codebook generation, and codeword assembly for high-resolution feature upsampling.

The fusion of multi-scale feature maps usually produces better performance. Take the feature maps of OS = 8, OS = 16, and OS = 32 in the encoder, first compress the channels of the three feature maps to 512 through 1×1 convolution and then downsampling the feature maps of OS = 8 and OS = 16 through channels Connect with OS = 32 to form m32.

Multi-scale fused feature maps m32 created for high-level and mid-level features, but their small resolution makes them lose many structural details of the scene. The small resolution of the multi-scale fusion feature map m32 results in the lack of details, so a series of unordered holistic codewords are generated from m32, which implicitly simulate the global context.

Holistic codewords can capture various global contexts of the input image, but most of the structural information has been removed during the codewords encode process. Therefore, the OS = 8 multi-scale fusion feature m8 is used to predict the linear combination coefficients of n codewords at each spatial position to create a high-resolution feature map.

3.4   SETR

SETR [15] is an image semantic segmentation model based on Transformer, which combines Transformer as an encoder instead of stacked convolution for feature extraction. To evaluate the effectiveness of the features encoded by SETR, three different decoders are designed. The composition of SETR consists of input preprocessing and feature extraction, conversion, and output. Unlike the fully-convolutional network (FCN) model based on the Encoder-decoder structure, SETR regards the semantic segmentation task as a sequence to sequence problem.

The transformer used as the encoder, the image must be converted to a format that Transformer can accept. SETR refers to the practice of ViT [17], slices the input image, and treats each 2D image patch as a 1D sequence, as the entire input into the network. To encode the spatial information of each slice, learn a specific embedding for each local position, and add it to a linear projection function to form the final input sequence.

Input the sequence into the Transformer architecture for feature extraction, which mainly consists of two parts: Multi-head Self-Attention (MSA) and Multilayer Perceptron blocks (MLP). The features extracted by the Transformer are consistent with the dimensions of their input and output. For semantic segmentation, it is necessary to reshape the original spatial resolution. SETR provides three methods: Naive upsampling (Naive), Progressive UPsampling (PUP), and Multi-Level feature Aggregation (MLA).

4   Method

In this study, object detection by Yolov4 or EfficientNetV2. At the beginning, we collected photos of different types and angles of pallets, and learned them through the object detection models, and then manually adjusted the accuracy of the models to complete pallet recognition. Then, the semantic segmentation models, such as EfficientFCN or SETR, using depth camera identifies the hole position of the pallet, and the algorithm calculates the angle and path of the self-propelled vehicle to achieve the automatic forking of the pallet.

4.1   Recognize the Pallet by Object Detection

In the implementation process, the first step is to collect color photos of the pallet by hand. The second step is to train the model by Yolov4 or EfficientNetV2 so that the model can recognize commercially available wood, plastic and various colors of pallets. The third step is to adjust the accuracy of the learning and then continuously optimize and improve the model so that the accuracy of the identification can be continuously improved, as shown in the Fig. 1.

4.2   Recognize the Fork Plate Hole by Semantic Segmentation

After the training of the object detection model is completed, the AGV automatic pallet picking system can automatically identify the pallet information by image acquisition

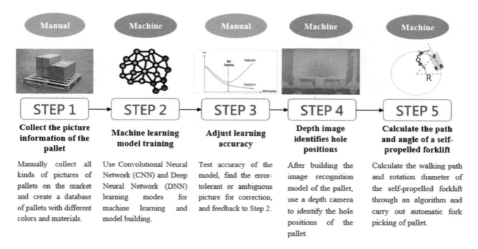

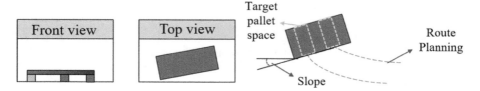

Fig. 1. The implementation process

after the self-propelled forklift reaches the target point, and then further analyze the pallet position and pallet hole angle through the algorithm. The depth distance of the target is sensed by the depth camera on the forklift, and the images of the front view and top view of the pallet are used to do semantic segmentation and to calculate the slope and relative position of the forklift and the target pallet, as shown in the Fig. 2.

Fig. 2. The pallet position and hole angle calculation

4.3 The Position Correction Algorithm

After knowing the pallet position and pallet hole angle, the steering wheel will control the rotation angle and make the self-propelled forklift travel to the planned path. The self-propelled forklift is driven by a single wheel, with the front wheel as the driving wheel, the rudder wheel with both walking and turning functions, and the last two driven wheels, whose turning radius exists, and the minimum radius of rudder wheel turning is calculated according to the planned path, meaning that the radius of self-propelled forklift rotation (R) is calculated under the limit angle of rudder wheel rotation, as shown in the Fig. 3.

The motion model of the single rudder wheel is established at the minimum diameter of the self-propelled forklift, as shown in the Fig. 4. The motion model of single tiller wheel is based on the initial position of the self-propelled forklift, and the

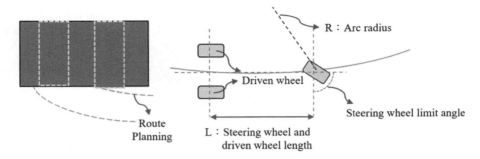

Fig. 3. Calculation of the minimum diameter of the rudder wheel

target rotation angle of δ tiller wheel can be solved, and a high-precision sensor is used to collect the rotation angle data and travel steps (at least 1000 pulse per revolution) to accurately calculate the diameter of each meter of the tiller wheel, and then the sensor feedback is corrected to obtain the current rotation angle of θ tiller wheel and accurately control it. Only when the AGV automatically forks the pallet system can enter the pallet without hitting the center of the pallet and finish forking the pallet automatically.

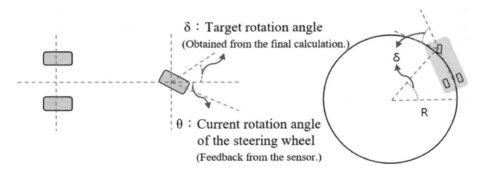

Fig. 4. The rudder wheel movement model

5   Conclusion

Through object detection and semantic segmentation, we can obtain effective object recognition results and calculate the travel path and rotation diameter of the forklift through algorithms to achieve automatic pallet forklift. This result can help logistics operators to solve the problem of manpower shortage, so that logistics personnel can reduce the workload and save a lot of operating time with the help and service of AGV automatic forklift pallet system, which can greatly improve the overall work efficiency.

References

1. Bochkovskiy, A., Wang, C., Liao, H.: YOLOv4: optimal speed and accuracy of object detection. arXiv abs/2004.10934 (2020)
2. Wang, C., Liao, H., Yeh, I., Wu, Y., Chen, P., Hsieh, J.: CSPNet: a new backbone that can enhance learning capability of CNN. In: 2020 IEEE/CVF Conference on Computer Vision and Pattern Recognition Workshops (CVPRW), pp. 1571-1580 (2020)
3. He, K., Zhang, X., Ren, S., Sun, J.: Spatial pyramid pooling in deep convolutional networks for visual recognition. IEEE Trans. Pattern Anal. Mach. Intell. **37**, 1904–1916 (2015)
4. Lin, T., Dollár, P., Girshick, R.B., He, K., Hariharan, B., Belongie, S.J.: Feature pyramid networks for object detection. In: 2017 IEEE Conference on Computer Vision and Pattern Recognition (CVPR), pp. 936-944 (2017)
5. Redmon, J., Farhadi, A.: YOLOv3: an incremental improvement. arxiv abs/1804.02767 (2018)
6. Liu, S., Qi, L., Qin, H., Shi, J., Jia, J.: Path aggregation network for instance segmentation. In: 2018 IEEE/CVF Conference on Computer Vision and Pattern Recognition, pp. 8759-8768 (2018)
7. Misra, D.: Mish: a self regularized non-monotonic activation function. BMVC (2020)
8. Tan, M., Le, Q.V.: EfficientNetV2: smaller models and faster training. arXiv abs/2104.00298 (2021)
9. Tan, M., Le, Q.V.: EfficientNet: rethinking model scaling for convolutional neural networks. arXiv, abs/1905.11946 (2019)
10. Touvron, H., Vedaldi, A., Douze, M., J'egou, H.: Fixing the train-test resolution discrepancy: FixEfficientNet. arXiv abs/2003.08237 (2020)
11. Liu, J., He, J., Zhang, J., Ren, J.S., Li, H.: EfficientFCN: holistically-guided decoding for semantic segmentation. arXiv, abs/2008.10487 (2020)
12. Mottaghi, R., et al.: The role of context for object detection and semantic segmentation in the wild. In: 2014 IEEE Conference on Computer Vision and Pattern Recognition, pp. 891-898 (2014)
13. Everingham, M., Gool, L., Williams, C.K., Winn, J., Zisserman, A.: The Pascal Visual Object Classes (VOC) challenge. Int. J. Comput. Vis. **88**, 303–338 (2009)
14. Zhou, B., Zhao, H., Puig, X., Fidler, S., Barriuso, A., Torralba, A.: Scene parsing through ADE20K dataset. In: 2017 IEEE Conference on Computer Vision and Pattern Recognition (CVPR), pp. 5122-5130 (2017)
15. Zheng, S., et al.: Rethinking semantic segmentation from a sequence-to-sequence perspective with transformers. arXiv abs/2012.15840 (2020)
16. Vaswani, A., et al.: Attention is all you need. arXiv abs/1706.03762 (2017)
17. Dosovitskiy, A., et al.: An Image is Worth 16x16 Words: Transformers for Image Recognition at Scale. arXiv abs/2010.11929 (2020)

Multichannel Convolutional Neural Network Based Soft Sensing Approach for Measuring Moisture Content in Tobacco Drying Process

Shusong Yu[1(✉)] ⓘ, Suhuan Bi[2], Xingqian Ding[1],
and Guangrui Zhang[1]

[1] College of Information Science and Engineering, Ocean University of China,
Qingdao, China
yushusong@ouc.edu.cn
[2] School of Information and Control Engineering, Qingdao University
of Technology, Qingdao, China

Abstract. To address the issue that the control timeliness of the drying process is affected by the unavoidable time delay of moisture content measurement, a soft sensing approach for moisture content in tobacco is proposed, which is based on a multi-channel convolutional neural network (MC-CNN). Firstly, the filtered parameter features are sampled within the lag time and converted into a two-dimensional matrix. Then the parameters of different categories are cyclically transformed to build the input image-like data with multiple channels. Through multiple convolution layers and pooling layers, the MC-CNN model extracts multi-channel features from the feature maps. Moreover, the model effectively perceives the time-sequential and state-spatial coupling characteristics from the raw production data. The proposed method is then validated based on the production data collected from the real production process in the cigarette factory. An online prediction application has been finally achieved to measure the moisture content. Therefore, the detection delay is eliminated and the response time for exceptions is greatly decreased. The amount of tobacco with abnormal moisture content is greatly reduced in the drying process. As the result, the MAE and RMSE of the data measured by the proposed method in a normal production batch are 0.0136 and 0.0257, respectively. Through comparison and analysis of a variety of prediction models, the proposed model has a greater improvement in performance.

Keywords: Soft sensor · Multi-channel · Convolutional neural network · Data-driven

1 Introduction

In industrial process control and optimization, it is important to keep efficiency, quality and energy consumption at the desired level. In practice, many of the associated variables are difficult to measure online due to technical and economic limitations, such as a severe measuring environment, large measurement delays, expensive costs, etc. [1] In the drying process of tobacco, the moisture content, as the feedback for the process

© The Author(s), under exclusive license to Springer Nature Singapore Pte Ltd. 2022
J. C. Hung et al. (Eds.): FC 2021, LNEE 827, pp. 130–142, 2022.
https://doi.org/10.1007/978-981-16-8052-6_14

control, is an important characteristic. Moisture measurement is difficult to be directly conducted especially when the tobacco is still in the dryer.

The soft sensing approach is to construct a mathematical model with measurable variables as input and estimated variables as output by selecting a series of related measurable variables. Data-driven soft sensors are recently applied to real-time prediction for process variables [2–5]. Yan et al. [6] proposed a novel soft sensor modeling method based on a deep learning network to estimate the oxygen content in flue gasses. A soft sensor based on DBN is applied for a debutanizer with unknown delay, which is a part of a refinery [7]. A spatiotemporal attention-based LSTM network is proposed for soft sensor modeling to predict the initial boiling points of heavy naphtha and aviation kerosene [8]. Literature proves that soft sensor modeling methods based on machine learning performed well in real-time predicting.

In the drying process of food and agricultural products, the moisture content is an important characteristic of product quality [9, 10]. Vieira et al. [11] used a hybrid neural model as part of a soft sensor for the online measurement of milk powder. The applications of ELM and ANN for predicting the moisture ratio (MR) were investigated in the drying process of black cumin seeds [12]. Li et al. [13] applied a recurrent self-evolving fuzzy neural network to predict the temperature and moisture content in red maple drying process. However, the process variables involved in the abovementioned online prediction methods are few, it is necessary to develop a novel soft sensor modeling method for the tobacco drying process to deal with more process variables.

This paper proposes a data-driven soft sensing method for measuring the moisture content of tobacco based on a multi-channel CNN (MC-CNN). The raw production data is time-continuous, which contains abundant useful information that represents dynamic changes in drying process. It is difficult to extract the time-sequential characteristics from the production data when used single time sampling as input data. Furthermore, the state-spatial coupling between process variables and manipulated variables is a feature that should not be neglected in predictive modeling. The proposed method can effectively extract the features related to time-sequential and state-spatial coupling characteristics, predict the moisture content in tobacco.

2 Methodology

2.1 Moisture Measuring with Large Delay

Generally, cut tobacco goes through warming, humidifying, drying and winnowing in the whole drying phase, as shown in Fig. 1. The moisture increase conditioner improves the consistency of tobacco by warming and humidifying it with hot steam. The rotary dryer with two drying zones can accelerate the dissipation of moisture by regulating working variables. The outlet moisture content is detected after the winnowing machine. The purpose of the process control is to keep the moisture content level within the required range. During the delay period (i.e., T), preventive and remedial actions are rarely conducted effectively and the tobacco is frequently over-dried or over-wet. Therefore, the quality of tobacco is always sacrificed while waiting for the time delay for moisture measurement in drying process control.

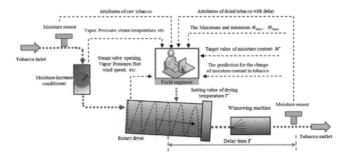

Fig. 1. Tobacco drying phase.

2.2 Data-Driven Soft Sensing Method

2.2.1 The Detection Delay Elimination Workflow

An MC-CNN-based data-driven soft sensing method is presented to simultaneously perform feature extraction and moisture prediction to overcome the above-mentioned problems. The flowchart of the method and the current control method is shown in Fig. 2. The control process mainly includes rotating drying, moisture measuring, anomaly detection, and taking corrective actions for the current method. The average time it takes to get measuring results is about 5 min, which has an impact on process control efficiency and performance. The method removes the delay T, improving the control timeliness and lowering the amount of improperly dried tobacco.

The data-driven soft sensing method consists of two parts: data processing and fully trained MC-CNN. Production-related data collection, which closely depends on the business requirements and preprocessing, is necessary for modeling the running state from raw data of the drying process [14]. The production-related linear sequential data is firstly transformed into two- dimensional data as the required input of the model. The novel MC-CNN model performs feature extraction and soft sensing for moisture content. The CNN-based model is fully trained based on the historical production data and is being used as a soft sensor for moisture online measuring.

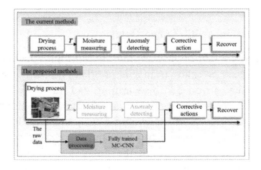

Fig. 2. Comparing regular control process and the proposed detection delay elimination method

2.2.2   Data Description and Conversion

The process data obtained is usually a huge, diverse time-series data set generated by industrial equipment, instruments or sensors. The data offers information about drying conditions and equipment status, both of which are critical for the future production process. It is possible to construct a time-series prediction model based on historical production data, for projecting the dynamic changes in the long drying process [15], as well as for providing a basis for the production process control and intelligent decision-making. Normally, production data includes dozens of factors and their changes over time, many of which are related. As a result, to effectively anticipate moisture content in the drying process, the prediction model should have the ability to extract state control information from both the time and space dimensions. Parameter filtration and data conversion are designed for the preprocessing of raw production data.

(a) Parameters filtration

The data set is taken from the manufacturing line before the outlet of the dryer and includes inlet tobacco properties, batch number, formula parameters and equipment parameters. There are more than 100 different parameters in the practical data acquisition system of the drying phase. For soft sensor modeling, the most important essential parameters must be chosen. First, based on practical expertise, the process variables that contributed nothing to the prediction (e.g. fixed-value frequencies) are discarded. Second, using the out-of-bag (OOB) [16] data error as an index, the features to be modelled are selected according to the impact weight of parameters. The Bagging classifier is built to predict the best training sample fraction according to the OOB data error. With the minimal OOB data error as the goal function, the type and number of parameters in the training set are optimized [17]. As a result, 42 parameters for soft sensor modeling are selected, as listed in Table 1.

Table 1. Specifications of the experiment data

Parameters	Number
Inlet tobacco properties	11
Moisture increase conditioner	11
Rotary dryer	20

(b) Data conversion

Data conversion of time series data before used as model input is an important step for predictive analysis using deep learning method. In the study, the selected production data are sequentially transformed into multi-channel two-dimensional image-like data that matches the proposed two-dimensional convolutional neural network. The conversion method for the raw data is as follows:

Step 1: Sequentially number each parameter. The parameters are numbered sequentially according to the different categories $x_1, x_2, x_3, \ldots, x_{11}$, $y_1, y_2, y_3, \ldots, y_{11}$ and $z_1, z_2, z_3, \ldots, z_{20}$, representing the parameters of inlet tobacco properties, moisture increase conditioner, and rotary dryer, respectively.

Step 2: Intercept data slice for modeling. The proposed method can perform online prediction when tobacco is still being dried in the dryer. The average time of tobacco moving from inlet to the moisture sensor is about 7 min and the sampling interval is 10 s. All data during this time period is taken as input data for modeling. A sliced 7min-length data set is presented as:

$$X = \{X_1, X_2, \ldots, X_{11}\} = \{\{x_{1,1}, x_{1,2}, \ldots, x_{1,42}\}, \{x_{2,1}, x_{2,2}, \ldots, x_{2,42}\},$$

$$\ldots, \{x_{11,1}, x_{11,2}, \ldots, x_{11,42}\}\} \tag{1}$$

$$Y = \{Y_1, Y_2, \ldots, Y_{11}\} = \{\{y_{1,1}, y_{1,2}, \ldots, y_{1,42}\}, \{y_{2,1}, y_{2,2}, \ldots, y_{2,42}\},$$

$$\ldots, \{y_{11,1}, y_{11,2}, \ldots, y_{11,42}\}\} \tag{2}$$

$$Z = \{Z_1, Z_2, \ldots, Z_{11}\} = \{\{z_{1,1}, z_{1,2}, \ldots, z_{1,42}\}, \{z_{2,1}, z_{2,2}, \ldots, z_{2,42}\},$$

$$\ldots, \{z_{20,1}, z_{20,2}, \ldots, z_{20,42}\}\} \tag{3}$$

Step 3: Transform the data sequence into the 2-D image-like data. The above-mentioned data sequences are converted into 2-D vector with size of 42×42 according to the sampling order.

Step 4: Convert the 2-D vector into multi-channel input vector. The input vector of soft sensing model is 2-D vector with three channels and the three channels can be presented as: $\begin{bmatrix} X \\ Y \\ Z \end{bmatrix}$、$\begin{bmatrix} Y \\ Z \\ X \end{bmatrix}$、$\begin{bmatrix} Z \\ X \\ Y \end{bmatrix}$.

The raw production data is transformed into multi-channel 2-D vector through the above steps, as shown in Fig. 3. All the original data in the data sequence N is firstly converted into a 2-D matrix according to the parameter numbers and sampling order, where $p(m, n)$ is the nth sample of the mth parameter. The size of preprocessed input vector is $M \times N \times 3$. The advantage of this kind of input vector transformation is that it provides a method to explore 2-D features of raw production data and quickly obtains the multi-channel vectors with enhanced feature without any predefined parameter or expertise.

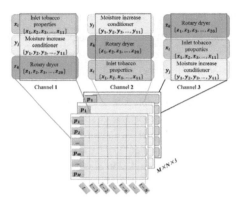

Fig. 3. The transformed 2-D input vector with three channels.

2.2.3   Multi-channel Convolutional Neural Network

The MC-CNN based soft sensing model is shown in Fig. 4. The input of model is the transformed multi-channel 2-D vector and the output is the prediction for moisture content of dried tobacco. The presented architecture includes three convolutional layers, two pooling layers and two prediction layers and the detailed configuration is listed in Table 2. The convolution kernel size of the convolution layers C1, C3 and C5 shows a decreasing trend successively, and the number of channels gradually increases, making the depth of the feature map gradually increase. In the pooling layers S2 and S4, the size of the pooling kernel is 2×2. The max-pooling strategy is applied for feature dimension reduction. The prediction layers perform the moisture prediction based on the multi-channel feature maps.

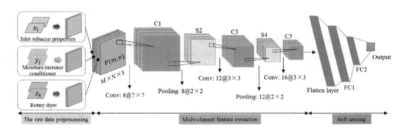

Fig. 4. The proposed MC-CNN model.

Three sources contribute to the multi-channel input vector: inlet tobacco characteristics, moisture increase conditioner, and rotary drier. Multiple convolution and pooling layers can improve the learning of meaningful information since parameters from the same channel have substantial spatial correlation. Convolution kernels can also be used to extract time continuity features from data sequences. As a result, using the acquired spatial and sequential characteristics, the proposed MC-CNN model could perform soft sensing for moisture content in tobacco.

Table 2. Detailed configuration of the proposed model

Layer name	Parameters and output channel size
Input	size: 42×42, channel: 3
Convolution C1	kernel: 7×7, channel: 8
Pooling S2	kernel: 2×2, stride: 2, channel: 8
Convolution C3	kernel: 3×3, channel: 12
Pooling S4	kernel: 2×2, stride: 2, channel: 12
Convolution C5	kernel: 3×3, channel: 16
Flatten layer	channel: 576
Fully-connected	channel: 256
Fully-connected	channel: 128
Output	channel: 1

3 Experimental Analysis

3.1 Experimental Establishment

The soft sensing model is constructed using production data from a cigarette manu-facturing factory's real-world production process. Tobacco is dried in batches, with each batch taking roughly 160 min to complete. As experimental data, 650 batches with approximately 480,000 records are randomly sampled. Among them, 500 batches are utilized as training sets, whereas 150 batches are used as testing sets. The prediction of moisture content is obviously a regression problem. Hence, the mean absolute error (MAE) and root mean squared error (RMSE) are employed to assess the proposed soft sensing model's performance.

3.2 Results and Discussion

The sequence of data remains unaltered during model training in order to retain the continuity and variance tendency in the time dimension. A comparison of the suggested CNN model, the traditional CNN model, and various alternative models is undertaken in order to evaluate the soft sensing model's actual performance.

3.2.1 Comparisons with the Standard CNN-Based Model

The prediction results of the suggested method and the typical CNN-based model are compared to verify the feasibility of the multi-channel feature extraction strategy. Three batches are randomly selected for analysis, one normal instance and two aberrant productions. Aberrant production refers to a production shutdown or other abnormal situations caused by unknown reasons in the manufacturing process. Effective moisture content prediction in dried tobacco is especially crucial for production guidance in abnormal situations. The input of standard CNN-based model is presented as $\begin{bmatrix} X \\ Y \\ Z \end{bmatrix}$ and

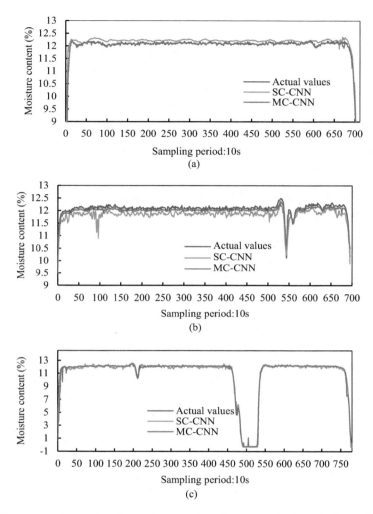

Fig. 5. The comparison of soft sensing results of two methods: (a) Normal instance; (b) Abnormal instance 1; (c) Abnormal instance 2.

the model is denoted as a single channel convolutional neural network (SC-CNN). The soft sensing results of two models are shown in Fig. 5.

In three cases, the MC-CNN model beats the SC-CNN model when the prediction curves are assessed, particularly in the abnormal production situation. Table 3 shows the prediction results of the two approaches in both normal and abnormal situations. When compared to the SC-CNN model, the MAE and RMSE of the MC-CNN model are much lower. The MAE and RMSE of the SC-CNN model were 0.0865 and 0.1135, respectively, in the normal instance (a). The MAE and RMSE of the MC-CNN model lowered by 0.0729 and 0.0878, respectively, after using the multi-channel feature extraction approach. The MAE and RMSE of the MC-CNN model were 0.0897,

Table 3. Prediction results of normal instance and abnormal instances

Methods	Normal instance (a)		Abnormal instance (b)		Abnormal instance (c)	
	MAE	RMSE	MAE	RMSE	MAE	RMSE
SC-CNN	0.0865	0.1135	0.1326	0.1801	0.5023	0.5326
MC-CNN	0.0136	0.0257	0.0897	0.1025	0.2026	0.2738

0.2026 and 0.1025, 0.2738 in the two anomalous cases, respectively. The causes of anomalous manufacturing conditions in the actual drying process are complicated. Although the model's prediction performance is not as good as the ordinary batch, it is better than the SC-CNN model. The enhancement of prediction accuracy for the soft sensor model has great practical significance for the optimization of drying process.

3.2.2   Comparisons with State-of-the-Arts Networks

The performance of proposed prediction approach is compared against that of other methods such as SVR, RNN, ARMAX, and ANN.

(a)   Soft sensing within prediction instance.

Figure 6 presents the moisture content prediction curve amongst prediction instances. Among the compared approaches, the kernel function of SVR is set to RBF kernel function, and the ANN model involves a four-layer neural network. The ARMAX model adopts system identification tool of MATLAB for data import and prediction, and other models are implemented on the TensorFlow platform.

As demonstrated in Fig. 6, SVR model performed the poorest, ARMAX and RNN model take second place, and the MC-CNN model performs best. The prediction based on the model is difficult to perceive the dynamic production state when the parameters are relatively fixed, and there is a certain deviation in the prediction of moisture content.

MAE and RMSR in various scenarios are illustrated in Fig. 7 to represent the overall prediction performance. In normal instance (a), the MAE and RMSE of SVR model are 0.3477 and 0.4687, respectively. However, the MAE and RMSE of SVR in instance (c) are 1.2006 and 2.7912 respectively, which far exceeding the acceptable range in actual production. The ANN model's prediction performance is fairly decent, but the prediction error is rather significant when the production condition is abnormal in instance (c), with MAE and RMSE of 0.4556 and 0.4866, respectively, which makes ANN difficult to meet the process control need for prediction accuracy. The input data of the proposed MC-CNN model is a multi-channel 2-D vector, which can more comprehensively extract spatial features and sequential features from the raw data. MAE and RMSE in normal instance are 0.0136 0.0257, respectively, and the MAE and RMSE in abnormal production are smaller than those of other models.

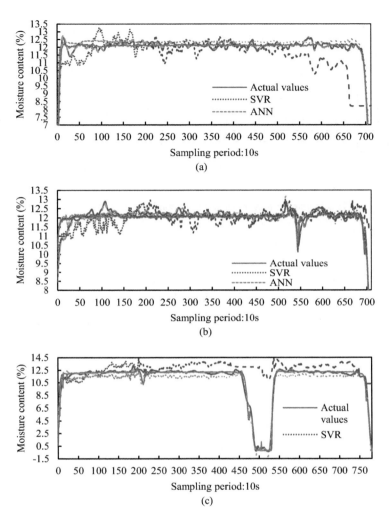

Fig. 6. Results of different prediction instances: (a) Normal instance; (b) Abnormal instance 1; (c) Abnormal instance 2

(b)   Results on testing dataset.

Table 4 and Fig. 8 illustrate the mean of the prediction outcomes of several models on the test dataset. In actual production, the proportion of normal batches is very high, and the proportion of abnormal production is relatively small. Therefore, the mean value of statistical results is close to the predicted result of normal batches. The mean values of MAE and RMSE of MC-CNN are the smallest, followed by ANN, ARMAX

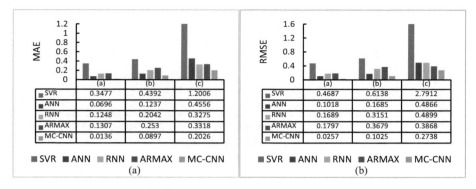

Fig. 7. Predicted errors under different conditions: (a) MAE; (b) RMSE

Table 4. The results on testing dataset.

Methods	MAE	RMSE
SVR	0.3571	0.4934
ANN	0.0740	0.1063
RNN	0.1276	0.1736
ARMAX	0.1339	0.1837
MC-CNN	0.0161	0.0291

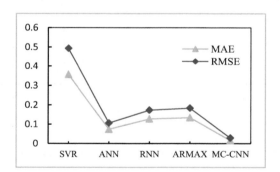

Fig. 8. Predicted errors on testing dataset.

and RNN, and SVR is the worst. The average MAE and RMSE of the MC-CNN model are 0.0161 and 0.0291 respectively. The prediction results can meet the requirements of the precision of moisture content in the actual process control, and can provide real-time predictions of moisture content for the optimization of tobacco drying process.

4 Conclusions

In this paper, a novel multi-channel CNN model was proposed to handle the time delay problem of moisture measuring by applying the multi-channel feature extraction strategy. The multi-channel feature extraction CNN-based structure was designed to learn spatial and sequential relationships automatically. Meaningful information contained in the raw production data was retrieved comprehensively by multiple convolution and pooling layers through resampling, converting, and multichannel vector constructing. The MC-CNN model's prediction performance is validated by comparison tests with other prediction models.

With the support of the data, the detection delay is eliminated and the response time for exceptions can be considerably reduced using soft sensing modeling and multichannel feature extraction. The proposed method introduces a new research strategy for addressing the issue of important production indicators being difficult to identify in the industrial process (e.g. lagging for the dryer in our case). Aside from soft sensing modeling for moisture content, the innovative coupling of control theory and deep learning for industrial process control optimization is an important and beneficial direction.

Foundation Item(s): Project supported by the National Key R&D Program of China (2017YFA0700601), and the Science and Technology Project of Qingdao West Coast New Area (No. 2019-2-3).

References

1. Chai, T.: Operational optimization and feedback control for complex industrial processes. Acta Autom. Sin. **39**(11), 1744–1757 (2013)
2. Murthy, T.P.K., Manohar, B.: Microwave drying of mango ginger (Curcuma amada Roxb): prediction of drying kinetics by mathematical modeling and artificial neural network. Int. J. Food Sci. Technol. **47**(6), 1229–1236 (2012)
3. Balbay, A., Avci, E., Sahin, O., et al.: Modeling of drying process of bittim nuts (pistacia terebinthus) in a fixed bed dryer system by using extreme learning machine. Int. J. Food Eng. **8**(4), 1–16 (2012)
4. Yu, H., Qin, S., Ding, G., et al.: Development of prediction model for machining precision of five-axis flank milling based on tool runout error. Comput. Integr. Manufact. Syst. **26**(12): 3359–3367 (2020)
5. Yao, L., Ge, Z.: Deep learning of semisupervised process data with hierarchical extreme learning machine and soft sensor application. IEEE Trans. Industr. Electron. **65**(2), 1490–1498 (2018)
6. Yan, W., Tang, D., Lin, Y.: A data-driven soft sensor modeling method based on deep learning and its application. IEEE Trans. Industr. Electron. **64**(5), 4237–4245 (2016)
7. Graziani, S., Xibilia, M.G.: Design of a soft sensor for an industrial plant with unknown delay by using deep learning. In: 2019 IEEE International Instrumentation and Measurement Technology Conference (I2MTC), pp. 977–982 (2019)
8. Yuan, X.F., Li, L., Shardt, Y.A.W., et al.: Deep learning with spatiotemporal attention-based LSTM for industrial soft sensor model development. IEEE Trans. Industr. Electron. **68**(5), 4404–4414 (2021)

9. Mozaffari, M., Mahmoudi, A., Mollazade, K., et al.: Low-cost optical approach for noncontact predicting moisture content of apple slices during hot air drying. Drying Technol. **35**(12), 1530–1542 (2017)
10. Liu, Z.L., et al.: Prediction of energy and exergy of mushroom slices. drying in hot air impingement dryer by artificial neural network. Drying Technol. (Online) (2019)
11. Vieira, G.N.A., Olazar, M., Freire, J.T., et al.: Real-time monitoring of milk powder moisture content during drying in a spouted bed dryer using a hybrid neural soft sensor. Drying Technol. **37**(9), 1184–1190 (2019)
12. Balbay, A., Kaya, Y., Sahin, O.: Drying of black cumin (Nigella sativa) in a microwave assisted drying system and modeling using extreme learning machine. Energy **44**(1), 352–357 (2012)
13. Li, J.S., Xiong, Q.Y., Wang, K., et al.: A recurrent self-evolving fuzzy neural network predictive control for microwave drying process. Drying Technol. **34**(12), 1434–1444 (2016)
14. Zhu, J.L., Ge, Z.Q., Song, Z.H., Gao, F.R.: Review and big data perspectives on robust data mining approaches for industrial process modeling with outliers and missing data. Annu. Rev. Control **46**, 107–133 (2018)
15. Administration, S.T.M.: Cigarette Making Process Specification. China Light Industry Press, Beijing (2016)
16. Ma, L., Fan, S.H.: CURE-SMOTE algorithm and hybrid algorithm for feature selection and parameter optimization based on random forests. BMC Bioinf. **18**, 1–18 (2017)
17. Martinez-Munoz, G., Suarez, A.: Out-of-bag estimation of the optimal sample size in bagging. Pattern Recognit. **43**, 143–152 (2010)

Influenza-Like Illness Patients Forecasting by Fusing Internet Public Opinion

Yu-Chih Wei[1(✉)], Yan-Ling Ou[1], Jianqiang Li[2], and Wei-Chen Wu[3]

[1] National Taipei University of Technology, Taipei, Taiwan
vickrey@mail.ntut.edu.tw, tl09ab8016@ntut.org.tw
[2] Beijing University of Technology, Beijing, China
lijianqiang@bjut.edu.cn
[3] National Taipei University of Business, Taipei, Taiwan
weichen@ntub.edu.tw

Abstract. Due to rapid change in influenza viruses, a prediction model for outbreaks of influenza-like illnesses helps to find out the spread of the illnesses in real time. In addition to using traditional hydrological and atmospheric data, popular search keywords on Google Trends are used as features in this research. Google Trends are popular keyword searches on the Google search engine. Popular keywords used in discussions of influenza-like symptoms at specific regions within specific periods are used in this research. Public holiday information in Taiwan, the population density, air quality indices, and the numbers of COVID-19 confirmed cases are also used as features in this research. An Ensemble Learning model, combining Random Forest and XGBoost, is used in this research. It can be confirmed from the actual experimental results in this research that the use of the ensemble learning prediction model proposed in this research can accurately predict the trend of influenza-like cases. The evaluation results show that the mean RMSLE of our proposed model is 0.2 in comparison with the actual number of influenza-like cases.

Keywords: Influenza-like illnesses · Monitoring and early warning

1 Introduction

The COVID-19 pandemic broke out at the end of 2019 and has spread all over the world at lightning speed. Past large-scale pandemics, such as SARS, H1N1, Influenza A and MERS etc. have drawn much global attention to the damages that pandemics may bring to the world. Due to the rapid development of modern technology and the convenience of transportation, viruses these days can be easily spread to every corner of the world. An influenza-like illness means any illness caused by a virus with symptoms similar to those that are caused by influenza viruses ("flu"), including symptoms such as fever, respiratory symptoms, muscle pain, and fatigue etc., and if they cannot be diagnosed to be caused by influenza viruses, they are called influenza-like illnesses. People who have influenza-like illnesses, at the same time [1]:

J. C. Hung et al. (Eds.): FC 2021, LNEE 827, pp. 143–151, 2022.
https://doi.org/10.1007/978-981-16-8052-6_15

1. have a sudden onset of an illness, a fever of an ear temperature of 38 °C or higher, and respiratory symptoms;
2. have muscle pain, headaches, or extreme fatigue; and
3. do not have a simple runny nose, tonsillitis, and/or bronchitis.

Many studies have shown the correlation between survival rates and outbreak periods of most viruses and seasonal climate changes. Prel et al. [2] explore effects of different climates on acute respiratory tract infections (ARI) and find that different viruses have different survival rates and prevalence to different temperatures, humidity and other climatic conditions and seasons. Chan et al. [3] observe Hong Kong from 1997 for 10 years, and find that there are two seasonal flu peaks every year and that although there are many seasonal factors of influenza, changes in weather are likely to play a key role. Taking Taiwan as an example [4], influenza outbreaks in Taiwan mainly occur in autumn and winter. The number of influenza cases gradually increases from November and peaks during the period between December and March. Influenza may cause acute respiratory infections in patients. Common symptoms include fever, runny nose, and muscle pain, etc.

In addition to meteorological factors that may affect the timing of virus outbreaks, Keyword searches have also been used by many studies to monitor epidemic virus outbreaks [5–10], including the number of times a keyword appears in posts and discussions on social media or search platforms, such as Facebook, Twitter and Wikipedia. Therefore, internet search volumes may immediately reflect how a virus is affecting people. However, as Google Trends can calculate the popularity of keyword discussions in real time, it is a better factor than the social media to be used to effectively predict outbreaks of illnesses [11]. Moreover, Google Trends has long been used by the Taiwan Centers for Disease Control (CDC) for tracking and predicting the spread of influenza [12].

This research proposes an ensemble learning multi-feature method, using the hydrometeorological data, the emergency infectious disease monitoring statistics - influenza-like illnesses, Google Trends keyword search volumes, the Taiwan public holiday information, the population data, air pollution indices, and the number of COVID-19 confirmed cases as features. Random Forest, XGBoost, SVR and ensemble learning are used to predict the number of influenza-like cases in Taiwan. This research uses RMSLE for model error evaluation. RMSLE is widely used to evaluate regression models and is an evaluation indicator used by many data sciences. It is similar to RMSE but uses logarithms for calculation. One of the advantages of using RMSLE as an indicator is that it has good robustness against outliers [13], and the closer it is to zero, the smaller the error rate is. After 6 weeks of evaluation, the RMSLE value is 0.2, which means that the prediction error rate between the predicted and the actual numbers of influenza-like cases for that period is close to the actual number of influenza-like cases.

Section 1 of this paper is introduction. Section 2 discusses literature contributed by scholars in similar fields in the past. Section 3 describes the methodology used in this paper. Section 4 contains our conclusion.

2 Literature Review

In this section, we will review past literature that have discussed factors affecting influenza-like illnesses and those that have discussed prediction models for outbreaks of influenza-like illnesses in machine learning.

2.1 Choices of Training Features

There are four possible modes of transmission of influenza viruses [14]. They are: (1) transmission through direct physical contact with an infected person; (2) transmission through mediums, usually inanimate objects (such as droplets on objects or surface); (3) transmission through droplets of an infected person produced through sneezing, coughing, etc., which are transmitted to the nasal cavity or oral mucosa of a recipient; and (4) transmission through particles of a radius of 2.5 μm propelled by coughing or sneezing into the air. Viruses can survive in the particles and float in the air for a long time and be transmitted through the particles.

The relative importance among the four transmission modes is a controversial issue. Lowen et al. [15] use guinea pigs as mammalian test objects to test the hypothesis that temperature and relative humidity affect the transmission rate of influenza viruses. They find that guinea pigs are very sensitive to influenza viruses that infect humans, and that the pups of guinea pigs exposed to the viruses are more likely to be infected. They use a variety of relative humidity and temperature conditions and various combinations of them to evaluate transmission rates of influenza viruses. They find that transmission speeds of influenza viruses depend on the temperature and relative humidity of the environment. Their findings support the hypothesis that meteorological conditions affect the spread of influenza viruses and help establish the link between meteorological factors and the spread and evolution of viruses, which was troublesomely uncertain in the past.

2.2 Machine Learning Models for Predicting Outbreaks of Influenza-Like Illnesses

Cheng et al. [16] use four machine learning algorithms, namely ARIMA, Random Forest, SVM and XGBoost, to establish a real-time national system to monitor influenza outbreaks and predict influenza-like cases in the succeeding four weeks for the Taiwan Centers for Disease Control (CDC). To combine the prediction results of the four different machine learning models, a stacking ensemble learning method is used to form the final prediction model. Its most accurate prediction result for the week when the prediction takes place has a MAPE of less than 0.75 and a hit rate 0.75. Darwish et al. [17] use machine learning and deep learning multiple algorithms to establish a model to predict the number of influenza-like cases in Syria. The lowest MAPE of its prediction results is 3.52% and the lowest RMSE 0.01662. Chen et al. [18] use the Seasonal Autoregressive Integrated Moving Average (SARIMA) to predict the outpatient rate of the influenza-like illnesses in Shenyang, China. The authors mention that the predicted values of influenza-like illnesses can be used as a reference for outbreaks of influenza-like cases in a short term, but other factors should be taken into consideration when forming strategies for influenza prevention and control.

3 Methodology and Evaluation

3.1 Prediction Framework

In this section, the methodology and techniques used in this research on prediction of outbreaks of influenza-like illnesses will be described. Hydrometeorological data taken from meteorological observation data, statistics on emergency infectious diseases - influenza-like illnesses, data on keyword search volumes on Google Trends, air quality indices, data on total population, population density and Taiwan public holiday information are used in this research.

The observation stations, where the hydrometeorological data is taken for this research, are the ground weather observation stations and the automatic weather/rainfall observation stations of the Central Weather Bureau, Taiwan. The observation data consists of several parts. The first part comprises the data taken from the "Data Bank for Atmospheric and Hydrologic Research" in Taiwan up to April 2020 and the data taken from the "Open Weather Data" in Taiwan, from that date up to the date of our study. This research focuses on predicting the number of influenza-like cases in counties and cities in Taiwan. However, as there are no ground weather observation stations of the Central Weather Bureau in some counties and cities, such as Miaoli County and Chiayi County etc., data is taken from the automatic weather/rainfall observation stations to fill in the missing data of these counties and cities.

Data on statistics on emergency infectious diseases - influenza-like illnesses is taken from the "Taiwan National Infectious Disease Statistics System" of the Taiwan Centers for Disease Control (CDC), which contains statistical data on the number of visits to emergency departments at hospitals by patients with influenza-like illnesses of every age in every county/city in every week of the year.

Data on keyword search volumes is based on Google keyword searches. This research selects various flu symptoms as keywords and collects their search volume values on Google. The search volume values are relative values and refer to the popularity of a search term in a specific area within a specific period. The value range is [0,100].

Monitoring data of the Environmental Protection Administration of the Executive Yuan in Taiwan is used as air quality indices in this research. Data on total population and population density is based on the statistical data of all counties, cities, towns and villages in Taiwan as provided by the Department of Statistics of the Ministry of the Interior, Taiwan. The total population is the statistical data of the statistical population, and the population density is the population indicator data. The Taiwan public holiday information is taken from the open government data platform at "data.gov.tw".

Datasets required for our study are first imported from their sources. They are then pre-processed using its applicable data processing method, and then all the processed data are grouped into its applicable training and testing datasets. Assuming prediction takes place this week (lag0) to predict the number of influenza-like cases in the following week, as the features used in this research to predict outbreaks of influenza-like cases do not predict outbreaks for the same week but they lag behind for a week or longer, data of the week before (lag1) are used this week for prediction. Machine learning is then used to predict the number of influenza-like cases for next week. The framework in this research for predicting influenza-like cases. The following subsection in this section will discuss the techniques used in and the reasons why they are chosen for this research (see Fig. 1).

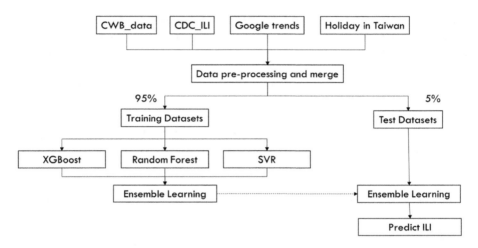

Fig. 1. A framework for prediction of influenza-like cases

3.2   Data Pre-processing

The pre-processed datasets discussed in this section are datasets publicly available as mentioned in Sect. 3.1 above. Features required for this research are selected from among the datasets. Features contained in Table 1 are the original features used in this research.

Lastly, consolidating the knowledge about influenza-like illnesses, we select the features introduced at Table 1 and perform different processing on different data according to the categories they belong to. The features of hydrometeorological data are combined and calculated according to the observation stations of the county and city, from which the data is collected. The minimum value of the PP data is 0 and there should be no negative value for RH data. As to the TX data, as only the mountainous areas at a high altitude may be subjected to a temperature below 0 °C in winter while the rest of the areas are above 0 °C, negative values of this set of data are also excluded. There should be no negative values for the WD set of data either. All anomalous negative values of these four sets of data are caused by instrumental or human factors (please refer to Table 2), and therefore, the negative values of the PP data are replaced by 0, and those of the RH, TX and WD data are excluded due to their characteristics. We also exclude outliers of a distance greater than three standard deviations. Weekly average values are calculated and log values (using one week as a

Table 1. Original features

Features	Measurement units	Descriptions
PP	Millimeter	Precipitation
RH	Percentage (%)	Average relative humidity
TX	Celsius (°C)	Average temperature
TD	Celsius (°C)	Daily temperature differences
WD	Meter per second (m/s)	Average wind speeds
ILI	Number of people	The number of influenza-like cases
ILI_D	Number of people	Differences in the numbers of influenza-like cases
GT_I	A value range of [0,100]	Keyword search volumes on Google Trends - influenza (flu)
HoC	Number of days	The number of public holidays in Taiwan per week

Table 2. Descriptions of negative values at the Atmospheric Hydrological

Negative Values	Descriptions
−9991	Instrument failures, to be repaired
−9996	Data accumulated later
−9997	No information available due to unknown reasons or malfunctions
−9998	Traces of rain

*Descriptions for the negative values at the Atmospheric Hydrological Observation Stations mentioned in this research.

unit) are taken for the PP and RH data. TD is the most special among the hydrometeorological features. TD represents the differences between the highest and the lowest temperature of TX, which are then calculated on a weekly basis to take logs. ILI is the emergency infectious disease monitoring statistics - the influenza-like illness, the aggregate data of the total number of emergency visits by patients of influenza-like illnesses for all age groups in each county/city. ILI_D represents the weekly changes in the number of emergency visits by patients of influenza-like illnesses by deducting the number of emergency visits in this week with the number of visits in the week before. ILI_D is the total number of emergency visits by patients of influenza-like illnesses from the week before. Each feature value of GT is the numerical data of a Google Trends search volume within 5 years. The numerical values are floating values, not absolute values. HoC is the number of public holidays per week, from Sunday to Saturday, in Taiwan within 5 years.

3.3   Ensemble Machine Learning Model

Random Forest (RF), eXtreme Gradient Boosting (XGBoost), Support Vector Regression (SVR), and Ensemble Learning, are selected as training models in this research. Random Forest affects proportions of features and facilitates the verification of hypotheses and ideas. XGBoost and RandomForest use similar principles and can more accurately predict results. XGBoost is eXtreme Gradient Boosting, a scalable machine learning system, used for tree boosting and based on the extension and improvement of Gradient Boosted Tree (GBDT), while retaining the original model. Random Forest consists of decision tree classifiers. Each of the classifiers is generated independently from random vectors in input vectors. A bagging algorithm is used for each feature or each feature combination, i.e. samples are randomly taken from the training data to train multiple classifiers. Gini coefficients are used to select features, to measure the impurity of the features to their categories, and to segment each feature.

SVR is different from the previous two. SVR is an extension of support vector machine (SVM). SVR can handle continuous prediction problems. Consistent with the classification method, SVR is characterized by the use of kernel functions, sparse solutions, VC marginal controls and the number of support vectors. One of the main advantages of SVR is that its complex calculation does not depend on the dimensionality of the input space. As the SVR prediction results are worse than the other three, we eventually only use the ensemble learning model of Random Forest and XGBoost as our prediction model. The SVR prediction results are not adopted. The final prediction results of our models are compared with the past data to obtain the accuracy rates. In this research, the values of the XGBoost and Random Forest models are combined by Ensemble Learning as the final prediction results of this research.

In this paper, the training data used in this research is the historical data taken from 2016 to 2021. The ensemble learning combining RandomForest and XGBoost is used as a predictive model, and the data of the previous week is used as the data of the current week, and the number of influenza-like cases for this week is predicted. For testing and evaluation, we predict the weekly number of influenza-like cases for 6 weeks of 2021. This research uses RMSLE to measure the effects of the models. RMSLE is RMSE in the log form. It considers relative errors in the same ways as MSPE and MAPE, but RMSLE error curves are asymmetry. The closer its values are to 0, the less the errors. The prediction error rate between the predicted and the actual numbers of influenza-like cases for that period, the RMSLE is 0.2, getting closer to the actual number of influenza-like cases.

4   Conclusion and Future work

Hydrological and atmospheric data is used in this research as features to predict influenza-like cases, adding popular search keywords on Google Trends as features. Search volumes on Google are put to better use to know the popularity of the keywords among Google discussions about influenza. Average temperature differences, public holidays, the population density, air quality indices and the number of COVID-19 confirmed cases are all used as features and taken into consideration in this research.

As to the construction of models, an Ensemble Learning model, combining Random Forest and XGBoost, is used as a prediction model. The weekly numbers of influenza-like cases are predicted for 6 weeks in 2021. The predicted results are compared with the actual numbers of influenza-like cases. The prediction error rate between the predicted and the actual numbers of influenza-like cases for that period, the RMSLE is 0.2, getting closer to the actual number of influenza-like cases. Predicted results produced by the model proposed in this research will be able to help the government, hospitals, pharmaceuticals and companies quickly understand the spread of influenza-like cases in the future, so that they can form informed decisions and preventive measures, as well as to help the general public to understand, through the government and hospitals, possible large scale outbreaks of influenza-like illnesses in the near future, so that they can take measures to protect their own health and safety.

As some previous researches have suggested that many features directly affect the number and the spread of influenza-like cases, and some question that they are not significantly related to outbreaks of influenza-like illnesses. In the future, we intent to evaluate different combination of features and do further investigations in order to improve the learning model to have better predict ability.

Acknowledgments. This work was partially supported by Onward Security (No. 209A136) and National Taipei University of Technology-Beijing University of Technology Joint Research Program (No. NTUT-BJUT-110-01).

References

1. Taiwan Centers for Disease Control: Practical Guidelines for Prevention and Control of Seasonal Influenza. Taiwan Centers for Disease Control (2020)
2. du Prel, J.-B., et al.: Are meteorological parameters associated with acute respiratory tract infections? **49**(6), 861-868 (2009)
3. Chan, P.K., et al.: Seasonal influenza activity in Hong Kong and its association with meteorological variations. **81**(10), 1797–1806 (2009)
4. Taiwan Centers for Disease Control: Severe Complicated Influenza (2020). https://www.cdc. gov.tw/Category/Page/HMC9qDI4FA-gDrbcnFlXgg
5. Wang, Y., et al.: Regional influenza prediction with sampling Twitter data and PDE model. **17**(3), 678 (2020)
6. Seo, D.-W., Shin, S.-Y.: Methods using social media and search queries to predict infectious disease outbreaks. Healthcare Inform. Res. **23**(4), 343 (2017)
7. Daughton, A.R., et al.: Comparison of social media, syndromic surveillance, and microbiologic acute respiratory infection data: observational study. JMIR Public Health Surveill. **6**(2), e14986 (2020)
8. Lampos, V., Zou, B., Cox, I.J.: Enhancing feature selection using word embeddings: the case of flu surveillance. In: Proceedings of the 26th International Conference on World Wide Web (2017)
9. Volkova, S., et al.: Forecasting influenza-like illness dynamics for military populations using neural networks and social media. PloS One **12**(12), e0188941 (2017)
10. Lee, K., Agrawal, A., Choudhary, A.: Forecasting influenza levels using real-time social media streams. In: 2017 IEEE International Conference on Healthcare Informatics (ICHI). IEEE (2017)
11. Huang, L.-H.: A deep learning based approach to forecasting influenza-like illness rate. In: Department of Medical Informatics. Tzu Chi University, Hualien County, p. 49 (2020)
12. Ginsberg, J., et al.: Detecting influenza epidemics using search engine query data. **457** (7232), 1012–1014 (2009)
13. Zeroual, A., et al.: Deep learning methods for forecasting COVID-19 time-series data: a comparative study. Chaos Solitons Fract. **140**, 110121 (2020)

14. Shaman, J., Kohn, M.: Absolute humidity modulates influenza survival, transmission, and seasonality. Proc. Natl. Acad. Sci. **106**(9), 3243-3248 (2009)
15. Lowen, A.C., et al.: Influenza virus transmission is dependent on relative humidity and temperature. PLoS Pathog. **3**(10), e151 (2007)
16. Cheng, H.-Y., et al.: Applying machine learning models with an ensemble approach for accurate real-time influenza forecasting in Taiwan: development and validation study. J. Med. Internet Res. **22**(8), e15394 (2020)
17. Darwish, A., Rahhal, Y., Jafar, A.: A comparative study on predicting influenza outbreaks using different feature spaces: application of influenza-like illness data from Early Warning Alert and Response System in Syria. BMC Res. Notes **13**(1), 1–8 (2020)
18. Chen, Y., et al.: Epidemiological features and time-series analysis of influenza incidence in urban and rural areas of Shenyang, China, 2010–2018 Epidemiol. Infect. **148** (2020)

Ubiquity of Aural Skills Development in Music Rhythm Through the Mobile Phone Mechanism

Yu Ting Huang[1] and Chi Nung Chu[2($\boxtimes$)]

[1] Department of Music, Shih Chien University, No. 70 Ta-Chih Street,
Chung-Shan District, Taipei, Taiwan, R.O.C.
yutingll@mail.usc.edu.tw
[2] Department of Management of Information System, China University of
Technology, No. 56, Sec. 3, Shinglung Rd., Wenshan Chiu, Taipei 116, Taiwan,
R.O.C.
nung@cute.edu.tw

Abstract. Music rhythm learning requires a critical ability with the immediate response to music staff reading and keeping up with the speed implied within the music notes. It is a difficulty of music rhythm learning for the learners but can be overcome by repeated practice. With the advent of the e-learning era, learning has transcended the choice of instruction beyond time and space. To this goal, this paper proposed a novel mobile learning system called Aural Skills Development in Music Rhythm through the Mobile Phone Mechanism. The application design is based on the learning theory of behavioral drill strategy to construct the music rhythm practice system. In order to achieve long-term effective practice, mobile devices can be introduced as a learning tool. Learners can use smart phones to practice music rhythms anytime and anywhere according to their personal needs. Learners would no longer be limited to music rhythm learning in traditional classrooms for instructors' correction on-side.

Keywords: Smart mobile phone · Music rhythm · Learning effect

1 Introduction

The aural skills include staff reading, music dictation, sight-singing and music rhythm, which is an extremely important professional foundation for music learning [8, 27]. The well-developed aural skills can create a deeper perception and understanding of music. The quality of aural skills can also affect the level of listening to music, it can enhance people's perception of music and make hearing more acute and enjoyable. With good aural skills, people can more easily understand and appreciate music, enhance their sensitivity to contact music [19, 28, 29].

Establishing the basic aural skills from music rhythm learning is a well beginning [18, 33]. Therefore, in all aspects of music learning, the establishment of music rhythm is critical, and the development of music rhythm needs time to form up [4, 16, 35]. Strengthening the music rhythm ability must extend the learning time. It is a long-term practice in the acquisition of music rhythm sensitivity. Learners can form a truly

J. C. Hung et al. (Eds.): FC 2021, LNEE 827, pp. 152–158, 2022.
https://doi.org/10.1007/978-981-16-8052-6_16

internalized rhythm, and then learn an advanced domain of music knowledge. The mastery of musical rhythm can effectively assist learners in the acuity of music dictation and the pitch of sight-singing. However, the difficulties in rhythmic learning come from the actual performing the length of music notes which are determined by the tempo of the music. The space symbol of music note is used to identify the pitch and time value of the sound. That is a problem in the transition of vision and hearing for students' difficulty in learning rhythm [22, 26]. Studies showed that rhythm teaching must stabilize learners on the stable tempo [7, 10]. Tradition rhythm exercises seem relatively drab. As the learners encounter rhythm exercises, the immediate reaction is to convert music notes into monotonous pithy formulas often in rhyme. Even the correct recitation or rhythm is confirmed by the instructor, it will limit the learners to develop their sense of rhythm with an instructor who must be on site guide.

With the advent of the digital learning trend, the information technology has been applied to learning-related research and design in recent years, including the technology in development of instruction systems, the platform planning of teaching platform, and the production of instruction contents [3, 13, 36]. The use of IT-assisted instruction including mobile learning has positive effects on the learning achievement [2, 17, 21]. Even for the courses with higher complexity and difficulty, digital learning tools have better learning efficiency than traditional instruction. Studies show that information technology can concretize abstract concepts of instruction contents, and achieve the effect of enhancing learning motivation and promoting self-learning [1, 6].

2 Learning Theory Behind the Design of Aural Skills Development in Music Rhythm

Learning theories provide educators with various education aspects for research and thinking on instruction strategies. As the human learning process is still in the exploratory stage, effective instruction strategies have absolute instruction benefits for learning. The design in this paper is based on the mobile learning with information technology and the application of practicing strategies of learning theory by the behaviorism.

2.1 Information Technology Convergence Strategy

With the development of information technology, acceleration in the communication, dissemination, and accumulation with information has been connected. Beyond the main resources of knowledge such as books or classrooms in the past, instead anyone can retrieve knowledge from everywhere as long as they turn on the mobile devices and surf the Internet [9, 12, 31, 34]. The impact of information technology is comprehensive. In education, it has affected teachers' instructions and students' learning methods, and even led to the transformation of traditional education patterns.

The rapid advancement of information technology has had a tremendous impact and influence on the teaching practice in the classroom. The popularization of smart phones, the Internet and the establishment of wireless networks have broken the time and space limitations of traditional classroom activities. Facing the advent of digital era, the world is close at hand with the connection of the Internet, and learning is even more ubiquitous because of the convenience of information technology. The modern learning model is gradually no longer centered on the instructor, but shifted to center on the learner. As a result, learners can not only learn with the aid of information technology, and even the scope of the classroom can also be infinitely extended due to the convenience of the Internet.

The learner-centered transformation and the combination of information technology with learning provide the possibility of exploring and improving instruction effectiveness [5, 14, 32]. As the information technology can integrate the digital media such as image, graphic, audio, text, and animation, various learning subjects can be vividly simulated and abstraction of domain knowledge can be presented concretely. The learning motivation can be greatly enhanced. Especially, information technology has gradually been developed into a new visual tool in art education. This is just the beginning. Information technology is no longer exclusive to the scientific professional field, but can be incorporated into other educational fields.

Therefore, the environment created by information technology is a satisfiable choice for learning needs of learners. It can not only facilitate learners' active operation choices to improve problem-solving and expand thinking skills. What digital learning intends to touch is not to merely adopt information technology for the sake of information technology, but to satisfy individual differences in learning. Therefore, instructors must bring learning to learners; not to bring learners to learn [11, 20, 25]. Information technology could allow learners to learn more actively, lively and interestingly.

2.2   Drill Strategy of Behaviorism

Behaviorism spanned from the field of psychology to the field of education in the early 1960s. The educational environment of the school was conducted to that if the correct stimulation is provided with the instructor then the knowledge would be got by the learners [15, 30]. Behaviorist believes that knowledge can be acquired passively, therefore emphasizing on the use of stimulus and feedback control makes meaning for learning. As one of the multiple spontaneous responses to stimuli makes sense to the learner, the stimulus and the response are connected to strengthen learning effects for the learner. Therefore the instruction strategy implements the most appropriate positive reinforcement to enhance the learner's need to show good behavior or provides a negative reinforcement to remove the behavior that learner should avoid.

The behaviorist view on learning emphasizes the task of instruction and should successfully guide learners to produce new connections of external behavior. The strategies of instruction design include designing different forms of exercises or questions to induce learners to respond and providing immediate feedback after doing practice questions, quizzes or homework by the learner.

3   System Development Strategy of Aural Skills Development in Music Rhythm

The acquisition of aural skills is a continuing process of practices that requires long-term training and continuous modifications [23, 24]. Time is a key factor in the existence of music forms. The abstract ion with various basic elements of music must be reflected in the course of time.

Learner drills these apps in conscious training to enhance learner's unconscious intuition and increase the effectiveness of music rhythm learning.

3.1   Steady Rhythm Maintenance

The metronome is an instrument for making sounds regularly. It is usually used as an auxiliary tool for practice to help learners maintain a steady rhythm and make sure to reach a precise rhythm during the entire practice. The tempo sign described in the text of Italian words, such as: Allegro, Andante, Moderato, etc. The music note sign is usually expressed in the form of "note = number". The learning interface is shown as Fig. 1.

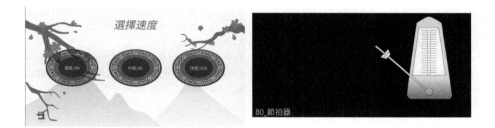

Fig. 1. System of Metronome

3.2   Drill in Music Rhythm

Music rhythm is mainly designed to help learners improve the ability of accurate rhythm in music, and its main goal is to provide learners with different types of music score exercises to improve their skills in stable rhythm.

There are two modes for practicing with smart phone: "tapping exercises" and "shaking exercises". The former one provides learners with choice to tap on the phone screen as the practice mode, the latter one provides learner with music rhythm practice by shaking. (Fig. 2).

Fig. 2. Functions of Drill Mode Selection and Drill

4   Conclusion

This convenient type of ubiquity of aural skills development in music rhythm through the mobile phone mechanism will "groove" with learners' performance anytime and anywhere. It can detect whether learners have tapped on or shaken with mobile phone in the correct rhythm key, and will respond to the learner until they accomplish the correct rhythm drill before continuing to play the next one. The learners just need to tap or shake to play, put their hands on/with the mobile phone, learners can concentrate on their performance, and they can know whether they are playing correctly through instant feedback—learners no longer need to stay with the tutors and be limited to the tutor's on-site guidance. From then on, there is no interference in the process of practicing music rhythm, only concentration and fluency.

The development of aural skills in music rhythm through the mobile phone mechanism received positive affirmation during the prototype trial test with potential users. This also strengthened further studies to bring this new technology to the majority of users.

References

1. Adekantari, P.: The influence of instagram-assisted project based learning model on critical thinking skills. J. Educ. Soc. Res. **10**(6), 315 (2020)
2. Ajabshir, Z.F., Sadeghi, K.: The Impact of Asynchronous Computer-Mediated Instruction (CAI) on EFL Learners' Vocabulary Uptake across Different Proficiency Levels. Teaching English with Technology **19**(3), 68–89 (2019)
3. Alam, M.M., Ahmad, N., Naveed, Q.N., Patel, A., Abohashrh, M., Khaleel, M.A.: E-learning services to achieve sustainable learning and academic performance: an empirical study. Sustainability **13**(5), 2653 (2021)
4. Bispham, J.: Rhythm in music: What is it? Who has it? And why? Music. Percept. **24**(2), 125–134 (2006)
5. Becker, H.J.: Pedagogical motivations for student computer use that lead to student engagement. Educ. Technol. **40**(5), 5–17 (2000)
6. Bond, M., Bedenlier, S.: Facilitating student engagement through educational technology: towards a conceptual framework. J. Interact. Media Educ. **2019**(1) (2019)

7. Calilhanna, A.M.: Teaching musical meter to school-age students through the ski-hill graph. In: Proceedings of Meetings on Acoustics 178ASA, vol. 39, no. 1, p. 025003. Acoustical Society of America (2019)

8. Condaris, C.: Correlating methods of teaching aural skills with individual learning styles. Athens J. Human. Arts **6**(1), 1–14 (2019)

9. Crompton, H., Burke, D.: The use of mobile learning in higher education: a systematic review. Comput. Educ. **123**, 53–64 (2018)

10. Dalby, B.: Toward an effective pedagogy for teaching rhythm: gordon and beyond. Music. Educ. J. **92**(1), 54–60 (2005)

11. de Brabander, C.J., Glastra, F.J.: The unified model of task-specific motivation and teachers' motivation to learn about teaching and learning supportive modes of ICT use. Educ. Inf. Technol. **26**(1), 393–420 (2020). https://doi.org/10.1007/s10639-020-10256-7

12. Dias, L.B.: Integrating technology: some things you should know. Learn. Lead. Technol. **27**(3), 10–13 (1999)

13. Giurgiu, L.: Microlearning an evolving elearning trend. Sci. Bull.-Nicolae Balcescu Land Forces Acad. **22**(1), 18–23 (2017)

14. Hall, A.B., Trespalacios, J.: Personalized professional learning and teacher self-efficacy for integrating technology in K–12 classrooms. J. Digit. Learn. Teach. Educ. **35**(4), 221–235 (2019)

15. Holland, J.G., Skinner, B.F.: The Analysis of Behavior: A Program for Self-instruction. McGraw-Hill, London (1961)

16. Iversen, J.R., Balasubramaniam, R.: Synchronization and temporal processing. Curr. Opin. Behav. Sci. **8**, 175–180 (2016)

17. Jdaitawi, M.: Does flipped learning promote positive emotions in science education? A comparison between traditional and flipped classroom approaches. Electron. J. e-learning **18**(6), 516–524 (2020)

18. Karpinski, G.S.: Aural Skills Acquisition: The Development of Listening, Reading, and Performing Skills in College-Level Musicians. Oxford University Press on Demand (2000)

19. Klonoski, E.: Improving dictation as an aural-skills instructional tool. Music. Educ. J. **93**(1), 54–59 (2006)

20. Kwon, B.R., Lee, J.: What makes a maker: the motivation for the maker movement in ICT. Inf. Technol. Dev. **23**(2), 318–335 (2017)

21. Li, J., Lin, J.T., Wu, C.H.: Deterministic factors influencing learners' online learning behaviors by applying IT-assisted music curriculum. In: Proceedings of the 2020 2nd International Conference on Modern Educational Technology, pp. 58–68 (2020)

22. Persellin, D.C.: Responses to rhythm patterns when presented to children through auditory, visual, and kinesthetic modalities. J. Res. Music Educ. **40**(4), 306–315 (1992)

23. Pesek, M., Suhadolnik, L., Šavli, P., Marolt, M.: Motivating students for ear-training with a rhythmic dictation application. Appl. Sci. **10**(19), 6781 (2020)

24. Pomerleau Turcotte, J., Moreno Sala, M., Dubé, F.: Factors influencing technology use in aural skills lessons. Revue musicale OICRM **4**(1), 1–16 (2017)

25. Rana, K., Rana, K.: ICT integration in teaching and learning activities in higher education: a case study of Nepal's teacher education. Malays. Online J. Educ. Technol. **8**(1), 36–47 (2020)

26. Reifinger Jr, J.L.: Skill development in rhythm perception and performance: a review of literature. UPDATE: Appl. Res. Music Educ. **25**(1), 15–27 (2006)

27. Rogers, M.: Aural dictation affects high achievement in sight singing, performance and composition skills. Aust. J. Music. Educ. **1**, 34 (2013)

28. Rohwer, D.: Predicting undergraduate music education majors' collegiate achievement. Texas Music Educ. Res. **45**, 52 (2012)

29. Schellenberg, E.G., Weiss, M.W.: Music and cognitive abilities (2013)
30. Skinner, E.A., Wellborn, J.G., Connell, J.P.: What it takes to do well in school and whether I've got it: a process model of perceived control and children's engagement and achievement in school. J. Educ. Psychol. **82**, 22–32 (1990)
31. Sharples, M., Taylor, J., Vavoula, G.: Towards a theory of mobile learning. In: Proceedings of mLearn, vol. 1, no. 1, pp. 1–9 (2005)
32. Scheurs, J., Dumbraveanu, R.: A shift from teacher centered to learner centered approach. Learning, **1**(2) (2014)
33. Song, A.: Alternative strategies for a collegiate aural skills classroom: an observational case study, Doctoral dissertation, Teachers College, Columbia University (2015)
34. Traxler, J.: Learning in a mobile age. Int. J. Mobile Blended Learn. (IJMBL) **1**(1), 1–12 (2009)
35. Vuust, P., Witek, M.A.: Rhythmic complexity and predictive coding: a novel approach to modeling rhythm and meter perception in music. Front. Psychol. **5**, 1111 (2014)
36. Yeung, C.L., Zhou, L., Armatas, C.: An overview of benchmarks regarding quality assurance for elearning in higher education. In: 2019 IEEE Conference on e-Learning, e-Management & e-Services (IC3e), pp. 1–6. IEEE (2019)

A Prospective Preventive Screening Tool-Pancreatic Cancer Risk Model Developed by AI Technology

Hsiu-An Lee[1], Kuan-Wen Chen[2], and Chien-Yeh Hsu[2,3(✉)]

[1] National Health Research Institutes, The National Institute of Cancer Research, Tainan, Taiwan
[2] Department of Information Management, National Taipei University of Nursing and Health Sciences, Taipei, Taiwan
cyhsu@ntunhs.edu.tw
[3] Master Program in Global Health and Development, Taipei Medical University, Taipei, Taiwan

Abstract. Pancreatic cancer is one of the cancers that are not easy to detect early due to the lack of obvious disease characteristics in the early stage, the tumor is mostly located in the posterior abdominal cavity, and the lack of early diagnosis tools. Therefore, when it is diagnosed, it is often approaching the late stage. In recent years, the studies of pancreatic cancer are mostly single-factor analysis and multi-factor analysis to summarize one or more risk factors, including past diseases, physical signs, family genes and long-term eating habits, etc., and for early evaluation models of pancreatic cancer is less. This study uses Logistic Regression (LR), Deep Neural Networks (DNN), ensemble voting learning (Voting), ensemble stacking learning (Stacking) and other methods to establish different models. Compare the performance between different models. It is hoped that based on the patient's past medical history, the high-risk group can be judged through the model, and whether it has a high probability of suffering from pancreatic cancer within one year, so that the public and doctors are aware of the risk of pancreatic cancer early.

In this study, the best model is 19 factors LR's model. The accuracy is 70%, the sensitivity is 70%, the specificity is 70%, and the AUC is 0.78. The contribution of this study is to use non-invasive factors to identify Chronic Kidney Disease, but it is a preliminary evaluation and ultimately requires doctors to diagnose.

Keywords: Pancreatic cancer · Machine learning · NHIR database

1 Introduction

In Taiwan, cancer has been the top cause of death for since 1982. According to statistics on the cause of death in 2020, pancreatic cancer ranks eighth. The incidence of pancreatic cancer is lower than other cancers, but its mortality rate is very high. The one-year survival rate of patients with pancreatic cancer in Taiwan is about 25.52%, the

J. C. Hung et al. (Eds.): FC 2021, LNEE 827, pp. 159–166, 2022.
https://doi.org/10.1007/978-981-16-8052-6_17

three-year survival rate is about 9.22%, and the five-year survival rate is about 6.6%, and the 10-year survival rate is about 4.71%. If calculated from the ratio of the number of new diagnoses to the number of deaths per year, the mortality rate is about 90% [1, 2]. Pancreatic cancer is one of the cancers that are not easy to detect early due to the lack of obvious disease characteristics in the early stage, the tumor is mostly located in the posterior abdominal cavity, and the lack of early diagnosis tools. Therefore, when it is diagnosed, it is often approaching the late stage. There will be certain difficulties.

Targeted drugs have always been one of the methods for the treatment of various cancers, but pancreatic cancer still lacks a wide range of effective targeted drugs. According to recent studies, the reason is that pancreatic cancer is not like lung cancer, breast cancer and other cancers that have found specific molecules. Variation, researchers even believe that there may be more than a single molecular variation in pancreatic cancer [3]. Compared with other diseases, pancreatic cancer tends to show higher non-specific symptoms. This also causes the disease to be diagnosed as other abdominal diseases at the beginning, making the initial treatment plan ineffective, delaying the timing of treatment, and adding pancreatic cancer. The lack of early diagnosis tools for visceral cancer has caused the overall treatment effect to fail to reach the expected level [4].

The estimated number of cases of pancreatic cancer will continue to grow from 2018 to 2040. Experts therefore call on the world to face up to the early prevention of pancreatic cancer and indicate that there is an urgent need for a screening method for early detection of pancreatic cancer. And effective treatment strategies [5]. The early prevention of pancreatic cancer must be an urgent task.

Early screening is one of the methods of disease management, sometimes even before the onset of disease symptoms, it helps to reduce the risk of death due to disease [6]. According to the World Health Organization (WHO) definition of cancer screening, early detection of cancer consists of two parts: early diagnosis and screening. The focus of early diagnosis is to find symptomatic patients as early as possible, aiming to reduce the proportion of patients diagnosed at an advanced stage, while screening involves testing healthy individuals and expecting patients to identify them before they develop any symptoms related to cancer [7]. In view of this, if the high-risk groups of pancreatic cancer can be identified in the population under the correct, efficient and low-cost conditions, it will help the early prevention of pancreatic cancer and improve the possibility of implementing high-risk population screening. In recent years, the studies of pancreatic cancer are mostly single-factor analysis and multi-factor analysis to summarize one or more risk factors, including past diseases, physical signs, family genes and long-term eating habits, etc., and for early evaluation models of pancreatic cancer is less.

This study uses Logistic Regression (LR), Deep Neural Networks (DNN), ensemble voting learning (Voting), ensemble stacking learning (Stacking) and other methods to establish different models. Compare the performance between different models. It is hoped that based on the patient's past medical history, the high-risk group can be judged through the model, and whether it has a high probability of suffering from pancreatic cancer within one year, so that the public and doctors are aware of the risk of pancreatic cancer early.

2 Method

2.1 Tool

In this study, SPSS 22 version was used for data pre-processing, chi-square independence test, and sampling matching division; R language version 3.5.2 was used for AIC standard verification such as Backward elimination in stepwise regression; Python 3.7.3 version was used for model establish which including: logistic regression, deep neural network, and integrated learning stacking method. Microsoft SQL Server 2014 was used for data management.

2.2 Data Resource

The National Health Insurance Research Database (NHIRDB) was used in this study. The data including health insurance outpatient details (OPDTE) and health insurance hospitalization details (IPDTE) in the sample files of 2 million people from 2000 to 2009. The ICD-9 coding format was used in the NHIRDB from 2000 to 2009.

The patients who suffered pancreatic cancer (first three codes of the ICD-9-CM main diagnosis are 157) were included in this study. First, exclude cases with fewer than two records of pancreatic cancer diagnosis, because they may only be a one-time diagnosis for an examination. Total number of cases was 738. Taking the time when the subject was first diagnosed with pancreatic cancer, the short-term medical history was followed up for one year. Then, according to the relevant literature recommendations of the case-control analysis, a control ratio of one to three was used to collect data from the control group [8]. A total of 738 people in the pancreatic cancer case group and 2,952 people in the control group without pancreatic cancer were finally adopted.

2.3 Data Pre-processing

In order to carry out short-term past disease counts, factor verification and subsequent model training of the research population, the ICD-9 codes in the records are summarized and processed. The processed ICD-9 are all performed in a three-code format. Then the data is transformed into a one-dimensional array of disease matrix through one-hot encoding. Add a field for each disease, and use category 1 to indicate that subject have suffered from this disease, and category 0 to indicate that subject have not suffered from this disease. Table 1 is an example of a one-dimensional array of disease matrix.

Table 1. An example of a one-dimensional array of disease matrix.

ID	CANCER	AGE	SEX	ICD-577	ICD-592	ICD-576	ICD-531	ICD-250
ruksuzdwfb	0	77	1	0	0	1	0	0
beacgcnrme	1	70	0	1	0	1	1	1
ayddayitna	1	64	1	1	1	1	1	1
rxyyskqehw	0	87	1	0	0	0	0	1

2.4   Factors Selection

We use chi-square test and Akaike information criterion (AIC) for factor selection. The model factors are divided into two groups. The first group is past diseases that are significantly related to pancreatic cancer. The second is that we use AIC's Backward Elimination to further filter significantly related factors. The method of AIC's backward elimination method is to discard variables item by item in a complete regression until the best solution is obtained when too much explanatory power is lost when any one variable is discarded. In this study, the P value set in < 0.01.

2.5   Model Training

We used eight different machine learning methods and compared accuracy, including Logistic Regression (LR), Deep Neural Networks (DNN), and Ensemble learning - integrated learning stacking method.

2.6   Model Evaluation

Confusion matrix and receiver operating characteristic curve (ROC curve), accuracy, sensitivity and specificity are used to evaluate the performance of different models.

3   Results

3.1   Demography

The sample control in this study was completed based on age and gender distribution. A total of 738 pancreatic cancer patients were used as the experimental group and 2,214 control groups (Table 2).

Table 2. Demographic of categories factors and p-value

Factors SUM	Description	Experimental group (With Cancer) 738	Control group (Without-Cancer) 2,214	p-value
Gender	Male	398(53.9%)	1194(53.9%)	.000
	Female	340(46.1%)	1020(46.1%)	
Age	0–17	1(0.1%)	1(<0.1%)	.000
	18–24	14(1.9%)	28(1.3%)	
	25– 34	21(2.8%)	64(2.9%)	
	35–44	55(7.5%)	141(6.4%)	
	45–54	95(12.9%)	234(10.6%)	
	55–64	144(19.5%)	421(19.0%)	
	>65	408(55.3%)	1325(59.8%)	

3.2   Factors Selection

This study uses a combination of two factors for model training. The first is a significant factor with a p-value of Chi-square test < 0.05. There are 33 items in total (hereinafter referred to as the first group of factors), which are cholangitis, pancreatitis, abdominal pain, Hepatitis, gastritis, flatulence, gastric ulcer, duodenal ulcer, peptic ulcer, abnormal gastric function, abnormal intestinal function, chest pain and cough, acute liver necrosis, hypertension, headache, low back pain, acute bronchitis, heart failure, urticaria, Kidney stones, conjunctivitis, arrhythmia, gout, hypertensive heart disease, acute nasopharyngitis, leg cellulitis, sleep disorders, anxiety, osteoarthritis, hypoglycemia, diabetes, coronary heart disease and bronchitis, these 33 items It will also be used as a candidate factor for stepwise regression to facilitate subsequent selection of factor combinations with higher explanatory power.

The second is to use the backward elimination method in stepwise regression, and select 19 factors with p value <0.05 (hereinafter referred to as the second group of factors), which are abdominal pain, peptic ulcer, flatulence, gastritis, and abnormal gastric function, Hepatitis, sleep disorders, cholangitis, pancreatitis, headache, chest pain and cough, urticaria, leg cellulitis, acute bronchitis, arrhythmia, acute liver necrosis, diabetes, gout and abnormal bowel function.

3.3   Model Evaluation

The evaluation results show that under the combination of two factors, the AUC of the LR model in the test set is higher than the other three models. The sensitivity evaluation is determined by the Stacking model of the first set of factors as the best, and the specificity is determined by the DNN of the second set of factors. The model is the best. In the validation set evaluation results, when the factor features are more complex (the first set of factors), the Voting, Stacking and DNN models are easy to overfit. However, after simplifying the model complexity (reducing the number of factors features and using the second set of factors), it has appeared Significant improvement. The detailed results are shown in the table below (Table 3, Figs. 1 and 2).

Table 3. Performance of different models.

First set of factors (33 Factors)	AUC	Accuracy	Sensitivity	Specificity
LR	**0.78**	**0.73**	**0.7**	**0.74**
Voting	0.87	0.77	0.76	0.77
Stacking	0.85	0.77	0.71	0.82
DNN	0.89	0.82	0.78	0.86
Second set of factors (19 Factors)	AUC	Accuracy	Sensitivity	Specificity
LR	**0.78**	**0.7**	**0.7**	**0.7**
Voting	0.83	0.73	0.74	0.72
Stacking	0.82	0.73	0.74	0.73
DNN	0.82	0.73	0.72	0.74

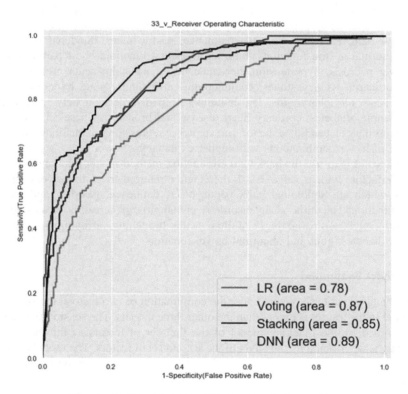

Fig. 1. Models (First Set of Factors) confusion matrix.

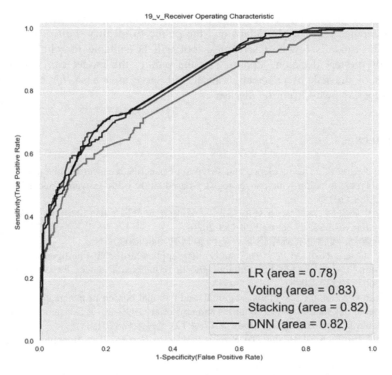

Fig. 2. Models (Second Set of Factors) confusion matrix.

4 Discussion and Conclusion

This study uses the disease diagnosis records in NHIRD to develop a pancreatic cancer risk identification and prediction model. The immediate detection of pancreatic cancer in the early stage of the disease is one of the key goals for evaluating the outcome of treatment. Based on the results of related research on the development of pancreatic cancer tumors, 6 to 12 months before diagnosis [9, 10], is a high-risk identification golden timing. This study explored the patient's disease diagnosis status within 12 months by means of statistics of real-world data. The evaluation results of the model showed that the short-term medical history has the ability to predict risk of pancreatic cancer.

This study uses the statistical presentation of Taiwan's real data as the final training factor test matrix. This will be closer to the real situation of patients in the country and will be more applicable to the local area. The previous literature pointed out that the main diagnoses related to the incidence of pancreatic cancer include pancreatitis, diabetes, peptic ulcer, cholecystitis, cholangitis, and hepatitis. Factors such as gastrointestinal-related inflammation, kidney stones, and chest pain and coughing. Although these features need further clinical verification in the future, they can also serve as potential future research candidates.

Through objective model development and comparison, this research has developed a reference tool for physicians and the public to identify the risk of pancreatic cancer. The model's low-cost screening tool will be suitable for clinical decision-making or pre-test decision support. For the public, the model can improve their awareness of the risk of pancreatic cancer and serve as a basis for changing their behavior to stay away from the disease.

References

1. Chang, J.S., et al.: The incidence and survival of pancreatic cancer by histology, including rare subtypes: a nation-wide cancer registry-based study from Taiwan. Cancer Med. **7**(11), 5775–5788 (2018)
2. 國家衛生研究院. 冰與火: 不死的癌王—胰臟癌治療展現生機 (2020). https://enews.nhri.org.tw/archives/3525. Accessed 20 Oct 2020
3. 好心肝會刊, 胰腺癌有新標靶藥 好心肝會刊第90期 (2020)
4. Chang, C.-L., Hsu, M.-Y.: The study that applies artificial intelligence and logistic regression for assistance in differential diagnostic of pancreatic cancer. Expert Syst. Appl. **36**(7), 10663–10672 (2009)
5. Pourshams, A., et al.: The global, regional, and national burden of pancreatic cancer and its attributable risk factors in 195 countries and territories, 1990–2017: a systematic analysis for the Global Burden of Disease Study 2017, **4**(12), 934–947 (2019)
6. National Institutes of Health: To Screen or Not to Screen? The Benefits and Harms of Screening Tests (2017)
7. Organization, W.H. Screening and early detection. https://www.euro.who.int/en/health-topics/noncommunicable-diseases/cancer/policy/screening-and-early-detection. Accessed 6 June 2021
8. Grimes, D.A., Schulz, K.F.: Compared to what? Finding controls for case-control studies. The Lancet **365**(9468), 1429–1433 (2005)
9. Canto, M.I., et al.: Risk of neoplastic progression in individuals at high risk for pancreatic cancer undergoing long-term surveillance. Gastroenterology **155**(3), 740-751.e2 (2018)
10. Yu, J., et al.: Time to progression of pancreatic ductal adenocarcinoma from low-to-high tumour stages. **64**(11), 1783–1789 (2015)

Data Security of Internet of Things Under Cloud Environment

Weiwen He[✉]

School of Information Engineering, Guangzhou Nanyang Polytechnic,
Guangzhou 510000, Guangdong, China

Abstract. With the increasing demanding of Internet of Things (IOT) and the improvement of information technology, the IOT service is facing greater challenges. The birth of cloud computing brings great opportunities to confront these challenges, but it also brings security problems. Considering the IOT data security problem under cloud service mode, this paper develops a trust framework and trust authentication scheme to solve problems including security from data network transmission, cloud data protection and acquisition terminal trust verification. It is believed that the framework will solve the data security problems to a certain extent.

Keywords: Internet of Things (IOT) · Data security · Trust framework · Trust authentication scheme

1 Introduction

With the development of global economy and the strategic objectives of "reading China" and "digital city", the adjustment of industrial structure in China has been gradually speeding up. Information technology, which is regarded as the key of enterprise development, is expediting the profound industrialized change. The services of Internet of Things (IOT) [1] based on cloud Computing Technology [2, 3] integrate Enterprise Resource Plan(ERP), supply chain, customer relationship, product data, financial and other information into an organic integration. It breaks the bondage of the enterprise resources, and makes links between management and public cloud platform. It also connects global industrial, reduces cost and improves enterprise competitive ability.

IOT is a new field of technology composing of the multi-disciplines form the perception of data processing. It can be divided into sensing layer, network layer, data layer and application layer. Sensing layer has the ability of transferring samples from the network layer to the data layer by two-dimension code, Radio Frequency.

Identification (RFID) and other sensor devices; application layer analyzes and mines dates from data layer to achieve different needs of enterprise services; IOT, based on the environment of technology of cloud computing, synthesizes cloud computing, grid computing, parallel processing and distributed computing, shifting data storage, data distribution, service response to the cloud services. However, IOT based on cloud service uses external network storage and sensor devices to automatically push regular

© The Author(s), under exclusive license to Springer Nature Singapore Pte Ltd. 2022
J. C. Hung et al. (Eds.): FC 2021, LNEE 827, pp. 167–172, 2022.
https://doi.org/10.1007/978-981-16-8052-6_18

data to the cloud, which inevitably brings many new unstable factors [4, 5]. The reliability of IOT network access and the security of data in cloud have become the primary concern.

There are still important problems to be solved in the research of IOT: 1. The credibility of network accessing can be guaranteed because of lacking the description of trust service specification. 2. Because the stored data in cloud is uncontrollable, it will face various types of threats, such as data leakage. The article has put forward a data trust scheme of IOT under cloud environment, it improves the security of IOT cloud data transmission and IOT cloud data storage.

2 Trust Framework

The development of communication technology and users' demand have determined the change from simplification to diversification on data collection terminal of modern IOT services. This paper puts forward a trust framework under cloud environment for data transmission of IOT, which ensures data reliable transmission and strong fault tolerance, it provides credibility monitoring, diagnosis and recovery to ensure transmission data movement. It also provides a security guarantee of stored data in the cloud. As shown in Fig. 1.

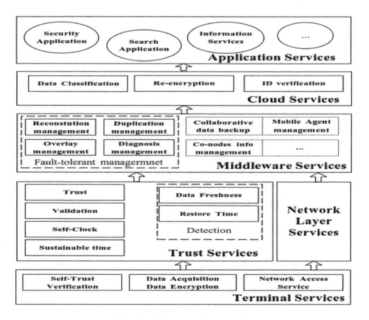

Fig. 1. The trust framework.

In design of reliable data transmission, the main idea is to create middleware layer to isolate network service layer and application layer, using the middleware layer network node management and data management. The key issues are: trust and cooperative relationships between the networks, mutual authentication of network object and so on. In terms of cloud storage, we use the re-encryption for data security over the middleware layer, which prevents cloud service provider's privilege to access the stored data in cloud [6, 7]. In fact, re-encryption method is performed by the cloud service agency, but it only can operate encrypted data after obtaining authorization. Although cloud service providers have data access privileges, they can't get any information from the encrypted data. When the final users receive the re-encryption of data and authorized decryption key, they can easily decrypt the data without extra computational overhead.

Considering many factors, this trust framework combines the trusted chains transmit and re-encryption [8]. In the bottom layer, it uses the TPM chip to ensure the self-trust of acquisition terminal; in the network layer, it uses dependable certification scheme to ensure reliable internet access; in the application layer, and it uses re-encryption to protect stored data in cloud [9]. From all above, it provides three-dimensional reliable protection.

3 Trust Authentication Scheme

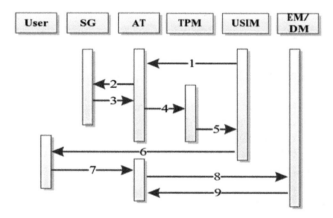

Fig. 2. The trust authentication scheme.

In dynamic IOT, the stored data in cloud often do increment operations, which is easy to implement re-encryption method. But the mobile acquisition terminal in the network will keep changing frequently, which complicates the cooperative relationship between network access terminal and acquisition terminal [10]. This article focuses on these problems and converts the relationship between network access terminal and acquisition terminal into measuring the dependence of self-trust authentication and mutual trust authentication. Combing with requirement cloud service, this paper propose the trust authentication scheme as shown in Fig. 2.

In Fig. 2, assumed that acquisition terminal of IOT holds the authority of terminal access. SG is sensor group, which collect data samples, AT represent the acquisition terminal, it mainly implement data summary. AT has the USIM module, which is a component of network. It is not only able to support multiple applications, butalso includesauthenticationof USIM tonetwork. EM/DM is an encryption/decryption module, which including many functions, such as password operation, digital certificate, sensitive information storage and so on. TPM module is the unit to ensure the credibility of sampling, it can implement self-trusted authentication. This scheme also contains its own private key, certificate TPMCert, and sharing SG keys Keyshare, it implements as following steps:

3.1   USIM→AT: r_1, $USIM_{ID}$, PR_1

Acquisition terminal of IOT needs to check its own platform integrity when power is on, so USIM module send random number r_1, USIM identification USIM $_{ID}$ and platform verification requests PR_1 to AT.

3.2   AT→SG: r_2, AT_{ID}, PR_2

After receiving PR_1 by the USIM platform, AT will transmit a random number r_2, AT's identification AT_{ID} and SG verification request PR_2 via the I/O bus.

3.3   SG→AT: EAC_{SG}

$$EACSG\ AC(E(Keyshare\ ,\ r2\ ),\ ATID\ \|\ SGHash\ ) \tag{1}$$

SG calculates Hash value after it receives integrity checking request PR_2, and then uses Key_{share} to encrypt these collected random numbers r_2.Finally, we can calculate SG authentication codes EAC_{SG} by formula (1). Then SG will send back authentication codes to AT for verification. Where AC and E represent the encryption algorithms, symbol | | represent cascade.

3.4   AT→TPM: r_3, TPM $_{Sig}$, ER

$$TPM_{Sig}Sig(TPM_{SK}, r_1\|r_3\|USIM_{ID}\|PCR) \tag{2}$$

AT needs to verify authentication codes after receiving EAC_{SG} from SG. Because SG integrity Hash value SG_{Hash} is stored within the TPM in advance, so the AT use random number r_2 generated by itself and SG_{Hash} to calculate EAC'_{SG} by formula (1), then do the cooperation. If $EAC_{SG} \neq EACS'_G$, it shows that SG information is not integrated or tampered. Thus, AT end its own credible verification process. If $EAC_{SG} \neq EAC'_{SG}$, it shows that the external acquisition is reliable and integrated, AT will go to next verification process. After that, AT will generate random number r_3, and use its own private key TPM $_{SK}$ to signPCR, transaction affairs ER and r_3 by formula (2), then send to the TPM.

3.5   TPM→USIM: TPM $_{Cert}$, r_3, TPM $_{Sig}$, ER

TPM only needs to do one operation after AT acquire TPM's authentication, it will send r3, TPM $_{Sig}$, ER and TPM certificate TPM$_{Cert}$ to USIM.

3.6   USIM→User: TS, r1, r3, TPM$_{CertID}$, NAI, ER, TPM $_{Sig}$, EAC$_{TPMID}$

$$EAC_{TPMID}AC(Key_{AU}, TS \, || \, TPM_{CertID}, \, || \, r_1 || r_3 || ER) \tag{3}$$

The certificate TPM$_{Cert}$ will be send by USIM to the mobile network access terminal, it needs timestamp TS, TPM certificate TPM$_{CertID}$ and Key$_{AU}$ to verify the legitimacy of the mobile terminal platform by EAC$_{TPMID}$ generate by formula (3). If it already has TPM$_{Cert}$, it will directly check the legitimacy of the signature TPM $_{Sig}$. If the signature verification fails, the certification process terminate. If not, it requests enduser password to obtain permissions to control the AT.

3.7   User → AT: (PR$_{re}$, Key$_{Usk}$)/(PR$_{se}$, Key$_{Upk}$)

If end user inputs the wrong password, login fails. If the user inputs the correct password, two types of actions can be performed. The first one is sending package, sending request PR$_{se}$ and public key Key$_{Upk}$. This action will ask AT to send encrypted data to the cloud for re-encryption operation. The second one is to receiving package. After sending request PRre and decryption key Key$_{Usk}$, AT will request to receive encryption data packets and do decryption operation. Considering the security of stored data in cloud, it will keep changing Key$_{Usk}$, so Key$_{Usk} \neq$ Key$_{AU}$.

3.8   AT → EM/DM: Data$_E$/Data$_D$

AT needs to determine the types of user's requirements after receiving them, then distributes encryption and decryption work to EM/DM module respectively according to different requirement.

3.9   EM/DM → AT: Data

$$EAC_{Data}AC(Key_{Data}, TPM_{ID} || TS) \tag{4}$$

AT must verify its integrity by formula (4) before data decryption. If integrity verification fails, AT sends verification results to the USIM to receive package again. If verification succeed, DM decrypts the data package, then sends the data to operational areas.

4 Conclusions

With the increasing demanding of IOT and the improvement of electronic information technology, the IOT services are facing more requirements and challenges. The birth of cloud computing brings great chances to confront these challenges, but it also brings security problems. To address the IOT data security problem under Cloud environment, this paper, investigated problems of security from data network transmission, cloud data protection and acquisition terminal respectively self-trust verification. It built a trust framework and it solved the data security problems to a certain extent.

References

1. ITU. The Internet of Things (2010). http://www.itu.int/internetofthings
2. Floerkemeier, C., Langheinrich, M., Fleisch, E., et al.: The internet of things. In: Sensor Applications in the Supply Chain: The Example of Quality-Based Issuing of Perishables, Chapter 9. LNCS, vol. 4952, pp. 140–154 (2008). https://doi.org/10.1007/978-3-540-78731-0
3. Wikipedia.Cloud Computing (2009). http://en.wikipedia.org/wiki/Cloud_computing
4. Shroff, G.: Enterprise Cloud Computing: Technology, Architecture, Applications. Cambridge University Press, England (2010)
5. Yu, X., Wen, Q.: A protect solution for data security in mobile cloud storage. In: Proceedings of SPIE, vol. 87841, pp. F1–F5 (2013)
6. Gu, X., Xu, Z., Xiong, L., Feng, C.: The security analysis of data re-encryption model in cloud services. In: Proceedings - 2013 International Conference on Computational and Information Sciences, ICCIS 2013, pp. 98–101 (2013)
7. Gu, X., Xu, Z., Wang, T., Fang, Y.: Trusted service application framework on mobile network. In: Proceedings - IEEE 9th International Conference on Ubiquitous Intelligence and Computing and IEEE 9th International Conference on Autonomic and Trusted Computing, UIC-ATC 2012, pp. 979–984 (2012)
8. Zhou, S., Liu, X.: Research on data security model of internet of things based on attribute based access control. J. Comput. Theoret. Nanosci. 13(12), 9596–9601 (2016)
9. Yue, J.H., Zhang, X., HD University: Research on the cloud services platform and service mode of intelligence community. Internet of Things Technologies (2013)
10. Carter, A.: Considerations for genomic data privacy and security when working in the cloud. J. Mol. Diagnost. 21(4), 542–552 (2019)

Impact and Application of Block Chain Technology on Urban Traffic Based on Artificial Intelligence

Zhenxing Bian[✉]

Software Engineering Department, Shandong Polytechnic College, Jining 272000, Shandong, China

Abstract. In recent years, with the development of artificial intelligence, blockchain technology has been used in many industries. The development of the financial industry is the fastest and most mature. Now in the transportation field, blockchain technology and big data technology are combined; it can solve some of the pain points in traffic. Blockchain technology is a decentralized technology. Decentralization means that each node in the network is independent of each other and carries out point-to-point information data transmission. In this process, there is no organization or individual pairing and transmission process. Using blockchain technology in traffic scenarios can put information and data on the chain to ensure that information and data will not be tampered with and can be traced back at all times. This article proposes a method to build a big data platform for urban intelligent transportation using blockchain technology. The method takes data as the core, eliminates the centralized computer management of multiple organizations, changes the methods of data collection, data processing and analysis, completes the data storage modules and methods, and realizes the complete platformization of the urban intelligent transportation system. The results show that the urban intelligent transportation big data platform architecture using blockchain technology overcomes the limitations of multiple organizations, discovers urban intelligent transportation data, and further solves the problem of insufficient data in existing network archives.

Keywords: Artificial intelligence · Blockchain technology · Urban transportation · Transportation big data platform

1 Introduction

With the rapid development of modern information and communication technologies, big data, Internet of Things, cloud computing, etc. have been widely used in modern urban rail transit. Among them, blockchain technology provides a basic guarantee for the intelligent and intelligent construction of urban rail transit through data collection, access, processing, analysis, mining and modeling [1]. At present, the intelligent development of urban rail transit involves a wide range of data sources and a huge amount of data, with the characteristics of multi-source, heterogeneous, and self-organizing. With the explosive growth of data volume in the later period, traditional data storage methods, data storage volume, data types and hardware architecture restrict the use of data value, and can no longer meet the needs of urban rail transit for the

J. C. Hung et al. (Eds.): FC 2021, LNEE 827, pp. 173–180, 2022.
https://doi.org/10.1007/978-981-16-8052-6_19

development of intelligence and intelligence [2]. More importantly, traditional data storage uses a centralized architecture. Once the data information center is destroyed, the entire information system will be paralyzed, seriously threatening the operational safety of urban rail transit [3]. Therefore, how to improve data sharing, storage and security has become a hot topic in the development of urban rail transit.

Blockchain is the supporting technology of Bitcoin. It has the characteristics of decentralization, non-tampering, and traceability. It is gradually applicable to the fields of finance, digital copyright, document storage, Internet of Things, and notarization, and has achieved great results. Technology comparable to artificial intelligence, big data, cloud computing, etc. [4]. Artificial intelligence is based on huge data and powerful computing power. The characteristics of blockchain technology are well integrated into artificial intelligence applications to promote the development of artificial intelligence. The blockchain is a distributed storage structure, all nodes of the blockchain are the same, there is no central manager, all users can upload the data information of the next node through the protocol process [5]. The biggest feature of blockchain is decentralization and uniform openness. Once the link is successful, the data is difficult to modulate, and the data of all nodes are consistent, which has a wide range of application prospects in the field of intelligent transportation [6].

Although there are a large number of people trying to reduce the traffic load, the situation will continue to deteriorate [7]. Now, thanks to the technology of the fourth industrial revolution, experts are optimistic that cities can eventually eliminate the traffic load [8]. According to a report from the World Economic Forum, there are two technologies that make the transformation of urban transportation possible: driverless cars and Bitcoin's blockchain technology. According to Thomas Birr and Carsten Stöcker, in order to realize this dream, blockchain technology is decisive: Travel expenses will be automatically withdrawn from passengers' digital blockchain wallets or loaded into their credit cards, and payments will immediately flow to the car [9]. Passes, ID cards and P2P transactions will make it easier and safer to share vehicles and infrastructure, such as charging stations, toll stations, and parking lots. Each user's identity, age proof, insurance and payment will be blocked ability to identify, while protecting the security of passenger names, travel information and payment mechanisms. Smart contracts that control such transactions are based on standard shopping malls. Tax records and visual reports can ensure accuracy no matter where the trip takes place sex [10].

2   Method

2.1   Data Statistical Algorithm

Statistical algorithms are used to obtain the main components of the signal matrix Y. In short, it is the process of decomposition and sampling:

The first is to decompose the covariance matrix, that is, the decomposition formula (1):

$$E(yy^{H}|G) = \rho GE(xx^{H})G^{H} + E(nn^{H}) \tag{1}$$

The above formula can be decomposed into:

$$E(yy^H|G) = \rho UD^2 U^H + I_M = [U_s \quad U_N] \begin{bmatrix} \rho D_S^2 + I_L & 0 \\ 0 & I_{M-L} \end{bmatrix} \begin{bmatrix} U_S^H \\ U_N^H \end{bmatrix} \tag{2}$$

When T is large, $\sum \approx E(yy^H|G)$ decomposes $\sum$, then:

$$\sum = [\hat{U}_S \quad \hat{U}_N] \begin{bmatrix} \hat{D}_S & 0 \\ 0 & \hat{D}_N \end{bmatrix} \begin{bmatrix} \hat{U}_S^H \\ \hat{U}_N^H \end{bmatrix} \tag{3}$$

In the base station antenna, $V^H X$ is unknown and the object to be estimated. From the derivation process of Eqs. (2) and (3), the PCA analysis method shows that:

$$V^H X = 1/2 \left(\sqrt{\hat{D}_S - I_L} \right)^{-1} \hat{U}_S^H Y \tag{4}$$

2.2 Multi-type Blockchain Collaborative Management

Urban traffic problems involve many aspects, and the channels to obtain traffic information involve all levels. However, all types of traffic information from collection to analysis, from release to update, the entire process is controlled by the transportation functional department, and the public and other government functional departments are rarely involved. Adopting a completely open information management model so that the public can be used as a node link information platform to read or publish traffic information at will. This approach is not desirable. Because only adopting completely open information management based on the public chain model will bring about two fatal problems. First, it is difficult to ensure that the traffic information uploaded by the public is true and accurate. The inability to tamper with the data of the blockchain technology determines that once the node is successfully linked, the published information is difficult to modify. Second, even if the information released is true and accurate, the complete transparency of the data may bring about a series of unpredictable traffic operation problems. To effectively use the co-financing traffic information provided by the public, we must not only delegate power to the public, allow them to participate in traffic management, improve management flexibility, but also strictly control the information released by the public to avoid unpredictability. Therefore, we adopt a multi-type blockchain collaborative traffic information management model to delegate power to the public while taking into account the reasonable management and control of information. The information management mode based on the private chain is adopted for the transportation functional departments, that is, the block chain is established in the functional departments with high trust, and the reading authority has a certain degree of restriction on the public.

2.3 Promote the Implementation of Blockchain Technology

The application mode of blockchain technology has three types: public chain, private chain and client chain. Blockchain technology has 7 types of open agreement, anonymity, trustlessness, immutability, decentralization, traceability and programmability. Although the development of international blockchain technology has undergone an obvious acceleration process, there is no substantial transportation application in the blockchain technology industry. The state should actively promote in-depth research on the level of blockchain technology and provide good research on the application of blockchain technology for urban transportation. Research and investigation will be conducted through a variety of methods. Blockchain technology needs a reliable network to drive. It needs a reliable identity, reliable ledger, reliable calculation and reliable storage technology to realize the application of blockchain technology in urban transportation. Therefore, urban transportation rights can be quickly and safely circulated and used on the blockchain, and quickly confirm its value in urban transportation. Through the application of blockchain technology, it not only promotes the safety of urban traffic, but also improves the mobility of urban traffic and improves the quality of urban traffic.

3   Experiment

3.1   Subject

In recent years, the urban economy of City A has developed rapidly, urban expansion, population surge, and motor vehicle ownership continue to rise, but traffic management problems such as traffic congestion, lagging public transportation facilities, insufficient parking space supply, and frequent traffic violations are also increasing. In order to study the impact and role of blockchain technology in urban transportation in the context of artificial intelligence, this article takes the actual urban traffic situation of City A as the research object, and analyzes the current status of traffic in City A and the experience with the support of blockchain technology. Compare the status quo of urban traffic after reform to understand the impact of blockchain technology on city A's urban traffic.

3.2   Experimental Method

The main research methods used in this experiment are as follows:

The first is comparative analysis. On the one hand, through horizontal comparison, that is, compared with the level of developed countries, we can clarify the gaps and shortcomings between the current urban traffic development of my country and City A and the international level. On the other hand, through longitudinal comparison, that is, time series comparison, we can clarify the development process and current situation of urban transportation.

The second is the case analysis method. When studying the development mode of urban transportation under the support of blockchain technology in foreign countries, case studies were carried out with typical developed countries such as Europe, the United States, and Japan, and provided valuable experience and useful enlightenment for the development of urban transportation in Province A.

The third is model analysis. Starting from the analysis of the consumption and urban traffic characteristics of City A, the relationship between the annual consumption, residents' transportation mode and GDP data of City A is decoupled and analyzed, and the current status of urban traffic and the passing blocks of City A are obtained. The current status and impact of city A's urban traffic after the support of chain technology.

The fourth is the questionnaire survey analysis method. By issuing a questionnaire to some residents of City A, the degree of satisfaction of the residents of City A with the urban traffic of City A was investigated, and thus the impact of City A on the urban traffic of City A after experiencing the blockchain technology on the urban transportation reform was analyzed.

4 Results

According to statistics from the Planning Bureau of City A, the city's land area is about 12,065 km^2, and the sea area is about 11,000 km^2. Beginning in 2017, the built-up area of the urban area has reached 255 km^2 (ranked 40th in the country and 3rd in the province), an increase of 6.0 km^2 compared to 2016. The city's total highway mileage is 14,700 km (5 national roads, 445 km, provincial roads 6, 322 km), and rural roads account for 68.7%. The total length of urban roads in the urban area is 1165.5 km, and the total area is 35.07 km^2. The length of the newly built road is 36.66 km and the total area is 1.21 km^2; the length of the renovated and modified road is 11.24 km, and the road area is 0.44 km^2.

According to the calculation of the urban permanent population of 3,003,500 in 2017, the per capita road area in the urban area was 12.21 m^2, an increase of 0.58 m^2 compared with 2016; the average road area of vehicles was 51.5 m^2, a decrease of 1.9 m^2 from the previous year. The indicators are shown in Table 1. Judging from the road conditions of City A, it is still in an orderly development, but the development trend is slowing down, and the density of the expressway network is far below the standard requirements.

Table 1. Urban road indicators of city a in 2017

Urban road	Length (km)	Area (km^2)	Average width (m)
Over 12 m	1153.0	34.95	30.3
Ancient city 7–12 m	12.5	0.12	9.6
Total	1165.5	35.07	30.1
Dongtou	121.73	1.60	
Total	1287.23	36.66	
Road area per capita		12.21	

According to statistics, in 2017, the average traffic congestion index (traffic index ranges from 0 to 10 during the working day of the congestion in urban area A, the higher the value, the more serious the traffic congestion) value is 5.91, which is an increase from 5.71 in 2016 0.20. In 2017, the urban congestion lasted about two hours during the working day, compared with an increase of 32 min in 2016, the degree of urban traffic congestion increased significantly.

According to the questionnaires issued and collected, regarding travel time, the urban area of City A is not large, but 19.03% of citizens still need to spend 30–45 min to travel, and even 12.01% of citizens need to spend more time. Regarding the degree of satisfaction with traffic conditions, 34.88% of people think that the traffic conditions are not very good and even 16.78% think it is very bad.

In the end, we have conducted a survey on the current status of urban traffic in City A after the blockchain technology reform, and the results are completely different from those before the reform. Most people are satisfied with the reformed urban traffic and only occupy one (Figs. 1 and 2).

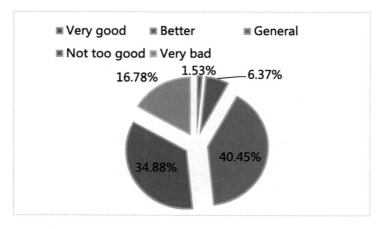

Fig. 1. Questionnaire survey on traffic satisfaction in City A

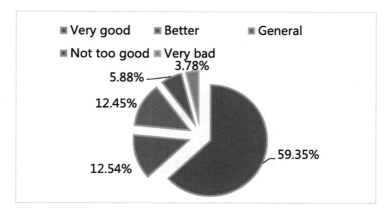

Fig. 2. Questionnaire survey on traffic satisfaction in City A

According to our Wenjuan investigation and analysis, it can be known that the urban transportation network of City A is relatively developed. Before the introduction of blockchain technology to reform the urban transportation of City A, the urban traffic of City A was crowded, and many residents were the city's traffic is dissatisfied, which affects their commuting and other life travel. However, after experiencing the reform of the blockchain technology on the urban traffic in City A, the urban traffic in City A has been greatly improved. It has basically been recognized by the residents. Most residents are satisfied with the urban traffic in City A. It can be concluded that the influence of blockchain technology in urban traffic is still great.

5   Conclusion

At present, the research of blockchain technology at home and abroad mainly focuses on the application of theoretical concepts and macrostructures, but the research on specific technologies and methods is not in-depth. With the active development of big data mining technology, artificial intelligence, intelligent identification technology, and network communication technology, integrated urban traffic management not only inherits traditional and standardized processes, but also changes creativity, innovation and development, intelligence, humanization, and flexibility and effectiveness. With the diversification of management modules, today with the rapid development of transportation informatization, blockchain technology clearly has the characteristics of decentralization, prevention of data modulation, data transparency and transparency, and will be integrated in all aspects of urban transportation And acceptance. Blockchain technology plays an important role in developing innovative solutions for developers and entrepreneurs to reduce the burden of traffic. The combination of blockchain technology, drive and driverless vehicles will significantly change the traffic situation. Our goal is to develop a series of previously incredible solutions and eliminate all traffic loads. On this basis, future research will focus more on specific technologies and methods. To promote the application of blockchain technology in big data, more blockchain technology and application cases will be proposed.

References

1. Lin, K., Li, C., Tian, D., et al.: Artificial-intelligence-based data analytics for cognitive communication in heterogeneous wireless networks. IEEE Wirel. Commun. **26**(3), 83–89 (2019)
2. Adams, R.C., Rashidieh, B.: Can computers conceive the complexity of cancer to cure it? Using artificial intelligence technology in cancer modelling and drug discovery. Math. Biosci. Eng. **17**(6), 6515–6530 (2020)
3. Yun, D., Xiang, Y., Liu, Z., et al.: Attitudes towards medical artificial intelligence talent cultivation: an online survey study. Ann. Transl. Med. **8**(11), 708 (2020)
4. Kzlta, M.A., Cankül, D.: Yyecek ecek letmelernde tedark zncr ve blokzncr teknolojs (supply chain and blockchain technology in food and beverage businesses). J. Gastronomy Hospital. Travel (joghat) **3**(2), 244–259 (2020)

5. Pandey, P., Litoriya, R.: Implementing healthcare services on a large scale: Challenges and remedies based on blockchain technology. Health Policy Technol. **9**(1), 69–78 (2020)
6. Darun, M.R., Adresi, A.A., Turi, J.A., et al.: Integrating blockchain technology for air purifier production system at FIM learning factory. Int. J. Control Autom. **13**(2), 1112–1117 (2020)
7. Cleophas, C., Cottrill, C., Ehmke, J.F., et al.: Collaborative urban transportation: recent advances in theory and practice. Eur. J. Oper. Res. **273**(3), 801–816 (2019)
8. Fernandes, D., Kanashiro, M.: Transportes urbanos e o paradigma assegurado por políticas públicas/Urban transportation and the public policy paradigm. Servio Social em Revista **23** (1), 143–159 (2020)
9. Mendili, S.E.: Big data processing platform on intelligent transportation systems. Int. J. Adv. Trends Comput. Sci. Eng. **8**(4), 1099–1109 (2019)
10. Beg, M.M.S., Hussain, M.M., Alam, M.S., et al.: Big data analytics platforms for electric vehicle integration in transport oriented smart cities: computing platforms for platforms for electric vehicle integration in smart cities. Int. J. Digit. Crime Forensics **11**(3), 23–42 (2019)

Data Mining Analysis Based on Cloud Computing Technology

Yizhi Li[✉]

College of Internet of Things, Jiangxi Teachers College,
Yingtan, Jiangxi, China

Abstract. With the rapid development of computer technology, Internet technology and artificial intelligence technology, the amount of global data has exploded, and the era of big data has come. This paper mainly studies the data mining analysis based on cloud computing technology. All the data needed in the experiment are put into the name node, which is responsible for copy management to other machines. All the intermediate data generated during the experiment are stored in the distributed file system of the cluster. Firstly, the data set to be mined is uploaded to Hadoop distributed file system (HDFS), and the program reads the data from HDFS when using it. At the same time, the output file name is set so that the system can store the calculation results in HDFS. The selection of indicators follows the scientific design principle to ensure that the indicators are comprehensive, scientific and operable, and the indicators are verified and selected by expert analysis. In the case of using complete 150g data, 72% of the single machine does not use Hadoop platform, and the accuracy of using Hadoop platform is 71%. The results show that the combination of cloud computing technology and data mining can significantly improve the operation effect of the algorithm.

Keywords: Cloud computing technology · Data mining · Hadoop cluster · Algorithm analysis

1 Introduction

Data mining involves a wide range of fields, which has attracted many experts, scholars and enterprises to participate in the research, innovation and improvement of data mining related algorithms. Huge data information has high requirements for algorithm, and the quality of an algorithm is directly related to the efficiency and timeliness of data mining.

Cloud computing is the development trend of information technology in the future. If association rules and related algorithms are brought into cloud computing, it can not only increase the service scope of cloud computing, but also improve the performance of association rules algorithm, and make it reach a new level [1, 2]. In addition, the virtualization technology of cloud computing also includes features and functions such as virtual machine replication and migration, which simplifies the management of large-scale computer cluster and improves the overall operation security of the platform [3, 4]. Single machine data mining algorithm performance is limited, the amount of data processing is also very limited, usually cannot exceed the memory size, unable to conduct in-depth analysis [5]. Application deployment and maintenance on multiple

© The Author(s), under exclusive license to Springer Nature Singapore Pte Ltd. 2022
J. C. Hung et al. (Eds.): FC 2021, LNEE 827, pp. 181–187, 2022.
https://doi.org/10.1007/978-981-16-8052-6_20

servers is very cumbersome, application monopolizes the server, and the waste of software and hardware resources is serious [6]. Therefore, it is necessary to introduce cloud computing, divide the calculation into several sub parts, each sub part is calculated by a server, and finally merge the calculation results of each sub part [7]. Cloud computing not only solves the problem of limited single machine memory, but also significantly reduces the computing time of data mining algorithm [8]. The data mining platform based on cloud computing is committed to providing data mining capability and big data storage capability services, which is convenient for other industries to build their own business system on this basis, which will greatly save the R & D investment cost of enterprises and improve the efficiency of the company [9, 10].

At present, many enterprises have built their own cloud computing platform. Facebook has the largest cloud platform system, which can store and process 100pb of data. It is one of the largest clusters in the world. All kinds of sources, such as social media, news channels, science labs, meteorological departments, produce a lot of data every day. These big data need the most effective storage and efficient analysis technology; these technologies can bring some knowledge, which is a huge challenge for the data world.

2   Cloud Computing Technology and Data Mining

2.1   Cloud Computing

Cloud computing has unlimited data storage, computing and mining functions to provide services to customers with "leasing" services. This can not only save the cost of software and hardware equipment purchase and maintenance, but also reduce the cost, and ensure the reliability and scalability of data storage, calculation and mining. The data collected on the Internet of things and computing devices are scattered, so the parallel data mining function of cloud computing must be used.

2.2   Data Mining

With the rapid development of science and technology, the amount of data obtained by all walks of life is increasing, and the information contained in these data is also growing by geometric multiples. In order to solve the above problems, discover useful information and knowledge from the huge data in time, and improve the utilization rate of data, data mining technology came into being, and has been vigorously developed, and gradually become one of the most cutting-edge research directions in the field of database and information decision-making.

Assume that each sample consists of multiple training attributes $A_k(K = 1, 2, \ldots, k)$ and prediction attributes. Divide N samples into a set of c different samples, and the number of samples in the category Ci is Ni. The initial amount of information in the decision tree is:

$$I(C_1, C_2, \cdots, C_C) = \Sigma_{i=0}^{C} - (\frac{N_i}{N}) \log_2(\frac{N_i}{N}) \tag{1}$$

Select the training attribute as the classification node. When each training attribute is set to $A_k(K = 1, 2, \ldots, k)$ and the number of samples belonging to the category C_i is n_{kji} for the n_{kj} samples of each branch, the expected entropy based on the training attribute A_K is as follows:

$$E(AK) = \sum_{j=1}^{J} \sum_{i=1}^{I} \left(\frac{n_{kj}}{N}\right) * \left(\frac{-n_{kj}}{n_{kj}}\right) * \log_2\left(\frac{n_{kji}}{n_{kj}}\right) \tag{2}$$

The information gain $Gain(S, D)$ can be expressed as:

$$Gain(S, D) = I(S_1, S_2, \cdots, S_m) - E(S, D) \tag{3}$$

In the expression, $E(S, D)$ represents the entropy weight of k subsets divided by the quotient of attribute D.

3   Data Mining Test

3.1   Experimental Environment

The test environment consists of one IBM server and six Dell desktops. The IBM server is configured with Xeon e5506 processor, 16g memory, dual Gigabit network card, 500g SATA hard disk. The Dell desktop is configured with i5760 processor, 4G memory and 500g IDE hard disk. Computers are connected by Huawei Gigabit switches, and all network cables are connected by gigabit network cables. All the data needed in the experiment are put into the name node, which is responsible for copy management to other machines. All the intermediate data generated during the experiment are stored in the distributed file system of the cluster.

3.2   Experimental Test

Firstly, the data set to be mined is uploaded to Hadoop distributed file system (HDFS), and the program reads the data from HDFS when using it. At the same time, the output file name is set so that the system can store the calculation results in HDFS. The selection of indicators follows the scientific design principle to ensure that the indicators are comprehensive, scientific and operable, and the indicators are verified and selected by expert analysis.

4   Discussion

4.1   Algorithm Classification Accuracy Comparison

The relationship between cluster size and runtime is shown in Fig. 1 when dealing with graph of fixed size. As can be seen from the figure, on the premise that the number of edges in the graph remains unchanged, increasing the number of computing nodes in the cluster can greatly shorten the running time of optimdm algorithm and improve the performance of the algorithm. The experimental results directly prove that the performance of optimdm algorithm is scalable and depends on the size of the cluster. In other words, optimdm algorithm can make full use of distributed computing resources and reduce the time cost of the algorithm. Compared with the original KNN algorithm, the classification accuracy of the improved KNN algorithm does not decrease when the selection of the limit nearest neighbor distance is more conservative, but for the classification task of large data sets, the classification efficiency of the algorithm has been significantly improved. If the limit nearest neighbor distance of the algorithm is further reduced, the operation efficiency is expected to be further improved. Of course, the classification accuracy of the algorithm will not be effectively guaranteed. When executing small data sets, the parallel algorithm consumes less computing resources, and the communication between parallel clusters has a fixed resource overhead, which makes the execution time of the algorithm longer when executing small data sets. If the data set size increases, and the communication between clusters is basically fixed, the ratio of computing cost and communication cost of the algorithm will be greatly increased, which can reflect the advantages of computer cluster computing, that is, the speedup ratio will also increase, and the performance of processing large-scale data will be better. Reasonable configuration of cluster nodes can improve the performance of the platform. In the experiment of changing support, it can be seen that the change of support also has a certain impact on the performance of the algorithm. The larger the selected support is, the faster the running speeds is, but it may have a certain impact on the accuracy of the generated rules. So choosing the appropriate number of support can speed up the running speed and get better rules.

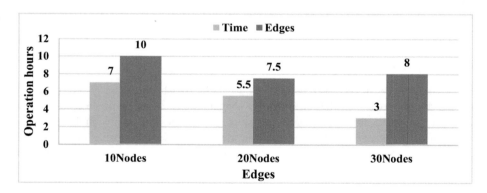

Fig. 1. The relationship between cluster size and running time

The cluster speedup results are shown in Fig. 2. It can be seen from the Fig that the cluster parallelization speedup is always less than the number of cluster computing nodes. With the increase of the number of cluster nodes, the speedup ratio is rising, and the overall computing time of the cluster is less and less, which reflects the strong overall computing power of the cluster. When the number of nodes is 2, the size of the sample number cannot see the performance difference, but with the increasing number of nodes, we can clearly see the trend of speedup under different samples. The more the number of nodes, the more linear the speedup of the algorithm, and with the increase of the sample size, the more obvious this trend is. Therefore, we can draw a conclusion that the improved Apriori algorithm greatly improves the computational efficiency and performance of the algorithm under the cloud platform, saves the time cost, and provides a guarantee for the high reliability of the system. The insertion and update performance of cloud storage system is much higher than that of traditional SQL, and the query performance is several times of that of traditional SQL. For the frequent insertion and query of massive GSM-R data, cloud storage system has incomparable advantages. Slicing will have a certain performance impact on the insert, update and query operations, but compared with the traditional SQL cloud storage system, the performance is still very strong after slicing, and it can support data expansion and solve the problem of storage capacity limitation.

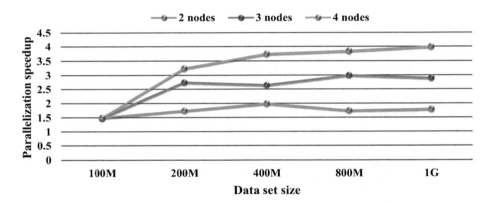

Fig. 2. Cluster speedup results

4.2 Algorithm Running Time Comparison

The execution time of aldck means algorithm in different clusters is shown in Table 1. When dealing with small dataset dataset1, the execution time in stand-alone mode is shorter than that in cluster mode. This is because dataset1 has a small data scale, and the cluster takes up few resources for clustering calculation. In Hadoop platform, certain resources are also needed for initialization, task allocation and communication between clusters. However, with the increase of the data set size, the communication tasks between clusters remain basically unchanged, while the proportion of computing tasks occupied by clustering algorithm will increase, which will show the advantages of parallel algorithm, and the parallel execution time will be far less than the serial execution time. In terms of the accuracy of logistic regression model, in the case of 60g data, the accuracy of single machine without using Hadoop platform is 73%, and the accuracy of using Hadoop platform is 69%. In the case of using complete 150g data, 72% of the single machine does not use Hadoop platform, and the accuracy of using

Hadoop platform is 71%. We can find that the accuracy of larger samples is slightly higher than that of smaller samples. The possible reason is that the training samples are segmented when the feature variables are selected to enter the regression model. In the case of large data scale, each node will get more data, so the accuracy will naturally increase. For fixed datasets of different sizes, the speedup increases linearly with the number of nodes. It shows that with the growth of the number of nodes, the processing time will be greatly reduced, and the parallel running speed of multiple hosts is higher than that of one host. At the same time, it is noticed that the experiment gets better speedup ratio with the increase of data set. The speedup of 10 m data set is closer to the ideal speedup than that of 100k when the number of nodes is large.

Table 1. ALDCK-means algorithm execution time in different clusters

Data set	Stand-alone	2	3	4	5	6	7	8
Dataset1	151	170	169	167	164	162	160	159
Dataset2	1013	972	917	840	810	789	761	750
Dataset3	2175	1740	1302	1037	917	830	801	771
Dataset4	4368	2834	1963	1687	1257	1069	972	942
Dataset5	8918	4652	3183	2739	2116	1792	1713	1646

5   Conclusions

This paper designs the virtualization and resource management scheme, and constructs the cloud computing data processing model, and does some research on its performance optimization and tests and compares the performance before and after optimization.

Data mining is the key technology of effective data management, the traditional data mining method is only suitable for the database with small amount of data, for massive data sets, parallel distributed processing is a good way, we will deploy it to the cloud computing platform, and we can quickly mine out the effective information.

Using the combination of data mining technology and social network, users' preferences are clustered and correlated in many aspects to achieve more accurate recommendation effect. It also verifies that cloud computing can not only improve the efficiency of data mining, but also provide more new ideas for the original data mining scenarios.

References

1. Du, J., Zhao, L., Feng, J., et al.: Computation offloading and resource allocation in mixed fog/cloud computing systems with min-max fairness guarantee. IEEE Trans. Commun. **66** (4), 1594–1608 (2018)
2. Wei, W., Fan, X., Song, H., et al.: Imperfect information dynamic Stackelberg game based resource allocation using hidden Markov for cloud computing. IEEE Trans. Serv. Comput. **11**(99), 78–89 (2018)

3. Hadjali, A., Mezni, H., Aridhi, S., et al.: Special issue on "uncertainty in cloud computing: concepts, challenges and current solutions." Int. J. Approximate Reasoning **111**(8), 53–55 (2019)
4. Cho, S.P.: Selected Peer-Reviewed Articles from 3rd International Conference on Big-Data, IoT, Cloud Computing Technologies and Applications (BICTA 2017), Daejeon, Korea, 9–11 November, 2017. Adv. Sci. Lett. **24**(3), 1942–1943 (2018)
5. Yang, J., Wang, C., Zhao, Q., et al.: Marine surveying and mapping system based on Cloud Computing and Internet of Things. Future Gener. Comput. Syst. **85**(8), 39–50 (2018)
6. Alkhanak, E.N., Lee, S.P.: A hyper-heuristic cost optimisation approach for Scientific Workflow Scheduling in cloud computing. Future Gener. Comput. Syst. **86**(9), 480–506 (2018)
7. Liu, X.F., IEEE, et al.: An energy efficient ant colony system for virtual machine placement in cloud computing. IEEE Trans. Evol. Comput. **22**(1), 113–128 (2018). Member S
8. Ning, J., Cao, Z., Dong, X., et al.: Auditable σ - time outsourced attribute-based encryption for access control in cloud computing. IEEE Trans. Inf. Forensics Secur. **13**(1), 94–105 (2018)
9. Senyo, P.K., Addae, E., Boateng, R.: Cloud computing research: a review of research themes, frameworks, methods and future research directions. Int. J. Inf. Manag. **38**(1), 128–139 (2018)
10. Helma, C., Cramer, T., Kramer, S., et al.: Data mining and machine learning techniques for the identification of mutagenicity inducing substructures and structure activity relationships of noncongeneric compounds. J. Chem. Inf. Comput **35**(4), 1402–1411 (2018)

Artificial Intelligence in Digital Media Technology

Jin Cai[(✉)]

Sichuan Vocational and Technical College, Suining, Sichuan, China

Abstract. With the advancement of society and the increase of computer, network and digital media technology, digital media technology has become an indispensable part of the modern service industry. The application of digital media technology in artificial intelligence has enabled the development of all aspects of life, such as the company's internal training, system development, update and maintenance from "quality" to "quantity". This article mainly introduces the application analysis of artificial intelligence in digital media technology. This paper uses the application analysis of artificial intelligence in digital media technology and proposes a visual media management model. Use artistic innovation methods to analyze and analyze the application of digital media technology in artificial intelligence to improve the standardization and accuracy of digital media. The experimental results of this paper show that the application of visual media and the technology of artistic innovation combined with artificial intelligence technology have increased the innovation rate of digital media by 23%, and improved the interactive synthesis of images and video materials.

Keywords: Artificial intelligence · Digital media · Visual media · Artistic innovation

1 Introduction

At present, digital media and film and television industries are cultural and creative industries produced by the combination of media technology and computer computing, but digital media technology based on artificial intelligence analysis has pushed digital media technology to a new level [1, 2]. The transformation of visual art in the context of digital media not only includes the creative reconstruction of new media art in terms of technology, form and concept [3, 4], but also includes the improvement of traditional visual media, the production of new visual media materials, and the old and new The cross-media of visual media [5, 6].

Visual communication has become the most important means of communication in the information society. In terms of concept and technology, the emergence of new media directly affects visual communication [7]. Levy believes that visual forms and characteristics, communication media and content, image delivery methods, manufacturers' creative methods and consumer experience have undergone tremendous changes [8]. In the in-depth construction of the sustainable development of cross-cultural, industrialization and sports under the cultural background, the transformation

J. C. Hung et al. (Eds.): FC 2021, LNEE 827, pp. 188–195, 2022.
https://doi.org/10.1007/978-981-16-8052-6_21

of visual art can also be seen [9, 10]. The transformation of visual art is not limited to the field of visual art creation, but a comprehensive transformation of the entire mechanism of visual art creation, the composition of visual communication, and the behavioral activities that consumers participate in [11, 12].

The innovation of this article is to propose an analysis of the application of artificial intelligence in digital media technology. According to the visual media technology and the secondary innovation of art, the analysis and application of digital media are strengthened. This article combines practical theory and uses standard case studies to conduct specific analysis, thus providing fresh materials and research for the application of digital media art.

2 Digital Media Innovation in Artificial Intelligence Environment

2.1 Picture Rendering for Digital Media

The image rendering module still occupies a dominant position in the digital entertainment system, and the development of image rendering technology is changing with each passing day. Based on the continuous development of computer image rendering technology, design and implement a graphics rendering subsystem with complete functions and good performance, and at the same time give secondary developers a good function configuration mechanism to facilitate the secondary development of the application system and improve the overall computing performance of the digital entertainment system. And user experience is an important part of the research content of this article.

Technologies such as artificial intelligence and network communication have become indispensable components of current digital entertainment systems. In order to achieve the integrity of the system functions, enrich the fun and interactivity of the digital entertainment system, and verify the results of the system scalability research in this article, this article will also discuss the artificial intelligence and physical simulation applied in the digital entertainment field. The technical methods are researched and realized, and a general method for applying these extended technologies to the overall system is proposed. Finally, the work of this article is a system-level research work. Based on the technical research of various sub-fields, these sub-technologies are effectively carried out. The system integration and application integration enable it to meet the immediate needs of reality. To this end, this article will combine actual project cases to verify the research work of this article. The technical characteristics of these specific projects have their respective focuses, which belong to different types of digital entertainment software products, which can fully prove the versatility and effectiveness of the system framework described in this article. As a graphics rendering engine, take the rendering of a sky as an example. At the top of the sky, you need to map the location of the flat square map to the hemisphere. When constructing a hemispherical dome, the longitude and latitude angles of the dome can be used as input parameters. Then, the local coordinates and latitude and longitude of any point in the dome have the following calculation relationship

$$\begin{cases} x = r \cos \beta \cos \alpha \\ y = r \sin \beta \\ z = r \cos \beta \sin \alpha \end{cases} \tag{1}$$

By setting static or dynamic light sources in the virtual scene, the color of illuminated objects can be changed in real time, or shadows can be cast on other objects. The formula is as follows.

$$L_D = \max(LdotN, 0.0) \tag{2}$$

If the object is only affected by Duse lighting and does not have a normal map, when there is no Specular reflection light, the simplest lighting Shader is used. The calculation method is as follows:

$$C_{Result} = C_{texture} * L_D \tag{3}$$

2.2   Digital Media Visualization Rendering

Rendering cycle: The graphics rendering system needs to go through a series of primitive drawing operations when drawing each frame of 3D scene and object picture. The whole system draws the same frame continuously, and finally forms a continuous animation effect. The series of drawing operations performed by the system in each frame is called a rendering cycle. System architecture design characteristics before proceeding with our system architecture design, it is necessary to clarify the unique functions that the system architecture design of this article needs to achieve. Compared with a pure game engine, our system architecture design should not only emphasize due resource management and object management functions, so as to facilitate developers to program descriptions of the virtual world, but also to integrate extension technologies well to reduce multi-party engines. Or the expansion of middleware. Object-oriented virtual world management includes two aspects.

The first is to use object-oriented program code development to realize the functional reuse of objective objects and logical objects in the virtual world and improve work efficiency. The second is to use the description mechanism between the agent-oriented objective and logical objects and the virtual world to make Agnet the basic element in the virtual world. These Agents can be used to describe the living "objects", such as virtual characters, or inanimate dynamic objects can also be used as packaging containers for some logical control units, such as event triggers. In this article, we will name these Aget packaging containers "Entity" between Entity and the system as a whole and between Eny the relationship between different technology types of development engines or functional packages (such as physics, AI, network, etc.) each has different initialization and release mechanisms, object description mechanisms, data representation methods and loop control procedures. These different engines must be used at the same time In a digital entertainment system, a lot of extra design, interface integration work and process control work are required, which not only increases the difficulty of development, but also affects the efficiency of the system. For this reason, the digital entertainment software system architecture in this article will place the digital entertainment system the required physics, AI, and network sub-modules are designed in a unified abstraction to standardize the control rules and operating procedures of the entire system. In the interface design of these sub-modules, this article will fully consider the current general physics, network and the design mode of the AI engine enables it to smoothly connect to these interfaces.

3 Visual Digital Media

3.1 Visualization System

Flexible rendering loop control as mentioned in the first chapter of this article, graphics rendering and rendering are the most fundamental function of digital entertainment software systems. It is responsible for presenting digital content to users from the most basic sensory way of vision. In the work of this article, the graphics subsystem is the only indispensable technical sub-module in the entire digital entertainment software system. The information processing results of other technical extension modules need to be finally presented through the graphics module, so they need to be surrounded by graphics Sub-module to do the work. The pre-built graphics sub-module, this article needs to make the graphics sub-module meet two requirements: one is the graphics drawing function, which can meet the needs of most application development; the other is the digital entertainment in real applications when there are special needs in software, physics, AI, and network modules often exist as auxiliary modules of the rendering engine. The fundamental reason is that the final processing results of the physics, AI and network communication modules need to be rendered with the help of graphics rendering functions.

3.2 Visual Media System Design

For physics simulation and AI operations, the internal architecture of these sub-modules is similar to the graphics rendering engine, which is composed of a procedurally described world model and abstract physics or AI individuals of objects that are constantly moving and changing in the world. Since their results are presented through a graphics engine, we can streamline the results presentation steps in the system design. We use a unified abstract interface IEXTENDMODULE to describe the management and control of all physics, AI and other sub-modules or third-party engine components. It is an overall manager. Positive EXTENDOBJECT is the object of specific physical simulation, a operation or network communication managed by IEXTENDMODULE, and generally corresponds to each Entity. Just like the initialization of XTENDMODULE, the released interface program initialization and the interface OnEngineInit, this interface is the interface of the sub-module engine manager, and the interface internally handles the related operations of the initialization of the sub-module engine. Under normal circumstances, the submodule engine is initialized together with the entire program system. Therefore, in the system of this article, the interface is called when the entire system is initialized. The program releases the interface ONENGINE DEIN. This interface is the interface of the submodule engine manager, and the interface internally processes the submodule. Operation related to engine release.

(1) Under normal circumstances, the submodule engine should be released together with the entire program system, so in the system of this article, this interface is called when the entire system exits.

(2) The operation interface of IEXTENDMODULE during scene loading and release. When the virtual scene is loaded and released, in addition to the relevant 3D rendering model data that needs to be changed, the agent Entity in the scene needs to be recreated, and the corresponding A, The description model of physics should also be changed accordingly. Therefore, the sub-module engine sets the corresponding operation interface scene loading interface On New WORLDLOAD for these times. This interface is the interface of the sub-module engine manager. It is called when a new virtual scene is read, used to update the new scene except 3D rendering data to describe the model, or to perform other related operations. The specific results are shown in Table 1.

Table 1. Terminal video media formats

Media format	Description
Sound	WAV, MP3
Static picture	JPEG, GIF, PNG, TIFF, PSD, BMP
Animated picture	GIF, M-JPEG
FLASH animation	FLASH
Web document	HTML, D-HTML, XHTML
Text	TXT, XML

4   Artistic Innovation of Digital Media

4.1   Digital Media Visual Communication Model

Visual communication is the behavior of people sharing visual information and presentation methods. "Visual communication" exists as an independent form of communication, and compared with other types of communication, it has become a more meaningful form of communication. Visual communication creates a form that exists as an important force in the content and provides a form with insignificant value in the content. The specific results are shown in Fig. 1: Using this model to solve, you can quickly find the optimal capacity allocation method. LR reaching 1.2 is the minimum value. When the category is relatively large, the calculation amount of the traversal search is very large, which is obviously undesirable. The meaning of solving the rate-distortion bound of the reversible information hiding under the non-uniform distortion metric is to quickly find the optimal allocation strategy.

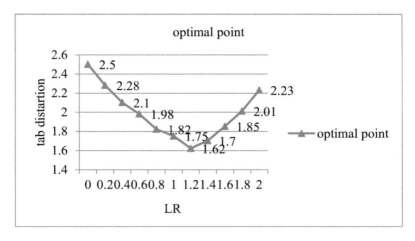

Fig. 1. The optimal capacity allocation

4.2   Artistic Innovation in Visual Media

The main purpose of the design and construction of the visual media material library is to provide various types of visual media material sharing and editing and reuse services for the digital content industry. Therefore, in addition to researching system architecture, efficient retrieval, material editing and other issues, breakthroughs are needed. The interactive technology of the web-based visual media material library system is how the system provides users with friendly interactive functions, such as real-time playback of video materials, visualization of human motion materials, real-time interaction and rendering of 3D models based on Web browsers, and similarity Material metadata comparison, etc. Systems and applications with reasonable design architecture, reliable systems, complete functions, and good user experience are the trend of future web-based visual media material library research. In the model browser, the user can realize the interactive operation of real-time rotation, scaling and movement of the three-dimensional model through the operation of the mouse or keyboard. The model browser also supports users to set the position and color of the light source and the color of the ambient light source to better display the model effect. The specific results are shown in Table 2.

Table 2.   The performance analysis of 3D model browser

	Windmill model		Airplane model	
Performance parameter	Before optimization	Optimized	Before optimization	Optimized
Memory footprint	17.82	12.56	30.48	26.52
Rendering time	6	4	42	30
Frame rate	24	32	8	12

McLuhan's view of "distributing information to the media" shows that the truly meaningful information is not what the media drives on people every time, but the media itself. He believes that the creation of each new medium, the new way of social life and social behavior, the form of visual communication and the meaning of value in the process of image reading have long been independent and transcends the value of content, which greatly affects people Understanding and mentality. In fact, as a term with good communication connotation, visual communication itself is the result of rich visual efficiency.

A large number of micro-cultural phenomena with micro-cultural features have appeared in the application of network technology in the new media environment. In the field of network economy, a long tail theory of small strokes has appeared, and the development of network technology. The crowdsourcing model (Crowd Sourcing) in Crowd Sourcing has created the accumulation of group wisdom to achieve development and innovation similar to the magic of gathering sand into a tower; the rise of microblogs in recent years has demonstrated the power behind seemingly simple gossip, and small mobile applications have dozens of apps Millions of downloads have helped the smart communication terminal represented by Apple mobile phones to become a novel and fashionable learning lifestyle for young people nowadays. The specific results are shown in Fig. 2. In the case of 1 thread, our algorithm is slightly slower. This is because the parallel mechanism introduces some additional overhead, which cannot be offset in the single-threaded case. These overheads are masked in the context of high parallelism.

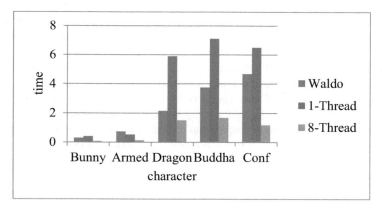

Fig. 2. Comparison of running time

5 Conclusions

Although this paper has made some research results on the application and analysis of artificial intelligence in digital media technology, there are still many shortcomings. Based on the analysis of the application of artificial intelligence in digital media technology, there is still a lot of in-depth content worth studying. There are many steps in the visual media process that cannot be covered due to reasons such as space and personal ability. In addition, the actual application effects of artistic innovation can only be compared with traditional models from the theoretical and simulation levels.

References

1. Li, Y., Tian, S., Huang, Y., et al.: Driverless artificial intelligence framework for the identification of malignant pleural effusion. Transl. Oncol. **14**(1), 100896 (2021)
2. Hassabis, D., Kumaran, D., Summerfield, C., et al.: Neuroscience-inspired artificial intelligence. Neuron **95**(2), 245–258 (2017)
3. Lu, H., Li, Y., Chen, M., et al.: Brain intelligence: go beyond artificial intelligence. Mob. Netw. Appl. **23**(7553), 368–375 (2017)
4. Ng, T.K.: New interpretation of extracurricular activities via social networking sites: a case study of artificial intelligence learning at a secondary school in Hong Kong. J. Educ. Train. Stud. **9**(1), 49–60 (2021)
5. Camerer, C.F.: Artificial intelligence and behavioral economics. NBER Chap. **24**(18), 867–871 (2018)
6. Hassabis, D.: Artificial intelligence: chess match of the century. Nature **544**(7651), 413–414 (2017)
7. Akerkar, R., Plantié, M., Trousset, F.: Editorial. Int. J. Artif. Intell. Tools **26**(02), 1702002 (2017)
8. Levy, F.: Computers and populism: artificial intelligence, jobs, and politics in the near term. Oxford Rev. Econ. Policy **34**, 393–417 (2018)
9. Dande, P., Samant, P.: Acquaintance to artificial neural networks and use of artificial intelligence as a diagnostic tool for tuberculosis: a review. Tuberculosis **108**, 1 (2018)
10. Douceur, J.R., Calligaro, M.P., Wood, R.C., et al.: Partitioned artificial intelligence for networked games (2016)
11. King, B.F., Editorial, G.: Discovery and artificial intelligence. AJR Am. J. Roentgenol. **209**(6), 1189 (2017)
12. Points, L.J., Taylor, J.W., Grizou, J., et al.: Artificial intelligence exploration of unstable protocells leads to predictable properties and discovery of collective behavior. Proc. Natl. Acad. Sci. U.S.A. **115**(5), 885 (2018)

Development of Virtual Reality and Computer Technology Application

Lixia Hou[✉]

College of Artificial Intelligence, Nanchang Institute of Science and Technology,
Nanchang, Jiangxi, China
wanly@ncpu.edu.cn

Abstract. Virtual reality technology, as a cutting-edge media technology, is widely used in various fields to enhance the user's autonomy and selectivity, make the user feel like in the real world, and realize the user's immersive experience. In this paper, through the related definition of virtual reality technology, at the same time, the integration of computer technology and virtual reality technology development research, through the survey, we can see that the number of people who choose to be optimistic is 71, accounting for 81.6% of the total number. The application of virtual reality technology in the development is inseparable from the help and promotion of computer technology, and the better integration of the two can achieve the purpose of common development.

Keywords: Virtual reality technology · Computer technology · Investigation and analysis · Common development

1 Introduction

Virtual reality technology is kinds of computer simulation system, which can create and let people, experience the virtual world. It is a high-tech that can use computer simulation to produce three-dimensional virtual world. The technology through the use of computer graphics technology, simulation technology, artificial intelligence and other computer related technologies, with the help of virtual reality wearable devices.

The birth of the world's first head mounted display in 1961 marks the beginning of virtual reality (VR) technology. In the past 50 years, VR technology has experienced the perfection from theory to technology, and finally ushered in the explosive stage of application in this century [1]. In 2014, social media giant Facebook acquired oculus, a VR technology equipment manufacturing company, with a huge sum of US $2 billion, which ignited a new round of virtual reality research and development boom. Google, Samsung, HTC and other technology companies have laid out the VR market one after another, and 2016 is even regarded as the "first year" of VR by the industry. VR technology has the advantage of "immersion" experience. It is famous for improving the sense of scene, participation and interactive experience. It attracts medical, tourism, education, games and other industries to explore the possibility of integrating new technologies. The development speed of computer technology is relatively fast. In the process of continuous progress of science and technology, the research and application

© The Author(s), under exclusive license to Springer Nature Singapore Pte Ltd. 2022
J. C. Hung et al. (Eds.): FC 2021, LNEE 827, pp. 196–204, 2022.
https://doi.org/10.1007/978-981-16-8052-6_22

of computer technology is more in-depth. A variety of new computers have been developed, and computer technology is also developing in the direction of informatization, intelligence and efficiency. For the traditional computer, its application performance has been unable to meet the current development environment, which requires the innovation of the application principle of computer technology to realize the effective application of a variety of new technologies [2]. In this development environment, virtual reality technology has been further developed. The development and application of a variety of science and technology, virtual reality technology has also been affected. In the process of continuous improvement of computer level, it provides a driving force for the development of virtual reality technology. Compared with the previous computer operation technology, the overall application advantage of this technology is more obvious [3].

To sum up, this paper mainly discusses and investigates the development of virtual reality technology and computer technology application, clarifies the relationship between the two by elaborating the relevant concepts and definitions, investigates the computer professional practitioners through questionnaire survey, and analyzes the development direction of virtual reality technology and computer technology application through research.

2   Definition of Related Concepts

2.1   Virtual Reality Technology

VR Chinese means virtual reality. Virtual reality is not real reality, but the combination of virtual and reality. It is a fictitious computer simulation system created by a series of corresponding computer technologies. Through VR technology, realistic real scenes can be created. In this virtual scene, virtual objects can be materialized through interactive devices, so as to make users feel comfortable. The image is brought into the real environment, the experience of the experience is more intuitive, and the feeling of being in the scene arises spontaneously. VR technology has greatly stimulated the interest of the majority of users. This kind of interactive environment, which is generated by VR technology on the computer and makes users feel more realistic, is called virtual environment [4].

2.2   Characteristics of Virtual Reality Technology

Virtual reality technology perfectly combines computer technology and media technology, brings users a highly realistic virtual experience, and provides an effective means for people to explore things and their development laws [5]. The sensing device and video implementation device are hardware technology, and the system application is software technology. The most important characteristics of virtual reality technology include immersion, interactivity and conceptualization. Specifically, by creating a virtual space which is highly similar to reality, users immerse in it and realize human-computer interaction through relevant hardware operation. Details are as follows:

Immersion: users can immerse themselves in the designed virtual environment when using virtual reality technology. They feel part of the virtual environment and participate in a variety of activities. Immersion is mainly reflected in the user's perception, such as the most basic visual perception, in addition to auditory perception, taste, smell perception and tactile perception [6].

Interactivity: this feature refers to the process in which users can interact with various elements in the virtual world and obtain system feedback after operation. Specifically, users can experience the same feeling as the real world through data helmets, data gloves and feedback devices. The interactivity of virtual reality technology is mainly reflected in two aspects: one is to emphasize the interactive feedback between users and the system in the virtual environment; the other is to emphasize the timeliness of interactive feedback [7].

Conceptual: the content of virtual situation is generated by the design and programming of developers, which mainly reflects the ideas that developers want to express. Therefore, the virtual situation designed in this way shows that a goal is conceptual. For example, the application of virtual reality technology in military, medical, education and other fields is to better solve the existing problems. Developers develop software according to the needs of users. In the field of education, bringing students into the imaginary virtual environment helps students improve their cognitive understanding and develop innovative development on the basis of understanding [8].

2.3 Application of Computer Technology

In view of the rapid development of computer technology, virtual reality technology also has a huge update. In the composition of virtual reality system, the application of 3D computer graphics technology plays an obvious role, which can simulate the real graphics, and people can get more intuitive and three-dimensional visual enjoyment. Moreover, computers produce high-performance and intelligent types, which provide many useful technical support for virtual reality technology. To build a virtual world is to achieve a visual, real-time interactive virtual environment state through computer operation of relevant data. In terms of operation and reality, it is significantly superior to the traditional computer [9].

2.4 Development of Virtual Reality Technology

The emergence and development of virtual reality has a strong inevitability, but also in the development of human society and production life has been a lot of exploration and practice. It is always a goal of human beings to imitate the real world objects and use them for people to achieve a certain purpose. Human's exploration and practice of "virtual reality" have been taking place all the time. The legends of shaman priests and Viking crazy soldiers in ancient times all decorate the animal's head and fur on their bodies in order to expect that they can obtain some ability of the corresponding animals. The paper-cut characters in shadow play and the wooden people in martial arts training are all human beings who achieve the established goal of "virtual reality" through the simple physical simulation of animals and human beings. With the advent of the Internet era and the continuous development of computer technology and its related technologies, human exploration and pursuit of virtual reality (Human-Computer Interaction) has shown explosive growth, reaching a new research level and height. Human beings are not only satisfied with simple physical simulation, but also open a new era of exploration and practice. After the technological revolution, virtual reality is the development of computer high-performance computing, computer graphics and image processing, human-computer interaction, artificial intelligence, communication and transmission technology, so that people's exploration and practice in simulating and fabricating the real world reach the latest level. Virtual reality has experienced several different stages from concept generation, technology practice to industrial development [10]:

2.4.1   Budding Stage

Industrial production began to emerge. A large number of physical simulation "virtual reality" are constantly explored and practiced. The concept of virtual reality is in the embryonic stage, and the corresponding concept principles and standards are not put forward. Physical simulation promotes the germination and exploration of virtual reality related concepts.

Taking the development of space technology as an example, before the emergence of simulation technology (the rudiment of Virtual Technology), the proportion of flight accidents caused by human operation errors in all accidents reached an amazing 90%. At the beginning of the 20th century, when the simulation technology was not developed, there was a saying in the aviation industry: "pilots were piled up with gold equal to their weight", which is enough to explain how expensive it was to train a qualified pilot at that time. In order to reduce the flight accident rate and train qualified pilots, a simple flight simulator appeared in Europe in 1920s. This kind of flight trainer has a simple simulated aircraft operating system, but it needs the help of natural wind, so it is not practical. Until 1929, Edwin link used the pneumatic parts of musical instruments to make a flight simulator (similar to today's children's rocking car), which was the earliest flight simulator. Interestingly, at that time, Edwin link sold his aircraft to the amusement park as a game machine, and even added a slot. The aircraft was more like an entertainment tool.

2.4.2   Exploration Period

In 1965, Ivan used his rich imagination to describe a kind of technology that can be used to simulate 3D scene by computer. Moreover, users can interact with virtual images as in real life, such as touching, sensing, smelling, controlling and moving virtual images, etc. In the early stage of Ivan Sutherland's research (1968), he developed a 3D display system: Sword of Damocles. This is the earliest augmented reality system, and its invention means opening a new chapter in virtual reality research.

2.4.3   Consumption Development Period

After the concept and principle of the second period were gradually improved and established, great breakthroughs and achievements were made in military, aerospace and other fields. Virtual reality technology has also ushered in the development stage of all technologies - from the professional field to the field of mass consumption, and finally achieved great breakthrough and success in the fields of film, games and so on.

Virtual reality technology has been widely popularized and gradually accepted by ordinary consumers, leading to a large number of capital entering the field of virtual reality, which has brought about the blowout development of the industry, It also brings the industry cold winter of technical bottleneck.

2.4.4   High Speed Development Period

Goldman Sachs Group (a famous investment bank in the United States) pointed out in the comprehensive report on the status of virtual reality industry released in 2016: the total output value of virtual reality industry is showing a rising trend year by year, and showing a huge potential. According to the prediction of Goldman Sachs, the total scale of virtual reality industry will reach 110 billion US dollars in 2025, which will officially surpass the scale of 90 billion US dollars set by the TV market. Thus, the future development potential of virtual reality technology can not be underestimated.

In short, the goal of human exploration of nature and social needs are the driving force of the emergence and development of virtual reality. The eternal pursuit of computer technology and artificial intelligence workers for faster, smarter and more harmonious computer and artificial intelligence system promotes the continuous development of virtual reality technology.

2.5   Related Formulas

Weight calculation:

$$CI = \sum_{i=1}^{5} W_i CI_i \tag{1}$$

$$RI = \sum_{i=1}^{5} W_i RI_i \tag{2}$$

$$CR_{all} = \frac{CI_{all}}{RI_{all}} \tag{3}$$

3   Development Survey of Virtual Reality and Computer Technology Application

3.1   Investigation Purpose

In recent years, the virtual reality industry has developed rapidly, whether it is military, medical, real estate, education or entertainment and many other industries are involved. In 2017, Jiangxi Province proposed to support the construction of Nanchang world-class virtual reality center, focusing on the construction of 10 artificial intelligence and intelligent manufacturing industrial bases. It can be seen that the development of virtual reality industry is the general trend. The following will be studied through questionnaire survey to analyze the development direction, problems and solutions of the application of virtual reality and computer technology, so as to pave the way for the future development of the technology.

3.2 Respondents

Through the questionnaire survey on the employees of virtual reality and computer related industries in our city, a total of 87 people were selected, including 67 people in the computer technology industry and 20 people in the virtual reality industry. Through the questionnaire results, the development of the industry and the prospects of the two technologies were analyzed.

4 Survey Results and Analysis

4.1 Development Prospect Survey

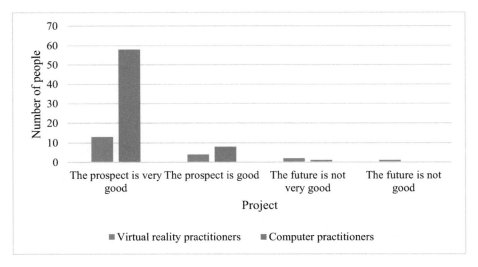

Fig. 1. Survey on the development prospects of virtual reality and computer technology applications

As shown in Fig. 1, through the survey of technology development prospects in the questionnaire survey, it shows that absolutely some employees in relevant industries are optimistic about the technology development prospects, and only a few employees are not optimistic about the development prospects. Basically, the daily work content of employees is related to the technology. They can well understand the development potential of the two technologies and represent the development expectation of the technology to a certain extent. The specific number of people is shown in Table 1.

Table 1. Survey on the development prospects of virtual reality and computer technology applications

	Virtual reality practitioners	Computer practitioners
The prospect is very good	13	58
The prospect is good	4	8
The future is not very good	2	1
The future is not good	1	0

It can be seen from the table that 71 people choose to be optimistic, accounting for 81.6% of the total number. Among them, 13 are engaged in virtual reality technology, 58 are engaged in computer industry, 12 are optimistic, accounting for 13.8% of the total number, 4 are not optimistic or not optimistic at all, accounting for 4.6%, and only 1 is not optimistic at all, accounting for 1.1%.

4.2   Technology Integration Survey

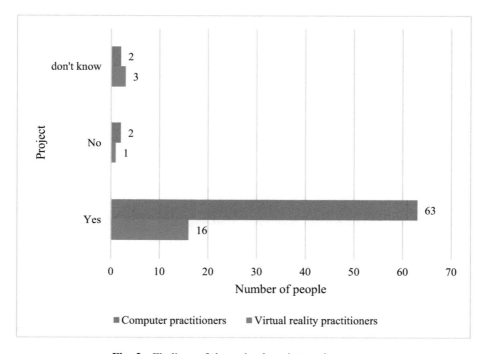

Fig. 2. Findings of the technology integration survey

The application of computer technology in virtual reality technology develops rapidly and gradually tends to mature, but inevitably there are many problems in the development. Therefore, whether the two technologies can be integrated more closely is investigated. As shown in Fig. 2, through the survey of technology integration in the questionnaire survey, absolute part of the employees in relevant industries said yes, only a few thought no.

4.3 Development Direction Analysis

The interactive experience of virtual reality technology provides similar services in life for the audience in the process of information release and acquisition. It not only makes the audience feel the intelligent experience, but also has the natural and real feeling. In addition, in order to make users have a better experience, it can also provide massage relaxation leisure activities. At the same time, the audience can browse the virtual reality news, and analyze the authenticity of the information on the basis of the virtual reality environment. Therefore, the interactive development of virtual reality technology is objectively the comprehensive utilization of time and space. Based on human sensory function, through the application of relevant calculation and simulation development, it can achieve stable and normal operation, and bring new experience and feeling to the audience.

5 Conclusion

With the rapid development of information technology, big data, computer technology, virtual reality technology and blockchain technology have been applied in many industries and fields. With the rapid development of foreign VR technology, in order to improve the application of VR technology, China needs to improve the level of relevant technology, strengthen the attention of the government and enterprises to this technology, use the mature computer technology to integrate the two, promote the development and construction of virtual reality technology, and realize common development.

Acknowledgements. This work was supported by Science and Technology Research Project of Jiangxi Provincial Department of Education under Grant no.GJJ202506 and the Nanchang Key Laboratory of VR Innovation Development & Application (No. 2018-NCZDSY-001).

References

1. Freeman, D., Reeve, S., Robinson, A., et al.: Virtual reality in the assessment, understanding, and treatment of mental health disorders. Psychol. Med. **47**(14), 1–8 (2017)
2. Rakhimova, A.E., Yashina, M.E., Mukhamadiarova, A.F., Sharipova, A.V.: The development of sociocultural competence with the help of computer technology. Interchange **48**(1), 55–70 (2016). https://doi.org/10.1007/s10780-016-9279-5
3. Ma, C.: Teaching application of computer virtual reality technology in international education of Chinese language. Educ. Sci.: Theory Pract. **18**(6), 1–3 (2018)
4. Liu, X.: Three-dimensional visualized urban landscape planning and design based on virtual reality technology. IEEE Access **8**, 149510–149521 (2020)

5. Żmigrodzka, M.: Development of virtual reality technology in the aspect of educational applications. Market. Sci. Res. Organ. **26**(4), 117–133 (2017)
6. Hyun, K.Y., Lee, G.H.: Analysis of change of event related potential in escape test using virtual reality technology. Biomed. Sci. Lett. **25**(2), 139–148 (2019)
7. Zhouping, Y.: Application and development of computer intelligent vision based on evolutionary computation. J. Comput. Theor. Nanosci. **13**(12), 9857–9863 (2016)
8. Koeva, M., Luleva, M., Maldjanski, P.: Integrating spherical panoramas and maps for visualization of cultural heritage objects using virtual reality technology. Sensors **17**(4), 829–830 (2017)
9. Zuev, A., Bolbakov, R.: On prospects of development of telecommunication systems and services based on virtual reality technology. Int. J. Adv. Comput. Sci. Appl. **9**(4), 18–21 (2018)
10. Liang, Z., Shuang, R.: Research on the value identification and protection of traditional village based on virtual reality technology. Bol. Tec./Tech. Bull. **55**(4), 592–600 (2017)

Use of Social Audit Network Intelligent System to Promote High-Quality Economic Development in the Era of Big Data

Bige Li[(✉)]

Fuzhou University of International Studies and Trade,
Fuzhou 350202, Fujian, China

Abstract. The audit system is an important part of the national supervision system. Promoting high-quality economic development is an important task put forward for audit work, and it is also a requirement for the positioning of audit work functions and audit business from the height of the new development concept. In the context of the era of big data, promoting and ensuring high-quality economic development is a major duty and mission of auditing. On the one hand, social audit should adapt to the requirements of the new situation and rely on big data technology to consciously serve the high-quality development of the economy. On the other hand, social auditing should also change ideological concepts, innovate auditing models, actively integrate into the country's modern governance system and governance capacity building, and act as a "consultant" for economic development in order to better provide information and basis for economic decision-making. In the context of the era of big data, this article analyzes the role of social auditing on high-quality economic development and provides corresponding countermeasures.

Keywords: Social auditing · High-quality economic development · Big data

1 Introduction

The high-quality development of the service economy is the responsibility and mission of auditing as an important part of the national supervision system, and it is also the need for auditing to adapt to the economic development of the new era [1]. China's audit supervision and management system includes national audit, social audit, and internal audit. These three have an inseparable connection, and each plays an important and irreplaceable role. Giving full play to the power of social auditing is an indispensable means to promote economic norms and orderly development [2]. In the context of economic globalization, the high-quality development of China's economy has put forward higher requirements on the depth, scope, specialization, and quality of auditing in my country. With limited human resources, information resources, and technical resources, the state gives full play to the power of social auditing to effectively integrate and maximize audit resources. This is also a trend in the development of the entire audit career [3]. Social auditing has the advantages of independence, objectivity, and impartiality. It can reveal problems in the process of high-quality

J. C. Hung et al. (Eds.): FC 2021, LNEE 827, pp. 205–212, 2022.
https://doi.org/10.1007/978-981-16-8052-6_23

economic development, analyze factors affecting high-quality economic development, and evaluate the effectiveness of economic transformation, thereby boosting high-quality economic development. At present, my country's economy has entered a stage of high-quality development. The quality, efficiency, and power changes of economic development require high-quality economic information as a basis for decision-making [4]. In addition to relying on government supervision for the high-quality development of the economy, it also needs to play the role of service supervision of social auditing to make up for the lack of power of the national auditing agency, thereby expanding the coverage of auditing. Compared with national audit institutions, social audit has its own advantages and characteristics: social audit is not subject to establishment restrictions, has great development potential, has extensive social relations, and can mobilize and organize many professionals to participate in professional audits; The rapid economic development has also brought about a large number of intricate relationships. Global economic integration, international capital flows, various economic vertical and horizontal integrations, and the "One Belt One Road Strategy" have promoted the internationalization of accounting information. These factors make various information users have higher and higher requirements for the objectivity and fairness of accounting information. Social auditing has obvious advantages in this respect, and these are inseparable from social auditing in accordance with relevant domestic regulations, international practices, or with reference to the laws and regulations of the country where the company is located or the third country [5, 6].

2   The Theoretical and Practical Significance of Big Data Technology for Social Auditing

In the context of the transformation of government functions and the high-quality economic development, the state must give full play to the role of certified public accountants and play the role of social audit as a bridge and link between the government and the economy. Lay a good foundation for improving economic efficiency, deepening reforms, and maintaining healthy and orderly economic development. Supervising economic activities through the intermediary function of social auditing can effectively achieve national macro-control [7]. Auditing practice has proved that social auditing has played an important role in promoting the development of multiple economic components and multiple business methods, straightening out the intricate economic relations between the state, enterprises, and individuals, and coordinating the normal operation of economic activities.

At the 2016 BRICS Conference, the top leaders of audit institutions clearly put forward suggestions on using big data technology to improve audit efficiency; in 2018, General Secretary Xi Jinping put forward suggestions on further enhancing audit informatization at the Central Audit Committee meeting. In the era of big data, audit institutions can provide more useful information for the operation and management of enterprises, investors, creditors and other relevant stakeholders through the classified collection and in-depth mining of big data, and then multi-angle processing and comprehensive analysis of relevant data. In the era of big data, the industrial chain of social audit institutions should be improved from the micro level of serving enterprises to the macro level of serving society and the country. In addition, social audit should enhance the sense of mission and responsibility of the audit industry, and establish a closer alliance with social and economic development. Social auditing uses data to guide enterprise production and operation activities, which can provide decision-making reference for national economic decision-making, thereby improving the status and voice of the social auditing industry;

In the era of big data, audit objects are showing diversified and informatized development, and traditional auditing models have been unable to adapt to the large amount of complex data information of enterprise development. In this context, audit work must change the audit mode and audit thinking, improve the audit risk management system, and use big data technology to organically integrate each process of the audit work, so as to better serve the high-quality economic development [8]. With the use of big data technology, auditors can continuously collect data from the audited unit and complete a large number of repetitive and tedious tasks in the audit work. Using big data analysis tools for analysis, the impact of big data and big data technology on audit work is shown in Table 1:

Table 1. The impact of big data and big data technology on audit work

Audit mode	Audit characteristics	Audit process	Audit evidence collection method	Audit scope
Traditional audit model	Static data	Judge selection based on audit experience	Request information and sampling from relevant responsible persons	Independent audit and key audit of each project
Information technology audit model	Dynamic data	Use data platform to contact analysis	Multi-dimensional correlation analysis using data platform	Comprehensive and continuous auditing using big data

3 Specific Suggestions for Playing the Role of Social Auditing to Promote High-Quality Economic Development

3.1 Improving Awareness of the Overall Situation

In the new era and new situation, social auditing must look at problems from a new height, be aware of the overall situation, and establish a concept of serving economic macro-control; understand the country's new development concept, and give full play to the three immune system functions of prevention, disclosure and resistance In the allocation of economic resources, strengthen its role in high-quality economic development; understand the important role of auditing in the national supervision system and economic operation, and find the best entry point for audit supervision services and high-quality economic development, so as to better provide counsel and basis for the formulation of national policies; Social auditing should also change its thinking in accordance with the needs of the development of the situation, have the consciousness of serving the interests of the whole society and economic activities, and assume the responsibility of social and economic services and supervision. At the same time, pay close attention to national and regional economic dynamics, improve the initiative to serve macroeconomic decision-making, summarize universality, tendencies and social concerns, and develop high-level information so that the audit results can be used by decision-makers [9].

3.2 Strengthening the Construction of Ideological Style

The state needs to strengthen ideological work and work style construction in the social audit industry. First, adhere to the audit in accordance with the law, strictly abide by the professional ethics of auditing, and establish a good image of auditors; secondly, strengthen the education of audit staff's integrity and dedication to improve the political quality and awareness of social auditors; third, strengthen party building in the social audit industry, carry out education on the theme of "not forgetting the original intention and keeping in mind the mission" in the social audit industry, and actively promote the experience and practice of effective integration of party building and business; finally, strengthen the industry's recognition of the professional spirit of integrity, adhere to honesty and integrity, and enhance social credibility, leading social audit institutions to be practitioners and guardians of integrity [10]. Audit institutions should face the society with high quality and credibility, and serve the society through strong legal status and effective practical actions, so as to truly maintain economic order and promote healthy economic development.

3.3 Innovativing Audit Ideas

Audit institutions must establish the awareness of advancing with the times, actively and timely change their thinking mode, adapt to the needs of the development of the big data era, rely on big data technology to form a proactive, professional, and overall view of auditing, and give full play to the interests of the country and society. The role of a defender of interests and promote high-quality economic development. At the same time, we must cultivate dialectical thinking, be good at looking at the essence through the phenomenon, looking at the macro through the micro, and making scientific analysis and judgment. It is necessary to cultivate innovative thinking, carry out work creatively, become a master at checking and analyzing problems, and explore new methods in practice; it is necessary to expand the breadth and depth of auditing, to lead by audit quality, and to highlight the orientation of audit results [11].

3.4 Strengthening the Construction of the Audit Team

In the era of big data, the scope of audit and supervision services is becoming wider and wider, and the subject of audit is becoming more and more complex. Therefore, the field of audit needs more high-level and compound audit talents to meet new demands and challenges. Social auditing must adapt to the requirements of economic development in the new era, build a professional and professional audit team that is proficient in business, and train auditors on advanced professional technologies such as intelligence, big data, Internet of Things, cloud computing, etc., in order to better adapt to the changes in science, technology, and markets [12]. Familiar with national laws, regulations and policies, study economic development plans and policies, earnestly understand the spiritual essence and rich connotations of the central reform decision-making arrangements, and take national economic development and the interests of the people at heart. Well versed in business, based on projects, fully researched, closely integrated with the characteristics and status quo of the region, industry, field, etc., study new situations and new problems in economic development, enhance the self-consciousness and initiative of audit work, and improve the service economy ability and quality.

3.5 Strengthening the Self-construction of Audit Institutions

In the era of big data, promote the establishment of an audit big data platform, develop audit software and audit analysis models that match cloud audit, and use big data technology as an important auxiliary means for audit work. It is necessary to collect audit evidence from various aspects and establish a big data platform integrating financial operation data of the audited entity, audit result data of audit institutions, and third-party information data. These measures can make the collection, analysis, processing, and update of audit data more timely and accurate. With the development of networks and terminal equipment, mobile office can be supported; to strengthen the maintenance of audit data security, authorization restrictions should be imposed on personnel contacting the data, and the data storage method and encryption level should be strict [13].Social audit institutions should also establish the concept of quality first. On the one hand, strengthen the awareness of risk responsibility and put quality control in a prominent position; on the other hand, strengthen the internal business and technical management, personnel management, information construction, and collaborative management of the headquarters and branches. At the same time, according to the needs of the development of the market economy, bold reforms in personnel management, salary distribution, quality management, financial management, etc. have made social auditing full of vitality.

3.6 Increasing Data Disclosure Through Legislation

Judging from the current open data, the sharing degree is limited, the structure is incomplete, and the overall quality is low. Neither the government, enterprises nor individuals have fully realized the effect of data on their own work and business activities. Increasing the publicity of these data will generate great value and will trigger huge changes in social, economic, cultural and other aspects. At present, there are not many platforms for government public data, and they are not widely involved. They focus on education, culture and sports, medical care, and the environment. The total amount of public data is low and the quality is not high. Enhance the data awareness of the whole society, develop a data culture, make the government, enterprises, and individuals fully aware of the role of data in their own work and production and business activities, and integrate big data awareness into everyone's daily work and life;

Big data, as data involving public interest, will have security problems in multiple links such as collection, processing, analysis, processing, storage, etc., so it needs to be regulated by legislation; currently there is no unified law and regulation to protect the development and use of big data, foreign countries such as the United Kingdom and the United States have established protective laws and regulations on data development and sharing; China should combine its own national conditions and absorb successful foreign experience, promulgate relevant laws and regulations as soon as possible, establish relevant laws and regulations on data disclosure, and clarify data disclosure and scope of use, to ensure the rights and obligations of governments, enterprises, individuals, etc. in the use and protection of data; improve the quality standards for publicizing various types of data. Since there is no data quality standard, the quality of data varies, and there is even a lot of redundancy data, only the high quality of all kinds of data publicly disclosed by the society can increase the value and effectiveness of data use [14].

3.7   Improving Relevant Laws and Regulations

The government should increase the practice quality inspections of social audit institutions, improve relevant laws and regulations, accounting standards and auditing standards, etc., block fraud loopholes in regulations and provide more operational guidelines for auditing practices; integrate administrative law enforcement with justice and introduce civil compensation, increase penalties for violations of laws and disciplines; increase the rectification of the industry order, and reverse the market phenomenon of "bad money driving out good money". Industry associations should guide social audit institutions to strengthen self-discipline management and optimize the practice environment; expand the scope of data collection allowed, formulate corresponding measures to stipulate the form and scope of data collection by audit institutions; revise relevant audit standards and supplement relevant regulations on big data audits. At present, big data auditing has been applied to a certain extent in social auditing, but there is no provision for big data auditing in the relevant auditing standards, so relevant content should be revised.

3.8   Strengthening the Guidance of Social Auditing

The new concept of promoting high-quality economic development is an important mission of the audit supervision system. Therefore, it is necessary to construct an audit supervision that is compatible with the new development concept and can effectively coordinate the audit forces of all parties from the perspective of political responsibility and audit business capabilities, so as to resolve various contradictions in the process of high-quality economic development and coordinate the interests of all parties. The national government should strengthen the attention, guidance and support of social auditing, fully delegate powers and coordinate relations, help develop new businesses, strengthen the information publicity of social audit work, and increase the social influence of social auditing so that it can better serve the market economy; guide social auditors to actively participate in economic decision-making, participate in government economic work conferences, and listen to their opinions and suggestions; social audits must be bold in pioneering, self-reliant, self-disciplined, and win social recognition and support with good reputation and quality services [15].

4   Conclusions

In the era of big data, multiple economic components still coexist, competition in domestic and international markets has become more intense, and business operations have become more complex and informatized. National audit institutions and the government should give full play to the role of social audit, guide and publicize its advantages, so that it can better serve the healthy development of the economy. In the context of big data, audit work is more closely related to high-quality economic development. As an important part of the national audit and supervision system, social audit must adapt to the development of the new era and have a deep understanding of the new characteristics, new missions, and new requirements of the new era. Social audit institutions should seize the opportunity to find integration points in the complex market environment, dare to undertake various businesses that are conducive to economic development, and give full play to the role of audit's immune system; at the same time, they should constantly change their thinking and concepts, establish scientific audit concepts, and constantly promote the integration of big data technology and audit work, change the audit model, innovate audit technical means, and improve audit effectiveness and efficiency. Through auditing methods, it is possible to reveal the difficulties and pain points in economic construction, improve the ability to audit and supervise guarantees in accordance with the law, the ability to advise and provide suggestions, and the ability to serve development, so as to promote the high-quality, sustained and healthy development of various economic undertakings.

References

1. Tao, Z.: Some thoughts on giving full play to the role of auditing to promote high-quality economic development. China Audit News, no. 005 (2019)
2. Zhang, M.: Collaborative utilization of government audit and CPA audit resources. Ind. Innov. **02**, 116–117 (2019)
3. Zhang, F.: Research on audit resource integration strategy under the current system. Ind. Sci. Trib. **04**, 228–229 (2017)
4. Cheng, L.: Vigorously improve audit quality and serve economic and social development. Financ. Account. **18**, 4–6 (2019)
5. Cao, W.: Several researches on the coordination between government audit and social audit. China J. Commer. **11**, 149–150 (2017)
6. Xu, X., Zhang, X., Xu, Y.: Research on the economic adaptability of government audit outsourcing and audit resources. J. Univ. Financ. Econ. **02**(01), 69–76 (2020)
7. Wu, C.: The mechanism and realization path of auditing to promote high-quality economic development. China Audit News, no. 006 (2018)
8. Yi, D.: Talking about CPA big data audit. Chin. Certif. Public Account. **08**, 69–71 (2020)
9. Ren, Y.: Planning audit work from the height of promoting economic development. Securities Daily, no. A03 (2013)
10. Wang, J.: Update ideas, highlight audit key points, and serve economic and social development with high quality. Mod. Audit. Econ. **02**, 8 (2014)
11. Lu, Q., Wang, H.: The evolution and logic of CPA auditing technology. Friends Account. **09**, 143–146 (2019)

12. Fu, D., Yang, J.: Auditing development trend and CPA audit risk under the Internet background. Chin. Certif. Public Account. **07**, 80–82 (2019)
13. Pei, C.: Audit risk prevention measures for accounting firms in the context of big data. Chin. Certif. Public Account. **04**, 79–81 (2020)
14. Hu, X.: Research on the construction of social audit information in the era of big data. China Townsh. Enterp. Account. **08**, 201–202 (2020)
15. Xu, Y., Feng, J.: Non-governmental audit participation in state governance in a social governance environment theoretical analysis and realization path of management. Financ. Account. Monthly **11**, 96–99 (2017)

Reform of International Economics and Trade Professional Course System in the Internet Age

Ping Wang[(✉)]

Henan College of Industry & Information Technology,
Jiaozuo 454000, Henan, China

Abstract. With the development of science and technology, Internet technology has become more and more mature. Now, we have entered the Internet era. International Economics and Trade has always been a popular major. Under the background of the Internet era, the traditional international economics and trade professional curriculum system can no longer meet the needs of the times. For this reason, this article expands the reform of the international economics and trade professional curriculum system in the Internet era. In the research, this article first conducted a questionnaire survey on the enterprises related to international economics and trade in our city, and analyzed the demand for talents in international economics and trade in the Internet era. Secondly, taking the international economics and trade major of our school as an example, researched its curriculum system and proposed reform measures for the international economics and trade professional curriculum system in the Internet era. The research in this paper finds that there are many factors that employers attach importance to when recruiting professionals in international economics and trade. Most employers value comprehensive quality. A total of 108 companies choose the factor of comprehensive quality, with an overall percentage of 93.91%. In addition to practical ability and professional knowledge, foreign language proficiency and social skills are also factors that companies pay more attention to when recruiting. In addition, in the analysis of the school's international economics and trade professional curriculum system, it is found that the current school pays more attention to general courses in the curriculum, while ignoring professional courses and practical courses. For this reason, we propose some reforms of the international economics and trade professional curriculum system The measures call on schools to pay attention to the education of professional knowledge and the cultivation of practical ability while paying attention to basic education, so as to cultivate better professionals in international economics and trade for the country.

Keywords: Internet era · International economics and trade major ·
Curriculum system · Talent demand

1 Introduction

The advent of the Internet era has caused tremendous changes in all aspects of the social environment [1, 2]. The advent of the Internet era has promoted the development of cross-border electronic business, and has provided more information exchange

J. C. Hung et al. (Eds.): FC 2021, LNEE 827, pp. 213–221, 2022.
https://doi.org/10.1007/978-981-16-8052-6_24

methods and trade methods for international trade, broke the traditional trade model and market area restrictions, and formed a globalization with information technology as the link Trade market [3, 4]. In addition, the advent of the Internet era has spawned many new international trade methods, such as online ordering, online payment, online sales, etc., both parties can use electronic information systems to complete commodity declaration, inspection, insurance, transportation management, foreign exchange settlement, etc. Work to improve trade efficiency [5]. In the Internet age, the society's requirements and needs for talents in various fields are constantly updated, especially for application-oriented talents who can meet market needs [6].

Nowadays, in the context of the Internet era, the traditional international economics and trade professional curriculum system can no longer meet the needs of the times [7, 8]. We know that the international economics and trade professional curriculum system includes a lot of content, and the quality of the curriculum system directly restricts or promotes the quality of teaching [9, 10]. In the Internet age, reform the international economics and trade professional curriculum system so that it can meet the needs of the development of the times and cultivate better professionals. In this way, it can effectively promote the development of international economics and trade.

This article provides a simple explanation of the Internet era and international economics and trade majors, and selects the city's international economics and trade-related enterprises to conduct questionnaire surveys and interviews, and analyzes the talent needs of international economics and trade in the Internet age. And taking the international economics and trade major of our school as an example, researched its curriculum system, and proposed reform measures for the international economics and trade major curriculum system in the Internet era.

2   Overview of the Internet Era and International Economics and Trade

2.1   Internet Era

The Internet age means a longer period of time with Internet technology as the core. Combined with the research views of many experts and scholars, the Internet age is generally divided into four stages, namely the CPU stage, Web1.0 stage, Web2.0 stage and In the big data stage, the development and use of Internet technology in each stage are different, and the manifestations are different, but all have a transformative impact on the government information management at that time.

The "Internet era" referred to in this article mainly refers to the fourth stage of Internet development, namely the big data stage under the new information technology environment. In the Internet age, mobile Internet, social networks, and emerging media have greatly expanded the boundaries and scope of the Internet, the information flood valve has been opened, all kinds of data have rapidly expanded, and the entire society has entered a period of information explosion, that is, the stage of big data.

2.2   International Economics and Trade Major

(1) Professional definition of international economics and trade

"International Economics and Trade Major" mainly includes two parts of international economics and international trade in professional learning, and it takes into account the cultivation of theoretical knowledge and practical ability in talent training. International economy and trade is a unit that cultivates applied talents that combine economics and trade.

(2) Talent training requirements for international economics and trade

The training of talents in international economics and trade must first emphasize the learning of applied economic management and trade theory. Should have the corresponding knowledge of international economics, international trade, management, accounting and other economic management and trade theories, and be familiar with relevant theoretical literature, which can provide scientific theoretical guidance for subsequent actual economic and trade operations. Secondly, we must pay attention to the cultivation of practical operation ability, execution development ability, communication and coordination ability and related professional skills. Emphasizing practice and hands-on operation is the most vivid interpretation of "application". Cultivating applied talents in international economics and trade is first to focus on the cultivation of their practical and operational capabilities, that is, to be able to apply theoretical knowledge and existing scientific research results in the field of international trade to practical international trade work. To become an excellent international trade application-oriented talent, we should focus on the training of market project execution ability and market development ability, so as to maximize the use of effective market information, grasp customer resources and do every international trade work with maximum efficiency At the same time, the ability of "execution and development" is also a concrete manifestation of emphasizing practice and application ability, and coordination ability is an important indicator of comprehensive quality.

(3) Problems in International Economics and Trade

 1) Talent training target is too single

In my country's universities, there is great convergence in the training objectives, curriculum systems, teaching plans, and teaching content of international trade talents, resulting in a lack of hierarchy and characteristics in the overall training of international trade talents. In the Internet age, international economics and trade majors the development of my country cannot meet the needs of our country's international trade, and reform is urgently needed.

2) Poor practical teaching effect

Practical teaching is an important part of teaching in colleges and universities, but the current practical teaching effect of international economics and trade majors is not significant, resulting in students not being able to use trade knowledge to solve practical problems. The main reason for the above phenomenon is that most of the international trade majors in our country currently implement passive practice methods, that is, the status of students' learning subjects is ignored, most of the practical routes and methods have been determined by teachers, and students cannot rely on their own knowledge system to learn. Complete teaching practice activities.

3 Research Ideas and Research Design

(1) Research ideas

This article adopts the literature analysis method to provide a theoretical basis for the research of this article by collecting and analyzing online and offline data. Secondly, we use a combination of questionnaire survey method and interview method to analyze the demand for international economics and trade professionals in the Internet era, and take the international economics and trade major of our school as an example to analyze the international economics and trade professional curriculum system, Thus proposed reform measures for the international economics and trade professional curriculum system in the Internet era.

(2) Research design

1) Analysis of talent needs

This article takes 120 international economic and trade-related enterprises in our city as the research object, and analyzes the current demand for international economic and trade professionals. The research was conducted by a combination of questionnaires and interviews. In this research, a total of 120 questionnaires were distributed and 115 valid questionnaires were returned. The questionnaire efficiency reached 95.8%, and each questionnaire corresponds to a company.

2) Course system analysis

Taking the international economics and trade major of our school as an example, the curriculum system of the international economics and trade major of our school was studied, and 200 students majoring in international economics and trade of our school were randomly selected to conduct a questionnaire survey. Inquiries ask students what they think about international economics and trade professional courses, and sort them into categories. Among them, the views on international economics and trade professional courses are divided into five types: very reasonable, reasonable, general, unreasonable, and very unreasonable.

4 Analysis and Discussion of Research Results

4.1 Analysis of the Demand for International Economics and Trade Professionals in the Internet Era

This article uses a combination of questionnaire surveys and interviews to analyze the demand for international economics and trade professionals in the Internet era. The results are shown in Table 1 and Fig. 1.

Table 1. Factors that employers attach importance to when recruiting professionals in international economics and trade

Value factors	Number of employers making selection	Overall percentage (%)
Professional knowledge	72	62.61
Gender	12	10.43
Comprehensive quality	108	93.91
Appearance	20	17.39
Social relationship	8	6.95
Foreign language level	56	48.70
Social skills	63	54.78
Academic performance	32	27.83
Actual ability	95	82.61
Knowledge	38	33.04

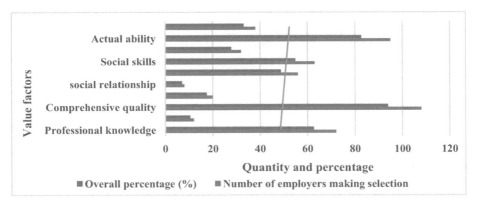

Fig. 1. Analysis of the demand for international economics and trade professionals in the Internet era

As shown in Table 1 and Fig. 1, there are many factors that employers attach importance to when recruiting professionals in international economics and trade. Among them, most employers value comprehensive quality. A total of 108 companies choose the factor of comprehensive quality. The overall percentage is 93.91%, followed by actual ability and professional knowledge. 95 companies choose the factor of actual ability, the overall percentage is 82.61%, 72 companies choose the factor of professional knowledge; the overall percentage is 62.61%. In addition, foreign language proficiency and social skills are also more important to companies when recruiting. Among them, 56 companies chose the foreign language factor, the overall percentage is 48.70%, and 63 companies chose the actual ability factor, the overall percentage is 54.78%. In addition to the appeal factors, some companies attach great importance to gender, appearance, social relations, learning ability and knowledge, but the proportion is not high. Only a few companies pay more attention to these factors. To sum up, most employers prefer to recruit graduates with high comprehensive quality, strong professional knowledge and ability, and outstanding practical ability.

4.2   The Reform of the International Economics and Trade Professional Curriculum System in the Internet Era

(1)   Analysis of the international economics and trade professional curriculum system
    Analyzing the analysis of our school's international economics and trade professional curriculum system, the credits, hours and ratios of some courses in the curriculum are shown in Table 2.

Table 2. Credits and proportions of some courses in the course

Course type	Credit	Percentage of total credits (%)	Class hours	Percentage of total hours (%)
General course	65	39.4	1206	39.0
Foundation course	21	12.7	336	10.9
Professional core courses	32	19.4	544	17.7
Professional development course	20	12.1	320	10.3
Concentrated practical teaching project	27	16.4	672	21.8

    It can be seen from Table 2 that in the international economics and trade professional courses, the general education courses have 65 credits, accounting for 39.4% of the total credits, and 1206 credit hours, accounting for 39.0% of the total credits. The credits of the basic courses are 21, accounting for 12.7% of the total credits, and the class hours are 336 h, accounting for 10.9% of the total credits. The professional core courses have 32 credits, accounting for 19.4% of the total credits, and the credit hours are 544 credits, accounting for 17.7% of the total credits. The credits of professional development courses are 20, accounting for 12.1% of the total credits, and the class hours are 320, accounting for 10.3% of the total credits. The intensive practice teaching project has 27 credits, accounting for 16.4% of the total credits, and 672 credits, accounting for 21.8% of the total credits. It can be seen that the international economics and trade majors of our school pay more attention to general courses, while the emphasis on other professional courses and practical courses is relatively low. This will result in poor professional basic knowledge and low actual ability of students.
    A questionnaire survey was conducted with students majoring in International Economics and Trade, and the results were shown in Fig. 2. A total of 200 questionnaires were distributed in this article, and 200 valid questionnaires were returned.

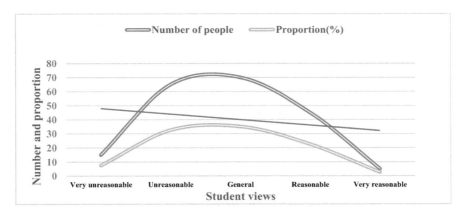

Fig. 2. Students' views on international economics and trade professional courses

As shown in Fig. 2, there are 15 people who think the course is unreasonable, accounting for 7.5%, 65 people who think the course is unreasonable, accounting for 32.5%, and 70 people who think the course is normal. The proportion is 35%, 45 people think the course is reasonable, accounting for 22.5%, and 5 people think the course is reasonable, accounting for 2.5%. It can be seen that students think that the international economics and trade professional courses are not very reasonable, and the school should reform the international economics and trade professional curriculum system.

(2) Reform of the international economics and trade professional curriculum system in the Internet era

In the Internet era, international economics and trade professionals should have the following characteristics: First, in terms of quality, they should have the necessary communication and cooperation capabilities in foreign trade activities, as well as a good sense of ethics, professionalism, and sense of responsibility. And other non-professional qualities. Second, in terms of knowledge, there should be a complete knowledge structure, and graduates of international trade majors should have a complete knowledge structure. Thirdly, in terms of ability, according to the needs of society, it has certain expression ability, learning ability, social ability, innovation ability and practical ability. For these kinds of abilities, the corresponding curriculum should focus on cultivating talents who meet these needs.

In order to adapt to the development of the times and cultivate more and better international economics and trade talents in the Internet age, we should reform the international economics and trade professional curriculum system. The specific reform measures are as follows:

(1) Curriculum setting of basic theory modules

The basic theory course module includes the related theory courses of the public basic course. Although there are many courses, the credits cannot be used for more. It is appropriate that the hours of basic courses account for one-fifth or even less of the total hours.

(2) Curriculum of professional ability module

The professional competence module is the core of the entire course structure of the international trade major; the professional competence module is a major compulsory course for the international trade major. It mainly learns major compulsory knowledge in economics, management, law, etc., and lays the foundation for future international trade work.

(3) The setting of capacity development module courses

The ability development module is a course that is not offered in the existing applied undergraduate international trade majors. According to the needs of the society, these abilities are social skills, communication skills, organization and coordination skills, etc. Based on these abilities, corresponding courses can be set up to lay the foundation for future international trade work. Ability development courses include website management and web production, Vietnamese, Thai, Korean, Japanese, Malay, speech and eloquence, international business negotiation, business English listening and speaking, social etiquette, ASEAN profile and customs, etc. In addition to classroom teaching, ability development courses can also set up some platforms, such as foreign language corners, so that students can strengthen the abilities they encounter in the classroom during exercise. In the process of teaching, teachers can also instill in students the ways to behave and do things, and better guide students with their own practical experience.

(4) The setting of the fourth practical ability module course

The practical ability module is a synthesis of the first two modules. Only by consolidating the foundation and highlighting the professional ability can the practice be carried out better. Practical course modules include customs declaration training, comprehensive international trade experiments, professional internships, employment and entrepreneurship internships, e-commerce, customs declaration practices, accounting computerization and other courses. The class hours of such courses can be extended, reducing classroom teaching and increasing practical training. Wait for practical lessons.

5   Conclusions

With the development of science and technology, the era of the Internet has quietly arrived. Nowadays, the Internet has had a huge impact on our lives. Under such a background, if the training of talents in international economics and trade remains unchanged, it will not be able to meet the needs of the development of the times. This article studies the reform of the international economics and trade professional curriculum system in the Internet era. In the research, this article analyzes the demand for talents in international economics and trade in the Internet era through a questionnaire survey of enterprises related to international economics and trade in our city, and takes the international economics and trade major of our school as an example. The system has been studied, and the reform measures of the international economics and trade professional curriculum system in the Internet era are proposed. This article believes that while the school attaches importance to basic education, it should also focus on the education of professional knowledge and the cultivation of practical ability, so as to cultivate better international economics and trade professionals for the country.

References

1. Shavina, E.V.: Formation and development of innovative growth points of economy: the experience of China and Russia. Int. Trade Trade Policy **4**, 107–120 (2018)
2. Krupenkov, V.V., Mikhailovich, R.N., Valerievna, G.Y., et al.: Development of small and medium trade enterprises in Russia and monotowns and worldwide in the digital economy. Int. J. Econ. Res. **14**(15), 341–352 (2017)
3. Mi, Z., et al.: Creating a better environment for foreign investment. Beijing Rev. **35**(v.60), 50 (2017)
4. Mei, X., et al.: Reform readjustment. Beijing Rev. **47**(v.60), 40–41 (2017)
5. Suryanto, N.A., Rusli, B.: Analysis of economy aspects in the policy on establishing housing and settlement in West Java, Indonesia. Int. J. Trade Glob. Mark. **10**(1), 91 (2017)
6. Fan, J.L., Wang, J.X., Li, F., et al.: Energy demand and greenhouse gas emissions of urban passenger transport in the Internet era: a case study of Beijing. J. Clean. Prod. **165**(nov.1), 177–189 (2017)
7. Greenhow, S., Hackett, S., Jones, C., et al.: Adoptive family experiences of post-adoption contact in an Internet era. Child Fam. Soc. Work **22**(S1), 44–52 (2017)
8. Popkin, P.: The effect of the Internet era and South Dakota v. Wayfair on the unitary business rule. Boston Coll. Law Rev. **60**(9), 8 (2019)
9. Chen, Z., Li, Y., Wu, Y., Luo, J.: The transition from traditional banking to mobile internet finance: an organizational innovation perspective - a comparative study of Citibank and ICBC. Financ. Innov. **3**(1), 1–16 (2017)
10. Mahajan, R., Spring, N., Wetherall, D., et al.: User-level Internet path diagnosis. ACM SIGOPS Oper. Syst. Rev. **37**(5), 106–119 (2017)

Visual Metaphor of the Short Video Eco-System

Xiaomin Zhang[✉]

Faculty of Journalism and Communication, Communication University of China,
Beijing, China

Abstract. The developments of short videos appear poised for growth. As image communication, short video presents an ecological chain development mode from content to platform, then to users, and finally to the formation of market influence. In this commercialized ecosystem, visual metaphor, which exists in artificial media using visual images, has become a set of strategies to stimulate insight. At the same time, visual metaphor is also a tool for users to think. Visual metaphor plays an important role in short video communication. This paper takes "happy country leaders program" on SnackVideo as the research object and applies the case analysis method to explore the generation mechanism and performance characteristics of metaphor mechanism in the theoretical perspective of visual metaphor, so as to provide useful references for the development of short videos.

Keywords: Short video · Ecological chain · Visual metaphor · Happy country leaders

1 Introduction

Recording self-growth and reflecting social transformation, short videos reshape the power of cultural communication and social innovation with richer and more sharing culture, knowledge and information. As a new media growing up in the Internet environment, the development of short video is challenged by both content and video-traffic.

Flow is the market vane, content is the cornerstone of development. In the era of new media, the scarcity and uniqueness of content are still of great value, and the new media based on channel needs the support of content. Visual metaphor exists in the artificial media using visual image, which is not only a strategy to stimulate insight, but also a tool for users to think. It is called "metaphor in metaphor", and its biggest advantage is to rely on the visual image constructed by visual practice for persuasion. These characteristics just fit the short video ecological chain development mode, from content, platform to users, influence and market. Therefore, under the theory of visual metaphor, it is beneficial to promote the development of short video to analyze how images act on the viewer, how to use visual symbols to express meaning, and how to grasp visual text through metaphor.

The "happy rural leader program" aims to enhance the leadership, business management ability and social responsibility of Chinese rural entrepreneurs, promote rural

J. C. Hung et al. (Eds.): FC 2021, LNEE 827, pp. 222–230, 2022.
https://doi.org/10.1007/978-981-16-8052-6_25

revitalization and talent cultivation, and help rural development [1]. Through the short videos display, local products in poor areas will be advertised, and produce economic benefits, which is an important goal of this kind of short videos communication. This demand is highly similar to the development model of "content-platform-users-influence- market" short video ecological chain that we are trying to explore. Therefore, this paper takes "happy rural leaders" as a case, selects four leaders, which are "laughing Shirley" "seven fairies of Dong family" "Yang Lili's wheat straw painting" and "PingWuGuanBa DuYong", as the research objects by random sampling method, and uses visual metaphor theory to analyze the generation rules and performance characteristics of metaphor mechanism, so as to provide useful reference for the development of short videos.

2   An Overview of Visual Metaphor Theory and Literature Review

Metaphor is an ancient way of thinking, which plays an extremely important role in human language, thought and culture. In ancient Greece, Aristotle's rhetorical study gave metaphor a high position and weight [2]. After metaphor entered the field of concept from the field of language, Victor Kennedy and John Kennedy found that metaphor has gone beyond the linguistic category of daily life and art, and become the most critical rhetorical device in visual art [2]. In the metaphor theory system of George Lakoff and Mark Johnson, the father of cognitive linguistics, metaphor is considered as a cognitive phenomenon and a basic cognitive way for people to know things and establish a conceptual system [2].

After the confluence of vision and reason, visual metaphor establishes its core position through the tradition of "visual centralism" [3]. As a common phenomenon in human thinking, metaphor is mediated by metaphorical words and is transformed between noumenon and vehicle. Its meaning is not fixed and needs to be explored by the observer. After entering the field of concept, visual metaphor becomes a set of visual images for communication and expression [3]. It is convinced that vision can objectively record the reality and accurately reflect the object, which is the psychological mechanism and structural basis of resorting to images, symbols and other visual images [3]. At the same time, visual metaphor is the source of thought, which is widely permeated in the knowledge system, ethical relations and power operation [3].

On the study of visual metaphor, Yan Gao combs the visual metaphor and spatial turn from the perspective of ideological history, and believes that focusing on the visual metaphor behind visual culture is the critical path to understand modernity, post-modernity and the modern life [3]. Tao Liu made a series of discussions from the perspective of visual rhetoric, such as *Metaphor Theory: Generation of Transferred Meaning and Analysis of Visual Rhetoric, Interactive Model of Metaphor and Metonymy: From Language to Image, Metonymy Theory: Image Reference and Visual Rhetoric Analysis, Sub Image Iconicity: Symbol Movement and Visual Metaphor of Pyles, etc.* Yiqing Hu pays attention to the metaphor of "entity or relationship", Danling Liu constructs the visual recognition framework of "image" of the national image, Wei Zhu analyzes the metaphor mechanism from the perspective of metafunction and discourse, Lang Chen constructs the concept of metaphorical discourse ability, Xiaoyun Li discusses the realization of metaphorical function from the perspective of media technology, and Jiang Chang expounds media from the perspective of media evolution Metaphor and imagination in ecology.

3 The Original Ecology is Chosen as the Source Domain, and the Target Domain Metaphor is Realized Through "Cross Domain Mapping"

According to the mapping theory, the thinking process of metaphor is reflected as the cross-domain mapping in the conceptual system. The occurrence mechanism of metaphorical practice is a kind of mapping or projection from the cognition of vehicle to the cognition of noumenon [2]. That is to say, when the interpretation process of a symbol system is in trouble, it is necessary to acquire another symbol system with universal cognitive basis, so as to approach and grasp the meaning system of the former along the meaning system of the latter [2]. In the image era, it is to use the image symbols of the original domain to explain the target domain through cross domain mapping.

The object presented in image text consists of "substantial part" and "element" [4]. The "substantial part" refers to the part that can leave the whole and self-maintain and present, with independence [4], Such as trees, mountains, rice fields, mobile phones, farm tools, the sun, children and so on. "Elements" refer to the parts that cannot be concrete objects and cannot be separated from the whole to which they belong, which are self-sustaining and independent, such as speed, tone, expression, posture, etc. The nature of the object in the visual text determines that the attribute recognition framework of the viewer in the cognitive process is completed by "joining" and "condensing" [4].

When watching, a user first confirms what is short video through the substantial part of video text, then identifies the elements that are attached to it, and discover and identify its feature. For example, Yang Lili's wheat straw painting represents the production and inheritance of wheat straw painting in Duolun County, Xilingele League, Inner Mongolia, PingWuGuanBa DuYong represents the daily life of forest protection in GuanBa village, PingWu County, Mianyang City, Sichuan Province, laughing Shirley and Romantic Seven Fairies of Dong Family are all set in Qiandongnan Miao and Dong Autonomous Prefecture, Guizhou Province. Among them, laughing Shirley mainly shows her daily life, while Romantic Seven Fairies shows Dong culture. These joining and condensing image features are called recording the world and recording you on the SnackVideo, that is, the emphasis and compliance on the original ecology.

The Four ID numbers are presented from the first perspective. They respect the original appearance of life, and use the method of true record to create and spread different cultures, showing the diversity and richness of life. Such as PingWuGuanBa DuYong's short videos show the placement of the infrared camera, the antelope jumping on the road, the life of workers fighting dams, Cattle and sheep on the mountain and so on. The most original life is exactly what users want to see. Here, the original ecology is affinity and penetration. The works of "Romantic Seven Fairies of the Dong Family" shows the food, scenery and ceremony of Dong village. From the perspective of visual metaphor, it mainly uses audio-visual language montage, relies on the time dimension of lens A and lens B to realize the construction and transmission of meaning, and guides the audience to form the metaphorical association that A is B. In this set of videos,A is every shot and every frame in the video, which exists in the context created by the author, and is concrete, vivid, sensible; B is built on the basis of cross- domain mapping, which needs to form a new understanding of the ontological attributes and connotation through the meaning construction of cognitive psychology. This process, in the Romantic Seven Fairies of the Dong Family, is to understand the logic and significance of the target domain along the meaning and logic of the original domain. Among them, the original domain is the content presented in the short video, and the target domain is to attract more people to pay attention to the Dong culture and travel to Shanbao village to get rid of poverty and become rich. At present, this goal has achieved initial results in gaibao village.

In poverty alleviation stories, visual metaphor builds a bridge between producers and users, satisfies emotional expression in coding and decoding, enhances social identity, and realizes communication value. Summing up the characteristics of these original stories, mainly in three aspects: first, the visual text has a real sense of reality. This sense of reality is similar to the situation or experience of the viewer, which easily leads to empathy or compassion. Second, the expression of visual practice is far from the experience of the viewer, which can arouse the curiosity, thinking and even action of the viewer. The third is that the point of view implied by visual discourse, whether agree or disagree, means profound, which can at least touch a certain point in the heart.

4 Encodes Concrete Spiritual Symbols and Obtains the Content Market Through Metaphorical Persuasion

Metaphor is called "metaphor in metaphor", and its importance lies in the pursuit of persuasive effect. One of the characteristics of visual metaphor is to rely on the visual images constructed by visual practice for persuasion, which is symbolic action in the sense of social construction [2]. The following is a sample of ID number "laughing Shirley", which has the highest number of fans and attention, to analyze the mechanism of its metaphorical effect.

Yuan Guihua is the author of "laughing Shirley". She comes from a big family of 14 people in leizhai village, Tianzhu County, Guizhou Province. Because of the need to bear the responsibility of supporting her family, Yuan Guihua chose to give up her studies after graduating from high school. Those short videos show her life,such as doing farm work, selling agricultural products and building houses,and so on. As Guy Debord said, almost every detail of life has been alienated into the form of landscape [5]. These Landscapes are visual and objective scenes displayed, which also means a subjective and conscious performance and show [5]. As Gérard Genette put it, this kind of display, performance and viewing are at different narrative levels. There are different rules of interpretation and different systems of partition between them.They are logically independent of each other [6]. Through representation and narration, the similarity between two parallel narrative layers is found, which makes the generation and flow of meaning possible [6].

In the "laughing Shirley", the land of idyllic beauty is constructed with materials such as melon bitten by insects, rose on the mountain, wild cherry, local bayberry, stone painting, girl catching fish, spring water, etc. Similarly, the image of tough girl also exists as a metaphor, and the noumenon is moving mountains's girl, challenging girl, female Rambo, poking horse honeycomb's girl, carrying a piece of sky, etc. In these two metaphors, that which are "paradise" and tough girl, and their corresponding construction materials are in two narrative levels, and their structural relevance is due to the similarity of spirit. The recognition and meaning of object image in visual text is because an image can be regarded as a non-material entity, a ghostly, phantom like appearance, depending on some material support to surface or obtain life [7]. It is an intentional structure given in experience, which integrates all appearances, perspectives and images [8]. It is the meaning that lies in what it expresses and hides behind them [8].

Shirley is a phenomenal product of platform for poverty alleviation. Short video creation not only solved the livelihood problem of Yuan Guihua's family, but also opened a market for local agricultural products. The achievement is built on the short video ecological chain of content platform user influence market. The key is the communication power and influence brought by visual metaphor. The emergence of visual metaphor also leaves the communication mechanism of CO production, participation and sharing bred by not short video soil. The advantages of this mechanism are mainly manifested in two aspects:

The first is that in the process of consuming products, users reproduce and redistribute products through individual forwarding, commenting, sharing and other behaviors, so as to create new product value. Laughing Shirley often interacts with users in short videos by asking questions. For example, in the short video of spring water published on June 24, 2017, the author's message is: what does mean of that "Cao jie"? The result is a heated discussion. Some say it is safe. Some say it is for health. Some say it is a ceremony to send water god. Some say it is to separate things from the water, or to tell others that the water is boiling water. Comments not only focus on the cursive knot, but also extend a lot of new topics, forming a new discourse. The old fellow described the comment as better than the video.

The second is the algorithm distribution mechanism based on platform and social network, which extends the diversified relationship between people, and breeds the soil for the production and dissemination of content. SnackVideo's Gini Coefficient mechanism is used to distribute the principle to avoid the polarization of attention resources and let the sun shine on more people. The figure below shows the relevant indicators of short video works produced by four happy rural leaders in the process of poverty alleviation.

These indicators directly show the inclusive communication based on the Gini Coefficient mechanism. Any person, even a naive person, can post short video on SnackVideo. And the system will match the number of basic users that may be interested in this video, at least 100 people. There is also a climbing mechanism. If the popularity index of 100 basic users is high, such as viewing time, comments, downloads, etc., the system will distribute it to more users again. On the other hand, for the head users, the platform takes measures to limit the flow. Compared with the popular video, Snack Video pays more attention to the distribution of long tail video. And at Snack Video, the head video accounts for only 30% of the platform traffic [9] (Fig. 1).

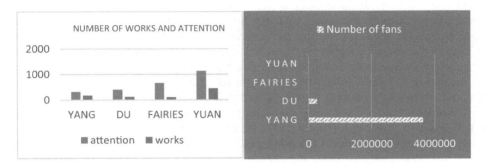

Fig. 1. Number of works, attention and fans of four happy rural leader

American sociologist Michael Collins defines an interactive ritual chain, which is a chain formed by the constant interaction between individuals in a certain situation [10]. Collins believes that the interactive ritual chain is the foundation of social structure. In the chain of interactive rituals, people's experience involves resources, status, exchange and other factors, so people's interaction has market characteristics [10]. In this interactive ritual market, people use emotional energy and capital symbols for social communication. The process of social communication is essentially a process of value production and exchange [11]. Therefore, short video can be spread in the metaphorical mechanism of content construction, and form a market extended by content.

5   With the Help of Allusion and Emotional Metaphor, the Meaning Outside the Painting Can Be Activated Through "Punctum Characterization"

Shirley has lost nearly 10000 fans in three hours after she interrupted the advertisement which does not match her image, and has not recovered to the previous level of fans in a short time. This mismatched advertisement, according to Roland Barthes' theory, is called punction in visual communication. It can pass through studium, burst into the center of visual attention, evolve into the cause of meaning increment, ignite the fire rope of emotion involvement, and activate the emotion recognition framework [4].

Roland Barthes put forward the concepts of studium and punctum to clarify what moves people in the images. Roland Barthes believes that "studium is a kind of extensive element, which aims to convey information, reproduce situations, surprise people, emphasize meaning. It is smooth and arouses casual desire, changeable interest and inconsistent tastes, or even half desire, half desire", and "enables me to approach a world" Some basic knowledge has provided me with a set of local conditions of objects and touched some fetishism in me. Prick point is a "detail" in the picture, a kind of "nameless" stimulus, where there is huge abnormality and destructiveness. Its existence always tempts people to ponder over some elusive meaning out of the picture [4].

Roland Barthes points out that the essence of studium identification is to touch the intention of the text producer, and the punctum is to "summon all possible meanings out of the picture". It emphasizes the foundational role of those infectious moments in the interpretation of the works, and the most authentic, natural and sensitive inner experience [4]. It can be seen that the punctum is the key to the interpretation of visual image in the visual metaphor mechanism. These punctums are usually expressed in two ways: space experience and time experience. Sherry's two short videos, "city boys and mountain girls" and "brother Niu are confused" can explain what is the punctum of visual metaphor and how it is expressed in time and space.

Shirley's short video focuses on everyday life, such as farming, cooking, catching fish, painting, carpentry, building a house, doing farm work, carrying wood, paving cement, etc. In the process of recording and sharing, What users feel is a Miao girl, who is sunny, kind, beautiful, cute, strong, and optimistic. However, in the two short videos, There are comments like "Can you change your clothes" "It's Keng duo duo all over the Internet" "Here's advertising again", etc. After the broadcast of "city boys and mountain girls", the number of fans did not fluctuate significantly, but after the broadcast of "PinDuoDuo", the number of fans decreased by nearly 10000 within three hours, and the mood and attitude of users fluctuated significantly. Therefore, what users perceive and understand is not only the image they see in front of their eyes, but also "the vision composed of potential and absence surrounds the actual presence of things" [4].If this potential perception is found to be unfavourable by the viewer, it will have a destructive effect.

As the punctum of time transcendence, it is manifested in the split, dramatic and conflict of image time expression [4]. For example, the user's evaluation of the tools in Shirley's video has changed on the timeline. At first, the videos show shovel, sack, rope, axe, etc. After that, there were famous chainsaw, cutting machine, UAV and so on. Old fellows felt that such a good tool must be very expensive,common people can't afford it, and there is conflict and split. This kind of punctum is opposite to the studium formed by poor family and independent girl. These emotions and attitudes eventually evolve into emotion recognition of users. In short videos, the number of fans decreased. Here, the culture, including experience, custom and knowledge, stipulates the habitus between the signifier and its implied meaning of the associated image symbols. Therefore, the specific objects, places, attitudes, references and frames of reference in the image have the potential to refer to other things and meanings [4].

6   Conclusion

Visual Metaphor mechanism is the driving force of short video ecological chainVisual metaphor is not only a figure of speech, but also a basic cognitive way for people to understand things and establish conceptual system [12]. In the age of image, visual metaphor has incomparable rhetorical ability and persuasive effect [12]. It shows great development potential. From the above analysis, we can see that in the publicity of poverty alleviation, short videos carry the technological advantages of large bandwidth, large connection and low delay, and promote the production and dissemination of visual images with unprecedented speed, power and effect. The platform mechanism, by calling individuals to participate in production, encourages grass-roots creation, and enables more original content to be displayed and transmitted through visual metaphor with the power of original nature. It provides a new framework for creating things, a new perspective for viewing things, and a new meaning for understanding things.

As an ordinary forest ranger, PingWuGuanBa Du Yong has no skillful shooting skills. His videos are often patchwork of photos. However, It is such a real expression that conveys the simple values of the protagonist, the forest protection, which is a dynamic teaching material to promote the concept of ecological and environmental protection. Wu Yusheng, once known as "the Secretary of the unorthodox poverty alleviation", reconstructs the communication time and space, expands the communication path of the Dong culture, and explores a way for the tourism transformation of Shanbao village through the operation of Romantic Seven Fairies of the Dong Family. Yang Lili's wheat straw painting and laughing Shirley have created a number of popular products on short video platforms, driving the consumption of new cultural landscape and new trends.

With the promotion of short video, the net red villages, scenic spots, products in poverty-stricken counties continued to emerge, which help the construction and dissemination of regional image, effectively drived the development of local tourism market, injected new momentum into economic development, and proved the feasibility of the short video ecological chain of "content-platform-users- influence-market" driven by visual metaphor mechanism effectiveness.

References

1. People's Daily Online. Candidate cases: happy country leader program. http://gongyi.pepple. com.cn/n1/2018/1112/c422231-30396322.html.2019.8.9
2. Liu, T.: Metaphor theory: generation of transferred meaning and analysis of visual rhetoric. J. Soc. Sci. Hunan Normal Univ. **6**, 141 (2017)
3. Gao, Y.: Visual metaphor and visualization of the world: on the ideological roots of contemporary visual culture. J. Sun Yat-sen Univ. **5**, 66–75 (2012)
4. Liu, D.: The way of watching: the visual recognition framework of "image" national image. Nanjing Soc. Sci. **10**, 121–128 (2018)
5. Debord, G.: Landscape Society, vol. 3, p. 10. Nanjing University Press, Nanjing (2006)
6. Liu, T.: Interactive model of metaphor and transformation: from language to image. Press **12**, 33–46 (2018)
7. Michel, W.J., Yongguo, C.: Trans. Gao Enthalpy. What Image Needs: Life and Love of Image, vol. 13. Peking University Press, Beijing (2018)
8. Sokolavsky, R.: Introduction to Phenomenology, vol. 20, p. 28. Wuhan University Press, Wuhan (2009). Translated by Gao Bingjiang and Zhang Jianhua
9. Yu, J.: Quick: Kwai Hui + Gini coefficient online community experiment. Media **3**, 19–21 (2019)

10. China's Autumn, China, TikTok, Li, Q.: Interactive ritual chain and value creation of. Chin. Ed. **9**, 70–76 (2018)
11. Zhao, Y.: Principles and Deduction of Semiotics. Nanjing University Press, Nanjing, vol. 164 (2016)
12. Liu, T.: Metaphor theory: generation of transferred meaning and analysis of visual rhetoric. J. Soc. Sci. Hunan Normal Univ. **6**, 140–148 (2017)

Self-deduction Training Method Based on Deep Reinforcement Learning and Monte Carlo Tree Search

Tongfei Shang[1(✉)], Weian Kong[1], and Bo Yang[2]

[1] College of Information and Communication, National University of Defense Technology, Xi'an, Shaanxi, China
[2] 31008 Troops, PLA, Beijing, China

Abstract. Aiming at the problem of how to construct the strategy and value function in reinforcement learning through deep neural network, this article first constructs the value function and strategy function under the framework of reinforcement learning through deep neural network, and improves the design of residual block to improve model performance. The feature extraction ability, and finally the Monte Carlo tree search method is introduced, and the decision model is intensively trained on the basis of supervised learning to improve its deduction ability. Simulation analysis proves the effectiveness of this method.

Keywords: Deep reinforcement learning · Monte Carlo tree search · Self-deduction training

1 Introduction

The decision space of wargaming is complicated and different from traditional expert systems. Taking into account the problem of position information description, through semantic segmentation modeling, the feature information and historical action information of the position in the position are integrated into the "image" tensor to compare and analyze the wargame position and Different Go positions, improve the residual network structure to improve the learning performance of the model, use the output of the deep neural network as the strategy and value function, introduce the Monte Carlo tree search method to design the model structure for training through self-deduction, in order to achieve Complete the exploration of the strategy space without relying on human deduction data [1–3].

2 Deep Reinforcement Learning

By semantically segmenting the operator information in the wargame deduction and enabling it to express the deduction process, the segmentation result can be used as the input of the neural network and has certain scalability. In order to enable the semantically segmented operator information to express the entire deduction process, the characteristic parameters of the operator are first fully mined, and then all the operator

J. C. Hung et al. (Eds.): FC 2021, LNEE 827, pp. 231–236, 2022.
https://doi.org/10.1007/978-981-16-8052-6_26

information at a certain moment is put into a matrix, which is used as an approximate image as a neural network [4–6].

The deep learning model constructed in this paper takes strategy and value as output, and aims to match the decision-making process of reinforcement learning, which includes two parts: strategy network and value network. Assuming x represents an input; the task of a mapping relationship $H(x) = x$ composed of a two-layer network is to fit a potential identity mapping function. This is difficult for deep networks, mainly because of the superposition of multiple non-linear layers. But if the network is designed as $H(x) = F(x) + x$, this will transform the problem into fitting a residual function $F(x) = H(x) - x$. If $F(x) = 0$, then it constitutes an identity mapping, and it can be seen that it is relatively easier to fit the residuals [7, 8]. The idea of residual error is to remove the same main part x and highlight small changes $F(x)$, so that the output becomes more obvious, the weight is easier to adjust, and the training effect is more ideal. The residual learning is integrated into the network layer. The two network layers are assumed to be two convolutional layers, and the residual structure is shown in formula (1).

$$y = F(x, \{W_i\}) + x \tag{1}$$

$$\frac{\partial loss}{\partial x} = \frac{\partial loss}{\partial y} \frac{\partial y}{\partial x} = \frac{\partial loss}{\partial y} \left(1 + \frac{\partial}{\partial x} F(x, \{W_i\}) \right) \tag{2}$$

Where x, y represents the input and output of the network, the function $F = W_2\sigma(W_1x)$ represents the residual mapping relationship, σ represents the activation function Relu, and the operation of $F + x$ is completed by "short-circuit chain" and element addition.

After passing through the Reproduce layer, Inception+ is a multi-scale and multi-channel structure block based on the improvement of GoogLeNet. The specific structure of Inception+ is shown in Fig. 1. First, after channel segmentation is performed on the output of the previous layer, two parallel twin convolutional layers are passed in, and the size of the convolution kernel is. The role of the twin convolutional layer and the Reproduce layer here is different, and no copy operation is performed. After the parallel convolution, the width of the network is expanded again by fusion, and the features of different levels are further mined. The result of the fusion is put into 3 asymmetric parallel convolutional networks as Fig. 1.

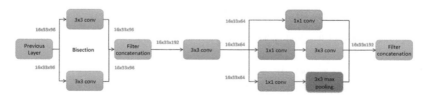

Fig. 1. Inception+ internal structure diagram

3 Self-deduction Model Based on MCTS

Wargame deduction is a turn-based game process, and the branch nodes of the game tree represent various situation states generated during the deduction process. Simulation based search is a common method to solve game decision problems. The simulation process is based on the reinforcement learning model for sampling to obtain sample data, but these data do not have a real effect on the environment; the search process uses the simulated sample results to guide the model to choose appropriate actions to maximize long-term benefits. The game tree is a top-down forward process, which expands all possible actions of a certain state node S_t. A subtree with the root node S_t can be constructed. This subtree can be regarded as an MDP. Solving the most valuable actions is the problem to be solved by forward search [9, 10].

Due to the large space of actions in wargaming, simple forward search takes a lot of time to complete, which is difficult to solve the requirement of decision-making time during wargaming. The Monte Carlo search method is based on a reinforcement learning model M_v and a simulation strategy π. For each possible sampled action $a \in A$, where A represents the action space, round sampling is performed to obtain the group K state sequence (episode) corresponding to the action, as shown in formula (3):

$$\left\{S_t, a, R^k_{t+1}, S^k_{t+1}, a^k_{t+1}, \ldots\ldots S^k_T\right\}^K_{k=1} \sim M_v, \pi \tag{3}$$

In the formula, R represents the reward in reinforcement learning. For each (S_t, a), the value function of its action and the choice of the optimal action are shown in formulas (4, 5):

$$Q(S_t, a) = \frac{1}{K} \sum_{k=1}^{K} G_t \tag{4}$$

$$a_t = \arg\max_{a \in A} Q(S_t, a) \tag{5}$$

In MCTS, the corresponding state sequence is shown in formula (6):

$$\left\{S_t, a^k_t, R^k_{t+1}, S^k_{t+1}, a^k_{t+1}, \ldots\ldots S^k_T\right\}^K_{k=1} \sim M_v, \pi \tag{6}$$

After sampling, a search tree of MCTS is constructed based on the sampling results, and the corresponding actions of and the maximum $Q(s_t, a)$ are calculated $Q(s_t, a)$ approximately, as shown in formulas (7, 8):

$$Q(S_t, a) = \frac{1}{K} \sum_{k=1}^{K} \sum_{u=t}^{T} 1(S_{uk} = S_t, A_{uk} = a)G_u \tag{7}$$

$$a_t = \arg\max_{a \in A} Q(S_t, a) \tag{8}$$

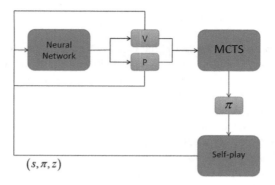

Fig. 2. Self-deduction structure diagram

After MCTS repeats the above search process N times, it selects the action branch of the root node, as shown in formula (9):

$$\pi(a|s) = \frac{N(s,a)^{1/\tau}}{\sum_b N(s,b)^{1/\tau}} \tag{9}$$

The input of the neural network is the description s of the current situation, and the output is the probability p of each feasible action in the current situation and the winning rate v of the current situation. What is used to train the strategy value network is a series of data (s, π, z) collected in the self-play process. According to the neural network training process in Fig. 2, the goal of model training is to make the action probability p output by the neural network closer to the output probability π by the MCTS, so that the position winning rate output v by the network can more accurately predict the real game result z.

4   Simulation Analysis

In order to verify the effectiveness of the improved residual network proposed in this article, select 200 rounds of deduction data from human players (the top 32 in the competition) in the "Urban Residential Area 8vs8" scenario, with a total of 64,000 position states, and train the improved network and Resnet respectively. The basic learning rate is set to 0.001, the maximum number of iterations is 80,000 times, the Batch size is set to 128, and the learning rate is attenuated by 0.1 times at each iteration of 15000 times. As is shown in Fig. 3, the parameters are updated by the gradient descent method, and the training of the two models is set according to the above hyperparameters.

Use the collected deduction data of 1027 rounds of "Urban Residential Area 8vs8" as a sample set to supervise and train the improved model. The training samples are the deduction data of the top 32 contestants in training and competition, and 100 rounds of data are randomly selected as the test Set, the training time is set to 7 days, and the

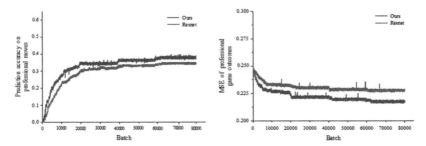

Fig. 3. Improved networks and Resnet training effect comparison diagram

learning rate is decayed every two days. At the same time, another network model with the same structure was trained through self-deduction, network parameters were initialized, and training was ensured from scratch. MCTS performed 500 simulations at each step, and 7 days of self-deduction produced 3125 rounds of deduction data. During the training process, the change of the loss function is shown in Fig. 4.

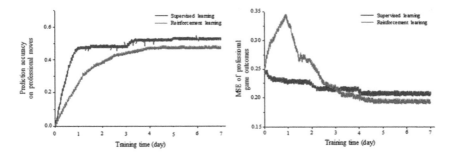

Fig. 4. Comparison of the effects of supervised learning and self-deduction training

5   Conclusion

In complex game problems, under the framework of reinforcement learning, DNN are used to construct strategies and value functions, and MCTS is used as the main reasoning process. The improved residual network is suitable for the learning of wargame situation information and the exploration of strategy space, can reduce the time overhead of simulation search, and has a certain auxiliary effect on the improvement of decision-making level.

References

1. Silver, D., Hubert, T., Schrittwieser, J., et al.: Mastering chess and shogi by self-play with a general reinforcement learning algorithm. Preprint arXiv:1712.01815 (2017)

2. Stanescu, M., Barriga, N.A., Hess, A., et al.: Evaluating real-time strategy game states using convolutional neural networks. In: Computational Intelligence and Games, pp. 1–7. IEEE (2016)

3. Barriga, N.A., Stanescu, M., Buro, M.: Combining strategic learning with tactical search in real-time strategy games. In: Magerko, B., Rowe, J.P. (ed.) Proceedings of the Thirteenth AAAI Conference on Artificial Intelligence and Interactive Digital Entertainment, pp. 9–15 (2017)

4. Heinrich, J., Silver, D.: Deep reinforcement learning from self-play in imperfect-information games. Preprint arXiv:1603.01121 (2016)

5. Mizukami, N., Tsuruoka, Y.: Building a computer Mahjong player based on Monte Carlo simulation and opponent models. In: Computational Intelligence and Games, pp. 275–283. IEEE (2015)

6. Tammelin, O., Burch, N., Johanson, M., et al.: Solving heads-up limit Texas Hold'em. In: International Conference on Artificial Intelligence, pp. 645–652. AAAI Press (2016)

7. Brown, N., Sandholm, T.: Safe and nested subgame solving for imperfect-information games. In: Advances in Neural Information Processing Systems, pp. 689–699 (2017)

8. Brown, N., Sandholm, T.: Reduced space and faster convergence in imperfect-information games via pruning. In: Proceedings of the 34th International Conference on Machine Learning, vol. 70, pp. 596–604 (2017)

9. Vinyals, O., Ewalds, T., Bartunov, S., et al.: StarCraft II: a new challenge for reinforcement learning. Preprint arXiv:1708.04782 (2017)

10. Vinyals, O., et al.: Grandmaster level in StarCraft II using multi-agent reinforcement learning. Nature **575**, 1–5 (2019)

Application Prospect of Film and TV Special Effects Synthesis Technology Under Digital Media Technology

Dong Wang[✉]

School of Digital Arts and Design, Dalian Neusoft University of Information, Dalian, Liaoning, China
wangdong_ys@neusoft.edu.cn

Abstract. In the era of digital media, new media art has been perfectly integrated with the film and television industry. Vigorously developing digital media technology synthesis technology will help improve the quality of film and television works, and film and television special effects synthesis technology will bring unprecedented visual shock to film and television production. This research mainly discusses the application prospects of film and television special effects synthesis technology under digital media technology. Through the impact of digital media technology and film and television special effects synthesis technology on film and television reputation, the economic benefits of digital media technology on the film and television industry, and the application of digital media technology and film and television special effects synthesis technology in the process of film and television production, discuss the film and television under digital media technology. Application prospects of special effects synthesis technology. During the production of "Wandering Earth", the proportion of special effects synthesis technology in the early film and television production process is 70%, and the proportion of film digital technology in the post film and television production process is 55%. This research helps to improve the quality of my country's film production.

Keywords: Digital media technology · Film television special effects synthesis technology · Application prospects · Film and television reputation digital media technology · Film and television reputation

1 Introduction

Digital media is the direction of media development. It has a profound impact on our lives and will continue to expand this impact in future development. Of course, digital media will not replace traditional media, but digitize traditional media and use digital means to serve the media.

The combination of the film and television industry and digital media technology has made the market economy a content that has attracted much attention. The industry can develop products related to film and television dramas according to market demand, promote the formation of the film and television industry chain, and expand the market economy, making online film and television Marketing, as a new growth

J. C. Hung et al. (Eds.): FC 2021, LNEE 827, pp. 237–243, 2022.
https://doi.org/10.1007/978-981-16-8052-6_27

pole of the cultural industry, combines the value of the market with P dramas to maximize economic benefits.

The film industry has participated in the production of special effects since its inception [1, 2]. With the rapid development of digital technology and the development of electronic computer technology, digital special effects synthesis has gradually been widely used in the film industry, which not only promotes the breakthrough development of the traditional film industry, but also opens up a new path for the future development of the film industry and direction [1, 3]. Digital technology was first widely used in the production of graphics and images, such as the production of advertisements, and then gradually introduced into the field of film and television. The maturity of digital special effects technology in the film industry can explain the problem. Digital technology has been successfully attempted to produce special effects for movies before this period, and digital technology gradually entered the field of film production during this period. Since then, special effects technology for film has been gradually promoted and applied to the various processes of film production. The control of the film's lens, post-production synthesis, etc. are inseparable from the participation of digital technology [4, 5]. At the same time, on the other hand, it is the most important feature of digital media technology, interactivity [6, 7]. In traditional movies, the story content, theme, and image style of a movie are basically fixed, reflecting the style of the director or the team, but under the influence of the era of digital media technology, the participation of the audience breaks all of this and makes The creators themselves have become more complicated [8, 9].

The film industry has participated in the production of special effects since its inception. With the rapid development of digital technology and the development of electronic computer technology, digital special effects synthesis has gradually been widely used in the film industry, which not only promotes the breakthrough development of the traditional film industry, but also opens up a new path for the future development of the film industry And direction.

2   Film and Television Technology

2.1   Film and Television Special Effects Synthesis Technology

In order to avoid putting the actors in a dangerous performance situation and state during the shooting process and to reduce the production cost of the film during the shooting process, it is more to use film special effects to achieve shooting scenes and movie plots that are impossible to achieve in the actual shooting process. As well as audio-visual effects, film special effects technology will be used to complete the film's picture when filming, making the audio-visual effects of the movie more exciting. In order to better link the relationship between the two in the film creation that combines reality shooting and digital CG, of course, it is inseparable from the camera motion control system.

2.2   Digital Media Technology

Although the media will influence people's way of thinking subtly, the media itself is also in constant evolution. The impact of digital media technology is first in the art of film, that is, it is changing the art of film. It has been more than a hundred years since the appearance of the Lumière brothers. For the definition of its identity, we are different from other arts. Film is an audiovisual art, its most important media attributes are audiovisual and audiovisual, but in the era of digital media technology, the nature of these two identity attributes cannot cover its characteristics, that is, the film in the era of digital media technology: not only an audiovisual art, It also has a variety of sensations such as touch, smell, cold and warm pain, and digital media technology has given many changes to traditional movies. With the development of digital technology, network technology and information technology, the trend of media digitalization has become more and more obvious. Since 2000, the emergence of web2.0 technology and the wide application of web2.0 platform have brought Internet technology to a new level. Under this platform, the interactivity is more prominent, allowing ordinary audiences who were originally in a disadvantaged position in the media communication process to participate more in the creation of Internet content. Web2.0 related technologies include social networking (SNS), P2P, RSS and so on. It is the interactivity provided by 2.0 technologies that provides a technical basis for ordinary users to comment, upload, and share, and also provide a technical basis for the birth and widespread dissemination of micro movies. Most of the special effects synthesis of movies is not placed in the post-production section, but the completed synthesis special effects are stored in the database before shooting. Through the pre-designed 3D photography system for real scene shooting and virtual shooting, the special effects synthesis database data is projected to synchronize with the shooting, which greatly reduces the special effects synthesis work in the later stage of film and television.

3 Experiment on the Development Prospects of Film and Television Special Effects

3.1 Three-Dimensional Special Effects and Film and Television Synthesis Special Effects in Digital Media Technology

The digital media technology of 3D special effects can make some scenes or even a single object be divided several times in the virtual scene, and the final synthesis can also be achieved by inputting 3D data and combining the ordinate. In an interactive three-dimensional special effects synthesis environment, the film can be freely changed angles, scenes, and three-dimensional deformations as needed.

Synthetic special effects are only used in the post-production and synthesis of movies. We can use a variety of special effects to build film and television scenes. The software for film and television post-production special effects synthesis learning includes Premiere, Photoshop, Maya (3DXMAX), Aftereffect, Shake, Difusion, Nuke, Avid, Luster, Boujou, Matchmove. The breakdown of traditional special effects and CG special effects is shown in Table 1.

Table 1. Breakdown of traditional special effects and CG special effects

Category	Ways of presenting	
Traditional special effects	Make-up, set scenes, firework special effects, early film	
CG special effects	Three-dimensional	Modeling, materials, lighting, animation, rendering
	Synthesis	Keying, wiping, coloring, compositing, scene view

3.2   Research Process

By studying the film "Wandering Earth", this research explores the impact of digital media technology and film special effects synthesis technology on film and television reputation, the economic benefits of digital media technology on the film and television industry, and the digital media technology and film special effects synthesis technology in the film and television production process. The application prospects of film and television special effects synthesis technology under digital media technology are discussed in three aspects.

4   Application Prospects of Film and Television Special Effects Synthesis Technology Under Digital Media Technology

4.1   Digital Media Technology and Film and Television Special Effects Synthesis Technology on Film Reputation

Digital media technology can bring a visual impact to movies. According to statistics, the output value of China's digital entertainment industry has grown from less than US$10 billion in 2018 to US$80 billion by the end of 2020, of which the film and television industry can account for about one-third. All kinds of online movies, high-end intelligent interactive video movies, smart phone movies, large arcade analog movies and other new forms of entertainment born under the application of a large number of new digital media technologies have profoundly affected our lives. The film "Wandering Earth" has a high proportion of its special effects synthesis technology in the early, mid and late stages, and it ranks first in the Douban movie score. Douban Koubei film reviews are shown in Table 2.

4.2   Digital Media Technology on the Economic Benefits of the Film and Television Industry

Digital media technology has brought huge economic benefits to the film and television industry. For example, the economic benefits of digital media technology in the movie "The Wandering Earth" are very impressive. According to statistics from Sina Yuyitong, between January 15th and 17th, 2019, the release of the first promotion song

Table 2. Douban's word-of-mouth film reviews

Rank	Douban ratings	Film
Top1	1259355 people	Wandering earth
Top2	1168718 people	I am not a medicine god
Top3	1121268 people	Farewell My Concubine
Top4	920232 people	Let the bullets fly
Top5	825812 people	The Great Sage of Journey to the West Marriage
Top6	685930 people	Operation Red Sea
Top7	681386 people	The richest man in Xihong City
Top8	679671 people	Infernal Affairs
Top9	676977 people	The unknown
Top10	667561 people	Journey to the West: Moonlight Treasure Box

"You Seed" MV caused the amount of information related to the film to rise rapidly, forming the first peak of public opinion during the monitoring period. As the release of the film approaches, the amount of information is also rising; as of February 2, 2019, the release of the second promotion song MV made its information volume reach another peak during the monitoring period. Netizens made this film work. Attention continues to rise, with the highest peak of the popularity index reaching 90% in the movie release. During January 2019, the amount of information on the entire network reached 3.96 million, and its related Weibo topic # The Wandering Earth # reached 3.222 million discussions and 480 million readings. Figure 1 shows the popularity index of the movie release.

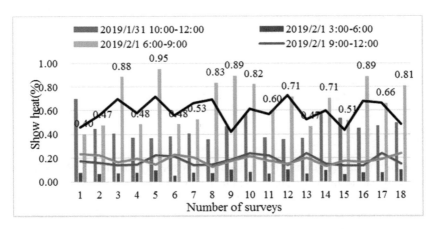

Fig. 1. Popularity index in movie releases

4.3   Digital Media Technology on Special Effects Synthesis Technology

Digital media technology and special effect synthesis technology are complementary in film production. Through the three-dimensional modeling method in digital media technology, the special effects of film and television in the film production process can be supplemented. As far as digital media technology and special effect synthesis technology are concerned, the ratio of film digital media technology to special effects synthesis technology in the early, middle and late stages of the film production process of "Wandering Earth" is shown in Fig. 2. In the early stage of film and television production, the proportion of special effects synthesis technology is 70%, and a large number of special effects synthesis is for better publicity of film and television works. In the post-production process, the proportion of film digital technology is 55%. In the post-production process, the filmmaker further modifies the film with special effects in order to improve the quality of film production.

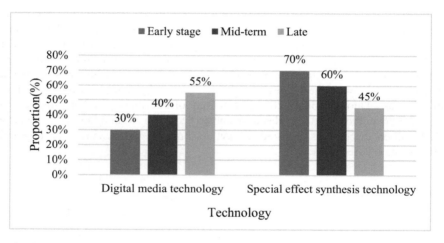

Fig. 2. The ratio of film digital media technology to special effects synthesis technology in the early, mid, and late stages of "Wandering Earth" film production

5   Conclusion

This research mainly discusses the application prospects of film and television special effects synthesis technology under digital media technology. Through the impact of digital media technology and film and television special effects synthesis technology on film and television reputation, the economic benefits of digital media technology on the film and television industry, and the application of digital media technology and film and television special effects synthesis technology in the process of film and television production, discuss the film and television under digital media technology. Application prospects of special effects synthesis technology. This research helps to improve the quality of film production in our country.

The rapid development of Chinese film brings opportunities as well as challenges for itself. The application of modern film production technology in the field of film special effects has improved the visual charm of commercial films, and is more convenient for the free expression of film forms. While bringing huge commercial benefits, it also promotes national culture to the world and promotes the international transformation of Chinese films.

References

1. Rosli, H., Kamaruddin, N.: Visitor experience's on digital media technology for the museum exhibition in malaysia: a preliminary findings. Int. J. Sci. Res. **7**(2), 245–248 (2020)
2. Chan, J.P.S., Yeung, J.H.Y., Wong, N.C.Q., et al.: Utilising digital media as enabling technologies for effective correctional rehabilitation. Safer Communities **18**(1), 30–40 (2019)
3. Yu, D.: Analysis of rural culture construction and innovation based on integration of digital media technology and paper material. Paper Asia **1**(9), 122–125 (2018). Smith, C.D., Jones, E.F.: Load-cycling in cubic press. In: Furnish, D., et al. (eds.) Shock Compression of Condensed Matter-2001, AIP Conference Proceedings 620, pp. 651–654. American Institute of Physics, Melville (2002)
4. Kamiyama, H., Takai, O.: Synthesis and film properties of carbon nitride film by unbalanced magnetron sputtering using pulse power supply. J. Surf. Finish. Soc. Japan **70**(3), 168–173 (2019). http://www.weld.labs.gov.cn
5. Lin, Z., Yunyun, L., Bin, C., et al.: Synthesis of antibacterial polyurethane film and its properties. Pol. J. Chem. Technol. **22**(2), 50–55 (2020)
6. Roggen, S.: CinemaScope and the close-up/montage style: new solutions to familiar problems. New Rev. Film Telev. Stud. **17**(7), 1–37 (2019)
7. Pratiwi, A.E., Elma, M., Rahma, A., et al.: Deconvolution of pectin carbonised template silica thin-film: synthesis and characterisation. Membr. Technol. **2019**(9), 5–8 (2019)
8. Saeedi-Jurkuyeh, A., Jafari, A.J.: Synthesis of thin-film composite forward osmosis membranes for removing organic micro-pollutants from aqueous solutions. Water Sci. Technol. **19**(3–4), 1160–1166 (2019)
9. Mazumdar, A.: Disney, others fend off copyright claims over motion capture tech. World Intellect. Prop. Rep. **32**(3), 20–21 (2018)

Upgrading Tourism Experience of Tourists by Using AR Technology

Zhen Gong[1], Danhong Chen[1(✉)], Zhaoxia Wen[1], Tianyu Yi[2], and Shiyu Zhang[1]

[1] College of Economy and Management, Shenyang Aerospace University, Shenyang, Liaoning, China
chendanhong@stu.sau.edu.cn
[2] Department of Investment and Financing, Shenyang DaSanHao Investment Co., Ltd., Shenyang, Liaoning, China

Abstract. Augmented reality (Augmented Reality, referred to as AR), is a technology that calculates the position and angle of the camera image in real time and adds the corresponding image. The goal of this technology is to set the virtual world in the real world and interact with each other on the screen. This technology was first proposed in 1990. With the improvement of the computing power of portable electronic products, augmented reality is more and more widely used. This paper will share and study the AR technology to enhance the tourist experience, and strive to do relevant technical support and reserve for upgrading the tourist experience.

Keywords: Augmented reality (AR) · Tourist experience · Image technology

1 Introduction

This is a new technology - enhances reality (AR), which can seamlessly integrate real-world information and virtual world information, and establish a related connection. It applies virtual information to the real world, through our well-known computers and other science and technology [1].

The reality is superimposed with virtual simulations so that virtual information can be perceived by human senses of real world. In order to transcend the realistic experience. In 2018, the number of investments in China's VR/AR industry was 109, and the investment amount was 10.977 billion yuan. From 2015 to 2016, China's VR industry "exploded", and the VR market once "let a hundred flowers blossom". In 2016, the number of VR/AR industry investment reached 299, which is the highest in history. However, from 2017 to 2018, the capital market gradually returned to rationality, and both the number and amount of investment in the industry declined. From the end of 2018 to 2019, with the approach of the 5G era, China's VR/AR industry capital market showed signs of recovery. From January to September 2019, the number of financing in China's VR/AR industry was 49, with the financing amount reaching 9.921 billion yuan [2–5].

According to online data and predictions by relevant agencies such as tourism agencies, the number of world tourists will grow at an annual rate of 3.8% from 2020,

J. C. Hung et al. (Eds.): FC 2021, LNEE 827, pp. 244–249, 2022.
https://doi.org/10.1007/978-981-16-8052-6_28

and it is estimated that by 2030, the number of world tourists will reach a staggering 1.8 billion. In addition, according to the data obtained by the World Tourism Association, In 2010, the number of global tourists and tourism revenue will grow at an annual rate of 4.3%, higher than the annual growth of world wealth (3%) in the same year. By 2020, the income of the tourism manufacturing will reach 16 trillion US dollars, countertype to 10% of the world's GDP; 300 million jobs will be provided, explaining for 9.2% of the world's total profession, further confirming the important position of the tourism manufacturing in the global thrift. However, there are few researches on AR technology in China [6].

In recent years, the use of AR technology in the tourism industry is also numerous, such as tourist path navigation, enhanced appreciation of AR scenic spots and other features, this paper will systematically analyze the related technologies [7].

2 Basic Principles of Augmented Reality

2.1 Augmented Reality

Early mobile augmented reality systems were based on the idea of adding images, sound, and other sensory enhancements to a real-world environment. For example, television networks use images to send messages around the world. Right? But all TV networks do is show still images, which cannot be adjusted as the camera moves, which is also one of its partial points. The augmented reality that this article is talking about is far more advanced than anything you see on TV, unlike traditional technologies like television. Because these systems can only display images that can be seen from one Angle. What VR is going to do is make the next generation of augmented reality systems capable of displaying images that everyone in the audience can see [8].

In all kinds of universities and high-tech corporations, augmented reality is still in the premier stage of research and development. Eventually, perhaps by the end of the decade, we will see the first augmented reality setups that have been put on the market in large quantities. One researcher called it "the Walkman of the 21st century". What augmented reality strives to achieve is not only to enhance images to the real environment in real time, but also to change these images to adapt to the rotation of the user's head and eyes, so that the image is always within the user's point of view [9].

2.2 Principle of AR Model

Assuming that u(n), x(n)is a stationary random signal, u(n) is white noise and the differentiation is Ó2, it is hoped that the correlation between the parameters of the AR model and the autocorrelation ak function of the x(n) can be established.

$$x(n) = -\sum_{k=c}^{D} c_k x(n-k) + u(n) \tag{1}$$

$$H(z) = \frac{1}{A(z)} = \frac{1}{1 + \sum\limits_{k=1}^{b} c_k z^{-k}} \tag{2}$$

$$P_x\left(e^{/c}\right) = \frac{\sigma^2}{\left|11 + \sum\limits_{k=c}^{y} c_k e^{-/\omega k}\right|^2} \tag{3}$$

Multiply both sides of the above Eq. (1) by x(n + m) at the same time, and find the mean, there are:

$$r_{1z}(m) = E\{x(n) * x(n+m)\} = E\left\{\left[-\sum_{k=1}^{0} c_k x(n+m-k) + u(n+m)\right] x(n)\right\} \tag{4}$$

So there are:

$$r_x(m) = -\sum_{k=\alpha}^{0} c_k E\{x(n+m-k)x(n)\} + E\{x(n+m)x(n)\} \tag{5}$$

So there are:

$$r_{1z}(m) = -\sum_{k=1}^{n} c_k r_{1z}(m-k) + r'_{1=c}(m) \tag{6}$$

There are:

$$r_{1z}(m) = E\{u(n+m)x(n)\} = E\left\{u(n+m)\sum_{k=0}^{\infty} h(k)u(n-k)\right\} \tag{7}$$

$$r_{1z}(m) = \begin{cases} -\sum\limits_{k=c}^{0} c_k r_{1x}(m-k)(\text{if} : m \geq 1) \\ -\sum\limits_{k=c}^{0} c_k r_{1x}(k) + \sigma^2(\text{if} : m = 0) \end{cases} \tag{8}$$

In the above derivation, the even symmetry of the autocorrelation function is applied, that is, so $r'_{1z}(m) = r_{1z}(-m)$, The above can be written as a matrix as follows:

$$\begin{bmatrix} r_{1x1}(0) & r_{1x}(1) & r_{1x}(2) & \cdots & r_{1x}(p) \\ r_{1x}(1) & r_{1x}(0) & r_{1x}(1) & \cdots & r_{1x}(p-1) \\ r_{1x}(2) & r_{1x}(1) & r_{1x}(0) & \cdots & r_{1x}(p-2) \\ \vdots & \vdots & \vdots & \vdots & \vdots \\ r_{1x}(p) & r_{1x}(p \cdot 1) & r_{1x}(p-2) & \cdots & r_{1x}(0) \end{bmatrix} \begin{bmatrix} 1 \\ a_1 \\ a_2 \\ \vdots \\ a_p \end{bmatrix} = \begin{bmatrix} \sigma^2 \\ 0 \\ 0 \\ \vdots \\ 0 \end{bmatrix} \tag{9}$$

3 AR Composition of Augmented Reality Technology

A set of closely connected real-time work hardware components and some related software systems constitute an enhanced reality system, which typically includes the following three forms [10].

3.1 Monitor-Based

In the AR implementation displayed by the computer display, first, by the captured actual image and enter the virtual scene generated by the computer, then generate the virtual scene generated in the computer graphics system database, finally output to the screen, display it to the screen.

The fulfillment of the monitor based augmented reality institution is shown in the following figure (Fig. 1).

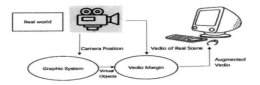

Fig. 1. Monitor-based schematic diagram

3.2 Optical Perspective

The method of displaying the wearable helmet is widely used in a virtual reality system, and the purpose is to enhance the user's visual imitation. According to the specific implementation principles, it has two forms: one is based on optical principles to penetrate HMD. The completion scheme of optical clairvoyant augmented reality system is shown in the following figure.

Simply, high-resolution and non-visual deviations are three advantages, high positioning accuracy, difficult delay matching, relatively narrow vision and high prices, etc., are the disadvantages of it (Fig. 2).

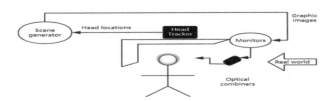

Fig. 2. Optical perspective schematic diagram

3.3   Video Perspective

Video perspective augmented reality system adopts penetrating HMD (Video See-through HMD) based on video synthesis technology, and the implementation scheme is shown in the figure (Fig. 3).

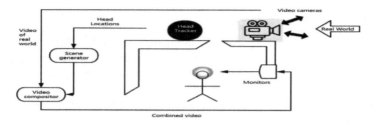

Fig. 3. Video perspective schematic diagram

4   Conclusion

The content of VR tourism will continue to increase with the pace of the times, which will also lead to continuous decline in investment and output equipment prices. Video display quality will gradually improve, and the practical application of software will become more powerful and easy to use. The application of AR technology will bring revolutionary changes to many fields. As an industry that uses this technology, tourism will inevitably have more possible development than other industries.

Acknowledgments. This research was supported by the college students' innovative entrepreneurial training plan of Shenyang Aerospace University under grant X202010143094. Danhong Chen is the corresponding author and instructor of this paper.

References

1. Cranmer, E.E.: Designing enhanced augmented reality tourism experiences: a multi-stakeholder approach. Int. J. Technol. Market. **13**, 3–4 (2020)
2. Kline, D.K., et al.: Arm motor recovery after ischemic stroke: a focus on clinically distinct trajectory groups. J. Neurol. Phys. Ther. **45**(2), 4–5 (2021)
3. Sven, B., Stefan, S., Maria, E.: Using augmented reality technology for balance training in the older adults: a feasibility pilot study. BMC Geriatr. **21**, 144 (2021)
4. Zheng, F.F.: Research on navigation system based on augmented reality technology. Softw. Navig. **9**, 57–59 (2016)
5. Zhang, L.B., Li, F.H.: Television technology of FPGA-based real-time video capture and remote transmission system (17), pp. 45–47 (2011)
6. Zhang, H.X., Jin, W.C., Zhou, G., Zhang, Y.M., Li, C.H.: Design of remote video transmission and real-time acquisition system, television technology (04), pp. 75–77 (2003)
7. Wu, Y., Wu, Y.Z., Cheng, J.F.: Real-time transmission and processing of video information over the network, minicomputer systems (08), pp. 78–81 (1999)

8. Huang, W., Chen, Y.M., Lu, B.H., Guo, J.: Real-time problem solving strategy, computer engineering and design (11), pp. 1920–1922 (2004)
9. Jiang, D.H., Tan, B.: Journal of the College of Real-time Display and Mapping of Terrain Scene in VR Technology **51**, 52–51 (2001)
10. Sun, L.Q., Li, R.F.: Real-time visibility of CAD models into VR models, computer engineering and applications (06), pp. 117–120 (2008)

Design of Cricket Control System Based on STM32

Jian Huang[(✉)]

Xijing University, Xi'an 710123, Shaanxi, China

Abstract. At that time, the 7670 plus warship with atom was used. It proved that it was difficult for warships to complete this topic. On the one hand, the 72 m dominant frequency was still slow for image processing. I tested to read and output a frame of QQVGA image, and it took nearly 30 ms after simple binary processing. In addition, the image read out from the FIFO of the 7670 is the image taken at the last time, which determines that the 7670 plus warship of STM32 is not suitable for cricket system.

Keywords: QQVGA · FIFO · STM32

1 Hardware

1.1 MCU

After my practice proved that the warship plus 7670 to do image processing is still very difficult, the completion of the design directly bought Apollo, main frequency 216 m, performance leverage drop. Everyone who used it said yes. If you change other cameras, warships can be competent. See below for cameras [1–4].

1.2 Camera

Using 7670, it takes a lot of time to read the image and has a headache in processing. It is not allowed to use the learning board in the competition. In addition, it is very easy for the camera to communicate with DuPont line, even if the atomic brother's camera extension line is used, the interference is also very serious. In general, the atomic camera is not suitable for cricket (personal opinion, if it is not appropriate, please give me some advice). Until the last day of the competition, my classmate found a camera called openmv. According to the seller of a treasure, it is dedicated to 2017 electronic games, and the output frame rate can reach 85 frames per second. I suddenly two eyes shine, but it's the last day, that kind of mood is like you look at hope in front of you humming a ditty slowly away, you can't catch [5–8].

Opemmv3 uses stm32f7 as the processing core, and the clock frequency can reach 216 m. Equipped with 7725, the output frame rate can be as high as 85 frames. In addition, the python interface is provided externally, which can complete a project with only a few lines of code. Finally, a variety of routines are provided, such as finding balls, color blocks, tracking, optical flow and so on.

J. C. Hung et al. (Eds.): FC 2021, LNEE 827, pp. 250–254, 2022.
https://doi.org/10.1007/978-981-16-8052-6_29

1.3 Motor

Before the game, a prediction post widely spread on the Internet said that this year's cricket control system is likely to come out, which is very accurate. However, the post recommended a motor - DC putter motor - which really made a lot of people suffer. The unit price is expensive, not to mention there is no feedback. PS: I'm also one of the victims (the key is that the list of components recommended by the organizing committee is also linear motor. I want to cry without tears!). The boss who bought the goods before the game was out of stock. After placing the order, he could receive a call from the boss to apologize that the goods were out of stock. Please return the order after the attempt, the final decision to use the steering gear, steering cycle for 20 ms, simple control, fast response. This time, I use mg946r metal gear steering gear instead of plastic. The metal has high torque and is stable [9, 10].

1.4 Little Ball

The ball is very important. Table tennis is too light and the center of gravity is not in the center of the ball. When we can't find a suitable ball, we can only use a kind of wooden ball similar to camphor ball to drive insects. (now Mr. Bi Shiyi bought a steel ball). By contrast, the steel ball is the most suitable choice, almost unaffected by the wind. But the steel ball is difficult to color, using the camera color block capture is not enough, can only use gray capture. Gray capture so play part of the fourth question do not know how to play, if you use color block capture, play part of the fourth question can use laser guidance, where to go. I use ordinary steel ball, and I want to draw a circle, but the effect is very poor, so I didn't do it. There is only one colored steel ball on a treasure, but it doesn't retail, so I can only ha ha.

1.5 Flat

It is suggested that you use acrylic plate, light and flat, glass is too thick. Acrylic board has colorless and white options. It depends on my personal preference. I take a transparent acrylic board, tear off the paper and dye the paper black with ink. In this way, I can track the ball by adjusting the gray threshold.

2 Mechanical Structure

Machinery is very important! Mechanical structure is very important! Mechanical structure is very important! How many teams are in the mechanical structure of the National Award. Firstly, the mechanical structure must be stable, and secondly, it must be flexible (Fig. 1).

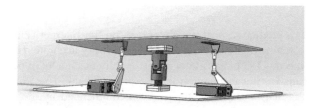

Fig. 1. Mechanical structure diagram

It's right to use a universal joint in the middle of this structure. I used it in the competition, and bisher uses it now. What I want to emphasize is the structure of connecting the two steering gears with plates. In this way, the connecting plates can only have one moving direction in this direction. As long as the moving plates of the two steering gears are deformed. If you change to three universal joints, you can link three. But the universal joint and acrylic are not well connected.

Part of the drill is inserted into a hollow wire or a screw. The fixed part is cut acrylic board and glued with acrylic glue. The viscosity of acrylic glue is awesome. In this way, when the board moves up and down, the board will not deform. My board area is not in line with the requirements of the topic, the surface did not draw circles in accordance with the requirements of the topic, this is mainly due to the small size, easy to move, big board is not interesting, in addition to occupying space!

3　Software

The software part is two PID parameter correction, here is the servo system, basically cannot use the micro component. PD is also very easy to find. I'm afraid this is the simplest control problem in the electric games in recent years. As long as we can set the ball in the middle, the basic part can be completed. Just change the setting position, and the play part 1 and 3 are no problem. It's interesting to ask the second question of the play part. Let's give you a detailed introduction to the second question of the play part. The second question is an automatic routing algorithm. My idea is to add four buffers to the nine areas required by the topic to accomplish the task.

When the serial port reads the camera output, the serial port assistant is used to observe the received data. If the camera finds multiple targets, the coordinate positions will be transmitted through, and the external bracket and bracket are used. The data length of each frame is inconsistent, so I compared the gray block found on the basis of the original code, and only the one with the largest area was sent. In addition, if the center coordinate of the color block is one or three digits, only one byte of the number will be sent in the jasson data string, so that the length of the received data will change when the serial port sees. I didn't understand how to use Jason decoding, but I just downloaded its keil codec library, which is in the attachment. I set up the area of interest to be 10–99.

```
int main (void)
{
    u8 key,i,buff[10];
    u8 *str=0;
    Stm32_Clock_Init(432,25,2,9);
    LED_Init();
    MPU_Memory_Protection();
    SDRAM_Init();
    LCD_Init();
    Remote_Init();
    LSS(30,50,200,16,16,"Apollo STM32F4/F7");
    LSS(30,70,200,16,16,"HaiNan university");
    LSS(30,90,200,16,16,"Cricket control system");
    LSS(30,110,200,16,16,"Writer:Liu Yifan");
    LSS(30,130,200,16,16,"Guide teacher:Yi Jafu");
  LSS(30,150,200,16,16,"KEYCNT:");
  LSS(30,170,200,16,16,"SYMBOL:");
    LSS(30,190,200,16,16,"Current Mode:");
    LSS(30,210,200,16,16,"Program State: ");
    LSS(30,230,200,16,16,"Current Rol position:");
    LSS(30,250,200,16,16,"Current Pit position:");
    LSS(30,270,200,16,16,"Set Rol position:");
    LSS(30,290,200,16,16,"Set Pit position:");
    LSS(30,310,200,16,16,"FPS:");
    for(i=0;i<10;i++)
switch(mode)
                {
                    case 0:    mode_0(); break;
                    case 1: mode_1(); break;
                    case 2: mode_2(); break;
                    case 3: mode_3(); break;
                    case 4: mode_4(); break;
                    case 5: mode_5(); break;
                    case 6: mode_6(); break;
                    case 7: mode_7(); break;
                    case 8: mode_8(); break;
                    default:break;
                }

                LSC(168,270,(u32)rol_setpos/10+0x30,16,0);
                LSC(176,270,(u32)rol_setpos%10+0x30,16,0);
    //              LSC(168,290,(u32)pit_setpos/10+0x30,16,0);
                LSC(176,290,(u32)pit_setpos%10+0x30,16,0);              //
                if(PASSBY_TIME >= 30)
                {
                    LSS(30,330,200,16,16,"OK!");
                }
                else
                {
                    LSS(30,330,200,16,16,"NO!");
                }
                LSC(30,350,PASSBY_TIME/1000+0x30,16,0);
                LSC(38,350,PASSBY_TIME/100+0x30,16,0);
                LSC(46,350,PASSBY_TIME/10+0x30,16,0);
                LSC(54,350,PASSBY_TIME%10+0x30,16,0);

}
```

4   Conclusion

After the analysis above, I think this question is no longer difficult for you? I hope that the small partners who are preparing for the electric games can achieve the ideal results. If a small partner wants to practice setting PID parameters with this problem, it is not recommended by individuals. PID parameters of this topic are very easy to set, as long as you understand the meaning of each parameter can be set out. To hone the ability of self-tuning PID parameters, it is suggested to do wind swing (15-year national race), inverted pendulum (13 years national race), flat pendulum (11 year national race). All of these topics are very good.

References

1. Siqi, Z.: Research on comprehensive improvement technology of Hall sensor motor speed measurement. Micro Spec. Motors **46**(5), 31–34 (2018)
2. Han, R., Guo, Y., Zhu, L., He, Q.: Improved speed measurement method of Brushless DC motor based on Hall sensor. Instrum. Technol. Sens. **10** 115–117 (2017)
3. Du, Y., Song, L., Wan, Q., Yang, S.: Accurate real-time speed measurement of photoelectric encoder based on wavelet transform. Infrared Laser Eng. **46**(5)1–6 (2017)
4. Wang, H, Hu, J., Wang, S.: Incremental photoelectric encoder angular displacement fitting velocity measurement method. Instrum. Technol. Sens. **10**, 99–101 (2014)
5. Zhao, S., Xiao, J., Guo, Y.: Motor speed data processing method based on Improved Kalman filter. Micro Spec. Motors **46**(9), 80–82 (2018)
6. Han, T.: Design of DC motor PWM closed loop control system based on MC9S12XS128. Mach. Tools Hydraul. **44**(7), 109–111 (2016)
7. Hua, Q., Yan, G.: Measurement of motor speed based on least square method. Electr. Drive **5**(12), 73–76 (2015)
8. Liu, Q., Zhang, R., Du, Y., Shi, L.: Research on speed measurement of long primary bilateral linear induction motor. Power Electron. Technol. **49**(5), 59–60 (2015)
9. Fu, Y., Sun, D., Liu, Y.: Speed signal estimation of low speed servo system based on Kalman filter. Appl. Motor Control **42**(5), 17–22 (2015)
10. Lei, W., Huang, C., Li, J.: Indirect speed measurement of DC motor based on morphological filter and center extreme difference. Meas. Control Technol. **34**(3), 17–20 (2015)

Legal Issues in the Development of Rural E-commerce in Chinese the Information Age

Qiuping Zhang[1($\boxtimes$)] and Leilei Li[2]

[1] Heilongjiang Academy of Agricultural Sciences Postdoctoral Programme,
Harbin 150086, Heilongjiang, China
[2] Harbin University of Commerce, School of Economics, Harbin 150028,
Heilongjiang, China

Abstract. The online sales of agricultural products are one of the important models in rural E-commerce, and it is also an important way to increase farmers' income and revitalize rural areas in China. However, there are many legal issues in the current E-commerce of online sales of agricultural products. These problems make it impossible to guarantee the various links in agricultural products. This article discusses relevant legal issues encountered in the process of online sales of agricultural products and on this basis, there are many suggestions for farmers to trade agricultural products on some trading platforms with effective laws and regulations. The purpose is to provide a certain reference for the smooth development of agricultural products in rural E-commerce.

Keywords: Rural E-commerce · Online sales of agricultural products · Legal issues

China's rural E-commerce has development rapidly like mushrooms after a spring rain. Two main forms of rural E-commerce have emerged in China: one is entry of industrial products that provide more convenience in rural areas, and the other one is the online sales of agricultural products that provide farmers with more opportunities to increase income. In the process of online sales of agricultural products relying on E-commerce, E-commerce companies with farmers as the main E-commerce operators encountered many legal problems. Therefore, it is meaningful to analyze the legal problems encountered in the process of online sales of agricultural products, such can promote the development of rural E-commerce.

1 Legal Issues Encountered in the Online Sales of Agricultural Products

1.1 The Legal Awareness of Rural E-commerce Operators is Weak

The main business entities of rural E-commerce are family-based farmers, rural cooperatives composed of farmers and rural enterprises [1]. Among them, farmers and rural cooperatives are different from rural enterprises in the operation of rural E-commerce. Because the farmers and rural cooperatives are not registered in the industrial and commercial department, and they lack legal awareness in the operation process, so they

J. C. Hung et al. (Eds.): FC 2021, LNEE 827, pp. 255–259, 2022.
https://doi.org/10.1007/978-981-16-8052-6_30

cannot use effective legal tools to maintain their own legitimate rights and interests. When legal disputes occur during the transactions, they cannot be reasonably resolved through legal methods. At the same time, since farmers and rural cooperatives do not have legal personality, it is also difficult for the industrial and commercial departments to effectively supervise them to ensure orderly operation of rural E-commerce [2].

1.2   Agricultural Products Lack Legal Protection

In rural E-commerce, some agricultural products traded are fresh, fruits and vegetables. These products are based on their small garden and cannot be effectively certified for quality. At the same time, due to the weak legal consciousness of some farmers, the phenomenon of "shitty good" and no trace-ability will appear in the process of agricultural products trading [3]. In addition, transaction products by E-commerce are surplus products from farmers'. These products do not have a unified price and a unified brand and so on. As a result, some farmers do not follow the market rules and it is easy to lead to chaos in agricultural product E-commerce platform transactions. Moreover, the relevant departments cannot effectively protect it.

1.3   The Agricultural Product Trading Platform Has Legal Risks

Compared with traditional agricultural products transactions, the use of E-commerce for transactions breaks the time and space restrictions between agricultural product sellers and consumers [4]. Therefore, in the process of rural E-commerce transaction process, in addition to the traditional thirty-party trading platforms such as Taobao, Pinduoduo, many proprietary trading platforms have begun to emerge. Meanwhile some other software applications also carry out online trading functions. However, the uneven quality of these trading platforms has caused some legal problems in the agricultural product E-commerce trading platform during the transaction. For example, If the platform is not compliant, it will not be able to effectively protect the rights and interests of both consumers and farmers [5]; there are too many malicious competitions among trading platforms; the payment methods used by various trading platforms are different, making the interests of some merchants unable to be protected; regarding the after-sales service after the transaction is completed, some platforms cannot provide effective services, such as delayed delivery, which has caused some legal disputes.

1.4   There is Few Laws Related to Rural E-commerce

The online sales of agricultural products in rural E-commerce is different from traditional E-commerce [6]. Because agricultural products have obvious characteristics, they are not included in the specific regulations in the "E-Commerce Law". For example: how to deal with the deterioration of agricultural products during transportation, how to conduct quarantine and inspection for agricultural and sideline products of poultry, and how to conduct quality control for the initial processing of some agricultural products [7]. However, some farmers and rural cooperatives mainly carry out quality control through non-standard forms such as word of mouth and trace-ability. In fact, the form cannot be unified. At the same time, there are a wide variety of agricultural products and the quality of agricultural products will be affected by many factors such as climate, soil, and water quality. This makes it more difficult to implement uniform laws and regulations for unified supervision.

2 Suggestions on Improving the Legal Protection of Agricultural Products

2.1 Improving the Legal Literacy of Rural E-commerce Employees

At this stage, most of the trading models adopted of the online sales of agricultural products in rural E-commerce are mainly sales farmer-based sales entities, enterprise-based trading platforms, and consumers urban residents-based consumers, that is, a C2C-based model. In this model, in addition to standardized E-commerce companies, neither farmers nor urban residents have the interpretation and understanding of the laws related to E-commerce in the process of E-commerce transactions. Coupled with the special nature of agricultural products, this kind of C2C model will encounter more legal disputes in actual operation. Therefore, it is imperative to improve the legal literacy of rural E-commerce personnel. In other words, farmers continue to increase their legal knowledge in the process of engaging in E-commerce [8], so as to effectively regulate their behaviors, standards, and procedures for engaging in rural E-commerce products. Continuously improve the legal awareness of urban resident when they purchase products through E-commerce to effectively protect their legal rights. And continuously enhance the standardization of the trading platforms through legal means to maintain the order of rural E-commerce [9].

2.2 Increasing the Legal Certification Method of Agricultural Products

For agricultural product quality certification methods, the traditional word-of-mouth certification should be discarded. Especially there is fierce competition brought by a large number of the current E-commerce platforms. In addition, due to the differences in agricultural products caused by the regions, climates and varieties, it is more necessary to standardize the certification methods of agricultural products [10]. Although there are certain difficulties, the certification of agricultural products can be completed through many methods such as the certification of farmer qualification, the certification of soil quality, and the trace-ability of the entire planting of agricultural products; the certification methods for the same types of agricultural products should be unified. Make the certification content and process transparent by formulating relevant regulations and rules. Only in this way, we can make the similar agricultural products comparable. Therefore, increasing the legal certification method for agricultural products is a key move to improve the online sales of agricultural products in rural E-commerce.

2.3   Regulate the Trading Platform Through Laws

Although different types of trading platforms have their own trading orders and rules, it is still necessary to regulate the qualifications of E-commerce farmers on each platform by laws. This can prevent shoddy farmers from entering the platform. In order to avoid malicious competition on the same platform, standardized regulations can be used to regulate the marketing model of merchants. Regulate the dispute settlement of merchants and consumers through relevant standards to improve consumer confidence. At the same time, self-operated E-commerce platforms and social media-based sales models should also strengthen legal regulations, so that the rights and interests of both farmers and consumers are protected.

2.4   Promulgating Relevant Laws and Regulations on Rural E-commerce

Different regions and departments should formulate laws or regulations related to rural E-commerce according to the characteristics of agricultural products in each region, and even refine them to different types of products. This standardized management can ensure product quality, expand sales, and reduce disputes. At the same time, relevant departments should also issue relevant laws and regulations on the online sales of agricultural products based on the type of platform. The ultimate goal is to protect the interests of farmers, increase their confidence in rural E-commerce, and truly achieve the purpose of increasing income through E-commerce.

3   Conclusions

In order to ensure that the online sales of agricultural products become an important magic way for farmers' income increase and rural revitalization, it is necessary to recognize the legal issues in the operation of rural E-commerce. Only the contents of farmers, trading platforms and agricultural products are regulated through legal methods, we can truly achieve the vigorous development of agricultural products upward model in rural E-commerce. Therefore, the current important measures to ensure the smooth development of the online sales of agricultural products are the introduction of relevant laws and regulations on rural E-commerce, improving the legal literacy of rural E-commerce employees, increasing legal certification methods for agricultural products, and standardizing trading platforms.

References

1. Sriyadi, Akhmadi, H., Yekti, A.: Impact of agrotourism development on increasing value added of agricultural products and farmers' income levels (a study in Karangtengah, Bantul, Yogyakarta). E3S Web Conf. **232**(232), 02–13 (2021)
2. Rui, S., Carolina, H., Ricardo, R., Elliot, R.: How to serve online consumers in rural markets: Evidence-based recommendations. Bus. Horiz. **63**(3), 351–362 (2020)

3. Cristobal-Fransi, E., Montegut-Salla, Y., Ferrer-Rosell, B., Daries, N.: Rural cooperatives in the digital age: an analysis of the Internet presence and degree of maturity of agri-food cooperatives' e-commerce. J. Rural Stud. **74**(74), 55–66 (2020)
4. Meyyappan, R.S., Nivash, T.N.G., Ariharan, J.: Rural e-commerce and last mile delivery: challenges and solutions in the indian context. Int. J. Manag. IT Eng. **9**(9), 164–170 (2019)
5. Jana, V., et al.: Digitizing of agricultural products sale towards the resilience of local communities to crisis situations. Agroeconomia Croatica **10**(01), 96–105 (2020)
6. Yoo, T.W., Oh, I.S.: Time series forecasting of agricultural products' sales volumes based on seasonal long short-term memory. Appl. Sci. **10**(22), 8169–8170 (2020)
7. Randy, K.A., Qibing, F.: Effect of good product design and packaging on market value and the performance of agricultural products in the Ghanaian market. Open Access Libr. J. **07**(09), 1–14 (2020)
8. Ibrahim, A.H., Purnomo, E.P., Malawani, A.D.: The most important agricultural products that sudan exports and the mechanisms to develop. Asian J. Agric. Ext. Econ. Sociol. **38**(08), 121–133 (2020)
9. Shrestha, A., Baral, S.: Consumers' willingness to pay for organic agriculture products: a case study of Nepalgunj city, Banke. Int. J. Agric. Environ. Food Sci. **3**(02), 58–61 (2019)
10. Proshchalykina, A., Kyryliuk, Y., Kyryliuk, I.: Prerequisites for the development and prospects of organic agricultural products market. Entrepreneurship Sustain. Issues **6**(04), 1307–1317 (2019)

Design of Electrical Automation Control System Based on Computer Technology

Lu Wang[✉]

School of Mechanical and Electrical Engineering,
Liaoning Jianzhu Vocational College, Liaoyang, Liaoning, China
wanglu@lnjzzyhyjdgchy.onexmail.com

Abstract. With the rapid development of globalization, information technology is also developing rapidly. Computer network is widely used in many fields. One of the key technologies is CT (Computer Technology). The application of CT to the design of EACS (Electrical Automation Control System) has been recognized by many experts and scholars, and many relevant personnel Work has begun on this issue. This paper discusses how to apply CT to the design of EACS, as well as the effect after application, analyzes the advantages of CT applied to EACS, and analyzes the design of EACS under CT by taking two power units of the same category as the research object Compared with the traditional technology and electrical automation system design advantages, and then the experiment shows that the former can improve the efficiency of the whole EACS. In the questionnaire about the satisfaction degree of employees in the two units, 35% of the participants in the experimental group are satisfied, and 40% of the participants are very satisfied, while the control group is not Only 25% of the group were satisfied. In this paper, in addition to expounding the influence of CT on the design of EACS, the development of this technology is also studied and discussed.

Keywords: CT · Electrical automation · Control system · Design research

1 Introduction

People's production and life are more and more inseparable from the power system, the power and social aspects of the contact is also more and more close, daily life on the power demand is also higher and higher. The main function of power system is to provide people with safe and sufficient power to meet the needs of people's daily life. With the advent of the Internet, followed by the emergence of CT. CT has also been applied to more fields. Relevant professionals should also apply CT to EACS, so as to transform its backward electrical system work. In the background of CT, EACS is becoming more and more intelligent, and the work efficiency and performance are also higher and better. Therefore, we discuss the design of EACS based on CT in order to better find a better way to improve work efficiency, improve production efficiency, enrich work forms, and meet the increasing demand for electric power in the current society [1].

Liu Ming believes that the development speed of all walks of life in our country is also very fast, the economy is constantly innovating, and the demand for industrialization is

J. C. Hung et al. (Eds.): FC 2021, LNEE 827, pp. 260–266, 2022.
https://doi.org/10.1007/978-981-16-8052-6_31

increasing, which also promotes the continuous improvement of the level of the whole social productivity, thus forming a virtuous circle. The design and application of EACS has a very positive impact on the improvement of industrialization level. Based on this technology, the EACS is highly integrated and more flexible [2]. At present, because of the increasing demand for electricity in the current society, local governments are also paying more and more attention to the power construction. Zhao libing believes that the current power enterprises should upgrade and optimize the existing EACS and technology, and update and improve the current equipment in the face of higher and higher power demand. With the application of CT in more and more fields, relevant professional and technical personnel should combine CT with EACS,to increase the production output, so as to meet people's demand for electric power [3]. The application of CT in the design of EACS can improve the efficiency and weight of EACS [4]. EACS can reduce the loss of power, but also can ensure the safe and stable operation of the power system, so as to bring people a better experience of using electricity and provide higher quality of power supply. Therefore, the optimization and upgrading of the existing EACS and the application of CT can improve work efficiency and quality, thus promoting the long-term and stable development of electric power enterprises [5, 6].

Using CT, we can manage and control the EACS, which is also an advantage of the EACS. The design scheme of EACS based on CT has also been significantly improved [7].The combination of CT and electrical automation has also achieved good results. Many industries have also begun to apply this technology. For example, many automobile factories now use this technology in the process of automobile production, which can carry out the simulation before production; so as to facilitate the staff to clarify in advance, timely measures should be taken to avoid the problems. At present, due to the application of computer in the design of EACS, the EACS is gradually becoming intelligent, which is conducive to all relevant industries to improve their work efficiency and quality, so as to further improve the social productivity level and raise the productivity level to a higher level [8–10]. It can not only increase the long-term production efficiency of the enterprise, but also improve the production efficiency of the production system.

2 Method

2.1 CT

CT not only refers to the general sense of technology, it contains technical methods and technical means, and these technologies and means belong to the field of computer. It also includes several technologies, first of all, software technology, and then hardware and application technology. CT and many disciplines are closely linked, so please have a strong comprehensive, and these disciplines include electronic engineering, applied material resources and modern communication technology, and also involve some knowledge in the field of mathematics. Its content is also very extensive, CT includes many aspects of technology, mainly four aspects of technology, one is computer system technology, the second is computer device technology, the third is computer assembly technology, and computer components technology. CT is a very complete system technology, and with its application in more and more fields, its development prospects and potential are very broad and huge.

2.2 Electrical Automation

Electrical automation is a new subject, and because of its strong practicability, this subject is also very popular. It includes electrical engineering and automation, which belongs to one of the fields of electrical information. Electrical automation is closely related to us and social production. Therefore, with the continuous improvement of industrialization, the development of electrical automation is also very rapid, and relevant departments and personnel have invested a lot of money and human resources in this research, so it is also relatively mature. Electrical automation in industrial production and people's daily life is also more and more, plays a very important role, has a very positive impact on China's economic development, has made a very great contribution. Electrical automation is now also applied to more and more industries. Because of its good development prospects, it also attracts a large number of talents to invest in this field, and its development will be more and more rapid.

2.3 Design Principles of EACS

The design principle of EACS is to be safe, reliable and easy to operate. Only under such a premise can we continue with our work. First of all, due to the development of technology, the EACS no longer uses the original traditional relay. Now the relay in the system is generally applied with new multi-functional relay, which can also make the program relatively simple. Then the good expansibility of the system is another principle of the EACS. When designing the EACS, we should consider all kinds of possible situations. For example, the scale of the system will be expanded, the development of the enterprise, the customer base is more and more, and the audience range is wider. All these possible situations require good expansibility. Finally, the EACS should have good compatibility, because there may be various situations in actual operation, and many facility components in the system are flexibly configured to facilitate the operation of the actual situation. Therefore, when designing the EACS, we should consider the principle of good compatibility of the system.

2.4 Formula Involved in the Experimental Part

From what we know above, we can know that compared with the traditional EACS, under the background of CT, EACS has more obvious advantages and better working effect. We use the methods of experiment and questionnaire survey when we compare the advantages and differences between the EACS and the traditional EACS combined with CT, and the results of these methods involve the following formulas:

$$D\overline{(X)} = \frac{D(X)}{n} \tag{1}$$

$$SD(\overline{X}) = \frac{SD(X)}{\sqrt{n}} \tag{2}$$

$$DX = \sum_{i=1}^{n} (x_i - EX)^2 pi \tag{3}$$

In the above formula, n is the number of samples, D is the sample variance, SD is the sample standard deviation, ex is the expected value, PI is the probability, Xi is the single sample, and X is the average number of samples.

3 Experiment

According to the previous content, we can apply CT to EACS, through data analysis experiments, to analyze and summarize the advantages of the design of EACS based on CT compared with the traditional EACS design. We can obviously find that the greatest advantage of applying CT to the design of EACS is to greatly improve the work efficiency, avoid the occurrence of errors as far as possible, save costs and bring greater benefits.

3.1 Selection of Experimental Objects

Taking two similar types of electric power enterprises as examples, this paper analyzes the situation of their EACS. Among them, the CT is adopted in the EACS of enterprise a, and the traditional technical method is used in school B.

3.2 Experimental Test Index

In this experiment, we conducted a questionnaire survey on the employees of the two enterprises by random interview, and compared and analyzed the results of the two groups of experimental data combined with the feedback.

3.3 Processing of Experimental Data

We usually use some mathematical methods in data analysis, such as the Bayesian formula of statistics. Bayesian formula is used to calculate the probability problem. Its mathematical significance is to estimate the probability of an event that you are not familiar with by virtue of the nature of things, so as to infer the probability of an unfamiliar thing. In mathematical language, it means that the greater the probability of something appearing, the greater the possibility of its occurrence. The formula is as follows:

$$P(A_i|B) = \frac{P(B|A_i)P(A_i)}{\sum j P(B|A_j)P(A_j)} \tag{4}$$

4   Result

4.1   Experimental Data Results

In this experiment, we analyze and compare the work results and quality brought by the EACS design of the two enterprises. The method adopted here is to mark the EACS work situation of the two enterprises in one year according to the relevant professional departments. The score is based on the percentage system. In this experiment, we record four groups of experimental results (every three of them) The final experimental results are shown in Table 1 and Fig. 1.

Table1. Comparison of variance results between the two enterprises

	Experiment 1	Experiment 2	Experiment 3	Experiment 4
Experiment group	80	83	79	86
Control group	71	63	60	72

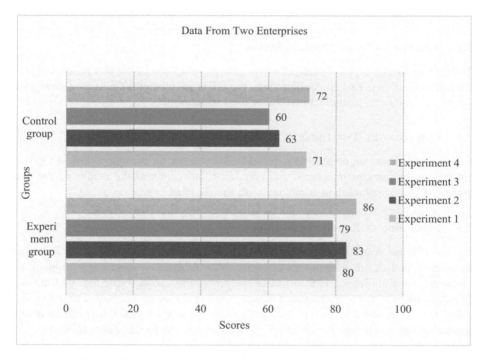

Fig. 1. Comparison of variance results between the two enterprises

From Table 1 and Fig. 1, we can see the working conditions of the EACS of the two power enterprises. We can see that the performance of enterprise a which adopts the EACS combined with CT is significantly higher than that of enterprise B which adopts traditional technical methods. Therefore, we can know that CT is used to design EACS It brings a lot of benefits. First of all, when designing, you can simulate the process first and make clear the situation of each process link, so as to avoid possible errors as far as possible, and greatly improve the efficiency and quality of enterprise power system work. It also shows that it is advisable to apply CT to the design of EACS, which is worthy of investment and research.

4.2   Employees' Satisfaction with the Two Technologies

In the questionnaire survey, we made statistics on the satisfaction degree of students in enterprise a who applied CT and enterprise B who applied traditional technology in the design of EACS. The results are shown in Table 2 and Fig. 2.

Table 2. How satisfied are the employees of the two enterprises with the two technologies

	Very satisfied	Quite satisfied	Dissatisfied	Unclear
Experiment group	35%	36%	17%	12%
Control group	22%	23%	30%	25%

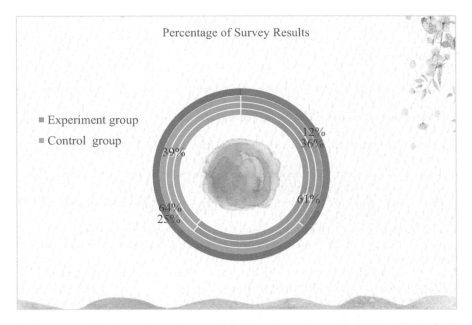

Fig. 2. How satisfied are the Employees of the Two Enterprises with the Two Technologies

From Table 2 and Fig. 2, we can see the satisfaction degree of the employees of the two enterprises to the two technologies. Among them, 35% of the staff in the experimental group were very satisfied, 36% of them were relatively satisfied, 17% of them were not satisfied, and 12% said that they were not clear. In the control group, 22% of the employees were very satisfied, 23% were satisfied, 30% were dissatisfied, and 25% were not clear. This shows that compared with the traditional EACS technology, employees prefer the experience of EACS combined with CT. In the combination of CT and EACS, the work efficiency and quality of enterprises are greatly improved, and at the same time, the work enthusiasm of employees is greatly improved and better work experience is brought. Therefore, the design of EACS based on CT is welcomed by enterprises and employees, and it also promotes the innovation and development of EACS. The development prospect of this area is also very broad.

5 Conclusions

In a word, with the continuous improvement of social level, people's demand for electric power is also increasing. The combination of CT and EACS helps to promote the intelligent development of EACS. When CT is applied to the design of EACS, it is conducive to the design scheme to be more perfect and the use of CT Advance link simulation can effectively clarify the process, improve work efficiency and avoid errors in actual work as much as possible. Therefore, it is of great practical significance to apply CT to the design of EACS, and the research of this area is also worthy of people's attention.

References

1. Zhang, S.: Design of EACS under the application of CT. Electron. Prod. **394**(08), 76–77 (2020)
2. Liu, M.: Design analysis of EACS under the application of CT. Software **039**(006), 170–173 (2018)
3. Zhai, L.: Design analysis of EACS under the application of CT. Digital World **174**(04), 249–250 (2020)
4. Chen, H., Lin, J., Luo, Y.: Research on synchronous control system of synthetic test. Electr. Drive Autom. **39**(003), 56–58 (2017)
5. Anjie: Design of EACS. In: Urban Construction Theory Research, Electronic edn. vol. 006, no. 008, pp. 3630–3631 (2016)
6. Yao, Z., Jiansu, G., Yuhan, Z.: Hardware design of dynamic closed loop autonomous tracking balance vehicle control system based on K60. Shandong Ind. Technol. **04**(4), 267 (2016)
7. Yixin, Z.: On the innovation of sports reality show under the background of Winter Olympics. News Res. Guide **010**(012), 97–99 (2019)
8. Ping, M., Kaichen, W., Zijun, L.: Design of temperature control system based on hardware in the loop simulation platform. Exp. Sci. Technol. **15**(005), 10–14 (2017)
9. Ming, T.: Design of laser beam self alignment electrical automation system based on FPGA. Comput. Meas. Control **025**(005), 240–243 (2017)
10. Zhang, Q., Zhang, P., Guan, Z.: New trend of media form change in 4G Mobile Internet era. Sci. Technol. Inf. 016(009), 19–20 (2018)

Islanding Detection Method of Active Phase-Shifting for Photovoltaic Microgrid Based on Fuzzy Control

Huaizhong Chen[(✉)], Jianmei Ye, and Riliang Xu

Zhejiang Industry Polytechnic College, Shaoxing, Zhejiang, China

Abstract. Island-detection is extremely important for grid-connection of photovoltaic microgrid and stable island operation. However, traditional island-detection methods have disadvantages of long detection time, low efficiency, low power quality, etc. An active phase shifting island-detection method based on fuzzy control is proposed in order to solve the problem. The advantages of fuzzy control and active phase-shifting island-detection method are combined in the method. Fuzzy control technology is utilized. The feedback parameters are adjusted by fuzzy self-adaptation based on the frequency deviation of the common connection point between the photovoltaic microgrid and the distribution network as well as the frequency deviation change rate. The effectiveness of the method is verified according to the simulation results. It can be used for accurately and quickly detecting the island state and reducing the influence of photovoltaic microgrid disturbance on the power quality.

Keywords: Microgrid · Islanding detection · Active phase-shift · Fuzzy control

1 Introduction

The island effect of photovoltaic microgrid refers that the public grid will be cut off due to various reasons after the distributed photovoltaic microgrid is connected to the grid, and the photovoltaic microgrid power supply still supplies a part of the load in the public grid at the time. Point of Common Coupling of microgrid and distribution network is also referred to as PCC. When it is switched from grid-connection to island, if the state of load change at the PCC switch cannot be detected quickly and effectively by the photovoltaic microgrid system, that is, the island state is detected. The microgrid is disconnected from the public grid within the specified time at the same time, personal safety of equipment and maintenance personnel will be endangered [1, 2].

The traditional active phase-shifting island-detection method can be applied to determine whether the island of the photovoltaic microgrid occurs or not by perturbing the relevant parameters in the inverter and detecting the changes in the parameters such as voltage amplitude, frequency, etc. The feedback coefficient is applied into the active phase-shifting island-detection method. It has the advantage of reducing the phase difference between the grid-connected current and the grid voltage, and improving the power quality of the grid-connected system. However, the value cannot be adjusted

J. C. Hung et al. (Eds.): FC 2021, LNEE 827, pp. 267–273, 2022.
https://doi.org/10.1007/978-981-16-8052-6_32

adaptively with the change of the voltage frequency of the common coupling point since the feedback coefficient must be a fixed value within the value range, thereby reducing the detection accuracy. An active phase shifting island-detection method based on fuzzy control is proposed in the paper, which has the advantages of fast response, high output control precision, etc. [3, 4].

2 Establishment of Mathematical Model of Island Mode

The circuit structure of the island mode of the photovoltaic microgrid is shown in Fig. 1 [5].

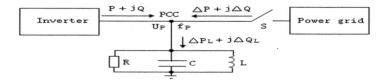

Fig. 1. Circuit diagram of photovoltaic microgrid island mode

In the Fig. 1, the local load is composed of resistor R, inductor L and capacitor C in parallel. R refers to the active load, while L and C represent the inductive and capacitive reactive loads respectively. The power grid supplies power to the local load through the common coupling point PCC. S refers to the PCC switch. The photovoltaic microgrid is operating in island mode when S is disconnected. The photovoltaic microgrid is operating in grid-connected mode when S is closed. Figure 1 shows the follows:

$$P_L = P + \Delta P \tag{1}$$

$$Q_L = Q + \Delta Q \tag{2}$$

$$P_L = \frac{V_P^2}{R} \tag{3}$$

$$Q_L = V_P^2 \left(\frac{1}{\omega L} - \omega C \right) \tag{4}$$

In the formula, V_P refers to the voltage at the common coupling point; P is the active power output of the inverter; Q is the reactive power output of the inverter power supply; P_L is the active power consumed by the load; Q_L is the reactive power consumed by the load; ΔP is the active power provided by the power grid; ΔQ is the reactive power provided by the grid. It is known from the formula that:

(1) Even if the distributed generation system is in an island operating state during power matching, the voltage and current waveforms do not change significantly at

PCC points due to $\Delta P = 0$ and $\Delta Q = 0$. The output current gradually increases when the active power P does not match. The output current frequency of the inverter decreases and the amplitude fluctuates when the reactive power Q does not match.

(2) The voltage at PCC is only related to the active power output by the photovoltaic grid-connected inverter and the R in the load after the occurrence of island, and the frequency at PCC is related to the reactive power output by the photovoltaic grid-connected inverter and the voltage at PCC.

3 Principle Analysis of Active Phase-Shifting Detection Method

The output current of the inverter is in the same frequency and phase with the power grid voltage when the photovoltaic grid-connected inverter and the power grid are in normal working state. However, when the active phase-shifting island-detection method is added, the phase thereof is not exactly the same because a small disturbance is added to the phase of the power grid voltage by the active phase-shifting algorithm. The voltage frequency at PCC point will continue to shift under the influence of inverter output current due to the existence of disturbance angle when island effect occurs. the disturbance angle can be expressed as follows at the point [6]:

$$\theta = \theta_m \left(\frac{\pi}{2} \bullet \frac{f - f_0}{f_m - f_0} \right) \tag{5}$$

In the formula: θ refers to the disturbance angle; f is the frequency at PCC point; f_m is the maximum frequency of the corresponding maximum phase Angle; f_0 is the rated frequency of the power grid; θ_m is the maximum phase-shifting angle.

The small change of current phase will make the voltage frequency decrease or increase continuously until it exceeds the frequency threshold due to the leading or hysteretic characteristics of the voltage waveform in the island state, therefore the island state can be detected.

It can be seen from the above formula that this algorithm needs one subtraction and three multiplication compared with a sinusoidal operation, the calculation process is complex and tedious, and the real-time performance is relatively low poor [7].

The method is further improved based on the method proposed in literature in order to further improve the detection speed and reduce the detection area [8]:

$$\theta = k(f - f_0) + \theta_0, f \geq f_0 \tag{6}$$

$$\theta = k(f - f_0) - \theta_0, f < f_0 \tag{7}$$

In the formula, k refers to the positive feedback coefficient so that the voltage phase angle difference at PCC changes in the same direction with the frequency drift; θ_0 refers to a small constant. It aims at introducing slight offset and accelerating the phase-shifting during network outage, thereby improving the detection efficiency to a certain extent. However, the method has the following defects: as a fixed value is adopted by the feedback coefficient within the value range, the value thereof cannot be adjusted adaptively with the change of the voltage frequency of the common coupling point, thereby reducing the detection accuracy. It is necessary to use fuzzy control to optimize and improve the existing island-detection technology [9].

4 Island Detection Method Based on Fuzzy Control

Slight changes in the current phase will lead to continuous increase or decrease of the voltage frequency until the frequency threshold is exceeded due to the leading or hysteretic characteristics of the voltage waveform when island occurs. A fuzzy optimization controller is designed to optimize the feedback coefficient k in the paper, and it can be used for quickly detecting the island state at PCC. It is ensured by the island detection method based on the fuzzy control technology that the frequency, voltage, phase angle and other relevant parameters remain normal and no distortion occurs when the island occurs in the photovoltaic microgrid, thereby realizing the seamless switching control of grid-connected/island mode [10, 11]. The working principle of island-detection based on fuzzy control technology is shown in Fig. 2:

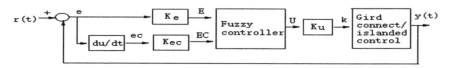

Fig. 2. Fuzzy control chart of islanding detection

In the Fig. 2, the deviation e of voltage frequency y(t) and the power grid voltage rated frequency r(t) at the PCC and the change rate du/dt of e are regarded as the input of the fuzzy controller. The feedback coefficient k is regarded as the output control quantity, Ke and Kec are regarded as quantitative factors of input, Ku is regarded as the output quantitative factor. The feedback coefficient k of the optimized fuzzy controller is used for detecting and monitoring the state of grid connection/island control.

The frequency deviation e ∈ [−0.5, 0.5] and the frequency deviation change rate ec ∈ [−50, 50]. The deviation value between the voltage frequency at PCC and the voltage frequency of the power grid changes positively due to the inductance of the load or negatively due to the capacitive change of the load within a power grid cycle are set. after the occurrence of island. The fuzzy domain of input e and ec is set as [−3, 3] and divided into seven symmetric fuzzy subsets, namely E = EC = {NB (negative big), NM (negative medium), NS (negative small), ZE (zero), PS (positive small), PM (positive medium), PB (positive big)}. The fuzzy theory domain is set as [0, 6] since the output quantity of the feedback coefficient K is a positive value correspondingly, which is also divided into 7 symmetric fuzzy subsets, namely, U = {ZE (zero), SS (relatively small), S (small), M (medium), BB (relative big), B (big) and VB (very big)}.

Whether the feedback coefficient k can be successfully detected in the island state or not in time in this algorithm is combined, the fuzzy control rule in the form of "if E and EC then U" is applied to obtain the output control rule of input E, input EC and output U based on the analysis of previous experimental data and expert experience.

Each fuzzy statement has a fuzzy relation corresponding to it. After the fuzzy statement is calculated, the fuzzy relation R corresponding to the whole control rule can be obtained. According to the fuzzy relation R, the fuzzy set output can be calculated, and then the weighted average method is used to deal with it. The fuzzy decision is transformed into a clear output, and then multiplied by the quantization factor, the output of the actual feedback coefficient k can be obtained.

Finally, the value is converted into analog quantity through the analog quantity module and D/A converter to detect and control the grid-connected/islet control unit, and quickly detect the island state at the photovoltaic microgrid PCC, thereby detecting the island state and executing corresponding control quickly and accurately.

5   System Simulation

A simulation model of island-detection is established in order to verify the effectiveness of the active phase-shifting island-detection method based on fuzzy control. The simulation model is mainly composed of an island-detection module, an fuzzy control module, etc. Model parameters are set as follows: input DC voltage 380 V, power grid voltage 220 V and frequency 50 Hz [12, 13]. The simulation of active phase-shifting island-detection based on traditional method is shown in Fig. 3, while the simulation of active phase-shifting island-detection based on fuzzy optimization control is shown in Fig. 4.

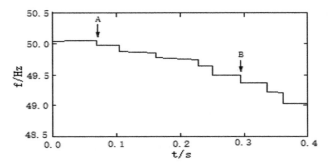

Fig. 3. PCC voltage frequency waveform based on traditional algorithm

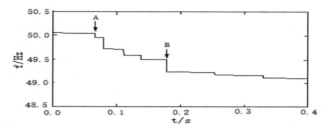

Fig. 4. Voltage frequency waveform of PCC based on fuzzy algorithm

In the figures, point A represents the time of islanding occuring, The time of islanding occuring is 0.065 s, and point B represents the time of islanding detection. It can be seen from Fig. 3 that the PCC frequency changes after the power is cut off. The traditional algorithm can be applied for detecting the transition of the photovoltaic microgrid from grid-connected state to island state in about 0.291s, and the detection time is 0.226 s.

It can be seen from Fig. 4 that after the optimization of fuzzy control technology, the optimization algorithm can be used for detecting the transition of photovoltaic microgrid from the grid-connected state to the island state in about 0.175 s, and the detection time is 0.11 s. It can be seen that the detection time is shortened aiming at the active phase-shifting islander detection based on the fuzzy optimization control algorithm, thereby improving the detection efficiency.

The detection time of the new islanding detection method is 0.116 s shorter than that of the traditional islanding detection method, that is, once the power grid is disconnected, this method can detect the islanding effect more quickly, and reduce the current harmonic distortion, so as to improve the output power quality of grid connected inverter.

6 Conclusion

An active phase-shifting islet detection method for photovoltaic microgrid based on fuzzy control is proposed in the paper in order to solve the problems of long island detection time and low efficiency of the traditional active phase-shifting algorithm, and the fixed feedback coefficient in the detection method is optimized into the adaptive adjustment coefficient. The feedback coefficient is adjusted adaptively according to the frequency change amplitude at PCC of the photovoltaic microgrid. The method not only has the advantages of traditional active phase-shifting algorithms in island detection, but also has the adaptability and flexibility of fuzzy control. The experimental results show that the method has good rapidity and stability. It can respond to the changes of grid-connected/island mode of photovoltaic microgrid as well as the external environment. The island state of photovoltaic microgrid can be quickly detected, and the overall performance of the system can be improved.

Acknowledgements. This research was supported by Zhejiang Provincial Basic Public Welfare Research Project of China under Grant No. LGG21F030001.

References

1. Zhou, L.H., Gao, F.: Islanding detection method based on MPPT. Dianji yu Kongzhi Xuebao/Electr. Mach. Control **22**(9),7–14 (2018). (in Chinese)
2. Shouxiang, W., Sijia, C.: Distributed generation hosting capacity evaluation for distribution systems considering the robust optimal operation of OLTC and SVC. IEEE Trans. Sustain. Energy **7**(3), 1111–1123 (2017). (in Chinese)
3. Jia, K., Zhu, Z.: Intelligent islanding detection method for photovoltaic power system based on Adaboost algorithm. IET Gener. Transm. Distrib. **14**(18), 3630–3640 (2020)
4. Liu, Z., Su, C., Høidalen, H.K., Chen, Z.: A multiagent system-based protection and control scheme for distribution system with distributed-generation integration. IEEE Trans. Power Deliv. **32**(1), 536–545 (2017)
5. Xie, Q.: Study on the anti-islanding detection based on energy storage inverter. Master thesis Submitted to University of Electronic Science and Technology of China, pp. 21–25 (2014). (in Chinese)
6. Fang, J., Sun, C.: An improved active phase-shift islanding detection method based on fuzzy control. Electr. Drive **47**(5), 71–74 (2017). (in Chinese)
7. Chen, Y., Zheng, S.: Improvement of active phase-shifting islanding detection algorithm in photovoltaic grid-connected systems. J. Syst. Simul. **25**(4), 748–752 (2013). (in Chinese)
8. Liu, F., Kang, Y.: Improved SMS is-landing detection method for grid-connected converters. IET Renew. Power Gener. **4**(1), 36–42 (2010)
9. Kim, D.-U.: Anti-islanding detection method using phase-shifted feed-forward voltage in grid-connected inverter. IEEE Access **7**, 147179–147190 (2019)
10. Fangde, W., Benben, Z., Fang, X.: Fuzzy optimization of islanding detection algorithm based on active frequency drift with positive feedback. J. Mech. Electr. Eng. **30**(2), 223–227 (2013). (in Chinese)
11. Chen, H., Xu, J.: Islanding detection method of distribution generation system based on logistic regression. J. Eng. **2019**(16), 2296–2300 (2019)
12. Xiaomei, F., Shuxiang, S.: A novel active phase-shifting islanding detection method based on fuzzy cotrol. Power Syst. Prot. Control **42**(20), 19–24 (2014). (in Chinese)
13. Haider, R., Kim C.H.: Passive islanding detection scheme based on autocorrelation function of modal current envelope for photovoltaic units. IET Gener. Transm. Distrib. **12**(3), 726–736 (2018)

Coping Strategy of University Counselors in "Microenvironment" Educational Background

Guoqing Chen[✉]

Department of Social Sciences, Baotou Iron and Steel Vocational Technical College, Baotou 014010, Inner Mongolia, China

Abstract. At present, the world has entered a two-way interactive microenvironment represented by micro-blogs, WeChat. Because the modern science and technology which includes the network, the digital information, the mobile communication and the intelligent mobile phone manufacturing is constantly updated and increasingly matured, everyone in the micro world finds or absorbs a variety of information. The micro environment brings many potential and far-reaching impact for contemporary university students' mind and body and changes their learning and life style. The advent of micro environment also brings new opportunity and challenge to university counselors' education. As the implementer of education, counselors need to have strong coping ability, so as to guide students positively. From the practical investigation, university counselor's "micro ability" is obviously insufficient. In order to effectively analyze the "micro environment" and summarize the coping strategy of counselor, firstly, analysis of university students' physical and mental characteristics in "micro environment" is proposed. Then, the current opportunities and challenges of university counselor's Education is discussed. In the above background, how to effectively launch university counselor's education is researched.

Keywords: Microenvironment · University counselor · Coping strategy

1 Introduction

After entering the universities, learning and living environment is relatively loose. University students have more disposable leisure time. With the rich content, abundance information, wide influence surface, fast transmission, opening concept and other characteristics, the network greatly satisfies the curiosity of university students to new things. At present, university students are more likely to accept new knowledge, new ideas, who have a strong thirst for knowledge and strong learning ability [1]. Under the micro environment, in the campus, classrooms, dormitories, buses, almost all students hold a mobile phone to read news, chat with WeChat, refresh micro-blog.

The contemporary university students have the active thought, who like make to public individual character and seek to prevail over others. At the same time, university students want to show true selves to the outside world. The Internet is a virtual world. Many contemporary university students show their talents and abilities through the

J. C. Hung et al. (Eds.): FC 2021, LNEE 827, pp. 274–281, 2022.
https://doi.org/10.1007/978-981-16-8052-6_33

network, thus creating a sense of achievement. But, the randomness and the lack of restraint of network causes that the contemporary university students are too self-centered, and excessively display the individuality. University students tend to apperceive things, rather than view and analyze problems rationally. Their heart will be carried along by the tide. The virtual identities and anonymous forms in network make many students speak freely on Internet [2, 3]. Individual students do not consider the phenomenon of "micro environment" and do not distinguish true from false. Sometimes, they make irresponsible comments or extreme comments, ignoring moral integrity. Contemporary university students have a strong attachment psychology to the network. The network has become an important platform in their lives, from learning to entertainment, from shopping to communication, which cannot be separated from the network. More and more time and energy are put into the network, so that the face-to-face communication and interaction of university students is reduced.

Consequently, university students' aloneness in the real world is more intense. In real lives, "frustration" makes students more dependent on the virtual space of "micro environment". Therefore, some university students ignore the interpersonal relationships in real lives, which is extremely unfavorable for the students' health growth [4].

2 Literature Review

2.1 Domestic Research Status

According to the data form Xiaoxiang Morning News, the number of micro-blog users reached 275 million in 2014.The "new media blue book in 2014" shows that WeChat users has reached 600 million, covering more than 200 countries and regions, publishing more than 20 kings of language versions. The number of active users at home and abroad is more than 270 million. University students are more likely to accept new things. They pursue individuality and freedom and have consciousness democratic rights, who become the main force of using micro media. Doing the job of public opinion on the Internet work well is a long-term task. We should improve and innovate the online publicity, and use the rule of network communication to highlight the theme of the times and excite the positive energy, and cultivate and practice the core values of Chinese socialism, and grasp the guidance of public opinion on the Internet. In the micro environment, university counselors need to understand and grasp physical and mental characteristics of university students accurately, and focus on opportunities and challenges faced by education of university students in micro environment, and actively explore the effective method, new ways and countermeasures of education [5].

2.2 Foreign Research Status

At abroad, there are many researches about countermeasures of university counselors in the microenvironment. Its point of penetration is often based on the system theory and psychological theory. American psychologist Bronfenbrenner divides the whole "micro environment" from the perspective of system theory [6, 7]. He believes that the environment is hierarchical, including microscopic level, mesoscopic level, facade layer and macro layer. This theory repeatedly stresses the most important characteristic of micro environment, which uses the characteristic to research coping strategy of university counselors. The psychologist Bronfenbrenner believes that micro environment of society should be a small environment in a sense, which is specific and sensible. Therefore, the research on the micro environmental problem should be at close range. The strategy that counselors should have is concrete and practical. But the biggest problem of these statements is that although it is theoretical, the practicality is not nearly enough.

3   Methods

3.1   Analysis of the Impact of "Micro Environment" on University Students

21st century is the era of information technology and digital technology. The network has spilled over into all fields of life. In this period, university students are always impacted by the network environment in the whole growth process. The network turns on their new cultural knowledge space, which influences their ideologies, life style and behavior [8]. During the years of counselor work, the author saw that a large number of students become excellent graduates because they have made good use of network resource, and also saw that a few students were kicked out of school due to their internet addiction. Therefore, in the process of enhancing ideological and political work of universities, the development of network technology puts forward new requirements for counselors in the new era. Recently, the author conducts questionnaire survey and individual interview and provides 500 questionnaires, so as to understand the relationship between university students and network from the online time, purpose, attitude and obtain the psychological demand and information demand of students on the network from data analysis.

Firstly, EViews7.0 is used to analyze and research the relationship between the network and the life time distribution of university students [9]. This paper uses multiple linear models for the hypothesis testing. The metrology model is:

$$SIT_i = \alpha_i + \beta_i CAIS_i + \gamma_1 S_1 + \gamma_2 S_2 + \varepsilon_i \tag{1}$$

In the formula: α_i is constant term; β_i and γ_i are the regression parameters. E_i is random disturbance term.

In this research, 495 questionnaires are recovered. The recovery rate is 99%. The major projects and data analysis of questionnaires are as follows Fig. 1:

As shown in Fig. 1, the randomness of surfing the Internet of students in three grades is high, where, the option "if you want, you can go" is the most choice.

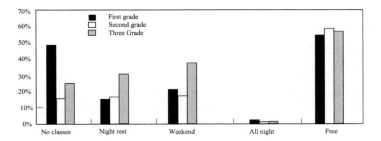

Fig. 1. Investigation of university students' online time

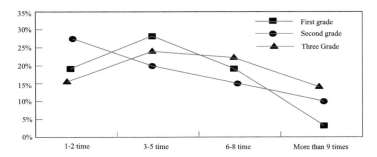

Fig. 2. The average number of online times per week of university students

Table 1. Students' time distribution statistics in "micro environment"

Time allocation	Entertainment of university students	The standard value of entertainment for university students	Entertainment in non university groups
Network game	75.27	23.98	11.94
micro-blog	66.69	18.35	24.20
WeChat	80.11	33.96	30.23
Dating software	65.82	37.4	33.24
Other	58.25	29.16	24.95

As shown in Fig. 2, the network has become a part of everyday life, and the number of surfing the Internet in three classes per week is most 35 times. The Students' time distribution statistics in "micro environment" is shown as Table 1.

3.2 Analysis of University Counselors' Challenges in "Micro Environment" Educational Background

3.2.1 The Overall Development of Students is Challenged

"All around development of moral, intellectual, physical, aesthetics and labor education" is always the requirement in the overall development of university students. But, the comprehensive quality of university students in the context of the micro environment is seriously influenced [10]. For most of the students, they spent all the spare time in WeChat, micro-blog, QQ, shopping and games. They are addicted to the Internet world. They are empty in spiritual world. At the same time, the overrun of network information rubbish can lead to the fuzzy sense of right or wrong of students and the weakening moral consciousness. Moreover, the long-term use of computer, tablet computer, mobile phone and other electronic equipment lead to the lack of physical exercise consciousness of student. Thus, the body diathesis decreases significantly.

3.2.2 The Traditional Education is Adversely Affected

The traditional thought generally adopts the top-down indoctrination approach, the students can only receive education passively and lack initiative. In the micro environment, this kind of education mode is facing great challenges. With the development of network, the time for university students to contact with the mass media is far more than that of the classroom education. In the presence of information transmission with multiple channels, one-way indoctrination education method is facing great challenges, moreover, the content and effect of education is also a severe challenge [11]. In the past, educators can grasp students' ideological change through words and behaviors in time, and make pointed references to carry out ideological education, thus helping students to solve problems. The advent of micro environment makes students increasingly express their thoughts through the network. Due to the concealment of network, educators cannot grasp the specific ideological trends in time and cannot carry out the work in a targeted manner.

3.2.3 The Quality of Science and Technology of University Counselor and Ideological and Political Educators Need to Be Improved

The advent of micro environment increases the technology content of the education work of university counselors. Counselors and ideological and political educators should not only have the excellent ideological and political quality and the extensive cultural knowledge quality, but also have high ability of network technology and information literacy. With the rapid development and application of new media, university counselors and ideological and political educators must adapt to its rapid change [12]. In the actual work, some counselors and ideological and political educators lack the attention on the network education, who stick to the tradition and follow a stereotype routine. Especially, some elder teachers stick to the traditional way of education, who lack the understanding of new media technology. This will cause that the university counselors cannot advance with the times. Thus, it is difficult to realize the transformation from the traditional way of education to the new media era. Therefore, the counselors and ideological and political educators must improve their consciousness and quality of science and technology in time.

4 Results

The survey result shows that nearly 20% of students regret wasting time and energy, and more than 70% of students hope for the occasional prompt and strict regulation of school. Therefore, the school should establish and improve the rule and regulation of network management to restrict university students' online behavior. The survey shows that the higher the grade, the higher the possession amount of individual laptop. Most of the laptops are used in the dormitory. In the work of counselors, we find that under the banner of learning, students get the parental permission, nut parents do not know the actual performance of students at school [13]. To introduce personal computers into dormitories is not helpful for learning, it is also a stumbling block for students to be an adult and to be a talent. In the actual work, our university use the application system of personal computer, then formulate the "requesting and using method of personal computer" and establish the oversight committee. Students need to apply for the use of personal computer. Some serious circumstances should be reflected to the parents.

Based on the above analysis, the countermeasures of counselors' education for university students in the context of micro environment are summarized:

Set up the micro media working platform of education work of university students. The use frequency of micro media is very high. When launching counselors' education work, universities must innovate the mode of education. They can try to establish the micro media platform, such as micro-blog union, WeChat public platform, and strengthen exchanges and interaction with students through these platforms [10, 14]. Through the "micro media", they can also spread the content of education to the network, so that students receive education and guidance at all times and places. The current hot issues concerned by university students can be discussed through the micro media, which helps students to be clear about what is right and wrong, thus achieving the purpose of education.

Set up the rapid intervention mechanism of public sentiment in network. In the context of micro environment, the students' education work must devote time and energy in the guide of public opinion. In the campus micro environment, the negative public opinion must be timely intervened. For some emergencies on campus, we should quickly respond to the hot topic on the micro media, and command a situation. Universities should pay attention to the use of micro media, and timely guide the public opinion in the party and the state policy guidance. The education of university counselors should grasp the characteristics and laws of university students' thought and behavior, and the habits of university students to use micro media, thus guiding students to use the network correctly.

Launch the rich and colorful cultural activities on campus. The universities should actively promote and organize the physical training after class and cultural activities, and encourage students to actively participate in various activities and social practices, and cultivate sense of teamwork and combatant spirit. Through the activities of student council, association activities and other activities, the university students who indulge in virtual network are attracted to the rich and colorful campus culture activities, which transfers or weakens students' internet addiction disorder, and enhance the body constitution of students. Thus, the comprehensive quality of students is improved. Students can develop in a comprehensive and healthy way.

Expand new positions in network ideological education. With the rapid development of Internet, new positions are continuing without end on the ideological front. WeChat group, circle of friends, and micro-blog have become an indispensable part of university life. It is unrealistic to keep students away from the Internet or out of the network. Only making full use of the network, and correctly guiding the students to use the network, and using the network to carry out the ideological education of the students can get twofold results with half the effort. The ideological educators of university students should master the functions of network, and use the network function to communicate with students. The ideological educators should establish the WeChat communication group in each class, and organize students to participate in in Liaoning university students online union, and expand the scope of communication, and understand the front situation of ideological education, so as to improve the comprehensive ability.

5 Discussion

The influences of micro environment on university students are not exactly negative. The micro environment also provides a new platform for education of university students. Counselors' education needs a certain carrier and platform to achieve their educational purpose. The ideological education of university students is launched mainly through forms of classroom education or special lecture, which is only realized within a certain range. The large network system under the micro environment has no space-time boundary, which is an interactive remote information exchange, thus providing a wide transmission route of education. With the advent of micro environment, the mobile phone is not only used for communication, but it also becomes the carrier of new media. Therefore, for many counselors and ideological and political educator, they can actively and rapidly disseminate correct theory, thought and consciousness through the micro media, which is not restricted by time and space, but also does not need tedious procedures.

6 Conclusion

To improve the micro environment of education has proposed a new scientific and technological quality requirement for university counselors and ideological and political educators. The traditional educational mode has been unable to satisfy and adapt to the education of contemporary university students completely. The ideological educators in universities should change the traditional educational concept, and deeply understand the importance of using micro media. Furthermore, they should learn and apply micro media technology into the education work of counselors. The ideological educators should pay attention to grasping the "micro language" of contemporary university students. In working, ideological educators should dexterously integrate the new language into the micro environment of university students, so as to reduce the barrier. Counselors can carry out the education work. Schools should carry out the network education training for ideological educators timely, and organize and carry out exchange activities, and constantly learn advanced experience to improve the quality of science and technology.

References

1. Zhang, Y.: Study on implicit education theory and reform in higher vocational universities. Creat. Educ. **06**(11), 1229–1232 (2015)
2. Zhou, W.F., Li, L.: The research on the function of university students' education to university students' employment. Open Cybern. Syst. J. **9**(1), 1806–1813 (2015)
3. Cheung, D.: The combined effects of classroom teaching and learning strategy use on students' chemistry self-efficacy. Res. Sci. Educ. **45**(1), 101–116 (2014). https://doi.org/10.1007/s11165-014-9415-0
4. Lu, F., Anderson, M.L.: Peer effects in microenvironments: the benefits of homogeneous classroom groups. J. Law Econ. **33**(1), 91–122 (2015)
5. Leach, D.A., Need, E.F., Toivanen, R.: Stromal androgen receptor regulates the composition of the microenvironment to influence prostate cancer outcome. Oncotarget **6**(18), 16135–16150 (2015)
6. Ma, T., Tsai, A.C., Liu, Y.: Biomanufacturing of human mesenchymal stem cells in cell therapy: influence of microenvironment on scalable expansion in bioreactors. Biochem. Eng. J. **108**(2), 4–50 (2015)
7. Linvill, D.L., Grant, W.J.: The role of student academic beliefs in perceptions of instructor ideological bias. Teach. High. Educ. **22**, 1–14 (2017)
8. Sun, G., Cui, T., Guo, W.: A framework of MLaaS for facilitating adaptive micro learning through open education resources in mobile environment. Int. J. Web Serv. Res. **14**(4), 50–74 (2017)
9. Jemielniak, D., Greenwood, D.J.: Wake up or Perish: Neo-Liberalism, the social sciences, and salvaging the public university. Cult. Stud. Crit. Methodol. **15**(1), 72–82 (2015)
10. Xue, L.: Exploration on construction of study style in application universities under the background of supply-side structural reform. Vocat. Tech. Educ. **37**(35), 56–78 (2016)
11. Qiao, X.Y.: Optimization of university students' educational resources integration. Comput. Simul. **34**(8), 239–242 (2017)
12. Yang, Y., Qi, T.: Research on the influence of internet on extracurricular learning and life of English major university students. Theory Pract. Lang. Stud. **7**(8), 695 (2017)
13. Wang, Y.: Big data era influence on university students' education and innovation strategy. In: Eighth International Conference on Measuring Technology and Mechatronics Automation, vol. 10, no. 2, pp. 126-128 (2016)
14. Zhu, J.L.: On the establishment of internal evaluation standard of university education quality. Heilongjiang Res. High. Educ. **32**(5), 39–56 (2017)

Construction of Data Monitoring System for Smart Classroom Based on Internet of Things

Yingjie Li[1] and Lianjun Chen[2(✉)]

[1] Information School, Shanghai Ocean University, Shanghai, China
[2] Information Technology School, Shanghai Jianqiao University, Shanghai, China

Abstract. The purpose of this system is to reduce the energy waste in the classroom of colleges and universities. At the same time, it also realizes the remote collection of sensor data and the intelligent control of the smart classroom. The system uses Arduino Mega 2560 MCU and sensor terminal to collect real-time temperature, humidity, smoke, carbon monoxide, PM2.5, light intensity and classroom equipment parameter information, and transmits the collected data to the cloud platform through ESP8266 WiFi module. Arduino ide tool realizes the design of WiFi module data format and data transmission. Python is used to develop the background application of cloud platform, control the underlying devices and local data processing. The results show that this system can effectively reduce the waste of resources and energy.

Keywords: Internet of Things · Cloud platform · Remote monitoring

1 Introduction.

Nowadays, because of the rapid growth of Internet of things, agricultural Internet of things, industrial Internet of things, smart city and other things related to Internet of things also develop rapidly [1–8]. After entering the 21st century, China vigorously develops higher education and cultivates high-quality talents. Colleges and universities begin to expand their enrollment on a large scale. The number of students in Colleges and universities increases greatly, and the teaching resources are seriously insufficient. As a result, colleges and universities must expand their teaching resources by establishing new campuses to meet the teaching needs of students [9]. The establishment of the new campus needs to spend a lot of manpower and resources to manage it. For the school, it greatly improves the consumption of human, material, financial and energy resources. How to effectively manage the energy resources in Colleges and universities has become one of the important issues in the construction of energy-saving campus [10]. In view of this situation, this paper develops a smart classroom data monitoring system, which reduces energy consumption and manpower.

J. C. Hung et al. (Eds.): FC 2021, LNEE 827, pp. 282–286, 2022.
https://doi.org/10.1007/978-981-16-8052-6_34

2   System Architecture

This system designed in this paper contains data acquisition, data monitoring and intelligent control functions. As shown in Fig. 1, the system consists of four layers.

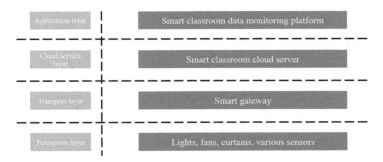

Fig. 1. System structure image

The first layer is the perception layer, which is deployed separately for different classrooms, mainly including power modules, sensor modules (temperature, humidity, smoke, carbon monoxide, PM2.5, light intensity, etc.), single-chip microcomputer (Arduino Mega 2560) and ESP8266 WIFI module. The Arduino Mega 2560 microcontroller mainly handles three parts of tasks. The first is to transform and integrate the data collected in the classroom; the second is to control the curtains and electric lights through the Bluetooth module; and the air conditioner and the projector are controlled through the infrared module. The third is to connect the physical entity to the transport layer via the ESP8266 WIFI module.

The second layer of the system is mainly composed of intelligent gateway. After the sensing layer collects the classroom data, it will pass the uplink and downlink service data via the smart gateway of the transport layer.

The third layer is the cloud server layer, which is in charge of gathering the relevant data of the smart classroom from the access network and storing it in the database according to different types of perception data. At the same time, in order to facilitate data acquisition, the interface is provided for the system.

The main functions of the fourth layer include calling the query interface to query platform data, and it can also send commands to the underlying control module to control the switches of curtains, lights and other equipment.

2.1   Perception Layer Hardware Design

The temperature and humidity measurement uses the SHT30 module, the working voltage is 2.4–5.5 v, the humidity measurement range is 0–100% RH, the temperature measurement range is −40–125 °C, and the I2C communication protocol is used. PM2.5 measurement uses the GP2Y1014AU sensor to monitor the tiny particles in the air, with a concentration range of 0–6000 mg/m^3. Use GY-30 module to measure light intensity, use 3–5 v power supply, light intensity range 0–65535 lx, use I2C communication protocol. Use the BMP180 air pressure module to obtain the air pressure data, the input voltage is 3.3–5 v, the measurable pressure range is 300–1100 hPa, and the I2C communication protocol is used. The combustible gas sensor uses the MQ-2 module, which can detect the concentration of carbon monoxide and methane in the air, with a working voltage of 5 v and a concentration range of 300–10000 pm.

2.2 Schematic Diagram of Node Structure

SHT30 temperature and humidity sensor module, GY-30 module and BMP 180 pressure module use I2C communication protocol, so they are connected together by bus. GP2Y1014AU sensor and MQ-2 combustible gas monitoring module can directly use the analog port to read data. The curtain control module and light control module use Bluetooth network, and Arduino Mega 2560 controls the light and curtain via bluetooth. The projector and air conditioner are controlled by infrared sensors, as shown in the Fig. 2.

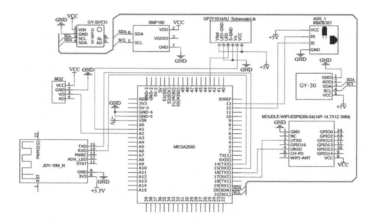

Fig. 2. Node structure image

2.3 Cloud Server Design

The server is built with a Raspberry Pi 4B development board. Its processor is an ARM Cortex-A72 quad-core processor with a main frequency of 1.5 GHz and a 4 GB of LPDDR4 memory. The server uses the Ubuntu Mate 20.04.1 LTS system, which uses the linux 5.4 kernel and natively supports the Raspberry Pi 4B development board. Raspberry Pi in the cloud server needs to use python3 socket to build TCP service, write the data uploaded by the sensor layer node to MySQL, and send the device control information issued by the monitoring platform to the main control board of the sensor node in the sensing layer.

2.4 Monitoring Platform Design

The monitoring platform uses Web development and adopts B/S architecture. It mainly includes two functions, one is data monitoring, and the other is equipment control. Data monitoring is mainly used to monitor the environmental data of the designated classroom. Use the Raspberry Pi cloud server built to read the latest data of the database on the server through HTTP get requests, including environmental data and equipment status data. Device control is mainly implemented through the post method of HTTP. After the device control information is submitted through the psot method, the cloud server forwards the control information, and the sensing layer node controls the device after obtaining the data.

2.5 Software Design

When the perception layer is turned on, initialize the device and complete the following steps:

Step 1 Use AT commands to set ESP8266 to connect to the classroom WIFI, and check the connection. Use AT commands to set the JDY-10M Bluetooth module for Bluetooth networking. If the connection fails, repeat step 1.

Step 2 After the connection is successful, read the switch status of the device, read the sensor data and perform A/D conversion.

Step 3 Integrate these data and upload them to the cloud server through ESP8266.

Step 4 Obtain the return value of the cloud platform and judge whether the upload is successful. If the upload fails, repeat step 3.

Step 5 Obtain the return value of the cloud platform and determine whether to modify the device opening and closing status. Modify the opening and closing status of the device according to the return value of the cloud platform. If no modification is required, repeat step 1.

3 Conclusions

The system consists of three parts, which are specifically divided into intelligent nodes at the perception layer, cloud servers and cloud platforms. The system combines single-chip microcomputer, wireless communication technology, and real-time monitoring the classroom environment online and managering classroom equipment remotely, which can reduce the energy consumption of the classroom to a certain extent and decrease a certain energy cost.

References

1. Shen, J., Cui, X.: Construction and application pattern analysis of English classroom intelligent teaching system. In: Proceedings of 2020 3rd International Conference on Education Technology and Information System (ETIS 2020), pp. 327–331. Clausius Scientific Press (2020)

2. Chao, D.: Application of wireless mobile communication and Internet of Things. In: Proceedings of 2019 Asia-Pacific Conference on Emerging Technologies and Engineering (ACETE 2019), pp. 491–495. Francis Academic Press, UK (2019)
3. Zhang, W., et al.: Internet of Things (IoT) enabled smart home safety barrier system. In: Proceedings of 2020 2nd International Conference on Computing, Networks and Internet of Things (CNIOT 2020), pp. 90–96. ACM (2020)
4. Zhou, M., et al.: Internet of Things (IoT) enabled smart indoor air quality monitoring system. In: Proceedings of 2020 2nd International Conference on Computing, Networks and Internet of Things (CNIOT 2020), pp. 97–101. ACM (2020)
5. Xia, W., et al.: Research on key agreement security technology based on power grid Internet of Things. In: Proceedings of the 10th International Conference on Computer Engineering and Networks (CENet 2020), pp. 1437–1444 (2020)
6. Shu, X., et al.: Deep reinforcement learning cloud-edge terminal computation resource allocation mechanism for IoT. In: Proceedings of the 10th International Conference on Computer Engineering and Networks (CENet 2020), pp. 1564–1570 (2020)
7. Wu, M., Liu, H.: Integration of Internet of Things and blockchain for chattel asset pledge financial service. In: Proceedings of 2020 19th International Symposium on Distributed Computing and Applications for Business Engineering and Science (DCABES 2020), pp. 172–175 (2020)
8. Xiang, H., Fu. X.: Design and application of wisdom classroom teaching mode in big data environment. In: Proceedings of 2019 7th International Education, Economics,Social Science,Arts,Sports and Management Engineering Conference (IEESASM 2019). Clausius Scientific Press,Canada, pp. 1311–1314 (2019)
9. Ru, W.: Research on flipped classroom teaching mode of college English based on intelligent course platform APP. In: Proceedings of the 2019 Northeast Asia International Symposium on Linguistics, Literature and Teaching (2019 NALLTS) (Volume B). New Vision Press, pp. 193–197 (2019)
10. Wei, X.: Research on practical teaching of intelligent manufacturing professional flip classroom based on modern educational technology. In: Proceedings of 2019 Asia-Pacific Conference on Advance in Education, Learning and Teaching (ACAELT 2019). Francis Academic Press, UK, pp. 1705–1709 (2019)

ICT Technology and Big Data Background Embedded Technology Professional Curriculum Optimization Research

Meifang Cai[✉]

School of Artificial Intelligence, Nanchang Institute of Science and Technology, Nanchang 330038, Jiangxi, China

Abstract. With the rapid development of science and technology, various new technologies have been proposed. Since the computer was proposed many years ago, in order to meet the professional needs of various professions, so we have developed an embedded system, the so-called micro-chip. It has helped us achieve richer results in various areas of expertise, so embedded technology is becoming more and more important. And now with the advent of the information age, information and communication technology and big data technology has developed rapidly and has become the cornerstone of modern life, now people's lives and information and communication technology has been inextricable. So, we can also apply these new technologies to our curriculum. Therefore, the purpose of this paper is to study the curriculum optimization scheme based on ICT technology and big data background embedded technology. This paper reviews the courses on embedded systems and embedded technology, and after looking up the development of big data and information and communication technology in recent years, we use the improved adaptive parameter DBSCAN clustering method to process and analyze the relevant data, model it, and then get the experimental results we need. The experimental results show that ict technology and big data technology can be used to optimize the curriculum system of embedded technology.

Keywords: ICT · Big data background · Embedded systems · System optimization

1 Introduction

Computer was a huge thing when it was first proposed, but later it was improved to the fourth-generation computer. And now the computer has become a symbol of convenience and fast, and has launched a series of intelligent products such as smart phones [1]. But the mainstream computers in the market are generally general-purpose computers, which mainly meet the needs of individuals and households, and cannot meet the needs of various professional fields [2]. Therefore, some people have built the embedded system to meet their own needs based on the general computer, which can help to solve the problems in the professional field. For example, medical computer and engineering computer, although they have many functions less than the mainstream computers in the market, they have functioned that other computers cannot surpass in one field. For example, operation, image processing and other aspects [3]. So, the development of embedded system is inevitable. In the future, with the development and subdivision of various professional fields, the research on embedded system will be more and more complex. Therefore, we need to develop the courses on embedded system [4].

But the learning of embedded system is very complex, because it involves many aspects, not only the construction of software, but also the installation of hardware, which brings a heavy burden to the learning of embedded system. Therefore, only undergraduate education is not enough for embedded technology [5]. And now the cultivation of embedded technology courses in Colleges and universities is far from the expected requirements. In the undergraduate era, embedded system mainly studies the theoretical knowledge of books, and then follows the tutor to install the practical system. However, teachers usually deliver the produced hardware and software to students for a rough installation, rather than complete installation processing, so some deviation may occur in learning [6]. And the subjects of embedded technology major are still traditional subjects. With the rapid development of technology, some subjects may not be needed. Some subjects need to be added. Therefore, we should optimize the curriculum system of embedded technology specialty to facilitate students to learn about embedded technology more conveniently, and gain knowledge and knowledge from them Meet the needs of the society for the embedded technology related talents. Therefore, we propose an improved adaptive parameter DBSCAN clustering algorithm to study this problem [7, 8].

Embedded system, it can also be regarded as a different kind of computer system. And it has hardware structure, so it is difficult to learn [9]. If it is only using offline teaching, after years of exploration, it is found that this is not enough. Because the books and knowledge points to learn are special, and some classes not enough to meet the teaching task needs, and the embedded system is better to carry out experimental operation to better understand. So, the course still uses online and offline double teaching methods to learn [10].

2   Improved Adaptive Parameter DBSCAN Clustering Method

Typically, in clustering analysis, the Average Square Integral Error Function (MISE) is generally selected to optimize bandwidth. The definition is as follows:

$$MISE(h) = E \int \left(\hat{f}(x) - f(x) \right)^2 dx \tag{1}$$

Under weak assumptions:

$$MISE(h) = AMISE(h) + o \left(\frac{1}{(nh)} + h^4 \right) \tag{2}$$

$$AMISE(h) = \frac{R(K)}{nh} + \frac{1}{4} m_2(K)^2 h^4 R(f'') \tag{3}$$

Among them:

$$R(K) = \int K(x)^2 dx \tag{4}$$

$$m_2(K) = \int x^2 K(x) dx \tag{5}$$

Minimizing MISE(h) is equivalent to minimizing AMISE(h), defletering and making the conductor equal to 0 has:

$$\frac{\partial}{\partial h} AMISE(h) = -\frac{R(k)}{nh^2} + m_2(K)^2 h^3 R(f'') = 0 \tag{6}$$

$$h_{AMISE} = \frac{R(K)^{\frac{1}{5}}}{m_2(K)^{\frac{2}{5}} R(f'')^{\frac{1}{5}} n^{\frac{1}{5}}} \tag{7}$$

Where m and R are determined according to the nuclear function.

3   Experiment

3.1   Experimental Process

We compare the advantages and disadvantages of the two modes of instruction by setting up two modes of instruction. We set up two classes to evaluate the results at the end of the semester by using different teaching methods. In order to make errors, we randomly disassociated students to reduce errors.

3.2   Selection of Experimental Data

We randomly selected 80 students from class 18 of our mechanical automation program, rehearsed them after the average upset, and then completed the experiment by using different teaching modes for the embedded system course.

4   Evaluation Results

4.1   Data Analysis

Table 1. Statistics on teaching performance results for both modes of instruction

	Excellent	Pass	Failed	Total number of people
Optimize system teaching	26	12	2	40
Traditional way of teaching	16	19	5	40

We use two different teaching methods to guide students on embedded technology courses. The exam is then performed at the end of the semester and the results are counted in a table, as shown in the table above. From it, we found that 65% of the people who had been assessed well in teaching after optimizing the system, 30% of those who had not, and only 4% of those who had not. However, in the assessment of achievement under traditional teaching, we found that only 40% of the good people were excellent and 12.5%. Therefore, after comparison, we find that the students' scores from the optimized system are better than 3ds from the traditional way of teaching. Of course, this is only in terms of performance, we also need to learn from the students of the two teaching methods of teaching evaluation to get a true reflection of the situation.

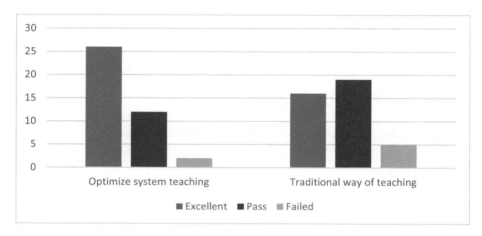

Fig. 1. Statistics of teaching achievement results for both modes of instruction

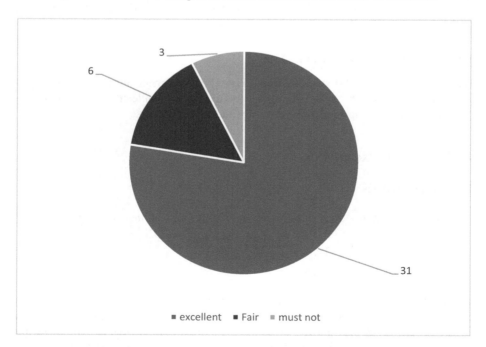

Fig. 2. About students' attitudes towards optimizing system teaching

Figure 1 is an intuitive diagram based on Table 1 to show the difference between the two modes of instruction more intuitively. Figure 2 is from these 40 students, the questionnaire issued to inquire about the optimization of the teaching system teaching evaluation. Figure 2 points out that 31 students think this new mode of teaching is very good, 6 students think the system is OK, but 3 students pointed out that no, our main inquiry is these three students think the optimized system is not good. After investigation, we found that the main reason why they thought it was bad was that they wanted to use traditional book teaching, which, suddenly, combined with communication and big data technology, made them so complex that they didn't keep up with the course schedule and felt unacceptable for the time being.

4.2 Embedded Systems

An embedded system is a device consisting of hardware and software that can run independently of itself. Its software is different from the computer's software composition, its software only includes the operating environment of the software and the operating system. But its hardware compared to the software is complex and diverse, there are signal processors, storage and other aspects of the content. Embedded systems are between different from ordinary computer processing systems. It cannot store as much information as a computer because it has only a small amount of storage media.

Embedded system is an application-centric, modern computer technology-based dedicated computer system. Its main purpose is to enable users to turn on the power can directly use the system's functions, basically do not need a second development operation or only a small amount of configuration to use. Its specificity lies in the fact that it is developed specifically for certain applications, so it is a system of software and hardware integration. Because this configuration can effectively reduce the cost of the system and increase the reliability of the system, so that users have a better experience. But its design and the current mainstream computer system design is basically the same, are using integrated circuits and the same system structure configuration. But compared to this, it adds embedded operating systems and real-time operating systems, as well as data analysis and processing systems for specific applications. Moreover, embedded systems can be used in multiple scenarios and flexibly replace hardware and software to implement the specific needs of users.

Embedded systems first appeared after microprocessors were discovered. Since the microprocessor was discovered in 1971, due to the widespread use of microprocessors, computer manufacturers have begun to provide microprocessors specifically for users to choose a dedicated embedded system suitable for their own use to embed between their own devices to meet their own needs. This is the birth of embedded systems. Then, in the 1980s, as the level of microelectronics improved, manufacturers of integrated circuits created a special microcontroller for io design, which we call a microcontroller. Later in the 21st century, with the rapid development of the network, embedded systems have also been greatly developed. But today's embedded systems are still isolated from the Internet, but with the development of the Internet and the Internet, in the future. Embedded systems will be more widely used in life.

Embedded system features a lot, we have a little bit of the following introduction. First, it's very specific. Because embedded systems are usually specific to an application. Second, it is smaller in size. Because it integrates many circuits into the chip so that embedded systems can be embedded in the targets we need. Third, real-time is good. Because the embedded system is mainly used in the production process control and transmission of signals and other occasions, it is mainly to control the host object, so his real-time requirements are relatively high, such as handheld computers. Therefore, in the later development of embedded systems, real-time good has also become a criterion for judging. Fourth, the tailorability is good. Because embedded systems make it easy for users to tailor hardware for different situations, the software is convenient for specific occasions. Fifth, high reliability. Because some of the tasks undertaken by embedded systems are related to confidential tasks and high-risk industrial environments, the requirements for reliability of embedded systems are also relatively high in the later development. Sixth, the power consumption is relatively low. Because embedded systems are typically targeted at small systems that do not have large power supplies, this results in embedded systems also using less power to save energy. Seventh, embedded systems cannot develop themselves. Unlike artificial intelligence, which can learn itself, it has to be revamped and developed with the help of a computer system, and often requires a dedicated set of tools to exploit it. The eighth point is that when making embedded systems, hardware software is designed at the same time. In the past, the main use of hardware first design, and then design software. However, this will lead to a relatively large defect, that is, if found in the later effect is not good, to push the whole experiment to redo. The current method is conducive to real-time modification of the accomplished goal at any time to obtain an optimized solution.

And the main purpose of embedded system is to embed CPU in the target to play a role, but the embedding method is also divided into two kinds. Generally divided into full embedding and semi-embedding. The full embedding method is characterized by a separate processor system that can act as a system independently. And the embedded system constructed in this way is generally used in small portable devices, and its working environment is generally in poor condition. And it can be designed by the system itself input and output circuit, can be suitable for a variety of voltages for power supply. And it is suitable for any general-purpose computer not suitable for the market, such as aerospace, fine operation and so on. The semi-embedded system generally needs to be combined with the computer system to work properly, it generally does not have a stand-alone processor, but borrows the processor in the target system to accomplish the required tasks. And such an embedded system can only be part of the system, cannot operate as a stand-alone system.

5   Conclusion

In summary, we can learn that because the embedded system curriculum learning needs to know a lot and complex knowledge, so we need to be as soon as possible to optimize the embedded technology professional curriculum system, so that students can learn and teachers to teach. Help students understand as quickly as possible what is needed in the profession and what needs to be accomplished in the future and the needs of society. In this way, they can have a clear plan for the future, which is in line with the professional needs for the development of talent. Therefore, we must optimize the curriculum of embedded technology.

Acknowledgements. Jiangxi Province Educational Reform Project: The development and practice of the embedded technology course system under the background of "class certificate financing" JXJG-20-27-10.

References

1. Sieniu, M.: Nadanie tytuu naukowego profesora. Kilka refleksji na tle regulacji zawartych w ustawie Prawo o szkolnictwie wyszym i nauce. Biaostockie Studia Prawnicze **25**(4), 137–157 (2020)
2. Lee, Y.J.: Defense ICT supply chain security threat response plan. J. Inf. Secur. **20**(4), 125–134 (2020)
3. Yulinetska, Y.V., Babii, O.Y., Hloviuk, I.V., Stepanenko, A.S., Pashkovskyi, M.I.: Implementing ICT into language and law classroom to develop law students' communicative competence. Inf. Technol. Learn. Tools **81**(1), 310–326 (2021)
4. Gross, K., Pawlak, F.: Using video documentation in out-of-school lab days as an ICT Learning and diagnostic tool. World J. Chem. Educ. **8**(1), 52–60 (2020)
5. Martin, J.: Pre-service TVET teachers' perceptions of their readiness to integrate ICT in the curriculum. Int. J. Sociol. Soc. Policy **1**(2), 1–15 (2020)
6. Wang, X., Zhou, X., Liu, H., Chang, J.: Optimal path for fault identification of marine communication network in the background of big data. Arab. J. Geosci. **14**(2), 1 (2021). https://doi.org/10.1007/s12517-020-06397-1
7. Zhao, Y., Zhang, Y.: Safety protection of E-commerce logistics information data under the background of big data. Int. J. Netw. Secur. **21**(1), 160–165 (2019)
8. Billah, M., Rashid, M., Bairagi, A.K.: Embedded system based on obstacle detector sensor to prevent road accident by lane detection and controlling. Int. J. Intell. Transp. Syst. Res. **18**(2), 331–342 (2019). https://doi.org/10.1007/s13177-019-00202-4
9. Mcmahon, M., Bornstein, S., Brown, A., et al.: Training for health system improvement: emerging lessons from Canadian and US approaches to embedded fellowships. Healthcare policy = Politiques de sante **15**(SP), 34–48 (2019)
10. Li, Z., Guo, K., Liao, M., et al.: Micro-hybrid energy storage system capacity based on genetic algorithm optimization configuration research. Int. Core J. Eng. **6**(2), 78–83 (2020)

The Collision Detection Technology in Virtual Environment

Yan Li[✉] and Wei Zhao

Changchun Institute of Technology, Mechanical and Electrical
Engineering College, Changchun, Jilin, China
0208088@ccit.edu.cn

Abstract. In-depth analysis of the development and advantages of virtual manufacturing technology, detailed discussion of current hot issues in virtual manufacturing technology, collision detection simulation status, overview of the development of collision detection algorithms, specific analysis of the current virtual manufacturing environment collision detection technology For existing bottlenecks, a parallel detection technology for optimized calculations is proposed, including: fast parallel collision detection based on graphics hardware and a divide-and-conquer strategy and pipeline technology to complete collision detection. Finally, the efficiency and superiority of the algorithm are demonstrated through experimental data.

Keywords: Virtual manufacturing technology · Collision detection simulation · Parallel detection

1 Introduction

Virtual manufacturing is a new idea first proposed by the United States in the late 1980s. It uses artificial intelligence technology and simulation technology to conduct a comprehensive simulation of people, objects, information and manufacturing processes in real manufacturing activities to discover the possibilities in manufacturing [1–5]. For problems that arise, preventive measures should be taken before the actual production of the product, so that the product is successfully manufactured at one time, in order to achieve the purpose of reducing costs, shortening the product development cycle, and enhancing the competitiveness of the enterprise. In virtual manufacturing, the entire production cycle of the product from initial appearance design, production process modeling, simulation processing, model assembly to inspection is simulated and simulated on the computer, without the need to actually produce the product to verify the mold design Reasonableness, therefore, can reduce the troubles caused by the early design to the later processing and manufacturing, and can avoid the scrapping of molds, so as to achieve the purpose of improving the primary yield of product development, shortening the product development cycle, and reducing the manufacturing cost of the enterprise [6–10].

J. C. Hung et al. (Eds.): FC 2021, LNEE 827, pp. 294–300, 2022.
https://doi.org/10.1007/978-981-16-8052-6_36

2   Collision Detection Simulation Technology

Collision detection is used to determine whether one or more pairs of objects occupy the same space area at the same time in a given time domain. It is an unavoidable research hotspot in the fields of human-computer interaction, robot motion planning, computer simulation, virtual reality, and computer graphics. With the increasing maturity of mathematical morphology, computational geometry, and algebraic geometry, as well as the rapid development of graphics processing units (GPUs), people are eager to perform fast real-time simulations of the real world. The key technology that needs to be solved urgently is real-time collision detection. At present, three-dimensional and spatial geometric models are becoming more and more complex, and the scene scale of virtual environments is getting larger and larger. People have higher and higher requirements for real-time interaction and scene authenticity. The strict real-time and authenticity requirements are increasing. Researchers in human-computer interaction, computational geometry, computer graphics, and virtual reality pose huge challenges, which make fast real-time collision detection once again a research hotspot.

3   Development History of Collision Detection Algorithm

In recent decades, researchers at home and abroad have conducted extensive and in-depth research on collision detection, and proposed a large number of efficient collision detection algorithms. From the perspective of the characteristics of the detection, collision detection technology can be classified into two categories. One is the use of the performance of the graphics processor (GPU) of the machine itself or the detection through algorithm programming. The advantages of this method are high efficiency and fast output of three-dimensional spatial geometry. However, this method will increase the hardware overhead of the machine and is not conducive to promotion. There is no good solution to the puncture phenomenon and missed imagination in the detection in this way. Another way is to use the method of constructing and decomposing geometric models for collision detection. Geometric models can be divided into two types: surface models and volume models. The surface model uses a patch to represent the surface of the object, and its basic geometric elements are mostly triangles; the volume model uses voxels to describe the structure of an object, and its basic geometric elements are mostly polyhedrons. The surface model is relatively simple, and the rendering technology is mature, and the processing is convenient, but it is difficult to carry out the overall form of the volume operation (such as stretching, compression, etc.), and it is mostly used for the geometric modeling of rigid objects. The body model has the internal information of the object, which can well express the body characteristics (deformation, splitting, etc.) of the model under the action of external force, but the time and space complexity of the calculation also increase correspondingly, and it is generally used for the geometric modeling of soft objects. Because this kind of method does not need to increase the machine hardware, the machine drawing simulation output is convenient, so it has been widely used.

In the past ten years, the University of North Carolina, the State University of New York, the University of Freiburg in Germany, the University of Bonn, the University of

Geneva in Switzerland, the University of Grenoble in France, etc., the Beijing University of Aeronautics and Astronautics, Zhejiang University, CAD &CG countries Researchers such as the Key Laboratory, National Key Laboratory of Parallel and Distributed Processing of National University of Defense Technology, Nanjing University of Aeronautics and Astronautics, Northwestern Polytechnical University, Xi'an Virtual Reality Engineering Technology Research Center, Xi'an Jiaotong University and other researchers have done a lot in the field of collision detection. Significant work, proposed some more mature algorithms, and developed corresponding software packages: PQ (sphere) SOLID (AABB), PQP (LSS), RAPID (OBB) and Quick-CD (k-DOP), etc.

4   Analyses of Technical Problems in Collision Detection

Although the research results on collision detection have been relatively rich, and some algorithms have been applied in practice, with the continuous development of human-computer interaction and virtual reality technology, the actual application system of large-scale and complex scenes has more requirements for collision detection technology. Come higher. At present, the following problems still exist in the field of collision detection, which urgently need to be studied and solved.

(1) Real-time collision detection between objects in large-scale complex scenes. With the development of graphics hardware technology, the system can process large-scale scenes in real time. These scenes often contain more than hundreds of thousands of patches, and even include a collection of objects whose data is so large that the memory cannot hold it. Carrying out collision detection in this kind of scene puts forward higher requirements on collision detection algorithms. Since the efficiency of the collision detection algorithm is inversely proportional to the complexity of the objects in the scene, although many methods have solved the collision problem between objects in a simple space (containing a small number of objects), as the complexity of the scene increases, the polygon surface The number of slices will also increase, and the efficiency of most algorithms will drop rapidly. In this way, the real-time and stability of multi-object collision detection in a complex virtual environment cannot be guaranteed. Therefore, the collision detection between multiple objects in the real scene in the complex virtual environment has become a bottleneck in the research of many application branches of virtual reality. It has once again become a hot spot in virtual reality research.

(2) Continuous collision detection of moving objects in complex scenes. At present, most algorithms are discrete collision detection algorithms, which refer to collision detection when an object is stationary at a certain point in time. This type of algorithm has two common shortcomings. One is that there is a puncture phenomenon. When the discrete detection step is too large, two objects may have a certain depth of puncture before they are detected to have collided, which cannot guarantee the object. The second is that collisions will be missed. For narrower objects, when the moving object is on both sides of the narrow object at adjacent discrete time points, the discrete algorithm cannot correctly detect the collision.

Although the continuous collision detection algorithm can solve these two problems, the calculation overhead of the continuous algorithm is too large, and it often becomes the computational bottleneck of the real-time system. Therefore, how to effectively combine the advantages of the discrete collision detection algorithm and the continuous collision detection algorithm has become a key issue to be solved in the current collision detection field.

(3) Collision detection between deformed objects in complex scenes. At present, the collision detection problems in some specific application fields, such as the collision detection of deformable objects such as clothing cloth, virtual surgery simulation, virtual plant modeling, etc., often have more special or even more demanding requirements, and most of them currently focus on rigid body collision detection. The algorithm is not yet able to meet its requirements. On the one hand, the structure of deformable objects such as clothing cloth, human skin, plant leaves, etc. is constantly changing in the simulation, and the structure of the object needs to be rebuilt. At present, most collision detection algorithms require a long preprocessing time to complete the structure of the object. It's usually difficult to achieve real-time reconstruction, and work such as virtual surgery simulation and flexible object deformation simulation also requires high precision. Therefore, the time consumption of collision detection between such objects is often several times that of object collision detection between rigid bodies. The general collision detection algorithm between rigid bodies is incapable of this. Therefore, it is necessary to find a better algorithm than existing methods to meet the needs of such applications.

(4) Collision response problems in complex scenes. When Surface-based algorithms process geometric feature (triangular patch) pairs of potential intersection models, this potential collision pair is often determined by bounding boxes. The contact force is calculated based on the proximity features of the geometry and the direction of the surface. When the collision detection or feedback is too fast or too large due to the speed of the model, it is possible that the vertex of one model penetrates the other model, which eventually led to the failure of the test. Moreover, due to the short reaction distance of the surface model, the simulation may not be able to repair this error state in time. Therefore, a long-term film can be used to avoid excessive puncture. In a complex virtual environment, the time slice requirement is very small, so this algorithm is difficult to simulate in real time. At present, there is no good algorithm that can solve the real-time collision response problem in complex scenes. Therefore, to solve this problem, a real-time fast response algorithm must be provided, which is also a key problem that must be solved in current collision detection research.

(5) Limitations of the method of handling non-convex objects. At present, some efficient collision detection algorithms are mainly limited to collision detection between convex bodies. For example, collision detection algorithms based on feature classes and image-based collision detection algorithms can only quickly detect collisions between convex bodies. In actual complex virtual scenes, most objects are non-convex objects. Although some improved algorithms propose to use bumps to organize non-convex objects, this will have a greater impact on the performance of the algorithm, and the detection effect will not be achieved.

5   Parallel Collision Detection Technology

With the rapid development of parallel machines, parallel technology has brought unprecedented opportunities and challenges to researchers. Parallel collision detection has been widely used in computational geometry and robot control. Some scholars have proposed parallel processing algorithms for voxel models of the same size, but it is difficult to deal with models that are too simple or complex. Some scholars have proposed a parallel collision detection algorithm based on convex polyhedrons, but the experimental model is convex The body has certain limitations. Later, a parallel collision detection algorithm based on MPI was proposed. Although the efficiency of the algorithm has been improved to a certain extent, the octree representation of the object is quite time-consuming, and the communication between the algorithm processes takes a certain amount of time.. Some scholars also proposed a parallel collision detection algorithm based on pipeline. Although the efficiency of the algorithm is greatly improved, the number of task processes p has a great influence on the performance of the algorithm, and it is difficult to find the best value.

5.1   Fast Parallel Collision Detection Based on Graphics Hardware

According to the advantages of the collision detection algorithm based on image space, this research proposes a fast parallel collision detection algorithm based on graphics hardware. The research scheme is: based on the collision detection of the image space, generally use the graphics hardware to sample the two-dimensional image of the object and the corresponding depth information to distinguish the intersection between two objects. The advantage of this type of algorithm is that it can effectively use graphics hardware acceleration technology to reduce the computational load of the CPU, so as to achieve the purpose of improving the efficiency of the algorithm. This project explores the use of programmable GPU to solve the real-time collision detection problem, maps the collision detection process of two arbitrary-shaped objects to the programmable GPU, and calculates the collision detection results in parallel through the real-time drawing process of the GPU, using GPU and CPU Multi-layer LDI is calculated in parallel in a combined way, and on this basis, two effective optimization algorithms are proposed. By comparing with the above parallel algorithms, it can meet the real-time and accuracy requirements of the interactive complex virtual environment.

5.2   Divide and Conquer Strategy and Pipeline Technology to Complete Collision Detection

The divide and conquer strategy in the parallel algorithm is used to establish a balanced bounding box tree for each object in the environment, and a task tree is formed by traversing every two bounding box trees, and the traversal of all task trees is evenly distributed to each processor, And then adopt the pipeline technology in the parallel algorithm to speed up the collision detection process by dividing the process to traverse the task tree, and at the same time apply multi-threading technology in the process, which can run on a single processor and a multi-processor. By quickly eliminating disjoint objects to speed up the algorithm, the OpenMP parallel model is used to traverse the hybrid bounding volume level in parallel to further accelerate the collision detection process. At the same time, using the typical algorithm in the symmetry breaking technology-coloring algorithm, each task tree is coded to generate different categories, different categories are assigned to different parallel machines, and multi-threading technology is used on the parallel machines to execute the same types of tasks. The tree is traversed to detect whether there is a collision. Experimental results show that compared with the existing classic I-COLLIDE and other algorithms, this algorithm has obvious advantages in efficiency and accuracy, and can meet the real-time and accuracy requirements of interactive complex virtual environments.

5.3   Experimental Data Analysis

The above two collision detection algorithms are compared with the classic serial algorithm and the spatio-temporal correlation algorithm. Table 1 shows the analysis data of the specific collision algorithm. From Table 1, it can be seen that the two collision detection algorithms mentioned in the article run speed Extremely fast, and the algorithm calculation frame rate is very high, easy to use.

Table 1.   Operational analysis of several classic collision detection algorithms

Algorithm type	Frame rate (sec.)	Time to run 1000 steps (ms.)
Classic serial algorithm	8.15	203
Spatiotemporal correlation algorithm	16.61	115
Open MP parallel algorithm	19.21	86
Detection line technology algorithm	46.67	51

6   Summary

(1) The first part of the article deeply analyzes the development process of virtual manufacturing technology and its application in reality, and discusses in detail the current hot issues in virtual manufacturing technology, and the current situation of collision detection technology simulation. At the same time, it summarizes and studies the specific collision detection algorithm. development path.

(2) The second part of the article specifically analyzes the bottleneck problems of collision detection technology in the current virtual manufacturing environment, and according to the actual situation, proposes parallel detection technologies for optimized operations, including: fast parallel collision detection and divide-and-conquer based on graphics hardware Strategy and pipeline technology complete collision detection.

(3) Finally, the article demonstrates the efficiency and superiority of the algorithm through experiments.

References

1. Bentley, J.L., Ottmann, T.A.: Algorithms for reporting and counting geometric intersections. IEEE Trans. Comput. (S0018–9340) **28**(9), 643–647 (1979)
2. Ahuja, N., Chien, R.T., Yen, R., Bridwell, N.: Interference detection and collision avoidance among three dimensional objects. In: Proceedings of the Annual National Conference on AI 1980. Stanford University, pp. 44–48 (1980)
3. Canny, J.F.: Collision detection for moving polyhedral. IEEE Trans. Pattern Anal. Mach. Intell. (S0162–8828) **8**(2), 200–209 (1986)
4. Moore, M., Wilhelms, J.: Collision detection and response for computer animation. ACM Comput. Graph. (S0097–8930) **22**(4), 289–298 (1988)
5. Bonner S, Kelley R B. A representation scheme for rapid 3-D collision detection. In: Proceedings of IEEE International Symposium on Intelligent Control 1988, Arlington, pp. 320–325 (1988)
6. Dobkin, D.P., Kirkpatrick, D.G.: A linear algorithm for determining the separation of convex polyhedral. J. Algorithms **6**, 381–392 (1985)
7. Chazelle, B.: An optimal algorithm for intersection three-dimensional convex polyhedral. In Proceedings of 30th Annual IEEE Symposium on Foundation Computer Science, pp. 586–591 (1989)
8. Lin, M.C., Gottschalk, S.: Collision detection between geometric models: a survey. In: Proceedings of IMA Conference on Mathematics of Surfaces, pp. 37–56 (1998)
9. Jimenez, P., Thomas, F., Torras, C.: 3D collision detection: a survey. Comput. Graph. **25**(2), 269–285 (2001)
10. Wong, T.H., Leach, G., Zambetta, F.: An adaptive octree grid for GPU-based collision detection of deformable objects. Vis. Comput. **30**, 729–738 (2014)

Application of Virtual Reality Technology in Interior Design Under the Background of Big Data

Ziyou Zhuang[✉]

School of Digital Arts and Design, Dalian Neusoft University of Information,
Dalian, Liaoning, China
zhuangziyou@neusoft.edu.cn

Abstract. Virtual reality technology is a new design method related to architectural art, widely used in interior space. It can make people experience different shapes or styles through vision and touch and has intuitive and visual characteristics. This paper aims to study the application of virtual reality technology in interior design under the background of big data, accelerate its integration with traditional interior design, and provide new ideas for the design industry's development and innovation. In the experiment, more than 80% of the respondents knew more about the application of VR technology in entertainment film and television but little about interior design. At the same time, 60% of the respondents were satisfied with the experience provided by the companies using VR technology.

Keywords: Big data · Computer graphics · VR technology · Interior design

1 Introduction

The combat simulation system of the U.S. military makes people begin to have the cognition of virtual reality technology (VR). It combines modern science and technology such as computer graphics technology, computer simulation technology, sensor technology and soon attracts many scientists' attention. In recent years, the application of virtual reality technology in various fields is endless [1]. With the advent of big data, diversified and huge data has promoted the continuous progress of modern science and technology, and the pace of society is accelerating. Compared with traditional technology, more efficient and convenient technology will replace it. More and more people have favored virtual reality technology. In the virtual reality world, users can experience all the feelings in real life most vividly, and even what we cannot experience in virtual reality can make a person immersive. Virtual reality can also imitate a kind of people who are confused with reality to make a variety of emotion and perception systems that human beings need to have, such as hearing, vision, touch, taste, smell, etc.; the most important thing is that its super-powerful simulation system really turns this human-computer interaction into virtual reality, People can operate autonomously and receive information and feedback from the environment. Its existence, a

J. C. Hung et al. (Eds.): FC 2021, LNEE 827, pp. 301–308, 2022.
https://doi.org/10.1007/978-981-16-8052-6_37

variety of perception, interaction, and other characteristics make virtual reality technology attracted many people's attention.

Once a new science and technology appear, it will be sought after by all countries. VR (virtual reality) technology is no exception. As the birthplace of virtual reality technology, the research of VR in the United States is at the leading level in the world. NASA has established a special VR training system and VR education system. Germany is committed to integrating VR and traditional industrial transformation, such as product design and product demonstration, which promotes the further development of virtual reality technology as a whole. There is still a gap between China's VR research and the developed countries globally, which also causes the country's great attention and puts the virtual reality technology in the key projects. With the continuous efforts of China's research institutions and universities, VR technology has gradually become the breakthrough point for various industries' development and innovation. Virtual palace museum tours, virtual zoos are all widely used in domestic virtual reality technology. With the continuous exploration of VR technology by scientists, VR technology will continue to meet higher-level needs and become more and more perfect.

The general interior design is mainly based on a two-dimensional plane. The maturity of 3D modeling software and technology has prompted interior designers to try three-dimensional modeling design boldly. However, only two-dimensional modeling software can demonstrate it, and the content and form of the final modeling design can only stay at the level of two-dimensional plane design [2]. The introduction of advanced virtual reality technology into interior design is conducive to the presentation and enhancement of space and a three-dimensional sense of interior design. It can help users quickly, efficiently, and accurately meet customers' needs, making the interior design more intelligent and humanized. At the same time, it is also conducive to the effective, economical, and flexible production organization of developers to shorten the development cycle and reduce the cost to the greatest extent, optimize the design quality, and maximize the production efficiency [3].

2 The Application of VR Technology in Interior Design Under the Background of Big Data

2.1 Big Data Era

Big data is also an emerging research hot spot. The "big" of big data is not only reflected in quantity but also various forms, including not only structured data such as pictures, video, audio, and even semi-structured data. With the popularity of mobile terminals, a large number of applications emerge endlessly, and the data generated is unpredictable. The data generated in one second can be calculated. Needless to say, the era of data everywhere is an era of big data construction. Classic cases such as beer and diapers, Obama's re-election success also make people understand that data has become an important production factor in this era [4].

2.2   Computer Graphics

The development of VR technology is inseparable from computer graphics (CG). Computer Graphics is a subject with computer computing technology, processing technology, and display technology to transform 2D and 3D graphics into computer grid form. Up to now, the research direction of computer graphics has included computer animation, graphics interaction, solid modeling, and virtual reality, and many computer technologies are inseparable from computer graphics, which are widely used in all fields of life [5, 6].

2.3   Concept of Virtual Reality

Virtual reality technology, whose essence is to simulate the virtual or real world. VR technology can also be said to be an advanced simulation system. The virtual environment it creates comes from real life or the virtual environment imagined by human beings. Its effect is immersion. You can not only see and hear it but also touch and smell it. It makes people feel the world existing in reality or human imagination through special electronic output devices.

2.4   Characteristics of Virtual Reality Technology

Multi-sensory refers to virtual reality that can let people experience vision, hearing, smell, touch, taste, and so on [7], which is to achieve the point of confusing the real with the fake.

The sense of existence is to make users feel that they exist in the virtual world. Interactivity means that users can get real feedback from all activities in the virtual environment. Autonomy means that objects' motion law in a virtual environment is consistent with the physical motion in reality. Conceivability refers to users creating a new environment and having surreal cognition when interacting with the simulated environment.

2.5   Virtual Reality Technology and Interior Design

The stigmatization of the Internet home decoration industry increases the cost of scheme selection for owners, and the VR immersion experience solves this problem greatly. Intuitive experience can not only help owners make decisions but also lead to becoming a new standard in the industry. Because of this advantage, many capitalists began to increase investment in VR technology [8].

VR technology breaks the traditional design mode of "plane, elevation, section, and 3D". Designers can use this technology to "walk-in" at the stage of improving the scheme design and intuitively feel the changes of light, space, and scale to make the design more perfect. In short, the application of VR technology in interior space design can be summarized as demonstration tools, design tools, promotion tools [9].

Demonstration tools. Professional design cannot be understood by everyone. No matter how professional the designer is, it may be difficult for customers to understand some professional words. Nevertheless, VR technology can let customers see and feel the design results intuitively and put forward more targeted requirements, promoting the communication between designers and customers [10].

Design tools. VR technology makes it possible for designers to roam in space. Designers can modify 3D models through VR experience and compare the feasibility of different schemes.

Promotion tools. VR has caught the attention of real estate developers since its birth. VR devices have advantages in a mobile and visual experience that traditional sand Table cannot match. It can be said that VR has become an effective tool for them to win customers and expand resources.

2.6   Defects of VR Technology in the Field of Interior Design

(1) Low Authenticity

The reality of immersive experience is the key standard to measure the virtual reality technology equipment. When customers use virtual devices to experience, sometimes their activities will be interfered with by the outside world, such as climbing stairs and sleeping in bed. These activities need to have the action of trampling and lying down, and the real scene of the experiencer often restricts these actions and even reduces the reality of the simulation environment.

(2) Poor User Experience

Using virtual reality technology to experience interior space design needs a special human-computer interaction interface, but the development of human-computer interaction is not very mature. In the process of wearing VR glasses, some users will feel dizzy and nauseous due to the operation of glasses. Sometimes the lens moves too fast and even affects the user's vision.

(3) High Cost

Time costs are high. There is no targeted interior design software in the market. The traditional design mode is mostly integrated with multiple software deployment designs, and the disadvantages of this model are complex operation, long cycle, and high cost. Although VR technology has improved these problems to the greatest extent, it is still necessary to cross-use other 3D software. The process of importing and exporting results in a long design time and many incompatibility defects.

The cost of investment is high. Although many developers are enthusiastic about virtual reality technology, they are in a wait-and-see state. The high cost of development and high price equipment make it difficult to popularize, leading to high yield and low return risk.

2.7   Innovation of VR Technology in the Field of Interior Design Practice

To solve the shortcomings of VR technology in the field of interior design and realize the further development of VR technology in the field of interior design, firstly, the design mode should be innovated, and the development of system software can simplify the design process. The most important thing is to strengthen the sense of reality. If the sense of reality reaches the standard, it will weaken the buffer between reality and the simulated environment. Generally, ray tracing algorithms and radiance algorithms are used in realistic rendering. The ray-tracing algorithm is faster and simpler than the radiance algorithm. When using this method, the octree data structure and the phone model are usually used, and the wllied model is used for the overall illumination, considering the regional characteristics of the building and the calculation speed.

3   The Experiment and Research of VR Technology in Interior Design

The objects of this experiment are two real estate development companies. Company A adopts virtual reality technology, while company B adopts traditional sand Table simulation.

The whole process of the sales of the two developers was tracked and investigated. The two companies used different technologies to select 100 ordinary residents to go to the two sales offices and conduct field experience and investigation. Then, the views of the 100 ordinary residents on the two companies were made. The form of the interview was a questionnaire.

The survey content is the respondents' feelings about the two companies' products and services, then points and sorts out their answers, and finally analyzes the results.

4   Analysis of Experimental Results of VR Technology in Interior Design

100 questionnaires were sent out, 93 questionnaires were effectively returned, and the effective recovery rate was 93%.

4.1   Public Understanding of VR Technology

In the questionnaire, we have an indefinite multiple-choice question of "what are the VR application examples you know" with 8 options including VR+smart city, VR smart medical, VR cultural education, VR entertainment, VR design, VR military, VR cultural protection, VR geography, etc. The data and results are as follows (Fig. 1):

Table 1.   The public's understanding of VR technology

	VR +smart city	VR smart medical	VR culture education	VR entertainment	VR design	VR military	VR culture protection	VR geography
Number of people	28	40	52	80	37	27	26	44
Percentage	30.1%	43.0%	55.9%	84.2%	39.8%	29.0%	28.0%	47.3%

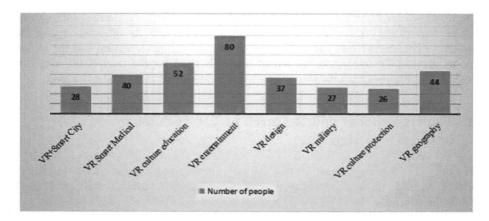

Fig. 1. The public's understanding of VR technology

From Table 1, we can see that VR involves all aspects of people's understanding, but the VR entertainment field has more understanding, like many games and film and television entertainment use VR technology. But in the smart city, VR military, VR cultural protection, but not much.

4.2 Which Kind of Experience and Service is More Preferred

In the survey, we interviewed the residents who participated in the survey about their feedback after visiting company A and company B. 48% of them were satisfied with company A, 25% were satisfied with company B, 15% were satisfied with company A and company B, and 12% were not satisfied with the both. The specific situation is as follows:

Fig. 2. Customer satisfaction comparison

It can be seen from the above Fig. 2 that company a, which uses virtual reality technology, brings a much better experience and service to customers than company B, which uses traditional sand Table simulation. VR technology allows customers to choose their own satisfying house without seeing the house on the spot. It can be said that VR technology realizes the everywhere movement of sales offices. It not only facilitates customers but also expands the source of developers' customers. However, the traditional sand Table simulation is not as vivid as VR technology and it is not easy to move around. Furthermore, the making of a sand Table model is also time-consuming and much more complex.

5   Conclusion

Virtual reality (VR) technology is an important branch of computer science and technology in the era of big data, and it has a bright future in this emerging technology era. It simulates an artificial environment through the information interaction between computers, which can be widely used in interior decoration. Its way of expression greatly innovates the level of traditional design so that artists and designers can transcend the limitations of time and space, have a conscious dialogue with the future, and inspire designers to create more inspiration. For customers, virtual reality technology provides us with more diversified choices, which helps to achieve reasonable and efficient space design and intuitive expression of space and realize the scientific control from the overall design to the final details. Interior space design is a large-scale data analysis and model reconstruction with the help of virtual reality technology, which breaks the barriers of user experience and human-computer interaction. There is no need to deny that VR technology is still insufficient in the field of interior design, and domestic technology is not particularly mature. However, with the progress and development of science and technology in China, it will become an inevitable trend in the development of interior design and other industries, and then combine other technologies such as artificial intelligence and the Internet of things. It will cause another revolution in interior design science and technology.

References

1. Żmigrodzka, M.: Development of virtual reality technology in the aspect of educational applications. Mark. Sci. Res. Org. **26**(4), 117–133 (2017)
2. Radley, L.: Interior design. Commer. Mot. **227**(5740), 32–37 (2017)
3. Meyrueis, V., Paljic, A., Leroy, L., et al.: A template approach for coupling Virtual Reality and CAD in an immersive car interior design scenario. Int. J. Prod. Dev. **18**(5), 395–410 (2017)
4. Wolfert, S., Ge, L., Verdouw, C., et al.: Big data in smart farming–a review. Agric. Syst. **153**, 69–80 (2017)
5. Peter, S., Thompson, W.B., Peter, W., et al.: Fundamentals of Computer Graphics. World Scientific Publishers Singapore, vol. 9, no. 1, pp. 29-51 (2009)
6. Rothbaum, B.O., Rizzo, A., Difede, J.A.: Virtual reality exposure therapy for posttraumatic stress disorder. Ann. N. Y. Acad. Sci. **208**, 126–132 (2019)

7. Lin, X., Song, S., Zhai, H., Yuan, P., Chen, M.: Physiological reaction of passengers stress metro fire using virtual reality technology. Int. J. Syst. Assur. Eng. Manag. **11**(3), 728–735 (2020). https://doi.org/10.1007/s13198-020-00991-y

8. Lee, Y., Choi, W., Lee, K., et al.: Virtual reality training with three-dimensional video games improves postural balance and lower extremity strength in community-dwelling older adults. J. Aging Phys. Act. **11**(3), 728–735 (2020)

9. Xue, W.: Virtual reality interior design based on paper material and fuzzy evaluation method. Paper Asia **2**(2), 186–189 (2019)

10. Liu, J.L., Zhu, L.: Research and implementation of virtual campus roaming system interaction function based on OSG. Appl. Mech. Mater. **687–691**, 2219–2223 (2014)

Earthquake Protection and Disaster Reduction Technology System Based on Optical Network

Chen Wu[✉], Junhao Qu, Hao Zhang, and Yang Zhao

The Seismological Bureau of Shandong Province,
Jinan 250014, Shandong, China

Abstract. The technical system of the new science and technology park for earthquake prevention and disaster reduction of Shandong Seismological Bureau includes: comprehensive wiring of the new office building, data center computer room, data communication network, basic information platform, intelligent property, etc. The technical system of this construction is a set of complex system, which not only includes computer system and other supporting equipment, but also includes redundant data communication connection, environmental control equipment, monitoring equipment and various safety devices. In the construction of the technical system of the central station, Shandong Seismological Bureau fully investigated the functional requirements and index requirements of each technical system according to the construction idea of "systematization, intelligence, modularization and integration". The requirements of functional and performance indicators follow the principle of "practical, advanced and forward-looking", which not only meets the needs of operation and management of existing business systems, but also meets the needs of business development for at least 10–15 years in the future. Combined with the actual needs of the industry and future business development planning, it provides reliable technical parameters, feasible technical solutions and valuable construction experience for similar projects in the same industry, and can also play a certain exemplary role in the industry.

Keywords: Virtualization · Information security · GPON technology · Data center · Intelligent property

1 Introduction

Advanced technologies such as virtualization technology, information security technology, GPON technology and cloud storage have been applied in the construction of technology system of Shandong earthquake prevention and Disaster Reduction Science and technology park. Virtualization technology greatly improves the efficiency and utilization of data storage devices. The research results of information security technology have been applied in the work of information security classified protection, and have obtained the three-level record qualification of classified protection in cooperation with the transformation and upgrading of the industry backbone network of our bureau. The research results of GPON technology and high availability of information center play a positive role in ensuring the operation and maintenance efficiency of the

J. C. Hung et al. (Eds.): FC 2021, LNEE 827, pp. 309–317, 2022.
https://doi.org/10.1007/978-981-16-8052-6_38

technical system of the central station. These technologies are the mainstream information technologies at present, and they are also the key technologies used by our seismic industry on the road of informatization [1]. The whole technology system has been built and operated for more than five years, It has fully undertaken the intelligent operation and maintenance of the whole Shandong earthquake prevention and mitigation Park, and all business transmission, storage, data analysis and processing, daily duty, emergency duty, guarantee and other tasks of Shandong earthquake industry, And the system has been applied to the daily business operation of the industry, and significantly improved the efficiency of the business system.

2 Project Background and Significance

With the development of earthquake prevention and disaster reduction and the pace of the information age, the business operation of the industry (system) has also opened a digital, network era. In this rapidly developing information age, only by applying advanced information technology to the work of earthquake prevention and disaster reduction, can our cause not lag behind the development of the times, and can better serve the society and national economic construction. At present, information system (technology system) plays a vital role in the industry, carrying the key business of the industry, providing us with timely and reliable data transmission, data analysis, data mining, high-performance computing and other services.

The earthquake industry information system of Shandong province carries the application support of industry network, Internet and internal office network. The application of industry network includes seismic measurement, strong earthquake, precursor, information, emergency command and other business systems; Internal office network applications include OA, document transmission, instant messaging and other services; Internet applications include external information release, e-mail and other services [2]. The strong motion system is responsible for the real-time data receiving and storage of 86 seismic stations and 135 strong motion stations in the whole province. It is also responsible for the earthquake quick report and earthquake catalogue in the whole province and serves for earthquake prediction and scientific research. The specific business system includes station data flow, database, real-time processing, monitoring and alarm, human-computer interaction positioning, earthquake quick report, station parameter synchronization and other seven services. Before the completion of the new system, all the services are in the mode of dual deployment and dual backup. The information system provides basic network services and communication links for other businesses, and carries 19 services including website, e-mail, DNS, e-government, instant messaging, intranet phone, document transmission, patch distribution, webpage tamper proof, government affairs publicity, digital library, and most of the businesses run on virtual machine.

3   Scientific Ideas and Technical Approaches

Cloud computing technology evolved from parallel computing and grid computing. After using cloud computing technology, people use information resources in more and more diverse ways and contents. In the Internet age, people get all kinds of information and services through the Internet. Our old information system was self-sufficient at that time. The utilization rate of original system resources (such as CPU, memory, etc.) is very low, but the new system cannot fully share and utilize the original system resources in the construction process. New resources must be purchased to meet the needs of new systems, which leads to a certain degree of resource waste and repeated investment [3]. The information system constructed by cloud computing technology only needs one-time construction of resource pool. The subsequent construction of new system can make full use of the resources in the original system resource pool, and carry out flexible scheduling and allocation of resources, so as to achieve the purpose of improving the use efficiency of resources and saving investment. Cloud computing architecture includes three basic levels: infrastructure as a service (IAAs), platform as a service (PAAS) and software as a service (SaaS). Infrastructure as a service (IAAs) refers to a service that provides IT infrastructure capabilities to users through the Internet [4], and charges according to the actual usage or occupancy of resources. IaaS: IAAs management platform includes two parts: resource management platform and business service management platform.

The key technologies to achieve infrastructure as a service IAAs are:

(1)   Host Virtualization: host virtualization applies system virtualization technology to servers, and creates multiple independent virtual machine servers on the server. Server virtualization provides the hardware resource abstraction that can support the virtual server, including virtual BIOS, virtual processor, virtual memory, virtual device and i/o, and provides good isolation and security for virtual machine (Fig. 1).

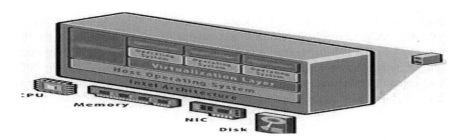

Fig. 1. Host virtualization

Host virtualization can accelerate application deployment: through host virtualization, virtual machines can be quickly configured, copied and started. Compared with traditional application deployment, it reduces manual intervention, shortens deployment time and reduces deployment cost. At the same time, it also has the following major functional advantages: improving service availability; realizing dynamic resource scheduling; reducing energy consumption.

(2) Storage Virtualization: a technology that abstracts, hides or isolates the internal functions of storage system from application, host or network resources. Its purpose is to manage storage or data independent of application and network [5]. Virtualization technology provides a simple and unified interface for the access of the underlying resources, so that users do not have to care about the complex implementation of the underlying system.

(3) Network virtualization: it is divided into the virtualization of external network environment and the network virtualization within the server host. Network virtualization is a process of integrating multiple hardware or software network resources and related network functions into a software available for unified management and control, and the realization of the virtual network environment is transparent (Fig. 2).

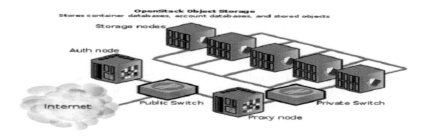

Fig. 2. Swift object storage architecture

In short, the IT system based on infrastructure as a service IaaS is highly advanced and forward-looking because of its strong functions. Its construction mode will become the mainstream of the IT system construction. The existing network, storage, host and virtualization technologies have basically become mature, and these technologies have been applied to the construction of this technology system, and the information of Shandong earthquake prevention and disaster mitigation is applied. Take the infrastructure construction as an opportunity, optimize the resource allocation [6], improve the service ability of the business system, and solve the existing problems.

4   System Construction

4.1   Construction Principle

(1) Reliability: the overall design of the system fully considers the safety and reliability, the availability of network equipment and network in the communication

network, the reasonable design of network architecture, and the redundant backup of core equipment and links.

(2) Security: the system deploys effective security policies to protect the seismic industry data with high security.

4.2   Technical Ideas

According to the development of technology and the needs of practical application in seismic industry, the mixed networking mode of Ethernet switch networking and PON equipment networking is adopted. The core layer adopts the dual machine mechanism to configure the core switch; In the access layer, the first floor command hall and the second floor duty center of the network center on key floors are connected to the core switch in the mode of access switch, PON devices are used for networking in other floors, OLT devices are placed in the core computer room, optical splitter is configured in each floor of each building according to the demand, and ONU devices are configured in each office according to the specific demand, so as to achieve fiber to desktop.

4.3   Advantages of the Scheme

The mixed networking mode of Ethernet switch and PON equipment not only ensures the security and stability of the core network, but also realizes the high reliability of key floors; The optical fiber instead of traditional transmission medium extends to the user terminal computer, which enables the user terminal to realize network access through the optical fiber [7]. In terms of core switch equipment: The industry network adopts cloud switch, which can not only meet the needs of the industry within 10–15 years, It is also well prepared to form a "double live" data center with the national network center in the future. In terms of PON networking equipment, OLT equipment is placed in the core computer room, the splitter is placed in each floor of each building, and the splitter is a passive device, which reduces the fault points (Fig. 3).

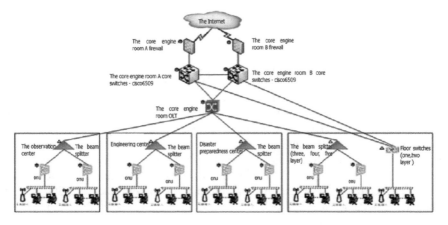

Fig. 3. Schematic diagram of Internet network topology

- Optical fiber can support transmission of further distance and higher bandwidth.
- Optical fiber is a non-metallic material, data transmission on optical wave can avoid external electromagnetic interference and radio frequency interference, and there is no crosstalk between fiber cores, and the signal will not be leaked, which plays a very good role in confidentiality.
- Optical fiber can be used in inflammable and explosive places with wide temperature range and no electricity for communication.
- Optical fiber is resistant to chemical corrosion and has long service life. Different sheath materials and internal structures can cope with harsh wiring environment.
- With the development of optical communication technology, the cost of raw materials has declined, and the installation and construction become more and more simple (Fig. 4).

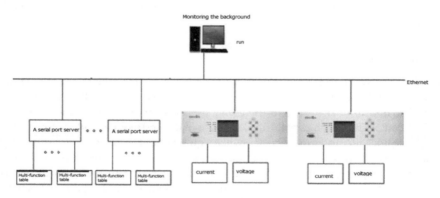

Fig. 4. Schematic diagram of energy consumption measurement

Data center plays an increasingly important role in the cause of earthquake prevention and disaster reduction. At present, data center can meet the new application requirements of cloud computing. The key technologies affecting the deployment and operation of the data center are analyzed one by one, It mainly includes: Network architecture design, virtualization technology, network convergence technology, security technology and green energy saving technology. The development mode of traditional data center has seriously hindered the development of data center. The convergence of transmission network, storage network and high-performance computing network is the development trend of data center network [8]. Through the convergence, we can reduce the cost, reduce the complexity of management, improve the security and other advantages. The goal of the network construction of the data center is: an information network system integrating office, command, decision-making and service, realizing office automation, information sharing and scientific management, and connecting with other public networks safely. At present, the running energy consumption and refrigeration energy consumption of the data center are very large, which adds too much burden to our users, and is not conducive to resource conservation and environmental protection. Therefore, one of the requirements of the new data center is to reduce energy consumption. When we designed the refrigeration system, we fully considered the N + 1 mode of power supply for the cooling equipment. It is the first time that this mode is adopted in the data center of the industry. We have conducted a more in-depth study on how to make the advanced power supply and refrigeration mode play the most effective role. In response to the call for energy conservation, the total power consumption of the main incoming box on each floor of the park is monitored to realize remote meter reading. The system platform has been built in the power equipment environment monitoring center of data center, and only the platform is expanded and debugged this time.

According to the code for design of intelligent property system, the system is designed according to A-level intelligent property system, and the main machine of power supply equipment, backstage support and air conditioner are designed according to redundancy to ensure uninterrupted communication link. The intelligent property system based on optical network is characterized by: the advanced system integrating high efficiency and convenient service can realize the scientificization of intensive management and safety connection with other public networks. The system can provide a high reliable and high-performance intelligent management platform for the earthquake prevention and Disaster Reduction Park of Shandong Province, and comprehensively consider the stability, compatibility and expansion ability of the network [2]. Efficiency and security are the core requirements of the system, so the construction of the system responds to the specific requirements and the comprehensive consideration of the development of network technology. In view of the development of information technology and the actual situation of the construction of earthquake prevention and Disaster Reduction Park of our bureau: large construction area, many information points, many specialties, as well as the actual application needs of the park and the construction of phase II and phase III, sufficient development space is reserved. The optical network adopts the mixed networking mode of Ethernet switch and PON equipment (Fig. 5).

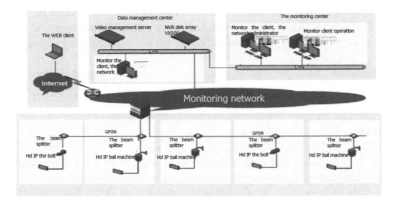

Fig. 5. Layout of security system

With the development of information technology, our bureau uses the existing mature technology to build the basic information platform supporting the technology system of Shandong earthquake prevention and Disaster Reduction Park. According to the construction concept of "advanced intelligent property", on the basis of original hardware server and storage resources [4], New hardware servers and storage resources are added, and the advanced optical network has been built, The establishment of a unified support platform for intelligent property background data processing and storage provides reliable infrastructure services, which is in line with the technical trend of the industry in the future. The advanced intelligent property system based on optical network has fully played its advantages in the earthquake prevention and Disaster Reduction Park of Shandong Province, This achievement through the research of the system's high-efficiency operation and maintenance, so that it can better serve our earthquake prevention and mitigation cause, and this achievement has been in the rescue training base (Park Phase II) weak current design and construction and has played a good effect.

5   Conclusions

The whole system adopts advanced design and construction concept, Such as: host room placed network switching equipment, storage, server, column head cabinet, distribution frame, etc.; The power battery room is equipped with UPS equipment and supporting battery pack, gas fire control, air conditioning and other equipment. It ensures the efficient and safe operation of data calculation, transmission and storage in the seismic industry of Shandong Province. Virtualization technology greatly improves the efficiency and utilization of data storage devices. The research results of information security technology have been applied to the classified protection of information security, and have achieved the three-level record of equal protection in cooperation with the transformation and upgrading of the industry backbone network of our bureau. The research results of GPON technology and high availability of information center play a positive role in ensuring the operation and maintenance efficiency of the technical system of the central station. At present, the whole system has fully undertaken the intelligent operation and maintenance of the whole Shandong earthquake prevention and mitigation Park, and all the business transmission, storage, data analysis and processing, daily duty, emergency duty, guarantee and other tasks of Shandong earthquake industry.

Acknowledgements. Fund projects: Key project of Shandong Natural Science Foundation (zr2020kf003).

References

1. Rongguo, F.: The emergency communication technology based on wireless Mesh network technology application. J. Disaster Prev. Mitig. Eng. **6**, 778–783 (2014)
2. Wu, C., Fang, X., Wang, P., et al.: Shandong seismic data center high availability study. Comput. Appl. Res. **32**(suppl.), 19–21 (2015)

3. Chen, W.: GPON optical fiber technology in the application of seismic information network in Shandong. J. Earthq. Def. Technol. **8**(1), 1–111 (2013)
4. Xin, G.: The earthquake emergency platform and emergency communication system research. J. Digital Commun. World **12**(7), 54–57 (2016)
5. Yong, H.: The design of the intelligent property management system and research. J. Henan Sci. Technol. **36**(10), 80 (2013)
6. Li, G., Sun, J., et al.: The MPLS VPN in Tianjin earthquake monitoring system of high speed LAN applications. J. Earthq. Def. Technol. **7**(1), 92–99 (2012)
7. Kan, Y.: The era of cloud computing data center construction and development. J. Inf. Commun. **166**(6), 100–102 (2011)
8. Jian, Q.: The light field network coding node model based on logic operation. J. Electron. **40**(7), 1304–1308 (2012)

Industrial Economics Analysis of Urban Agglomeration Effect in the Information Age

Jinfeng Wang[✉]

Department of Economics and Management, Yunnan Technology and Business University, Kunming, Yunnan, China

Abstract. With the rise and rapid development of mobile communications, smart terminals and information technology, we are entering the era of mobile information technology. The urban agglomeration effect, as the name implies, is the unified planning of towns and villages. Specifically, it is to coordinate the relationship between urban and rural areas, the relationship between industry and agriculture, and to agglomerate the city. This article mainly introduces the industrial economics analysis of urban agglomeration effects in the information age. Optimize resource allocation, and gradually clear the gap between urban and rural areas. Urban agglomeration is to adhere to the guidance of the scientific development concept and carry out integrated construction of cities and rural areas. The experimental results of this paper show that the industrial economics analysis of the urban agglomeration effect in the information age has increased the urban agglomeration by 16%. The limitation of the industrial economics of the urban agglomeration effect in the information age is very important for the urbanization of local industries. The planning methods and approaches are analyzed, discussed and summarized, thereby enriching the academic research results.

Keywords: Informatization · Urban agglomeration · Industrial economics · Local industry

1 Introduction

The effect of urban agglomeration in the information age. In the process of continuous improvement of the social security system in urban and rural areas, my country should adhere to the practical method of "urban agglomeration" in accordance with my country's basic national conditions in order to finally realize the urban-rural integration of the social security system [1, 2]. Urban-rural integration involves all aspects including social culture, national economy, natural environment, etc., and needs to realize the common development of society and economy of urban agglomeration [3, 4]. Complete the free circulation of resources in urban and rural areas and industrial interaction, promote market integration and promote industrial urban-rural integration, and promote the development of urbanization and urban-rural integration through industrialization [5, 6].

With the advancement of science and technology and the rapid development of the Internet, Zhou J. divides the location factor into local factors, agglomeration factors and

© The Author(s), under exclusive license to Springer Nature Singapore Pte Ltd. 2022
J. C. Hung et al. (Eds.): FC 2021, LNEE 827, pp. 318–323, 2022.
https://doi.org/10.1007/978-981-16-8052-6_39

scattered factors. The so-called local factors mainly refer to transportation costs and wage factors [7]. Ren C thinks that the so-called aggregation factor refers to the reduction of labor or sales costs due to a certain amount of production concentrated in a specific area [8]. Formulate a reasonable and effective entrepreneurial industrial economic agglomeration policy. Gathering is of decisive significance for the improvement of the city's core competitiveness [9, 10]. However, there are errors in their experimental process, which leads to inaccurate results.

The innovation of this article is to put forward an industrial economic analysis of the effect of urban agglomeration in the information age. A good environment is the fertile soil for industrial economic agglomeration, and it is also the basis and guarantee for the industrial economic agglomeration effect. The emergence of the industrial economic agglomeration effect requires a sufficient material basis, including advanced equipment, comfortable working environment, and reasonable material income. The city should ensure the adequate supply of these material conditions. The purpose of this research is to find a new path for the development of industrial economics that is suitable for the current urban agglomeration effect in the information age.

2 Industrial Economy Under Informatization

2.1 Domestic Industrial Economy

Since 2005, many domestic scholars have done a lot of research on industrial economic theories, and further expanded Western industrial economic theories based on the characteristics of China. The content mainly includes the concept and property rights of industrial economy, the status quo and form of China's industrial economy, and the relationship between industrial economy and economic growth. Scholars divide aggregation factors into two types: one is the aggregation of "large-scale production benefits" brought about by the expansion of urban production levels.

$$X_\mathrm{t} = \sum_{j=0}^{q} \theta_i \varepsilon_{i-j} \tag{1}$$

The calculated value of the algorithm is as follows;

$$3I_{01} = I_{A1} + I_{B1} + I_{C1} = 3U_0\mathrm{j}\varpi C_{01} \tag{2}$$

The test shall use the following formula:

$$X_\mathrm{t} = \sum_{i=1}^{p} \mathrm{a}_i x_{t-i} + \varepsilon_t \tag{3}$$

2.2 Urbanization Industry

According to the city's characteristics and positioning, combined with its own resources and location advantages, it can formulate reasonable ways and directions to attract entrepreneurs to invest according to local conditions. Formulate reasonable tax and land preferential policies. Profit is the main goal pursued by entrepreneurs. Preferential taxation and planning policies have a certain incentive effect on entrepreneurs, and are conducive to attracting entrepreneurs to gather in the industrial economy. However, this kind of policy lacks a certain degree of sustainability and has strong negative effects. It is more suitable for the initial use of urban development. Improve relevant service guarantee policies for enterprises, including enterprise registration, loan processing, tax payment, and normal production and operation. Gradually establish a market environment that is conducive to fair competition, and use rule-based and related policies to guide the realization of the internal interdependence between enterprises and cities, so that enterprises have a strong dependence on the city, forming a city that enterprises rely on, and cities rely on enterprises, and each other A win-win situation of dependence and common development.

3 Local Industries Under Informatization

3.1 Local Economic Analysis

The concentration of local industrial economy, changes in spatial distribution, and the formation of industrial economic agglomeration, can generate greater energy and create results that cannot be achieved under the dispersed state of a single industrial economy. Talents are generally one-skilled, and those who have specialties in two or three areas are great talents, or "generalists", and those who have specialties in more areas are called "all-rounders". Generalists and omnipotent talents are not only rare but also limited, and "all-round" and "all-round" talents, strictly speaking, do not exist at all. In order to enhance the core competitiveness of a country or city and promote the rapid and sound development of the economy, there are many things to be done, and talents from all aspects are needed. Therefore, it is necessary to gather talents from all aspects to form the advantages of the industrial economy and form a "reasonable structure." Rely on "joint force" to win. The practice of economic development in developed countries has confirmed this point. It causes a decrease in the individual value of the industrial economy, increases transaction costs, and produces uneconomic effects of industrial economic agglomeration.

3.2 Status Quo of Local Industry

The economic situation of local industrial entrepreneurs is mostly innovative. This can be done by changing the organization, market, and product. In the post-urbanized industries, most people with industry skills are relatively lagging behind in allocating resources, discovering new opportunities, insight into the business development process, and management and interaction capabilities. As an industrial economy with scarce resources, they yearn for a better living environment. Clean air, high-quality environment, and convenient transportation have become indispensable considerations in the flow of industrial economy. The specific results are shown in Table 1.

Table 1. Comparison of several wireless data transmission methods

Number of hidden layer nodes	2	3	4	5
Relative error percentage	6.34%	4.57%	2.64%	0.24%
Mean square error	0.0095	0.0125	0.0016	0.0007

4   Economic Analysis in the Effect of Urban Agglomeration

4.1   Urban Agglomeration System

The laboratory project in this article puts forward a hypothesis of urban agglomeration, and briefly describes that the competitiveness of industrial economy individuals and groups have been improved, thereby promoting technological innovation within the industry and technological upgrading of the entire industry. If it plays a role in the enterprise it will greatly increase the profitability of the enterprise and produce huge benefits; if the agglomeration effect of the industrial economy is formed in the industry, the advantage of the industry will be established to realize the agglomeration and development of the industry; if the agglomeration effect of the industrial economy is brought into play in the city, will greatly enhance the value of the city. The specific results are shown in Fig. 1. Urbanized cities have more industrial advantages, especially first-tier cities can integrate more resources.

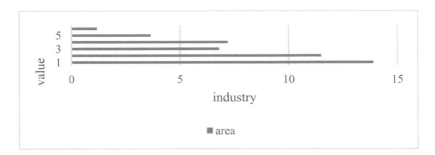

Fig. 1. Especially first-tier cities

4.2   Economic Analysis of Urban Agglomeration Industries

The core competitiveness of a city is closely related to the industrial economy. In other words, the industrial economy is a necessary condition for the formation of the core competitiveness of a city. Including natural resources and social resources. Although the stocks and types of resources in many cities are relatively similar, the development of cities is quite different. Some cities have even experienced "resource tragedies". The reason is that there are differences in the ability of resource integration and transformation. The acquisition of these capabilities mainly comes from the development of the enterprise. Entrepreneur capital is an indispensable factor for the development of the enterprise. The specific results are shown in Table 2.

Table 2. Statistical table of sample library

	Normal	Traditional algorithm matching time	Improved algorithm matching time
Number of transformers	17	1025	638
Total sample	305	1214	352
Training samples	204	1179	386
Validation sample	101	1233	429

Formulate a proactive industrial economic flow policy. The aggregation of industrial economy is conducive to the full display of industrial economic value and produces a series of economic effects that are conducive to urban development. The industrial economy directly participates in production and value formation. With rapid growth, the city's production capacity has been enhanced, and the industrial economy has high cultural qualities and good moral cultivation. Its gathering will inevitably form advanced values, mature social psychology, and a good humanistic environment, thereby enhancing the culture of the entire society. Quality and spiritual level. The specific results are shown in Fig. 2. Compared with the urban industrial economy, there is more trade in the economic system.

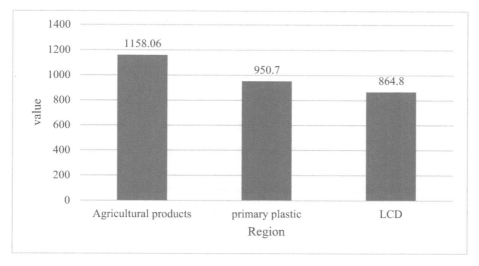

Fig. 2. Economic effects

5   Conclusions

Although this article has a lot of deficiencies in the industrial economics analysis of urban agglomeration effects in the information age. Cities should formulate industrial economic strategies based on their own resource advantages and characteristics, including what type of industrial economy to attract, and then formulate various preferential policies, including salary, housing, facilities, The design of indoor path planning algorithms for complex environments not only requires extensive theoretical knowledge, but also a solid theoretical foundation and competence. In the information age, the industrial economics analysis of urban agglomeration effects still has a lot of in-depth content worthy of study. There are still many steps to study the interior path design analysis because of space and personal ability, etc., which are not covered. In addition, the actual application effects of the related experiments of algorithm design can only be compared with traditional models from the level of theory and simulation.

References

1. Yu, M., Liang, Q.: Research on the construction of Guangdong, Hong Kong and Macao Bay Area in the new pattern of comprehensive opening up. Int. Trade **445**(01), 4–11 (2019)
2. Chen, W., Li, Z., Xie, W.: Discussion on the urban spatial development strategy of Dongguan under the strategy of Guangdong, Hong Kong and Macao Bay Area. Planner **35** (287, 11), 69–74 (2019)
3. Huang, H., Wen, Y., Qiu, W., et al.: Prospects and countermeasures of the "sustainable development plan" for the development of small and medium-sized enterprises in China: a survey of 16 cities in Yunnan, Guangxi, Guangdong (Guangdong, Hong Kong, Macao, Dawan District), Fujian and other places. Natl. Circ. Econ. (32), 2–3 (2019)
4. Kuang, H., Wu, S.: Discussion on the development trend of Higher Vocational exhibition education under the background of Guangdong, Hong Kong and Macao. Mod. Comm. Trade Ind. **041**(006), 162–163 (2020)
5. Zhai, X.: Research on the path of promoting rural revitalization in Zhuhai under the background of Guangdong, Hong Kong and Macao. Natl. Circ. Econ. **2259**(27), 96–98 (2020)
6. Chuan, T., Jinwei, Z., Wang, X.: CEPA and regional economic and trade integration of Guangdong, Hong Kong and Macao under e-commerce environment. Mall Modernization **21**, 161–162 (2005)
7. Zhou, J.: Survey on the living conditions of small and medium-sized electronic traders in Huaqiangbei, Shenzhen. Electr. Comp. Inf. (011), 56–57 (2009)
8. Ren, C., Hangrong: Analysis of influencing factors and Countermeasures of e-commerce development in Shenzhen. Rural Staff **628**(16), 231–232 (2019)
9. Hong, J., Lu, X.: Development status, problems and Countermeasures of cross border e-commerce in Shenzhen. Spec. Econ. **347**(12), 98–100 (2017)
10. Jingbo, F.: E-commerce helps SMEs develop international trade. E-commerce **06**(6), 37 (2009)

Application of Video Teaching Platform Based on Internet in Guzheng Teaching in Information Age

Yinghong Fu[(✉)]

Hunan Yiyang Normal School, Yiyang 413000, Hunan, China

Abstract. As an ancient musical instrument loved by people, the teaching and practice of Guzheng are paid more and more attention. It is of great significance to develop students' personality, expand their knowledge structure, improve their comprehensive quality and activate campus culture. This paper discusses the development opportunity of Guzheng art in the network era and the significance of constructing Guzheng network teaching, and puts forward the role of network public platform in Guzheng teaching. The results show that: from 2018 to 2020, the scale of online education market will increase from 12.132 million yuan to 34.154 million yuan in three years, and the growth rate of market scale will increase from 17% to 47%.

Keywords: Information · Internet · Teaching platform · Guzheng teaching

1 Introduction

With the attention to the traditional culture of the nation, more and more people begin to learn this instrument, and its teaching and artistic performance are gradually concerned by the society. Guzheng performance and artistic performance play an important role in shaping personality, optimizing knowledge structure, improving personal accomplishment and comprehensive quality, and have important significance and influence on promoting traditional Chinese culture.

With the development of science and technology, many experts have studied the teaching of guzheng. For example, some Chinese teams have studied the teaching contents of zither textbooks. Guzheng can inherit and develop the excellent cultural heritage of the nation. From the perspective of Guzheng playing techniques, the development and development of Guzheng performance techniques are described. According to the characteristics of modern formation, the paper analyzes the cultural connotation of each school, discusses the different views existing in the current teaching syllabus of each school, puts forward its own views, and makes a thorough and detailed study on the formation mode and psychological causes of Guzheng performance. This paper discusses the specific performance skills and methods of Guzheng concerto from the perspective of teaching, and designs specific teaching practice methods, aiming at improving the performance level of students' guzheng Concerto. The advantages and disadvantages of traditional zither teaching are analyzed in detail, and some suggestions are put forward, which are hoped to be helpful to the reform of

J. C. Hung et al. (Eds.): FC 2021, LNEE 827, pp. 324–332, 2022.
https://doi.org/10.1007/978-981-16-8052-6_40

Guzheng teaching. Through the analysis of the current situation of Guzheng teaching mode, the paper summarizes the corresponding countermeasures [1] some experts have studied the teaching of Guzheng education, and put forward the feasibility of developing guzheng in Colleges and universities, so as to make the guzheng art better inherit and better move to the world cultural and artistic stage. Through the discussion of teaching skills, sitting posture, hand posture and basic technology of guzheng, a simple learning method of guzheng is proposed. This paper discusses the inquiry teaching of guzheng. Inquiry teaching method is an effective teaching method which conforms to the practice of teaching reform. The purpose of introducing situational teaching method into guzheng teaching is to provide some reasonable suggestions for improving the teaching level of Guzheng in Colleges and universities in China [2]. Some experts have also studied the teaching practice of collective lessons of guzheng, expounded the basic situation of Guzheng teaching, and analyzed the mode of Guzheng education in Colleges and universities outside the school. Through the discussion and summary of the advantages, disadvantages and development trend of the teaching of the national instruments, the importance of the teaching of the national instruments is reflected, and the development of the national instruments is promoted. Based on the teaching mode of Guzheng art and appreciation, through the analysis of the classical music of guzheng, this paper discusses the performance and teaching of Guzheng from the aspects of performance skills, work forms, artistic conception and melody connotation, so as to improve the aesthetic grasp and artistic processing ability of the music, and accurately inherit and develop music works [3]. Although the research results of Guzheng teaching are quite abundant, there are still some shortcomings in the video teaching platform of Guzheng teaching in the information age.

In order to study the video teaching platform of Guzheng teaching in the information age, this paper studies the guzheng teaching platform and guzheng teaching, and finds the quantitative formula. The results show that the network teaching is beneficial to the development of Guzheng teaching.

2   Method

2.1   Guzheng Teaching

(1) The teaching concept of Guzheng

The inner ear ability in guzheng teaching refers to the individual understanding and internal performance of each teacher in the teaching process [4]. It is a kind of ability that does not depend on acoustics, it can form the musical image thinking of the soul accurately through memory and memory [5]. This kind of inner hearing ability is a kind of signal theory guzheng quality teaching [6]. The cultivation of inner hearing ability can be carried out through skills training, color experience, connotation improvement, role play, scene presentation and other ways [7]. Music score is a symbol of recording music and a basic tool for learning music [8]. Music appreciation, performance and creation are inseparable from the ability of music reading [9]. For children who first know music scores, teachers can use music familiar to students to teach music [10]. Although the teaching of folk music is mostly simplified, the quintessence is also the most popular and scientific method of recording music. Today, many performance levels and ensemble tracks require five line music to record [11]. Learning five element spectrum can make students learn music more deeply and lay a foundation for music study in the future. So the collective teaching of Guzheng should not only let students read Zheng spectrum, but also let students read five elements.

(2) The function of Guzheng Teaching

Guzheng belongs to the plucking instrument, which is accomplished by the movement of fingers and the contact of strings. Guzheng teaching is a kind of skill training, and the intensive training of fingers promotes the cultivation and development of students' tactile thinking. Scientific experiments show that if we want to strengthen the strength of the fingertips, our finger nerves can make our senses sensitive and also play an extraordinary role. In zither performance, fingers need a lot of finger training and playing to achieve the best musical performance effect, which plays an exciting and stimulating role in the development of brain and thinking. In addition, in learning the zither playing skills, the use of hands and frequent transformation and cooperation not only promote the development of left brain, but also develop the right brain, thus promoting the comprehensive development of intelligence. Through the study and understanding of these zither music, on the one hand, it has brought positive influence to the students in the process of subtlety, on the other hand, it also learned the content of national cultural knowledge, and expanded the history and literature knowledge of the Chinese nation by using music skills, and then improved the students' cultural cultivation.

(3) Problems in the teaching of Guzheng

Imitation is the most popular traditional teaching method in the teaching of performance, and it is no exception in the teaching of guzheng. Guzheng teachers usually demonstrate the complete music first, then start the imitation teaching of "you sentence, I sentence", and cultivate students' relevant performance skills. Such teaching method only teaches students to complete a piece of music according to their own rules and regulations, so that all students play music in a constant way, without their own ideas, they lose the ability to think independently and the ability to play zither. At the same time, it also wiped out the students' creativity and imagination in music. Some teachers lack enough knowledge of instrumental teaching, think that classroom instrumental teaching is only a flash in the pan, can not get long-term development, so the attitude is indifferent, but instrumental teaching has also been recognized by many teachers in classroom teaching, they think it is difficult to implement it. Learning instrumental music is different from learning singing, because it has a strong dependence on the performance basis of students. Because of the different students' acceptance ability and the recognition degree of instrumental music, the students who are based on practical instruments will enter the state quickly and keep up with the progress of learning. Those students with relatively weak musical instrument foundation will be difficult to keep up with the rhythm. Therefore, it is difficult to carry out classroom teaching smoothly in the face of students with large differences in the classroom. When listening to all kinds of music played by zither, we should try to expand the ability of feeling in the ears and understand the differences between the characteristics of zither art and other music works. It can also listen to guzheng and other music versions of the same music repeatedly, especially when practicing certain music; it can listen to famous music or famous performer. While increasing the music memory, the author can absorb the characteristics of each celebrity, effectively grasp the overall feeling of music, make the auditory impression integrated, and make the brain form a complete sound whole. For a long time, we can strengthen the accumulation of musical emotional experience, master the artistic language and expression means of zither, and have certain musical emotional experience, and gradually improve the perception and understanding ability of Guzheng art.

2.2   Teaching Platform

(1) Video teaching

To a certain extent, the setting of a single skill course and the compiling method of teaching materials separate the horizontal connection of language skills. In the actual classroom teaching operation level, it is difficult to only listen but not read, only speak but not write. As a result, some students' interest in second language learning will be reduced, which will inevitably lead to misunderstandings. Learning language through video is not to study the meaning of language itself from the internal structure of language, such as pronunciation, vocabulary and grammar, but to observe language examples from the outside and study the real meaning of speech activities in context from the discourse level. The integration of video will make the classroom more vivid and eye-catching, increase the interest of classroom teaching, help to mobilize the enthusiasm of students, and let students seek the accurate understanding and appropriate expression of discourse meaning in a specific context. On the network video teaching platform, students can fully consider their own needs and learning interests when selecting courses. The combination of video teaching with online discussion and online examination helps students consolidate their knowledge.

(2) Network video teaching

At present, there are three basic interactions in the learning environment of online video teaching: the interaction between students and teaching resources (i.e. online learning materials), the interaction between students and teachers, and the interaction between students and students. The communication and interaction between teachers and students is one of the components of distance video teaching. The quality of their interaction directly affects the quality of distance teaching, which is reflected in the speed and degree of interaction. It mainly depends on the speed, whether the interaction is timely and smooth, whether the students' questions can be answered in time, whether the students' learning situation can be checked and guided in time; to a certain extent, it mainly depends on the controllable degree of the interaction mode and the breadth and depth of the interaction content.

2.3   Quantitative Formula

In the video data compression algorithm, discrete cosine transform (DCT) is used to eliminate the spatial redundancy of video data. The definition of 8×8 DCT formula is as follows (1)

$$F(u, v) = \frac{1}{4} C(v) \sum_{x=0}^{7} \sum_{x=1}^{7} f(x, y) \tag{1}$$

However, no matter what kind of resources, there will be detailed description information in the database. Therefore, the project characteristics of network teaching system are described by teaching resources. The logarithmic information between content sets rel (q) is shown in formula (2)

$$\log MI(w_i, rel(Q)) = \log\left(\frac{P(w_i | w_i \in rel(Q))}{P(w_i)}\right) \tag{2}$$

Assuming that student I's online video course evaluation set is Ni, the formula for predicting the preference for any online video $t(t \notin I_N)$ course is shown in Eq. (3):

$$predictiojn = \bar{i} + \frac{\sum_{j=1}^{n} sim_j \times (cour_j - \bar{u})}{\sum_{j=1}^{n} sim_j} \tag{3}$$

3   Experience

3.1   Extraction of Experimental Objects

The data access layer is mainly responsible for the interactive operation of database access. The system uses Java language to access the database and Dao technology to process the database. Dao technology is a collection of database operations. Different Dao is used to operate different database resources. Logic layer is the core of the whole system. Responsible for handling the core functions of the system. It is mainly based on the needs of users to find the corresponding operation, to achieve the functional requirements of users. The main logic layer function of the system is realized by java language. In baseaction, different operations are implemented by different functions. Among them, open is open, update is update, remove is move, delete is delete and list is list function. In different class libraries, some of them rewrite and implement some functions according to their own functional requirements. In addition, some class libraries also add new functions to achieve different functions.

3.2   Experimental Analysis

In the whole system structure, it is divided into three levels: representation layer, logic layer and data layer. Data exchange and transmission are realized by data transmission module. The whole system adopts springmvc framework mode. It includes two important data transmission modes. One is the data transmission channel from the presentation layer to the logical layer. In this case, the form based transmission is used. That is, the user input data through the generated JSP page form, and the JSP page submits the form data to the corresponding action operation of the logic layer through hibernate for processing. The other is the data transmission channel from the logical layer to the presentation layer. In this case, the input is used for transmission. An entry is a data object whose internal type and content can be defined by the user. The data interaction between logical layer and data layer is realized by bean object. Bean object plays the role of media, and realizes the data exchange task between Dao and logical layer actions of database.

4 Discussion

4.1 Online Practice Mode

In the current guzheng learning, parents are puzzled by after-school guidance. Parents who have not received professional study do not know how to guide their children to practice. It's not just the simple practice and supervision of music that makes children tired of the company. As parents gradually realize the network education mode, online software can solve the problem that there is no right or wrong standard in the process of piano practice, the market demand of network accompaniment training is gradually emerging. As shown in Table 1.

Table 1. Market size of China's online education industry from 2018 to 2020 (ten thousand yuan)

	Market size	Growth rate
2018	1213.2	17%
2019	2637.5	36%
2020	3415.4	47%

It can be seen from the above that in 2018, the scale of online education market is 12.132 million yuan, with a market growth rate of 18.754 million yuan; in 2019, the scale of online education market is 26.375 million yuan, with a market growth rate of 25.874 million yuan; in 2020, the scale of online education market is 34.154 million yuan, with a market growth rate of 36.733 million yuan. The results are shown in Fig. 1.

From the above, the online education market increased from RMB 12.213 million to 3.4154 million yuan in three years from 2018 to 2020, and the growth rate of market scale increased from 17% to 47%.

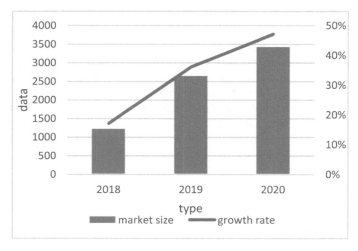

Fig. 1. Market scale of China's online education industry from 2018 to 2020 (10000 yuan)

4.2 Survey Results of Teaching Effect of China Guzheng Net

Through the survey of the teaching effect of Guzheng net in China, we can see the users' desire and demand for guzheng net teaching. Of course, this is also the inevitable trend of the times. Therefore, the popularization and promotion of Guzheng art must combine network teaching with traditional teaching. As shown in Table 2.

Table 2. The influence of online teaching column of Guzheng on the dissemination of Guzheng Art

Online teaching column of Guzheng	Proportion
Protect Guzheng Art	45.7%
Inheritance of Guzheng Art	23.8%
The development of Guzheng	30.5%

It can be seen from the above that the proportion of users' love for guzheng art teaching columns is 45.7%, the proportion of users' love for inheriting guzheng art teaching columns is 23.8%, and the proportion of users' love for promoting the development of Guzheng art teaching columns is 30.5%. The results are shown in Fig. 2.

From the above, the user's favorite proportion of Guzheng art teaching column is 45.7%, which indicates that more and more people attach great importance to the teaching of Guzheng art.

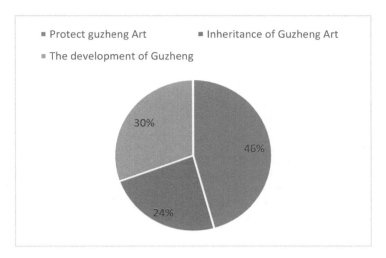

Fig. 2. The influence of online teaching column of Guzheng on the dissemination of Guzheng Art

5 Conclusion

In recent years, the guzheng ensemble training in various professional colleges is booming, so it is imperative to set up an independent guzheng ensemble course for guzheng major in normal universities. While improving students' performance skills, it can also exercise students' team spirit, let more students on the stage, help students accumulate practical experience, and promote the development of students' comprehensive ability. In order to improve the performance level of students, in guzheng teaching, we should increase the learning time of basic performance skills, enrich the teaching forms, strengthen the psychological adjustment of students, and improve the demonstration time of teachers.

References

1. Granjo, J., Rasteiro, M.G.: LABVIRTUAL—a platform for the teaching of chemical engineering: the use of interactive videos. Comput. Appl. Eng. Educ. **26**(5), 1668–1676 (2018)
2. Schwade, F., Schubert, P., et al.: The ERP challenge: developing an integrated platform and course concept for teaching ERP skills in universities. Int. J. Human Cap. Inf. Technol. Prof. **9**(1), 53–69 (2018)
3. Aungaroon, G., Vawter-Lee, M.: Teaching Video NeuroImages: Ictal vomiting in a child. Neurology **91**(19), e1836–e1837 (2018)
4. Schoenfeld, A.H.: Video analyses for research and professional development: the teaching for robust understanding (TRU) framework. ZDM: Int. J. Math. Educ. **50**(8), 1–16 (2018)
5. Stafford, A., Osman, C.: Teaching video neuroimages: a 20-year-old man with distal paresthesia. Neurology **92**(2), e170–e170 (2019)

6. Lee, S.Y., Lo, Y., Lo, Y.: Teaching functional play skills to a young child with autism spectrum disorder through video self-modeling. J. Autism Dev. Disord. **47**(8), 2295–2306 (2017). https://doi.org/10.1007/s10803-017-3147-8
7. Mendez-Guerrero, A., Lopez-Blanco, R., de Urabayen, D.U.P.: Reader response: teaching video neuroimages: olivary enlargement and pharyngeal nystagmus. Neurol.: Off. J. Am. Acad. Neurol. **90**(16), 754–754 (2018)
8. Lohmann, S., Frederiksen, L.: Faculty awareness and perception of streaming video for teaching. Collect. Manag. **43**(2), 101–119 (2018)
9. Xiong, C., Ge, J., Wang, Q., et al.: Design and evaluation of a real-time video conferencing environment for support teaching: an attempt to promote equality of K-12 education in China. Interact. Learn. Environ. **25**(5–8), 596–609 (2017)
10. Zkul S, Ortaçtepe, D.: The use of video feedback in teaching process-approach EFL writing. Tesol J. **8**(4), 862–877 (2017)
11. Yawiloeng, R.: Second language vocabulary learning from viewing video in an EFL classroom. Engl. Lang. Teach. **13**(7), 76–83 (2020)

Optimization of New Media Marketing Methods for Apparel Brands Based on Information Technology

Lin Li[✉]

Jilin Engineering Normal University, 3050 Kaixuan Road,
Changchun 130052, Jilin, China

Abstract. With the rapid development of information technology, great changes have taken place in the dissemination of information and the way people communicate. New media marketing has become a link that cannot be ignored in the marketing of apparel brands. Based on the analysis of the concepts and characteristics of new media marketing, taking the Chinese market as an example, the main methods and development status of new media marketing for clothing brands under the information technology environment are studied, and the existing problems are analyzed from positioning, content, interactive methods, and transformation functions. In terms of exploring new media marketing optimization strategies for apparel brands.

Keywords: Information technology · New media marketing · Clothing

With the rapid development of information technology, communication methods represented by new media are widely used in people's lives. The combination of new media and marketing can effectively change the past marketing methods and communication concepts, and realize a marketing model without interaction barriers. It can not only achieve the purpose of promoting corporate brand promotion and expanding product sales, but also dig out new business opportunities. New media marketing has become a link that cannot be ignored in the marketing of apparel brands.

1 The Connotation of New Media Marketing

1.1 The Concept of New Media Marketing

The "new media" is relative to traditional media. New media is a medium for information dissemination based on digital technology and the Internet as a carrier [1]. New media marketing refers to the use of new media platforms, such as portals, search engines, Weibo, WeChat, Douyin, blogs, forums, APPs, etc., based on the conceptual appeals and problem analysis of specific products, to conduct targeted psychological guidance to consumers. A marketing model is the realization of a company's soft penetration business strategy in the form of new media. It uses media expression and public opinion dissemination to make consumers agree with certain concepts, opinions, and analysis ideas, so as to achieve the purpose of corporate brand promotion and product sales [2].

J. C. Hung et al. (Eds.): FC 2021, LNEE 827, pp. 333–339, 2022.
https://doi.org/10.1007/978-981-16-8052-6_41

1.2   Features of New Media Marketing

New media marketing has the characteristics of intuitiveness, precision, interactivity, quickness and creativity [3, 4]. Through new media marketing, time and space boundaries can be broken and a platform for direct communication between consumers and enterprises can be built. For consumers, they can learn about the relevant information about the products they need more quickly and widely, and make purchase choices quickly and accurately. For enterprises, by browsing the big data information of the marketing background, consumers can be accurately positioned in time, and they can deeply understand the needs of consumers and make targeted adjustments. Due to the rapid information dissemination and the high degree of topicality, it can quickly arouse the attention and resonance of consumers, and will quickly place related companies on the same arena to compete, which will affect the entire industry's interest chain.

2   The Status Quo of New Media Marketing of Clothing Brands Based on Information Technology

Competition in the apparel industry is extremely fierce. With the advent of the new media era, various apparel companies have used Internet resources to accelerate the construction of a new sales model centered on users, driven by data + content, and the integration of consumption and experience scenarios. According to the "Statistical Report on China's Internet Development Status" released by the China Internet Network Information Center (CNNIC), as of December 2020, the number of Chinese Internet users has reached 989 million, and the Internet penetration rate has reached 70.4%. In 2020, China's online retail sales reached 11.76 trillion yuan, and it has become the world's largest online retail market for eight consecutive years. In the competitively stimulated Chinese market, e-commerce platforms are no longer the only outlet for clothing sales. With the emergence of a series of new media such as WeChat, Weibo, Douyin, and Xiaohongshu, the younger generation is more inclined to use these new media channels to make purchases.

2.1   Wechat

As a social app with a large user base, WeChat is an important position for online marketing of apparel brands. Currently, the monthly active users of WeChat Mini Programs have exceeded 1.1 billion. Various clothing brands have used the WeChat platform's public account, subscription account, community, circle of friends and other functions to promote brand image or product sales. For example, for the Uniqlo brand, when people follow Uniqlo's official WeChat official account, they will automatically receive a message prompt to add members to the "Uniqlo Pocket Mall", and users can follow them immediately with just one click. On the basis of LBS, geographic location is shared to realize online shopping and offline pickup. In different seasons, the brand will recommend some different styles of clothing to users, and regularly push some designer cooperation models and benefits to users, thereby attracting more fans.

The information dissemination of WeChat presents the characteristics of divergent fission dissemination. Many clothing brand companies use WeChat official accounts to push product information, brand stories, discounts, etc. to the friend's information column and circle of friends in the form of subscription accounts. This one-to-one follow and push notification method has further strengthened Marketing efforts, and the content of communication is more vivid and intuitive. Second, the social and inter-active nature of WeChat is very obvious. Clothing brands use the same interests of WeChat community members to establish common interests, increase the sense of belonging and recognition of community members during interactions, and meet the needs of members or achieve sales goals. Third, the convenience of WeChat is very prominent. Apparel brands rely on QR codes to achieve O2O marketing [5]. Merchants use the QR code scanning function in WeChat to generate a QR code from the clothing brand, membership card system, product discounts and other information produced by their company. Customers can clearly obtain the relevant information of the product by scanning this QR code., And can follow the development of the brand at any time, WeChat Pay has become one of the most frequently used payment methods.

2.2 Short Video

The short video platform has developed rapidly in recent years. Short video applications such as Kuaishou, Douyin, Miaopai, and Watermelon Video have formed a very large market and achieved a strong flow of traffic [6]. As of December 2020, the number of short video users in China is 873 million, accounting for 88.3% of the total netizens [7]. According to data from iiMedia Research, in the monthly active rankings of China's short video platforms, Douyin and Kuaishou ranked the top two with 47.723 million and 37.8293 million respectively.

With the support of high-quality content, video sites continue trying to optimize their business models. The short video platform achieves a better content balance and traffic continuity through the PGC + UGC (professionally produced content + user-produced content) operation model, and relies on the algorithms of Internet information technology. Cooperate with e-commerce to launch shopping modules or jump links. Bloggers can display products in videos and recommend them to accurate people, or directly place product links corresponding to their fan groups under daily videos. For example, on the Douyin platform, the "new product launch" under the Dior account can directly jump to the Dior purchase interface. CHANEL has released 12 short videos in 12 days through the Douyin account "Good Life Images", and users can jump directly from Douyin to Chanel's official website. Many brands also use a series of "effective interactions" on short video platforms such as following, likes, shop visits, comments, and conversion rates to tap in-depth users of their products [8].

2.3 Sharing Community

Many consumers are keen to see real shopping experiences and experiences in the sharing community. The sharing community platform mainly uses UGC as the main and PGC (professional production content) and PUGC (expert production content) as a supplementary model to continuously output high-quality content. Every experience sharing plays its unique role. Sharing communities improve the efficiency of content matching through personalized distribution, relying on e-commerce or advertisers to monetize traffic, and complete the closed loop of "content + community + consumption".

Take the "Xiaohongshu" app, which is popular among post-90s generations in China, as an example. The content posted by "Xiaohongshu" users comes from real life. Through the sharing of text, pictures, and video notes, it triggers "community interaction" and promotes other users to spend offline. These users in turn will do more "Online sharing" eventually, which forms a positive cycle. Many readers leave an impression or have a good impression of the clothing brands they share or make purchases through the content of pictures and texts of ordinary people. Many clothing brands have put out graphics and texts that fit their own brands and have achieved great success through various direct or indirect publicity. For example, a batch of new fashion brands such as Lost in echo, simple piece, and 7 or 9 completed the rapid growth after establishment in "Little Red Book" in only 2–3 years. In the future, the involvement of more brands and the investment of capital will make the content on the sharing platform richer and more varied, and bring more influence to apparel brands.

2.4   Live Broadcast

According to the statistics of iimedia, as of March 2020, the number of live broadcast users in China has reached 560 million, of which live broadcast e-commerce users are about 270 million, and about half (48.2%) of live broadcast users have buying behavior. In the first half of 2020, the number of live broadcast products of China's apparel-related products reached 8.549 million, accounting for 37.6%, which is the largest transaction volume in live streaming.

For clothing brands, the two common methods are self-broadcasting by merchants and cooperation with external anchors. Brands adopt different live broadcast operation strategies according to the different characteristics of each platform. For example, the practice of the down jacket brand "Gofan" is mainly in two aspects. On the other hand, it is to determine the co-host selection mechanism, based on brand positioning and specific products, to match suitable co-hosts. This aspect mainly depends on the host's fan audience and the host's own characteristics, that is, whether the host's fan portrait, consumption characteristics and brand products match, and whether the host's delivery style and past delivery status are consistent with the brand positioning.

The development of live broadcast e-commerce in the apparel industry will present four major development trends: first, all participants in the industry chain have deployed virtual anchors, and virtual anchors may be widely used in the field of apparel; Secondly, the value of traditional culture is demonstrated, and the live broadcast of national style elements such as Hanfu and cheongsam has attracted attention; in addition, the second-child policy has been fully implemented, and the scale of the silver-haired population has increased. Fourthly, the C2M model of the apparel industry will become the mainstream, and apparel live broadcast e-commerce will tend to be personalized, quantitative, and data-based.

2.5 Other Forms

In addition to WeChat, short videos, sharing communities, and live broadcast platforms, there are still many new media positions suitable for marketing and promotion of apparel brands. For example, on Weibo, a large number of executives of apparel companies, such as Zhou Shaoxiong, Chairman of Seven Wolves, Chen Nian, Chairman of Vancl, and Fang Jianhua, CEO of Inman, are all active; Even Chanel CHANEL also entered Weibo. These corporate senior officials and official microblogs can not only use Weibo to provide new product introductions and customer service, but also to conduct market research, detect needs and find crises. In addition, Toutiao, Youku, iQiyi, Bilibili and other entertainment and leisure platforms, Douban, Zhihu, Tianya and other Q&A platforms, and online shopping platforms such as Vipshop and Mogujie are also new media marketing platform options for clothing brands.

3 Problems with New Media Marketing Methods for Apparel Brands

3.1 Unclear Brand Positioning

Brand positioning affects the overall development direction of the brand. With the diversification of brands in the apparel market, the homogeneity of many brands in design, promotion, and marketing has become more and more obvious. The main reason is that the positioning of the clothing brand is not clear, the target consumer group cannot be accurately classified, and the special new media marketing style cannot be created according to their age, interest, aesthetics, and hobby needs.

3.2 Dissemination Channels are Chaotic

In the new media environment, many apparel brands are facing difficulties in the selection of communication media in the process of realizing the expansion of communication channels. Many companies will ignore the degree of agreement between the brand and the media, blindly choose multiple channels and diversified communication methods, resulting in the same clothing brand expressing different brand concepts and demands in different communication media, and the voices conveyed to consumers are confused. This in turn affects the overall communication effect of the clothing brand.

3.3 Dissemination Content Lacks Innovation

Creativity is the life of the current fashion brand communication. Without creativity, it is difficult to attract consumers' attention, and it will not be able to achieve the ideal marketing promotion effect. The promotion content of many clothing brands is limited to the combination of models and clothing, lack of innovative awareness, unable to fully display the personality characteristics of clothing brands, difficult to resonate with the audience, and even false propaganda, causing consumers' boredom and aesthetic fatigue, which is extremely unfavorable to brand's long-term development.

4   Optimization Measures for New Media Marketing Methods of Clothing Brands Based on Information Technology

4.1   Positioning Optimization

In the development process of the clothing brand itself, positioning affects the overall development direction of the brand. Clear positioning and precise marketing can save companies from taking many detours. When companies conduct new media marketing for apparel brands, they should give full play to their own advantages, differentiate similar products, show individuality, form unique brand emotional appeals in the hearts of consumers, and accurately targeting customer groups. According to the needs and characteristics of different consumer groups, provide corresponding marketing data support.

4.2   Content Optimization

In the process of new media marketing, clothing brands mainly meet the preferences of consumers, but they also need to ensure that the brand itself has certain characteristics and integrate communication content on this basis. Based on the connection between new media technology and information terminals, clothing brands should convey differentiated brand concepts from competitors, win the favor of consumers through distinctive brand differences, At the same time, note that no matter which media method is used, the dissemination of information must maintain consistency in content.

4.3   Optimization of Interactive Mode

New media can use convenient information technology to establish an effective communication and interaction platform between clothing brands and consumers [9]. Good interaction with consumers can not only gain increased popularity, but also enhance consumer reputation and promote re-orders. Clothing companies can establish contact with consumers through pictures and texts, use live broadcasts to enhance stickiness, online and offline linkages, so that consumers can personally participate in clothing design, choose clothing fabrics, colors, and styles according to their preferences and needs, and then use the new media platform to communicate and place orders, to combine services and after-sales to create a complete optimized closed loop. This can not only improve the consumer's brand experience, but also make consumers form loyalty to the brand in a subtle way.

4.4   Conversion Function Optimization

Order conversion is the most critical step in the complete marketing process, which determines the profitability of the apparel company in the entire process. With the advent of the era of 5G and artificial intelligence, people's pace of life is gradually accelerating, and more convenient shopping experience is needed. Clothing brands must integrate various channels of new media marketing, improve various functional configurations of APP, and provide a complete and smooth technical support and service system for consumers to browse. And focus on "people, goods, field" effective allocation of resources, to achieve an omni-channel operation model of mutual drainage between online and offline [10].

5 Conclusions

The development of information technology has promoted the development of new media, and has promoted the diversification, precision, visualization and interaction of clothing marketing methods. In the context of information technology, clothing companies seize the opportunity for development and use new media to carry out corresponding marketing activities, which will surely bring good development opportunities for the company.

Acknowledgement. This work was supported by Research Project of Vocational Education and Adult Education Teaching Reform in Jilin Province 2020ZCY367.

Project Title: Research on the Countermeasures for the High-quality Development of Garment Vocational Education in Jilin Province.

Project Manager: Shuchang Zhu.

References

1. Dibb, S.: New media needs new marketing: social network challenging traditional methods. Strateg. Dir. **6**, 24–27 (2012)
2. Huo, C., Zhou, L.: The influence of new media marketing strategies on the communication of apparel brands. Workplace Mod. **01**, 84–85 (2018)
3. Pham, P.H., Gammoh, B.S.: Characteristic of social media marketing strategy and customer-based brand equity outcomes: a conceptual model. In: Kim, K.K. (ed.) Celebrating America's Pastimes: Baseball, Hot Dogs, Apple Pie and Marketing? DMSPAMS, pp. 433–434. Springer, Cham (2016). https://doi.org/10.1007/978-3-319-26647-3_87
4. Mathur, M.: Leveraging social media-based determinants to build customer-based brand equity of a retailer. Int. Rev. Retail Distrib. Consum. Res. **28**, 554–575 (2018)
5. Lu, X.: WeChat marketing strategy of clothing brand from the perspective of new media. Chin. Mark. **28**, 124–125 (2019)
6. Liu, P.: The development process and trend of my country's short video platform. News Writ. **01**, 81–84 (2019)
7. Statistical Report on China's Internet Development. China Internet Network Information Center, Beijing, 03 February 2021. http://cnnic.cn/hlwfzyj/hlwxzbg/hlwtjbg/202102/t20210203_71361.htm
8. Li, X., Sun, C., Shen, L.: The optimization strategy of apparel enterprises driven by short video. Western Leather **01**,17–18+20 (2021)
9. Lei, R., Wang, X., Li, J.: Research on clothing brand communication in the new media environment. Text. Rep. **09**, 32–33+40 (2019)
10. Zhu, Z.: Research on omni-channel marketing strategy of retail industry under O2O mode. Mod. Mark. **05**, 92–93 (2019)

Radar-Based Activity Recognition with Deep Learning Model

Han Zhang[(✉)]

Faculty of Information Technology, Beijing University of Technology,
Beijing, China
zhanghan1218@emails.bjut.edu.cn

Abstract. Over the past decades, Human action recognition is very significant for human action analysis, which is a strongly active research area. The relevant target detection and classification technologies have become research hot topics, with a very wide range of applications, including security protection, disaster relief, smart home and other fields. In the academia, radar has gradually become the focus of research in the field of target detection and classification. Compared with the other sensors, radar is more advanced in non-environmental impact and good data integrity in human body detection and classification. While the current studies have combined machine learning with radar signals, the performance of machine learning methods such as SVM and decision tree is not well. To address this issue, this work utilized convolutional neural networks, combined with micro-Doppler image features extracted from radar signals. The experimental demonstrated that CNN model achieved a higher accuracy in human activity recognition task.

Keywords: Human activity recognition · Deep learning · CNN model · Radar data

1 Introduction

With the development of the modern society, the demand for human activity recognition is increasing rapidly, which plays an increasingly important role in various application scenarios, from the ubiquitous cameras, the discovery and rescue of the wounded in the disaster site, to smart homes. Since some countries in the world have entered an aging society, the remote health monitoring of the elders based on sensors has been widely concerned and applied. By employing human activity recognition to obtain real-time human motion information, a remote monitoring system can respond quickly to ensure the safety of the elderly. Simultaneously, for the elderly who are recovering after surgery, this method enables doctors to remotely obtain the patient's real-time exercise, sleeping habits and other health related data, which will significantly improve the health of patients and alleviate the pressure of the health care industry. The untouchable monitoring also applied in dealing with the COVID-19 era.

From an analysis of the certain studies, there are already some works on human activity recognition. A wearable home device is proposed to obtain the physical health parameters of the wearer [1]. However, since the health parameters acquired by the device are sparse and incomplete, it is unable to identify and feedback the information in time when the patient is in an emergency situation. Moreover, wearing devices will

© The Author(s), under exclusive license to Springer Nature Singapore Pte Ltd. 2022
J. C. Hung et al. (Eds.): FC 2021, LNEE 827, pp. 340–348, 2022.
https://doi.org/10.1007/978-981-16-8052-6_42

bring extra burdens to patients, elderly people at home often forget to wear devices, and wearing these devices for a long time can make people feel uncomfortable and burdensome. The camera is also proposed to conduct human activity recognition [2]. This is a very mature technology, but the privacy at home that people are paying more and more attention to hinder the application of cameras. Besides, the camera is very sensitive to the environment, such as the light conditions and the occlusion of surrounding objects. WiFi channel state information is proposed to be used for human activity recognition [3]. Nevertheless, as the wavelength of WiFi is longer than that of K-band radar, it causes the poor performance of WiFi system in human motion detection. At the same time, WiFi-based human activity recognition system has a poor directivity, which means that the system is vulnerable to interference. A comprehensive evaluation on the system parameters and operating conditions of the micro-Doppler characteristics is conducted in [4]. The result shown that radar has the advantages of both high accuracy and long detection distance in comparison with wearable devices. Compared with the camera, radar does not involve privacy issues, is not affected by the light, can work normally in the dark environment, and has the penetrating ability. Compared with the WiFi system, radar has a good directivity. Therefore, it can be found that radar has great more advantages in addressing human activity recognition. Radar is not only insusceptible to interference, but also can protect the privacy of users.

Due to the small movement of the target (for example, rotation and vibration), the echo signal of the radar in use will produce additional frequency shift, which is called the micro-Doppler effect [5, 6]. This effect reflects the modulation effect of moving target on radar signal. The micro-motion state of targets with different structures will be different when performing different actions. In this paper, by obtaining the physical characteristics of the targets such as structure, motion details, etc., from the micro-Doppler features, we comprehensively employed deep learning algorithms and machine learning algorithms for human activity recognition.

The experiments are conducted on a public data set [7]. To be specific, the dataset includes radar signatures of different indoor human activities performed by different people in different locations. The data set was collected from December 2017 to March 2019, during which 1754 data sets were generated. We also use both the machine learning and convolutional neural network models to learn the extracted data. Experiments show that the accuracy of convolutional neural network learning is superior than those methods based on machine learning. The accuracy of linear SVM is 0.863, the accuracy of XGBoost is 0.874. By a sharp contrast, the accuracy of convolutional neural network is 0.949 for activity classification.

2   Problem Description

The data [7] was collected using an off-the-shelf FMCW radar operating at C-band (5.8 GHz) with bandwidth 400 MHz and chirp duration 1 ms, delivering an output power of approximately +18 dBm. The radar is connected to transmitting and receiving Yagi antennas with a gain of about +17dB and is capable of recording micro-Doppler signatures of the people moving within the area of interest. Those involved in data collection are asked to repeat three to five different activities such as walking, sitting down, standing up, picking up an object, drinking water, and falling.

In this study, we further divide the data set into three subsets with a ratio of 8:1:1, namely the training data set, the validation data set, and the test data set. There are six categories for activity recognition and 72 people for human identity recognition.

3   Models

3.1   Feature Engineering

In this section, we introduce the conversion of the original FMCW data into micro-Doppler signals. The original data are illustrated in Fig. 1.

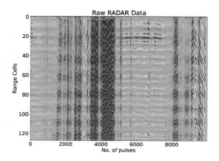

Fig. 1. An illustration of original data

We first reshape the data matrix to a matrix of 128 * M. 128 is the number of time samples for each scan, which is also the size of Fast Fourier Transform (FFT). M is the linear frequency modulation number of an active sample. We then apply FFT to each column of the data matrix, and calculate the spectrum and form the distance distribution map, as shown in Fig. 2(a). Subsequently, we adapt infinite impulse-response 9th-order high-pass Butterworth notch filter with the cut-off frequency of 0.0075 Hz to remove the interference, and keep the data of the target distance from 1.875 m to 9.375 m. (please see Fig. 2(b)).

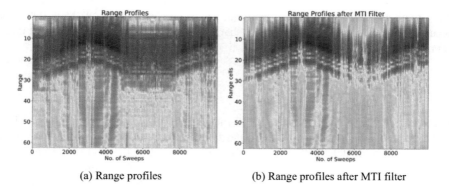

(a) Range profiles   (b) Range profiles after MTI filter

Fig. 2. Comparison between Range profiles and Range profiles after MTI filter

Then, STFT is adopted to extract the signal of micro-Doppler feature on the distance-time matrix, as shown in Fig. 3. It can be observed that the micro-Doppler feature maps of six activities, and find that the micro-Doppler maps of each activity are quite different and have the characteristics of action. For example, the micro-Doppler feature map of walking is shown in Fig. 3(a). The Doppler map shows periodic changes due to the natural and regular movement of the limbs when a person is walking. The activity of sitting down is shown in Fig. 3(b). Note that when a person is sitting down, as the movement range is not large, the value of the Doppler map is not high and irregular, and also since the person is static, there is no value in the micro-Doppler feature map that occupies the majority. The activity of standing up is shown in Fig. 3(c). When a person stands up, the arms and legs will have a force action, and they are in one direction, which makes the Doppler map converge in one direction and vary greatly. The activity of picking up an object is shown in Fig. 3(d). When a person picks up an object, it is a continuous and back-and-forth movement, thus the micro-Doppler map lasts for a long time and fluctuates. Since the movements of bending and raising the waist are not very regular, the waveform of the micro-Doppler map is non-periodic. The activity of drinking water is shown in Fig. 3(e) When a person drinks water, there is a movement of holding and placing a cup, and there is little difference in limb changes, so the waveform of the micro-Doppler map is periodic. However, since the movement range of the hands for drinking water is smaller than that for walking, the amplitude of drinking water is less than that of walking. The activity of falling is shown in Fig. 3(f). Because the activity of falling is very fast and completes in an instant, the micro-Doppler map is concentrated in the starting area, and there is no follow-up. From the above analysis, we find that the micro-Doppler maps of different activities are very different, we thus put pictures as features into the convolutional neural network for training. At the same time, we observe that there are samples of 5 s and 10 s in the data set. We use the sliding window to extract a picture every 5 s, so the number of pictures as input data is more than the number of original data.

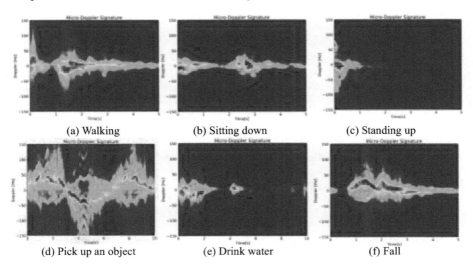

| (a) Walking | (b) Sitting down | (c) Standing up |
| (d) Pick up an object | (e) Drink water | (f) Fall |

Fig. 3. Signal of micro-Doppler feature by utilizing STFT

3.2   Activity Feature Extraction

After obtaining the micro-Doppler map, we extract characteristics further. We obtain several sequence data, and extract characteristics on each sequence by using the TSFRESH package one by one. A total of four sequences are obtained as follows:

- Maximum value sequence: directly calculate the maximum value of each bin in the Doppler map;
- Minimum value sequence: directly calculate the minimum value of each bin in the Doppler map;
- Mean value sequence: directly calculate the mean value of each bin in the Doppler map;
- Energy sequence: directly sum up each bin in the Doppler map.

We use the TSFRESH package to calculate 35 characteristics for each sequence, such as entropy, quadratic sum, etc. We get 140 values in total as the vector input.

3.3   Recognition Models

We try linear SVM, XGBoost [8], and convolutional neural network (CNN) [9], and finally decide to adopt the CNN model, which also achieves a state-of-the-art performance in a series of problems [10–13]. By changing the number of convolution kernels used in the first and second convolution layers and the third and fourth convolution layers of the convolutional neural network, we modify the model structure and compare the results. To combine different input features, we adopt the following model structure shown in Fig. 4, with a concatenation layer for the image and vector inputs.

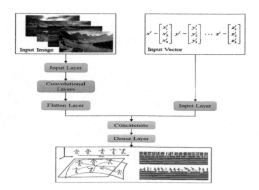

Fig. 4. Human activity recognition model structure

4   Experimental Results

4.1   Experiment Settings

All experiments are conducted on a computer with 1.80 GHz Intel i5-8265U processor, 8 GB RAM, and the Windows 10.0 operating system. The experimental codes of data preprocessing and modeling are written with Python 3.8.5. The main package used is the deep learning package of TensorFlow, scikit-learn and time sequence feature extraction package TSFRESH.

Different models are evaluated with the following metrics:

- Test accuracy: It is used to evaluate the generalization ability of the final model. It refers to the percentage of the total number correctly predicted by the model using the testing set.
- Train accuracy: It is used for the samples of fitted data by the model. It refers to the percentage of the total number correctly predicted by the model using the training set.
- Validation accuracy: The validation set is composed of separate training sets. It refers to the percentage of the total number correctly predicted by the model using the validation set. It is used for adjusting hyper parameters and monitoring whether the model has been fitted.
- Train loss: It is mainly used to quantify the quality of training set classification.
- Validation loss: It is mainly used to quantify the classification of validation set.

4.2   Results

The basic CNN models we use have four convolutional layers, one fully connected layer, and one output layer. For the convolutional layers, the filter size is (3, 3). For the output layer, the activation function used is softmax and for other layers, the activation function used is relu. After the second and the fourth convolutional layers, we also use a maxpooling layer with a pool size of (2, 2) and a dropout layer with a probability of 0.25. We alter the CNN structure by altering two numbers, which represent the filter number for the first two convolutional layers and the filter number for the last two convolutional layers, respectively. The corresponding experimental result is listed in Table 1.

Table 1. Accuracies of different CNN models for activity classification

Model_types	Only image input			Image input + Vector input		
	train_acc	val_acc	test_acc	train_acc	val_acc	test_acc
2–4	0.973	0.885	0.882	0.962	0.904	0.909
4–8	0.963	0.902	0.917	0.971	0.912	0.922
8–16	0.992	0.934	0.936	0.972	0.938	0.940
16–32	0.993	0.918	0.921	0.989	0.942	0.949
32–64	0.992	0.928	0.933	0.980	0.904	0.937
64–128	0.995	0.926	0.935	0.977	0.908	0.935

The accuracy and loss changes during the training process of the 16–32 CNN are shown in Fig. 5.

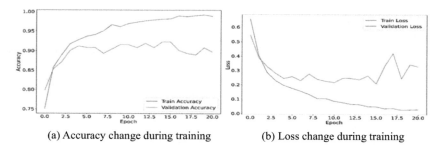

(a) Accuracy change during training   (b) Loss change during training

Fig. 5. Accuracy and loss changes during the training process of the 16–32 CNN

From the accuracy curve we see that with the increase of training times, train accuracy and validation accuracy gradually increase, which indicates that the model has not over-fitted. From the curve of loss value, we can see that the train loss and the validation loss gradually decrease, which indicates that the model has gradually learnt the data information, and the output results are approaching the ground truth. We also draw the confusion matrix of the CNN model which has the best performance, on the test set, as shown in Fig. 6. We note that the classification effect is superior, and its classification accuracies of most activities are significantly high, while the activities which are easily misclassified are picking up an object and drinking water.

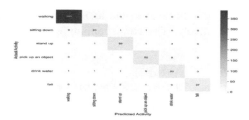

Fig. 6. Confusion matrix of the best CNN model

Table 2. Accuracies of different CNN models for human identity classification

Model_types	Only image input			Image input + Vector input		
	train_acc	val_acc	test_acc	train_acc	val_acc	test_acc
2–4	0.796	0.152	0.156	0.671	0.191	0.221
4–8	0.837	0.141	0.150	0.750	0.191	0.206
8–16	0.870	0.176	0.180	0.782	0.162	0.165
16–32	0.951	0.186	0.180	0.814	0.202	0.219
32–64	0.921	0.170	0.144	0.556	0.327	0.309
64–128	0.918	0.152	0.158	0.642	0.335	0.325

Moreover, we find that it is far from enough to rely solely on micro-range signature for more complex person identity classification, in particular when there are as many as 72 different participants. The results obtained by different CNN models are given in Table 2. The accuracy and loss changes during training process of the 64–128 CNN model are plotted in Fig. 7.

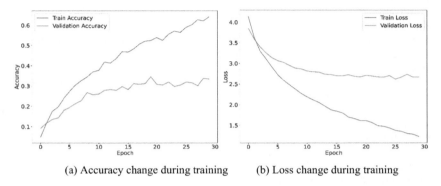

(a) Accuracy change during training   (b) Loss change during training

Fig. 7. Accuracy and loss changes during training process of the 64–128 CNN model.

Clearly, we can see that notwithstanding the training accuracy increases steadily, the increase of validation accuracy is very slowly, and while the train loss decreases, the validation loss increases, which seems to show that the model is seriously over fitted.

5 Conclusion

In this paper, we adapt the micro-Doppler feature map converted from FMCW data to extract features, and put these features into the deep learning model (CNN model) for training and testing. The model based on deep learning achieved a superior recognition performance in the field of activity recognition by learning features. It is worth remarking that the recognition ability can be applied to certain smart home equipment to identify and record the action of the detection target. However, the performance in human identity recognition is not advanced at present, and it cannot distinguish the participants effectively. In the future, a series of significant studies can be conducted. For instance, we would like to verify a quantitative research on the relationship between micro-Doppler and movements, and use the research results to re-extract features and build models, therefore as to extract deeper features, so that the deep learning model can identify the differences between people and improve the accuracy of human identity recognition in an effective manner.

References

1. Clawson, J., Pater, J.A., Miller, A.D., et al.: No longer wearing: investigating the abandonment of personal health-tracking technologies on craigslist. In: Proceedings of the 2015 ACM International Joint Conference on Pervasive and Ubiquitous Computing, pp. 647–658 (2015)
2. Zhao, J., Snoek, C.G.M.: Dance with flow: two-in-one stream action detection. In: Proceedings of the IEEE/CVF Conference on Computer Vision and Pattern Recognition, pp. 9935–9944 (2019)
3. Zhang, J., Wei, B., Hu, W., et al.: Wifi-id: human identification using wifi signal. In: 2016 International Conference on Distributed Computing in Sensor Systems (DCOSS), pp. 75–82. IEEE (2016)
4. Gürbüz, S.Z., Erol, B., Çağlıyan, B., et al.: Operational assessment and adaptive selection of micro-Doppler features. IET Radar Sonar Navig. 9(9), 1196–1204 (2015)
5. Chen, V.C., Li, F., Ho, S.S., et al.: Micro-Doppler effect in radar: phenomenon, model, and simulation study. IEEE Trans. Aerosp. Electron. Syst. 42(1), 2–21 (2006)
6. Chen, V.C., Li, F., Ho, S.S., et al.: Analysis of micro-Doppler signatures. IEE Proc.-Radar Sonar Navig. 150(4), 271–276 (2003)
7. Li, X., Li, Z., Fioranelli, F., et al.: Hierarchical radar data analysis for activity and personnel recognition. Remote Sens. 12(14), 2237 (2020)
8. Chen, T., Guestrin, C.: XGboost: a scalable tree boosting system. In: Proceedings of the 22nd ACM SIGKDD International Conference on Knowledge Discovery and Data Mining, pp. 785–794 (2016)
9. LeCun, Y., Bengio, Y., Hinton, G.: Deep learning. Nature 521(7553), 436–444 (2015)
10. Zhou, X., Liang, W., Kevin, I., et al.: Deep-learning-enhanced human activity recognition for Internet of healthcare things. IEEE Internet Things J. 7(7), 6429–6438 (2020)
11. Jiang, W., Zhang, L.: Geospatial data to images: a deep-learning framework for traffic forecasting. Tsinghua Sci. Technol. 24(1), 52–64 (2018)
12. Jiang, W.: Applications of deep learning in stock market prediction: recent progress. arXiv preprint arXiv:2003.01859 (2020)
13. Bhattacharya, A., Vaughan, R.: Deep learning radar design for breathing and fall detection. IEEE Sens. J. 20(9), 5072–5085 (2020)

A Joint Approach Based on Matrix Factorization for Multi-view Clustering

Bailin Chai[✉]

School of Economics and Management, Beijing Jiaotong University,
Beijing, China
18711081@bjtu.edu.cn

Abstract. In recent years, multiple view clustering tasks has attracted sustained concern as numerous realistic data are composed of distinct expressions. The point is to adequately integrate knowledge from multiple perspectives for facilitating the clustering tasks. The non-negative matrix factorization (NMF) technique is extensively applied in multiple perspective clustering on account of possessing the ability of dimensionality reduction. This paper introduces an innovative NMF multi-view clustering approach, which can extract complementary and compatible information presented in multiple perspective data. Besides, for solving the non-convex optimal issue of the objective function presented in this paper, an available iterative renewing approach is introduced. Experiments on two standard data sets indicate the advantages of our raised algorithm contrasted with basic methods.

Keywords: Multiple view clustering · Non-negative matrix factorization · Iterative algorithm

1 Introduction

With the speedy development of computer technique, data can usually be collected from multiple sources [1, 2]. For example, webpages can be expressed by words or hyperlinks. Text can be translated into different languages [3]. Different views may contain distinct information, that is to say, each view may include critical information not involved in other views. Hence, multi-view clustering [4–7] has become an important research area, which aims at dividing diverse information into different groups. A typical multiple view clustering method is to fuse distinct views directly, then separate them into different categories using traditional clustering methods. Nevertheless, this approach ignores the inherent nature of each view.

To solve the above problems and get better clustering performance, numerous of multiple view clustering techniques [8] have been raised in past few years. For example, the work [9] propose an method which studies from both marked and unmarked samples, where kernel-based spectral clustering approach regarded as an important framework and use regularization terms integrate the information of labeled data points. Multi-view subspace clustering [10] is proposed to utilize a shared cluster construction to ensure the accordance among disparate views. Unluckily, these approaches may yield undesirable results when the dimension of different views varies

© The Author(s), under exclusive license to Springer Nature Singapore Pte Ltd. 2022
J. C. Hung et al. (Eds.): FC 2021, LNEE 827, pp. 349–356, 2022.
https://doi.org/10.1007/978-981-16-8052-6_43

greatly. Therefore, the works [11–13] are proposed to handle this matter. NMF plays an essential role in dimensionality reduction, and some multi-view clustering methods based on NMF [14–16] come into being. For example, the work [17] is put forward to adopt semi-NMF to study the semantic level of multiple view data and graph regularization are offered to recognize the intrinsic manifold structure in every view. The work [18] introduced a multi-view NMF-based model on patch alignment with view consistency, its purpose is to learn the potential representation shared by various views.

Despite these advancement on multi-view clustering, there still exists defects owing to computational loss of high-dimensional data and combination problem among diverse views. To deal with the above points, in this work, we come up with a joint model NMF-based for multiple view clustering task with the purpose of exploiting consistent and supplementary from multi-view data. Specifically, we drive an efficient algorithm based on the iterative updating tactics to solve the proposed model. In brief, the chief highlights of this work can be summed up as below:

We introduce a joint approach based-NMF for multiple view clustering, which endeavors to realize this goal based on the decomposed coefficient matrix.

The proposed model can capture consistent and complementary information from multi-view coinstantaneous by the coefficient indicator to benefit clustering.

We deduce an iterative alternating updating tactics to solve the non-convex optimal problem of model, the corresponding convergence analysis is explained.

The remanent part of this work is set as below. In Sect. 2, we depicted the establishment process of our model and its optimization scheme. In Sect. 3, tests on actual datasets evaluate the validity of our designed method. Section 4 summarizes this paper with certain probe.

2   Methodology

2.1   Problem Definition

Multi-view data $\{X^1, X^2, \ldots, X^V\}$, where $X^v = \{x_1^v, x_2^v, \ldots, x_n^v\}$ includes n items. It is easy to think that apply all views directly to the NMF definition, which can be shown below:

$$\min_{\substack{W^v, H^v \\ v=1,2,\ldots,V}} \sum_{v=1}^{V} \| X^v - Z^v H^v \|_F^2 \tag{1}$$

Where Z^v is the basis matrix of the vth view, H^v is the coefficient matrice of the vth view, V stands for the view.

2.2   The Proposed Model

In this work, a restriction is used on the coefficient vector factorized by NMF and makes the latent expression of views be a common consensus representation. To measure the unequal between coefficient matrix and consensus matrix, the loss function is described below:

$$\min_{H^*} \sum_{v=1}^{V} \alpha_v \parallel H^v - H^* \parallel_F^2 \tag{2}$$
$$s.t. \quad H^* \geq 0$$

Where α_v is the weight corresponding to the vth view. Considering H^v from different views which might be incomparable at the equal size, diagonal matrix Q^v is used to simplify the calculation. Incorporating the above thought into the NMF, we can achieve the nether loss function:

$$\min_{W^v, H^v} \sum_{v=1}^{V} \parallel X^v - Z^v H^v \parallel_F^2 + \sum_{v=1}^{V} \varpi_v \parallel H^v Q^v - H^* \parallel_F^2$$
$$v = 1, 2, \ldots, V \tag{3}$$
$$s.t. \quad Z^v \geq 0, \quad H^v \geq 0, \quad 1 \leq v \leq V, H^* \geq 0$$

The above model can entirely extract consistent information from different perspectives. Nonetheless, in real-world applications, different views may fluctuate drastically in quality. We expect that the coefficient matrix received from two perspectives can complement each other in the procedure of factorization. Thus, the overall loss function can be described as below:

$$\min_{W^v, H^v} \underbrace{\sum_{v=1}^{V} \parallel X^v - Z^v H^v \parallel_F^2}_{NMF\ loss\ function} + \underbrace{\sum_{v=1}^{V} \varpi_v \parallel H^v Q^v - H^* \parallel_F^2}_{consistency\ term} + \underbrace{\sum_{v,t} \beta_v \parallel H^v - H^t \parallel_F^2}_{complementary\ term}$$
$$v = 1, 2, \ldots, V$$
$$s.t. \quad Z^v \geq 0, \quad H^v \geq 0, \quad 1 \leq v \leq V, H^* \geq 0, \quad H^t \geq 0 \tag{4}$$

Where β_v is the balance parameter corresponding to the complementary term, v and t represent the distinct view state. In this way, when facing multi-view problems, the proposed joint framework can handle the accordant and supplementary message at the same time, the clustering evaluate indicator can also be further improved.

2.3   The Optimization Process

For simplicity, Eq. (4) is equal to make the following function minimized:

$$O = \sum_{v=1}^{V} \parallel X^v - Z^v H^v \parallel_F^2 + \sum_{v=1}^{V} \varpi_v \parallel H^v Q^v - H^* \parallel_F^2$$
$$+ \sum_{v=1}^{V} \sum_{t=1}^{V} \beta_v \parallel H^v - H^t \parallel_F^2 + \sum_{v=1}^{V} (tr(\phi^v W^{vT}) + tr(\varphi^v H^{vT})) \tag{5}$$

where ϕ^v and φ^v are the Lagrange multipliers for Z^v and H^v, individually.

Keeping H^v and H^* fixed, update Z^v: Taking the partial derivatives of O concerning Z^v, we can get the formula below:

$$\frac{\partial O}{\partial Z^v} = Z^v H^{v^T} H^v + \varpi_v R^v + \phi^v - X^v H^v \tag{6}$$

Where k stands for the clustering number and R^v is shown as below:

$$R^v = Q^v_{k,k}(H^{v^T}H^v)_{k,k} - (H^{v^T}H^*)_{k,k} \tag{7}$$

Using the Karush-Kuhn-Tucker (KKT) condition, we can receive the updating equation:

$$Z^v_{i,k} = W^v_{i,k} \times \frac{(X^v H^v)_{i,k} + \varpi_v (H^{v^T}H^*)_{k,k}}{(Z^v H^{v^T}H^v)_{i,k} + \varpi_v Q^v_{k,k}(H^{v^T}H^v)_{k,k}} \tag{8}$$

Keeping Z^v and H^* fixed, update H^v: Taking the partial derivative in the same way that is applied in Eq. (6), we get the updated equation in connection with H^v as below:

$$H^v_{j,k} = H^v_{j,k} \times \frac{(X^{v^T}Z^v + \varpi_v H^* + \beta_v \sum_{t=1}^{V} H^t)_{j,k}}{(H^v Z^{v^T}Z^v + \varpi_v H^v + \beta_v \sum_{t=1}^{V} H^v)_{j,k}} \tag{9}$$

Keeping Z^v and H^v fixed, update H^*: Taking partial derivation concerning H^*, we can get the following updating formular:

$$H^* = \frac{\sum_{v=1}^{V} \varpi_v H^v Q^v}{\sum_{v=1}^{V} \varpi_v} \tag{10}$$

Base on the descriptions above, an alternating tactic is outlined in Algorithm 1.

Algorithm 1: The solve procedure for the presented model
___
Input: Multiple views samples $\{X^v, \forall 1 \le v \le V\}$, parameters $\{\varpi_v, \beta_v\}$
  for $v = 1 : V$ **do**
      Initialize Z^v, H^v by using NMF.
  end
  while all variables not converged **do**
    while Z^v, H^v not converged **do**
      for $v = 1 : V$ **do**
         Keeping H^v and H^* fixed, renew Z^v by using Eq. (8)
         Keeping Z^v and H^* fixed, renew H^v by using Eq. (9)
      end
    end
    Keeping Z^v and H^v fixed, renew H^* by using Eq. (10)
  end
Output: Z^v, H^v, H^*
___

3   Experiments

In this part, several tests are implemented on two benchmark samples to assess the advantages of the suggested approach. Firstly, the features of data samples are listed in Table 1. Then the baseline methods and evaluation indicators are introduced. Finally, we summarized the clustering evaluate indicator of the suggested method on all data samples, moreover, parameter sensitivity and astringency analysis are displayed in detail.

Table 1. The datasets utilized in the experiments

Datasets	Samples	Cluster	Views	Dimensions
BBC	685	5	4	4659, 4633, 4665, 4684
BBCSport	554	5	2	3183, 3203

3.1   Compared Methods and Evaluation Metrics

Several baseline methods have used. MCDNMF applies the semi-NMF to fetch the hidden knowledge of the input multiple view data and present graph regularized to get the representation studing. MultiNMF is proposed by sharing latent lower dimension space from multiview and developing a normalization scheme motivated by the connection between NMF and PLSA. SC is a typical graph-based approach, we experiment with best-view data and describe the corresponding consequences. GNMF attempts to construct an association graph to encode the intrinsic geometrical information and explore a matrix factorization which considers the graph structure.

Three metrics, accuracy (*ACC*), normalized mutual information (*NMI*), and purity (*PUR*) are applied to assess the clustering ability, which is extensively utilized for clustering and is positive corrections.

3.2   Clustering Results, Parameter Sensitivity and Convergence Analysis

3.2.1   Clustering Results

For these experiment outcomes in terms of ACC, NMI and PUR in Figs. 1 and 2 on multi-view clustering works, we can observe that our method exceeds baseline methods

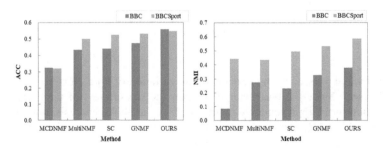

Fig. 1.   Accuracy and NMI results on two standard datasets for all approaches

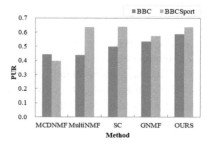

Fig. 2. Purity results on two standard datasets for all approaches

in most conditions. However, in some situations, single view methods, i.e. SC, are even slightly better than our model on PUR, which means that extracting multi-view information still requires better techniques.

3.2.2 Parameter Sensitivity Analysis

In this section, we chiefly discuss the sensitivity analysis on parameters used in our model. Tables 2 and 3 illustrate the impacts of the variation of ϖ on clustering for all datasets. As for the parameter ϖ that commands the smoothness from consensus term. For the BBC dataset, as the ϖ increases, the maximum difference between ACC is 0.026, which is 0.048 for NMI and PUR. The same analysis applies to the BBCSport. On the whole, our model is insensitive to ϖ parameters.

Table 2. Clustering performance of ϖ changes on BBC dataset

α	0.01	0.05	0.1	0.15	0.2	0.25	0.3
ACC	0.521	0.547	0.537	0.526	0.530	0.529	0.526
NMI	0.369	0.376	0.384	0.369	0.380	0.336	0.373
PUR	0.581	0.577	0.571	0.558	0.566	0.533	0.561

Table 3. Clustering performance of ϖ changes on BBCSport dataset

α	0.01	0.05	0.1	0.15	0.2	0.25	0.3
ACC	0.546	0.515	0.535	0.533	0.529	0.537	0.535
NMI	0.445	0.341	0.369	0.367	0.374	0.379	0.380
PUR	0.636	0.555	0.577	0.575	0.572	0.585	0.583

Tables 4 and 5 describe the impacts of the variation of β on clustering performance for all datasets. Since the parameter β controls the smoothness of the complementary regularization in the representative matrix. Follow the above analysis, noticing that our model on ACC, NMI, and PUR obtain well-pleasing results. Therefore, we can verify that the proposed method is immunity to the value of β.

Table 4. Clustering performance of β changes on BBC dataset

β	1	2	3	4	5	6	7
ACC	0.547	0.510	0.552	0.552	0.558	0.514	0.553
NMI	0.376	0.349	0.380	0.381	0.380	0.373	0.377
PUR	0.577	0.512	0.580	0.581	0.585	0.583	0.581

Table 5. Clustering performance of β changes on BBCSport dataset

β	1	2	3	4	5	6	7
ACC	0.546	0.544	0.518	0.520	0.544	0.544	0.546
NMI	0.445	0.380	0.364	0.361	0.382	0.380	0.385
PUR	0.636	0.586	0.561	0.559	0.585	0.586	0.588

3.2.3   Convergence Analysis

In this subsection, a minimization mechanism is adopted to deal with the optimization task, and we test the stypticity of the iterative algorithm on BBC and BBCSport datasets. As we can see from Fig. 3, the objective function value fell sharply while the number of iterations increases. Moreover, it's not difficult to see that convergence can be achieved within the five steps of the iterations.

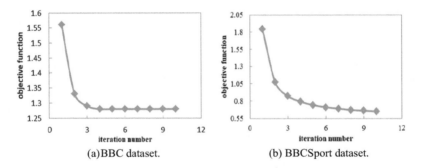

(a)BBC dataset.   (b) BBCSport dataset.

Fig. 3. Objective values of our model on two datasets

4   Conclusions

In this paper, we raise an innovative NMF method for clustering works over data from multiple views. The presented method is a joint framework that can both learn the underlying representation of each view and deal with consistent and complementary information concurrently. In addition, a repeating alternation tactic is intended to handle the non-convex optimal task and validate the astringency of our work.

However, the proposed method holds promising performance in clustering tasks. In future work, we intend to extend it into other learning tasks such as classification, retrieval and so on.

References

1. Yang, Y., Wang, H.: Multi-view clustering: a survey. Big Data Min. Anal. **1**(02), 3–27 (2018)
2. Li, Y., Yang, M., Zhang, Z.M.: A survey of multiview representation learning. IEEE Trans. Knowl. Data Eng. **31**(10), 1863–1883 (2018)
3. Kan, M., Shan, S., Zhang, H., Lao, S., Chen, X.: Multi-view discriminant analysis. IEEE Trans. Pattern Anal. Mach. Intell. **38**(1), 188–194 (2016)
4. Cai, X., Nie, F., Huang, H.: Multi-view K-means clustering on big data. In: International Joint Conference Artificial Intelligence, pp. 2598–2604 (2013). (in Chinese)
5. Bisson, G., Grimal, C.: Co-clustering of multi-view datasets: a parallelizable approach. In: International Conference on Data Mining, pp. 828–833. IEEE (2012)
6. Zhan, K., Niu, C., Chen, C., Nie, F., Zhang, C., Yang, Y.: Graph structure fusion for multiview clustering. IEEE Trans. Knowl. Data Eng. **31**, 1984–1993 (2018)
7. Sharma, A., Kumar, A., Daume, H., Jacobs, D.W.: Generalized multiview analysis: a discriminative latent space. In: 2012 IEEE Conference on Computer Vision and Pattern Recognition, pp. 2160–2167 (2012)
8. Hou, C., Nie, F., Tao, H., Yi, D.: Multi-view unsupervised feature selection with adaptive similarity and view weight. IEEE Trans. Knowl. Data Eng. **29**(9), 1998–2011 (2017)
9. Mehrkanoon, S., Suykens, J.A.K.: Multi-label semi-supervised learning using regularized kernel spectral clustering. In: International Joint Conference Neural Networks, pp. 4009–4016 (2016). (in Canada)
10. Gao, H., Nie, F., Li, X., Huang, H.: Multi-view subspace clustering. In: Proceedings of the IEEE International Conference on Computer Vision, pp. 4238–4246 (2015)
11. Perera-Lluna, A., Kanaan-Izquierdo, S., Ziyatdinov, A.: Multiview and multifeature spectral clustering using common eigenvectors. Pattern Recogn. Lett. **102**, 31–36 (2018)
12. Ayesha, S., Hanif, M.K., Talib, R.: Overview and comparative study of dimensionality reduction techniques for high dimensional data. Inf. Fusion **59**, 44–58 (2020)
13. Kang, Z., Wen, L., Chen, W., Xu, Z.: Low-rank kernel learning for graph-based clustering. Knowl. Based Syst. **163**, 510–517 (2019)
14. Wang, H., Peng, J., Fu, X.: Co-regularized multi-view sparse reconstruction embedding for dimension reduction. Neurocomputing **347**, 191–199 (2019)
15. Hidru, D., Goldenberg, A.: EquiNMF: graph regularized multiview nonnegative matrix factorization. Comput. Sci. (2014)
16. Liu, J., Wang, C., Gao, J., Han, J.: Multi-view clustering via joint nonnegative matrix factorization. In: Proceedings of the 2013 SIAM International Conference on Data Mining, pp. 252–260 (2013)
17. Wang, J., Wang, X., Tian, F., Liu, C.H., Yu, H., Liu, Y.: Adaptive multiview semi-supervised nonnegative matrix factorization. In: International Conference Neural Information Processing, pp. 435–444 (2016). (in Japan)
18. Ou, W., Yu, S., Li, G., Lu, J., Zhang, K., Xie, G.: Multi-view non-negative matrix factorization by patch alignment framework with view consistency. Neurocomputing **204**, 116–124 (2016)

Management Accounting System of Commercial Bank Based on Big Data

Hui Zhang[(✉)]

Shandong Commercial Vocational and Technical College,
Jinan 250103, Shandong, China

Abstract. Although the current management accounting model has also made contributions in profit, it has not been well combined with the actual situation of the enterprise, and there are many problems in the concrete application. This paper mainly studies the management accounting system of commercial bank based on big data. The project is technically developed in JAVA language. With the help of the existing management accounting system application platform and system architecture, the existing management accounting technology platform is a service-oriented architecture (Service-Oriented Architecture, SOA). It solves the problem of business integration between different business applications in the Internet environment and completes a specific function by connecting loosely coupled coarse-grained services. After the application of the management accounting system, the fine and accurate management ability of the daily operating cost of commercial banks is significantly enhanced, which provides accurate cost data for the profitability analysis of institutions, business lines, products, customers and account managers, and provides decision basis for performance appraisal, business analysis and product pricing.

Keywords: Big data · Commercial banks · Management accounting · Service-oriented architecture

1 Introduction

Management accounting and financial accounting are two branches of modern accounting theory and practice. Financial accounting is mainly the accounting form of confirming, measuring, recording and disclosing the economic events of an enterprise, which we call external accounting. Management accounting is mainly aimed at the analysis of enterprise production and operation, and gives relevant information for managers' reference in the aspects of prediction, decision-making, budget, evaluation and planning, so as to promote the improvement of internal management level and the growth of economic benefits. We call it internal accounting. The financial accounting construction of domestic commercial banks has been in line with the international level, but in the theory and practice of management accounting, it is far behind the international level. Due to the influence of a series of drastic changes in the external operating environment of commercial banks in China in recent years, such as marketization of interest rate, strengthening of financial supervision, promotion of accounting

J. C. Hung et al. (Eds.): FC 2021, LNEE 827, pp. 357–364, 2022.
https://doi.org/10.1007/978-981-16-8052-6_44

authorities and financial disintermediation, commercial banks in China have generally recognized the importance of management accounting [1].

Because of the development of western management accounting in the past hundred years, management accounting is becoming more and more perfect, and the research field is widening in depth. However, because of the inconsistency with the practical application of some fields, the theoretical research is not very mature, which makes management accounting not very ideal in the practical use of western countries, especially since the end of the last century, the problems and new challenges to be solved are endless. At present, the western management accounting researchers try to solve the practical problems by introducing a new "empirical research" method. It can be seen that management accounting still needs to be improved, there is still room for development in application, or a new knowledge in accounting. However, it cannot be completely denied that the prospect of management accounting is partly adapted to the current social and economic development, especially the external environment of enterprises, the prospect of management accounting is still very optimistic [2].

This paper attempts to provide relevant experience guidance and reference for the better construction and application of management accounting system by commercial banks in China, so as to provide targeted strategies for banking management accounting. Through analysis, we can see that both in theory and in the solution of practical problems, this paper has a very far-reaching significance.

2   Management Accounting System

2.1   Big Data

(1) Big Data Overview

With the rapid rise of the Internet of things and the Internet of things, its information technology has been widely used in various fields of people's lives, resulting in massive data. Especially in recent years, the emergence of various Internet devices and APP has produced a lot of data, data scale use TB and PB are difficult to express [3]. How to make good use of these data to play a good role has been highly concerned by enterprises and schools. Through the mining and analysis of big data, we can give full play to the decision-making role of management. But in order to store and analyze data better, many enterprises need to invest a lot of money to buy storage equipment and hire a large number of personnel to manage to promote the use of massive data in enterprises and schools. Because the traditional data analysis method can not meet the needs of Internet information, the big data era has come, a large number of enterprises and university researchers have begun to engage in massive data mining research, Under big data technology, all the data are analyzed and processed fast [4].

Big data technology has been rapidly expanded and developed under the common impetus of Internet of things, artificial intelligence, machine learning and cloud computing, and has become the focus of attention of enterprises and colleges and universities. Through the Internet of things technology, artificial intelligence technology and so on to collect massive data, and then on the big data platform using machine learning and cloud computing algorithm analysis, so that they promote mutual development. At present, enterprises and universities and related research departments are increasing the of research and development and use of big data technology [5].

2.2   Contents of Management Accounting System of Commercial Banks

In short, the function of management accounting is to participate in the operation and management of enterprises. Combined with the circular characteristics of enterprise economic activities, the functions of management accounting in enterprise operation and management can be divided into three parts: pre management, in-process management and post management. Prediction and analysis in advance, supervision and control in the event, and assessment and evaluation after the event [6]. When the commercial bank constructs the management accounting system, it must first satisfy the demand of the management accounting function, and then conform to the actual situation of the commercial bank. Through the support of information means, build an information management accounting platform of data integration [7].

Different from the common office software and independent business system, the management accounting system of commercial banks needs to connect with the core business system of the bank and almost all the peripheral business systems and management information systems. The data involved covers all the account level data and transaction level data in the process of bank operation, and all the data are integrated, calculated and transformed through certain procedures to meet the needs of all the customers There are different management and decision-making needs. Because the bank management accounting system is affected by many factors, such as wide range, large amount of data, complex application requirements and so on, the bank management accounting system is a huge "system group" composed of many subsystems. In the case of meeting the functional requirements of management accounting, bank management accounting system mainly includes comprehensive budget management system, internal fund transfer pricing system, cost allocation system, credit rating system, risk asset measurement system, performance management system, etc. [8].

(1) Comprehensive budget management system

Comprehensive budget management system is an important strategic implementation tool for enterprises. By integrating the budget management data scattered in different subsystems and channels, it improves the depth and breadth of budget management information, realizes the symmetry of information, improves the scientificity and feasibility of budget, and achieves the expected refined budget management objectives [9].

(2) Internal fund transfer pricing system

Internal fund transfer pricing system is the core of dividing and measuring the cost of capital and income. It can calculate the internal fund transfer (FTP) price of each business, that is, the opportunity income of capital source and the opportunity cost of capital utilization. According to the data requirements of internal fund transfer pricing, the internal fund transfer pricing system obtains the basic pricing data in each source business system, and "processes" the required pricing result data one by one by configuring various pricing rules.

(3) Cost sharing system

Cost allocation system is like the basis of the division of operating costs in management accounting. Through the determination and selection of allocation factors, the operating costs from public management departments and business departments are reasonably allocated to different dimensions such as institutions, business lines, customers and products according to certain rules, so as to obtain the operating costs that should be borne by each dimension.

(4) Credit rating and risk assets measurement system

The measurement of customer credit rating, risk and risk cost is usually completed automatically by the corresponding professional system background. For example, the customer credit rating system can automatically obtain all the business data of each customer in the bank, and automatically count them into the rating score. The bank can obtain the rating of each customer through the customer credit rating system; the risk cost can be measured through the customer credit rating and credit management system, and the economic capital and its cost can be measured through the risk assets System or economic capital management system.

(5) Performance management system

Performance management system is one of the core tools for commercial banks to use management accounting for operation and management, and it is also one of the internal conditions for commercial banks to achieve healthy development. In the performance management system, the development strategy of banks is in the core position, and the management system decomposes the overall development strategy, and implements it to each staff member of each department, through each business outlet and management system The specific work arrangements of employees, the realization of the bank's development strategy, improve the performance level [10].

3   System Test

After the design and implementation of the system is completed, the function and performance of the system need to be simulated. Testing is very important for a system that is about to go online. The system can run normally and ensure the smooth development of bank cost sharing. Through the system test and development, we can find the loopholes and short board of the system, and also confirm whether the target set in the early stage of development is met, whether it meets the customer needs and ensure the quality of the system.

3.1   Test Environment

The test environment of this system is mainly reflected in the server side and the client. The server side includes processor, host, storage, standard interface and other parts, while the client takes PC facilities as the core. The specific configuration is as follows:

Processor: Intel Xeon 3.5 GHz.
Memory: 2 DDR4 memory, 16 g.
Physical storage: 4 hard disks, 4T.

Network configuration: 1000 Mbps rate network card.

Interface: standard USB interface *4, evenly distributed in front and rear.

Operating system: Windows 7, Linux CentOS dual system.

3.2   Test Method

Software testing methods include static test, dynamic test, white box test and black box test, and so on. The white box and black box test are commonly used.

When testing software, the whole system is regarded as a closed box. The programmer operates the software according to the function in the requirement, and finds the problem of software. This test method is called black box test. Black box testing is functional oriented, regardless of how it is implemented within the program. In the test, the appropriate test cases are often designed according to the requirements specification, and the conclusion is drawn through the typical test results. Black box test is widely used in software development, and is suitable for most projects. In black box test, the design of use cases is very important and key.

When testing software, the whole system is regarded as an open box. The programmer operates the software according to the code logic, and finds the problem of software. This test method is called white box test. It pays more attention to code and operates the software according to the requirements of design specification. By designing use cases, the process of the code is tested to see if it is correct. The white box test attempted to detect all logical paths. By comparing the detected program state with the expected, the correctness of the program is judged.

System performance test formula:

$$F = \frac{N_{PU} \times R}{T} \tag{1}$$

$$C^{\mu} = C + 3\sqrt{C} \tag{2}$$

4   System Test Results

This paper uses the system throughput performance test tool and the calculation formula (1) and formula (2) to test the system performance. The test results are shown below (Fig. 1).

4.1   Interface to Docking Server Performance

According to the data response test results of system processing capacity, when the total number of transactions is 8520, the average transmission time is 0.0259 s, the average interface time is 0.006 s, the average front-end time is 0.0176 s, and the average post-end time is 0.0023 s; when the total number of transactions is 24152, the average transmission time is 0.0268 s, the average interface time is 0.007 s, the average front-end time is 0.0167 s, and the average post-end time is 0.0031 s; when the total

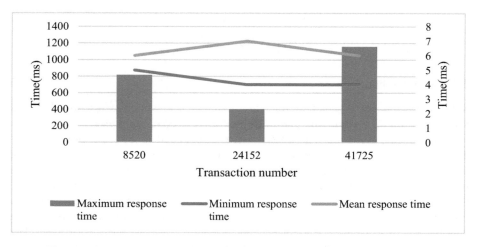

Fig. 1. The database records data in response to the application interface data

number of transactions is 24152, the average transmission time is 0.0268 s, the When the number of transactions is 41725, the average transmission time is 0.0207 s, the average interface time is 0.006 s, the average front-end time is 0.0129 s, and the average back-end time is 0.0018 s. According to the benchmark results, when sending requests continuously, each request completes the whole process from receiving to distributing in 0.03 s on average, meeting the second response requirements.

4.2 Application Interface Data Response Performance

Table 1. Application interface data response records data

Time (min)	Transactions	Maximum response time	Minimum response time	Mean response time
1	759	373 ms	43 ms	78 ms
3	2284	425 ms	53 ms	77 ms
5	3851	424 ms	53 ms	77 ms

As shown in Table 1 and Fig. 2, when the total number of transactions is 759, the average transmission time is 0.02875 s, the average interface time is 0.078 s, the average front-end time is 0.165 s, and the average post-end time is 0.0382 s; when the total number of transactions is 2284, the average transmission time is 0.3266 s, and the average interface time is 0.077 s, The average time consuming of front-end is 0.198 s, and that of post-end is 0.0416 s; when the total number of transactions is 3851, the average time consuming of transmission is 0.3032 s, the average time consuming of interface is 0.077 s, the average time consuming of front-end is 0.178 s, and the average time consuming of post-end is 0.041 s. According to the benchmark test results, when the requests are continuously

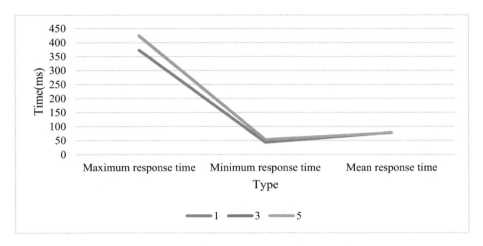

Fig. 2. Application interface data response records data

sent, the average time of each request is 0.033 s from data transmission request to successful transmission, which meets the second transmission requirements.

5   Conclusions

Based on the analysis of the research background and the research status at home and abroad, the management accounting system for commercial bank management accounting application is divided into four business modules: cost collection, cost reduction, cost allocation and report platform based on business analysis. A complete system requirements analysis system is formed through functional requirements analysis. In the part of system outline design, from the perspective of system internal and external overall architecture and technical scheme, through the design of system business process and data process, the functional architecture corresponding to system requirements is established, and the design results of system operation environment and network architecture are introduced. In the detailed design part, the structure of UAP platform, basic setting and cost sharing are introduced through class design, and the data structure and peripheral system interface are designed. In the process of system implementation and testing, the front-end application interface of each module is introduced, and the realization effect of system function is elaborated.

References

1. Kaur, N., Sood, S.K.: Efficient resource management system based on 4Vs of big data streams. Big Data Res. **9**(3), 98–106 (2017)
2. Qiu, W.: Enterprise financial risk management platform based on 5G mobile communication and embedded system. Microprocess. Microsyst. **80**(4), 103–104 (2021)

3. Wang, F., Ding, L., Yu, H., et al.: Big data analytics on enterprise credit risk evaluation of e-Business platform. Inf. Syst. E-Bus Manage. (2), 55–56 (2019).

4. Pana, E., Vitzthum, S., Willis, D.: The impact of internet-based services on credit unions: a propensity score matching approach. Rev. Quant. Financ. Acc. **44**(2), 329–352 (2013). https://doi.org/10.1007/s11156-013-0408-2

5. Khemka, B., Friese, R., et al.: Utility functions and resource management in an oversubscribed heterogeneous computing environment. IEEE Trans. Comput. **64**(8), 2394–2407 (2015)

6. Tobias, B., Stephan, T., Heiko, N.: On event-based optical flow detection. Front. Neurosci. **9**(12), 137–138 (2015)

7. Yang, W.A., Sui, X.B., Qi, Z.C.: Can fintech improve the efficiency of commercial banks?—an analysis based on big data - ScienceDirect. Res. Int. Bus. Financ. **11**(2), 55–58 (2020)

8. Santamaría-Bonfil, G., Reyes-Ballesteros, A., Gershenson, C.: Wind speed forecasting for wind farms: a method based on support vector regression. Renew. Energy **85**(32), 790–809 (2016)

9. Jlh, A., Wu, H.B., Js, C.: Big data analytics for supply chain relationship in banking - ScienceDirect. Ind. Mark. Manage. **86**(1), 144–153 (2020)

10. Wakkee, I., Sleebos, E.: Giving second chances: the impact of personal attitudes of bankers on their willingness to provide credit to renascent entrepreneurs. Int. Entrep. Manag. J. **11**(4), 1–14 (2015)

Stride Length Estimation Model Based on Machine Learning Algorithms

Xueling Zhao[1], Zhiyong Chen[1(✉)], and David K. Yang[2]

[1] Department of Precision Instrument, Tsinghua University, Beijing 100084, China
[2] Villanova Preparatory School, Ojai, CA 93023, USA

Abstract. The accuracy and feasibility of the stride length estimation model are important for pedestrian dead reckoning (PDR). In the current algorithms, the linear frequency models or nonlinear models consider few variable factors, and need to determine the coefficients previously for different pedestrians, so these models are not universal. The paper proposes to use machine learning algorithm (MLAs) to try to establish a general stride length estimation model by using a consumer-grade micro inertial measurement unit (MIMU) to collect data. Two MLAs including back propagation artificial neural network (BP-ANN) and least squares support vector machine (LSSVM) are adopted for this model. The relationship between input parameters and output values is trained by data of 13 subjects, then 3 new test subjects' stride length estimation using the trained net is discussed. The experiment shows that the stride length estimation algorithm using the two MLAs both can meet the requirements of pedestrian distance estimation, and does not need to determine the coefficient in advance, which verifies the feasibility of the proposed methods.

Keywords: Stride length estimation · BP-ANN · LSSVM · MIMU

1 Introduction

The principle of satellite communication is to launch a satellite to the geostationary orbit, use the communication transponder to receive the signal, and the signal is amplified and converted and forwarded to other ground stations to complete the transmission between two ground stations. It has high frequency bandwidth, large communication capacity, low bit error rate, and wide coverage. However, satellite signal is very poor in indoor environment such as shopping mall or museum. In this case, the application based on inertial positioning system can play a great role in obtaining azimuth and distance [1, 2]. Furthermore, according to the data of micro inertial measurement units (MIMUs), pedestrian dead reckoning (PDR) can be applied to tracking the position of firefighters [3, 4]. Because the acceleration will change regularly during the walking, the number of walking steps can be obtained by analyzing the periodic changes of the data. PDR not only depends on the number of steps, but also on the average stride (or step) length. The stride length is complex because it is not only very different among persons, but also related to the motion patters of walking [5]. Modeling stride length is the key to accurately estimate the distance of walking [6].

© The Author(s), under exclusive license to Springer Nature Singapore Pte Ltd. 2022
J. C. Hung et al. (Eds.): FC 2021, LNEE 827, pp. 365–375, 2022.
https://doi.org/10.1007/978-981-16-8052-6_45

In addition, in many sports, it is very important to master the stride (or step) length of athletes in different states, which can improve their performance. MIMU is small in size, light and cheap and it can be easily fixed on athletes' shoes, so it is very suitable for use in daily sports practice [7, 8]. From a practical point of view, it is attractive to estimate the stride length with the MIMU on the foot. However, the stride length cannot be obtained directly from the signal of MIMU [9].

The simplest way is to take each person's step length estimation as a constant model [10], but the constant model can't estimate the step length well due to different motion patters [11]. P. D. Groves proposed to use the pedestrian height to calculate the step length, but ignored the step length change during walking [12]. J. E. Bertram found that there is a close relationship between the step frequency and the step length [13]. The most common way is to approximate the step length and the step frequency into a linear relationship [14], but the step length also has some relationship with acceleration variance and the vertical velocity [15, 16]. Just considering the step frequency will lead to a larger error, and different people will have different model parameters.

Many scholars have also studied the nonlinear models [17]. Fang proposed a nonlinear step length estimation model with only one parameter [18], which is easy to implement in a real-time estimation algorithm, but for different pedestrians, the coefficients need to be determined in advance. Grejner-Brzezinska proposed to use artificial neural network (ANN) and fuzzy logic to build the step length estimation model through combining the heading information, the navigation accuracy has been greatly improved [19]. But due to the use of a backpack system carrying multiple sensors, the structure is complicated and the cost is relatively high.

Walking can be roughly divided into the standing phase and the leg swinging phase [20]. Each stage corresponds to one step, and the two steps correspond to one stride. When only one MIMU is tied to the foot, one cycle corresponds to one stride after processing the output data of the accelerometer. The accurate stride length cannot be obtained directly based on the acceleration, but the stride length can be estimated by some statistical characteristics of acceleration data [21].

When walking, some statistical characteristics of acceleration are closely related to the change of step length. The relationship between step frequency and step length is approximately linear [12, 22, 23]. In order to obtain the relationship between the two, SUN Zuo-lei allowed the tester to walk a fixed distance along a straight path, recorded the number of steps and the walking time of the walker in two cases of slow walking and fast walking, and calculated the average step length and step frequency for fitting the relationship between the two in real time [23]. Timothy H. Riehle statistically analyzed 52 parameters in a step cycle, including the time of each step, acceleration variance and maximum value, acceleration derivative, etc., and found that the acceleration variance and the maximum acceleration value were the dominant factors [24]. However, this literature only made qualitative analysis, and did not give the functional relationship between these parameters and step length.

In this paper, a consumer-level MIMU is used to collect the acceleration data of the testers in four states: slow walking, normal walking, fast walking, and running. After acceleration data preprocessing, the stride length estimation models based on back propagation artificial neural network (BP-ANN) and least squares support vector machine (LSSVM) are respectively established, then the established models are put to use to estimate the new tester's stride length. The purpose of work in this paper is to discuss how to build universal stride length estimation models based on machine learning algorithms (MLAs).

Here is the structure of this paper. The Sect. 2 is the introduction of the two MALs and discussing the preprocessing methods of acceleration data. The Sect. 3 verifies the models with experimental data. The Sect. 4 is the conclusion.

2 Methodology

2.1 Data Collection

In this paper, a consumer-grade MIMU consisting of MPU6050 is used to collect experimental data. The serial rate is 2400 bps–921600 bps and the measuring range of accelerometer is ±16g. During the experiment, the mobile application receives data in real time through Bluetooth, and the sampling frequency is 100 Hz. The data obtained are quite different when the MIMU is fixed on different parts of the body. The lower the MIMU is placed, the more sensitive it is to the different stages of stride [24]. The MIMU fixed on foot can better reflect the law of pedestrian movement, and the step detection is more reliable [25]. Therefore, as shown in Fig. 1, the MIMU is fixed on the front foot for data collection.

Fig. 1. MIMU for experiment and its fixed position

The stride length discussed in this paper is the distance that the same foot leaves the ground and then touches the ground. As shown in Fig. 2, generally speaking, the stride length is about twice the step length.

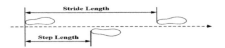

Fig. 2. Difference between stride length and step length

This paper measures the data of 13 people in four walking patterns including slow walking, normal walking, fast walking, and running. The subjects are aged between 22 and 29. The experimental data sampling rate is 100 Hz, and two sets of data are collected in each walking pattern. The experiment location is in the hallway of one school building, and the movement distance was set to 30 m by laser rangefinder.

2.2   Data Analysis

The three-axis acceleration data output of MIMU are a_x, a_y, and a_z. Figure 3 reflects the change of a_x, a_y, and a_z during a tester's normal walking. The x-axis corresponds to the forward direction, the y-axis is the left side of it, and the z-axis is the cross direction of x-axis and y-axis. It can be seen that a_x, a_y, and a_z all have a certain periodic rule, and the acceleration of the z-axis is more obvious. This is because the vertical acceleration will change regularly with the foot lift and landing when one person walks [26].

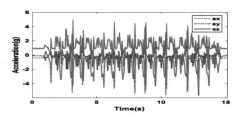

Fig. 3.   The data changes of the three-axis accelerometer during norm walking.

However, due to the installation error and the body swinging during walking, a_z does not truly reflect the change of acceleration in the vertical direction, and a_x and a_y will also couple the acceleration in the vertical direction. If the acceleration vector a_{sum} (where $a_{sum} = \sqrt{a_x^2 + a_y^2 + a_z^2}$) are used for stride detection and stride length estimation, the influence of sensor tilt on the acceleration signal can be avoided [22]. Figure 4 (a) is the changing of acceleration vector a_{sum} when a tester walks. Figure 4(b) is a partially enlarged view. It can be seen that a walking cycle can be roughly divided into the swing phase and the stance phase. It can be seen that during the movement, the acceleration a_{sum} also exhibits periodic characteristics, so this paper chooses a_{sum} as the data processing object.

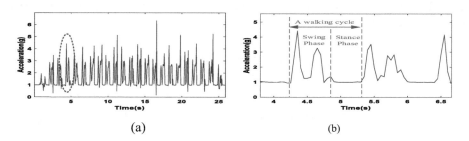

| (a) | (b) |

Fig. 4.   (a) Variation of acceleration in normal walking (b) local enlarged view.

In this paper, a total of 104 sets of three-axis acceleration data of 13 people are collected, and then the data are processed to obtain the maximum acceleration a_{max}, the stride frequency f_{stride}, the standard deviation σ_a and the average acceleration a_m during the stride period of each set of data. The calculation formulas of the stride frequency f_{stride} and average stride length L_m are as follows.

$$f_{stride} = N/t \tag{1}$$

$$L_m = D/N \tag{2}$$

Where, it is the time of walking distance, D is the walking distance, and N is the number of strides.

2.3   Overview of BP-ANN and LSSVM

The neural network performs function conversion on the input and output data. The BP-ANN has a wide range of applications and it continuously approximates the desired input and output mapping relationship by continuously adjusting the weights and thresholds. Figure 5 reveals the structure block diagram of BP-ANN.

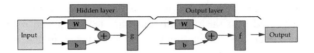

Fig. 5. Structure block diagram of BP neural network.

The hidden layer has a non-linear function. The sigmoid function shown in (3) is commonly used, where x is an independent variable.

$$g(x) = \frac{1}{1 + e^{-x}} \tag{3}$$

The input of the hidden layer is

$$v_i = \sum_{i=1}^{n_0} w_{ij}x_j + b_i \tag{4}$$

Where v_i is the input of the ith neuron in the hidden layer, w_{ij} is the connection weight, and x_j refers to the output value, n_0 is the number of neurons, b_i is the threshold.

The output of the ith neuron in the hidden layer is

$$x_i = g(v_i) \tag{5}$$

The final result is

$$\hat{y} = \sum_{i=1}^{n_1} w_i x_i \tag{6}$$

Where $\hat{y}$ is the network output value, n_1 is the number of hidden layer neurons, w_i is the connection weight, and x_i is the output value of the ith hidden layer neuron.

Next, this paper discusses the other machine learning algorithm, LSSVM, which is different from BP-ANN, but also has similar applications. The LSSVM is a machine learning algorithm proposed by Suyken, which applies the least square linear system as the loss function, not quadratic programming in the case of standard SVM [27]. LSSVMs have been successfully used in pattern recognition and function estimation [28, 29].

Fig. 6. The principle of SVM

The principle of LSSVM is illustrated in the Fig. 6. The training data $\{x_i y_i\}_i^n$, where x_i is the n-dimensional input vector, $x_i \in R_n$, y_i is the desired output value. The data in input space are projected into the high-dimensional feature space by nonlinear mapping, and then the linear estimation function is constructed in the high-dimensional feature space to obtain the nonlinear regression effect in the original space. As shown in Eq. (7)

$$f(x) = w \cdot \phi(x) + b \tag{7}$$

Where, $\phi(x)$ is the nonlinear mapping between input space and high-dimensional feature space, w is weight vector, b is constant offset.

Using the structural risk minimization (SRM) principle to find w and b, which can be estimated by minimizing the regularized risk function

$$\text{Minimize } J(w, e) = \frac{1}{2} \|w^2\| + c \sum_{i=1}^{n} e_i^2 \tag{8}$$

$$\text{s.t. } y_i = \phi(x_i) \cdot w + b + e_i \, i = 1, 2, \cdots, n$$

Where, e_i is the introduced error variable and c is an adjustable parameter. The optimization problem is solved by Lagrange method showed in (9).

$$L(w, b, e, a) = J(w, e) - \sum_{i=1}^{n} a_i(\phi(x_i) \cdot w + b + e_i - y_i) \tag{9}$$

Then the mapping relationship between input space and high-dimensional feature space can be solved by (9).

2.4   Data Fitting of Two MALs

The input vector X is composed of 4 variables, which are $x_1 = f_{stride}$, $x_2 = a_{max}$, $x_3 = \sigma_a$, $x_4 = a_m$, and the target value of output is Lm. 104 sets of data are randomly divided into training dataset and test dataset. The data from the training set is to generate a mapping network, and the data from test set is to verify the performance of the trained network. The purpose of training is to make the output as close as possible to the target value by continuously adjusting the parameters of mapping function. It can be seen from Fig. 7(a) that the two MALs both can fit the data well. From the Fig. 7(b), the predicted stride length values are close to the actual stride length values using the two MALs. The mean value of prediction errors are 0.007 m and −0.06 m, and the standard deviation (SD) are 0.11 m and 0.08 m respectively using BP-ANN and LSSVM, which indicates the feasibility of establishing stride length model based on MALs.

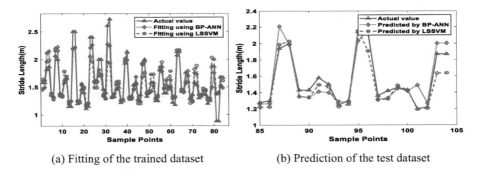

(a) Fitting of the trained dataset   (b) Prediction of the test dataset

Fig. 7. The effect of fitting and prediction of stride length using BP-ANN and LSSVM

3   Experimental Verification

To verify the feasibility of the proposed stride length estimation algorithm, further experiments are conducted in the playground. Three new testers are selected. Judging the effect of the algorithm by estimating the walking distance. The smaller the error, the more reliable the proposed model. As shown in Fig. 8, the testers walk along the playground one circle at one time, and the movement patters included slow walking, normal walking, fast walking and running. The actual distance traveled is 500 m. Figure 9 shows the variation of accelerometer data during the experiment.

Fig. 8. Playground experiment

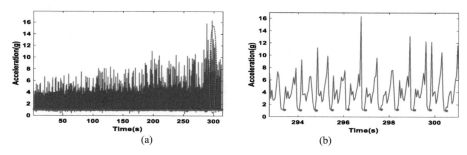

Fig. 9. (a) Acceleration change during movement; (b) Partial enlarged drawing (running).

In each stride cycle ΔT, the stride frequency f_{stride} can be obtained.

$$f_{stride} = 1/\Delta T \tag{10}$$

Furthermore, the maximum value a_{max}, average value a_m and SD of acceleration σ_a can be obtained, and then the estimated stride length corresponding to each footstep can be obtained by formula (11).

$$L = f\,(net, X) \tag{11}$$

Where, L is the estimated stride length, X is the input vector, which is composed of four variables: stride frequency f_{stride}, maximum acceleration a_{max}, standard deviation σ_a and average value a_m in the stride period; The parameter net is the trained network by BP-ANN or LSSVM.

This paper studied the stride length estimation based on inertial sensor. In the experiment, the positioning error e was used to evaluate the performance of the proposed model.

$$e = \frac{|s_c - s_t|}{s_t} \times 100\% \tag{12}$$

Where s_c is the total distance calculated by the stride length estimation algorithm, s_t is the real total distance, and e is the estimation error.

The estimated values and errors of the estimated distance of the three subjects are shown in Table 1. Experimental results show that the two proposed MALs are both feasible and universal in stride length estimation. This verifies the rationality of the parameters selected in this paper, and the feasibility and effectiveness of MALs in stride length estimation.

Table 1. The calculated distance values and the errors

Tester	A	B	C
BP-ANN model	484.11	506.44	494.06
Error	3.18%	1.29%	1.12%
LSSVM model	487.89	498.83	483.19
Error	2.42%	0.23%	3.36%

4   Conclusion

After carefully and strictly conducting the test, all the results are concluded in this paper. As shown here, a consumer-grade MIMU is used to collect data when the subjects are walking, and two MAL models, including BP-ANN and LSSVM, are used to fit the relationship between stride length and the other four walking parameters including stride frequency, maximum acceleration, standard deviation and average value. The test results show that the estimation error of walking distance can meet the requirements of pedestrian distance estimation. According to the data arranged from the experiment, the proposed MALs both can train and learn the statistical characteristics of the acceleration data during walking, which verifies the potential application of the proposed stride length model in estimation of stride length.

In the future, with more attention from the academic circle and the increase of the need in the society, the estimation of pedestrian stride length needs to be further studied. Moreover, since data is the foundation of all the scientific guess and conclusion, the data of testers should be collected, because the larger the data, the more reliable the model can be established. Due to the limitation of conditions, this paper proposes an idea to estimate the stride length, but it is not the final solution. The accuracy of stride length estimation needs to be further improved. The application of multiple sensors or stride length estimation in complex environment needs further research.

References

1. Qian, C., et al.: An integrated GNSS/INS/LiDAR-SLAM positioning method for highly accurate forest stem mapping. Remote Sens. **9**(1), 3 (2017)
2. Dag, T., Arsan, T.: Received signal strength based least squares Lateration algorithm for indoor localization. Comput. Electr. Eng. **66**, 114–126 (2018)
3. Nilsson, J.-O., Rantakokko, J., Händel, P., Skog, I., Ohlsson, M., Hari, K.: Accurate indoor positioning of firefighters using dual foot-mounted inertial sensors and inter-agent ranging. In: 2014 IEEE/ION Position, Location and Navigation Symposium-PLANS 2014, pp. 631–636. IEEE (2014)

4. Bousdar Ahmed, D., Munoz Diaz, E., García Domínguez, J.J.: Automatic calibration of the step length model of a pocket INS by means of a foot inertial sensor. Sensors **20**(7), 2083 (2020)

5. Weinberg, H.: Using the ADXL202 in pedometer and personal navigation applications. Analog Devices AN-602 Appl. Note **2**(2), 1–6 (2002)

6. Libotte, G.B., Lobato, F.S., Neto, F.D.M., Platt, G.M.: Adaptive second order step length algorithm for inverse reliability analysis. Adv. Eng. Softw. **146**, 102831 (2020)

7. Ammann, R., Taube, W., Wyss, T.: Accuracy of PARTwear inertial sensor and Optojump optical measurement system for measuring ground contact time during running. J. Strength Cond. Res. **30**(7), 2057–2063 (2016)

8. Falbriard, M., Meyer, F., Mariani, B., Millet, G.P., Aminian, K.: Accurate estimation of running temporal parameters using foot-worn inertial sensors. Front. Physiol. **9**, 610 (2018)

9. de Ruiter, C.J., van Dieën, J.H.: Stride and step length obtained with inertial measurement units during maximal sprint acceleration. Sports **7**(9), 202 (2019)

10. Ho, N.-H., Truong, P.H., Jeong, G.-M.: Step-detection and adaptive step-length estimation for pedestrian dead-reckoning at various walking speeds using a smartphone. Sensors **16**(9), 1423 (2016)

11. Harle, R.: A survey of indoor inertial positioning systems for pedestrians. IEEE Commun. Surv. Tutor. **15**(3), 1281–1293 (2013)

12. Groves, P.D.: Principles of GNSS, inertial, and multisensor integrated navigation systems, [Book review]. IEEE Aerosp. Electron. Syst. Mag. **30**(2), 26–27 (2015)

13. Bertram, J.E., Ruina, A.: Multiple walking speed–frequency relations are predicted by constrained optimization. J. Theor. Biol. **209**(4), 445–453 (2001)

14. Díez, L.E., Bahillo, A., Otegui, J., Otim, T.: Step length estimation methods based on inertial sensors: a review. IEEE Sens. J. **18**(17), 6908–6926 (2018)

15. Jing, Z., Miaohong, C., Haojie, W.: An indoor positioning system based on bluetooth RSSI and PDR for a smartphone. DEStech Trans. Comput. Sci. Eng. (iccis) (2019)

16. Zhang, Y., Li, Y., Peng, C., Mou, D., Li, M., Wang, W.: The height-adaptive parameterized step length measurement method and experiment based on motion parameters. Sensors **18**(4), 1039 (2018)

17. Ayane, R., Hamdaoui, A., Braikat, B., Tounsi, N., Damil, N.: A new analytical formula to compute the step length of Padé approximants in the ANM: application to buckling structures. Compt. Rendus Mécan. **347**(6), 463–476 (2019)

18. Fang, L., et al.: Design of a wireless assisted pedestrian dead reckoning system-the NavMote experience. IEEE Trans. Instrum. Meas. **54**(6), 2342–2358 (2005)

19. Toth, C., Grejner-Brzezinska, D.A., Moafipoor, S.; Pedestrian tracking and navigation using neural networks and fuzzy logic. In: 2007 IEEE International Symposium on Intelligent Signal Processing, pp. 1–6. IEEE (2007)

20. Hoseinitabatabaei, S.A., Gluhak, A., Tafazolli, R., Headley, W.: Design, realization, and evaluation of uDirect-an approach for pervasive observation of user facing direction on mobile phones. IEEE Trans. Mob. Comput. **13**(9), 1981–1994 (2013)

21. Zou, H., Chen, Z., Jiang, H., Xie, L., Spanos, C.: Accurate indoor localization and tracking using mobile phone inertial sensors, WiFi and iBeacon. In: 2017 IEEE International Symposium on Inertial Sensors and Systems (INERTIAL), pp. 1–4. IEEE (2017)

22. Leppäkoski, H., Collin, J., Takala, J.: Pedestrian navigation based on inertial sensors, indoor map, and WLAN signals. J. Signal Process. Syst. **71**(3), 287–296 (2013)

23. Wu, Y., Zhu, H.-B., Du, Q.-X., Tang, S.-M.: A survey of the research status of pedestrian dead reckoning systems based on inertial sensors. Int. J. Autom. Comput. **16**(1), 65–83 (2019)

24. Riehle, T.H., Anderson, S.M., Lichter, P.A., Whalen, W.E., Giudice, N.A.: Indoor inertial waypoint navigation for the blind. In: 2013 35th Annual International Conference of the IEEE Engineering in Medicine and Biology Society (EMBC), pp. 5187–5190. IEEE (2013)
25. Yuan, X., Liu, C., Zhang, S., Yu, S., Liu, S.: Indoor pedestrian navigation using miniaturized low-cost MEMS inertial measurement units. In: 2014 IEEE/ION Position, Location and Navigation Symposium-PLANS 2014, pp. 487–492. IEEE (2014)
26. Chen, G., Meng, X., Wang, Y., Zhang, Y., Tian, P., Yang, H.: Integrated WiFi/PDR/Smartphone using an unscented kalman filter algorithm for 3D indoor localization. Sensors **15**(9), 24595–24614 (2015)
27. Alharbi, N., Hassani, H.: A new approach for selecting the number of the eigenvalues in singular spectrum analysis. J. Franklin Inst. **353**(1), 1–16 (2016)
28. Lin, W.-M., Tu, C.-S., Yang, R.-F., Tsai, M.-T.: Particle swarm optimisation aided least-square support vector machine for load forecast with spikes. IET Gener. Transm. Distrib. **10** (5), 1145–1153 (2016)
29. Xing, H., Hou, B., Lin, Z., Guo, M.: Modeling and compensation of random drift of MEMS gyroscopes based on least squares support vector machine optimized by chaotic particle swarm optimization. Sensors **17**(10), 2335 (2017)

Application of Artificial Intelligence in the Design of Animation Special Effects

Bing Zhang, Qian Ma$^{(\boxtimes)}$, and Haidong Li

School of Digital Media and Creative Design, Sichuan University of Media and Communications, Chengdu, Sichuan, China

Abstract. Modern people live in a social environment of ubiquitous intelligence, science and technology. If they do not have a sufficient understanding of the concept of intelligence on a philosophical level, even if they have greater enthusiasm for mechanical thinking, they will still be in a state of inadequacy. In this article, a simplified beam curve model is used to represent the structure of curved tree trunks. The vibration frequency and damping ratio are used as the parameters of the time-domain motion equation to input the motion equation of each main vibration function and the time-domain equation and space-time equation of a simple solution analysis method. The motion of each part of the curved branch is integrated to obtain the total motion of the curved branch, and the motion of each curved branch is integrated to obtain the total motion of the tree. The results show that the motion tree simulation system used in this article can interact with wind direction and wind speed, and can transmit smooth motion tree motion at a frame rate of about 25–30 fps, which is more effective; the motion pattern of the tree varies with the wind and wind direction. The long branches in the middle layer have obvious elastic deformation, and the thin branches of the crown move with the movement of the mother branch. The simulation movement of trees conforms to the characteristics of the leafless magnolia tree in winter; in addition, the movement simulation of the uniform point of the cloud and the model of the ground laser point cloud show that the simulation system in this paper is universal.

Keywords: Artificial intelligence · Machine thinking · Animation effects · Smart environment

1 Introduction

The emergence of information civilization has created the rapid progress of human society. It is not only the result of continuous innovation in human cognition, but also an important manifestation of the realization of human intelligence [1]. As the driving force of the information civilization, the concentrated expression of computer algorithms, the Internet, and artificial intelligence, intelligent science and technology will inevitably become the research core with the most attention and exploration potential. When scientific researchers and technical experts feel all kinds of doubts, troubles, and obstacles to the nature of intelligence in the wave of intelligence theory exploration and

J. C. Hung et al. (Eds.): FC 2021, LNEE 827, pp. 376–383, 2022.
https://doi.org/10.1007/978-981-16-8052-6_46

practice, philosophical reflection and analysis highlights the extremely high importance. In fact, the theory of intelligence has been legendary from the very beginning. It is almost the same as the natural science and philosophy of ancient Greece. It is destined to be the ultimate dialogue between philosophy and science and technology. As the research on the theory of intelligence continues to move forward, The philosophical discussion on the understanding of the concept of intelligence has never stopped [2]. For the realization of the concept of intelligence, the relationship between computer science and intelligence is self-evident. The most concentrated expression is intelligent computer algorithm and intelligent information processing. Computer science is the basic driving force for the realization of intelligence. The relationship between neuroscience and intelligence is divided into two levels, namely the system structure level and the operating mechanism level. The most concentrated expression is in intelligent behavior and intelligent knowledge [3]. The constant interaction and stimulation between neurons brings unlimited sensory knowledge and provides an excellent model for intelligent knowledge.

Chinese scholar Zhong Yixin believes that current artificial intelligence research has made a lot of progress, but there are important shortcomings of "shallow depth in depth, fragmentation in breadth, and closed system". This is not a problem that can be solved by improving algorithms or improving hardware performance, but to find the root cause in the scientific view and methodology [4]. Zhong Yixin also believes that the methodology of scientific research is not only the source of theoretical research, but also the leader of theoretical research. Therefore, in the face of the current situation of artificial intelligence development, the top priority should be to reform and innovate the scientific methodology of artificial intelligence research, grasp the leader of artificial intelligence innovation research, and achieve breakthroughs in the basic theory of artificial intelligence at the source, surpassing and leading [5]. Guo Yunzhong believes that animation special effects have become an important part of most of today's film and television advertisements, and even the main component or main part of some film and television advertisements. A good film and television advertisement is mostly inseparable from the support and technology of animation special effects [6].

The research direction of this article is the application of artificial intelligence in the design of special animation effects. This article focuses on the dynamics of trees as the research content. By improving the performance of software and hardware and the rapid development of virtual reality technology, a new virtual technology method is being provided for scientific research on plant objects. In recent years, the establishment of three-dimensional dynamic visualization models of plants on computers based on virtual plants, research on efficient and realistic plant reconstruction and motion simulation technologies, has attracted the attention of researchers at home and abroad. However, due to the complex morphological structure and physiological characteristics of plants, high-precision, high-realistic plant three-dimensional modeling and motion simulation are still hot and difficult issues in research [7–10].

2 Analysis of Artificial Intelligence in the Design of Animation Special Effects

2.1 Artificial Intelligence Environment Analysis

Since the birth of modern computers in the 1940s, information technology has experienced more than 70 years of development, and now it has passed the key turning point based on increasingly powerful computing capabilities, and information technology has begun to have an impact on all aspects of society. Artificial intelligence is a branch of computer science and a comprehensive technical discipline, including the intersection of thought science, computer science, biological science, psychology and brain science. Artificial intelligence research includes not only robots, language recognition, image recognition, but also natural language processing and professional systems. At present, it has achieved world-renowned achievements in the fields of pattern recognition, natural language processing, games, expert systems, intelligent robots, knowledge processing, automatic programming, knowledge bases, etc., and has created a diversified, multi-level situational development direction. This new technological science will become the new technological trend of this era. In fact, in terms of artificial intelligence, it is not far from our lives, nor is it far from our imagination. This article focuses on the application of artificial intelligence in motion characteristics and special effects trends.

2.2 Application Analysis of Animation Special Effects

As an indispensable part of our life and entertainment, special effects animation is an art that reflects human emotions in real life, and is an ideology that is older than language. The rapid development of science and technology has brought many new elements to animation special effects, and has gradually changed the way people create, execute, produce and appreciate animation. In recent years, the active efforts and major progress of artificial intelligence in the field of special effects animation have been amazing. The rapid development and progress of modern science and technology have brought huge challenges and enlightenments to the field of animation, brought material and technical changes to the development of the field of animation special effects, and improved the special effects of animation from the perspective of animation. Design innovation plays an important role in promoting growth that cannot be underestimated. The introduction of different theories in the field of vitality, the formation of new ideas, the improvement of creative methods, expression methods and methods have brought positive changes and influences. Based on the latest results of domestic and foreign research, this article deeply explores the possibilities and trends of the development of artificial intelligence technology in the field of film, so that readers have the understanding and new knowledge of macroeconomics on the combination of the two.

2.3   Research on Tree Animation

The realistic animation of the tree is very important to increase the appeal and immersion of the game. In recent years, in order to improve the reality of tree animation simulation, data-driven cartoon animation simulation methods have become a research hotspot. However, the current database method has expensive data acquisition equipment and complicated procedures, and there is a big gap between the tree and the actual model. Simulation problems, such as restricted movement of trees. In order to solve the above problems, this paper uses cheap Kinect somatosensory equipment to study the method of extracting dynamic tree parameters based on video motion and depth data, and the method of constructing a 3D tree model based on cloud points and then inputting the dynamic tree parameters. Build a realistic tree model on the Power Tree Scientific model, and create an animation simulation system based on how to perform realistic tree motion simulation based on data.

3   Application Research Methods of Artificial Intelligence in Animation Special Effects Design

3.1   The Purpose of the Study

Artificial intelligence is an emerging science and technology that is used to research, simulate, develop and expand the methodology and application systems of human intelligence. The state attaches great importance to and has carried out work related to artificial intelligence, such as the "Technology Innovation 2030 China Project". The central government will create special projects for this purpose. Local governments have also begun to introduce enhanced support policies for the artificial intelligence industry. At present, about 30 cities in my country regard robotics and artificial intelligence technology as objective research objects for local development. At the same time, the national government is building and has even built more than 40 robotic industrial parks. In the next five years, Beijing, Shanghai and Shenyang are expected to become top cities in the artificial intelligence industry. This article focuses on the combination of artificial intelligence and animation special effects design. Its purpose is to provide others with detailed scientific reports based on the status quo of international and domestic development to understand the field of artificial intelligence and vitality. They have also been widely recognized and verified. Strong scientific and technological achievements will have a positive, stable and far-reaching impact on the establishment of the animation industry. We hope that through research, we can make small academic research contributions to the combination of the two, so that others can contribute to this field. And with reference prices, with high-tech achievements to better serve the animation industry.

3.2   Research Process

The purpose of this paper is to study the method of obtaining trunk motion data and analyzing trunk dynamic parameters based on Kinect, the reconstruction technology of 3D realistic tree model based on cloud point, and the simulation method of realistic motion tree based on data. The basic idea is to use cheap Kinect equipment to capture real-time color image data and single-sided depth image data when pulling trees, and combine video surveillance to watch color images. Branch according to the color characteristics of the branch mark. Define the two-dimensional trajectory and local coordinate system of the dry point to calculate the time deflection angle of the marked point. Then, the time zone distortion angle signal is converted into a vibration signal in the frequency domain through fast Fourier transform, and the vibration frequency and attenuation rate of the tree are calculated in the frequency domain. Tree movement generates a 3D model based on tree cloud reconstruction. By inputting the parameters (natural frequency and damping ratio) obtained from the actual captured motion analysis into the motion equation of the dominant tree mode, the displacement displacement of the motion equation can be solved and converted into the rotation angle of space to create a realistic tree.

3.3   Data Processing

Through the use of SPSS22.0 calculation software, the following calculation formulas are used to calculate and process the experimental data:

Quadratic formula:

$$x = \frac{-b \pm \sqrt{b^2 - 4ac}}{2a} \tag{1}$$

Fourier series:

$$f(x) = a_0 + \sum_{n=1}^{\infty} \left(a_n cos \frac{n\pi x}{L} + b_n sin \frac{n\pi x}{L} \right) \tag{2}$$

4   Analysis of Research Results on the Application of Artificial Intelligence in Animation Special Effects Design

4.1   Natural Frequency and Damping Ratio of Different Branches

The vibration deflection angle of each branch is obtained by fast Fourier transform to obtain its frequency range. The amplitude of the first-order vibration frequency is selected as the standard for calculating the damping ratio, and the tree damping ratio is calculated. The data obtained is shown in Table 1 and Fig. 1.

Table 1. Natural frequency and damping ratio of different branches

Branch number	Natural frequency (HZ)	Damping ratio-greater tension	Damping ratio-less tension
A	2.4	7.088	7.808
B	2.4	7.780	8.228
C	2.4	7.508	7.987
D	2.4	7.544	8.122

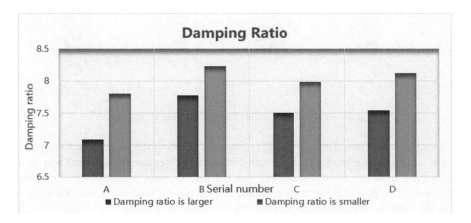

Fig. 1. Natural frequency and damping ratio of different branches

Combining Table 1 and Fig. 1, comparing the damping ratios of different tensile forces, it can be seen that under the condition of larger tensile force, the damping ratio is also larger, and the damping ratio of high-level branches is lower than that of low-level branches.

4.2 Reconstruction Parameters and Time-consuming

The parameters and time-consuming for skeleton extraction and 3D reconstruction of experimental data collected. The data obtained is shown in Table 2 and Fig. 2.

Table 2. Reconstruction parameters and time-consuming

serial number	Influence radius	Delete threshold	Number of point clouds	time consuming/s
1	0.8	0.2	193	10.972
2	0.6	0.2	143	11.624
3	1.2	0.3	97	7.745
4	0.4	0.1	85	8.610

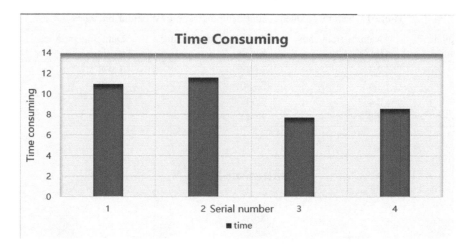

Fig. 2. Reconstruction parameters and time consumption

It can be seen from Table 2 and Fig. 2 that by deleting the skeleton points without growth space in the skeleton generation algorithm, the number of skeleton points involved in each iteration is reduced, and the convergence speed of the iteration is improved. Therefore, for the measured dense tree point cloud in the natural environment, the implementation method in this paper can achieve skeleton extraction and tree model reconstruction with higher efficiency.

5   Conclusions

In this paper, based on the space colonization algorithm, the tree point cloud skeleton extraction obtained by Kinect is implemented. In order to retain the characteristics of the bending of the branches, a skeleton reduction method based on the curvature of the skeleton is proposed. In order to realize the tree hierarchy construction and the final geometric entity model rendering, design and the process of rendering from tree skeleton to tree hierarchical structure and geometric entities is realized. The following results are obtained through experiments: In the space colonization algorithm of iteratively generating tree skeletons, by deleting the skeleton points without growth space in the skeleton generating algorithm, the number of skeleton points participating in each iteration is reduced, and the external efficient data structure is not used at the same time, improve the speed of iterative convergence.

References

1. Wu, Y., Si, G., Luo, P.: Application of artificial intelligence technology in cyberspace security defense. Appl. Res. Comput. **32**(008), 2241–2244 (2015)
2. Wei, X.: Analysis of interactive animation special effects production techniques in visual communication design. Nongjia Staff **648**(05), 275–275 (2020)

3. Hu, X.: Discuss the application and expressiveness of animation special effects technology in animated films. Sci. Technol. Innov. Herald **17**(510(06)), 122–123 (2020)
4. Zhong, Y.: Mechanismist artificial intelligence theory-a general artificial intelligence theory. J. Intell. Syst. **13**(69(01)), 2–18 (2018)
5. Zhong, Y.: From "mechanical reduction methodology" to "information ecological methodology"-the successful road to the source innovation of artificial intelligence theory. Philos. Anal. **8**(005), 133–144 (2017)
6. Guo, Y.: Application analysis of animation special effects technology in film and television advertisements. Western Radio Telev. **41**(487(23)), 116–119 (2020)
7. Du, C., Zhang, M.: The important role of 3D animation special effects technology in film and television works. Rep. Cradle **588**(12), 65–66 (2019)
8. Cui, Y.: On the interactive animation special effects production method of visual communication design. Art Technol. **032**(012), 159–160 (2019)
9. Li, Y.: Analysis of interactive animation special effects production techniques of visual communication design. Prose Baijia **390**(12), 256–256 (2019)
10. Gao, Y.: Analysis of the application of special effects of three-dimensional animation in "ten thousand years before history". News Res. Guide **010**(023), 126–128 (2019)

Smart Community Management Based on Internet of Things

Yue Zhao[✉]

Department of Economics and Management, Yunnan Technology and Business University, Kunming, Yunnan, China

Abstract. With the introduction and evolution of the concept of "smart city", with the development of technologies such as the Internet of Things, cloud computing, and big data, information technology has had a greater impact on the construction, management and social life of cities, and has gradually become one of the main bodies of a smart city, how to do a good job in efficient management of smart communities, effectively use the resources of smart communities, ensure the safe and orderly progress of people's lives and work, and provide people with a good living environment, which becomes crucial to the construction of smart communities an important issue for development. The purpose of this article is to study the management of smart communities based on the Internet of Things. This article mainly combines the current development situation of the Internet of Things and the development status of urban community management in my country, using relevant theories of public management, combined with the development of a certain community, and applying the Internet of Things technology and concepts to the innovative urban community management model. In the application, specific implementation suggestions are put forward in order to achieve the purpose of efficient community management and improvement of the quality of life of citizens. Experimental results show that the smart community will exceed 100 billion in the future market development, and its growth rate has been maintained at more than 50%.

Keywords: Internet of Things · Smart community · Efficient management · Integrated research

1 Introduction

In today's era, the development of the "Internet of Things" cannot be ignored. Now, with the continuous advancement of technology, the "Internet of Things" technology has become more and more mature, and the "Internet of Things" can be seen in ordinary people's homes [1, 2]. The "Internet of Things" refers to connecting things with things, where people live, and even the cities they live in through the Internet, and ultimately through the use of "Internet of Things" to control [3, 4]. In view of the trend of the Internet of Things, the use of Internet of Things technology as a key method and means to build an intelligent community, the management of innovative Internet of Things technology is an important way to achieve intelligent community management [5, 6].

J. C. Hung et al. (Eds.): FC 2021, LNEE 827, pp. 384–391, 2022.
https://doi.org/10.1007/978-981-16-8052-6_47

The practice of using the Internet of Things in community management in foreign countries. As Western developed countries, Europe and the United States are relatively at the forefront of technological development and maintain a continuous innovation growth trend. Many efforts and explorations in the field of community information. Take the United States as an example, it is too early to participate in community media research. Participate in the practice of Chinese IoT community management. With the rapid growth of my country's national economy, information network technology is rapidly popularized in this context. The Internet penetration rate continues to increase, which makes this article an indispensable part of work, study and life [7, 8].

This article uses bibliographic research methods, empirical analysis methods and grey forecasting model methods. Through reading a large number of Chinese and foreign classics and books, combined with the content of this research, the theoretical results related to this research field have been studied, and the research results for reference have been extracted and classified. Discuss and use its uniqueness to provide an effective theoretical basis for research, and then combine IoT-specific practices in the management community to propose an innovative community management model system, and then summarize it. The article combines the theory and practice of this article and makes it more practical [9, 10].

2 Smart Community Management Based on the Internet of Things

2.1 Application of the Internet of Things in the Management of Smart Communities

In smart communities, Internet of Things management services have been applied to smart homes, security monitoring and other scenarios. However, with the construction of smart communities and the development trend of accelerating social informatization, the scope of smart community management services is becoming wider and wider. The types and quantity of information are increasing. Therefore, it is necessary to highly integrate the information in management, use modern information technology to realize automated comprehensive management, and continuously improve management methods and systems in order to meet the increasing demand for management services of the people and society and conform to the development of the times. The following analyzes the management functions that the key technologies of the Internet of Things can achieve in the smart community from the functional structure of the smart community.

(1) Comprehensive management of community properties

In the integrated property management of the smart community, RFID cards can be issued to users, and the non-contact identification of RFID card readers can be used for authority management to realize smart access control and smart parking management; security personnel in security patrols can be assigned patrol rods, Built-in RFID reader, install RFID electronic tags at designated locations on the security patrol route. When security personnel patrol to the designated locations, use the RFID reader to identify the RFID electronic tags and record the information to prove that the security personnel are patrolling normally. Supervise security work; use image acquisition equipment to implement a full range of video surveillance of the community and video calls between residents' homes and building access control and community management departments.

(2) Community smart home

Research on smart homes has been going on for many years. A number of Internet of Things technologies have been successfully applied to the construction of smart home projects. Among them, there are more researches on the four key Internet of Things technologies: Wi-Fi, ZigBee, RFID, and Bluetooth. Wi-Fi and ZigBee are used for wireless networking management of home appliances and other devices, RFID is mainly used for access control, and Bluetooth is mainly used for continuous communication between small devices with low data volume. In the construction of a smart community, the research on these four technologies should focus on the integration of resources. Indoor environment monitoring and other low-data-volume devices use ZigBee network management and information transmission uniformly, saving each device using its own link to transmit data the resulting repeated consumption of hardware resources. Smart phones are already a popular device, ranging from senior citizens to children, almost everybody has one. In smart phones, Bluetooth is a necessary hardware, you can try to develop smart phone applications and use Bluetooth as the key to open the door. In addition, you can also try to develop the NFC (Near Field Communication) function of the smart phone, and use the phone as an access card to improve the intelligent level of management services.

(3) Community housekeeping services

In housekeeping services, interest classes, academic counseling, and child care generally have fixed places. RFID readers can be installed in these places, and teenagers and children and RFID cards can be used to achieve safety supervision. In the washing service, the RFID electronic label with the information of the clothes owner can be scanned to automatically print the labels registered when the clothes are washed, eliminating the cumbersome manual registration and realizing electronic automatic management.

(4) Community medical and health

The RFID access control system can be used to integrate the health file management of the medical and health system, and the RFID card can be used to realize the on-site number-taking function of appointment registration, which is convenient for the management of the medical and health system and the residents' visits.

(5) Community logistics services

Express delivery service, express delivery status inquiry service and warehouse management can use community RFID access card to realize self-service delivery delivery, express delivery status inquiry and warehouse management identity recognition, which is convenient and fast.

The above are the main management functions that can be realized by the key technologies of the Internet of Things in the smart community, and the remaining functions in the functional structure of the smart community are mainly realized by software. Through the above analysis, it is known that through the Internet of Things technology, certain functional systems of the smart community can be integrated, the reuse of software and hardware can be improved, and resources can be saved to facilitate the residents and managers of the community. For example, through the expansion of the RFID access control card function, multiple sub-functions in comprehensive property management, smart home, housekeeping services, medical and health, and logistics services can be integrated to achieve one card for multiple purposes; security equipment and smart home in community public areas the security equipment and other devices that can be managed through the ZigBee network can use the ZigBee network to achieve unified management.

2.2 Management Methods

(1) Constructing a big data system environment in the community
The big data software system environment in the community should involve the community data operating system, management system, exchange and integration system, network antivirus system, various intermediate systems, and various auxiliary support software in the community. Only when the software and hardware big data system is fully established and exerts its support and auxiliary role in community management, the community can serve the community people more conveniently and efficiently.

(2) Establish a community big data resource warehouse
The community street office authorizes technology-based private enterprises with core technologies and high-quality talents to develop and utilize the community's big data information system through bidding or cooperative development, and at the same time establish a data resource warehouse belonging to the community. Private companies rely on their own high-tech and other advantages to develop and utilize the community's big data resources in depth and fully, excavate potentially valuable information, and store it in the resource data warehouse in order to provide services for users who need community big data. And the community big data resource warehouse and the private enterprise big data resource warehouse can exchange and share the data resources they own.

(3) Build a data resource sharing platform
Communities and private companies publish the resources in the warehouse publicly on the data resource platform for community big data users to use. Community big data users include enterprises, residents, voluntary organizations, etc. The community or private enterprise provides the required data resources to the users for free or at a cost based on the usage of the community big data users' use of data resources, that is, for individuals or non-profit institutions. The data needed for non-profit purposes can be used for free by applying to the community street office or private enterprise. For those enterprises or individuals that have repeatedly developed and used community data resources to obtain commercial benefits for the purpose of profit, the community street Offices or private companies should charge a certain fee for the big data they obtain.

2.3   Research on the Application of the Internet of Things in Smart Communities

It can be seen from the basic structure of the smart community that the eight functional systems of the information display unit include applications in the application layer of the Internet of Things, the network in the information transmission unit includes the network in the network layer of the Internet of Things, and the information collection unit includes the perception of the Internet of Things perception function in the layer. Therefore, from the perspective of system functions, the management functions implemented by the smart community IoT are integrated together as a subsystem of the smart community. The functions realized by the management system are divided into three categories according to the main application technologies, RFID application system, ZigBee application system, and video application system.

The video application system includes video security monitoring and video intercom, which are realized by image acquisition equipment. The communication links are both bidirectional, and the ZigBee device and the RFID reader are connected via UART. Taking the transmission from the bottom layer to the top layer as an example, the signals of small data volume devices such as RFID readers at the bottom layer first pass through the ZigBee network and gateway, and then are transmitted to the top layer through a wired LAN. The data of large data volume devices such as high-speed network cameras pass through the wired LAN. In addition, there are RFID readers that are directly connected to the background management system computer through a serial port or USB to realize RFID card registration and other operations.

3   Experimental Research on the Management of the Internet of Things in Smart Communities

3.1   Research Objects

In order to make the research results more scientific, comprehensive and effective, this time we will investigate the functions realized by smart community management. This article will randomly invite 50 relevant engineers for consultation, of which the proportion of men and women is equal. Their working age is more than 3 years, which ensures the reliability of the experimental results, and then predicts the smart community market in my country. This study uses a questionnaire survey, using a ten-point scoring system, "1" means disapproval, "10" means approval, the degree of recognition from 1 to 10 is from low to high, and the data obtained is obtained by using the analytic hierarchy process. More accurate results.

3.2   Questionnaire Survey Method

The questionnaire used in this study is based on consulting a large number of documents, according to the purpose and content of the research, after many consultations with experts, the research indicators are selected, and the principles and requirements of the questionnaire are followed. The questionnaire uses semi-open and closed answer methods, and its purpose is to promote correct filling.

3.3 Mathematical Statistics

Use software to perform statistical processing on relevant data and analyze relevant data.

3.4 Analytic Hierarchy Process

In judging and judging the relative importance of the existing plan, and judging the implementation degree of the existing plan based on the final calculation result.

$$a_j = a/ \sum_{j=1}^{m} (a_{j_i}) \quad j = 1, 2, 3\ldots\ldots, m \tag{1}$$

In order to meet the data usage habits, the data is normalized

$$a_j = a_j/ \sum_{j=1}^{m} (a_j) \tag{2}$$

4 Experimental Research and Analysis of the Internet of Things in the Management of Smart Communities

(1) The management system is the center of a smart community management, and its function will directly affect the operation of the smart community. Its functions are divided into three categories according to the main application technologies: RFID application system, ZigBee application system, and video application system. The following table is developed The person is the satisfaction with the main technology (1–10) (Table 1).

Table 1. Management system satisfaction

	RFID system	ZigBee system	Video application system	Others system
Man	8	7	7	3
Woman	7	9	7	2

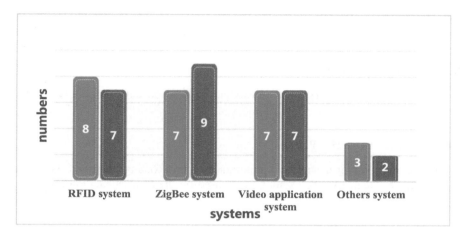

Fig. 1. Management system satisfaction

It can be seen from Fig. 1 that the three system management technologies are not much different, but ZigBee technology is relatively used a little more, which verifies the high efficiency, short distance and convenience of ZigBee technology.

(2) Through the analysis of the collected data, the prediction results of the market size of my country's smart community from 2014 to 2017 are finally calculated, as shown in Table 2:

Table 2. 2014–2017 smart community market scale forecast value and growth trend

	2014	2015	2016	2017
Predictive value	264	319	517	982
Growth rate (%)	53	63	65	66

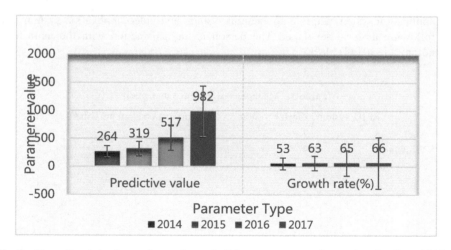

Fig. 2. The estimated value and growth trend of the smart community market size from 2014 to 2017

As shown in Fig. 2, through the prediction of the market scale of the application of the Internet of Things in the community field from 2014 to 2017, it can be seen from a quantitative level that the smart community will exceed 100 billion in the future market development, and its growth rate Has been maintained at more than 50%.

5 Conclusions

The construction of smart communities is the requirement of the development of the times, and it is also an important part of the current construction of smart cities in various regions. To build a smart community, we must proceed from reality, combine the development of new technologies such as the Internet of Things, and in accordance with the top-level design and planning of relevant national policies, through innovative business models, the joint influence of the economic and social benefits of the entire industrial chain has been brought into play. Upgrade and transformation of the entire national economy and social management, improve the service management level of the community, put people first, and realize the "Benefiting People Project" in the true sense. The smart community is the basic unit of a smart city. With the continuous development of technology, the form of smart communities will continue to change; life will be more convenient, social management will be more humane, intelligent, and refined. With the introduction of the Internet of Things, people's lives will enter a space that transcends space constraints.

References

1. Huang, Y., Ali, S., Bi, X., et al.: Research on smart campus based on the internet of things and virtual reality. Int. J. Smart Home **10**(12), 213–220 (2016)
2. Huang, L., Yuan, X., Zhang, J., et al.: Research on internet of things technology and its application in building smart communities. J. Phys.: Conf. Ser. **1550**(2), 022029 (4pp) (2020)
3. Mital, M., Pani, A.K., Damodaran, S., et al.: Cloud based management and control system for smart communities: a practical case study. Comput. Ind. **74**(C), 162–172 (2015)
4. Guo, B., Dong, B., Zhang, X., et al.: Research on home healthcare management system based on the improved internet of things architecture. Int. J. Smart Home **9**(9), 51–62 (2015)
5. Du, M.: Research on smart park information system design based on wireless internet of things. Int. J. Online Eng. **13**(5), 134 (2017)
6. Liu, M., Ma, J., Lin, L., et al.: Intelligent assembly system for mechanical products and key technology based on internet of things. J. Intell. Manuf. **28**(2), 271–299 (2017)
7. Li, R., Kido, A., Wang, S.: Evaluation index development for intelligent transportation system in smart community based on big data. Adv. Mech. Eng. **7**(2), 541651 (2015)
8. Guo, C., Zhang, C., Wang, G.: Research on Construction of smart community Based on GIS. In: IOP Conference Series: Earth and Environmental Science, vol. 580, no. 1, 012067 (4pp) (2020)
9. Li, Y., Wen, Z., Cao, Y., et al.: A combined forecasting approach with model self-adjustment for renewable generations and energy loads in smart community. Energy **129** (Jun.15), 216–227 (2017)
10. Abeysiriwardhana, W., Wijekoon, J.L., Nishi, H.: Smart community edge: stream processing edge computing node for smart community services. IEEJ Trans. Electron. Inf. Syst. **140**(9), 1030–1039 (2020)

Sentiment Analysis of Film Reviews Based on Text Mining—Taking the Shawshank Redemption as an Example

Li Tang[⊠]

School of Foreign Studies, Anhui Sanlian University, Hefei, Anhui, China

Abstract. Based on the user text evaluation data of Douban and IMDB, this paper realizes the judgment of movie tendency and the mining and analysis of hidden information by judging the emotional tendency in movie reviews, so as to provide guidance and Suggestions for other movie viewers. This paper mainly USES the emotional tendency to analyze and explore whether this can become a way to understand China's foreign cultural exchanges which will be the content of active exploration and research.

Keywords: Affective computing · Semantic computing · Cross-cultural research · Cultural communication

1 Introduction

Under the framework of the "Belt and Road" initiative, we must strengthen China's foreign cultural exchanges, which is an effective way to gain strong public support. The cultural exchange between Chinese and foreign cultures inevitably involves learning and understanding of each culture, which can not only promote the improvement of cross-cultural communication skills, but also enhance people's cognition and interest in Western culture. The popularization of computers and the birth of the Internet have narrowed the space and distance between people. In the process of global economic integration, countries with different cultures have more frequent exchanges, and people are faced with more complex intercultural communication context. Under the cultural differences of different languages, people are bound to have communicative conflicts [1]. Although the space for communication and exchange between various cultures is reduced, the distance between cultures and the psychological distance is not shortened but enlarged.

In the classification of text sentimentality, a text with emotional color is used as the research object, and the sentiment orientation of the text is determined by processing, classification and analysis [2]. The research on emotional orientation based on digital media has more advantages in the way of obtaining materials: first, the selected data can be updated in time; second, the speed of selecting data is faster; third, the scope is wider. The user's real emotion and behavior on the network platform represent their cultural background and cultural differences in real life. Through the study of the emotional orientation of film reviews, the cultural background and civilization level expressed by the emotional orientation can be analyzed.

J. C. Hung et al. (Eds.): FC 2021, LNEE 827, pp. 392–398, 2022.
https://doi.org/10.1007/978-981-16-8052-6_48

2 Related Theories and Research Foundation

Since the beginning of 2000, affective analysis has grown to be one of the most active research fields in natural language processing. It has been researched extensively in data mining, Web mining, text mining and information retrieval. In fact, it has spread from computer science to management science and social science [3], such as marketing, finance, political science, communication, medical science, and even history, which have attracted widespread attention from the whole society due to their important commercial nature.

At present, the research methods of text emotion analysis are mainly based on supervised learning and unsupervised learning. Research in this field was carried out earlier abroad. Beineke et al. (2004) combined traditional machine learning with manual annotation to improve the accuracy of emotion analysis in English texts. Fei et al. (2004) used machine learning method to conduct emotion analysis on the study of English sports commentary on Yahoo website. So far, the research on the analysis of the emotional tendency of English text comments has achieved rapid development [4–7].

The research on emotion analysis in China started late due to the complex structure of Chinese words and their many meanings. Based on semantic similarity and semantic correlation field, Zhu Yanylan et al. (2006) proposed two kinds of lexical semantic tendency computing methods based on HowNet to calculate the degree of praise or criticism by using the similarity between words [8]. Zhang Ziqiong, Ye Qiang, Li Yijun et al. (2010) explored and implemented the theory and method of emotion analysis based on Chinese environment, preliminarily established the method of emotion mining and analysis based on semantics in Chinese environment, and carried out relevant experimental verification, demonstrating the correctness of the method [9]. From the perspective of textual granularity, Yang Ligong Zhu Jian, Shang Shiping (2013) reviewed text sentiment analysis literature from five aspects, namely, extraction of sentiment words, construction of corpus and sentiment dictionary, analysis of evaluation object and opinion holder, discourse level sentiment analysis and practical application, made necessary comments as well [10, 11].

What is the fundamental source of the cultural difference between China and the West? Wang Shanshan, Jiang Xingjun ect. agreed that due to history, geography, different regional cultural backgrounds, life custom, and various social factors, there exist many differences between Chinese and western culture in the aspects of language, knowledge, beliefs, outlook on life, mode of observation and thinking, ethics, and customs. Furthermore, this difference is also the cause of the Chinese and western cultural communication obstacles [12, 13].

In view of the research status at home and abroad, it has become an urgent problem for us to analyze emotions through comments and understand the cultural differences behind emotions. At present, computers have provided more valuable reference for the market and businesses, but there are few studies on Chinese and Western cultural studies and cross-cultural communication under different backgrounds. At present, in China, very few studies analyze the emotional tendency of film criticism based on emotion calculation, and study the background information hidden behind Chinese and

Western culture. Through our study, we will have a clearer understanding of the cultural consciousness and cultural level of the western people, so as to better carry out cultural and civilized exchanges and integration [14, 15].

3   The Research Process

Based on the existing research results, the following aspects also need to be considered: first, due to the complex network environment, the messy acquired text data, many invalid data and a large amount of filtering, the workload should be reduced and efficiency should be improved as much as possible; Second, Chinese and English are in different contexts, so they cannot be measured and compared with the same standard. They need to be transformed instead. Third, a lot of technical words and emoticons used in film reviews also need to be converted.

Based on the above questions, we converted the valid English data text into Chinese version, and then calculated the emotional tendency of all the key words in the sentences to judge their emotional color (Fig. 1).

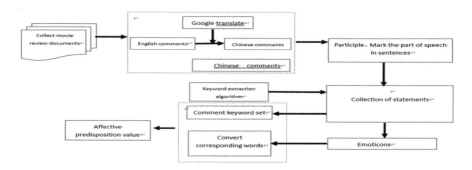

Fig. 1. Sentiment orientation calculation process

3.1   Data Collection

The first choice is to select appropriate text data: The text data studied in this article mainly comes from Chinese movie playback platform- Douban and English movie playback platform - IMDB. By collecting relevant film reviews through web crawler technology, valid data applicable to this topic are filtered out. Python web crawler technology is a program script that automatically crawls network information according to certain rules. This article uses Python web crawler technology to obtain movie review data in both Chinese and English. Through various settings for data collection, a total of 5 indicators are captured: one is the (primary key) movie name, the other is the number of movie reviews, and the third is the comment users' username, the fourth is the star rating of each comment (star rating from 1 to 5, corresponding to very poor, poor, okay, recommended, highly recommended), and the fifth is the users' comment content. Due to the large number of comments, we put the data in a single Excel table.

Among them, the fields contained in the data table in the data capture process are shown in Table 1:

Table 1. Web crawls data field table

Field names	Field type	Note
filmname	String	primary key, Crawl the movie name of the webpage
num	Int	Comment number of likes
usrname	String	username
identity	String	Comment on the authentication status of the user, can be empty
comment	String	User comments

By collecting the Chinese and English film reviews of the same movie The Redemption of Shawshank, and analyzing and processing the data of 35 XML files in Chinese and English, a preliminary data format suitable for this study was obtained. The result file is automatically processed by Perl to delete invalid data, including unrated comments, advertising comments, comments with links, and some empty text. The generated form is shown in Table 2 (Table 3):

Table 2. Parsed result file style sheet (Chinese version)

usrName	Star rating	Evaluation
rhinoceros	Five stars (highly recommended)	......Whenever reality makes me feel so exhausted, I will regain my strength when I turn out this disc...... Top 3 must-see movies for men
Jidi	Five stars (highly recommended)	I like it so much. If you don't watch it, your life will not be complete
Nine-tailed black cat	Two stars (poor)	I really don't like it, it doesn't look good, it doesn't feel

Table 3. Parsed result file style sheet (English version)

Usrname	Star rating	Evalution
blairgallop7	highly recommended	One of my favorite movies ever, The Shawshank Redemption is a modern day classic as it tells the story of two inmates who become friends and find solace over the years in which this movie takes place....This film is easily considered one of the greatest films of all time and I definitely agree with this fact
aziz-737-972827	highly recommended	I saw this film in 1995 and still vividly felt the euphoria of the final half hour which stayed with me for a couple of weeks

After preliminary sorting, 35648 movie reviews were obtained. Generally speaking, comments containing links have only promotional purposes and do not involve emotion, so they can also be ignored. After further screening, there are 20471 comments suitable for the research content of this topic, including 13,520 comments in the Chinese version and 6,951 comments in the English version. Finally, the content of the comments is stored in a text file, and 3,000 comments are extracted as sample data for sentiment oriented classification. 3,000 comment samples are categorized by hand. The categories include comments with positive, negative, and neutral sentiment tendencies as well as noise comments.

3.2   Data Analysis

Due to the different data sources used, the comparison of emotional tendencies brings many difficulties. In order to be compared, it is necessary to switch to the same context. We use Google Translate, a machine translation system with relatively high accuracy in the process of sentence translation, to translate English comments into Chinese comments so that the comparison of the emotional orientation of the two in a unified context is logical.

First, select the emotional dictionary. By selecting a subset of the sentiment analysis word set in HowNet, it includes a total of 6 subsets: positive and negative emotional words, positive and negative evaluation words, level words and proposition words. Secondly, a new emotional dictionary is constructed. Emotional words cannot satisfy our subject research. We need to accommodate multiple aspects of content such as film industry professional vocabulary, movie title dictionary, movie star dictionary, popular internet words and so on based on HowNet. In addition, the film review website also provides the function of expressions, which need to be converted and stored in the database, such as [Applause], [laugh], etc. Third, preprocess the sample set we collected (feature value extraction, stop words removal, part-of-speech tagging). Finally, the keywords of all sentences are extracted through the keyword extraction algorithm to calculate the sentiment value to determine the sentiment tendency.

3.3   Data Result Analysis

According to the experimental analysis of the sample set data extracted in the third stage, we can get two groups of different data results: Table 4 Proportion of emotional analysis tendency.

When the value we calculate is infinitely close to 1, it indicates that the emotion of the word tends to be positive, good or positive. When the value we calculate is infinitely close to zero, it indicates that the emotion of the word tends to be negative, depressed or negative, which is a negative word. Based on the above data, we can perceive the differences of civilization level and cultural origin. We have different views and opinions on the same thing, but basically maintain the same emotional expression. In the Shawshank Redemption, which has a critical rating of more than 90%, everyone has the same vision for a good movie, but different countries and nations form different cultural backgrounds and its derivative products, such as the way of thinking, pragmatic rules and ways, are inconsistent.

Table 4. Proportion of emotional analysis tendency

Classification rules		Positive	Negative	Neutral	Total number of comments
Sentiment value ≈ 1 Express positive; ≈ 0 Express negative; The rest means neutral (Source text is Chinese)	Number of comments	2343	408	249	3000
	Percentage of ratings	78.10%	13.60%	8.30%	100%
Sentiment value ≈ 1 Express positive; ≈ 0 Express negative; The rest means neutral (Source text is English)	Number of comments	2067	523	410	3000
	Percentage of ratings	68.90%	17.43%	13.67%	100%

According to the film, we can learn that such differences are mainly reflected in the following aspects: First, Chinese and Western words are different in terms of describing the plot and characters. In film reviews, Chinese culture is more euphemistic and implicit. The expressions and words of Western culture are direct and calm, and the evaluation is warm, which can be clearly felt as the charm of the movie plot. Secondly, Chinese audience are more likely to show a strong sense of involvement in the performance of the protagonist of the film. In turn, they judge the quality of the actor based on the character image performed by the actor, which is particularly likely to cause discomfort to other audiences. The audience under the influence of Western culture is more intuitive and calm, and can give objective evaluation through the performance of the actors. Of course, there are also "crazy" netizens. At the same time, the culture behind the language teaches us the most critical values, which are a core part of society. In China, the golden mean has always been pursued in interpersonal communication, while in western countries, individualism is more advocated and independence, freedom and equality more pursued. These are two basically different values, and these factors will bring potential barriers to cross-cultural communication, and may even lead to cultural conflicts.

4   Conclusion

By the end of March 2019, the Chinese government had signed 173 cooperation documents with 125 countries and 29 international organizations. China and other countries and regions including central and Eastern Europe and ASEAN have jointly held activities for the Year of Culture, forming more than 10 cultural exchange brands such as the Silk Road Tour and the China-Africa Cultural Focus. At the same time, 17 Chinese cultural centers have been set up in countries along the Belt and Road, 153 Confucius Institutes and 149 Confucius classrooms have been set up in 54 countries along the Belt and Road. These data tell us that with the economic globalization, China's communication with the world has become increasingly close, and cultural exchanges have been continuously deepened. Therefore, the demand for the integration of Chinese and Western cultures has become higher.

This paper obtains the data by collecting and analyzing the film reviews of The Shawshank Redemption. Through a series of data, we get a different language and culture background of emotional expression on the same thing, and this kind of emotional expression will be the difference between the cross-cultural communication ability and the key point of cognitive, which is also the significance of our research theory and practice.

Acknowledgment. This paper is a key project of the Humanities and Social Sciences of the Department of Education of Anhui Province: Research on English Academic Writing Citation Ability (Project Number: SK2019A0761) and the Key Research Project of Social Sciences of Anhui Sanlian College: Emotional Tendency of English Film Criticism Based on Digital Media (Project Number: SKZD2019009) It is one of the phased results.

References

1. Zhang, C.: The enlightenment of Chinese and western cultural differences on college English vocabulary teaching in the cross-cultural context. J. Zunyi Norm. Univ. **05**, 126–130 (2016). (in Chinese)
2. Tang, L.: Research on the emotional tendency classification of online film criticism. J. Zunyi Norm. Univ. **06**, 160–164 (2018). (in Chinese)
3. Jin, X.: Sentiment analysis of Chinese short text based on deep learning. Shenyang University of Technology (2018)
4. Beineke, P., Hastie, T., Vaithyanathan, S.: The sentimental factor: improving review classification via human-provided information. In: Proceedings of ACL (2004)
5. Turney, P.D., Littman, M.L.: Measuring praise and criticism: inference of semantic orientation from association. Trans. Inf. Syst. **21**, 315–346 (2003)
6. Fei, Z.C., Liu, J., Wu, G.F.: Sentiment classification using phrase patterns. In: Proceedings of the Fourth International Conference on Computer and Information Technology (CIT 2004). IEEE, Wuhan (2004)
7. Zheng, S., Wang, F., Bao, H., et al.: Joint exitraction of entities and ralationgs based on a novel tagging schenem (2017)
8. Zhu, Y., Min, J., Zhou, Y., Huang, X., Wu, L.: Calculation of lexical semantic orientation based on HowNet. Chin. Inf. J. (2006). (in Chinese)
9. Zhang, Z., Ye, Q., Li, Y.: Review of research on sentiment analysis of internet product reviews. J. Manag. Sci. (2010). (in Chinese)
10. Yang, L., Zhu, J., Tang, S.: A review of text sentiment analysis. Comput. Appl. (2013). (in Chinese)
11. Hua, X.: A linguistic study on the emotional polarity of internet comment texts. Hebei University (2012). (in Chinese)
12. Wang, S.: A brief talk on the influence of Chinese and western cultural differences on translation. Overseas Engl. **04** (2010). (in Chinese)
13. Zhao, L.: The influence of Chinese and Western cultural differences on English translation and countermeasures. Sci. Technol. Wind **26** (2019). (in Chinese)
14. Jiang, T., Wan, C., Liu, D.: Evaluation object-sentiment word pair extraction based on semantic analysis. Chin. J. Comput. **4** (2016). (in Chinese)
15. Zhu, L., Xu, J.: Research on the key techniques and application of sentiment analysis of internet comments. Inf. Theory Pract. **1** (2017). (in Chinese)

CPU-GPU Heterogeneous Parallel Algorithm for Metal Solidification Molecular Dynamics Simulation Based on Artificial Intelligence

Huhemandula[1(✉)], Jie Bai[2], and Wenhui Ji[1]

[1] School of Physics, Ji Ning University, Jining 012000, Inner Mongolia, China
111990046@imu.edu.cn
[2] School of Foreign Languages, Ji Ning University, Jining 012000,
Inner Mongolia, China

Abstract. AI (Artificial Intelligence) technology is one of the most popular technologies in today's society, and its development and application are getting more and more attention. Especially in recent years, AI technology has been greatly developed, and its shadow can be seen everywhere in our lives. Now its application range is very wide. In order to explore the influence of AI technology on CPU-GPU heterogeneous parallel algorithm of metal solidification molecular dynamics simulation, we set up an experimental group and a control group as the experimental research object, and then the experimental group applied big data technology in its CPU-GPU heterogeneous parallel algorithm of metal solidification molecular dynamics simulation, while the control group still used the conventional method. Then the experimental results show that the calculation efficiency and accuracy of the experimental group are much higher than those of the control group, and the calculation efficiency and accuracy of the experimental group can reach 97.80% and 99.51% respectively, while the corresponding data results of the control group are 85.10% and 91.80% respectively.

Keywords: Artificial Intelligence · Metal solidification · Molecular dynamics simulation · Parallel algorithm

1 Introduction

At present, AI technology is one of the most cutting-edge technologies in the current era. Its rapid development has also brought great changes to our lives and has a great impact on our way of life and work [1, 2]. At present, the researchers try to apply AI technology to the CPU-GPU heterogeneous parallel algorithm of metal solidification molecular dynamics simulation, so as to improve the computing power of the algorithm. So how much effect can AI technology bring to the algorithm, and what kind of influence can it have? These are the purpose of this study [3, 4].

Molecular dynamics simulation of metal solidification is a method to study metal solidification. This method is derived from computer technology. It can not only eliminate some unavoidable interference factors in some physical experiments, but also solve some headache problems in physical experiments, such as some experimental

J. C. Hung et al. (Eds.): FC 2021, LNEE 827, pp. 399–405, 2022.
https://doi.org/10.1007/978-981-16-8052-6_49

conditions are difficult to achieve [5, 6]. Nowadays, with the rapid development of computer and the emergence of AI technology, especially the wide application of GPU in computer, these have brought new development opportunities and new research methods for metal solidification molecular dynamics simulation, including CPU-GPU heterogeneous parallel algorithm for metal solidification molecular dynamics simulation [7, 8].

The rise of AI technology has brought new changes to various industries and fields, especially for the research of metal solidification. This technology has brought important development opportunities and favorable development conditions for the research and application of this field, and relevant researchers should pay enough attention to it [9, 10].

2   Method

2.1   AI

AI is a new technology science. This technology science can be used to simulate and extend the theory and method of human intelligence, and the technology can also expand the research of these contents, so as to establish a more intelligent application system. AI technology is used to understand the essence of intelligence, in fact, it is still a computer science. People use AI technology to develop and research some fields, including robot, language and image recognition, as well as expert system and natural language recognition processing. People use it to try to produce a machine similar to human intelligence, so as to help people complete some complicated and boring work. With the development of AI technology, its theory and technology have been relatively mature, and with the passage of time, its application field is also more and more extensive, and our life and work rely on it more and more. This technology can simulate our human consciousness and thinking, so it can replace human beings to do some work, and help people reduce the pressure of work. AI technology also involves many subjects' knowledge content like computer technology. These disciplines include computer science first, then philosophy, linguistics and psychology. All disciplines of social and natural sciences are almost involved. AI technology also involves mathematical theory. Mathematics can be said to be the theoretical basis of many disciplines It provides thinking and computing tools for various disciplines, and also provides theoretical basis for logical thinking. Because of the strong theory, AI technology can develop so fast. It is also known as one of the greatest technologies in the new era. From its characteristics, AI technology belongs to an application branch of thinking science. However, from the perspective of thinking, AI technology is not only limited to logical thinking, but also includes image thinking and inspiration thinking. Only by integrating these three thinking perfectly, can AI technology make breakthrough progress in the future.

2.2   Development Trend of AI

The appearance of AI technology not only brings great convenience to people's life, but also enriches people's life, making people's life better and colorful. It is a kind of technical means used by people to produce intelligent machines. This technology method has human thinking ability to some extent, and can also be used to simulate human intelligence and to make this kind of intelligent Apply to real life. People always try to build an intelligent system with human intelligence, so that computers can be applied to higher level applications. With the change of the times and the rapid development of society, computer technology and AI technology have been widely used. The role played in our life is becoming more and more important. Even some industries will not work normally once they leave their technical support. Because the computer technology and AI technology have penetrated all aspects of our lives. They are no longer necessary. As a kind of cutting-edge science and technology, AI technology has been paid more and more attention in the world. In addition, with the changes of the times and the rapid development of economy, some existing computer technologies have gradually failed to meet the needs of the present people. Therefore, we must study more advanced technology to meet the growing needs of people. The emergence of AI technology is the problem The solution of the system provides the solution ideas and technologies.

2.3   Solidification of Metals

Solidification of metal refers to the process of metal changing from liquid to solid state. Generally, except for some liquid alloy metals, when they are solidified, they will become amorphous metal without fixed structure. Most of the metals will become similar to crystal or grain during solidification, and this process is a gradual process, which is like a process of small crystals or grains growing up gradually. Metal solidification process also accompanied by a series of physical and chemical changes, including volume change, gas desolvation and element segregation. Most of the metal materials are made of liquid components and impurities and degassing. Then people are casting and processing these metal materials, and then they can be made into the required metal materials or directly made into some parts. The solidification of metals involves a wide range of materials. This solidification process not only determines the structure and properties of metals, but also plays a decisive role in the structure of metals and some alloys. The solidification of metal plays an important role in the ingot and the macro structure of the casting, and it can also help people to repair the defects of some castings and to the macro segregation of the castings. The thermodynamics, phase diagram and diffusion of metal are the theoretical basis for the metal solidification analysis by researchers. Now heat and mass transfer and phase transformation are also included in the theoretical basis. The homocrystallization and the differentiation crystallization are two kinds of metal crystallization processes. The crystal crystallized in the solidification process of metal is identical with the chemical composition of the metal mother liquor, and vice versa, if the chemical composition of the crystal formed in the process is not consistent with the metal mother liquor, it is called the differentiation crystallization. In reality, most of the gold composite crystals belong to the heteromorphic crystallization.

2.4   CPU-GPU

CPU and GPU are two different things, CPU refers to the central processor of computer, and GPU refers to the graphics processor of computer. The main function of CPU is to control and calculate computer system, while GPU is a kind of centralized circuit, and this integrated circuit has a very large scale.

2.5   Relevant Algorithm Formula

When we check the data in the experiment, we will generally apply some mathematical formulas, and then we can use these mathematical calculation formulas to make statistics and analysis. At the same time, we can find the characteristics of experimental data through some calculation methods. The following are the algorithms involved in the experiment:

$$M = \frac{x_1 + x_2 + x_3 + \ldots + x_n}{n} \tag{1}$$

$$s^2 = \frac{(M - x_1)^2 + (M - x_2)^2 + (M - x_3)^2 + \ldots + (M - x_n)^2}{n} \tag{2}$$

$$P(X = k) = \frac{\lambda^k}{k!} e^{-\lambda} \tag{3}$$

3   Experiment

3.1   Selection of Experimental Research Objectives

In order to explore the influence of AI technology on CPU-GPU heterogeneous parallel algorithm of metal solidification molecular dynamics simulation, we set up two groups of experiments, one group was set as the experimental group, the other group was set as the control group, and then the experimental group applied AI technology to CPU-GPU heterogeneous parallel algorithm of metal solidification molecular dynamics simulation, while the other group was set as the control group. The control group still used the original algorithm.

3.2   Specific Implementation Steps of the Experiment

At the same time, in order to ensure the authenticity and objectivity of the experimental data results, we carried out five groups of experimental tests on the two groups, then recorded and counted the five groups of data results of the two groups, and drew charts according to the data, and then analyzed and summarized the results. What kind of influence and effect does intelligent technology bring to CPU-GPU heterogeneous parallel algorithm of metal solidification molecular dynamics simulation of AI.

4 Discussion

4.1 Investigation of Computational Efficiency of Two Experimental Groups

In order to ensure that the experimental data results are more accurate and convincing, each experimental group has carried out five experimental tests, and then we check and count the calculation efficiency. The final calculation efficiency is as follows (Table 1):

Table 1. The investigation of the calculation efficiency of two experimental groups

	Computational efficiency	
	Experience group	Control group
Test one	86.5%	82.9%
Test two	89.3%	85.1%
Test three	92.1%	78.9%
Test four	95.6%	84.7%
Test five	97.8%	81.6%

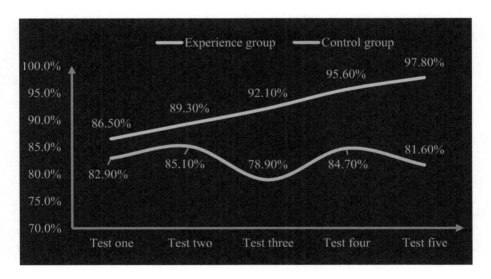

Fig. 1. The investigation of the calculation efficiency of two experimental groups

From the above chart, we can see that the computational efficiency of the experimental group is 86.5%, 89.3%, 92.1%, 95.6% and 97.8% respectively, while the computational efficiency of the control group is 82.9%. The efficiency of the second group was 85.1%, the third group was 78.9%, the fourth group was 84.7%, and the fifth group was 81.6%. From the results of these experimental data, we can know that the computational efficiency of the experimental group is much higher than that of the control group. The computational efficiency of the experimental group is high and has been rising steadily, while the computational efficiency of the control group is low and unstable. The state is good and bad, up and down, ups and downs, ups and downs. We can also see this clearly in Fig. 1. The above experimental results show that the computational efficiency of the control group is relatively low, and the computational ability of the control group needs to be strengthened.

4.2   Investigation of Calculation Accuracy of Two Experimental Groups

In order to ensure the authenticity of the experimental data and reduce the influence of subjective factors as much as possible, we also conducted five groups of accuracy test surveys on the two groups. The accuracy results are shown in the following chart:

Table 2. Investigation on the accuracy of calculation of two experimental groups

	Survey of calculation accuracy	
	Experience group	Control group
Test one	93.38%	88.92%
Test two	95.71%	88.53%
Test three	96.98%	89.24%
Test four	98.76%	91.80%
Test five	99.51%	87.62%

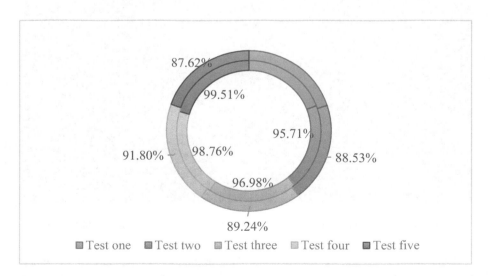

Fig. 2. Investigation on the accuracy of calculation of two experimental groups

First of all, by roughly comparing the data results in Table 2 and Fig. 2, we can find that there is a certain gap in the calculation accuracy of the two experimental groups. We can see that the calculation accuracy of the first group is 93.38%, the second group is 95.71%, the third group is 96.98%, the fourth group is 98.76%, the fifth group is 99.51%, while the control group is 99.51%. The accuracy results were as follows: 88.92% in the first group, 88.53% in the second group, 89.24% in the third group, 91.80% in the fourth group and 87.62% in the fifth group. We can see from these data results that the calculation accuracy of the experimental group is also higher than that of the control group. The highest calculation accuracy of the experimental group is 99.51%, close to 100%, while the highest calculation accuracy of the control group is 91.80%, which is lower than that of the experimental group.

5 Conclusions

To sum up, AI technology can have a positive effect and influence on CPU-GPU heterogeneous parallel algorithm of metal solidification molecular dynamics simulation. This technology can improve the computational efficiency and accuracy of the algorithm. Therefore, it is very correct to apply AI technology to the research of the algorithm. We also look forward to the better integration of AI technology and the algorithm in the future together.

References

1. Song, P., Zhang, Z., Liang, L., et al.: Study on parallel efficiency optimization of MOC neutron transport calculation based on CPU-GPU heterogeneous parallel. At. Energy Sci. Technol. **053**(011), 2209–2217 (2019)
2. Xiao, H., Li, C., Li, Q., et al.: Research on matrix transpose algorithm of CPU + GPU heterogeneous parallel. J. Northeast Norm. Univ. (Nat. Sci. Edn.) **51**(04), 75–82 (2019)
3. Liu, Y., Xiong, R., Xiao, Y.: Research on CPU/GPU heterogeneous hybrid parallel algorithm for MT Occam inversion. Petrol. Geophys. Prospect. **57**(3), 470–477 (2018)
4. Zhu, W., Yu, W.: Application of CPU-GPU heterogeneous parallel computing in numerical core linear elastostatics finite element simulation. Progr. Geophys. **31**(004), 1783–1788 (2016)
5. Wang, H., Wei, X.: Research and application of remote sensing image preprocessing algorithm based on GPU. Mod. Electron. Technol. (003), 47–50 (2016)
6. Qu, Y.L., Lan, C., Ren, Z.G.: Review of CPU/GPU heterogeneous parallel systems. Autom. Instr. (004), 25–26 (2016)
7. Tang, Y., Zhou, H., Fang, M., et al.: Research and implementation of hyperspectral remote sensing image data processing based on CPU/GPU heterogeneous mode. Comput. Sci. **43** (02), 47–50 (2016)
8. Qiu, H.: Research on heterogeneous computing programming based on CPU + GPU. Sci. Technol. Innov. (001), 74–75 (2020)
9. Luo, Y., Chen, Z., Tang, R., et al.: Parallel acceleration of PIC particle simulation based on intelmic coprocessor and comparison with CPU/GPU. Chin. Sci. Pap. **13**(08), 118–123 (2018)
10. Zhai, J., Gao, X.: Research on mixed programming mode of CPU + GPU heterogeneous system. Inf. Record. Mater. **17**(004), 31–32 (2016)

Cost-Benefit Analysis of VPP Considering Environmental Value Under EV Access

Yingying Hu[1(✉)], Di Wu[2], Jianbin Wu[1], Wei Song[1], and Peng Wang[1]

[1] Economic and Electrical Research Institute of Shanxi Electrical Power
Company of SGCC, Taiyuan, Shanxi, China
[2] State Grid Taiyuan Power Supply Company of Shanxi Electric Power
Company of SGCC, Taiyuan, Shanxi, China

Abstract. VPPs can effectively promote new energy consumption and reduce environmental pollution by gathering a large number of distributed energy sources and other load-side resources such as demand response. EVs have the dual attributes of power and load, which will affect the cost-effectiveness of the VPP when it is connected to the VPP. This paper first analyzes the VPP environment and system stability value under the EV access; then constructs a VPP multi-objective optimization scheduling model considering the environmental value; finally, the impact of five EV access schemes on the cost-effectiveness of VPP is analyzed. The results show that there is an optimal access scale for EV access to VPP.

Keywords: Electric vehicle · Virtual power plant · Cost-effectiveness · Environmental value · Multi-objective

1 Introduction

In recent years, under the goal of carbon neutrality, reducing the emission of pollutants by promoting the use of clean energy has become the main way to achieve the goal of carbon neutrality. On the one hand, Virtual Power Plants (VPPs) have high environmental value due to the accumulation of distributed and other clean energy, but due to the deviation of distributed power output, the system cannot achieve real-time balance; on the other hand, EVs (EVs) with power and load characteristics are similar to storage. The energy system can effectively adjust the real-time deviation of the VPP. Therefore, the clustering of EVs in the VPP has become a research hotspot [1].

The participation of EV VPPs in the grid operation can effectively promote the safe and stable operation of the grid, including: promoting the consumption of renewable energy [2], reducing the grid loss [3] and increasing the stability of the grid frequency. Literature [4] constructed a VPP containing wind turbines and EVs. Data analysis shows that the VPP has high economic benefits. Literature [5, 6] established a multi-party electricity price linkage game model consisting of EVs and wind power, traditional generators, users, and grid companies participating in a multi-party electricity price linkage game model, and used the collaborative genetic evolution algorithm to solve the problem. Literature [7] proposed a single EV V2G model, on which the V2G response capability of an EV VPP was evaluated. The literature [8, 9] aimed at the best AGC frequency modulation effect and the highest net income expectation, and

© The Author(s), under exclusive license to Springer Nature Singapore Pte Ltd. 2022
J. C. Hung et al. (Eds.): FC 2021, LNEE 827, pp. 406–415, 2022.
https://doi.org/10.1007/978-981-16-8052-6_50

established a VPP including thermal power units, hybrid energy storage systems and EVs to participate in AGC frequency modulation decision-making models, and the analysis of examples showed that the model can Significantly improve the AGC frequency modulation effect. There are many existing studies on the evaluation of the response of EVs to the VPP and the benefit evaluation under a certain scale of connection, but the cost and benefit of different scales of EV access to the VPP under the environmental value have not been studied.

Based on the existing research, this paper analyzes the environmental and system stability value of EVs connected to the VPP, constructs a VPP multi-objective optimization stochastic dispatch model considering the environmental value, and analyzes a case study.

2 Environment and System Stability Value Analysis of EV Connected to VPPs

2.1 Analysis of Pollutant Emissions

The distributed wind power system, distributed photovoltaic system and pumped storage system included in the VPP belong to renewable energy power generation, and its pollutant emissions are basically zero. The gas turbine power generation system will produce a certain amount of sulfur dioxide, carbon dioxide, nitrogen oxides and other pollutants during operation. Since the electricity in the large power grid is mainly derived from coal-fired thermal power generation, a certain amount of pollutant emissions will also be generated during the purchase of the large power grid by the VPP. After the EV is connected to the VPP, due to its charging and discharging characteristics and the connection of different EV scales, it will affect the output of the gas turbine unit and the purchased electricity. Therefore, the pollutant emissions generated during the operation of the VPP are also inconsistent. The different pollutant emission coefficients during the operation of the VPP are shown in Table 1:

Table 1. Pollutant emission coefficient of VPP operation (g/kWh)

Pollutant emission source	SO_2	NO_x	CO_2	CO	Soot
Distributed gas generator set	0.0093	0.6185	184.0827	0.1708	–
Coal-fired generator set	6.4830	2.8840	623	0.1183	2.583

Based on the pollutant treatment cost in each region, the environmental treatment cost of the VPP can be obtained as shown in formula (1):

$$C_s^{env} = \sum_{m=1}^{N^{env}} \left[\rho_m^{env} \times \sum_{t=1}^{T} (\sum_{i=1}^{N^{FC}} D_{s,m}^{FC} G_{i,t}^{FC} + D_{s,m}^{in} L_t^{in}) \right] \tag{1}$$

Where, C_s^{env} is the environmental treatment cost when the EV is connected to the scale of s; N^{env} is the type of pollutants, the pollution referred to in this report is the five pollutants SO_2, NO_x, CO_2, CO and soot; ρ_m^{env} is the environmental treatment cost of pollutants; $D_{s,m}^{FC}$ is the pollutant emission value of the gas-fired distributed unit when the EV is connected to the scale of s; $D_{s,m}^{in}$ is the pollutant emission value of the external large grid electricity when the EV is connected to the scale of s. b and c respectively represent the two ends of the moderate interval $[b, c]$.

2.2 Analysis of Environmental Loss Value

Distributed wind power and distributed photovoltaic power generation in VPPs are clean energy sources and have significant positive externalities in terms of environmental value. The full-load power generation of these units is conducive to giving full play to the environmental benefits of the VPP, so when the phenomenon of abandoning wind and light occurs, it can be regarded as a loss of environmental value of the power plant. When EVs are charged and discharged disorderly, discharge at valley time and charge at peak time, the consumption value of renewable energy such as wind power and photovoltaic will decrease; on the other hand, EVs of different scales connected to VPPs will affect wind power and photovoltaic power. The consumption space of renewable energy causes different degrees of positive and negative impacts, and therefore the environmental loss values are also inconsistent. Based on the concept of opportunity cost, this paper defines the environmental value loss C_s^{loss} caused by abandoning wind and abandoning light as the product of its reduced clean energy generation and the penalty coefficient, expressed as Eq. (2):

$$C_s^{loss} = \rho^{out} \sum_{t=1}^{T} \left[\sum_{i=1}^{N^m} (G_{i,t}^{w^*} - G_{s,i,t}^{w}) + \sum_{i=1}^{N^m} (G_{i,t}^{pv^*} - G_{s,i,t}^{pv}) \right] \quad (2)$$

Where, ρ^{out} represents the penalty coefficient of abandoning wind and light, taking the average power generation cost value of the coal-fired power plant; $G_{i,t}^{w^*}$ is the maximum output value of wind turbine i at time t; $G_{i,t}^{pv^*}$ is the maximum output of photovoltaic generating set i at time t Value; $G_{s,i,t}^{w}$ is the output value of wind generator set i at time t when the EV is connected to the scale s; $G_{s,i,t}^{pv}$ is the output value of photovoltaic generator set i at time t when the EV is connected to the scale s.

2.3 System Stability Analysis

The charging and discharging of renewable energy generator sets such as wind power and photovoltaic in the VPP and EVs are all uncertain and random. The addition of these uncertain resources will affect the peak load and valley load of the VPP. The increase in the load difference between peak and valley periods will increase the reserve capacity of units during peak periods, which will greatly increase the investment cost, and the low load during valley periods will cause the units to be idle and reduce the utilization efficiency of the units. Therefore, the product of the peak-valley difference and the penalty coefficient represents the loss of the VPP due to the enlarged peak-valley difference, as shown in formula (3):

$$C_s^{pvd} = \rho^{pvd} \sum_{t=1}^{T} (L_{t,s}^{\max} - L_{t,s}^{\min}) \tag{3}$$

Where, C_s^{pvd} is the peak-valley deviation penalty cost when the scale of the EV connected to the VPP is s; ρ^{pvd} is the peak-valley deviation penalty cost coefficient; $L_{t,s}^{\max}$ and $L_{t,s}^{\min}$ are the scale of the EV connected to the VPP for the time period t at s Peak and valley load.

3   Multi-objective Optimization Stochastic Dispatch Model of VPP Considering Environmental Value

3.1   Objective Function

Based on the analysis of the environmental value of the VPP, this section will establish a multi-objective optimization model of the VPP that considers the environmental value. After considering the environmental value, the objective function of the VPP optimal dispatch is as follows:

(1) Maximize the operating income of the VPP
     The formula for maximizing the operating revenue of a VPP is shown in formula (4):

$$
\begin{aligned}
\max TR^{all} &= I_{sell}^{PB} - \sum_{i=1}^{N_{IB}} C_{i,s}^{IB} - \sum_{i=1}^{N_W} C_{i,s}^{W} - \sum_{i=1}^{N_{PV}} C_{i,s}^{PV} - \sum_{i=1}^{N_{FC}} C_{i,s}^{FC} - \\
&= \sum_{i=1}^{N_{BT}} C_{i,s}^{BT} - \sum_{i=1}^{N_{PEV}} C_{i,s}^{PEV} - C^{CW} - C^{in}
\end{aligned} \tag{4}
$$

Where, I_{sell}^{PB} is the income of the VPP from selling electricity to end users in a specific area; $\sum_{i=1}^{N_{IB}} C_{i,s}^{IB}$ is the economic compensation to the user after the system call when the EV access scale is s can interrupt the load; $\sum_{i=1}^{N_W} C_{i,s}^{W}$ is the EV access scale is s When the power generation cost of distributed wind power; $\sum_{i=1}^{N_{PV}} C_{i,s}^{PV}$ is the power generation cost of distributed photovoltaic power when the scale of EVs is connected to s; $\sum_{i=1}^{N_{FC}} C_{i,s}^{FC}$ is the cost of gas distributed power generation when the scale of EVs is connected to s; $\sum_{i=1}^{N_{FC}} C_{i,s}^{FC}$ is the scale of EVs connected to s Hour battery charging cost; $\sum_{i=1}^{N_{FC}} C_{i,s}^{FC}$ is the pumping cost of the pumped-storage system when the EV is connected to the scale of s; C_s^{in} is the external power purchase cost when the EV is connected to the scale of s.

(2) The environmental management cost of the VPP is minimal

The minimum environmental governance cost of the VPP is shown in formula (5):

$$\min C_s^{env} = \min \sum_{m=1}^{N^{env}} \left[\rho_m^{env} \times \sum_{t=1}^{T} (\sum_{i=1}^{N^{FC}} D_{s,m}^{FC} G_{i,t}^{FC} + D_{s,m}^{in} L_t^{in}) \right] \tag{5}$$

(3) The environmental value loss of the VPP is minimal

The minimum environmental value loss of the VPP is shown in formula (6):

$$\min C_s^{loss} = \min \rho^{out} \sum_{t=1}^{T} \left[\sum_{i=1}^{N^m} (G_{i,t}^{w^*} - G_{s,i,t}^{w}) + \sum_{i=1}^{N^m} (G_{i,t}^{pv^*} - G_{s,i,t}^{pv}) \right] \tag{6}$$

(4) The peak-valley deviation cost of the VPP is minimal

The peak-valley deviation cost of the VPP is shown in formula (7):

$$\min C_s^{pvd} = \min \rho^{pvd} \sum_{t=1}^{T} (L_{t,s}^{max} - L_{t,s}^{min}) \tag{7}$$

3.2   Constraints

The optimal scheduling of the VPP should satisfy conditions such as power balance constraints, distributed power output constraints, interruptible load constraints, and battery operation constraints.

(1) Power balance constraint

$$\sum_{i=1}^{N_W} \mu_{i,t}^{W} G_{i,t}^{W}(1 - \varphi_W) + \sum_{i=1}^{N_{PV}} \mu_{i,t}^{PV} G_{i,t}^{PV}(1 - \varphi_{PV}) + \sum_{i=1}^{N_{FC}} G_{i,t}^{FC}(1 - \varphi_{FC}) + \sum_{i=1}^{N_{BT}} \mu_{i,t}^{dis} G_{i,t}^{BT,dis}$$
$$+ \mu_i^{CWg} G_i^{CWg} + \mu_t^{in} L_t^{in} = \sum_{i=1}^{IB} L_{i,t}^{PB} - \sum_{i=1}^{IB} \mu_{i,t}^{IB} \Delta L_{i,t}^{IB} + \sum_{i=1}^{N_{BT}} \mu_{i,t}^{chr} G_{i,t}^{BT,chr} \tag{8}$$

Where: φ_W, φ_{PV}, and φ_{FC} are the plant power consumption rates of wind power, photovoltaic power generation and gas power generation.

(2) Distributed power output constraints

Distributed wind power, distributed photovoltaic and gas distributed power generation should meet the upper and lower limits:

$$\begin{cases} 0 \leq G_{i,t}^{W} \leq \mu_{i,t}^{W} \overline{G}_t^{W} \\ 0 \leq G_{i,t}^{PV} \leq \mu_{i,t}^{PV} \overline{G}_t^{PV} \\ \mu_{i,t}^{FC} G_{i,min}^{FC} \leq G_{i,t}^{FC} \leq \mu_{i,t}^{FC} G_{i,max}^{FC} \end{cases} \tag{9}$$

Where: $\overline{G}_t^W$, $\overline{G}_t^{PV}$ are the maximum power generation output power of distributed wind power and distributed photovoltaic i; $G_{i,\min}^{FC}$, $G_{i,\min}^{FC}$ are the maximum and minimum power generation output power of gas-fired distributed unit i, respectively.

Gas-fired distributed power generation should also meet the ramp rate constraints and minimum start-up and shutdown time constraints, as shown in Eq. (10):

$$\begin{cases} \mu_{i,t}^{FC} \Delta G_{\min,t}^{FC} \leq G_{i,t}^{FC} - G_{i,t-1}^{FC} \leq \mu_{i,t}^{FC} \Delta G_{\max,t}^{FC} \\ (T_{t-1}^{FC,\mathrm{on}} - T_{\min}^{FC,\mathrm{on}})(\mu_{i,t-1}^{FC} - \mu_{i,t}^{FC}) \geq 0 \\ T_{t-1}^{FC,\mathit{off}} - T_{\min}^{FC,\mathit{off}})(\mu_{i,t-1}^{FC} - \mu_{i,t}^{FC}) \geq 0 \end{cases} \tag{10}$$

Where: $\Delta G_{\min,t}^{FC}$ and $\Delta G_{\max,t}^{FC}$ are the upper and lower limits of the climbing power of the gas-fired distributed unit i; $T_{t-1}^{FC,\mathrm{on}}$ and $T_{t-1}^{FC,\mathit{off}}$ are the continuous operation time and continuous shutdown time of the unit at $t-1$; $T_{\min}^{FC,\mathrm{on}}$ and $T_{\min}^{FC,\mathit{off}}$ are the minimum start-up time of the unit And minimum downtime.

(3) Constraint on callable capacity of interruptible load

$$\Delta L_{\min,t}^{IB} \leq \sum_{i=1}^{N_{IB}} \mu_{i,t}^{IB} \Delta L_{i,t}^{IB} \leq \Delta L_{\max,t}^{IB} \tag{11}$$

Where: $\Delta L_{\min,t}^{IB}$ and $\Delta L_{\min,t}^{IB}$ are the upper and lower limits of the interruptible load of the VPP dispatching at time t.

(4) Restrictions on battery charging and discharging

In order to ensure the service life of the battery, it cannot run the two working modes of charging and discharging at the same time during the scheduling period, as shown in Eq. (12):

$$0 \leq \mu_{i,t}^{dis} + \mu_{i,t}^{chr} \leq 1 \tag{12}$$

The charge and discharge power of the battery should meet the upper and lower limits as shown in formula (13):

$$\begin{cases} 0 \leq G_{i,t}^{BT,dis} \leq \mu_{i,t}^{BT,dis} \overline{G}_i^{BT,dis} \\ 0 \leq G_{i,t}^{BT,\mathrm{chr}} \leq \mu_{i,t}^{BT,\mathrm{chr}} \overline{G}_i^{BT,\mathrm{chr}} \\ SOC_t^{\min} \leq SOC_{i,t} \leq SOC_t^{\max} \end{cases} \tag{13}$$

Where, $\overline{G}_i^{BT,dis}$ and $\overline{G}_i^{BT,\mathrm{chr}}$ are the maximum discharge power and maximum charging power of battery i, respectively; $SOC_t^{\min}$ and $SOC_t^{\max}$ are the upper and lower limits of the state of charge of battery i, respectively.

4 Example Analysis

4.1 Basic Data

In this paper, the improved 2-unit system data is used to verify and analyze the model. G1 and G2 are VPPs VPP1 and VPP2, respectively. The detailed parameters of these two units are shown in Table 2 [10].

Table 2. Basic data

	A(¥/ MW²h)	B(¥/ MWh)	C(¥)	∂ (¥/ MW²h)	β (¥/MWh)	γ (¥)
G1	-0.0004	252	781	0.00023	72.28	324
G2	-0.0006	235	973	0.00021	73.32	366

The VPP G1 consists of two 50 MW wind turbines, six 20 MW gas turbines and 50 MW interruptible loads. The VPP G2 includes three 50 MW wind turbines, two 10 MW photovoltaic units, five 20 MW gas turbine units and 40 mW interruptible load. In the two VPPs, the output power of a single wind turbine is 50 MW under the rated wind speed of 8 m/s, the cut in wind speed is 1 m/s, and the cut out wind speed is 10 m/s. The area of each photovoltaic power generation system is about 60000 square meters, the photoelectric conversion efficiency is 16%, and the maximum output power is 10MW when the light intensity reaches 105. The upper and lower limits of climbing rate of a single gas-fired generator unit are 60 to 200, the minimum start-up and stop time is 1H, the upper limit of output is 25 MW, and the lower limit is 10 MW. The unit deviation penalty price of the actual output of the VPP is 500 yuan/MW, and the rated power of the EV is 2 kW. The load demand forecast values of EV users in typical days are shown in Fig. 1 below:

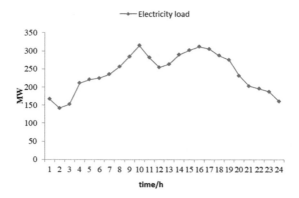

Fig. 1. EV user load demand forecast value on a typical day

4.2   Result Analysis

Set the number of EVs to be 21,100. In order to analyze the impact of different EV intervention scales on the revenue of VPPs, the following scenarios are set up:

Scenario 1: The number of EVs connected to the VPP is 80% of the predicted number, that is, 168,100 vehicles.
Scenario 2: The number of EVs connected to the VPP is 90% of the predicted number, that is, 189,100 vehicles.
Scenario 3: The number of EVs connected to the VPP is 100% of the predicted number, that is, 21,100.
Scenario 4: The number of EVs connected to the VPP is 110% of the predicted number, that is, 231,100.
Scenario 5: The number of EVs connected to the VPP is 120% of the predicted number, that is, 252,100.

Based on the above scenario analysis, combined with formulas (4)–(7), the net income of G1 and G2 VPPs, environmental governance costs, environmental value loss, and peak-valley deviation costs under different scenarios can be obtained as shown in Table 3 and Table 4:

Table 3. Benefits and costs of G1 VPP under different scenarios

	Scenario 1	Scenario 2	Scenario 3	Scenario 4	Scenario 5
PV power generation (MW)	–	–	–	–	–
Wind power generation (MW)	651.32	648.41	644.00	631.29	627.53
Gas turbine output value (MW)	1714.56	1702.34	1700.70	1688.45	1672.83
Interruptible load (MW)	0	0	0	0	0
EV output value (MW)	18.91	16.81	21.01	23.11	25.21
Environmental governance costs (Ten thousand yuan)	41.33	39.25	37.12	32.98	31.16
Environmental damage costs (Ten thousand yuan)	51.29	46.72	41.24	36.45	34.88
Peak-to-valley deviation cost (Ten thousand yuan)	51.14	57.93	60.73	62.18	69.91
Net income (Ten thousand yuan)	140.52	149.24	151.29	153.44	150.82

The net income of VPPs, environmental governance costs, environmental value losses, and peak-valley deviation costs under different scenarios are shown in Table 4:

Table 4. Benefits and costs of G2 VPP under different scenarios

	Scenario 1	Scenario 2	Scenario 3	Scenario 4	Scenario 5
PV power generation (MW)	66.09	65.67	63.24	60.12	57.30
Wind power generation (MW)	1060.01	1054.85	1045.25	1021.56	1012.73
Gas turbine output value (MW)	4348.51	4309.83	4294.03	4229.78	4180.88
Interruptible load (MW)	0.00	0.00	0.00	0.00	0.00
EV output value (MW)	28.84	25.63	32.04	35.24	38.44
Environmental governance costs (Ten thousand yuan)	34.24	30.88	47.81	41.47	38.73
Environmental damage costs (Ten thousand yuan)	41.50	44.09	35.26	37.45	34.87
Peak-to-valley deviation cost (Ten thousand yuan)	82.05	92.54	96.49	98.34	110.39
Net income (Ten thousand yuan)	225.43	240.99	244.96	248.79	246.35

It can be seen from Table 3 and Table 4 that as the proportion of EVs connected to the G1 VPP increases, the wind power generation and gas turbine output value decrease; due to the decrease in gas turbine output value, environmental governance costs and environmental loss costs are reduced, but the peak The cost of valley deviation has risen because the increase in the proportion of EVs connected has an impact on the safety and stability of the system, which has increased the cost of peak and valley deviation to a certain extent. But for the G1 system, the net income of the VPP increases first and then decreases as the scale of EV access increases. Specifically, when the EV access ratio is 80%, 90%, 100%, and 110%, the income is 2,254,300 yuan, 2,409,900 yuan, 2,449,600 yuan, and 2,487,900 yuan respectively; but when the EV access rate is 120% Hourly income was reduced to 2,463,500 yuan. This shows that the connection scale of EVs is not as high as possible, and there is an optimal connection ratio.

Compared with the G1 VPP, the G2 VPP's wind power and photovoltaic power generation are higher than G1, and the gas turbine value is also higher than G1, which causes the peak-to-valley deviation cost to increase. However, due to the increase in the power of renewable energy, both the cost of environmental governance and the cost of environmental losses have decreased, resulting in an increase in the net income of the G2 VPP.

5   Conclusions

In this paper, by studying the cost-benefit of the VPP under the connection of EVs, it can be obtained that when the proportion of EVs is 110%, the income is 2,487,900 yuan, which is higher than the income of 120% when the proportion of EVs is connected. It shows that the connection scale of electric vehicles is not as high as possible, and there is an optimal connection scale and proportion. Therefore, the connection scale of electric vehicles should be set reasonably in the subsequent process of connecting electric vehicles to virtual power plants.

References

1. Vaya, M.G., Andersson, G.: Self-scheduling of plug-in electric vehicle aggregator to provide balance services for wind power. IEEE Trans. Sustain. Energy **7**(2), 886–899 (2016)
2. Liu, W., Hu, W., Lund, H., et al.: Electric vehicles and large-scale integration of wind power-the case of Inner Mongolia in China. Appl. Energy **104**, 445–456 (2013)
3. Liu, X., You, J., He, Y., et al.: Research on loss reduction of electric vehicle charging and swap station connected to grid based on experimental design. Electr. Power Autom. Equip. **38**(09), 70–76 (2018). (in Chinese)
4. Vasirani, M., Kota, R., Cavalcante, R.L.G., et al.: An agent based approach to virtual power plants of wind power generators and electric vehicles. IEEE Trans. Smart Grid **1**(3), 1311–1322 (2013)
5. Song, X., Wang, Y.: Microgrid economic and environmental protection dispatch based on co-evolution genetic algorithm. Power Syst. Protect. Control **42**(5), 85–89 (2014). (in Chinese)
6. He, L., Wang, X., et al.: Wind power and electric vehicles form a virtual power plant to participate in a multi-agent game in the market. Renew. Energy **38**(12), 1686–1692 (2020). (in Chinese)
7. Tian, L., Shi, S., Jia, Z.: Statistical modeling method of electric vehicle charging power demand. Power Syst. Technol. **34**(11), 126–130 (2010). (in Chinese)
8. Yuan, G., Su, W.: Research on virtual power plant participating in AGC frequency regulation service considering the uncertainty of electric vehicles. Power Syst. Technol. **44**(07), 2538–2548 (2020). (in Chinese)
9. Tian, L., Cheng, L., Guo, J., et al.: A review of research on virtual power plant's management and interaction mechanism for distributed energy. Power Syst. Technol. **44**(06), 2097–2108 (2020). (in Chinese)
10. Ying, F., Xu, T., Li, Y., et al.: Research on day-ahead dispatch optimization strategy of commercial virtual power plant including electric vehicle charging station. Power Syst. Protect. Control **48**(21), 92–100 (2020). (in Chinese)

Prediction Model of Internationalized Talent Training Based on Fuzzy Neural Network Algorithm

Zhong Wu[1] and Chuan Zhou[2(✉)]

[1] Physical Education School, Wuhan Business University, Wuhan, Hubei, China
[2] Institute of Mechanical Engineering, Wuhan Institute of Shipbuilding Technology, Wuhan, Hubei, China
20150459@wbu.edu.cn

Abstract. Since entering the 21st century, the network information age, China's rapid integration into the global economy, the rapid development of international economy, science and technology level unceasing enhancement, the computer communication network technology, computer network virtual simulation technology, cloud computing technology, the Internet of things technology and multiple wireless sensor network technology under the background of the fast development. The cooperation and competition between countries promote the cultivation of international talents, and the prediction model of national talent cultivation plays an important role in talent cultivation. At the same time, with the support of technological development, the international talent training mode based on fuzzy neural network algorithm appears in a creative image, which is of great significance for the accuracy of the prediction model of international talent training. In this paper, based on the fuzzy relation analysis, using the BP algorithm based on fuzzy neural network algorithm for internationalized talents cultivation model and internationalization of traditional personnel training efficiency and accuracy of forecasting model to calculate, it is concluded that based on fuzzy neural network prediction model of internationalized talents training algorithm at 5.3% and 3.1% respectively in terms of accuracy and efficiency higher than traditional internationalized talents cultivation prediction model.

Keywords: Fuzzy neural network · Algorithm · International talent training · Training prediction model

1 Introduction

The 21st century is a country of constant exchange and integration among countries in the world. Under this background, the integration and development of the world economy require more national talents to deal with various phenomena and problems in the national economic and political exchanges. The cultivation of national talents is an important field of talent cultivation today. The cultivation of international talents should not only realize the educational purpose of teaching and educating people, but also meet the special requirements of the country and society for such talents. How to

J. C. Hung et al. (Eds.): FC 2021, LNEE 827, pp. 416–423, 2022.
https://doi.org/10.1007/978-981-16-8052-6_51

realize the high efficiency and high quality of national talent cultivation is an important issue of social concern. The prediction model of international talent cultivation appears in this context, but the traditional model of international talent cultivation has some problems, and it needs to be combined with more disciplines to create a more efficient and high quality prediction model of international talent cultivation.

In 1965, L.A. Zadeh created the fuzzy idea and realized the fuzzy expression of accurate data through membership function. Later, neural network was introduced to make up for the defect that it could not express fuzzy information [1]. Fuzzy neural network has been used in fuzzy control system, fault prediction, risk assessment and other fields. In order to further improve the effectiveness of FNN, Tekin A Kunt made full use of the flexibility of FNN through A combination of various algorithms to complete talent training evaluation [2]. Jiao Aihong et al. constructed a new prediction model by using fuzzy rough set and clustering algorithm [3]. Neural network can constantly improve the adaptability of the algorithm through learning, and can adjust its parameters according to environmental changes, but its output expression form is often abstract and difficult to be understood by people. However, the fuzzy control algorithm focuses on the expression of prior knowledge and is very suitable for dealing with various influencing factors and other non-deterministic information in international talent training [4, 5].

Combining the above analysis and research, this paper applied fuzzy neural algorithm in internationalized talents cultivation model, the internationalization of constructing algorithm based on fuzzy neural network prediction model of personnel training, and based on the fuzzy relation analysis, using the BP algorithm based on fuzzy neural network algorithm for internationalized talents cultivation model and internationalization of traditional personnel training efficiency and accuracy of forecasting model to calculate.

2   Fuzzy Neural Network and Prediction Model of International Talent Cultivation

2.1   Fuzzy Neural Network

Fuzzy neural network is a linear complex control system with the help of a large number of processing units. To some extent, it is similar to the digital information processing, memory and search function of human brain, and has intelligent processing ability such as learning and memory. Fuzzy neural network is a fusion technology based on neuroscience, mathematics, statistics, physics, computer science and other disciplines [6]. Fuzzy neural network is a combination of fuzzy control and neural network. It is generally a multi-layer feed forward network, which is divided into the first layer, the first layer and the last layer. The former layer realizes fuzziness, the middle layer realizes fuzzy reasoning, and the latter layer realizes anti-pattern dextrinization. It combines the advantages of fuzzy control and neural network, can use the limited information of fuzzy rules to do fuzzy logic reasoning, and has a good approximate ability to nonlinear systems, and is effective in solving nonlinear problems with large hysteresis.

The calculation process of the fuzzy neural network algorithm is as follows: first, the neural network structure is input as the basis to determine the level and number of neurons, and on the basis of the input characteristic parameters, appropriate membership function is established to establish the corresponding fuzzy setting decision rule base. According to the neuron conduction model, the input and output of each layer network are calculated [7]. The weight of conductive network at all levels is calculated according to the model built by the collected sample data. In the calculation, the weight value of the artificial neural network that needs to be modified is adjusted continuously, and the error value of network output is set.

2.2   Prediction Model of International Talent Cultivation

International talent training is the inevitable trend of economic development with the time development and the inevitable requirement, but the current our country the development of international talent and develop some problems, such as: the lack of international talent development and training system construction, the internationalization of talents development and training of long-term goals and short-term goals are not clear, lack of systemic concept design internationalized talents training work, internationalization development and training the lack of professional management talents and internationalized talents development and training the lack of scientific management methods [8]. For the sustainable development of economy in China, with the deepening of the foreign cooperation, urgently needs to cultivate international talents, cultivating international talents cannot blindly, should be combined with the present science and technology and the discipline research results, build internationalized talents cultivation model, lay a foundation for internationalization talented person's raise is effective [9]. Internationalized talents cultivation model is the internationalization of talents cultivation in China, the various problems and build, on the basis of the forecast, we can know the development direction of internationalized talents cultivation, the progress of the work requirements and scheduling, to make internationalized talents cultivation based on the model of long-term goals and short-term goals, provide internationalized talents cultivation of the overall work of scientific and effective management and arrangement, for China's internationalization talented person troop transport firm power [10].

3   Research on Prediction Model of International Talent Training Based on Fuzzy Neural Network Algorithm

3.1   Influence Mechanism of Fuzzy Neural Network Algorithm on Prediction Model of International Talent Cultivation

The internationalization of China's traditional talent training than in other developed countries there are still some gaps: for internationalized talents cultivation education awareness is low, the internationalized talents cultivation idea is not clear, hard to combine the basis for the development of our country and advantages to develop talent, transmission applied to the development of economic internationalization, skills talents.

There are also some problems in the traditional prediction model of international talent cultivation built on this basis. Most of the prediction is based on qualitative analysis, which makes it difficult to comprehensively consider all the influencing factors and make accurate and reasonable prediction of China's international talent cultivation.

Fuzzy neural network uses a large number of processing units to constitute a nonlinear complex control system. Fuzzy prediction is based on the fuzzy dynamic model. Through the processing of original data, the fuzzy model is constructed, the development rules of the system are found and the future state of the system is scientifically and concretely predicted quantitatively. Fuzzy neural network algorithms to predict the presence of the props to solve in time and make up for the defects in the traditional algorithm, which based on fuzzy prediction system to reflect the various influence factors to predict, so fuzzy sequence forecast on the basis of related numerical, the biggest characteristic, used to predict the object itself data and other related factors, to predict fuzzy object scientific budget.

Based on the fuzzy neural network algorithm, to the students' thinking quality, learning methods and ideas, with the goal to pursue hobbies specialty by the deepest level of analysis, on the basis of the focus of the data and information for accurate prediction of talent training, to provide personalized training objectives for the cultivation of the talent, improve the comprehensive quality of talents with professional skills, to become truly international talents, in order to meet the demand for international talents and social development in our country.

3.2   Data Calculation Formula

Based on fuzzy relation analysis, the index system and quantity of internationalized talents cultivation as a whole system, calculation of internationalized talents cultivation quantity of each index correlation degree, according to the correlation order, sorting indicators for internationalized talents training again forecast system has the characteristics of poor information, high dimension and its index system directly dimension reduction method is optional. On this basis, the specific calculation formula can be obtained as follows:

$$P_{ci} = P_{cmax} - (P_{cmax} - P_{cmin}) * T(F_i + F_{min}) \tag{1}$$

The learning process of BP algorithm consists of forward propagation and back propagation. In the forward propagation process, the input layer is processed layer by layer through implicit layer and transferred to the output layer, and the state of neurons at each layer only affects the state of neurons at the next layer. If the desired output cannot be obtained in the output layer, the error signal is backward propagated and calculated to the connection path, and the weight of neurons in each layer is adjusted by gradient descent method to reduce the error signal. Therefore, the calculation function of efficiency can be expressed as:

$$e(k) = y(k) - y_n(k) \tag{2}$$

4 Data Analysis of the Prediction Model of International Talent Cultivation Based on Fuzzy Neural Network Algorithm

4.1 Advantages of Fuzzy Neural Network Algorithm on National Talent Training Prediction Model

Fuzzy neural network can imitate human brain neural system, set up similar to the model based on a single nerve cell network, fuzzy neural network algorithm based on fuzzy set theory, the rules of the fuzzy language variables and fuzzy logic reasoning, the combination of computer technology, relying on the internationalization collected during the process of personnel training of the students' thinking quality, learning methods, ideas and hobbies specialty and target the pursuit of all sorts of factors as a parameter, such as by using fuzzy neural network control algorithm internationalized talents cultivation of model calculation, so as to improve the accuracy of internationalized talents cultivation model calculation and efficiency, optimized training and development of international talents through feedback and calculation.

By investigating a region to the internationalization of traditional forecasting model of training teachers and internationalized talents cultivation model based on fuzzy neural network algorithm of satisfaction, we can see the teacher for two kinds of international talent training model has different attitude, which can reflect based on fuzzy neural network algorithm for the role of internationalized talents cultivation model, specific circumstances such as Table 1:

Table 1. Teachers' satisfaction with the traditional prediction model of international talent cultivation and the model of international talent cultivation based on fuzzy neural network algorithm

Model	Teachers		
	Teachers A	Teachers B	Teachers C
Traditional forecast model of international talent cultivation	65.8%	78.9%	77.9%
International talent training model based on fuzzy neural network algorithm	66.3%	79.8%	82.4%

On the basis of theoretical analysis and data statistics, the prediction model of international talent cultivation based on fuzzy neural network algorithm can clarify the goal of international talent cultivation, improve the model of international talent cultivation, promote the construction of the curriculum system of international talent cultivation, and help the construction of the teaching material system of international talent cultivation.

4.2 Analysis of Calculation Results

Based on fuzzy correlation analysis and BP algorithm, we calculate the accuracy and efficiency of international talent cultivation prediction under different models.

According to the experiment and calculation, students of international talent training class in a certain region were selected as data samples to calculate the accuracy of international talent training prediction under different models. According to the relevant calculation data, Fig. 1 can be drawn as follows:

At the same time, corresponding formulas were used to calculate the predicted efficiency of international talent cultivation under different models for relevant samples, and the data were compared and analyzed, as shown in Fig. 2:

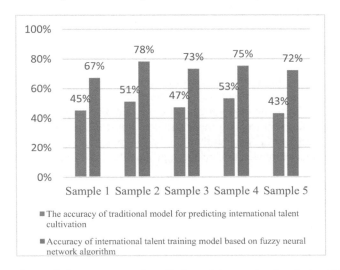

Fig. 1. Comparison of the accuracy of prediction of international talent cultivation under different models

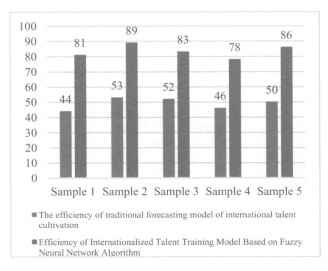

Fig. 2. Comparison of predicted efficiency of international talent cultivation under different models

Comparison of According to the calculation data, the prediction model of international talent cultivation based on fuzzy neural network algorithm is far superior to the traditional prediction model of international talent cultivation in terms of accuracy and efficiency. With the help of its own characteristics, fuzzy neural network algorithm is of great significance for the prediction of international talent cultivation. We should make full use of this advantage to promote the development and progress of the prediction model of international talent cultivation in China, and improve the quality and efficiency of international talent cultivation.

5 Conclusions

To sum up, using a fuzzy neural network algorithm of today's technology and the discipline research results, combined with internationalized talents cultivation model, can the analysis from the data information of the internationalization strategy of directional cultivate talent training more targeted, boost the further development of the national internationalized talents cultivation, on the maximum continuous conveying to the nation, the internationalization of the high quality, high quality talent.

Acknowledgements. This work was supported by the Provincial University Teaching Research Project of Hubei Province (2018466), Teaching and Research Program at City-owned Universities in Wuhan (2019102), and the Construction of Specialty Features of Sports Economics and Management Major in Wuhan Business University (2019N013).

References

1. Zadeh, L.A.: Nonlinear systems modelling based on self-organizing fuzzy neural network with hierarchical pruning scheme. Appl. Soft Comput. J. (3), 95–97 (2020)
2. Kunt, T.A., Le, T.-L., Huynh, T.-T.: Self-evolving function-link interval type-2 fuzzy neural network for nonlinear system identification and control. Neurocomputing (9), 275–278 (2019)
3. Chen, X., Wang, L., Huang, Z., et al.: Principal component analysis based dynamic fuzzy neural network for internal corrosion rate prediction of gas pipelines. Math. Prob. Eng. (23), 67–68 (2020)
4. Tang, B., Xiang, K., Pang, M., et al.: Multi-robot path planning using an improved self-adaptive particle swarm optimization. Int. J. Adv. Robot. Syst. (5), 17–19 (2020)
5. Ayoola, R., Larion, S., Koya, R.H., et al.: Clinical factors associated with high-risk patients who have not undergone hepatitis b screening: data from two large academic hospitals in an urban setting. Gastroenterology (15), 152–154 (2019)
6. Kim, D.-H., Yoon, H.B., Sung, M., et al.: Evaluation of an international faculty development program for developing countries in Asia: the Seoul intensive course for medical educators. BMC Med. Educ. (1), 15–18 (2021)
7. Weaver, S.R., Heath, J.W., Ashley, D.L., et al.: What are the reasons that smokers reject ENDS? A national probability survey of U.S. adult smokers. Drug Alcohol. Dependence (12), 211–213 (2020)

8. Shawe, R., Horan, W., Moles, R., et al.: Mapping of sustainability policies and initiatives in higher education institutes. Environ. Sci. Policy (8), 99–101 (2019)
9. Tómasson, B., Karlsson, B.: The role of households in Nordic national risk assessments. Int. J. Disaster Risk Reduct. (24), 45–47 (2020)
10. Maraseni, T.N., Poudyal, B.H., Rana, E., et al.: Mapping national REDD+ initiatives in the Asia-Pacific region. J. Environ. Manag. (21), 269–271 (2020)

Path to Improve the Intelligent Level of Rural Governance in the Era of Big Data

Danni Shi[✉] and Qinyi Chen

Department of Marxism, Jilin Agriculture and Science University,
Jilin, Jilin, China

Abstract. In order to adapt to the new requirements of big data, China puts forward the corresponding guidance and action plan technology. The rapid development of big data technology has a fundamental impact on rural governance. Now it can be seen that big data technology has become the key to rural governance innovation. This paper analyzes the problems in rural governance. At the macro level, combined with the development of the current big data era, we should take a reasonable innovation path of rural environmental governance, and analyze the resource value and platform value of big data from the perspective of promoting the modernization of rural governance. The results show that the proportion of left behind women is 36%, and the proportion of left behind elderly is 5% higher than that of left behind children.

Keywords: Big data · Rural governance · Intelligence and rural revitalization

1 Introduction

With the continuous progress of science and technology, big data technology has become an indispensable part of our life. The application of big data technology is conducive to the construction and governance of rural information. From single supervision to multiple governance, from experience decision to data decision, from passive disposal to active prediction, from refined management to refined management service. But in the process of big data embedded in rural governance, there are still obstacles.

With the continuous development of computer technology, many experts have studied rural governance in the era of big data. For example, some domestic teams have studied the modernization strategy of rural governance from the perspective of targeted poverty alleviation, sorted out the relationship between various data, predicted the trend, solved problems through collection and analysis, built a comprehensive information network, and built a mufti information sharing and processing platform. In this paper, using the methods and algorithms of data mining and the quality index of rural governance project constructed through large-scale research, we respectively discussed the mobile sensing and cloud computing in detail, and combined the two concepts to form a unique concept of mobile cloud sensing. This paper will give an intuitive description of mobile cloud sensing architecture and discuss its building modules [1]. Some experts have studied the efficiency of village rules and regulations in the new period of rural governance. Using the concept of overall planning and coordination and the open and transparent evaluation model of big data is an inevitable perspective to

J. C. Hung et al. (Eds.): FC 2021, LNEE 827, pp. 424–431, 2022.
https://doi.org/10.1007/978-981-16-8052-6_52

realize the scientific grasp of rural governance and construct the consensus of mufti-agent participation. This paper discusses the intelligent building system based on related technologies, and uses these key technologies to promote the innovative application system and user, network and service system of building physical model, so as to make buildings become perception [2]. Some experts have also studied the participation of rural sages in rural governance under the background of rural revitalization, constructed a theoretical framework, and analyzed a series of effects of the participation of new rural cooperative economic organizations in rural governance. This paper reviews the history of rural governance; reexamines the concept of rural governance, and attempts to reconstruct the possible paradigm of rural governance research, that is, the comprehensive study of rural political, economic, social and cultural development. On this basis, it gives a preliminary overview of the methods and stages of constructing the research paradigm of rural governance. On the basis of quantitative analysis, some effective suggestions are put forward. This paper summarizes the characteristics of e-government driving rural governance, reveals the driving force and development trend of e-government driving rural governance from both inside and outside the village, and thinks that the adjustment of interest structure and the upgrading of e-commerce will reconstruct the rural internal governance and external governance respectively, so as to provide enlightenment for rural construction and planning. From the perspective of social resources and land variables, the relationship between class division and rural governance is different. Different rural governance has different interests, social status, political attitude and social performance, and different ability and motivation to maintain rural social order. According to the current social stratification, rural governance should change the mode, strengthen rural social management, adjust class interest relations, maintain social stability, and promote rural political development. Agricultural and sideline products are the middle class and the mainstay of rural society. Harmonious class interest relations, maintain rural social order, promote rural democracy, and realize agricultural modernization must be strengthened Cultivate and support it [3]. Although the research results of rural governance in the era of big data are quite abundant. But in the era of big data, there are still some deficiencies in the research of rural governance intelligence.

In order to study the intelligence of rural governance in the era of big data, this paper studies the intelligence of big data, information and rural governance, and finds the Governance Dilemma of rural community public affairs in the environment of big data. The results show that big data technology is conducive to the intellectual of rural governance.

2   Method

2.1   Big Data and Informatization

(1) Big data

Big data technology pushes human beings to the era of measurement [4]. Universal computing and social media enable us to measure the physical and social environment of a city, making it a reasonable core of urban and rural planning [5]. With the development of information technology, urban and rural planning survey is developing towards relational database, data warehouse, multi-dimensional online analysis, data mining, visualization, etc. [6]. At the same time, urban and rural planning survey is also facing data challenges [7]. Big data encourages all subjects to participate in joint governance, and promotes decision-making, data opening, information sharing, information sharing, and privacy protection And so on. Enterprises, institutions, social organizations and other disciplines at the level of rural society have mastered certain data resources, such as economy, environmental protection, education, etc. [8]. The application of big data technology has inspired these social subjects to realize the importance and significance of data, and actively participate in the process of rural social governance through big data technology [9].

(2) The function characteristics of information platform

Information platform. The platform adopts information technology, integrates information resources, establishes information network, builds interactive platform for rural informatization construction, and solves the problems of long distance, difficult communication and poor time limit in rural governance [10]. Through the transmission of mobile terminal network, we can effectively and timely process and collect all kinds of information, and shorten the digital gap between urban and rural areas, rural areas, rural areas and villagers [11]. The platform is an intelligent tool with mobile terminal as the main body, which changes the original way of rural grassroots governance. As an information platform, based on modern communication and network technology, it develops rapidly, connects the developed information network, establishes an accurate and efficient database, combines the reality and characteristics of rural governance, deeply integrates the needs of villagers, and becomes a dynamic platform for information integration and information release.

(3) Big data platform

Big data platform. The platform collects, manages and processes all kinds of information data of each village through the background data storage center. In view of the analysis of complex data such as age, geographical distribution, village hot spots, active plates and public opinion monitoring under the background, the platform adopts big data means to process data resources, optimize the platform construction, accurately grasp the dynamic and moving track of certified villagers, give play to the advantages of "let the data run" and "let the masses walk less", and provide a lot of real valuable information for rural governance Effective data. The platform has a large and fine database. Through data analysis, it has formed a big data link from collection, processing, storage to result application. Through data analysis, it can understand the law and key nodes of rural governance, find the most important part of villagers' attention, improve the accuracy of information analysis, and reduce the duplication and waste of information resources, it is helpful to the innovation of rural information governance.

2.2 Rural Governance

(1) Rural Governance

Rural governance is an important factor in the development of rural society. It can also promote the development of society. Harmonious social capital is an explanatory concept, which reflects the social structure and social state governance. Therefore, it is a major breakthrough to improve the level of farmers' organization and reconstruct the social basis of rural governance. The concept of governance focuses on the diversification of governance subjects and the multi dimension of governance power. It is a kind of top-down dialogue, consultation and cooperation, rather than a relationship of command and obedience, emphasizing the multiple interaction between the governance subjects. Community governance is the application of governance theory in the field of community, which means to meet the needs of community residents, improve the community governance mechanism, and promote the development and progress of the community through the coordination and cooperation of multiple governance subjects.

(2) Rural governance mechanism

The word "machine" originated from Greek, which refers to the structure and motion principle of machine. Biology and medicine use this concept to explain how organs interact and regulate when physiological or pathological changes occur. Later, the word was introduced into social science research, referring to the way in which various parts of a social organism interact. Governance mechanism is a dynamic description of governance system, formal informal system operation mode and interaction process of subsystems. The rural governance mechanism refers to the specific operation mode of multi governance subjects' joint governance and integration of rural governance resources. It is about the rural governance subject cooperation mechanism, resource integration mechanism and other specific mechanisms collectively. Rural governance emphasizes the effective integration of governance resources to maximize the efficiency of resources. It mainly includes rural human resources, information resources, cultural resources, internal idle funds, enterprise capital investment, government financial support, etc. Rural governance emphasizes the dynamics of governance process and the openness of governance purpose. Rural governance aims to solve the problems of rural public affairs, resolve rural social contradictions, provide public services for rural society, coordinate the main body of interest conflicts, and promote rural public interests, so as to maintain the stability of rural society, build a good rural order, and realize the dynamic and orderly development of rural society.

2.3 Governance Dilemma of Rural Community Public Affairs Under Big Data Environment

Assuming that the probability r of all participants choosing to provide public goods is the same, then the probability of n $-$ 1 participants except villager B choosing not to provide public goods is r^{n-1}, and the probability of all participants choosing to provide public goods is $1 - r^{n-1}$. Using the payment equilibrium method of equilibrium calculation, the pure strategic payment of farmer B is set as equilibrium, then the formula (1–3) is as follows:

$$Y_{\text{VillagersA}}(\text{Choose to offer}) = 7 = Y_{\text{VillagersA}}(\text{Choose not to offer})$$
$$= 7 \times r^{n-1} + 7 \times \left(1 - r^{n-1}\right) \tag{1}$$

$$r^{n-1} = 0.2 \tag{2}$$

$$r = \frac{1}{0.2^{n-1}} \tag{3}$$

3 Experiences

3.1 Extraction of Experimental Objects

The survey object is a village in a city, which will become one of the pilot areas of rural community governance in 2020. Since then, Shuangluan district actively promotes the construction of rural communities, and promotes the development of rural communities according to the overall idea of strengthening Party building, deepening service and villagers' satisfaction. Meanwhile, we will promote the city to drive the countryside, realize complementary advantages, improve the service environment and optimize the economic environment. The survey method is mainly questionnaire survey, and interview is the auxiliary. This paper uses random questionnaire survey to investigate the age structure, education level, occupation division and other basic situation of the respondents. The current situation of village a governance is also prospected. The development direction of the future governance is prospected, so as to understand the actual needs of rural residents of different ages and occupations for village a governance, so as to provide the basis for the development path of new rural community governance According to. The interview survey mainly summarizes and extracts the relevant content of the interviewers' answers through individual interviews, and obtains the corresponding investigation results to support the study.

3.2 Experimental Analysis

First, it can process and analyze the big data of various data sources, combine various big data with artificial intelligence algorithm, provide the tourists with intelligent planning before travel, intelligent push of various destination information in travel, and intelligent extension service after travel. Secondly, by comparing the analysis results of various big data, the tourists can be intelligently marked with the possible congestion sections or scenic spots or the latest global tourism hotspots, and give suggestions for guidance and connection to help tourists make judgments and decisions.

4 Discussions

4.1 The Change of Rural Population is Obvious

The most important factor affecting rural social structure is the change of population. Over the years, there have been large-scale population movements in China. The main trend is the migration of rural population to urban areas. The most direct impact is the obvious change of rural social population structure. As shown in Table 1.

Table 1. The proportion of Rural three left behind population in 2020

Type	Data
Aged	26%
Children	21%
Woman	36%
Other	17%

From the perspective of age structure, the rural population has changed in age level and gender structure. Due to the influence of work, study, urban settlement and other factors, most of the rural floating population concentrated in the young and middle-aged stage. At present, the rural population is mainly young people, children and the elderly, and the number of rural women is more than that of men. As shown in Fig. 1.

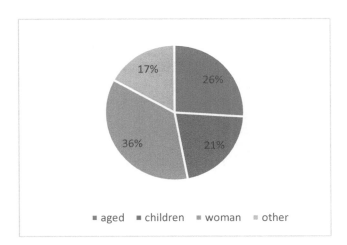

Fig. 1. Rural three left behind population in 2020

It can be seen from the above that the number of women left behind in rural areas accounts for 36% at most, and the number of elderly left behind in rural areas is 5% more than that of children left behind.

4.2 Rural Construction Investment in China

In some new rural construction led by the government, large-scale demolition and farmers' resettlement blindly copy the form of urban residential areas, which is not in line with the villagers' living habits; the abuse of modern urban building materials and morphological elements destroys the traditional rural cultural environment and natural style, and brings great disaster to rural society and ecology. This is far from the idea of systematically studying the natural laws of rural areas, analyzing the harmonious mechanism of environmental growth, discovering the human factors of rural evolution, and finally realizing rural modernization. The investment in rural construction in China is shown in Table 2.

Table 2. Rural construction investment in China from 2018 to 2020 (100 million yuan)

	2018	2019	2020
Investment in village construction	2573	2865	3267
Township Construction	2644	2795	3046

It can be seen from the above that in 2018, China invested 257.3 billion yuan in village construction and 264.4 billion yuan in township construction; in 2019, China invested 286.5 billion yuan in village construction and 279.5 billion yuan in township construction; in 2020, China invested 326.7 billion yuan in village construction and 304.6 billion yuan in township construction. As shown in Fig. 2.

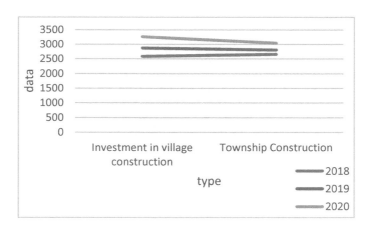

Fig. 2. Rural construction investment in China from 2018 to 2020 (100 million yuan)

It can be seen from the above that China's investment in village construction and township construction is an increasing trend. In 2020, China's investment in village construction will reach 326.7 billion yuan, and the investment in township construction will reach 304.6 billion yuan.

5   Conclusions

With the development of modern network technology and the increasing amount of information, the society has gradually entered the era of big data. In the big data environment, how to use big data to implement management strategy and how to strengthen the use of big data make the construction of "intelligent rural governance" become the focus of people's thinking. To promote the modernization of rural governance, we must innovate the way of embedding big data in rural governance, including building multi governance big data system and building embedded rural governance big data security system. On the basis of summarizing the theoretical research results of rural governance in China at the present stage, this paper analyzes the difficulties, current situation and path of big data application in rural governance in China, so as to provide useful reference for exploring and refining the ways suitable for rural social governance in China.

References

1. Sajid, A., Hussain, K., Shah, S., et al.: History and buffer rule based (forward chaining/data driven) intelligent system for storage level big data congestion handling in smart opportunistic network. J. Ambient. Intell. Humaniz. Comput. **10**(7), 2895–2905 (2018). https://doi.org/10.1007/s12652-018-1030-x
2. Spyratos, N., Sugibuchi, T.: HIFUN - a high level functional query language for big data analytics. J. Intell. Inf. Syst. **51**(3), 529–555 (2018). https://doi.org/10.1007/s10844-018-0495-6
3. Sun, H., Liu, Z., Wang, G., et al.: Intelligent analysis of medical big data based on deep learning. IEEE Access **PP**(99), 1 (2019)
4. Chergui, H., Verikoukis, C.: Big data for 5G intelligent network slicing management. IEEE Netw. **34**(4), 56–61 (2020)
5. Uddameri, V.: JAWRA: fifty-five years of sustained contributions for improved water resources management—ongoing decade (2015–2019). JAWRA J. Am. Water Resour. Assoc. **55**(6), 1367–1369 (2019)
6. Coyne, E.M., Coyne, J.G., Walker, K.B.: Big data information governance by accountants. Int. J. Account. Inf. Manag. **26**(3) (2018)
7. Chen, M., Ye, C., Lu, D., Sui, Y., Guo, S.: Cognition and construction of the theoretical connotations of new urbanization with Chinese characteristics. J. Geog. Sci. **29**(10), 1681–1698 (2019). https://doi.org/10.1007/s11442-019-1685-z
8. Westfall, J.M.: Rural health more than just big data. JAIDS J. Acquir. Immune Defic. Syndr. **74**(3), e84 (2017)
9. Arora-Jonsson, S.: The realm of freedom in new rural governance: micro-politics of democracy in Sweden. Geoforum **79**(FEB.), 58–69 (2017)
10. Min, Z., Qin, L.I.: Restoration and regeneration of traditional villages' culture under rural governance: a case study of Zhentou village in Ji'an county. J. Landsc. Res. **01**, 116–119 (2017)
11. Boyle, D.: We need better rural governance. Town Ctry. Plann. Q. Rev. Town Ctry. Plann. Assoc. **87**(8), 315–316 (2018)

Support Load of Tunnel Construction by Crossing Broken Fault and Water-Rich Geological Mine Method by Using MIDAS Software

Jingang Fang[✉]

Guangzhou City Construction College, Guangzhou, Guangdong, China
fangjingang@gzccc.edu.cn

Abstract. For tunnel engineering through faults or through faults and other bad geology, the comparison method of tunnel construction is adopted, and reasonable mining method construction is selected to analyze the load of the adopted mining method construction support structure. Using MIDAS software to analyze the internal force of the secondary lining of the mining method tunnel, and determine the surrounding rock pressure of the ultra-shallow, shallow, and deep buried tunnels, the verification results are that the structure of the single-track tunnel and the double-track tunnel can meet the design force requirements.

Keywords: Broken fault · Mining method · Supporting load

1 Project Overview and Difficulties

1.1 Project Overview Analysis

The section from Xili Railway Station to Shigu Station on Shenzhen Metro Line 13. The location of the project is between Xili Railway Station and Baoshi Road Station in Nanshan District, Shenzhen. After leaving Xili Railway Station, this section of the line went northward through Vanke Yuncheng Project Department, Dashi 1st Road, and finally entered Shigu Station. The route mainly passes through the urban roads of Shenzhen City and the planned plots of Vanke Cloud City. The section line structure is a single-layer three-span box structure and a single-hole single-line horseshoe-shaped tunnel. The length of the left line of the section is about 300 m, and the starting end of the left line (about 230 m) is parallel to the stop line and turn-back line. 70 m after the left line is the left line parallel to the parking line; the right line is about 300 m long and laid separately. The strata crossing the section are strongly weathered granite (soily), strongly weathered granite (mass), moderately weathered granite, lightly weathered granite, strongly weathered fault breccia, and developed boulders. According to regional geological data, the underground section of the site passes through a structural fault. The cave body is all rock strata through the strata, the rock mass around the fault is broken, the joints and fissures are very developed, and it has good connectivity. It is a water-rich zone of groundwater. When the tunnel is under construction, water inrush and water inrush are prone to occur.

© The Author(s), under exclusive license to Springer Nature Singapore Pte Ltd. 2022
J. C. Hung et al. (Eds.): FC 2021, LNEE 827, pp. 432–440, 2022.
https://doi.org/10.1007/978-981-16-8052-6_53

1.2   Impacts and Difficulties of Bad Geology on Projects

Tunnel engineering crossing faults or passing through faults and other bad geology will cause tunnel tunnel collapse, mud and water inrush, coal and gas outburst problems. In addition, the construction tunnel is located in a water-rich area, and measures must be taken to protect the groundwater from the flow of water, prevent surface settlement, protect the underground ecological environment, and unify the benefits of environmental protection, society, and economy [1].

2   Comparative Selection of Tunnel Construction

2.1   Comparison of Construction Methods

Shield tunneling or TBM construction is used for the left line of the interval, and nonstandard sections cannot be realized, and the interval is short, and the cost performance is not high. The starting end of the left line (approximately 230 m) is parallel to the stop line and the turn-back line. If the mining method is used, the span is large and the construction risk is high. In the 74-m section behind the left line, the under crossing the Dashi 1st Road construction with cut and cover method requires interruption of the road and the relocation of pipelines, which has a great social impact. After comprehensive comparison and selection, the starting end of the line (approximately 213 m) is parallel to the double wiring, and the cut and cover method is adopted for construction. The structure is a single-layer three-span box structure. The 85-m section after the left line is parallel to the left line and the single line. The mine method is adopted for construction. The lining structure is a single-hole double-line horseshoe-shaped structure. In the construction of the unde crossing the road, advanced support measures are adopted to strengthen monitoring to avoid affecting the safety of pipelines and driving [2].

The right line of the section is laid separately. If the shield method is used to cross faults, the upper soft section and the lower hard section are risky, and the tunneling efficiency of the shield machine in the hard rock section is not high. Moreover, the right line is about 300 m long, and the use of the shield method is not economical. If the cut and cover method is adopted, the Dashi 1st Road must be interrupted, which will have a great social impact. Comprehensive comparison and selection, the right line adopts the mining method to construct, crosses the fault section, adopts advanced support measures, and uses curtain grouting to reinforce the ground when necessary.

2.2   Determination of Tunnel Mining Method Construction Method

The mileage of the left line (69.987 m) is parallel to the left line and one wiring. The mining method is adopted, and the lining structure is a single-hole double-line horseshoe-shaped structure [2]. The right line tunnel is constructed by the mining method, and the structure is a single-hole single-line horseshoe-shaped tunnel [5].

Tunnel excavation construction method: Interval tunnels are single-hole single-line tunnels, which are excavated manually or mechanically [1]. The soil surrounding rock of the single-track tunnel is excavated in sections by circular steps, and the stone surrounding rock is excavated by steps. After excavation, see the stability of the tunnel face, if necessary, use spray concrete to seal the tunnel face and the bottom of the steps, and set longitudinal connection channel steel or lock foot anchors (pipes) at the arch feet to control the sinking [4] (see the Table 1).

Table 1. Tunnel construction method

Section type	Length	Surrounding rock grade	Section size (width*height)	Construction method
A-1	58	III	6.1 * 6.5	Step method
A-2	201.485	IV/V	6.5 * 6.9	Circular step method
A-3	40	V	6.5 * 6.9	Circular step method
B	73.94	III/IV/V	15.1 * 10.3	CRD method

Auxiliary construction measures: deep hole pre-grouting in the cave; advanced support of large pipe shed; advanced (grouting) small conduit pre-support; temporary sealing of the excavation face with sprayed concrete [6].

3 Analysis of the Loads of the Supporting Structure in the Construction of the Mining Method

3.1 Load Analysis

The loads acting on the underground structure are classified as shown in Table 2 for the classification of loads acting on the underground structure. When determining the value of the load, the changes that have occurred during the construction and service life should be considered. According to the current national standard "Building Structure Load Code" and the most unfavorable conditions that may occur in relevant codes, the combination coefficients for different load combinations are determined [3].

Table 2. Classification of loads acting on underground structures

Load type		Load name
Permanent load		Structural weight
		Formation pressure
		The pressure of facilities and buildings on the upper part of the structure and damage to the prism range
		Hydrostatic pressure and buoyancy
		Prestress
		Shrinkage and creep of concrete
		Equipment weight
		Formation resistance
Variable load	Basic variable load	Ground vehicle load and its dynamic effect
		Lateral earth pressure caused by ground vehicle load
		Metro vehicle load and its dynamic effect
		Crowd load
	Other variable loads	Influence of temperature change
		Construction load
Accidental load		Earthquake load
		Air defense load

Formation Pressure Combined with Engineering Analogy, Calculated and Determined According to Relevant Formulas.

The water pressure acting on the underground structure should be based on the changes in the groundwater level during the construction phase and the use phase, distinguish different surrounding rock conditions, and calculate it according to the hydrostatic pressure. The calculation of the hydrostatic pressure shall be based on the characteristics of the stratum in Shenzhen. The method of water and soil separate calculation and water and soil calculation shall be used in the construction stage, and the water and soil calculation shall be used in the long-term use stage.

The numerical value and distribution law of lateral stratum resistance and foundation reaction force should be determined according to the structural form and its deformation under load, construction method, backfilling and grouting conditions, stratum shape characteristics and other factors.

3.2 Vertical Load and Horizontal Load

Vertical load—According to the "Code for Design of Railway Tunnels", the pressure formula of loose surrounding rock is used [7].

Horizontal load—According to the relationship between the wall displacement and the ground during the stress process of the structure, it can be calculated according to active earth pressure, static earth pressure or passive earth pressure [8].

During the construction period, the horizontal pressure of the enclosure structure should be calculated according to Langkin's formula. According to the geological conditions, the method of water and soil separate calculation or water and soil calculation shall be adopted [9]. The main structure should be calculated according to the static earth pressure during the use stage, and the water and soil classification should be adopted [10].

The load combination sub-factors are shown in load combination sub-factors (see theTable 3).

Table 3. Load combination sub-factors

Serial number	Load combination check calculation condition	Permanent load	Variable load	Accidental load	
				Earthquake load	Air defense load
1	Strength calculation of basic composite members	1.35 (1.0)	1.4		
2	Component crack width calculation	1.0	1.0		
3	Component deformation calculation	1.0	1.0		
4	Checking calculation of member strength under seismic load	1.2 (1.0)	0.5 * 1.2	1.3	
5	Checking calculation of member strength under civil air defense load	1.2 (1.0)			1.0
6	Checking calculation of anti-floating stability of components	1.0			

4 Mine Method Construction Support Structure Model Calculation

4.1 Basic Assumptions

The internal force analysis of the secondary lining of the mining method tunnel is carried out with the load-structure model and the MIDAS software [9]. Only the main loads are considered, that is, ground building load (or ground overload), surrounding rock pressure and structural self-weight [11].

The surrounding rock pressure adopts the vertical uniform pressure and the horizontal uniform pressure. The specific process is as follows: firstly, it should be judged whether the tunnel is deeply buried, shallow buried, or super shallow buried. The surrounding rock pressure is calculated according to the formula provided by the "Code for Design of Railway Tunnels" and "Tunnels". When the buried depth of the tunnel is $h \geq 2.5$ ha, it is deep buried, when 2.5 ha $> h \geq$ ha, it is shallow buried, and when $h <$ ha, it is super shallow.

4.2 Determination of Surrounding Rock Pressure of Deep-Buried Tunnels

When calculating the lining of a deep-buried tunnel, the surrounding rock pressure is considered as loose pressure, and the vertical uniform pressure is

$$q = \gamma h a \tag{1}$$

$$h_a = 0.45 \times 2^{S-1} \omega$$

ω-Width influence coefficient, $\omega = 1 + i(B - 5)$

B-tunnel width (m)

i-the rate of increase or decrease of surrounding rock pressure when B increases or decreases by 1 m. When B < 5 m, take i = 0.2; when B > 5 m, take i = 0.1. The horizontal distribution pressure can be determined in accordance with the "Railway Tunnel Design Code" and the "Tunnel" specification.

4.3 Determination of Surrounding Rock Pressure for Shallow Tunnels (Design Code for Railway Tunnels, Appendix E)

Vertical pressure:

$$q = \gamma h (1 - \frac{\lambda h t g \theta}{B}) \tag{2}$$

$$\lambda = \frac{tg\beta - tg\varphi_c}{tg\beta[1 + tg\beta(tg\varphi_c - tg\theta) + tg\varphi_c tg\theta]}$$

$$tg\beta = tg\varphi_c + \sqrt{\frac{(tg^2\varphi_c + 1)tg\varphi_c}{tg\varphi_c - tg\theta}}$$

B-tunnel width (m)

γ-weight of surrounding rock (kN/m^3)

h-buried depth (m)

θ-the friction angle on both sides of the roof soil column (°), which is an empirical value

λ-side pressure coefficient

ψc-calculate the friction angle.

$$\text{Horizontal pressure: } ei = \gamma h i \lambda \tag{3}$$

4.4 Determination of Surrounding Rock Pressure of Ultra-Shallow Tunnel

$$\text{Vertical pressure: } q = \gamma h \tag{4}$$

$$\text{Horizontal pressure: } ei = \gamma h i \lambda \tag{5}$$

4.5   If the Overlying Rock Mass is Relatively Loose or is a Soil Layer, the Vertical Pressure on the Top of the Tunnel is

$$P_v = \frac{\gamma}{Ktg\phi}[b + h_t tg(45^o - \frac{\phi}{2})] \ (kPa) \tag{6}$$

If it is sandy soil, K = 1, then

$$P_v = \frac{\gamma}{tg\phi}[b + h_t tg(45^o - \frac{\phi}{2})] \tag{7}$$

The horizontal pressure at any point on the side wall is

$$P_h = (P_v + \gamma dh_t)tg^2(45^o - \frac{\phi}{2}) \tag{8}$$

5   Computational Verification Results Summary

5.1   Calculation Results for Single-Track Tunnels

Select YCK8+670 section for analysis and calculation. The section is buried at a depth of 15 m, the surrounding rock is fully weathered granite and strongly weathered mixed rock, and the surrounding rock is grade V. The calculation results are shown in Fig. 1 Tunnel Bending Moment (KN.m), Fig. 2 Tunnel Axial Force (KN), and Fig. 3 Tunnel Shear Force (KN).

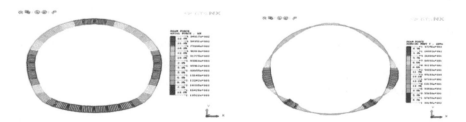

Fig. 1. Tunnel bending moment (KN.m)   **Fig. 2.** Tunnel axial force (KN)

Fig. 3. Tunnel shear force (KN)

From the data analysis of the calculation results in Fig. 1, Fig. 2, and Fig. 3, it can be seen that in the grade V surrounding rock section, the maximum positive bending moment of the secondary lining of the tunnel (tension on the inner side of the tunnel) is 137 kN·m, and the corresponding axial force It is 1100 kN; the maximum negative bending moment (tension on the outside of the tunnel) is −130 kN·m, and the corresponding axial force is 1185 kN. The amount of reinforcement for the secondary lining of the tunnel designed this time is $\varphi22@200$. After calculation, the structure can meet the design force requirements.

5.2   Calculation Results for Two-Lane Tunnel

Select YCK8+670 section for analysis and calculation. The section is buried at a depth of 15 m, the surrounding rock is fully weathered granite and strongly weathered mixed rock, and the surrounding rock is grade V. The calculation results are as follows.

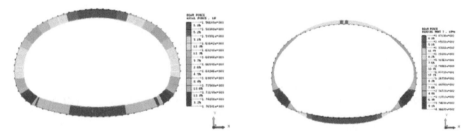

Fig. 4. Tunnel bending moment (KN.m)   **Fig. 5.** Tunnel axial force (KN)

Fig. 6. Tunnel shear force (KN)

From the data in the load calculation result graph in Fig. 4, 5 and 6, it can be seen that in the grade III surrounding rock section, the maximum positive bending moment of the secondary lining of the tunnel (tension on the inner side of the tunnel) is 426 kN·m, and the corresponding axial force is 1765 kN; The maximum negative bending moment (tension on the outside of the tunnel) is −307 kN·m, and the corresponding axial force is 1650 kN. In this design, the reinforcement of the secondary lining of the tunnel is $\varphi 25@200$, the steel bars at the arch feet are mutually anchored, and a row of steel bars are added at the invert. After calculation, the structure can meet the design force requirements.

Acknowledgments. This work was funded by the provincial key platform and scientific research project (natural science) of Guangdong universities and the (2018GKTSCX095) Guangzhou Science and Technology Plan Project (20202030310).

References

1. Shang, C.L.: Key techniques for construction of high pressure and water-rich faults in Qiyueshan tunnel of Yiwan railway. J. Tunnel Constr. **30**(03), 285–291 (2010)
2. Wang, D.M.: Research and application of the disaster mechanism of water bursting and mud bursting in tunnels in mud fault fracture zones. Degree Thesis of Shan Dong University (2017)
3. Li, D.X.: Analysis of composite control grouting reinforcement technology for tunnels in water-rich fault fracture zone. J. Sichuan Build. Mater. **42**(02), 109–111 (2016)
4. Zhu, H.T.: Qiyueshan tunnel lining water pressure characteristics and karst treatment technology. Degree Thesis of Beijing Jiaotong University (2011)
5. Yang, J.: Discussion on construction technology of high pressure and rich water fault in railway tunnel. J. Technol. Inf. (17), 106 (2011)
6. Zhao, J.: Research on treatment measures for structural compression fractured water-rich zone of Xiushan tunnel on Yumeng railway. J. Railw. Constr. (06), 54–57 (2015)
7. Zhang, L.S., Weng, X.J.: Treatment technology for water bursting and mud bursting in the water-rich fault fracture zone of Zhongjiashan tunnel. J. Jiangxi Build. Mater. (18), 178–179 (2016)
8. Zhang, M.Q., Sun, G.Q.: Study on methods and standards for inspection and evaluation of grouting effect on pingan water-rich faults. J. Railw. Eng. **26**(11), 50–55 (2009)
9. Xiao, W.: Curtain grouting construction technology for tunnel crossing the lower fault fracture zone of the reservoir. J. Railw. Constr. Technol. (08), 48–53 (2012)
10. Wen, W.Z.: Fault construction technology for the high pressure and rich water section of the north Tianshan tunnel of Jingyihuo railway. J. Railw. Stand. Design (06), 90–92 (2009)
11. Li, Y.Q.: Three-bench construction of soft surrounding rock with large cross-section in shallow buried section of tunnel. J. Enterp. Technol. Dev. **29**(01), 75–77 (2010)

Model Method of Intelligent Decision Support System Based on Machine Learning

Yiqiang Lai[(✉)]

South China Business College, Guangdong University of Foreign Studies,
Guangzhou 510545, Guangdong, China

Abstract. Intelligent decision support system is the inevitable product of the combination of real decision support system and modern artificial intelligence technology. It introduces and applies the traditional concept of mutual processing between information representation and actual knowledge in traditional artificial intelligence technology to the decision support system. It is unique one of the scientific research theoretical methods, extensive technological development and future prospects has successfully emerged, and it has become a technical hotspot and main technology development research direction of decision support system scientific research. This article aims to study the model method of intelligent decision support system based on machine learning, through the literature research method and quantitative research method to gain an in-depth understanding of the intelligent decision support system, introduce the decision tree into the intelligent decision support system, and make relevant decisions on this basis Experiments verify it. Experimental research shows that the RABDT algorithm is faster than the FP-growth algorithm. As the amount of data increases, the advantage of the algorithm is more obvious, and the calculation speed increases more. It proves the feasibility and practicability of the improved algorithm.

Keywords: Machine learning · Intelligent decision support system · Decision tree · FP-growth algorithm

1 Introduction

Machine learning computing methods have received more and more attention in various fields by virtue of their powerful computing capabilities that can perform automatic deep learning of complex data models, automatically extract complex computing models and make various intelligent management decisions on them. The effective solution of decision support problems in these fields provides a new way [1, 2]. Data analysis sample variables can be said to be used not to be regarded as or only to be used as a sample data analysis sample that reveals the interrelationship between two sample variables obtained by observer analysis [3, 4]. The main focus of research in machine learning is to conduct deep learning on the automatic collection of user data, identify the complex thinking patterns in it, and make more intelligent decisions on it [5, 6]. Therefore, how to effectively construct an efficient classification machine learning data model that exhibits strong general visual analysis capabilities and provide efficient,

© The Author(s), under exclusive license to Springer Nature Singapore Pte Ltd. 2022
J. C. Hung et al. (Eds.): FC 2021, LNEE 827, pp. 441–449, 2022.
https://doi.org/10.1007/978-981-16-8052-6_54

scientific and reasonable classification decision data support for classification decision-making methods is undoubtedly a big problem to be solved at present [7, 8].

The intelligent decision support system has made considerable research progress and has been widely used in many industrial fields. It has become a modern decision support system tool that is very impossible or lacking in operation and management in many different industries [9]. When the decision support system personnel have richer theoretical knowledge, strong professional knowledge and information processing ability, it is possible to provide more effective business decision management support to the major decision makers. The development, construction, and promotion of the use of process intelligent decision support systems can continuously enhance the comprehensive ability of system knowledge resource development, management and utilization, and improve the process intelligent management level of system decision-making [10].

Based on the structure and problems of the intelligent decision support system of machine learning, this paper designs the scheme of the intelligent decision support system based on machine learning, and implements the proposed decision tree classification algorithm in the implemented decision system platform, which further verifies the proposed decision tree classification algorithm. The effectiveness and practicality of the algorithm and design of the decision support system platform.

2   Research on the Model Method of Intelligent Decision Support System Based on Machine Learning

2.1   Research Methods

(1) Literature research method
    The literature research method refers to obtaining research materials by consulting books, newspapers, books, etc., and generating inspiration in the research. The advantage of doing so is that you can understand the development process of the research object from the source and understand the situation. By comparing the data, we have a more comprehensive understanding of the research objects.
(2) Quantitative analysis method
    Qualitative analysis is related to quantitative analysis. Quantitative analysis refers to the analysis of mathematical hypothesis determination, data collection, analysis and testing. Qualitative analysis refers to the process of conducting research through research and bibliographic analysis based on subjective understanding and qualitative analysis.

2.2   Basic Structure of Decision Support System

Auxiliary function decision support function: This is one of the main features of the decision support system. It refers to the advanced management information system management technology system that is gradually produced as the management technology information system develops to a higher level. It provides a virtual environment for enterprise decision makers to analyze decision-making problems, establish decision models, simulate enterprise decision-making processes and solutions, call various enterprise information technology resources and decision analysis technology tools, and help enterprise decision makers effectively improve enterprise decision analysis Level and quality of service. The structure of the decision support system is shown in Fig. 1:

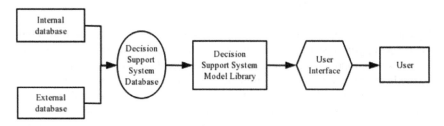

Fig. 1. Basic structure of decision support system

Internal information database: applications such as enterprise management internal information data system, transaction dynamic processing information system or knowledge base-based work management system, etc.

External information database: Collect relevant information from outside relevant organizations, such as other relevant organizations or corporate markets.

Decision support system model database: The database is a set of data collections of decision mathematics system models and decision analysis system models, including analysis history, summary, and analysis excerpts.

Decision support system model library: It is the database, model library and user interface for managing DSS to ensure that users can interact easily with the database and model library of the decision support system.

User interface: graphical and interactive.

2.3   Problems of Intelligent Decision Support System

Although the research of intelligent decision support system has made great progress, due to technical reasons, for most intelligent decision support application systems, some problems remain to be solved, such as:

(1) Vulnerability and difficulty in acquiring basic knowledge points are the main reasons why our traditional artificial intelligence decision support system is difficult to develop and apply quickly.

(2) Information closure, knowledge design and maintenance and knowledge update are difficult, that is, information system users can only make full use of local information resources; in addition, once the knowledge design work of the system is completed, it is more difficult to add local resources to users in the information system, because The information connection between the various information modules is relatively close, and a little change may directly affect the entire information system.

(3) The application of various intelligent reasoning process auxiliary analysis support for the entire traditional decision-making plan reasoning analysis process is low, because the current reasoning analysis for traditional description decision-making is completely procedural tradition. Decision analysis and reasoning process analysis and calculation methods cannot analyze the process results of the entire traditional decision analysis and reasoning according to various new descriptive decision reasoning analysis methods to provide a variety of intelligent auxiliary reasoning support.

(4) Operational flexibility and poor learning adaptability. When traditional intelligent decision support systems are provided with traditional intelligent decision support in the core components of the traditional intelligent decision support system, most of the learning adaptation behaviors are static and passive, rather than according to actual conditions. The demand in the learning environment makes a dynamic learning strategy by itself, and lacks an active learning adaptation mechanism.

(5) The system integration is poor. It is difficult to truly solve the system versatility in various aspects such as the mutual integration of different human-machine artificial intelligence, the mutual integration of different human-machine software operating systems, and the mutual integration of different artificial intelligence hardware technologies. The contradiction between system efficiency and system efficiency.

2.4 Intelligent Decision Support System Based on Machine Learning and Its Discussion

Machine language learning has been widely used in natural language semantic understanding, non-monotonic reasoning, machine learning vision, pattern recognition and many other fields, especially in enterprise knowledge management systems and enterprise decision management scientific research. Using the latest research results of artificial intelligence technology, the introduction of a variety of machine deep learning technologies into the artificial intelligence decision support system is for this main purpose. This needs to be achieved through the integration of the following ways:

(1) Through the establishment of a design decision-making problem model, it provides solutions for the automated construction, composite and manual reuse of design models, and establishes a set of highly adaptable design problem model solving strategies; the system's design model processing components will design. The calculation method that the model needs to call is combined with the encapsulation of various design data types required by the design to realize the structure of the design model.

(2) It is necessary to establish a set of decision-making knowledge system structure that can fully represent the observation and practical experience of the past few years, and support the establishment and comprehensive application of basic knowledge in the decision-making field. For some repetitive decision-making application scenarios, establish a set of relatively standardized. The decision-making example analysis template for solving the key steps directly helps to solve similar indirect problems in subsequent decision-making by using observation experience knowledge.

2.5 Optimization Algorithms for Machine Learning

Since the training parameters of the data model using machine intelligence learning are usually relatively large, and the training data is also large, we cannot find a 21-order deep optimization calculation method that is expensive to use model calculations, while first-order deep optimization calculations the model training data efficiency of the method is usually relatively low. In order to make full use of some mature and efficient local optimization research methods in convex local optimization research theories, many typical machine deep learning research methods will select the appropriate basic model and the lowest loss convex function model to require the solution of the convex function as a local optimization research. However, many basic models such as mathematical neural networks require non-convex local optimization research goals, and only require finding a locally optimized solution. In machine learning, the simplest and most commonly used optimization algorithm is the gradient descent method, that is, first initialize the parameter 0, set the risk function as the objective function, and then iteratively calculate the minimum value of the risk function on the entire training set. The formula is as follows:

$$\theta_{t+1} = \theta_t - \alpha \frac{\vartheta R(\theta)}{\vartheta \theta} \tag{1}$$

$$= \theta_t - \alpha \frac{1}{N} \sum_{n=1}^{N} \frac{\vartheta \mathcal{L}(y^n, f(x^n; \theta))}{\vartheta \theta} \tag{2}$$

Among them, θ_t is the parameter value at the t iteration, and α is the learning rate. This is the most primitive form of the gradient descent method, which is equivalent to approximately calculating the expected risk by the empirical risk of samples independently sampled from the real data distribution. This method is easy to calculate in parallel and can converge to the global optimal solution, but when training samples when the number is large, the computational overhead of each iteration is high, and the training process will be very slow. To solve this problem, the stochastic gradient descent method only calculates the gradient of a randomly sampled sample during each iteration update and uses it for parameter update. At the t iteration, the update method is as follows:

$$\theta_{t+1} = \theta_t - \alpha \frac{\vartheta \mathcal{L}(y^n, f(x^n; \theta))}{\vartheta \theta} \tag{3}$$

3 Experimental Research on Model Method of Intelligent Decision Support System Based on Machine Learning

3.1 Research Purpose

At present, association rules are widely used in intelligent decision support systems. However, as the complexity of data increases, it becomes more and more difficult for association rule algorithms to mine frequent itemsets, and multiple database scans have an impact on computer performance. Put forward higher requirements. In order to reduce the difficulty and the number of database scans, this paper studies and analyzes an improved method of association rules based on decision trees.

3.2   Data Collection

(1) The system automatically extracts relevant data used for data analysis from the theme database according to the needs of enterprise theme analysis. For this reason, in the process of secondary data extraction, it is necessary to perform secondary classification, summation, statistics and other comprehensive processing of various original data types. The processing process of extraction is actually the regeneration and organization of the original data.

(2) In the data processing process of the original data extraction, a purification of the existing database is completed, that is, the data extracted from the incomplete use of qualified original data must be completely removed, and these serious defects must be corrected when necessary. The original data system is optimized to replenish it.

(3) When you need to change the relevant subject of business analysis and decision-making, you can directly perform related database query and file access according to the subject's requirements.

(4) The system adopts a multi-level data storage structure model that supports offline coexistence, large-capacity disk storage, online shared disk capacity storage, and memory disk storage, to solve the problem of the huge amount of system data and the complex data organization structure divided according to data topics and granularity.

4   Instance Verification

According to the algorithm flow of the fast association rule algorithm based on decision tree.

Table 1. Example classification results

Instance	Included items
T1	I1, I2, I5
T2	I1, I2, I3
T3	I2, I3
T4	I1, I2, I3, I5
T5	I1, I2, I4
T6	I2, I3
T7	I1, I3
T8	I2, I4

Construct the decision tree of Table 1: First, create the root node of the tree, mark it with "Root", and create an empty table of the node index table. Then, scan the instance database row by row, and the items in each instance are arranged in the order of I1, I2, I3,`, In, create a branch for each instance, and update the count in the node index table. Until the scan is complete the entire database generates a decision tree and node index table. The details are as follows:

Create an empty table for the "Root" root node of the tree and the node index table. Then, scan the instances in the table in turn. The first instance contains items {I1, I2, I5}. Therefore, first add 1 to the support counts of items I1, I2, and I5 in the node index table, and then construct the first branch of the tree {(I1: 1), (I2:1), (I5:1)}. This branch has 3 nodes, among which I 1 is the child node of the root, I2 is linked to I1, and I5 is linked to I2.

Using the decision tree method described in this article to mine association rules has several important advantages: one is that the structure of the decision tree method is simple, and it can generate a rule that is easy for researchers and people to master; the other is the comparison of the work efficiency of the decision tree model. It is relatively high, and the computational data capacity is relatively small. It is more suitable for the case of large data capacity in the training set; third, the decision tree method generally does not require additional technical domain knowledge, and the self-organization ability is relatively strong.

When all the contents of a database based on the instance structure are added again and then restarted for data mining of all association structure rules, as long as the decision tree is read out from it and stored in other memory, then it can already be used for data mining. Data mining is carried out based on the content of all the association rule increments of the instance structure database. Let's verify it:

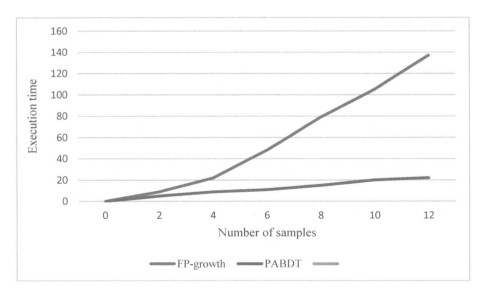

Fig. 2. Simulation diagram of algorithm execution time

Table 2. Comparison experiment of the execution time of the two algorithms

Experimental data	Number of instances	FP-growth algorithm execution time	RABDT algorithm execution time
Test.1 date	312	0.28	0.12
Test.2 date	4929	2.48	1.61
Test.3 date	42286	18.79	9.15
Test.4 date	124201	136.55	22.6

From the experimental results in Fig. 2 and Table 2, the RABDT algorithm is faster than the FP-growth algorithm. As the amount of data increases, the advantage of the algorithm is more obvious, and the computing speed increases more. It proves the feasibility and practicability of the improved algorithm.

5 Conclusions

With the development of science and technology, the application of machine learning and intelligent decision support systems has become more and more extensive, involving all aspects of life. The predecessors have also done a lot of related research and achieved great results. This paper proposes an intelligent decision support system based on machine learning, and verifies the characteristics and usability of the designed system framework through a practical use case, and performs algorithm design on the functional part of the system. Finally, the strength system is tested. The test results show that this article the availability and rationality of the system framework involved.

References

1. González Rodríguez, G., Gonzalez-Cava, J.M., Méndez Pérez, J.A.: An intelligent decision support system for production planning based on machine learning. J. Intell. Manuf. **31**(5), 1257–1273 (2019). https://doi.org/10.1007/s10845-019-01510-y
2. Zhu, A., Hao, D.: Research on decision support system of ship planning management based on data warehouse. J. Phys.: Conf. Ser. **1651**(1), 012082 (4pp) (2020)
3. Nie, X.: Research on economic function data and entrepreneurship analysis based on machine learning and computer interaction platform. J. Intell. Fuzzy Syst. **39**(4), 5635–5647 (2020)
4. Gharehbaghi, A., Lindén, M., Babic, A.: A decision support system for cardiac disease diagnosis based on machine learning methods. Stud. Health Technol. Inform. **235**, 43–47 (2017)
5. Nijeweme-D'Hollosy, W.O., Velsen, L.V., Poel, M., et al.: Evaluation of three machine learning models for self-referral decision support on low back pain in primary care. Int. J. Med. Inform. **110**(FEB.), 31–41 (2018)
6. Woldaregay, A.Z., Arsand, E., Walderhaug, S., et al.: Data-driven modeling and prediction of blood glucose dynamics: machine learning applications in type 1 diabetes. Artif. Intell. Med. **98**(Jul.), 109–134 (2019)

7. Singh, G., Vadera, M., Samavedham, L., et al.: Machine learning-based framework for multi-class diagnosis of neurodegenerative diseases: a study on parkinson's disease. IFAC Papersonline **49**(7), 990–995 (2016)
8. Doko, F., Kalajdziski, S., Mishkovski, I.: Credit risk model based on central bank credit registry data. J. Risk Financ. Manage. **14**(3), 138 (2021)
9. Yang, C.: Evaluation of maker space index system based on machine learning and intelligent interactive system. J. Intell. Fuzzy Syst. **39**(4), 5941–5952 (2020)
10. Chen, Y., Wu, B., Qi, Y.: Using machine learning to assess site suitability for afforestation with particular species. Forests **10**(9), 739 (2019)

The Application of Data Visualization in Economic Law Under the Background of Big Data

Manxia Huang[✉]

Wuhan Institute of Shipbuilding Technology, Wuhan, Hubei, China

Abstract. In recent years, the information resources of economic law formulated in order to solve the problems in the process of economic construction and development in China are innumerable. In the face of massive economic law data, it is necessary to strengthen the analysis and application of these data. And provide intuitive and easy to understand visual query results. This paper mainly discusses the application of data visualization in economic law under the background of big data. In the process of compiling this paper, this paper reviews and summarizes the relevant books and research results of economic law and the relevant laws of various countries, and summarizes the basic principles and basic system of economic law. When studying the application of data visualization under the background of big data, this paper collects the relevant research results of big data technology and information visualization technology, summarizes the technology related to big data and analyzes all kinds of visualization technology in detail. This paper summarizes their visualization methods and theoretical formulas. Through the research, it is found that in the process of visualizing the actual case data of economic law, in order to ensure the credibility of the data, more than 50% of the data originate from the government departments, and choose a variety of visualization methods according to the nature of the data itself.

Keywords: Big data · Economic law · Data visualization · Case analysis

1 Introduction

The enhancement of information sharing ability has caused the exponential growth of information, which has caused inconvenience to understand the relationship between various kinds of information and their characteristics. The use of visualization technology can show the relationship and characteristics of these massive information well [1, 2]. With the continuous improvement of the educational commitment of the masses, the demand for legal knowledge is increasing day by day, including economic law, because in the current era, economic production is still the primary task of various countries, especially China [3, 4]. The application of data visualization in economic law conforms to the need of judicial transparency in the era of big data and ensures the fairness of justice. It enriches the communication mode of economic law and meets the needs of the public to understand the law [5, 6].

J. C. Hung et al. (Eds.): FC 2021, LNEE 827, pp. 450–457, 2022.
https://doi.org/10.1007/978-981-16-8052-6_55

Because of the importance of economic law to various countries and the time factors of big data, many scholars have analyzed the integration between them, some of which have brought theoretical and technical support to this study. For example, Li Cai thinks that visualization technology can directly reflect the changing trend of space-time data and help to mine the of knowledge and law in data [7]. In order to study the main points of contemporary data visualization design in practical application, Wang Wenyi discusses the development of data visualization design to contemporary common forms and visual styles by analyzing many cases of visualization application. In addition to showing the significance and value of data itself, we should also attach importance to real-time data transmission and interactive experience between users and data visualization applications [8, 9].

This paper mainly introduces the application of data visualization in justice under the background of big data, with emphasis on the application in economic law. In this paper, the analysis technology of big data and the technology used in visualization are compared and discussed, the related concepts and characteristics of data visualization technology are introduced, and the overall design and execution flow and function module of some visualization platforms are analyzed. It provides some ideas for the development of judicial-based visualization system. Through consulting a large number of documents, this paper summarizes the basic principles and basic embodiment of economic law, based on the practical case data visualization example of economic law, and summarizes the development strategy of data visualization in economic law.

2 Research on Data Visualization Application in Economic Law under the Background of Big Data

2.1 Basic Principles and Systems of Economic Law

(1) Basic principles

Concept of basic principles of economic law refers to the guiding ideology and principle which runs through the whole process of economic law from legislation to use and serves as the basis of economic law rules [10]. Although many scholars have put forward their own views, some scholars have refuted these views with dialectical thinking, saying that these views cannot describe the basic principles of economic law comprehensively and succinctly. According to these studies, this paper puts forward the following two views:

1) Principle of appropriate intervention

Appropriate intervention refers to the effective but reasonable and prudent intervention of the state or economic autonomous group on the premise of full respect for economic autonomy, including two aspects, namely, legitimate intervention and prudent intervention. As a basic principle of economic law, proper intervention runs through the whole process of legislation and law enforcement of economic law.

2) Rational competition principles

Competition does not mean pure laissez-faire. Therefore, while playing a positive role in competition, we should also pay attention to controlling competition within a reasonable range and restrain and eliminate the negative role of unfair competition. In order to ensure that economic law can promote reasonable competition in the market, it is necessary to achieve orderly competition and effective competition.

(2) Basic system

The economic law system refers to a systematic and unified whole according to the current economic legal norms of a country according to different classification. In the current concept of economic law system, the most supported view is Li Changqi's viewpoint, which can be summarized as: market subject law, market order law, macro-control law, economic supervision law and so on.

2.2 Data Visualization

(1) Classification of data visualization

Data visualization is to show the connection and characteristics of target data in a certain way, so data visualization should include two aspects: the scientific visualization of interpreting structured data and the information visualization of solving unstructured data.

(2) Classification of data visualization techniques

1) Classification based on data objects

According to the data visualization technology, it can be divided into low-dimensional data, multidimensional data, hierarchical structure data, text or hypertext data. Among them, low-dimensional data usually refers to data with only one or two elements, and multidimensional data usually refers to data greater than or equal to three elements. Hierarchical structure data usually refers to data containing and containing relationships between each other. Visualization of text or hypertext data can be divided into two categories: visualization of a single document itself and visualization of a large set of documents.

2) Keim classification methods

Keim, all kinds of data visualization techniques are divided into three categories according to the standard of mutual orthogonal, that is, view deformation technology, visualization technology and interaction technology.

3) Card classification methods

Card and others divide the data visualization according to the different applications of information visualization technology, which are the visualization tools of information set, the visualization tools of data display model, the visual knowledge tools and the visualization tools of mining the representation of data objects.

According to the visual structure, the visual knowledge tools can be subdivided again. Typical visual structures include three-dimensional low-dimensional structures, three-dimensional multidimensional structures and tree structures.

2.3   Application of Data Visualization in Economic Law

(1) Issues

In the course of the practical case data visualization application research of economic law, the following problems are faced:

1) Data visualization talent shortage

Generally speaking, the authors in the judicial visualization platform should master more relevant professional knowledge so that they can interpret some behaviors in the judicial process to the masses. However, it is impossible for relevant departments to let all high-level personnel compile work, especially in the process of studying the visual application of practical case data of economic law. Although the number of economic law departments is large, but the high level of business is not many. Not only compilation staff shortage, data collection talent is also scarce. Judicial data visualization is based on a large number of data collection and analysis, but due to the limited professional level of data collectors, it is often impossible to collect all information comprehensively, and sometimes it may be affected by subjective factors. The authenticity of data needs to be considered, which leads to the effectiveness of data analysis.

2) Data monopoly

Governments, as the largest gatherers of raw information, sometimes force people to provide data without having to obtain it by persuasion or payment. But only a small fraction of the data are published, and most of the other data are eventually hidden for private or other reasons. Whether at home or abroad, it takes complicated procedures and time to obtain unpublished data from the government, and sometimes is rejected. The applicability of the data is the same for both government departments and private enterprises. Only through further analysis can the potential value of most of the data be released, but because the government is not completely public after obtaining the data, The value of the data is reduced.

3) Invasion of privacy and confidentiality

Under the background of big data, judicial visualization may involve the problem of personal privacy. In the process of judicial visualization, data is inevitably needed, and data is the key to visualization. Informing and licensing, ambiguity, anonymity, these three privacy protection strategies before the big data era is difficult to play its role, in the big data and network background, even if only a back, can quickly know the details of this back. In addition to personal privacy issues, trade secrets also face a certain degree of threat. In visual presentation, data maps are a good form of expression. However, this way of using maps may reveal the secrets of some technology research and development companies.

(2) Prospects

1) Promoting the Improvement of Economic Law

In the process of visual application of practical case data of economic law, it is found that many people have expressed their views on the actual case in the visualization platform. Some of these views are suggestions for the improvement of economic law, and some of them are worthy of reference.

2) Simultaneous development in terms of ease and depth
   There has been a debate in the field of data visualization, whether data visualization should make data easier to read or more complex. These two arguments represent the two directions of the development of data visualization.
3) Leading the masses to understand the new model of economic law
   In the process of making data visualization, whether it is data mining analysis or final visualization, its purpose is to convey information more accurately and to serve the audience's reading experience.

2.4   Interactive Adjoining Management Algorithm—Bezier Curves

Bezier curves are derived from polynomial mixed functions, usually defining a n polynomial n+ one vertex. Its mathematical expression is:

$$P(t) = \sum_{i=0}^{n} P_i B_{i,n}(t), \qquad t \in [0, 1] \tag{1}$$

Where P is the position vector of each vertex, $B_{i,n}(t)$ Bernstein function, defined as:

$$B_{i,n}(t) = \frac{n!}{i! \cdot (n-i)!} \cdot t^i \cdot (1-t)^{n-i}, \qquad t \in [0, 1] \tag{2}$$

3   Experimental Research on the Application of Data Visualization in Economic Law Under the Background of Big Data

3.1   Research Subjects

On the basis of the theory of economic law and data visualization under the background of big data, this paper takes data visualization as the research object and the practical case data visualization application of economic law as the investigation object. This paper explores the data sources and main presentation methods of data visualization in economic law.

3.2   Research Methodology

(1) Literature Comprehensive Law
   From the focus of the research, through the literature search and retrieval tools to consult the relevant monographs, papers and so on, comprehensive views, learn from the valuable content, and put forward their own views.
(2) Case study
   Taking the financial fraud case of Wanfusheng as a research case, this paper analyzes the application of data visualization in economic justice, analyzes its production concept, data source and presentation mode, and summarizes the

feasibility suggestions of data visualization in the process of making and implementing the law.

(3) Comparative analysis

By drawing lessons from the application of data visualization in other fields, it provides ideas for the application of data visualization in judicial field.

4 Experimental Research and Analysis on the Application of Data Visualization in Economic Law Under the Background of Big Data

4.1 Visual Data Source Analysis

For visualization technology, data is the core part, and the accuracy of data will affect whether the final visualization results are credible. Therefore, how to obtain data and the degree of credibility of data sources is also one of the concerns of data visualization. In the process of data visualization of financial fraud cases in Wanfu Department, the source and credibility of visual data are summarized, as shown in Table 1:

Table 1. Visual data source analysis

	Government	NGO	Enterprise	Network data
Percentage	55.48%	25.1%	14.01%	5.41%
Reliability (5 is the highest)	4.8	3.9	3.5	1.5

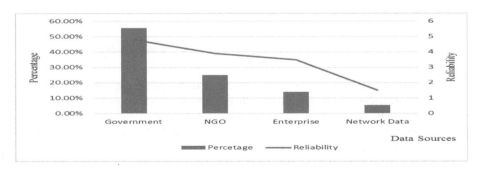

Fig. 1. Visual data source analysis

As shown in Fig. 1, the government accounts for half and a large proportion of all data sources, NGOs are also important data sources, accounting for more than 20% of all data. The rest of the data is basically from the enterprise network, which adds up to less than 20% of all data.

4.2   Presentation Analysis of Visualized Works

The ultimate purpose of data visualization is to spread the data with visual elements and present the massive data in the form of simple charts so that readers can understand more easily. Therefore, it is necessary to study the presentation form of visual works. In the process of data visualization of financial fraud cases in Wanfu Department, the following visualization methods are used and their readability is investigated, such as Table 2:

Table 2. Analysis of presentation form of visualized works

	Timeline	Coordinate system axis	Data map	Bubble chart
Use ratio	43.21%	8.48%	13.81	34.5%
Readability (5 highest)	4.5	1.6	2.5	3.1

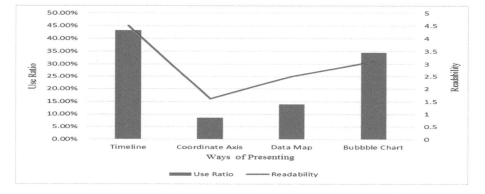

Fig. 2. Analysis of presentation form of visualized works

As shown in Fig. 2, in the research process of data visualization of financial fraud cases in Wanfusheng, more use of time line and bubble chart presentation, a small number of coordinate axis and data map presentation, and in the masses. The visual presentation of time lines is considered to be more readable. However, the presentation of data visualization is more determined by the nature of the data itself.

5   Conclusions

Under the background of big data, the visualization of a large number of judicial data presents a form of data vision after data integration and analysis. In the past, the spread of justice has often deterred audiences from lacking legal knowledge with intricate data and obscure legal terms. Nowadays, with the development of Internet of things, Internet of things and new media, information is pouring in, and human society is gradually entering a visual data age. Both legal professionals and ordinary audiences involved in the legal field are more eager to see concise, practical, vivid and novel judicial data, so the visualization of judicial data has practical needs, especially in the era of economic development as the focus. The visualization of economic judicial data meets the needs of the times. Although judicial data visualization is one of the trends of judicial development in the future, we need to pay attention to it, but we can not ignore the important development of other aspects of justice, such as making more perfect law is more important than anything.

References

1. Prommaharaj, P., Phithakkitnukoon, S., Demissie, M.G., et al.: Visualizing public transit system operation with GTFS data: a case study of Calgary, Canada. Heliyon 6(4), e03729 (2020)
2. Ruan, Z., Miao, Y., Pan, L., et al.: Visualization for big data security—a case study on KDD99cup data set. Digit. Commun. Netw. 3(4), 250–259 (2017)
3. Park, N.: Data protection in the TPP: more emphasis on the "use" than the "protection". In: Chaisse, J., Gao, H., Lo, C. (eds.) Paradigm Shift in International Economic Law Rule-Making. Economics, Law, and Institutions in Asia Pacific, pp. 363–370. Springer, Singapore (2017). https://doi.org/10.1007/978-981-10-6731-0_21
4. Yu, C.H., Reimer, D., Lee, A., Snijder, J.-P., Lee, H.S.: A triangulated and exploratory study of the relationships between secularization, religiosity, and social wellbeing. Soc. Indic. Res. 131(3), 1103–1119 (2016). https://doi.org/10.1007/s11205-016-1290-9
5. Mastropaolo, E.M.: Economic violence: which economy law instruments are more effective. Manage. Res.: Engl. Edn. 007(003), 214–221 (2019)
6. Cai, L., Zhou, Y., Liang, Y., He, J.: Research and application of GPS trajectory data visualization. Ann. Data Sci. 5(1), 43–57 (2017). https://doi.org/10.1007/s40745-017-0132-1
7. Brigdan, M., Hill, M.D., Jagdev, A., et al.: Novel interactive data visualization: exploration of the ESCAPE trial data. Stroke 49(1), 193–196 (2017)
8. Wang, W.: Application research of data visualization design. Design 032(007), 48–50 (2019)
9. Wong, J.C., et al.: Pilot study of a novel application for data visualization in type 1 diabetes. J. Diabetes Sci. Technol. 11(4), 800–807 (2017)
10. Brath, R., Banissi, E.: Using typography to expand the design space of data visualization. She Ji: J. Design Econ. Innov. 2(1), 59–87 (2016)

Design and Application of New Economic Power Index Based on New Energy and New Economy

Shanshan Wu[✉], Xiang Wang, and Lili Zhang

State Grid Energy Research Institute, Beijing, China

Abstract. This article starts from studying the basic concepts of the new economy, reviews relevant domestic and foreign research, and then designs the new economic power index from two parts: the new energy power development index and the power new economic development index. And use power data to analyze the current status and trends of the new economy. It is found that the overall trend of the New Economy Electricity Index is relatively stable, showing greater volatility under major events than NEI.

Keywords: New economy · Power index · New energy

1 Introduction

In the past 40 years of reform and opening up, after China has experienced rapid economic growth in the 20 years from 1991 to 2010, economic growth has begun to enter a bottleneck, from 10.4% in 2010 to 6.6% in 2018. Facing the pressure of economic growth, cultivating new kinetic energy and developing a new economy has undoubtedly become an inevitable choice. With the rapid development of the Internet, big data, artificial intelligence and other fields, China's economic system is in the process of transforming from a traditional economy to a new economy. In this context, this article starts from studying the basic concepts of the new economy, reviews related research, and then uses index design to reflect the current status and trends of the new economy with electricity data.

In the 1990s, the United States had about three views on the new economy. The first is the long-term growth perspective, which believed that strong productivity growth at that time could enable the US economy to grow at a high speed without the pressure of inflation. The second is the business cycle view, which believes that the short-term substitute relationship between inflation and unemployment has changed, and low unemployment and low inflation can coexist. The third is the new growth source viewpoint, which believes that the characteristics of the information age have led to the network economy, increasing returns, and knowledge spillover effects, which have changed the pattern of economic growth (Lin Ling (2000) [1]).

The research on the new economy in China in the early stage mainly analyzed and summarized the characteristics and causes of the new economy in the United States (Chen Baosen (1998) [2], Wu Jiapei (2000) [3], Fan Gang (2000) [4]). In recent years, Chinese scholars have interpreted more from the microscopic, structural and

J. C. Hung et al. (Eds.): FC 2021, LNEE 827, pp. 458–464, 2022.
https://doi.org/10.1007/978-981-16-8052-6_56

morphological perspectives. Bai Jinfu (2015) proposed that the new economy mainly refers to new industries, new services and new business formats based on modern information technology from the perspective of the "four new" economy [5]. Ma Jiantang (2016) believes that the new economy refers to the "five new" integration of new products, new services, new industries, new business formats, and new models created by the new round of technological and industrial revolutions [6]. Li Guojie (2017) believes that the new economy is essentially a transition from an industrial economy to an information economy (digital economy) [7]. Xu yunbao (2018) believes that the current China's new economy is supported by information technology, with big data, cloud computing, Internet of everything and artificial intelligence as basic means, and new energy, new manufacturing and new retail as important promoters, so as to realize the integrated development of primary, secondary and tertiary industries and breed new industries [8]. It can be seen that the key to understanding the new economy lies in what "new" is and where it is. At present, China's new economy is closely integrated with digital technology and focuses on institutional innovation and industrial upgrading.

Combining the connotation of the new economy at home and abroad, this article believes that the new economy refers to a new economic form. In different historical periods, the new economy has different connotations. The current new economy refers to a smart economy in which innovative knowledge dominates knowledge and creative industries become the leading industry.

2 New Economic Power Index Design

Considering that the research purpose of this article is to directly reflect the development trend of the new economy related to electricity through power data, and provide a new method for measuring the development of the new economy. Therefore, when constructing the index, research is carried out from two parts: the new energy power development index and the power new economic development index.

2.1 New Energy Power Development Index

New energy mainly refers to wind power and photovoltaics, which are significantly different from traditional energy sources such as thermal power, nuclear power, and hydropower due to their technical characteristics and social benefits. Not only has it become an important part of China's energy system, but it is also a concrete manifestation of the new economy in the energy field. Constructing a new energy power development index can keep abreast of China's new energy power generation infrastructure construction and the overall level of new energy power generation.

The New Energy Power Development Index is a comprehensive quantitative evaluation of China's wind power, solar and other new energy power development levels, covering three first-level indicators of development level, development effectiveness, and development potential. (1) The level of development is mainly used to reflect the market scale and market structure of the development of new energy power, in which indicators such as new energy power generation and the proportion of new energy power generation in the total power generation are used to measure the market size, the proportion of wind power generation in new energy generation are used to measure market structure. (2) The development effect reflects the resource allocation of new energy power development and the improvement of people's livelihood from the three aspects of market efficiency, stability and social welfare. Among them, indicators such as the cumulative average utilization hours of wind power equipment nationwide and the average transaction unit price of wind power are used to measure market efficiency, indicators such as transaction price volatility measure the stability of transactions, and the average salary indicators of wind power and photovoltaic industries measure social welfare. (3) The development potential is used to reflect the innovation driving ability of the new energy industry. The number of authorized new energy patents is selected as the second-level index under this first-level index. According to the availability of data, some three-level indicators have been modified and deleted.

2.2 Power New Economic Development Index

(1) Industry definition and adjustment

The Power New Economy Development Index reflects the development of new economy industries that are highly correlated with power from the perspective of power consumption. Based on the availability of electric power data, combined with the typical characteristics of new economy, such as high human capital investment, high-tech investment, light assets, sustainable and rapid growth, international experience of industrial upgrading and expert opinions, the new economy industry with high correlation and rapid development speed with electric power is selected. It covers industry, information transmission, software and information technology service industry, financial industry, leasing and business service industry, public service and management organization industry, 5 major industries and 14 small industries. It should be noted that due to the fourth revision of the national economic industry classification in 2017, the statistical caliber of various industries in the national economy has changed in the years after 2017, and relevant data adjustment is needed to unify the data caliber of 2017–2020 and make it comparable. By comparing the "Classification of National Economic Industries" before and after the 2017 reform, the industry classification in Chapter 4 was adjusted, and the original 14 small industries were merged and adjusted into 9 industries.

(2) Indicator construction

Industry Growth Index: Reflects the growth of electricity consumption in this industry. The electricity consumption growth index of this industry is obtained by comparing the electricity consumption of the industry in the current period to the previous period. The formula is as follows:

$$InG^i = \frac{Q_t^i}{Q_{t-1}^i} \times 100\% \tag{1}$$

Where Q_t^i represents the electricity consumption of industry i in the current period, Q_{t-1}^i represents the electricity consumption of industry i in the base period, and InG^i represents the electricity consumption growth index of industry i.

New economic growth index: reflects the growth of electricity consumption in the new economy industry, and the new economic growth index is obtained by weighted average of the electricity consumption growth indexes of various industries in the new economy. The formula is as follows:

$$InG^{new} = \sum_{i=1}^{n} \alpha_i \times \frac{Q_t^i}{Q_{t-1}^i} \times 100\% \tag{2}$$

Where Q_t^i represents the electricity consumption of industry i in the current period, Q_{t-1}^i represents the electricity consumption of industry i in the previous period, α_i represents the index weight of the industry i electricity consumption growth index, InG^{new} represents the new economic growth index, and $InG^{new} - 1$ represents the new economic growth rate year-on-year.

(3) Determination of weight

Taking into account the characteristics of the industry's electricity consumption indicators, the weight is directly measured by the proportion of the industry's electricity consumption in the new economy industry's electricity.

$$\alpha_i = \frac{Q_t^i}{\sum_{i=1}^{n} Q_t^i} \tag{3}$$

The index weight of the industry electricity consumption growth index is the ratio of industry i's electricity consumption to the new economy industry's electricity consumption. The industry with large electricity consumption has a larger weight, and the weight changes dynamically with time.

3 New Economy Power Index Application

3.1 Application and Analysis

(1) New energy power development index

In index synthesis, first standardize the index data, then synthesize three first-level indicators with equal weights, and finally use the average of the three first-level indicators as the new energy power development index. From the development trend of new energy power development index, the overall development of new energy power in 2017–2018 is relatively stable, and the new energy power development index continues to rise after August 2019, mainly due to the increase of new energy power generation and the improvement of new energy scientific research level. By March 2020, there will be a downward trend, mainly due to the effect and level of new energy development (Fig. 1).

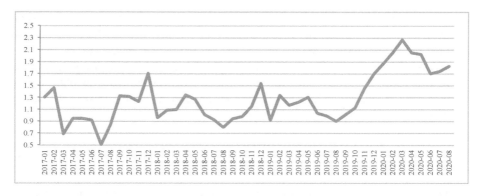

Fig. 1. New energy power development index calculation results

(2) Power New Economic Development Index

According to the above industry classification, the calculation of the new economic development index of electric power is carried out. It can be seen from Fig. 2 that from 2017 to the present, the overall development of the new power economy development index has been in a relatively stable trend. Among them, in 2017, 2018, 2019 and February of 2020, there was a significant decline in the new economic development index of electricity, mainly because of the slowdown in the growth of electricity consumption in various industries, which has a great relationship with the Spring Festival holiday. The largest decline in February 2020 is mainly due to the delay in resuming production and work under the epidemic, which has greatly affected the electricity consumption of the new economy industry. From the perspective of the development of each year, the power new economic development index shows the same cyclical trend, and the annual average of the power new economic development index in each year is about 1.02.

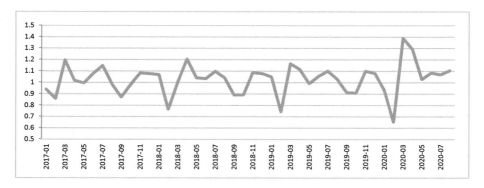

Fig. 2. Trend map of power new economy development index

From the perspective of weighting, among the big industries, industry will occupy the first place in the development of the new power economy from 2017 to 2020, followed by leasing and business services, residential services and other services and information transmission, computer services and software industries. Among the small industries, the electrical and electronic equipment manufacturing industry, metal products industry, and general and special equipment manufacturing industries have higher weights in the composition of the new power economy development index, indicating that they have made greater contributions to the development of the new power economy at this stage. The cumulative weight of these three industries exceeds 60%, which is about twice the cumulative weight of the other six industries.

(3) New economic power index

The new energy power development index and the mean value of the power new economic development index are regarded as the new economic power index. It can be seen from Fig. 3 that the overall trend of the new economic power index is relatively stable, and there has been a relatively obvious rise after 2020, mainly due to the impact of the development of new energy power, and the improvement of the development level and development potential of new energy power.

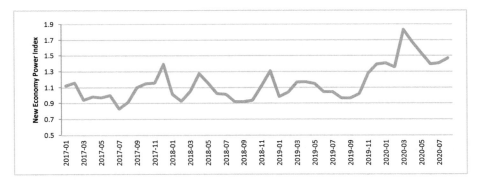

Fig. 3. Calculation results of the new economy power index

3.2   Comparative Analysis with Other New Economic Indexes

In order to verify the applicability of this power index, compare it with other new economic indexes. The most authoritative one is the China New Economy Index (hereinafter referred to as NEI) developed by Caixin Think Tank and Shulian Mingpin (BBD). NEI estimates the importance of the new economy in the entire economy, that is, when the Chinese economy generates a dollar of output, how much comes from the new economy. The changing trend of NEI reflects the degree of activity of the new economy relative to the traditional economy, and is an important indicator for judging the growth and decline of the new and old economies in the process of China's economic transformation.

Plot the new economy power index and the NEI index in the figure below, you can find that the overall trend of the new economy power index and the NEI index are roughly the same. Compared with NEI, the New Economy Electricity Index has shown greater changes. This may be because the changes in the new economy industry are more sensitive and direct to the impact of the new economic development index of power, while NEI more reflects the proportion of the new economy. And judging from the changes in the index in 2020, the impact of the new economic power index under major events has shown greater fluctuations (Fig. 4).

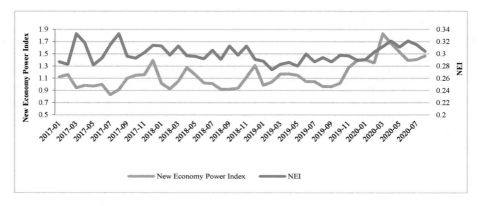

Fig. 4. Comparison of new economic power index and Nei

Acknowledgments. This work was financially supported by the State Grid Science & Technology Project No. SGNY0000GXJS2100065 (Research on bidirectional data value mining technology and typical application of "power economy" in SGCC).

References

1. Lin, L.: Questioning the New Economic Theory of the United States. J. World Econ. **23**(05), 38–45 (2000)
2. Chen, B.: On America's "New Economy". J. World Econ. (06), 3–5 (1998)
3. Wu, J.: Economic globalization and new economy. Mark. Econ. Res. (06), 13–14 (2000)
4. Fan, G.: "New economy" and China—also on the traditional industry in China is still promising. Knowl. Econ. (11), 32–34 (2000)
5. Bai, J.: Lead the new normal with the new economy. Only the new economy can strengthen China. China Econ. Weekly (14), 22–23 (2015)
6. Ma, J.: Accelerate the development of the new economy and cultivate new growth momentum. Adm. Manage. Reform (09), 4–6 (2016)
7. Li, G., Xu, Z.: Looking at the new economy from the development trend of information technology. Chin. Acad. Sci. **32**(03), 233–238 (2017)
8. Xu, Y., Zeng, G.: Exploration on the innovation path of China's creative industry format under the big data strategy – based on the perspective of new economic connotation evolution. Theor. Discuss. (06), 108–114 (2018)

Analysis on the Frame Model of Poor Student Identification and Funding Management Based on Financial Information Sharing System

Haiyan Guo[(⊠)] and Zhibin Jia

Xi'an Urban Architectural College, Xi'an 710114, Shaanxi, China

Abstract. In our country's major colleges and universities, the special group of poor students exists widely. How to effectively solve their economic difficulties and make them complete their studies smoothly, so as to ensure the stability of colleges and even the whole society, has become a country, society and society. The focus of the school's attention. This article first studies the financial information sharing system, and uses computer technology to design the system; then it designs the design of the overall function of the management system for the evaluation of poor students and the design of the public system database. Finally, the basic functional modules of the system are designed and implemented, and simple functions and performance tests are performed on the system. This study on the identification and funding management system of poor students in this article can not only provide experience to the staff related to the management of poor students, but also provide help for the identification and funding of poor students in China in the future.

Keywords: Financial information sharing · Identification system · Management framework · Case model

1 Introduction

In the process of achieving strategic goals for enterprises, how to solve financial dilemmas and how to transform financial management models are very important research directions. Innovative research on financial management is of great significance [1]. As we all know, big data, Internet, artificial intelligence and other technologies have been applied in various fields of society today. In terms of economy, society has entered the era of network economy. Most European and American companies have established shared service systems, especially in financial shared services. The most important thing for financial shared services is to have a fully functional financial sharing system [2]. Nowadays, resources are becoming more and more optimized. Under the challenge of many problems, researchers from various large enterprises have been trying to create a high-level financial sharing system. The financial sharing system is a technical concept that emerged in the context of economic globalization. It transforms computerized accounting into financial integration, allowing the system to automatically complete basic and repetitive accounting tasks. All the staff need to do is simple verification and confirm the work [3].

© The Author(s), under exclusive license to Springer Nature Singapore Pte Ltd. 2022
J. C. Hung et al. (Eds.): FC 2021, LNEE 827, pp. 465–472, 2022.
https://doi.org/10.1007/978-981-16-8052-6_57

Chinese scholar Wu Middle School believes that financial information disclosure and information sharing have been included in the matter of government information disclosure. As a unit of an economic department, colleges and universities need to accelerate the improvement of the quality and efficiency of financial information disclosure in colleges and universities. Logistic services and other business-related economic activities use the Internet and modern information technology to innovate management models, and establish a highly unified data management database with information classification and standards [4]. Xiong Hang believes that from the perspective of today's multinational companies or enterprises, and based on the informatization of management accounting, the in-depth promotion of the establishment of financial shared services can reduce production costs to a large extent, thereby greatly improving production efficiency. It can help the management to strengthen the control of the enterprise, and it can also realize the efficient allocation of resources in all aspects of the enterprise and obtain greater profits [5]. Liu Wei believes that in recent years, my country's economy has developed rapidly, and science and technology and network information technology have also made certain progress. Under such an environment, it not only provides opportunities for enterprise development, but also brings more challenges. The issue of financial sharing has become a key topic of research by enterprises. The application of the corporate financial sharing management model based on the information environment is not only the need for corporate management innovation, but also the key to the construction of corporate financial management information [6].

In our country's major colleges and universities, the number of students enrolled and the cost of learning are increasing every year, leading to the emergence of more impoverished students with insufficient family financial ability. Providing assistance to students with insufficient financial ability has become the education industry. An important question of for the development of education science in our country, it is necessary to do a good job in supporting the poor students, which is also an important means to serve the people and maintain the results of education. This paper designs a financial information sharing system through the use of related technologies, and applies the system to the construction of the identification of poor students and the construction of the funding management framework model.

2 Method of Analyzing Poor Student Identification and Funding Management Framework Model Based on Financial Information Sharing System

2.1 Related Technologies Used

SHL system: SHL system is an integrated process management system. It is an electronic process system based on V3 workflow engine software development, using database and C language technology. The platform has been redesigned and packaged at the bottom of V3 to build the platform. The workflow platform adopts the SBM architecture, which is implemented in layers and has good reusability; it adopts distributed growth and has good scalability. The service interfaces are smoothly matched and have good scalability; at the same time, based on the graphic designer provided by the engine, various work and approval procedures can be quickly implemented on the platform. In order to handle the approval process, SHL provides many auxiliary functions similar to OA, which can be effectively integrated with external systems in a variety of ways. KOP system: KOP system is the image data center [7]. The main function of the KOP system is to send the image of the account during the financial distribution process, transmit the image data to SHL and display its actual location in real time. It is deployed on the NET platform and exists in the unified database with SHL. The KOP system involves the development of interfaces: high-speed scanner interface, SHL system interface for image transmission. For technical applications, the scanner is called through the web adder, and the web service is used for communication, and the image data is transmitted to the SHL system. Separate business receipts, create corresponding one-dimensional barcodes, and collect those using CODE codes. Generate the account index number in the simulated account, and logically associate the online process with the simulated account through the barcode and index number [8].

2.2   Design of the Financial Sharing System

The basic mode of the financial sharing process in the financial sharing system is that students fill in the electronic application form in the SHL system. Basic data such as the student's name, class, home phone number, and bank account information are automatically obtained by the financial sharing system from the database of each subsystem. Students only need to fill in simple financial information, complete the electronic application form and click submit. The application form is automatically transferred to the student's teacher agent. After the teachers of each grade approve the approval by means of digital signatures, the electronic application form is transferred to the financial sharing center. After the accountant of the financial sharing center reviews the form and confirms the accounting voucher automatically generated by the system, the electronic application form flows to the cashier. After the cashier of the financial sharing center further reviews the form information, the submission process goes to the cashier review link, and the payment information is confirmed in the bank-enterprise direct connection system [9]. If it is a student reimbursement business, after the process is over, the bank will transfer the approved financial amount actually reimbursed by the student to the student's bank card within 1 to 3 days. In the process of using the financial sharing process, data support and interaction of multiple systems are involved, mainly task information and banking information in the SAP system, image information in the KOP system, student information in the student system, and organizational structure information. The interaction between the SHL workflow platform and the SAP system is realized through the interface service layer, and the master data synchronization between the two systems is realized through SSIS [10]. Both the SHL system and the KOP system adopt the BS architecture, provide a rich service structure for the system in a unified manner, and achieve effective integration with the school's own student system, mail system, and short message system. The server-side code is developed on the server based on Windows Server2003, and all use relational database.

2.3 Database Design

The database design of the financial sharing system takes the SHL system as the main body, and the image function as the auxiliary. The image is transmitted to the SHL system to realize the integration of the image and the electronic approval process. Through the interface of SHL system and SAP system, the financial accounting processing function is realized. According to the design requirements, the database of the SHL system is divided into 5 basic database logic levels in complex application scenarios with various transaction types, allowing the DBA to design redundant data structures to improve the performance of the database. For the database design of process forms, considering the diversity of form logic. Data storage faces huge challenges. In order to solve this problem, the SHL system does not store business data in a way that corresponds to the traditional form and relational database table, but stores the corresponding control of each business data and the serialized key value, which avoids the process of The problem of infinitely expanding relational data tables with increasing complexity has emerged, adapting to the ever-changing business logic. The back-end database deployment of the SHL system adopts a cluster mode, that is, multiple DB Servers can run independently, but share storage, perform hot backup of the storage medium, and regularly store it in different places. When a DB Server is stopped for some reason, the other server can be started immediately, but the front-end users will not notice it. Hot backup and off-site storage of storage media can cope with various emergencies and disasters as much as possible, minimize system losses, and restore system operation as soon as possible. These deployment strategies ensure the SHL system's requirements for database failure recovery and high availability.

3   The Construction of Poor Students' Identification and Funding Management Framework Model

3.1 Build Factor Set

Regarding the identification of poor students, the basic assessment is to assess each indicator of the student's family economy or poverty. Therefore, it is necessary to create an urgent factor about the economic situation of the student's family. This set of factors mainly includes indicators and weights. The calculation method is as follows:

$$A1 = \{a_1, a_2, a_3, \ldots .a_n\} \tag{1}$$

Among them, ak(k = 1, 2, ... n) represents the weight of the index, and satisfies $\sum_{i=1}^{n} a_i = 1$.

Each text is represented by a vector space model, expressed as (W1, W2, …, Wn) as the sample to be tested, and Ci is a given category; what is obtained is the objective function:

$$C_{MAP} = \arg \max \ P(C_i d) \tag{2}$$

3.2   Construction of Index System for Poor Student Identification System

According to the established index factor set, after performing related calculations, you can know the evaluation conclusion for the students. In this system, the first step is to analyze some related problems such as opacity, irregularity, and lack of comparison in the process of assessing the poverty level of students in the past. Then, based on this, a poor student identification index system was created, and the related functions of the system were constructed to form an evaluation model to realize the evaluation of the comprehensive index system for poor students, compared with the previous evaluation work. It is more accurate, transparent, scientific and easy to operate. After the system model is created, the data needs to be processed in a scientific and standardized manner, especially the index data in it, which needs to be obtained through certain channels. This article adopts the form of online questionnaire survey to conduct online questionnaire surveys on certain schools, including various types of schools, so that more accurate and reliable indicator data can be obtained.

3.3   Modeling the Evaluation Hierarchy of Students with Family Financial Difficulties

After obtaining the indicator system for student poverty identification, the evaluation level of student poverty identification can be established, and the evaluation level of students with family financial difficulties can be modeled according to the mathematical model designed above. In the process of modeling, the indicator system is divided into There are three levels, and the three levels are target W, first-level index Ui, and second-level index Vi. In the above three levels, there are 4 first-level indicators collected, and different first-level indicators are combined into a first-level indicator set, which is expressed as U = {U1, U2, U3, U4}, Ui (i = 1, 2, 3, 4). Each first-level index can be further divided into a second-level index. For the second-level index Vij, it is expressed as V = {Vi1, Vi2, Vi3, Vi4}. For the regional evaluation of student poverty identification, the value can be expressed as S(S = 1, 2......q), W(s) represents the comprehensive evaluation value of the s-th assessee.

4   Data Analysis

4.1   The Numerical Value of the Fractional Segment is Shown in Table 1 and Fig. 1

Analyze the indicators and systems of student poverty determination. The most important thing in the determination process is the construction of the indicator system

Table 1. Percentage of fractional segments

Poverty level	Fractional segment	Percentage
Very poor	8–10	10%
General poverty	5–8	30%
Not poor	0–5	60%

and weight model. Therefore, it is very important for the construction of the indicator system. It is necessary to build a more scientific and reasonable model to complete the evaluation of the indicators. Design and use, and apply the constructed poor student judgment index model to specific applications. According to Table 1 and Fig. 1, it can be concluded that the proportion of students from very poor families is 10%, the proportion of generally poor student families is 30%, and the proportion of non-poor student families is 60%.

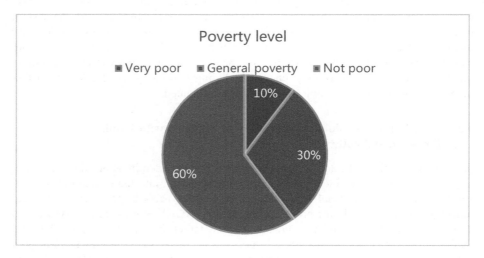

Fig. 1. Percentage of fractional segments

4.2 The First-Level Indicator Economic Resource Indicators Are Evaluated and Analyzed, and Each of the Sub-projects is Evaluated. The Results Are Shown in Table 2 and Fig. 2

Table 2. Evaluation of economic indicators

	Very poor	Very poor	Not poor
Source of income	5	8	3
Fixed assets	3	5	7
Expenditure item	2	4	5

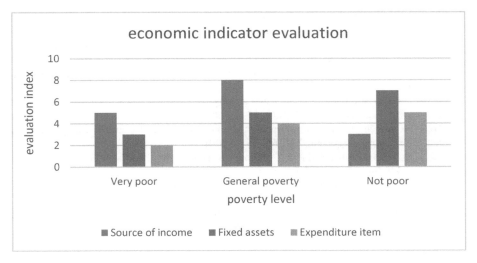

Fig. 2. Evaluation of economic indicators

In the above-mentioned judgment vector data, as a reference standard and vector, it is necessary to make a comprehensive evaluation according to different students and compare and reference with the judgment vector, and finally get the result. According to Table 2 and Fig. 2, compared with poor and non-poor families, student families with very poor families have fewer sources of income, the least fixed assets, and the least expenditure items.

5   Conclusions

This article is mainly to design a framework model for the identification of student poverty. In the process of design, for the problems and deficiencies in the traditional evaluation system, the system is digitized and standardized, so as to solve the opacity and intransparency of the evaluation process. Issues such as specifications. The first is to create factor sets and judgment sets, calculate these sets, and then establish a poor student identification indicator system based on the calculated results. Finally, conduct a case study of this indicator system and use relevant algorithms to further confirm the feasibility of the system Sex and effectiveness.

References

1. Liang, B., Zhang, X.: Thoughts on the influence of financial information sharing on accounting profession and subject education. J. Shanxi Financ. Taxat. Coll. **017**(001), 71–73 (2015)
2. Fu, G.: Effective implementation of financial accounting informatization under financial information sharing services. Contemp. Account. **97**(13), 76–77 (2020)

3. Xin, X.: Optimization and implementation of financial information sharing process for commercial circulation enterprises. Bus. Econ. Res. **759**(20), 108–110 (2018)
4. Wu, Z.: Research on the financial information sharing service mode of colleges and universities. Financ. Econ. (Acad. Edn.) **504**(03), 124 (2019)
5. Xiong, H.: From the perspective of financial shared services. Resour. Regener. **191**(06), 39–42 (2018)
6. Liu, W.: Research on corporate financial sharing management mode under the information environment. Account. Study **18**(126), 43+45 (2015)
7. Luo, P.: Research on financial information sharing and financial management optimization. Chin. Certif. Public Account. **232**(09), 95–98 (2018)
8. Zhang, X.: Financial information sharing in equipment manufacturing. Chin. Agric. Account. **345**(04), 10–12 (2020)
9. Lu, Y., Gong, L.: Research on hospital financial information sharing under big data. Econ. Res. Guide **419**(33), 114–115 (2019)
10. Zhao, H.: Discussion on the construction of Shougang group's financial information and financial sharing center. Taxation **13**(09), 45–46+49 (2019)

A Method for Measuring and Calculating Low Temperature Air Humiture

Huan Zhang[✉], Cheng Zhang, Jing Huang, and Shouchuan Wang

Hefei General Machinery Research Institute, Hefei, Anhui, China
zhanghuan86@keyaninfo.com

Abstract. In the performance test of refrigeration and air-conditioning products, the measurement of air humiture in a wide temperature range is often involved. Usually measurement with wet-and-dry-bulb thermometer is used to measure air humidity at room temperature, and moisture-sensitive humidity sensor method is used to measure air humidity at low temperature. However, they all have measurement flaws. With the continuous development of society, various types of low-temperature heat pump units and other new products have been developed and applied. The humidity parameter at low temperature also needs to pass the test by using the air enthalpy difference method to join in the calculation of the basic performance of the product. The existing moisture-sensitive humidity sensor has poor stability and is difficult to meet the accuracy requirements of low-temperature humidity measurement, while the cost of high-precision dew point temperature sensor increases multiples, and the device is complicated to achieve accurate measurement and control of the dew surface temperature. Therefore, there is an urgent need for a new type of measurement device to achieve accurate and stable measurement of low-temperature air humidity, and then achieve a high-precision measurement target of air temperature and humidity in a wide temperature range. In view of the shortcomings of the existing traditional humidity measurement methods, this application aims to propose a new air humidity measurement method to meet the requirements for wet bulb temperature measurement of −30 °C to 80 °C.

Keywords: Low temperature air · Humiture · Measurement · Performance test

1 Introduction

Air humidity is used to characterize the water vapor content in the air or the humidity of the air. It is a very important physical quantity in the air condition parameter that is difficult to accurately measure. Currently, air humidity measurement methods mainly include the following: First, using fiber, hair and other substances that expand and contract with the change of air humidity. It has the advantages of simple and low cost. However, it is poor at measurement accuracy [1]. Second, measurement with wet-and-dry-bulb thermometer. Its advantage lies in high accuracy. When the air temperature is 20–30 °C and the measurement error of the dry and wet bulb temperature does not exceed 0.1 °C, the wet and dry meter with constant ventilation device can be regarded as the second-class standard for humidity measurement, and the measurement error won't be more than ±2%RH (relative humidity) [2]. However, since the saturated vapor pressure of water is an exponential function of temperature, the saturated vapor pressure at low temperature is much lower than the corresponding data at room temperature, resulting in the extremely low measurement resolution of the dry and wet bulb method under low temperature conditions, and its measurement scope is 5–40°C generally. Third, dew point temperature method. This method measures the humidity indirectly by measuring the dew point temperature of the air, which is mainly composed of a cooling device, a dew surface and a temperature sensor [3]. This method has high measurement accuracy under low temperature and low humidity conditions, but the device is complicated, the cost is high, and it is difficult to achieve precise control of the condensation surface temperature. Fourth, the moisture-sensitive humidity sensor method. It has fast response, low hysteresis, and wide measurement range. The nominal applicable range is −40 °C−80 °C. The measurement accuracy of the middle and low humidity section is ±2%RH, and the measurement accuracy of the high humidity section is ±4%RH [4]. However, the third-party test results show that it is still very difficult to achieve this accuracy even in a ventilated state of 20–30 °C, and the measurement error in the low temperature section and high humidity section is even more difficult to guarantee. In addition, because the humidity sensor needs to be exposed to the environment to be tested for a long time, it is easy to be polluted and affect its accuracy and stability [5].

In the performance test of refrigeration and air-conditioning products, the measurement of air temperature and humidity in low temperature and wide temperature range is often involved. Usually measurement with wet-and-dry-bulb thermometer is used to measure air humidity at room temperature, and the moisture-sensitive humidity sensor method is used to measure air humidity at low temperature. Because the moisture-sensitive humidity sensor method test cannot meet the measurement accuracy problem, the air humidity parameter test by the method at low temperature is only used as the operating condition parameter, not as the cooling and heating capacity calculation.

2 The New Air Humidity Measurement Method

2.1 Measuring Principle

Aiming at the shortcomings of various existing measurement methods, this paper innovatively designs a new air humidity measurement method, which is mainly used to meet the needs of wet bulb temperature measurement of −30 °C−80 °C. This method innovatively uses the reheat constant humidity method to heat the measured air at low temperature, measures the air wet bulb parameters after heating, so as to realize the calculation and determination of the measured air humidity at low temperature by combining the principle of constant absolute humidity during the isohumid heating process and the parameters mentioned above. The system principle of this measurement method is shown in Fig. 1.

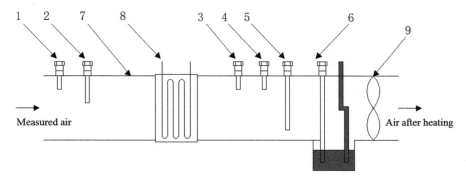

Fig. 1. Principle diagram of measuring low temperature air humidity by reheating and constant humidity method

1-The pressure measure point of the measured air; 2-The dry bulb temperature measure point of the measured air; 3-The air velocity measure point after heating; 4-The air pressure measure point after heating; 5-The measure point of air dry bulb temperature after heating; 6-The measure point of air wet bulb temperature after heating; 7-Measuring air duct; 8-Adjustable electric heater; 9-Frequency conversion fan.

2.2 Measurement Procedure

The measurement of the measured air humidity at low temperature and the measurement of other state parameters is achieved through the following technical solutions. The method for measuring temperature and humidity of low-temperature air, which is characterized in that the low-temperature air is heated and measured by the reheating and humidity-constant method, and the measurement steps are as follows:

Step 1: Set the pressure measure point of the measured air pressure at the inlet 1 and the dry bulb temperature measure point of the measured air 2; The adjustable electric heater 8 is arranged on the measuring air duct 7. The arrangement of the adjustable electric heater 8 should not affect the measurement of the low temperature measured air parameters and the heated air parameters; Near the outlet, set the air velocity measure point after heating near the outlet 3, the air pressure measure point after heating 4, the measure point of air dry bulb temperature after heating 5, and the measure point of air wet bulb temperature after heating 6; Set an frequency conversion fan at the outlet of the heated air 9.

Step 2: Adjust and control the adjustable electric heater 8 according to the temperature of the low-temperature measured air to realize the iso-humid heating process of the low-temperature measured air, so that the heated air parameters can reach the wet bulb temperature range measurement of the dry and wet bulb temperature sensors. At the same time, frequency conversion fan 9 is used to adjust and control the wind speed, so that the air flow rate at the measure point of air wet bulb temperature after heating 6 is within the range of 5 m/s ± 1 m/s.

Step 3: Measure the measured air pressure p1 and the measured air dry bulb temperature t1 through the pressure measure point of the measured air 1 and the dry bulb temperature measure point of the measured air 2; Through the air pressure measure point after heating 4, the measure point of air dry bulb temperature after heating 5, the measure point of air wet bulb temperature after heating 6 and the matching wet bulb water cup and wet bulb gauze, etc., the heated air pressure p2, the heated air dry bulb temperature t2 and heating After air wet bulb temperature t2′ can be measured.

Step 4: Calculate the absolute humidity x2 by the heated air parameters p2, t2 and t2′.

$$x_2 = \frac{(2501 - 2.340t_2')x_{s2} - 1.005(t_2 - t_2')}{1.846t_2 + 2501 - 4.186t_2'} \tag{1}$$

$$x_{s2} = 0.6220p_2'/(p_2 - p_2') \tag{2}$$

$$p_2' = 10^{B2} \tag{3}$$

$$B_2 = -7.90298(A_2 - 1) + 5.02808\log A_2 - 1.3816 \times 10^{-7}\left\{10^{11.344(1-1/A_2)} - 1\right\} \\ + 8.1328 \times 10^{-3}\left\{10^{-3.48149(A_2-1)} - 1\right\} + \log 101325 \tag{4}$$

$$A_2 = 373.15/(273.15 + t_2') \tag{5}$$

Thereinto:

x2: Absolute humidity (kg/kg) of humid air after heating, measured value;

t2: Dry bulb temperature (°C) of the heated air, measured value;

t2′: Wet bulb temperature (°C) of the heated air, measured value;

p2: Air pressure (Pa) after heating, measured value;

xs2: Saturated water vapor absolute humidity of the heated air (kg/kg), measured value;

p2′: Saturated water vapor pressure (Pa) corresponding to the heated air wet bulb temperature t2′, calculated value;

A2, B2: constant coefficients.

Step 5: Using the principle of x1 = x2 in the equal heating process, combined with the measured parameters p1 and t1 of the measured air at low temperature, calculate its wet bulb temperature t1′:

$$t_1' = \frac{(1.846\ t_1 + 2501)x_1 + 1.005\ t_1 - 2501\ x_{s1}}{4.186\ t_1 + 1.005 - 2.340\ x_{s1}} \tag{6}$$

$$x_{s1} = 0.6220\ p_1'(p_1 - p_1') \tag{7}$$

$$p_1' = 10^{B1} \tag{8}$$

$$B_1 = -7.90298(A_1 - 1) + 5.02808 \; \log A_1 - 1.3816 \times 10^{-7} \left\{ 10^{11.344(1-1/A_1)} - 1 \right\}$$
$$+ \, 8.1328 \times 10^{-3} \left\{ 10^{-3.48149(A_1-1)} - 1 \right\} + \log 101325$$

$$\tag{9}$$

$$A_1 = 373.15 / (273.15 + t_1') \tag{10}$$

Thereinto:

t1′: The wet bulb temperature (°C) of the measured air, calculated value;

x1: The absolute humidity (kg/kg) of the measured air, calculated value;

t1: The dry bulb temperature (°C) of the measured air, calculated value;

p1: Measured air pressure (Pa), measured value;

xs1: Saturated water vapor absolute humidity (kg/kg) of the measured air, calculated value;

p1′: Saturated water vapor pressure (Pa) corresponding to the measured air wet bulb temperature t1′, calculated value;

A2, B2: constant coefficients.

Step 6: The enthalpy value h1 of the measured air at low temperature can be directly calculated from the absolute humidity x2 of the heated air and the dry bulb temperature t1 of the measured air [6]. Other parameters can also be calculated, such as the relative humidity and dew point temperature of the measured air at low temperature.

$$h_1 = 1.005 \, t_1 + (2501 + 1.846 \, t_1) x_2 \tag{11}$$

Thereinto:

h1: The enthalpy value of the measured air (kJ/kg), calculated value;

t1: The dry bulb temperature (°C) of the tested air, measured value;

x2: Absolute humidity (kg/kg) of humid air after heating, calculated value;

2.3   Measurement Characteristics

The feature of the invention is to establish a new low-temperature air temperature and humidity sampling measurement method, which innovatively heats the low-temperature air to be tested, measures the dry bulb temperature, wet bulb temperature and air pressure parameters after heating, and performs absolute humidity calculation. It uses the principle of constant absolute humidity in the isohumid heating process and combines with low temperature dry bulb temperature and air pressure parameters, the humidity and enthalpy value of the measured air at low temperature are calculated, so as to achieve accurate humidity measurement under low temperature conditions.

2.4   Outstanding Results of the Measurement

Compared with the existing measurement methods, the present invention has the following outstanding results:

(1) This method realizes the integrated accurate measurement of air temperature and humidity under low temperature conditions;

(2) This method uses conventional instruments to measure low-temperature air humidity, which greatly reduces investment and maintenance costs;

(3) This method uses the reheat constant humidity method to measure the parameters after heating the low-temperature air. There is no need to directly measure the amount of electric heating, which simplifies the measurement process and reduces the measurement links that produce errors;

(4) This method only involves the measurement of two state point parameters, namely the dry bulb temperature and air pressure of low-temperature air, and the dry and wet bulb temperature and air pressure of heated air. The measurement method is relatively simple and easy to implement. Under the low temperature environment, it has the advantage of high humidity measurement accuracy;

(5) This method realizes the isohumid heating process of air through adjustable electric heating [7], and its theoretical range is only limited by the range of the low-temperature temperature sensor;

(6) The method of the present invention can realize the measurement of other state parameters of the measured air at low temperature.

3   The Implementation Case

According to the layout of the temperature measurement points in the figure, after the operating conditions are stable, the measured values of the dry bulb temperature and the air pressure of the low-temperature measured air at a certain moment are: t1 = -10 °C, p1 = 100510 Pa; The measured values of the dry bulb temperature, the wet bulb temperature and the air pressure are: t2 = 23.4 °C, t2' = 9.1 °C, p2 = 100360 Pa. The measurement accuracy of the sensors at each dry and wet bulb temperature measurement point is ± 0.1 °C. At the same time, the air flow velocity at the wet bulb temperature measurement point 6 of the heated air is about 5 m/s.

According to the measured temperature and pressure values, the method and steps to calculate the wet bulb temperature and various state parameters of the measured air state point at low temperature are as follows:

(1) According to the measured values of the dry and wet bulb temperature and pressure parameters of the heated air, calculate the absolute humidity of the heated air x2:

$$x_2 = \frac{(2501 - 2.340\, t_2')x_{s2} - 1.005\,(t_2 - t_2')}{1.846\, t_2 + 2501 - 4.186\, t_2'} \qquad (12)$$

$$x_{s2} = 0.6220\, p_2'/(p_2 - p_2') \qquad (13)$$

$$p_2' = 10^{B2} \tag{14}$$

$$B_2 = -7.90298(A_2 - 1) + 5.02808 \ \log A_2 - 1.3816 \times 10^{-7}\left\{10^{11.344(1-1/A_2)} - 1\right\}$$
$$+ \ 8.1328 \times 10^{-3}\left\{10^{-3.48149(A_2-1)} - 1\right\} + \log 101325$$

$$\tag{15}$$

$$A_2 = 373.15/(273.15 + t_2') \tag{16}$$

Thereinto:

x2: The absolute humidity of the heated air (kg/kg), calculated value;

t2: The dry bulb temperature (°C) of heated air, measured value;

t2': The wet bulb temperature (°C) of heated air, measured value;

p2: Air pressure (Pa) of heated air, measured value;

xs2: Saturated water vapor absolute humidity of heated air (kg/kg), calculated value;

p2': Saturated water vapor pressure (Pa) corresponding to the heated air wet bulb temperature t2', calculated value;

A2, B2: constant coefficients.

Calculate the absolute humidity of the air state point after heating by formula (12)–formula (16):

$$x2 = 0.001431 \ \text{kg/kg};$$

(2) Using the principle of x1 = x2 in the equal heating process, combined with the measurement parameters p1 and t1 of the measured air at low temperature, calculate the wet bulb temperature t1' of the measured air [8]:

$$t_1' = \frac{(1.846\,t_1 + 2501)x_1 + 1.005\,t_1 - 2501\,x_{s1}}{4.186\,t_1 + 1.005 - 2.340\,x_{s1}} \tag{17}$$

$$x_{s1} = 0.6220\,p_1'/(p_1 - p_1') \tag{18}$$

$$p_1' = 10^{B1} \tag{19}$$

$$B_1 = -7.90298(A_1 - 1) + 5.02808 \ \log A_1 - 1.3816 \times 10^{-7}\left\{10^{11.344(1-1/A_1)} - 1\right\}$$
$$+ \ 8.1328 \times 10^{-3}\left\{10^{-3.48149(A_1-1)} - 1\right\} + \log 101325$$

$$\tag{20}$$

$$A_1 = 373.15/(273.15 + t_1') \tag{21}$$

Thereinto:

t1': The wet bulb temperature (°C) of the measured air, calculated value;

x1: The absolute humidity (kg/kg) of the measured air, calculated value;

t1: The dry bulb temperature (°C) of the tested air, measured value;

p1: The air pressure (Pa) of the measured air, measured value;

xs1: Saturated water vapor absolute humidity (kg/kg) of the measured air, calculated value;

p1': Saturated water vapor pressure (Pa) corresponding to the measured air wet bulb temperature t1', calculated value;

A2, B2: constant coefficients.

Calculate the wet bulb temperature of the measured air by formula (17)–formula (21):

$$t1' = -10.64\,°C;$$

(3) From the absolute humidity x2 of the heated air and the dry bulb temperature t1 of the tested air, the enthalpy value h1 of the tested air can be directly calculated:

$$h_1 = 1.005\,t_1 + (2501 + 1.846\,t_1)x_2 \tag{22}$$

Calculate the enthalpy value of the measured air state point by formula (22):

$$h1 = -6.493\,kJ/kg;$$

Other parameters of the measured air: relative humidity is 80.71%, dew point temperature is −12.18 °C.

4   Conclusion

Aiming at the shortcomings of the existing low-temperature air humiture measurement methods, this paper designs a new calculation method through experiments, which not only realizes the integrated and accurate measurement of low-temperature air humiture, but also ensures the safety and convenience of low-temperature operation of the device. Accordingly, the reliability of the measurement method is greatly increased, the investment cost is reduced. Its practical application value is very high.

Acknowledgements. The work described in this paper is supported by a youth science and technology fund project established by Hefei General Machinery Research Institute (2019010374).

References

1. Qin, H., You, J., Liu, Z., et al.: Research on a kind of high precision humidity sensor. Heilongjiang Sci. Technol. Inf. **13**, 34–35 (2015)
2. Liang, X., Zhang, J., Lin, Z., et al.: Some discussions on the measurement of humidity by wet-and-dry-bulb thermometer. Meas. Tech. **07**, 68–69 (2007)
3. Tong, M., Liu, Z.: Iterative algorithm of apparatus dew-point temperature for air conditioning systems and its application. Heat Ventilating Air Cond. **37**(08), 159–160 (2007)

4. Liu, Y., Liu, D.: A method of weaking the temperature and humidity on the impact of change on gas sensor. Inf. Technol. Netw. Secur. **38**(01), 36–40, 48 (2019)
5. Tao, Z., Wang, Z., Yuan, S.: Study on properties of ZrO2/poly (sodium-p-styrenesulfonate) composite humidity sensor. Electron. Compon. Mater. **33**(05), 48–52 (2014)
6. Qiu, W.: The application of enthalpy control and PID adjustment in the subway intelligent temperature control system. Installation **09**, 26–28 (2020)
7. Wang, S., Mei, Z., Yu, H.: Continuously adjustable phase shifting circuit from module integration of high frequency induction heating power. Electron. Design Eng. **20**(11), 132–134 (2012)
8. Ning, D., Pang, F., Yan, C., et al.: Application of a novel instrument for wetness measurement in water-air flows. J. Harbin Eng. Univ. **27**(06), 825–829 (2006)

The Hidden Troubles and Countermeasures of Perfect Education System for College Students Based on Big Data Analysis

Yan Zhang[(✉)] and Xinwen Bai

Department of Psychology, Chinese Academy of Science University,
Beijing 100049, China

Abstract. Perfectionist education refers to the excessive pursuit of the best and perfection by parents and teachers in their children's growth education. Although the original intention of this idea is good, the pursuit of perfection is often counterproductive. On this basis, this article discusses and analyzes the hidden dangers of perfectionism education in the context of big data. In order to further analyze the impact of the perfectionist education concept on students, this article visited some schools in our city. Through the survey, we learned about the basic situation of the students affected by the concept of perfectionism education, and tested and evaluated them. The results showed that the score of emotion test of students who were not affected by the concept of perfectionism education was 19.37, the score of interpersonal relationship was 8.53, the score of learning objective was 15.76, and the score of adaptability was 21.56. The scores of students affected by perfectionist education concepts were 15.42, 5.27, 11.32, and 11.32, respectively. It can be seen that the ideal of perfectionism education will have a certain negative impact on students' psychology.

Keywords: Perfectionism education view · Hidden danger analysis · Psychological influence · Big data

1 Introduction

Since the beginning of the study of perfectionism, many researchers have advocated that perfectionism is a multi-dimensional personality trait, which has duality, negative side and positive side [1, 2]. As a personality trait and mode of thinking, perfectionism is considered to be closely related to mental health [3, 4]. The perfectionism mentioned in a large number of studies in the past mainly points to its negative aspects, which is associated with pure negative personality characteristics and is classified as one of neurosis problems. Its basis is mainly based on some mental disorders and maladjustment symptoms of perfectionists, such as obsessive-compulsive disorder, depression, anxiety, mania, periodic headache, type a behavior, low self-esteem and fear of negation, suicidal ideation, etc., the most significant ones were anxiety, mood disorder and depression [5, 6].

Perfectionism, as a national characteristic of China, has been one of the qualities pursued by successful people since ancient times [7, 8]. But parents and teachers' excessive pursuit of perfect behavior and thought will bring invisible pressure to

J. C. Hung et al. (Eds.): FC 2021, LNEE 827, pp. 482–490, 2022.
https://doi.org/10.1007/978-981-16-8052-6_59

children, causing physical and mental harm. Under the cultivation of perfectionism education, those so-called "excellent students" constantly pursue the "first" plot. They are competitive, vain and have poor frustration. They are easy to cause a series of psychological problems such as emotion, cognition and behavior. Even when they can't reach the goal in mind, they will have excessive behaviors such as self-abuse and suicide [9, 10]. Therefore, under the background of fierce social competition, the hidden danger behind the perfectionism education is worth pondering. It is of great significance for the education and growth of contemporary students to deeply explore and analyze the hidden dangers of perfectionism education.

This article first expounds the concept of perfectionism and the concept of perfectionism education in the context of big data. Using the method of literature analysis, it summarizes and analyzes the influence of perfectionism education concept. In addition, in order to further analyze the impact of perfectionism on students, we visited many schools and tested and evaluated the conditions of some students. The research results show that the ideal of perfectionism education will have a negative impact on students' psychology.

2 Review of Perfectionism Theory

2.1 The Concept of Perfectionism

In the past studies, scholars have different definitions of perfectionism because of their different understanding of it. There are three types of perfectionism: Perfectionism and perfectionism. In the early research, researchers generally regarded perfectionism as a concept of one-dimensional structure. Some people think that perfectionism refers to the individual's excessive pursuit of order and strict requirements on himself, and he puts forward that perfectionists believe that their own standards in intelligence, achievement and other aspects are higher than others, and have a sense of excellence. Others believe that perfectionism means that individuals set unrealistic goals for themselves. Others believe that perfectionism is a natural drive to pursue perfection, which helps individuals adapt to social development. In the later research, more and more researchers advocate the multidimensional structure of perfectionism. Other scholars believe that perfectionism is multi-dimensional. They think that perfectionism refers to individuals' pursuit of high standards to achieve their goals. However, in the process of achieving the goals, they will worry about mistakes, care about parents' expectations and evaluation, act too hesitantly, and overemphasize order and organization. Some scholars believe that perfectionism includes interpersonal and intra individual components, which can be divided into others oriented perfectionism, social determined perfectionism and self oriented perfectionism. In some studies, perfectionism is multi-dimensional, divided into high standards, order and difference. The non adaptive component is difference, while the adaptive component is high standard and order. It can be seen from the above that the concept of perfectionism has no unified definition because of different research propositions.

2.2   The Concept of Perfectionism Education

There are many forms of perfectionism in the current educational practice in China, and the most typical slogan is: "there are no bad students, only bad teachers". Other performances are as follows: "everything is decided by the students, everything is decided for the sake of the students"; "everyone can succeed and everyone can become a talent"; "a perfect education does not need any coercion. All compulsions in education are unjust.", "let the children grow up independently", "there are only wrong teachers and no wrong students" and so on. In a word, perfectionism education view is such a view that school education can make students become perfect individuals no matter what kind of students they are facing, and make them have perfect knowledge and moral quality.

2.3   The Hidden Dangers of Perfect Education

(1) Increased pressure on teachers

The concept of perfectionism puts teachers on a road that must improve the quality of education infinitely, otherwise they will have to bear the name of unqualified teachers. This has seriously increased the pressure of teachers, almost unlimited to improve the professional difficulty of teachers. The imposed perfectionism poses a potential threat to teachers' mental health.

(2) Set up the right and wrong view of students' mistakes

There is an amazing theoretical premise in the concept of perfectionism education, that is, all students are good, and students have an inevitable trend to become moral individuals. If there is any deviation in this process, it is the teacher's education that has gone wrong. This is a typical romantic philosophy of education, which has been proved to be totally wrong in practice. This kind of educational view actually pushes students to an indeterminate path. With its natural development, students may become individuals with good character and learning, or they may become individuals with bad quality and no learning meta skills. But in the view of perfectionism, the former is the result of individual development, while the latter is the performance of teachers' incompetence.

(3) It leads to the perfection, simplification and utilitarianism of education evaluation

The popularity of perfectionism education in the society leads to the judgment tendency of simplifying the originally complicated educational theory, idealization of educational practice and utilitarianism of educational evaluation. Since in theory, qualified teachers can educate all students well, and education is omnipotent, then school teachers have the obligation and responsibility to educate every student well. If there is any problem in the process, it is inevitable that there is something wrong with education and teachers, rather than the responsibility of students, parents and society, It must be the reason for the low ability of the school and teachers, thus ignoring the real cause of the problem.

(4) It leads to the continuous tension of the whole school

In reality, it is the perfectionism education theory that leads to the society's devaluation of pedagogy. It thinks that the educational theory is simple and everyone can speak, which leads to the society's imposing on school education the completely successful educational goal which cannot exist in the perfectionism education theory itself, thus forming the continuous tension and impetuousness of school education. "The ideal should be regarded as reality, and the possibility must be regarded as necessary. The dislocation between ideal and reality brings about the deviation of educational practice and theoretical cognition".

(5) Forming compulsive personality

Children with perfectionism education tend to have compulsive personality. They seem to be cautious, hesitant, pursuing 100% accuracy and hoping that everything is under control or they won't feel safe. These students require too much order and order, pay attention to details, and pursue perfection. They are always dissatisfied with their achievements. Therefore, they are more likely to procrastinate and fail to complete tasks on time, which has a certain impact on their study and life.

(6) With anxiety and depression

A large number of studies have shown that in the Perfectionism Scale, attention to errors and doubt about actions are highly positively correlated with depression and anxiety. In other words, perfectionists tend to have negative emotions and low life satisfaction. Under the requirements of extremely high standards, these students blindly pursue excellence and do not consider their own abilities. In this way, they will bear too much pressure, resulting in lower scores. In this way, they will be more worried about failure, worried about test pressure, and can not correctly treat failure. In addition, long-term depression will produce depression, and even suicidal behavior.

(7) Low self-evaluation

Theoretically speaking, the idea of perfection is to pursue perfection and avoid imperfection. Therefore, the perfection of education often ignores the advantages of children, and clings to the shortcomings and pursues perfection. This leads to excessive accusations and denials in the teaching process, which makes students live in the shadow of disappointment for a long time, and their self-evaluation gradually weakens, which easily leads to the decadent thoughts of shame and self blame and dim future. In the long run, this will lead to their consistent negative evaluation mode and lack of self-confidence.

2.4 Suggestions for Improvement of Perfect Education

(1) Starting from family education, change the expectations of parents. Promote the damage caused by high expectations, support proper parenting, and don't let the love of parents cause a heavy psychological burden on the child. Create a harmonious, warm and democratic family atmosphere. Parents and children need to respect and understand each other. Parents need to transform from decision makers to supporters, allowing their children to explore and create their own lives independently.

(2) For school education, it is necessary to emphasize the value of "people-oriented comprehensive development". Although our education system has weaknesses, teachers must bear in mind the purpose of education and adopt correct teaching

methods. From the perspective of long-term development, teachers must Caring for the lives of students, caring for the spiritual level of students, based on the personal development of students, caring for students, respecting students, listening to students, encouraging and praising students.

(3) For the students themselves, they must give full play to their individual subjective initiative and avoid the negative impact caused by the unfavorable objective environment. No matter how parents and teachers evaluate themselves, they must decide and learn to accept and treat them objectively. nobody is perfect. Accept that you make mistakes and set achievable goals for yourself. In life, it is necessary to communicate frequently with parents and teachers to inform them of their true situation and avoid psychological burdens caused by their high hopes on themselves.

3   Research Design

This paper first uses the method of literature analysis to provide a certain theoretical basis for the study of this paper by consulting relevant materials. Then, through the way of interview and investigation, the students affected by the concept of perfectionism education are investigated and interviewed.

(1) Research object
    This paper visited many schools in our city, and investigated the students affected by the concept of perfectionism. In this paper, from the visited schools, 40 students were randomly selected as the research object. Among them, 20 students were affected by the concept of perfectionism education, which was set as the test group, while the other 20 students were not affected by the concept of perfectionism education as the control group.

(2) Test content
    The content of the test in this paper is mainly divided into emotional test, interpersonal relationship test, learning and goal test and adaptability test. In order to prevent data error caused by individual reasons, this paper takes the average value of each test data as the final experimental analysis data.

4   On the Hidden Danger of Perfectionism Education

4.1   Analysis of Hidden Dangers Brought by Perfectionism

By using the method of literature analysis and consulting relevant materials, this paper summarizes and analyzes the hidden dangers caused by the concept of perfectionism education. The results are shown in Table 1 and Fig. 1.

This paper summarizes the hidden dangers caused by the concept of perfectionism education, and analyzes the consequences. It can be seen from Table 1 and Fig. 1 that the perfectionism education concept will increase the pressure on teachers, increase the difficulty of teachers' occupation, and deprive teachers of their professional rights,

Table 1. Hidden dangers brought by perfectionism

Hidden danger	Effect
Increased pressure on teachers	It improves the difficulty of teachers' profession and deprives them of their professional rights
Set up the right and wrong view of students' mistakes	This leads to the lack of sense of responsibility and discipline
It leads to the perfection, simplification and utilitarianism of education evaluation	Neglect the causes of the problem
It leads to the continuous tension of the whole school	It is not conducive to the sustainable development of the school
Forming compulsive personality	Hesitation in doing things
With anxiety and depression	Lead to students' pressure is too high
Low self-evaluation	Lead to students' inferiority complex

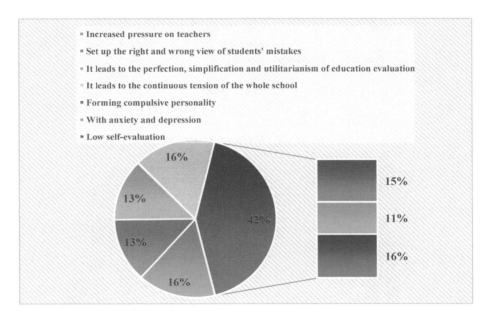

Fig. 1. Hidden dangers brought by perfectionism

accounting for 15.6% of the total. Secondly, the concept of perfectionism will make students establish a wrong view of right and wrong, which leads to the lack of sense of responsibility and discipline, accounting for 13.1% of the students. Moreover, the perfectionism education view will also lead to the perfection, simplification and utilitarianism of the social evaluation of education, thus neglecting the causes of the problems, accounting for 12.7% of the total. 3.3% of the total school development is not conducive to the continuous development of the school. Perfectionism education also makes students form compulsive personality and hesitant to do things, which accounts for 15.2% of the total. Students affected by perfectionism are often accompanied with anxiety and depression, which leads to excessive pressure of students, accounting for 11.4%. Finally, the students affected by perfectionism tend to have low self-evaluation, which leads to their inferiority complex, accounting for 15.7%.

4.2   Analysis of the Influence of Perfectionism Education Concept on Students' Psychology

This paper visited many schools in our city for investigation, and tested some students who had been influenced by the concept of perfectionism education, and compared the test data of the students who had not been influenced by the concept of perfectionism education. The results are shown in Table 2 and Fig. 2.

Table 2. The influence of perfectionism on students' Psychology

Dimension	Grouping	N	Mean	t
Emotion	Control group	20	19.37	0.128
	Experience group	20	15.42	0.124
Interpersonal relationship	Control group	20	8.53	0.653
	Experience group	20	5.27	0.642
Learning objectives	Control group	20	15.76	2.122
	Experience group	20	11.32	2.107
Adaptability	Control group	20	21.56	2.357
	Experience group	20	14.37	2.334

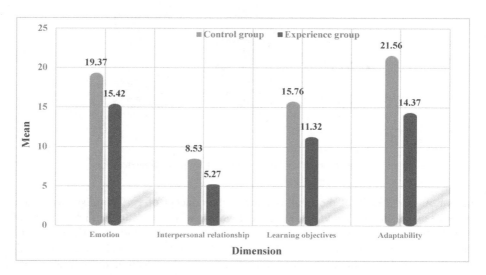

Fig. 2. The influence of perfectionism on students' Psychology

In order to analyze the influence of perfectionism education concept on students' psychology, this paper conducts relevant tests, including: emotion test, interpersonal relationship test, learning goal test and adaptability test. It can be seen from Table 2 and Fig. 2 that there are obvious differences in the test results between the students who have received the influence of the concept of perfectionism education and those who have not. Among them, the students without the influence of perfectionism education concept scored 19.37 on emotion test, 8.53 on interpersonal relationship, 15.76 on learning goal and 21.56 on adaptability. The scores of the students who had been influenced by the concept of perfectionism education were 15.42, 5.27, 11.32 and 11.32 respectively. In each score test, the students who have not been influenced by the concept of perfectionism education are more than those who have been influenced by the concept of perfectionism education.

5   Conclusions

In modern social life, many parents and teachers have high hopes for their children, which leads to the phenomenon of the pursuit of perfection. There is nothing wrong with the pursuit of perfection. However, the blind pursuit of perfection without considering the actual situation will have a significant impact on children's physical and mental health. Therefore, this article explores and analyzes the hidden dangers of perfectionism education in the context of big data. This article analyzes the impact of perfectionism education on students through literature analysis and interview investigation methods, and highlights the shortcomings of perfectionism education through comparison. This article believes that the ideal of perfectionism education has a certain negative impact on students' psychology, and parents and teachers should consider the actual situation and educate students in accordance with the actual situation.

References

1. Seeliger, H., Harendza, S.: Is perfect good? – dimensions of perfectionism in newly admitted medical students. BMC Med. Educ. **17**(1), 206 (2017)
2. Farvis, J., Hay, S.: Undermining teaching: how education consultants view the impact of high-stakes test preparation on teaching. Policy Futures Educ. (14), 147821032091954 (2020)
3. Chen, I.J., Hu, M., Zhang, H., et al.: The effect of parenting behavior on the obsessive-compulsive symptom of college students: the mediating role of perfectionism. Creat. Educ. **09**(5), 758–778 (2018)
4. Kim, Y., Jang, et al.: The mediating effects of achievement motivation and self-consciousness between university students' type of perfectionism and speech anxiety. Asian J. Educ. **18**(3), 441–462 (2017)
5. Prichard, R.: Redefining the ideal: exquisite imperfection in the dance studio. J. Dance Educ. **17**(2), 77–81 (2017)
6. Cho, H.H., Kang, J.M.: Factors influencing clinical practice burnout in student nurses. Child Health Nurs. Res. **23**(2), 199–206 (2017)

7. Hartman, J., Härmark, L., van Puijenbroek, E.: A global view of undergraduate education in pharmacovigilance. Eur. J. Clin. Pharmacol. **73**(12), 1–9 (2017)

8. Ztrk, Z.B.: A theorical view to the relation of design training and design of education. Turk. Online J. Design Art Commun. **9**(1), 22–27 (2019)

9. Wilkinson, A.: U.K. English teaching—an American view. Engl. Educ. **3**(3), 102–105 (2017)

10. Mutale, J.: Educational challenges: issues with power engineering education [in my view]. IEEE Power Energ. Mag. **16**(5), 120–127 (2018)

Analysis of Uniqlo Enterprise Profitability Under the Supply Chain Dual Distribution Channel Marketing Model Based on Big Data

Yang Jiao[(✉)]

School of Economics and Management, Lanzhou University of Technology,
Lanzhou, Gansu, China
18734871802@139.com

Abstract. In the era of knowledge economy and the post-epidemic environment, the role of information technology and big data has been highlighted. More and more companies regard big data as their business helpers and strategic core. This phenomenon is not only obvious. In addition to technology companies, the fast fashion industry, which is closely integrated with users, is particularly prominent. Among them, Uniqlo, the leader of fast fashion brands, has the most outstanding performance. This article combines the case of Uniqlo to explain how companies use big data to integrate channels in the supply chain to form a complete system. Then extract data from the wind database to conduct an empirical analysis of profitability. The results show that the combination of big data development will lead to partial financial improvements in the short-term, which is beneficial to the development of the enterprise in the long-term. Finally, relevant suggestions are put forward based on the empirical pair, which can provide reference for the better development of related enterprises.

Keywords: Big data · Dual-channel integration of supply chain · Profitability

1 Introduction

In the context of the post-epidemic era, information technology and human production and life have been further integrated to form a closer situation. The Internet has gained a lot of popularity in various countries and industries, and the sinking development has also opened up a broader market for enterprises. In the daily operation and business life of enterprises, a large amount of data is generated every moment [7]. For example, Facebook users provide more than 2.5 million pieces of data per minute, Google's search data volume reaches an astonishing 4 million pieces per minute, and domestic companies such as Alibaba have an astonishing amount of data on products, users, and transactions each year. The number, hidden behind the double eleven hundred billion turnover is the support of massive data.

The aggregation of big data has also contributed to the rise of technical means such as cloud computing, artificial intelligence, the Internet of things, and blockchain. In the context of data support, traditional marketing models have also ushered in a new round of growth and development [8]. The company combines the collection and transfer of user information and data, modular management and connection, new technology analysis and community management, etc., through the construction and mining of offline and online channels, and transmits the company's own values and characteristics to the audience [1]. Finding is not just the actual offline experience. Brands and products form a more connected state with people through machines and data [10]. People's information perception and actual perception are complementary, participating behaviors are more convenient and fast, and the sense of participation is stronger. In fact, the core value of new marketing combined with big data is the user dividend between the enterprise and the user. Consumers' browsing and purchasing preferences are recorded in real time [4]. The recorded data reflects the enterprise's understanding and mastery of users. The degree is a direct connection with the customer [2].

How to create such a connection, how to use and meet the needs contained in the connection, help realize the self-worth of customers, and formulate corresponding strategies are the problems that companies have been facing and seek to solve, and it is also the value of big data research and development [6]. At present, many companies have already taken the steps to use big data, exploring the application of big data in different ways in order to realize a new era marketing road with corporate characteristics and integrated corporate value [3]. In this regard, in addition to platform-based companies that can rely on the advantages of corporate attributes to apply big data to corporate marketing and operations more effectively, more companies are also exploring their own big data marketing roads. Among them, the more brilliant and effective brands are GAP, Zara, MUJI and Uniqlo. They have formed a unique SPA (private brand clothing professional retailer) model on the road of exploring big data marketing, in addition to their own business lines These fashion brands have also cooperated with major platforms to further broaden their channels, and at the same time obtain more information and data about customers. These brands have increasingly realized that big data and the Internet have the importance of business development. Combined with big data, advanced brand design promotes the enhancement of corporate value [5]. Next, we will take Uniqlo, which is most well known to us, as an example to analyze its profitability and explore the development path of big data for modern enterprises.

2   Overview of Uniqlo's Marketing Model

In fact, Uniqlo's dual-channel marketing includes two modes. One is SPA (Professional Retailer of Private Label Clothing), which specifically refers to the direct connection between the production end of the clothing brand and the retail end, so that the product can directly face the customer and save money. In addition to the cost of the enterprise and the customer, the timeliness of the product can be grasped, and business opportunities and profits can be taken into account. Under this model, companies make overall revisions to business processes, including planning, R&D and design, circulation, and marketing, and strive to establish a business model with the highest degree of matching with consumers, and consider issues from the actual needs of consumers, and Put it into production practice. The second mode is BOPS (online purchase and offline pickup), which has become a hot issue in the current supply chain management field. Existing studies have shown that the BOPS model can realize the conversion of online and offline user traffic and the sharing of information between e-commerce and physical stores. It improves the dual-channel linkage capabilities of enterprises and also realizes a convenient shopping experience for consumers [9]. Uniqlo has adopted this model to achieve the upgrade of consumer consumption patterns. While the consumers' offline try-on experience is perfectly integrated with online service content, it is also of great help to solving inventory problems and shortening logistics delivery time. We can clearly understand the BOPS model through the supply chain model in Fig. 1:

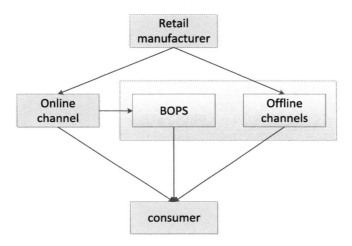

Fig. 1. Dual-channel supply chain structure under BOPS mode

Combining the two models, Uniqlo has achieved a successful O2O transformation, combined with big data to transform and integrate the supply chain, forming a unique O2O model, as shown in Fig. 2. This model can also be called omni-channel marketing and retail [11]. Consumers have changed the situation of directly facing offline physical stores. Between the store and customers, a trinity of official website, official APP and Tmall online flagship store has been formed. At the same time, customers are doing experiential shopping in physical stores. The staff will also guide customers to carry out related online operations and actively introduce customers to the experience of online platforms. Although the intermediate links in shopping may be increased, the distance between the brand and the consumers has been further shortened, and the customers have gained with more considerate services and convenient online channels, combined with store shopping, the shopping experience is more diverse and flexible. While grabbing customers and enhancing the stickiness between the company and customers, companies have obtained first-hand user data, channels and Users are better matched and a win-win situation is achieved. In summary, we put forward the hypothesis that the supply chain dual-channel marketing model based on big data can improve the business capabilities and performance of enterprises.

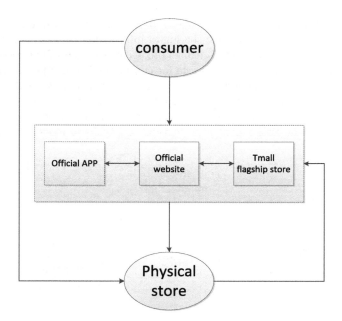

Fig. 2. UNIQLO dual-channel mode indication

3   Empirical Design

3.1   Selection of Research Samples and Variables

This article draws on the research of previous scholars and uses financial indicators as indicators to measure profitability. The data comes from the Wind database, and Uniqlo's financial data from 2008 to 2020 is collected for analysis. Uniqlo entered the Tmall flagship store in 2009, and developed its own official APP and put it into use in 2013. The collection of data covers these two important time nodes, and the 2013 node is the focus, because after this year Uniqlo The supply chain dual-channel marketing system based on big data is formally formed. In our research, we will treat the formation of a dual-channel supply chain model based on big data as a dummy variable, and the formation mark is 1 and the unformation mark is 0.

Profit performance mainly includes two aspects: operating efficiency and operating results. Operating efficiency refers to the performance generated by the company's operations, reflecting the company's operational and operational efficiency after adopting a big data strategy for dual-channel supply chain integration, mainly through total asset turnover and Measured by inventory turnover rate. Business results are what we usually call profit performance, which reflects the profit or loss of a company after adopting a big data strategy, which is measured by the return on total assets and the net sales margin. These data can generally be directly obtained by reading the company's financial statements. If there is no directly available data, it can be calculated by the method in Table 1.

Table 1. Variable selection and calculation

Dimension	Metrics	symbol	Calculation
Operating efficiency	Inventory turnover rate	ITR	Operating cost/[(Inventory balance at the beginning of the period + Inventory balance at the end of the period)/2]
	Turnover rate of total assets	TAT	Operating income/[(total asset balance at the beginning of the period + total asset balance at the end of the period)/2]
Operating results	Net sales profit	NSP	Net profit/operating income
	Return On Total Assets	RTA	(Total profit + financial expenses)/[(Total asset balance at the beginning of the period + Total asset balance at the end of the period)/2]

3.2 Determination of Research Methods

There are three main research methods for the hypotheses of this article. The first is to use SPSS to test the reliability and validity of the data to verify the possibility of empirical analysis of the data. The second is the parameter test method. The t value is used as the test parameter and the t value obtained from the data before and after the introduction of big data marketing is compared to study the significant changes in profitability. The third is the non-parametric test. In order to avoid the deviation of the research results due to the sample data, the rank sum test is used to analyze the changes in the profit performance of the supply chain dual-channel model under Uniqlo's big data from the quantitative level.

3.3 Empirical Analysis

3.3.1 Reliability and Validity Test

Using SPSS 25 to test the reliability and validity of each dimension data, it is found that KMO = 0.648, greater than 0.6, factor loading = 0.673, which meets the reliability and validity requirements required for empirical analysis.

3.3.2 Parameter Test

Using SPSS25 to test the parameters of the sample data, the results are shown in Table 2:

Observing the results of the t test, the inventory turnover rate (ITR) t value is increasing year by year under the dual-channel mode of supply chain integration after the introduction of big data. The test results are significant in the two years, five years, and the overall comparison. Observing the total asset turnover rate (TAT) data, after the introduction of big data supply chain integration, the t value increased rapidly during the two to five years period, and also increased during the five-year to the overall period, and from the two-year period The insignificant status of "has changed to remarkable within five years". The t-value of the net profit rate (NSP) has not changed much, and the test results are all insignificant. The t value of return on total assets

Table 2. Test result

Index		Operating efficiency		Operating results	
		ITR	TAT	NSP	RAT
Comparison two years before and after 2013	Average value	3.48	1.21	0.06	0.16
	t	−50.20	−2.42	−1.53	−26.08
	p	0.00**	0.14	0.26	0.00**
Comparison five years before and after 2013	Average value	3.34	1.11	0.06	0.18
	t	−9.14	−2.46	−2.03	−3.13
	p	0.00**	0.04**	0.08	0.01**
Overall comparison before and after 2013	Average value	3.28	1.02	0.06	0.16
	t	−6.22	−2.73	−1.71	−4.11
	p	0.00**	0.02**	0.11	0.00**

(RAT) increased significantly from two to five years, but there was a slight decrease in the overall test, and the test results were all significant.

3.3.3 Non-parametric Test
Use SPSS25 to perform non-parametric test on the sample data, and the test results are shown in Table 3:

Table 3. Rank sum test

Index		Operating efficiency		Operating results	
		ITR	TAT	NSP	RAT
Comparison two years before and after 2013	z	−1.63	−1.55	−1.55	−1.55
	p	0.33	0.33	0.33	0.33
Comparison five years before and after 2013	z	−2.62	−2.19	−2.40	−1.78
	p	0.01**	0.03**	0.02**	0.10**
Overall comparison before and after 2013	z	−2.93	−2.63	−1.90	−2.34
	p	0.00**	0.01**	0.07	0.02**

From the inspection results, the inventory turnover rate (ITR), total asset turnover rate (TAT), net interest rate (NSP) and return on total assets (RAT) did not change significantly in the first two years of introduction, and showed the same The inspection results showed significant changes in the five-year comparison, but the net profit rate was not significant enough in the overall inspection.

4 Conclusions and Recommendations

Based on the above research and analysis, it is found that in the short term, the integration of two channels in the supply chain under big data will have a significant positive impact on operating performance, and will have a significant impact on the return on total assets in profitability, but it will have a significant impact on the net profit margin. The impact is fluctuating. We infer that it is because the company needs to invest more in the application of big data technology within a certain period of time. The improvement of hardware and the normal and continuous operation require a large amount of early investment, and the introduction and introduction of relevant professionals Cultivation will also occupy corporate funds, which will have an impact on corporate profits in the short term. However, in the long term, the application of big data and effective channel integration will undoubtedly affect the corporate integration capabilities, business process level, customer management and operational efficiency. To improve and create benefits, the time delay response is inevitable.

Based on the research conclusions, we put forward the following suggestions:

(1) From managers to employees, companies should establish awareness of the era of big data and knowledge economy, realize that the era of data has arrived and is developing rapidly, and take a positive attitude towards the positive effects of big data on the company, and take the initiative to learn Understand and accumulate relevant experience.
(2) Enterprises should take a long-term perspective and get rid of the impact of short-term interest fluctuations. In particular, corporate executives should be good at discovering new data business footholds and develop and occupy the market as soon as possible.
(3) While introducing big data talents and technologies, enterprises should also pay attention to the cultivation of local relevant talents and technological breakthroughs, and strive to form their own unique business process framework, combine their own business with big data more flexibly, and have a plan, Targeted implementation of big data strategy to achieve all-round optimization of channels and businesses.

References

1. Gao, F., Su, X.: Omnichannel retail operations with buy-online-and-pick-up-in-store. Manage. Sci. mnsc.2016.2473 (2016)
2. Gao, F., Su, X.: Online and offline information for omni-channel retailing. Oper. Res. **58**(4), 385–387 (2018)
3. Yan, R., Zhi, P., Sanjoy, G.: Reward points, profit sharing, and valuable coordination mechanism in the O2O era. Int. J. Prod. Econ. **215**(SEP.), 34–47 (2018)
4. Timoshenko, A., Hauser, J.R.: Identifying customer needs from user-generated content. Mark. Sci. **38**, 1–20 (2019)
5. Boyd, D.E., Kannan, P.K., Slotegraaf, R.J.: Branded apps and their impact on firm value: a design perspective. J. Mark. Res. **56**(1), 76–88 (2019)

6. Hu, M.M., Dang, C.I., Chintagunta, P.K.: Search and learning at a daily deals website. Mark. Sci. **38**(4), 609–642 (2019)
7. Trabucchi, D., Buganza, T.: Data-driven innovation: switching the perspective on big data. Eur. J. Innov. Manag. **22**(1) (2018)
8. Liu, X., Lee, D., Srinivasan, K.: Large-scale cross-category analysis of consumer review content on sales conversion leveraging deep learning. J. Mark. Res. **56**(6), 918–943 (2019)
9. Zhang, X., Chen, H.R., Zhi, L., et al.: Strategies of pricing and channel mode in a supply chain considering showrooming effect. Kongzhi yu Juece/Control and Decision (2020)
10. Rehman, N.: Information technology and firm performance: mediation role of absorptive capacity and corporate entrepreneurship in manufacturing SMEs. Technol. Anal. Strategic Manage. **32**, 1049–1065 (2020)
11. Li, Z., Yang, W., Jin, H.S., et al.: Omnichannel retailing operations with coupon promotions. J. Retail. Consum. Serv. **58**, 102324 (2021)

Influence of Network Opinion Leaders on College Students' Tourism Behavioral Intention in Weibo Marketing Based on Big Data Analysis

Fengqi Zhang[✉]

School of Economics and Management, Lanzhou University of Technology, Lanzhou, Gansu, China

Abstract. In recent years, big data analysis technology has become more and more mature, and it is of great value to apply big data analysis technology in microblog marketing to promote enterprises to tap potential users. The article analyses the characteristics of big data, the network opinion leaders traits and according to the characteristics of the tourism industry and college students' group's unique features, big data analysis in the application of the microblogging marketing value, and how to use big data analysis technology, etc., to solve these problems and to explore the network opinion leaders about what impact on university students' tourist behavior intention, using the results of the study, can be targeted to provide guidance for the weibo marketing of tourism development.

Keywords: Microblog marketing · Network opinion leader · College students' travel behavior intention · Big data analysis

1 Introduction

With the rapid development of society and the rapid development of science and technology, more and more scientific products have entered our lives. Big data is the product of this high-tech era. Big data has the following characteristics: (1) The size of the data is large and can be close to all the data. (2) There are many types of data and different types of data. (3) Increasingly unstructured data. (4) Low value density [1–4]. With the concept of big data, enterprises' judgment of consumer behavior and precise marketing scope have been comprehensively improved and optimized. With the continuous development of social economy and the continuous improvement of industrial structure, China's tourism industry has gradually developed into an indispensable part of the overall income of the national economy. And the state keeps issuing policies to support the growth of tourism. Mobile travel users have gradually become the rising star of tourism consumers, among which Weibo users occupy a very important main position. With the rapid expansion of we-media content, the opinions of "big Vs" in the vertical field have more and more influence on fans, and decision-making "KOL" has become an aspect that the development of various industries have to pay attention to. In recent years, the tourism market of college students in China has been sought after, and

J. C. Hung et al. (Eds.): FC 2021, LNEE 827, pp. 499–505, 2022.
https://doi.org/10.1007/978-981-16-8052-6_61

the market size is expanding continuously. Therefore, the economic benefits of these tens of millions of college students will not be underestimated. Based on the above background, college students and opinion leaders are selected as the subjects of this study.

In the previous researches on microblog marketing, most scholars only constructed the network opinion leaders as a study of information characteristics, while ignoring the specific research on the internal characteristics of network opinion leaders. In addition, among the related researches on opinion leaders, the number of researches on the tourism industry is relatively small, and there are few direct studies on the impact of online opinion leaders on tourism behavior. A detailed study on the different influences of promoting tourism products with their different characteristics on the variable of college students' tourism behavioral intention is conducted, and the research on the characteristics of network opinion leaders as a single trait resource is a supplement to the research on tourism behavioral intention as well as the research on microblog marketing.

2 Theoretical Basis and Research Hypothesis

The author combined with the technology acceptance model and S - O - R model, university students' tourist behavior intention of stimuli is network opinion leaders, based on this, through the use of network opinion leaders traits (professional, innovative, information quality, visual environment, product involvement) release travel-related information makes the college students perceive the quality of information, which in turn affect their travel behavior intention (Fig. 1).

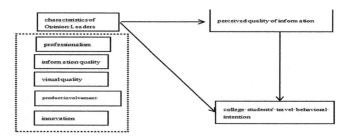

Fig. 1. The influence model of network opinion leader traits on college students' tourism behavioral intention in microblog marketing based on big data analysis

Gilly (1998) thinks that the professional ability of information source is an important measure of information credibility [5]. Zeng (2013) has proved that the purchase intention of female consumers is affected by the gender, professionalism and product involvement of online opinion leaders, and has a significant positive correlation. Turcotte et al. (2015) proved that people behavioral; purchase intention is more easily influenced by Internet opinion leaders. The form of information recommended by Internet opinion leaders is positively related to the quality of information [6]. Alba

(1987) concluded that the stronger the personality of network opinion leaders, the higher their innovation and influence [7]. Fischer (2014) concluded that the characteristics of Internet opinion leaders have a positive correlation with perceived quality. The author puts forward the hypothesis by studying the previous literature.

H1: In Microblog Marketing in the Era of Big Data in the Era of Big Data, network opinion leaders have a significant positive influence on college students' tourism behavioral intention.

H2: In Microblog Marketing in the Era of Big Data, network opinion leaders have a significant positive impact on the perceived quality of information.

H3: In Microblog Marketing in the Era of Big Data, the perceived quality of information has a significant positive impact on college students' tourism behavioral intention.

3 Questionnaire Design and Data Analysis

This questionnaire is designed and formulated by the author on the basis of summarizing relevant literature and combining with the research purpose of this paper [8–11]. The first part is the demographic characteristics of the respondents. This includes gender, educational background, monthly income, and whether they use microblogs (used to screen eligible respondents). The second part is the description of Weibo user behavior of college students, and statistics are made on the number of years, frequency and duration of using Weibo of the respondents. The third part is the most important part of the questionnaire. In order to make the data of respondents more accurate, the author explains the definition of opinion leader in the explanatory part. To see whether college students browse microblog of opinion leaders, and to measure whether the characteristics of network opinion leaders in microblog have an impact on the tourism behavior intention of college students. The third part mainly uses Likert scale to measure each variable. It includes usefulness, professionalism, information quality, innovation, visual quality, product involvement, information perceived quality, and college students' tourism behavioral intention.

This questionnaire is about the influence of network opinion leaders in Microblog Marketing in the Era of Big Data on college students' tourism behavior intention. According to the research needs, the research objects should meet the following conditions: first, college students; second, they should have experience in microblog; third, they should have tourism intention. The questionnaire was distributed through the Internet, among which 162 online questionnaires were collected and 148 were valid.

There are more girls than boys in terms of research objects, with boys accounting for 28.4% and girls 71.6%. According to the previous investigation and research literature, girls are more likely to be influenced by online opinion leaders or social media environment, so it is more reasonable for them to collect tourism information. On the other hand, because there are more female students in the author's social circle and the ratio of male and female students in finance and economics school is different, it is a normal phenomenon that there are more female students than male students. From the perspective of monthly disposable income, the majority of people are below 1000–

2000 yuan, which is related to the fact that the respondents are mainly college students. Those who have used Weibo account for 91.36%, and those who have the idea of traveling account for 96.3%, which indicates that most college students use Weibo and have the intention of traveling. It also indicates that the research topic of this paper is meaningful. 77.78% of respondents have used microblogs for more than one year, and they are experienced in microblogs.

Nearly 60% of them log on to microblogs every day. More than 80% of respondents will browse the microblogs of opinion leaders. To sum up, the samples collected in this study meet the requirements of the research topic (Table 1).

Table 1. Questionnaire overall reliability analysis table

	Cronbach's alpha	Number of items
Questionnaire as a whole	0.987	31

It can be seen from the table that the overall reliability coefficient of the questionnaire is 0.987, which is greater than 0.9, indicating that the reliability quality of the research data is very high (Table 2).

Table 2. Tests of kmo and Bartlett

Sampling the Kaiser-Meyer-Olkin measure of adequacy		0.937
Bartlett's test for sphericity	The approximate chi-square	4897.077
	df	465
	Sig.	0.000

As can be seen from the analysis table, the Kaiser-Meyer-Olkin measure is 0.937, which is obviously greater than 0.9;In addition, Bartlett's sphericity test Sig is infinitesimally small, so it can be concluded that the data collected in this questionnaire are valid.

This part tests whether there is a significant correlation between college students' travel behavioral intention and the characteristics of network opinion leaders (professionalism, visual quality, information quality, product involvement, and innovation). The results are shown in the table below (Table 3).

It can be seen from the table that the characteristics of network opinion leaders, such as professionalism, visual quality, information quality, product involvement, and innovation, are significantly positively correlated with college students' tourism behavioral intention and the quality of information perception, so it can prove the validity of H1and H2 hypothesis (Table 4).

It can be seen from the above table that the perceived quality of information is significantly positively correlated with college students' travel behavioral intention. So the H3 hypothesis is true.

Table 3. Correlation analysis

Factor		Tourism behavior intention of college students	Information perception quality
Speciality	Pearson correlation	0.956	0.949
	Sig.	0.004	0.001
Product involvement	Pearson correlation	0.965	0.949
	Sig.	0.001	0.002
Visuality	Pearson correlation	0.952	0.958
	Sig.	0.002	0.000
Information quality	Pearson correlation	0.963	0.976
	Sig.	0.003	0.003
Innovation	Pearson correlation	0.917	0.976
	Sig.	0.000	0.001

Table 4. Correlation analysis

Factor		Tourism behavior intention of college students
Information perception quality	Pearson correlation	0.966
	Sig.	0.000

4 Hypothesis Discussion and Conclusions

According to the results of correlation analysis, there is a significant positive correlation between professionalism and information perceived quality, and a significant positive correlation between professionalism and college students' travel behavioral intention. As a group of highly educated people, college students have already acquired a considerable degree of judgment ability, and their reference opinions tend to be specialized. The more professional the information released by network opinion leaders is, the more useful the information will be for college students, and the greater the influence of this information on their travel behavior intention.

According to the results of correlation analysis, the visual quality is significantly positively correlated with the perceived quality of information, and the visual quality is significantly positively correlated with college students' tourism behavioral intention. Visualization is a special hobby of college students. They prefer dynamic pictures, short videos, forms similar to cartoons and music backgrounds to pure text. When the microblog of opinion leaders has various forms and caters to their preferences, they will be more willing to spend time to browse the information and have a deeper understanding of the published content, which will make them feel that the information is useful and more likely to guide and suggest their behavioral intentions.

According to the results of correlation analysis, product involvement has a significant positive correlation with information perceived quality, and product involvement has a significant positive correlation with college students' tourism behavioral intention. Few people travel by heart, most college students before travel, must be in advance to do the strategy, and must refer to the views of experienced tourists. When the products of network opinion leaders are more involved, college students will find their information more useful, and their behavioral intentions will also be affected.

According to the results of correlation analysis, information quality is significantly positively correlated with information perception quality, and information quality is significantly positively correlated with college students' tourism behavioral intention. College students receive longer education, they not only have their own characteristics of young people in the age, in the judgment of things are more rational. When receiving information, they also pay attention to the quality of information, and have high expectations for the credibility and timeliness of information quality. The higher the information quality is, the more college students will perceive the usefulness of the information, and it will certainly have a corresponding impact on their travel behavior intention.

According to the results of correlation analysis, there is a significant positive correlation between innovation and information perceived quality, and a significant positive correlation between innovation and college students' tourism behavioral intention. As a young generation, college students yearn for new things, are easy to accept new ideas and follow new trends. "New" can be said to be a new mark given to them to distinguish themselves from others' uniqueness. On the way of pursuing individuality, they never slow down. The information released should be novel enough to stay ahead of the trend and do something unexpected by others. Only in this way can we become a pacesetter and an opinion leader to attract college students who are eager for new things and further influence their behavioral intentions.

According to the results of correlation analysis, there is a significant positive correlation between the perceived quality of information and college students' travel behavioral intention. In the technology acceptance model, perceived quality is the influence on behavioral intention. The research results of this paper also confirm this result again. There is a significant positive correlation between the perceived quality of information and several traits of opinion leaders, indicating that the perceived quality of information indeed acts as the mediating variable of the model in this study.

References

1. Rahul, K., Banyal, R.K., Goswami, P.: Analysis and processing aspects of data in big data applications. J. Discrete Math. Sci. Cryptogr. **23**(2), 385–393 (2020)
2. Alaei, A.R.: Sentiment analysis in tourism: capitalizing on big data. J. Travel Res. **58**(2), 175–191 (2019)
3. Fang, A.-D., Xie, S.-C., Cui, L., Harn, L.: Research on the structure and practice of internet environment of things based on big data analysis. Ekoloji Dergisi **28**(107), 4239–4247 (2019)

4. Chen, H., Wang, S., Liang, D., Su, Y.: Big data analysis research of power saving in consumer side. Dianwang Jishu/Power Syst. Technol. **43**(4), 1345–1353 (2019)
5. Gilly, M.C., Graham, J.L., Wolfinbarger, M.F., Yale, L.J.: A dyadic study of interpersonal information search. J. Acad. Mark. Sci. **26**(2), 83–100 (1998)
6. Turcotte, J., York, C., Irving, J., et al.: News recommendations from social media opinion leaders: effects on media trust and information seeking. J. Comput.-Mediat. Commun. **20**(5), 520–535 (2015)
7. Alba, J.W., Hutchinson, J.W.: Dimensions of consumers expertise. J. Consum. Res. **13**(I), 411–454 (1987)
8. Chen, Y.-C.: A novel algorithm for mining opinion leaders in social networks. World Wide Web **22**(3), 1279–1295 (2019)
9. Walter, S., Brüggemann, M.: Opportunity makes opinion leaders: analyzing the role of first-hand information in opinion leadership in social media networks. Inf. Commun. Soc. **23**(2), 267–287 (2020)
10. Qiu, L., Gao, W., Fan, X., Jia, W., Yu, J.: Detection of opinion leaders in social networks using entropy weight method for multi-attribute analysis. J. Nonlinear Convex Anal. **20**(7), 1377–1390 (2019)
11. Liu, Q., Xiao, R.: An opinion dynamics approach to public opinion reversion with the guidance of opinion leaders. Complex Syst. Complex. Sci. **16**(1), 1–13 (2019)

Innovative Development Based on Educational Technology in the Era of Big Data

Yan Li[(⊠)]

Xi'an FanYi University, Xi'an, Shaanxi, China
duduma2010@sohu.com

Abstract. Modern educational technology is not only the result of the development of modern education, but also the product of the era of big data. The advent of the era of big data has also played an important role in promoting the development of educational technology, and has also triggered changes in the technology and thinking of the entire society. It is the introduction of big data and its characteristics that explains the impact of big data on traditional education, and explores the way that education technology can promote education development in the context of big data.

Keywords: Big data · Educational technology · Teaching

1 Introduction

Today, Global data is exploding. People's understanding of the objective world is also constantly improving. Every day, many kinds of data are generated. General Secretary Xi Jinping said: "Science and technology in today's world is advancing with each passing day. Modern information technologies such as the Internet, cloud computing, and big data have profoundly changed the way human beings think, produce, live, and learn, and deeply demonstrate the prospects for world development" [1].

In the era of big data in the "Internet + " environment, exploring the integration of big data technology and education technology is of great significance, because there is a natural relationship between big data technology and education technology. Research institutions, experts and scholars have conducted research on the impact, influence and changes of big data in the education field, and put them into practice, and achieved good results. Big data is an issue that must be considered by Chinese university educators.

2 Overview of Big Data

2.1 The Concept of Big Data

Big data is a collection of massive data. Big data (such as infrastructure) has become a basic resource. It is not sampled data, but all data. Relying on the development of technologies such as data storage, statistics, and analysis, people will gain subversive understanding from massive amounts of data, and even be able to understand and

J. C. Hung et al. (Eds.): FC 2021, LNEE 827, pp. 506–510, 2022.
https://doi.org/10.1007/978-981-16-8052-6_62

predict the development trend of things. Therefore, big data is not only a resource, a technology, but also a new discipline and a new way of thinking [2].

2.2 Big Data Technology

Big data technology is far beyond the scope of traditional databases [9]. Big data technology cannot be simply regarded as big "data" and big "database technology" [3]. Big data technology is mainly for professional processing of meaningful data, rather than mastering huge data information. Its innovation is an improvement on the existing technology system, not abandoning the original technology system.

2.3 The Impact of Big Data on Traditional Education

The connotation of "Internet + education" is to use digital technology to realize the efficiency and convenience of knowledge dissemination and update; expansion is to accept knowledge through digital tools and online virtual classroom and dissemination of changes to teachers and audiences [4, 5]. A schematic diagram of big data education is shown in Fig. 1.

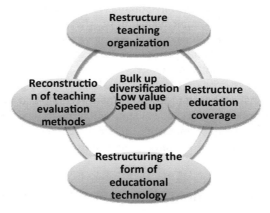

Fig. 1. Schematic diagram of big data education

(1) Reconstruction of teaching organization
   Without the support of all data, it can only plan the teaching process based on experience and assumptions. This can usually only be a subjective, arbitrary, idealized design, and may even run counter to actual teaching. Online education using big data can collect almost all sample data. On this basis, a comprehensive, systematic and objective analysis is helpful to the reorganization of the teaching organization. Online learning tools will record the relevant performance of each student's learning, can evaluate the teaching process, clarify how to obtain electronic resources, which types of problems are most likely to make mistakes,

which knowledge is not mastered, determine appropriate learning methods, and develop appropriate teaching designs, Allocate appropriate teaching resources, and truly teach students in accordance with their aptitude.

(2) Reconstruction of teaching evaluation methods

The focus of teaching evaluation should be mainly on the whole process of teaching. Teaching evaluation is divided into formative evaluation and summative evaluation. Formative assessment mainly analyzes students' learning behaviors, such as online time, number of clicks, number of exchanges, number of speeches, etc., so that teachers can accurately and quickly obtain relevant data of students. For example, the teacher found that 40% of students watched micro-videos, and the ruminating rate was 220%, which means that nearly half of the students watched 2.3 times the video. This shows that students generally think this knowledge point is more difficult. Therefore, teachers can use these big data analysis to understand students' learning dynamics, so as to better improve teaching methods and supervise students' learning behaviors.

(3) Reconstructing the form of educational technology

In the era of big data such as flipped classrooms, micro-courses, and MOOC, changes in the education field have also followed, involving teaching concepts. The concepts of resources and resources have also changed accordingly, which will make informatization teaching continue to face new technical challenges. For example, the design and implementation of MOOC has realized the interactive feedback function and efficiency that the previous open courses, resource networks, and ordinary learning platforms could not achieve.

(4) Reconstruction of education coverage

If education data is open and transparent, it will have a greater impact on real education and make education decisions more credible and more persuasive. At that time, the roles of teachers and students will be assimilated and communicated. New degrees of freedom of interaction and iterative modes will continue to appear in the construction of new teaching relationships.

3 The Development Path of Educational Technology in the Context of Big Data

3.1 Transition from "Digitalization" to "Digitalization"

Digitization is the norm in today's social development. It uses two digital symbols "0" and "1" to describe things; digitization and digitization are closely related, but digitization is not digitization, digitization is the product of digital development and upgrading, and it is also a product of informationization. It can be said that digitization is the foundation of digitization, and digitization is a new product of digitization. Big data technology can make educational technology develop towards intelligence. In the environment of intelligent education technology, teachers can obtain real-time data in the learning process of students, and can grasp the learning behavior and psychological state of students in time, and then can fully understand the differences in students' learning.

3.2 Modern Education is Developing Towards Science

Modern education urgently needs to improve its scientific level, which is a necessary condition for modern education to improve itself. Various new educational technology methods represented by MOOC have also emerged. Big data can enable teachers to more objectively and accurately grasp the characteristics and impact of students' learning, predict students' knowledge level, through the results of data analysis, and use this student who finds learning difficulties, teach students to teach students in accordance, and reduce student losses. The introduction of big data-based learning analysis technology into the construction of large-scale online open courses can help each student choose the most suitable learning method and process, and achieve the purpose of providing students with personalized analysis, push and service.

3.3 The New Category of Educational Technology Development-Analysis

Today, learning analysis technology will become a new hot spot of people's attention. Analytical technology has brought the development of educational technology into a new development field. Therefore, it can be considered that the new category of educational technology research is "analysis". With the informatization of learning resources and the popularization of online learning, students' learning behaviors, processes and results can be recorded comprehensively and truthfully. The learning analysis process can be decomposed into several stages of data collection, analysis, data application and intervention; learning analysis technology can automatically track and record students' learning and statistical analysis, and display the analysis results in an "obvious" visual way. Present complex learning data to teachers and students in a more intuitive and expressive form. Therefore, through the learning of analytical skills, teachers can better grasp the learning process of students, and then implement teaching in accordance with their aptitude.

3.4 The New Mission of Educational Technology Development-Wisdom Education

In the era of big data, students' learning methods are different from digital learning based on multimedia and the Internet, and are no longer limited to books and classroom learning. Compared with the traditional step-by-step learning method with limited time and space, smart learning can make full use of various smart mobile devices to seamlessly connect the learning process of students. Provide a smarter and more humane learning environment [6]. In addition, intelligent learning is an intelligent, anytime, anywhere learning method that can greatly improve students' autonomy and learning efficiency, and is of great help to students' information technology practice and innovation ability.

4   Concluding Remarks

The trend of education development in the information age is big data analysis and education reform. The application of big data makes educational decision-making more efficient and accurate [7]. Teachers can dynamically grasp the learning process of students in real time and run through the entire learning process; make teaching and learning more intelligent, provide more personalized services, and provide personalized resource recommendations and services for teachers and students. In the era of big data, through in-depth analysis and mining of educational data, personalized education is carried out in a digital environment to improve students' thinking style, independent learning and innovation ability, so that students' growth will be better developed in the direction of individualization [8–10].

Acknowledgements. This work was supported by T1901.
Xi'an FanYi University Interpretation "Provincial First-Class Major in E-commerce".

References

1. An, T., Zhao, K.: The development orientation of educational technology in the era of big data. Mod. Educ. Technol. **02**, 27–32 (2016). (in Chinese)
2. Zhang, Z.: Research on the development of online education in the era of big data. J. Jiangsu Open Univ. **02**, 48–53 (2016). (in Chinese)
3. Yang, X., Jiang, Q., et al.: Big data analysis and education reform—reflections on the 15th international forum on educational technology in 2016. Mod. Dist. Educ. **03**, 69–79 (2017). (in Chinese)
4. Yang, X.: Research on the path of modern educational technology promoting educational development under the background of big data. J. Heilongjiang Inst. Educ. **12**, 61–68 (2016). (in Chinese)
5. Hu, Z., Guo, H., et al.: Research on modern education technology under the environment of "Internet+" and big data technology. China Educ. Inf. **08**, 30–33 (2016). (in Chinese)
6. Chen, Y.: Research on the effect evaluation of ideological and political education in colleges and universities based on big data technology. J. Shandong Inst. Agric. Eng. **06**, 127–128 (2017). (in Chinese)
7. Cen, J.: Challenges and innovations faced by higher education in the era of big data. J. Liaoning Inst. Sci. Technol. **10**, 67–70 (2016). (in Chinese)
8. Lei, W.: The path analysis of modern educational technology promoting education development under the background of big data. In: Proceedings of 2018 International Conference on Education and Cognition, Behavior, Neuroscience (ICECBN 2018) (2018)
9. Atiquzzaman, M., Yen, N., Xu, Z. (eds.): Big Data Analytics for Cyber-Physical System in Smart City. Springer, Singapore (2020). https://doi.org/10.1007/978-981-15-2568-1
10. Yu, S., Yang, D., Feng, X.: A big data analysis method for online education. In: 2017 10th International Conference on Intelligent Computation Technology and Automation (ICICTA) (2017)

Urban Landscape Design Based on VR Immersive Interactive Experience App Platform Research and Practice

Lanjian Zeng[✉]

Jiangxi University of Applied Science, Nanchang, Jiangxi, China

Abstract. Urban landscape design is a major content of urban construction. With the development of modern science and technology, design software with different functions has emerged on the market. Based on VR technology, this article makes an idea for the development of urban landscape design App, aiming to provide a new concept for the development of urban landscape design and develop an APP that allows customers to have an immersive interactive experience. This article first expounds the concept of VR technology, then focuses on the immersive interactive experience, and then analyzes the current situation of urban landscape design, including the design principles and deficiencies of urban landscape design. Through questionnaire surveys and interviews, we also explored the market demand and functional requirements of immersive interactive experience landscape design apps. The experimental results show that 86% of people think it is necessary to develop immersive interactive experience landscape design apps, and there are 130 people. I think this App should have the function of rehearsing effects.

Keywords: VR technology · Immersive interactive experience · Urban landscape design · App design

1 Introduction

Although the core concept of the design is the harmony and unity of man and nature, in architectural design and planning, the overall surrounding environment must be controlled. The level of modern urbanization is constantly improving, and people are gradually transitioning from material pursuits to spiritual pursuits. Urban landscape design is particularly important at this time. It can embody people's spirit in environmental construction, and designers also pay more attention to landscape design. The meaning of performance [1]. The current landscape design still uses field surveys, field surveying and mapping, scheme design and then modeling with other design software. In the display of works, it is also displayed in physical venues. Such a model not only takes a long time and wastes resources, but its interactivity and immersion are not well reflected. The emergence of VR technology provides a new development direction for landscape design. Combined with Internet technology, it allows experiencers to break through the limitations of time and space, and they can feel the surrounding environment without going to the venue in person. The immersive interactive experience of

J. C. Hung et al. (Eds.): FC 2021, LNEE 827, pp. 511–519, 2022.
https://doi.org/10.1007/978-981-16-8052-6_63

VR allows users to experience the overall atmosphere of the environment and improve their understanding and experience of the hidden meaning of the design works. This enables landscape design to carry out operations based on the most authentic experience, and allows experiencers to intuitively experience landscape design works, discover deficiencies and formulate solutions, which not only saves resources and costs but also ensures the quality of landscape design works.

Yang yan, lin shanghai and others systematically sorted out the literature closely related to "interactive experience", using systematic literature method, induction method and systematic case method to systematically describe the different specific ways of "interactive experience" in the literature. Sorting out, analyzing, and summarizing, in-depth exploration of the basic connotation of "interactive experience" from the meaning of the literature, and in-depth exploration of the level of its knowledge points from the perspective of social psychology, and through induction to draw their theme, participation, and Dynamic, dramatic and innovative characteristics are the five important characteristics of interactive experience landscapes [2]. HUANG, Ting, JIANG integrate the idea of "immersive theater" and use multi-threaded drama narrative mode as the design method, adding a lot of interactive flow lines in the building, and focus on the design of the sensory experience of the environment, the purpose is to make Traditional regional culture can really touch the audience [3]. Shen Yinghua, Pang Ying and others studied the factors affecting the landscape design of the experience area based on the exclusive landscape characteristics of the experience area. Finally, they summarized the key points of the experience area landscape design, and provided help to showcase the characteristics and advantages of the project [4]. Ziping Y U took the Ming Dynasty exterior wall from Xianhe Gate to Qilin Gate as an example, and proposed a landscape design from the perspective of experience tourism, highlighting the city wall, and absorbing the beautiful environment and rich history and culture of the Ming Dynasty. It also pointed out that the outer wall is an indispensable experience element. The cultural leisure corridor with the theme of "Experiencing Ming Dynasty Culture" can provide visitors with an unforgettable experience. The quality of tourism, the popularity of the Ming outer wall and its cultural heritage will be further extended [5].

The current VR applications are mostly reflected in the game field, but the needs of the market and life will also promote the development of more valuable VR applications, and urban landscape design is a more valuable field. In traditional urban landscape design, due to the obstacles of time and space, it is not possible to achieve a truly immersive interactive experience. Through VR immersive interactive experience, the shortcomings of traditional urban landscape design can be solved, and concepts and thoughts that cannot be described in words can be completely conveyed. This article provides a theoretical basis for the development of urban landscape design App by discussing the theory of VR immersive interaction, as well as a theoretical research foundation for the development of VR applications.

2   Method

2.1   VR

VR, or virtual reality technology, is a particularly hot emerging computer technology at the moment. Refers to the simulation of the real world through computer technology, and the experiencer can use professional equipment to experience and operate this virtual world. VR has the characteristics of multi-sensing, immersive, interactive, and conceptual. In the VR system, people as the subject can experience the virtual world through touch, smell, taste, and hearing.

2.2   Immersive Interactive Experience

The most prominent feature of VR is that it allows the experiencer to be immersed in it and to get feedback from the virtual environment. In addition, it is also possible to freely construct a virtual environment by unfolding imagination. The immersion of VR is mainly achieved by simulating the unique perception of vision, touch, taste, smell, and hearing [6].

Vision. The resolution of the external device can create a sense of visual immersion. VR devices have a PC terminal and a mobile terminal. The resolution of the mobile terminal is generally higher than that of the host terminal, and the VR headset has a very high resolution. But this does not mean that users can have a higher-end experience. In the case of hardware conditions not reached, high resolution can easily make users feel dizzy. To solve this problem and achieve a deceptive effect, the refresh rate of the screen must be accelerated and the resolution must be at least nanometer level [7]. Light often plays a big role in reflecting the realism of vision. In the field of design, VR generally uses ray tracing algorithms and radiosity algorithms to draw realism. The calculation speed of the ray tracing algorithm is fast, and the whole is simpler than the radiosity algorithm. When using this method, taking into account the regionality of the structure and the calculation speed, usually the octree data structure and the Phone model are used, and the overall lighting uses the Wllited model. The formula is as follows:

$$I = I_c + k_s I_s + k_t l_t \tag{1}$$

$$I_c = k_a I_{pa} + [k_d I_{pd}(N_o \cdot L_o) + K_s I_{ps}(N_o \cdot H_o)n] \tag{2}$$

Hearing. For users to experience the immersion of sound in a virtual environment, the sound must be able to adjust the size of the sound, the level of pitch, and the change of timbre according to the actual situation of the user in the real world. The effect of the currently used sound equipment is not ideal, and only changes are made based on the change of the user's position. The sound changes of the sound presented in the virtual reality environment are as complicated as the real world. For example, different media emit different sounds, and the size of the space will also change the size of the sound and the speed of the echo produced.

Taste. If vision and hearing can be simulated and imaginable, then simulated taste is definitely beyond the cognition of many people. New At present, there is a kind of taste simulation technology called Digital Taste Interface in foreign countries. Taste simulation is mainly to simulate the taste and mouthfeel of food. The five flavors of "sour, sweet, bitter, spicy, and salty" are used as the taste attributes of food, and then these attributes can be combined like toning to achieve different tastes. Simulation. The simulated taste is transmitted to the human brain through electronic sensors to achieve the effect of deceiving the tongue. Need to rely on mechanical feedback equipment to realize the simulation of taste. The taste of chewing is achieved through feedback of the skeletal force. There is no technology to simulate the taste of the tongue at present.

Smell. The sense of smell can stimulate the human brain, thereby causing people to imagine the scene of the picture. The sense of immersion brought by smell is stronger, and it can even affect people's mood. For example, the smell of asphalt will reveal asphalt roads and vehicles, which will undoubtedly make people suffering from motion sickness feel uncomfortable. Japan has developed an olfactory assist device. The user wears it on his head, and the switch of the virtual scene triggers a mechanism to make the chemical combination and atomization in the device so that the user can smell the smell. Taste does not want color and taste to be superimposed, so the sense of smell in the virtual environment must be based on the visual counterpart, otherwise the existence of this smell will be very abrupt.

Touch. Tactile sensation can be achieved through the combination of body tracking technology and mechanical feedback technology of wearable devices. At present, wearable devices have partial types such as tactile gloves and finger cots, as well as full-body wearable devices. In addition, the user's hand neurons are stimulated by ultrasonic waves, air pressure fields, and high-frequency sound waves, so as to achieve tactile simulation [8]. At present, wearable products have appeared on the market, but these products do not feel comfortable to wear, which hinders the tactile experience. If you can sense the sense of touch without these external devices, it will become a milestone in the immersive interactive experience.

2.3   The Concept of Urban Landscape Design

The concept of humanistic thought. Each city has its unique regional culture and historical culture. In urban landscape design, humanistic thought is an element that designers often consider [8]. A city landscape rich in regional characteristics can make foreign tourists feel the charm and unique cultural heritage of the city strongly [9]. For example, the mosques in Xinjiang, the Forbidden City in Beijing, the Bund in Shanghai, and the Diaojiaolou in Xiangxi. For areas with a large number of ethnic minorities, the urban landscape design must take into account the local humanistic ideas, including historical culture, myths and legends, and religious beliefs. The injection of humanistic ideas not only allows the traditional Chinese culture to be inherited and carried forward, but also makes this city unique and mysterious, and enhances the attractiveness and competitiveness of this city.

Ecological harmony concept. With the acceleration of urbanization, natural ecology is facing even greater problems. The air quality in the city has decreased, the temperature has increased, and the vegetation has decreased [10]. In order to ensure the sustainable development of all mankind, mankind must attach importance to environmental protection. Therefore, one of the most important functions of urban landscape design is to promote ecological harmony. Urban parks, rivers, and forests must be protected. Many cities will design afforestation and try their best to preserve the original natural landscape during urban planning [11]. In order to improve the natural environment of the city, the coverage of green plants can be increased, and large-scale green parks and green buildings can be established. Applying the concept of ecological harmony to urban landscape design can not only optimize the urban environment, but also affect the development direction of a city. For example, Suzhou City has the reputation of being a garden city, and it is well-known throughout the country.

Functional value concept. Cities are the living places of modern humans, and they bear the heavy responsibility of providing living space and comfortable environment for humans. Therefore, urban landscape design must take into account the basic living environment of humans [12]. For example, increase parks and pedestrian streets in residential areas, and expand green areas in industrial areas.

2.4   Insufficiency of Traditional Urban Landscape Design

The display of traditional landscape design works is the display of objects, so the current problems in its preservation and display also come from objects. The interaction and experience problems arising from this are in the final analysis the many shortcomings in the display of objects. This part analyzes the current situation of preservation and display of landscape design, explores its specific problems, and proposes new functions of landscape design works. Changes and new requirements for interaction and experience.

Current status of preservation of landscape design works. Landscape design is a comprehensive discipline that integrates art, architecture, social behavior, psychology, anthropology and other disciplines. Its professional integration determines the diversity of the display form of landscape works, so a single display form Unable to meet the requirements of its performance, a large number of different forms of work display methods have been widely used in the display of landscape works, such as hand-drawn sketches, physical models, etc. Pictures and texts can no longer meet the immersion of landscape design works to convey profound concepts. Interactive display research. With the development of design software, the form of expressing design works has been extended to three-dimensional models, roaming videos, etc. Different types of landscape design works are displayed in different ways. After investigation, the form of exhibits is usually divided into physical models, posters and videos. With the increase in the number of excellent works, the requirements for space are getting greater and greater. Due to the constraints of space and manpower, it is gradually difficult to preserve the exhibits. Increase. Physical models can be divided into three types: large, medium and small. Small and medium models are generally stored in indoor warehouses. Large physical models are generally stored in the open air. As new works are born in the school every year, the space is limited. The works are replaced by new ones, and even processed by disassembly. Only award-winning or extremely outstanding works can be preserved for a long time. Display posters are usually stored in the warehouse for a short time after the exhibition hall is dismantled, and then destroyed.

Documents in electronic form such as videos are generally copied and sealed, which also brings great inconvenience to later viewing and browsing.

Present situation of landscape design works display. At present, the display of landscape design works is mainly divided into two display methods, physical and network. The physical display is mainly on-site physical display, which can be divided into two types. One is to display design works in designated exhibition halls. For example, the New York School of Visual Arts uses multimedia projection equipment to display student works on each floor of the college. In China, professional colleges and universities generally hold exhibitions of students' graduation design works during the graduation season, and they also use physical exhibition halls for exhibition; another physical display is to use landscape design works as practical signs for on-site production, usually in urban parks., Vacant land and some creative bases for the production and display of landscape works, such as Shenzhen Overseas Chinese Town Creative Park. The network display of landscape design works is mostly web-based display platforms, such as Landscape China, Zhuzhu.com, and foreign websites such as LANDZINE, PLACES, Places journal, etc., which mainly explain the design ideas and design details of the works in the form of pictures and texts And show. The content is mostly excellent cases. Most of the works of landscape design students on display are award-winning works.

3   Experiment

3.1   Experimental Content

In order to design an app that better meets the needs of users, we conducted an online questionnaire on the users of the urban landscape design app. We distributed 150 questionnaires, and finally received 140 valid questionnaires.

3.2   Experimental Process

We first set up 36 questions about the landscape design app, and then carried out a questionnaire survey on the Internet. After the questionnaire survey, we selected focus groups for in-depth interviews. The focus group members are 8 characteristic and representative urban landscape architects. In this interview, the author carried out an urban landscape design app demand symposium in the form of a host. Participants put forward their own opinions and opinions based on the topic and discussed with each other.

4   Discussion

4.1   Market Demand for Urban Landscape Design App

In the questionnaire, we designed the question "Do you think it is necessary to develop an urban landscape design app?" And set up the three options of "necessary, unnecessary, and okay", of which 120 people think it is necessary, accounting for 86% of the total number; 15 people who think it is unnecessary, accounting for 11% of the total; There are 5 people who think it's okay, accounting for 3% of the total;

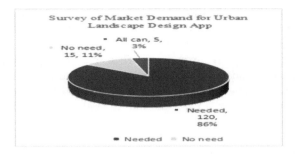

Fig. 1. Survey of market demand for urban landscape design app

As shown in Fig. 1, most of the interviewees believe that it is necessary to design an App specifically for urban landscape design. In the following interviews, we learned that many factors should be considered in urban landscape design. If there is a fully functional App to assist designers in design, such as surveying, calculation, and effect preview, it will greatly improve the quality and quality of landscape design. effectiveness. Some people feel that it is not necessary to develop apps because urban landscape design must consider many factors such as ecological sustainable development, transportation, and aesthetic design. It is difficult for an app to consider all of these aspects.

4.2 Functional Requirements of Urban Landscape Design App

Table 1. Urban landscape design app functional requirements

	Number
Preliminary investigation	115
Data mapping	120
3D modeling	120
Material selection	90
Cost budget	125
Finished preview	130
Late feedback	85

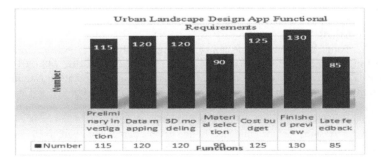

Fig. 2. Urban landscape design app functional requirements

It can be seen from Table 1 and Fig. 2 that for the urban landscape design App, the demand for its effect preview function is the highest. We learned in the following interview that for the preview of the effect, the designer hopes that this App can give people an immersive interactive experience. There are many presentation tools. The traditional multimedia rehearsal does not achieve the ideal effect in the mind of the designer, and it only stays in the video presentation. If there is an auxiliary design platform that allows designers to feel the traffic, light, and overall aesthetics of the actual site, this will greatly reduce design costs, and provide designers with different design ideas, and provide customers with more suitable solutions. After the design is completed, designers and customers can perceive the final effect of the design through virtual reality technology, and propose improvements to the deficiencies of the design, which is conducive to saving resources.

5 Conclusions

Based on the research of VR immersive interactive experience, combined with the basic concept of urban landscape design, this paper analyzes the market demand and functional demand of urban landscape design App. If VR immersive interactive experience is to be applied to urban landscape design apps, it will face huge challenges. First of all, VR devices are not portable. If you want to combine them with apps, you must overcome this difficulty. Secondly, regarding VR immersive interactive experience, the current technology is not very mature, so the experience effect is not ideal.

Acknowledgements. Project: Science and technology research project of Jiangxi Province Department of Education in 2020: "Research on city landscape design app platform based on VR technology", compere: Zenglanjian. (Project Number: GJJ203007).

References

1. Matovnikov, S.A., Matovnikova, N.G.: Innovative urban planning methods for the urban landscape design in the Volgograd agglomeration. Proc. Eng. **150**, 1966–1971 (2016)
2. Yang, Y., Lin, S.: Discussion on interactive experience landscape design strategy. Heilongjiang Agric. Sci. (007), 101–105 (2018)
3. Huang, T., Jiang, S., et al.: New perspective of architectural design based on immersive theatre thinking: a case study of rural living room design in Majia marsh. J. Landsc. Res. **06** (v.9), 61–63 (2017)
4. Shen, Y., Pang, Y.: Landscape design of residential real estate experience zone under the impact of "experience economy". Shanxi Archit. **042**(022), 190–191, 248 (2016)
5. Ziping, Y.U.: Landscape design of the Ming Dynasty outer city wall in Nanjing from the perspective of experience tourism. Landsc. Res. Engl. Edn. **010**(006), 134–136 (2018)
6. Chang, L.: "Immersive experience" in the visual field of media communication and application analysis. Design Art Res. **008**(001), 93–96 (2018)
7. Kim, W.S.: A study on the influence of life sports participants perceived organizational culture on their immersive experience and exercise addiction. Korean J. Sports Sci. **27**(2), 47–60 (2018)

8. Zhang, C.: The why, what, and how of immersive experience. IEEE Access **8**(1), 90878–90888 (2020)
9. Zhou, J.: VR-based urban landscape artistic design. Landsc. Res. Engl. Edn. **012**(001), 114–116 (2020)
10. Kim, S.H., et al.: Effect of resident participation methods on the urban landscape-design - focused on 'turtle-market' in Suwon and 'specialized animation-street' in Bucheon. J. Korean Soc. Design Cult. **23**(2), 83–99 (2017)
11. Xin, B.: Discussion on the urban landscape design and garden plant protection. Manage. Technol. Small Medium Enterp. (008), 108–109 (2019)
12. Xia-Li, Z.: Discussion on the problems and optimization measures in urban landscape design. J. Liaoning Teach. Coll. (Nat. Sci. Edn.) **020**(004), 106–108 (2018)

Based on Big Data the Influence of Chinese Words in Korean on Korean Chinese Teaching

Lin Li[(⊠)] and Qinyi Chen

Department of International Exchange, Jilin Agriculture and Science University, Jilin, Jilin, China

Abstract. The Chinese characters in Korean system are not only the witness of history, but also the wealth of language and culture exchange. It is a very effective way to make full use of this nearly 60% Chinese characters, whether to teach Korean or Chinese to learn Korean. Based on the historical background of the existence of Chinese words, this paper discusses the differences between Chinese and Korean words, and explores the positive and negative transfer of Chinese words in the process of learning Chinese, and puts forward some strategies for teaching Korean Chinese words. In order to make rational use of Chinese characters to improve the efficiency of Korean vocabulary teaching.

Keywords: Big data · Korean teaching · Influence and characters

1 Introduction

In recent years, with the increasing popularity of Chinese and Korean, both Korean and Korean learning Chinese are increasing day by day, and because of the existence of nearly 60% of Chinese characters in Korean, it undoubtedly provides learning advantages for learners of both languages. However, it is obvious that in the actual learning process, this advantage is either ignored or negative transfer occurs, and few learners really use this advantage to learn quickly. This paper aims to explore this phenomenon and study how to use this advantage to improve the efficiency of Korean vocabulary teaching.

1.1 Historical Origin of Chinese Characters in Korean

Chinese is isolated language, Korean is sticky language, in essence, this belongs to two language systems. However, due to the close geographical relationship between China and the Korean Peninsula and the profound cultural influence, Korean is deeply influenced by Chinese. There are three main sources of Chinese characters in Korean. First of all, they naturally come from Chinese. This part of Chinese characters is not only a large number of words, but also widely used, and even many of them have been integrated into the daily life of the common people. Secondly, the Chinese characters introduced from Japan for historical reasons, because Japan belongs to the cultural circle of Chinese characters, and since modern times, because of the expansion of

J. C. Hung et al. (Eds.): FC 2021, LNEE 827, pp. 520–527, 2022.
https://doi.org/10.1007/978-981-16-8052-6_64

Japan's war and the influence of modern economic factors, Japanese has also become an important factor in modern Korean. The last part is the self-created Chinese characters in Korean, the plasticity of Chinese characters is very strong, Koreans use the characteristics of Chinese characters and their own national situation, create a part of the Chinese characters, mainly with names, place names, life terms and other words.

Chinese characters from Chinese.

Chinese characters 1. introduced from traditional classics.

The Korean Peninsula has been influenced by Han culture since early, and since the first century BC, Chinese cultural classics have been introduced into the Korean Peninsula. 'According to the 'Biography of Koguryo in North History', Koguryo' books have five classics and three history, three Kingdoms, and Jinyang Qiu'. "Five Classics" is "Shang Shu", "Yi Jing", "Book of Songs", "Yi Li", "Spring and Autumn", "Three History" is "Historical Records", "Han Shu", "Later Han Shu", "Three Kingdoms" and "Jin Yang Qiu" [1]. During the period of the Korean dynasty, its relationship with the Ming Dynasty was more close, political, economic, cultural, military and many other aspects were deeply influenced by the culture of the Ming Dynasty. Confucian classics became the mainstream thought of the Korean Peninsula.

1.2   Chinese Characters Introduced by Buddhism

(1) The influence of vernacular writing

Compared with classical Chinese, vernacular Chinese is a written form formed on the basis of spoken language, which is between spoken language and classical Chinese. Originally originated from popular literature, such as strange novels, Song and Yuan dialect books, to modern times, gradually evolved into modern Chinese. The vernacular language in Korean system comes from both the words of Song and Yuan Dynasty and the novels of Ming and Qing dynasties, as well as from the modern vernacular words after the May 4th Movement.

(2) Chinese characters derived from Japanese

After the Meiji Restoration, Japan continuously learned from the western society and introduced a large number of western civilizations, which made the national strength gradually strong. In the process of learning from the West, Japanese was influenced by Western culture, constantly introduced western vocabulary, and made full use of the word-making function of Chinese characters to form a number of new Japanese Chinese characters vocabulary, in the process of Japanese colonial expansion. Then this kind of Japanese characteristic Chinese character word brought into the Korean peninsula. In the current Korean system, most of the Chinese words absorbed since modern times belong to this column, mainly nouns, adjectives and verbs, and other parts of speech are less.

(3) Creating Chinese characters

Chinese is an isolated language, an ideographic character, master its word-formation rules, you can use Chinese characters to create a new vocabulary of Chinese characters, which is also a major advantage of Chinese. The Korean Peninsula has used this characteristic of Chinese characters to create many new Chinese characters. This kind of vocabulary is mainly living language, including a large number of names, place names. Because this kind of vocabulary creation follows the rules of Chinese character word formation, even if it has never been contacted, it can understand its meaning as long as it understands Chinese characters, which belongs to a relatively convenient and flexible class of Chinese characters with Korean style.

(i) Synonyms

Homomorphic synonyms refer to the same characters and meanings in Korean and Chinese. There are many such words. According to the research results, homomorphic synonyms account for 73% of the whole Korean Chinese characters [2]. The synonyms, which account for 73% of the Chinese characters in Korean, are an excellent bridge for both Korean and Chinese to learn Korean, such as Li , Kim, State , Research , students and so on. In the process of teaching Korean Chinese, this kind of vocabulary is the easiest to master and the most difficult to produce errors. Therefore, reasonable utilization can greatly reduce the difficulty of learning and improve the efficiency of learning.

(ii) Synonyms

China and South Korea have experienced their own different historical processes. In this long historical process, the language representing their culture will naturally change. Taking modern Chinese as an example, in comparing modern Chinese with ancient Chinese, We will find that the same phenomenon of word meaning change, which is the normal evolution of language development. Compared with ancient Chinese, modern Chinese is the same, and it is inevitable to change between the two language systems.

1. Word meaning expanded

Refers to the meaning of Chinese words in Korean more than its corresponding Chinese meaning. There are two main reasons for the formation of this kind of words. On the one hand, modern Chinese vocabulary has shrunk compared with ancient Chinese, but its semantics in Korean has not changed, so it leads to more Korean semantics than Chinese semantics. On the other hand, Korean semantics has expanded, adding extension items in line with Korean historical development or national development, making its meaning richer.

For example:

Extracurricular: Modern Chinese [3];Modern Korean: Amateur; Extracurricular Activities; Extracurricular Tutoring; Extracurricular Teaching [4];: Take medicine, as opposed to external application (different from external application) [3] Contemporary Korean: underwear; oral [4]

2. The meaning of words

In this case, the meaning of the word in Korean is less than that in modern Chinese. The reason for this situation can be understood as two aspects. On the one hand, the meaning of modern Chinese itself has expanded in the process of development, while Korean still retains the original meaning of Chinese characters in ancient Chinese, which leads to the change of meaning in modern Chinese than in Korean.

For example: Wife: wife and children in modern Chinese; formal spouse of a man [3].

Modern Korean: Wife and Children [4]: In and Out; Income an Expenditure [3] Modern Korean: Advance, Advance, Enter [4]

3. The same part
The so-called word meaning part is that Chinese characters retain the same meaning in modern Chinese and modern Korean, but because of the evolution of the two languages, they have developed different meanings. This makes this part of the vocabulary in the two language systems have the same meaning, but also have their own different meanings.

For example:

Article modern Chinese: originally refers to the text, now refers to the length of not very long and independent text.A general reference to a work.The metaphor is the meaning of twists and turns.A matter; a procedure [3]; Modern Korean: article (with modern Chinese meaning). Sentences.A(essayist) phrase [4].

(iii) Synonyms of near form
In addition to the above, there are a class of Chinese characters with similar glyph and the same meaning. Some of these Chinese characters contain common morphemes, some are morphemes reversed, and some add morphemes to Korean. There are similarities in word formation, but not exactly the same, but the semantics are consistent.

The weather is—sunny; the advantages—longer

To call—glory—glory

Products—products; translators—translators

This part of the synonym of word order inversion is usually called Chinese and Korean reverse order words.

2 The Influence of Chinese Words in Korean on Chinese Vocabulary Teaching

2.1 Positive Migration

1. Synonyms and reduce memory difficulty
Homomorphic words account for 73% of Korean Chinese words, while Chinese words account for nearly 58.5% of Korean words. That is, nearly half of the words in Korean can be translated directly with Chinese words, which greatly reduces the difficulty of Korean learning Chinese. And in this part of the vocabulary, although the Korean writing style is widely used in modern Korean, which makes the characters different from the Chinese characters, in fact, Koreans are no stranger to

the Chinese characters that can correspond to Korean, especially the names and place names, so it is relatively easy to remember this part of the synonyms.

2. Lexical similarity enhances semantic comprehension

   For non-homomorphic words, whether homomorphic or near-shaped synonyms, because of their similar word-formation method and the ideographic function of Chinese characters, it is easy to understand the semantics of Chinese characters and reduce the deviation of understanding. It can also reduce the difficulty of learning.

3. Speech is similar, improve recognition

   Many Chinese characters have similar pronunciation in the two language systems, such as the protagonist every household and so on. This kind of vocabulary increases the recognition degree in pronunciation. Enable learners to actively correspond to Korean words and be willing to incorporate them into the language system.

2.2 Negative Migration

Chinese characters bring convenience to Korean learners in the process of learning and promote Korean learners' Chinese learning. However, Chinese characters will also bring negative transfer to Korean learners and become a stumbling block for Korean learners to learn Chinese. If learners cannot break through this effect, their learning progress may be worse than that of learners without negative transfer of mother tongue in Europe and the United States.

1. Differences in parts of speech

   Different from Chinese, Korean's part of speech identification is very obvious, verbs, adjectives often need to add as a sign, [5]. But Chinese does not, and Chinese vocabulary often has a variety of parts of speech, specifically need to contact the context to distinguish. This is difficult for Korean learners who are used to dividing parts of speech by identification. Make friends make friends make friends make friends make. In Korean Most of the parts of speech in Korean Chinese characters only take the noun word, but in order to do the verb, we need to use auxiliary, to do the adjective need auxiliary, which will lead to errors in Chinese.

2. Semantic differences

   As mentioned earlier, there are quite a number of different meanings in modern Chinese and modern Korean, among which the meaning of Chinese characters is enlarged, the meaning of words is reduced, and the meaning of words is partly the same. Even there are completely different situations. Although in the process of learning, the similar meaning of words can play a positive role in transfer, so that learners can quickly replace their mother tongue, but also produce negative transfer, that is, hope for literary meaning. Using the meaning of your mother tongue to understand Chinese semantics is likely to lead to errors. For example, the above mentioned "article", Korean text also has the meaning of text, but more often used to express the meaning of "sentence", and it is obviously wrong to understand the meaning of "figurative zigzag concealment" in Chinese. But for similar to "turbulence (Chinese refers to wave ups and downs, instability, not calm; Korean refers to (face) rich and beautiful)", "consultation (Chinese refers to exchange of views; Korean refers to the amount of thought", "physique (Chinese refers to body and energy; Korean refers to zombies)" [6]. This kind of Chinese words with completely different meanings are more prone to errors.

3. Emotional Color Differences

Emotional color refers to the emotional attitude of existence or approval or opposition to an objective thing contained in the vocabulary, which can be divided into three types: positive, neutral and derogatory. The emotional color of Chinese character vocabulary itself will change with the development of the times and society. There are also great differences in the two language systems of modern Chinese and modern Korean. If we turn a blind eye to it, the light will make jokes, and the heavy will even affect the feelings of both sides of the conversation [7].

For example: dereliction of duty, expressed in modern Chinese as failure to fulfill their duties, is derogatory. Modern Korean only refers to the loss of work, as a neutral word. the middle finger of modern Chinese shelved, placed; in Korean meaning placed, shelved, regardless of, do not care about the meaning, implied derogatory meaning.

3 Strategies for Teaching Han Chinese Vocabulary by Using Chinese Characters

3.1 Comparative Teaching

The application of contrastive teaching method in the teaching of Korean Chinese characters is both interesting and useful. In modern Chinese and modern Korean, Chinese characters have both "same" and "different". Taking "same" and "different" can effectively promote positive migration and reduce negative migration.

(1) In the teaching of Korean Chinese, we should make full use of the Chinese characters of "synonymous form" to help students understand the internal relationship between Korean and Chinese. Since modern Korean is written in Korean, it is very different from Chinese characters. Therefore, while teaching this part of the vocabulary, teachers need to guide Korean learners to learn Korean and Chinese characters, although the writing method is different. But Chinese characters correspond to Korean characters, such as flowers, Li, week. For this part of Chinese characters, not only synonymous with form, but also similar pronunciation, can be used to activate the corresponding Korean Chinese characters, [8] and then through the mother tongue system to master Chinese characters, so as to quickly grasp and understand Chinese characters, Reduce learning difficulty.

(2) Context teaching

Language teaching cannot be one-sidedly understood as vocabulary and grammar teaching, especially in the teaching of Korean Chinese, the existence of 58.5% of Chinese words makes the teaching of Korean vocabulary relatively easy, but only increases vocabulary, but neglects the nuances of Chinese words in the two languages systems, still does not achieve the purpose of teaching.

Chinese characters have great differences in part of speech, emotional color and stylistic color, and there are more or less differences in the meaning of words. It is difficult to grasp these differences in simple vocabulary teaching, which needs to be combined with the situation and be taught in pragmatics [9]. Therefore, teachers need to provide students with specific contexts for vocabulary use in order to help students understand the differences between the two languages, especially the problems of emotional color and stylistic differences, and need context to help teaching.

(3) Morpheme teaching

Chinese and Korean have strong similarities in word formation, Korean and Chinese word formation converge, and can fully understand and use the method of Chinese morpheme word formation [10]. And the corresponding way of Chinese characters is basically morpheme correspondence, and many Korean corresponding Chinese characters exist in one-to-many way, such as mentioned above According to morpheme teaching, students can understand can correspond to "words, disasters, transformation, fire, flowers", and find the corresponding Chinese characters in the mother tongue system can be quickly understood. This teaching method can make learners form the ability to derive their own vocabulary. Even if new words appear, according to word formation and the same morpheme, learners can quickly substitute their mother tongue and learn new words.

4   Conclusions

Chinese characters are the bridge to Korean Chinese teaching and the link. The rational use of Chinese characters can promote the positive transfer of mother tongue, reduce the negative transfer of mother tongue, and achieve the purpose of learning Chinese efficiently. The existence of Chinese characters makes the difference between Korean Chinese teaching and non-Chinese circle Chinese teaching. Therefore, we should be more targeted in the process of thinking about Korean Chinese teaching in order to highlight the advantages of Chinese characters. Bring learning convenience to learners.

Acknowledgment. This work was supported by Jilin Province Education Science Planning Project "Study on the Influence of Chinese-Korean Language Contrast on Second Language Acquisition" (ZD19063).

References

1. Bai, L., Cui, J.: A Comparative and Common Error Analysis in Chinese. Educational Science Press, Beijing (1991)
2. Chen, Y.: DongGai Language: A Study on Korean Chinese Characters, Words. Liaoning Normal University Press, Dalian, p. 2 (2007)
3. Pu, J.: Chinese character words in Korean and its teaching method exploration. J. Anhui Univ. Technol. **01**, 125 (2010). http://xh.5156edu.com/. Online Xinhua Dictionary

4. Hong, M., Li, S., Li, F.: Chinese and Korean Dictionary, p. 8. Heilongjiang Korean National Publishing House (2012)
5. Wang, Y.: A study on Chinese words in Korean and an analysis on the teaching strategies of Chinese vocabulary in Korean; and Liaoning University, p. 5 (2016)
6. Liu, P.: Korean. Grammar Commercial Press (2017)
7. Wang, Y.: A contrastive study and teaching strategies of homomorphic words in Chinese and Korean, p. 5. Nanchang University (2020)
8. Gold lotus. A study on the acquisition of homomorphic words in Han, p. 5. Nanjing University (2018)
9. Any meeting. A study on the teaching strategies of Chinese character words in Korean Chinese teaching.1 ± dissertation. Lanzhou University (2012)
10. Wang, L.: Comparison and teaching of Korean homomorphic words. To Korean vocabulary overseas Chinese education (2008)

The Effect of New Type Smart Classroom Teaching Based on Structural Equation Model

Shuzheng Zhao[⊠]

School of Economics and Management, Yunnan Technology and Business University, Kunming, Yunnan, China

Abstract. Students are the main body of learning. How to mobilize students' autonomy in teaching and provide them with a good and harmonious development environment is particularly important. In order to improve the quality of classroom teaching and the level of class management, it is necessary to continuously improve the ability of interaction and communication between all teachers and students in the class and their participation in the classroom to enhance their sense of autonomy and creativity; this article first deals with the relevant situation of the new smart classroom introduction, including its functions, changes brought to traditional classrooms, and existing shortcomings; secondly, the structure and construction steps of the structural equation model are introduced; finally, the structural equation model is constructed to study the teaching effect of the new smart classroom, and the research shows students are the biggest factor affecting the teaching effect of the new smart classroom. The comprehensive influence coefficient of students on teaching effect is 0.88, the influence coefficient of teachers on teaching effect is 0.81, and the influence coefficient of external environment on teaching effect is 0.77.

Keywords: Structural equation model · Smart classroom · Effect research · Teaching mode

1 Introduction

Gerbing DW, Hamilton JG, Freeman EB, etc. studied and analyzed the construction of a large-scale structural equation model with 57 projects, thereby analyzing the relationship between management's participation in strategy formation and organizational plan revenue. The results show that there is a strong causal link between management participation and the two types of strategic interests, which indicates that management participation can enhance the effectiveness of the strategic process [1]. Elferink-Gemser MT, Roos ID, Torenbeek M and others studied the training volume and competition performance of young speed skaters by constructing a structural equation model, and pointed out that the psychological structure of young speed skaters is very important for training volume and competition performance [2]. Badri M, Rashedi AA, Yang G built a structural equation model based on causality and determinants, and studied the willingness of college students to learn online courses. The results showed that the intention of using e-learning, the perception of ease of use, and the perception of usefulness As well as other factors (such as user characteristics and support) there is a

J. C. Hung et al. (Eds.): FC 2021, LNEE 827, pp. 528–536, 2022.
https://doi.org/10.1007/978-981-16-8052-6_65

significant connection [3]. Yuan K H, Bentler PM et al. found that in structural equation modeling (SEM), parameter estimates are usually calculated by Fisher scoring algorithm, which is usually difficult to obtain a convergent solution. Even for simulated data with correctly specified models, non-fusion replication is frequently reported in the literature. However, research on the ridge method in SEM shows that adding a diagonal matrix to the sample covariance matrix can also improve the convergence speed of the Fisher scoring algorithm. By using statistical and numerical analysis, they clarified why both methods can improve the convergence speed of SEM and provided suggestions on how to improve the speed and convergence speed in parameter estimation [4].

2   Method

2.1   New Smart Classroom

The construction of smart classrooms mainly uses three technologies: cloud, internet and terminal. The function of the cloud is mainly to collect, store, manage and use resources; while the network is a miniature cloud server, which can guarantee the communication between teachers and students without a network; the terminal refers to the mobile internet device [5]. Combining the characteristics and teaching mode of smart classrooms, we can define smart classrooms as follows: Smart classrooms are the use of "Internet +" thinking, the use of Internet of Things technology, artificial intelligence technology, big data technology, cloud computing, etc. to transform and update traditional classroom teaching the environment, with the help of real-time data analysis system, provides a full range of technical support for the whole process of teaching, such as teacher preparation before class, student preview, teacher teaching in class, and review and expansion after class, making teaching decisions more science, teacher-student interaction is more three-dimensional, teaching feedback is more timely, and teaching resource sharing is more comprehensive. The new smart classroom is a new teaching model that is conducive to the overall development of students [6].

2.2   The Changes Brought by the New Smart Classroom to the Traditional Classroom

(1) Updated the teaching philosophy of teachers
     Many people think that smart classrooms just use the form of "Internet + classroom", and use some modern information technologies on the basis of traditional classrooms, such as multimedia, teacher's tablet and student's tablet, so that the work of preparing lessons, expanding and reforming homework becomes convenient efficient, not very different from traditional classroom [7]. In fact, it is not the case. The use of traditional teaching to develop quality education still has great limitations. The new smart classroom provides new teaching methods and learning methods. Students are the main body of classroom teaching activities, and students' independent learning ability is improved by cultivating students' independent learning ability. Their innovative ability, using more targeted teaching methods to improve the comprehensive quality of students. The new type of smart classroom

makes the classroom more interesting and closes the teacher-student relationship. Through the use of modern teaching technology, teachers have also changed the traditional teaching concepts, which is conducive to formulating plans to teach students in accordance with their aptitude.

(2) Enriched teaching methods and learning methods

The teaching process in traditional classrooms is a process in which teachers teach students to passively receive knowledge. Some teachers have tried to change teaching methods, hoping to let students go deep into classroom learning and experience the fun of the classroom. However, due to the limitations of teaching conditions, methods such as discussion, raising hands, and questioning have not significantly improved the initiative of students in learning. The new smart classroom provides many different teaching methods and learning methods. Functions such as taking photos to answer questions and like scoring arouse students' curiosity and increase the interest of the classroom. In the new smart classroom, the teacher's teaching is no longer based on the text, and the students have also changed from passively receiving knowledge to actively learning knowledge [8].

2.3 Insufficiency of Smart Classroom

(1) The application of technical functions is not mature enough

The construction of the smart classroom technology system has basically covered all teaching activities including pre-class preparation (preparation), in-class teaching (listening), and post-class extension (review), but the actual data shows that there is such a phenomenon: when teachers use smart classroom related equipment for teaching, they will most commonly choose functions such as whiteboard comments, electronic textbooks, resource sharing, assignment and correction of homework, video recording on the same screen, and class questioning and awarding. Functions such as projectors, courseware production, and class space are rarely used, and the utilization rate of human-computer dialogue and one-on-one tutoring is almost zero. Students use the functions related to classroom interaction and completion of homework the most, and use the pre-class preview function, the after-class expansion function, and the wrong question collection function less. As for the use of functions related to micro-class review and non-meeting with the teacher, etc. It is very few.

2.4 Structural Equation Model

Structural Equation Model (SEM) is a statistical method that uses the covariance matrix of variables to study the relationship between variables. This model is a theoretical model for abstract phenomena that are difficult to explain simply and more complex, and then analyzes and processes them using certain statistical analysis techniques. First determine the model of the research object, and then optimize and analyze the data obtained from the model, and finally achieve the purpose of scientific and effective research on the actual problem studied [9]. SEM is actually a model for studying the causal relationship between variables. Customer satisfaction is difficult to directly measure products or services due to human subjective factors, so it generally needs to be measured indirectly through some other external variables [10].

2.5 The Structure of the Structural Equation Model

The structural equation model is mainly used to measure the error and study the structural relationship of the latent variables in the model. The structural equation model is mainly composed of variables and variable relations. According to the relationship between variables, three equations can be used to express the structural equation model [11, 12].

Measurement model (external model)

$$X = \Lambda x \xi + \delta \tag{1}$$

$$Y = \Lambda y \eta + \varepsilon \tag{2}$$

Λx represents the regression coefficient matrix of x versus ξ, Λy represents the regression coefficient matrix of y versus η, δ represents the dominant variable, and ε represents the measurement error.

Structural model (internal model)

$$\eta = B\eta + \Gamma \xi + \varsigma \tag{3}$$

In this equation, η and ξ are expressed as an endogenous latent variable vector and an exogenous latent variable vector; B and Γ are the structural coefficient matrix of η and ξ, respectively; ζ is the error vector of the latent variable model.

2.6 Structural Equation Model Construction Steps

(1) Model construction

To construct a theoretical model, we must first clarify the relevant professional theory and the purpose of research. Structural equation model is used to analyze the relationship between variables. Therefore, to establish a structural equation model of high school class self-management satisfaction, it is necessary to clarify the individuality and role of each variable in the model. Clarify the latent variables of the model, and then use the data obtained from the survey questionnaire to verify whether the theoretical model constructed is reasonable.

(2) Model recognition

The key to determining the structural equation model is that the free parameters of the model must not exceed the sum of the variance of the observation data and the variance of the universe, otherwise the model may not be able to identify and obtain estimates of the free parameters.

(3) Model estimation

The parameters should be estimated after the model is set. Through research, we can know that there are actually many parameter estimation methods in the estimation of structural equation models. However, each estimation method has its own applicable situations and advantages. Usually parameter estimation methods include maximum likelihood method and generalized least square method.

(4) Model evaluation

If you want to know how well the constructed model fits, you should evaluate the model, that is, whether the survey model can adequately explain the observed data. In order to evaluate the model, it may be necessary to perform multiple tests on the model throughout the evaluation process. When evaluating, it is necessary to check the various parameters in the model, but also to check the structural equation, and more importantly, to evaluate the fit of the entire model.

(5) Model modification

In order to improve the fitting degree of the initial model, a variety of tests in the structural equation model are used for correction and evaluation. When the analysis result finds that the degree of fit is not ideal, it is necessary to find a way to correct the model and test a reasonable fit index.

3 Experiment

3.1 Experiment Content

Through the analysis and summary of the relevant literature, we found that the factors affecting the effectiveness of the new smart classroom teaching are external factors, teacher factors and student factors. This experiment selects these three indicators to construct a structural equation model, and analyzes its impact on the effect of new smart classroom teaching.

3.2 Experimental Hypothesis

(1) Teachers are participants in the new smart classroom and maintainers of order, which have a great influence on the effectiveness of classroom teaching. This experiment sets up three variable indicators of "teacher attitude", "teacher ability", and "teacher quality", and makes a hypothesis: "teacher factor" has a very significant positive path influence on the new smart classroom.

(2) Students are the main body of the new smart classroom and the main carrier for showing the pros and cons of teaching effects. This experiment sets up three variable indicators of "student attitude", "student ability", and "student quality", and makes the hypothesis: "student factor" has a significant positive influence on teacher factors, and students have a significant effect on teaching.

(3) External factors are factors other than teachers and students. In the new smart classroom, there are many external factors that affect the teaching effect, such as "device software", "device hardware", "family problems", etc. These external factors are summarized into three variable indicators: "equipment factor", "resource factor" and "family factor". Based on these variable indicators, we propose a hypothesis: external factors have a significant positive path impact on teacher factors, and external factors have a very significant positive path impact on student factors.

According to the hypothesis, we can draw a conceptual model of the structural equation.

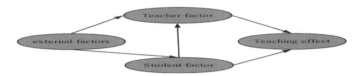

Fig. 1. Conceptual model of influencing factors of new smart classroom teaching effect in colleges

3.3   Experiment Procedure

We have set up a questionnaire on the effect of new smart classroom teaching. The questionnaire covers four aspects: "teacher factor", "student factor", "external factor" and "teaching effect". We selected a university as the experimental research object and distributed 270 questionnaires, of which 255 valid questionnaires were returned, and the questionnaire response rate was as high as 94%. Then we used SPSS22.0 as a data analysis tool to test the reliability of the questionnaire. The result was that the Cronbach's coefficient of the total table reached 0.952, and the Cronbach's coefficient of each component was also between 0.829 and 0.901. In addition, the result of the spherical test shows that the KMO value of the total table is 0.921, which can be used for factor analysis. We use the model fitting tool AMOS21.0 to fit the model. In order to verify the fit of the model, we choose standard fitting index, comparative fitting index and other commonly used model fitting indexes to measure, among which the NFI value is 0.922 The TLI value is 0.974, the CFI value is 0.981, and the RMSEA value is 0.041. The results show that the indicators of the model have reached an acceptable level, and the model fits well on the whole.

4 Discussion

4.1 Output Path Normalization Coefficient After Model Fitting

Table 1. Structural equation model coefficients for evaluating the effect of smart classroom teaching in colleges and universities

	Student factor →			Teacher factor →			External factor →			Teaching effect →			
	Attitude	Ability	Quality	Attitude	Ability	Quality	Equipment	Family	Resources	Achievements	Experience	Knowledge	Skills
	0.71	0.82	0.73	0.58	0.15	0.74	0.63	0.78	0.67	0.73	0.77	0.83	0.82
→ Student factor	#			#			0.80			#			
→ Teacher factor	0.44			#			0.60						
→ External factor	#			#			#						
→ Teaching effect	0.52			0.81			#						

From Table 1, we can see that the direct influence coefficient of student factors on teachers is 0.44, and the direct influence coefficient on teaching effects is 0.52; the direct influence coefficient of teacher factors on teaching effects is 0.81; the direct influence coefficient of external factors on students is 0.80, the influence coefficient of team teachers is 0.60.

4.2 Comprehensive Influence Coefficient of Each Module

According to the relevant content in Fig. 1 and Table 1, we can calculate the comprehensive influence coefficient of each indicator variable on the teaching effect of the new smart classroom (Table 2).

Table 2. The comprehensive influence coefficient of each index variable on the teaching effect

	Coefficient
Student factor	0.88
Teacher factor	0.81
External factor	0.77

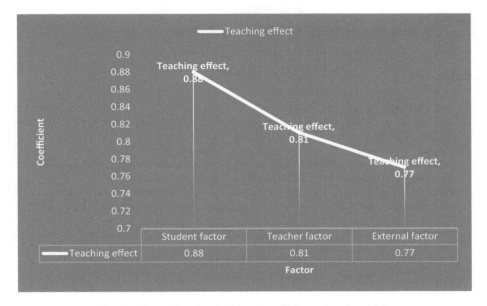

Fig. 2. Comprehensive influence coefficient of each module

From Fig. 2 we can see that on the whole, student factors have the greatest impact on the teaching effect of the new smart classroom, with an impact coefficient of 0.88; followed by teacher factors, with an impact coefficient of 0.81; external factors indirectly affect the teaching effect of the smart classroom by influencing students and teachers, and its influence coefficient on the teaching effect is 0.77.

5 Conclusions

Through the exploration of this article, we can know that in the new type of smart classroom, the main body of the classroom has changed from teacher to student, so the student factor has the greatest impact on the teaching effect of smart classroom. This requires students to improve their own learning initiative, actively participate in the classroom, and consolidate relevant knowledge and expand in time after class. In the new smart classroom, although the workload of teachers is reduced, it is still a key element that affects the quality of teaching. This not only requires teachers to master the use of modern teaching equipment, but also to maintain classroom order and guide students to learn independently. External factors have also had a great impact on the teaching results of smart classrooms. Colleges and universities have also made every effort to improve teaching equipment. Families should also increase the importance of students' learning, and at the same time supervise students to use the Internet rationally after class.

Acknowledgements. Supported by the Scientific Research Foundation of Department of Education of Yunnan Province, China, Item number: 2021 J0894

References

1. Gerbing, D.W., Hamilton, J.G., Freeman, E.B.: A large-scale second-order structural equation model of the influence of management participation on organizational planning benefits. J. Manage. **20**(4), 859–885 (2016)
2. Elferink-Gemser, M.T., Roos, I.D., Torenbeek, M., et al.: The Importance of psychological constructs for training volume and performance improvement. A structural equation model for youth speed skaters. Int. J. Sport Psychol. **46**(6), 726–744 (2016)
3. Badri, M., Rashedi, A.A., Yang, G., et al.: Students' intention to take online courses in high school: a structural equation model of causality and determinants. Educ. Inf. Technol. **21**(2), 471–497 (2016)
4. Yuan, Ke-Hai., Bentler, P.: Improving the convergence rate and speed of Fisher-scoring algorithm: ridge and anti-ridge methods in structural equation modeling. Ann. Inst. Stat. Math. **69**(3), 571–597 (2016). https://doi.org/10.1007/s10463-016-0552-2
5. Wentao, H., Yaping, W., Gang, M.: An analysis of the interactive features of collaborative learning in smart classroom: based on IIS map and social network analysis% the perspective of network analysis. J. Dist. Educ. **036**(003), 75–83 (2018)
6. Li, X.: Smart classroom and its effective generation. US-China Educ. Rev. B (1), 32–38 (2020)
7. Zhang, L.: Smart classroom design on the basis of internet of things technology. Mod. Electron. Technol. (Engl.) **003**(001), P.31–P.35 (2019)
8. Faritha, B.J., Revathi, R., Suganya, M., et al.: IoT based cloud integrated smart classroom for smart and a sustainable campus. Proc. Comput. Sci. **172**, 77–81 (2020)
9. Beaumelle, L., Vile, D., Lamy, I., et al.: A structural equation model of soil metal bioavailability to earthworms: confronting causal theory and observations using a laboratory exposure to field-contaminated soils. Sci. Total Environ. **569–570**(nov.1), 961–972 (2016)
10. Brandmaier, A.M., Prindle, J.J., Mcardle, J.J., et al.: Theory-guided exploration with structural equation model forests. Psychol. Methods **21**(4), 566–582 (2016)
11. Yang, L., Wu, Q., Hao, Y., et al.: Self-management behavior among patients with diabetic retinopathy in the community: a structural equation model. Qual. Life Res. **26**(2), 1–8 (2017)
12. Jacobucci, R., Grimm, K.J., Mcardle, J.J.: A comparison of methods for uncovering sample heterogeneity: structural equation model trees and finite mixture models. Struct. Eqn. Model. Multidisc. J. **24**(2), 1–13 (2017)

Research into Network Consumers' Luxury Value Perception in E-Shopping Environment: Scale Development and Validation

Zeyun Li[(⊠)]

School of Management, Tianjin University of Finance and Economics Pearl River College, Tianjin, China
lizeyun86@xueshumail.cn

Abstract. E-shopping have become an essential part of modern life. The combination of fashion and technology is a key business strategy of many luxury brand. This investigation developed a luxury value perception scale of network consumers by using grounded theory method. Results showed that the scale of value perception is composed of three dimensions and each dimension having their own components. A SEM method was used to prove the high reliability and validity of the scale. This paper also provide reference and inspiration for further research in Internet consumption field.

Keywords: Online shopping · Scale development · Luxury brand · Value perception

1 Introduction

With the development of technology, the functions of Internet services increasingly powerful, while the consumer demands become more complicated. Traditional offline consumption does not satisfy all today's consumers. Many of the luxury brands has been making a strong push into online promotion. And as luxury market has been the most growth market segment, both the marketing industry and academia have been paying more attention to its potential and growth [1]. Although the global luxury market still faces a great deal of challenges, with the further maturity and development of consumers, total luxury consumption amount has reached a record $9.8 trillion which rise at record levels in 2019. And, it's worth noting that, due to the COVID-19 outbreak, the explosion of online economy is a new business form in the luxury industry.

Since the end of the last century, computer and network has become universal in China. Not only this can help people in every aspects of their daily life, but also a consumption fashion is created from this as well. For example, a new luxury product may doesn't fit some customer group, but these consumers will be still purchase it online which they could not afford. Therefore, it will be necessary for researchers to explore the reason for luxury online shopping.

The present article makes two contributions. First, Chinese luxury consumption need a fully specified understanding of Chinese context, identifying value perception

© The Author(s), under exclusive license to Springer Nature Singapore Pte Ltd. 2022
J. C. Hung et al. (Eds.): FC 2021, LNEE 827, pp. 537–544, 2022.
https://doi.org/10.1007/978-981-16-8052-6_66

of Chinese consumers is essential. Second, this new scale of luxury value perception offers further direction and ideas for benefiting luxury online marketing.

2   Literature Review

2.1   Online Luxury Consumers in China

Due to the rapid development of economic and large luxury expense population in China, China's luxury market has become a rising concern to luxury companies all over the world. As for the characteristics of online luxury consumption in China, many scholars' conclusions basically include three aspects.

Firstly, Chinese consumers like having a good face. Face is a rich and ancient concept in Chinese vocabulary, and it is also one of the most common psychological and behavior in Chinese social activities. With the development of cross-cultural research in recent years, scholars have found that consumer behavior in Chinese society cannot be understood through western marketing theory which was born in individualist societies. The reason for that may be the different cultural values and the speed of network technology development in China. Secondly, the need of social communication. Undeniably, the development of online consumption platforms such as Taobao and Jingdong Mall, as well as the widespread online communication methods such as WeChat and QQ, have made people's lifestyle more open. "WeChat Moments" have become a common platform to show consumption power. Morever, gift giving can be seen as a means of maintain social relations.The Internet has made gift-giving more convenient [2]. Thirdly, the cultural taste of Chinese. Online shopping platforms provide more choices of goods, allowing consumers to buy luxuries in line with their taste, which may not be seen in physical stores.

2.2   Luxury Value Perception

Earlier research on luxury goods, some scholars present "brand luxury index", which consists of personal value perception and the non-personal value perception [3]. Hedonism value and extended-self form personal value, non-personal value perception refers to conspicuousness, uniqueness and quality. Subsequent scholars developed this structure as four dimensions, which consist of personal value, price value, social value and function value [4]. In addition to all the above, Tynan et al. (2010) expand the theoretical source of luxury value, consider a new type of value which offered by customer-brand relationships and brand community [5]. He asserts that this value is especially important for those high-value luxury goods which need more supplier service.

The aforementioned research shows that the discussion on the value perception of luxury goods had been well developed in the theory. But it is not enough for the days of information and network. As network consumption shows new consumer characteristics and value demand, new theories are needed to develop and serve marketing. On the basis of literature review, relevant qualitative and quantitative studies are carried out in turn.

The following research contents are arranged as follows. Study 1 develop the model of online luxury value perception by grounded theory. In Study 2, the validity and reliability of the model was tested empirically. After that, a comprehensive discussion and analysis of the previous study are summarized. This paper also present conclusions, research limitations and the further research directions.

3   Scale Development and Test

3.1   Study 1: Grounded Theory and Conceptualization

Grounded theory was chosen to develop the conceptual model of online luxury value perception in this paper. The method originated from a field observation by Glass and Strauss on medical staff dealing with dying patients in a hospital. It mainly includes three steps: open coding, spindle coding and selective coding.

The data collection process of this study was during the COVID-19 epidemic period, so face-to-face interviews were not adopted. Instead, semi-structured and in-depth interviews were conducted through voice calls and WeChat communication with interviewees. Interviewees must be those who have online luxury shopping experience in the past. All interviews were conducted with individual interviewees. A total of 14 interviewees, including 8 males and 6 females, ranged in age from 20 to 60. The interview usually lasts about 30 min.

Software NVIO was used to process the voice data and dialogue data. After the open coding and spindle coding, we found 190 labels, 12 categories. 3 main categories formed for selecting coding. Their relationship is present in Table 1.

Table 1. Axial coding

Main category	Categorical relation	Category (A)
Functional value perception	Price	A1: Very expensive
	Investment	A2: Preservation and appreciation
	Quality	A3: Higher price, higher quality
	Uniqueness	A4: Highlight different types
Personal value perception	Hedonic	A9: Life become pleasant and comfortable by online shopping
		A10: More choice about aesthetic, cultural and taste
	Materialistic	A8: Buy it online and own it earlier
Social value perception	Conspicuous	A5: Online to show off
	Status symbol	A6: Demonstrate one's social status
		A7: Online social
	C2C	A11: Relationship between customers to each other
	B2C	A12: Relationship between the brand service provider and customers

The process of selective coding is to connect the main categories together by selecting the core categories with strong generality. In this study, the core category of "network consumers' luxury value perception" was determined, which was composed of three main categories: functional value, personal value , and social value. In addition, in order to confirm the superiority of the model structure, another competitive model was proposed which is based on Tynan et al. (2010) [5]. These two theoretical framework models are shown in Fig. 1 and Fig. 2.

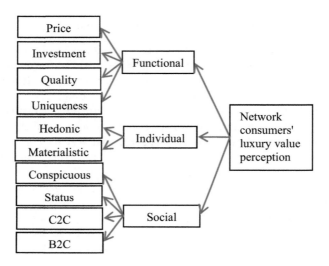

Fig. 1 Proposed model.

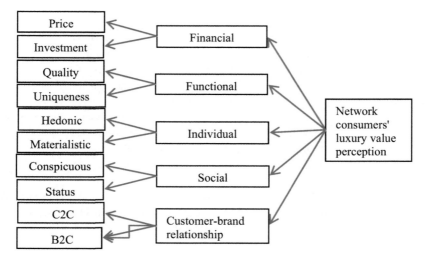

Fig. 2. Competing model

3.2   Study 2: Model Test

This study generated items based on the reviews of past literature and grounded theory. Measurement of price value, uniqueness value, hedonic value, materialistic value was largely based on Shukla (2012) [6]. Investment value was taken from Choo et al. (2012) [7], who modified the scale measurement of economic value. Quality value adapted the scale used by Hennigs et al. (2012) [8]. In addition, measurement items for relationship value between the customers to each other used by Stokburger (2010) [9] in their study of brand community integration. Measurement of relationship value between the brand service and customers was according to Mcalexander et al. (2002) [10], Veloutsou and Moutinho (2009) [11]. Their scales were closely compared with those in the present study, with some necessary adjustments.

In an effort to test the measurement model, a structured questionnaire prepared by the author was used to collect the data. The data collection procedure yielded 416 subjects for the analysis. Of the respondents, 60.6% were female. About 44.7% of the respondents belonged to the age group of 20 to 29, 36.5% were in the 30 to 39 age group, the age group of 40 to 49 were 10.3%, the remainder (8.4%) were over 50 and under 20. In addition, 90.6% had a bachelor's degree or higher, indicating that participants had some cognizance of luxury brands and products. Most of the respondents were currently employed (96.4%).

This research first examine the applicability of the data for factor analysis. The Kaiser–Meyer Olkin (KMO) value of 0.930 and a significant Bartlett sphericity ($P < 0.001$) showed that some common factors between matrix and data are suitable for factor analysis. For further to determine the reliability of the scale, the study calculated the correction item-to-total correlation coefficient (CITC) and Cronbach's alpha coefficients. All items' CITC values ranged from 0.314 to 0.727, the Cronbach's alpha for the scale was 0.950, each one of ten dimensions were higher than 0.857. So the data could be further examined in the next analysis.

To assess the measurement model, this study conducted confirmatory factor analysis by AMOS 21.0, on the basis of which the maximum likelihood method is used to estimate the parameter. The results show good fitting degree of observed data and confirmatory factor analysis model of the good, with $\chi2 = 1074.060$ (df = 450, p = 0.000), normed $\chi2 = 2.387$, GFI = 0.863, CFI = 0.940, IFI = 0.940, TLI = 0.929 and RMSEA = 0.058. Table 2 present the convergent validity of the measurement model which reflect by examine the statistical significance of AVE and composite reliability. All factor loading were statistically significant ($P < 0.01$) and were greater than 0.7. Besides that, AVE factors were all >0.5 and the CR for each construct >0.8 in all cases. The results show that the scale possesses good reliability.

Table 2. Confirmatory factor analysis

Construct	AVE	CR
Price	0.6735	0.8915
Investment	0.6910	0.8701
Quality	0.7740	0.9102
Uniqueness	0.7436	0.8968
Hedonic	0.7152	0.8827
Materialistic	0.6809	0.8646
Conspicuous	0.6608	0.8859
Status symbol	0.6700	0.8903
C2C	0.6958	0.8726
B2C	0.6676	0.8576

To examine the dimensionality of the proposed construct in study 2, this study performed second-order confirmatory factor analysis. The results show the the specified model has satisfactory model fit ($\chi2$ = 1377.096, df = 482, p = 0.000, normed $\chi2$ = 2.857, GFI = 0.822, CFI = 0.913, IFI = 0.914, TLI = 0.905 and RMSEA = 0.067. For confirm the superiority of proposed model, corresponding second-order factor analysis of the competing models also been conducted. But the result show that covariance matrix of the five dimensions in competing model is not positive definite. The reason for this issue may be the number of first-order factor which belong to each second-order factor is less than 3 in competing model, that led to the problem of model recognition.

Furthermore, this study view customer luxury value perception as a third-order factor. More specifically, the three second-order constructs of functional value perceptions, personal value perception and social value perception would sum algebraically to give the third-order customer luxury value perception. These two hierarchical factor structure was shown in Fig. 3.

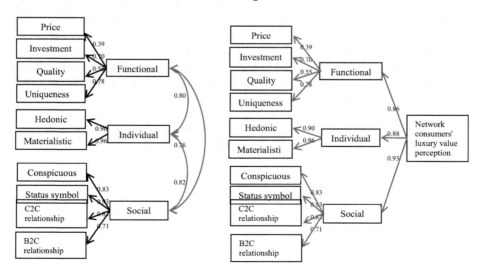

Fig. 3. Second-order factor model and third-order factor model

4   Discussions and Conclusion

Through qualitative and empirical research, the following conclusion can be reached.

Study 1 built two theory framework model through the method of grounded theory. In the proposed model, functional values perception, individual value perception and social value perception are the main factors which affect the Network consumers' luxury value perception.

Study 2 examine the reliability and validity of first-order model. Statistical analysis results shows that the scale has a satisfactory measurement quality. Standardized regression weights are all greater than 0.7, indicating that the measurement scale generated is reasonable for further research. Results of Confirmatory Factor Analysis provide satisfactory support for the three-value structure model consisted of functional, individual and social values. Functional value includes price, investment, quality and uniqueness, whereas individual value encompasses hedonic and materialistic value. Social value is reflected in conspicuous value, status symbol value, relationship value between customers to each other and relationship value between the brand service and customers.

In Study 1, some respondents argue that the value of luxury goods may increase in the future, especially those classic luxury accessories. Specifically, the expected price not only don't fall in the future, even will rise to some extent. And the possibility of resell could also occur. In addition to these, customer-brand value and customer-customer value also should be added to the luxury value framework [7]. It is important for consumers to feel positive behavior and attitude from the brand service. Consumers would create a bond among brand, customer and the company [12]. When consumers feel a brand long for learning the views expressed by consumers about their system of values, they would tend to support the brand and purchase its products [13].

This research provides some helpful implications. The implication for academics is confirmed the third-order structure of Network consumers" luxury value perception in this study. The findings of new dimensions indicated that Hofstede's theory of cultural dimensions maybe not fit all cultures. The findings also suggest some valuable managerial hints. The customer value perception scale is helpful for luxury managers in determine the characteristics of the target market and selecting customers to establish a good interactive relationship. This study reveals that consumers in Chinese context not only perceive status symbol and conspicuous values for luxury brand, but also pay attention on the relationship value of brand. Thus, establishing and maintaining a strong long-term brand relationship with customers is the ultimate task for brand service. Not only refers to the relationship between B to C, relationship between customers to each other, is also very important.

The present study was limited to Chinese network consumer, the generalisability of our results to other cultures requires careful treatment, a cross-cultural studies on the scale can be further discussed in future research. Moreover, the number of items used in the scale is a little bit more. A more simplify version of the scale could be explored from different aspects.

Acknowledgements. Supported by the major project of Tianjin University of Finance and Economics Pearl River College (ZJZD20-06).

References

1. Sung, Y., Choi, S.M., Ahn, H., et al.: Dimensions of luxury brand personality: scale development and validation. Psychol. Mark. **32**(1), 121–132 (2015)
2. Joy, A.: Gift giving in Hong Kong and the continuum of social ties. J. Consum. Res. **28**(2), 239–256 (2001)
3. Vigneron, F., Johnson, L.W.: Measuring perceptions of brand luxury. J. Brand Manag. **11**(6), 484–506 (2004)
4. Wiedmann, K.P., Hennigs, N., Siebels, A.: Value-based segmentation of luxury consumption behavior. Psychol. Mark. **26**(7), 625–651 (2009)
5. Tynan, C., McKechnie, S., Chhuon, C.: Co-creating value for luxury brands. J. Bus. Res. **63**(11), 1156–1163 (2010)
6. Shukla, P.: The influence of value perceptions on luxury purchase intentions in developed and emerging markets. Int. Mark. Rev. **29**(6), 574–596 (2012)
7. Choo, H.J., Moon, H., Kim, H., et al.: Luxury customer value. J. Fashion Mark. Manag. **16**(1), 81–101 (2012)
8. Hennigs, N., Wiedmann, K.P., Klarmann, C., et al.: What is the value of luxury? A cross-cultural consumer perspective. Psychol. Mark. **29**(12), 1018–1034 (2012)
9. Stokburger-Sauer, N.: Brand community: drivers and outcomes. Psychol. Mark. **27**(4), 347–368 (2010)
10. Mcalexander, J.H., Schouten, J.W., Koenig, H.F.: Building brand community. J. Mark. Q. Publ. Am. Mark. Assoc. **66**(1), 38–54 (2002)
11. Veloutsou, C., Moutinho, L.: Brand relationships through brand reputation and brand tribalism. J. Bus. Res. **62**(3), 314–322 (2009)
12. Thomson, M., MacInnis, D., Park, W.: The ties that bind: measuring the strength of customers' attachment to brands. J. Consum. Psychol. **15**(1), 77–91 (2005)
13. Kates, S.: Out of the closet and out on the street: gay men and their brand relationships. Psychol. Mark. **17**(6), 493–513 (2000)

On the Value Demand and Implementation Strategy of Innovative Education in Colleges and Universities in the Information Age

Yang Yang[✉]

School of Jilin Agricultural Science and Technology College, Jilin, Jilin, China

Abstract. With the development of Information Technology, university innovation education based on information technology has gradually attracted great attention of colleges and universities, with the overall improvement of the quality of the educated as the main means of implementation, to train them to become an all-round development of the new era of talent and all-round development of human values. Therefore, in the era of information technology, colleges and universities should establish a sound talent training mechanism, improve the construction of curriculum system, stimulate students' learning motivation, encourage students to explore learning, and deal well with the relationship between innovative education and basic education, truly realizing the well-being of mankind brought about by innovative education [1].

Keywords: Information technology · Higher education · Innovative value · Implementation strategy

1 Introduction

In the present era, innovative ability and spirit have become the key driving force of national development and social progress. As the training base of innovative talents, universities shoulder the heavy responsibility of realizing the national mission, higher Education should focus on the value of innovative education in colleges and universities, and on this basis, deeply explore how to carry out innovative education and how to achieve good results [2].

2 Value Demand of Innovative Education in Colleges and Universities

2.1 Innovative Education and the Improvement of Students Comprehensive Quality

Innovation is the soul of national development and the source of social progress. The primary goal of innovation education in colleges and universities is to cultivate high-quality talents with all-round development. Here the emphasis on high quality refers to the students in the ideological and moral character, wisdom enlightenment, strong physique and aesthetic ability in all-round improvement. This is consistent with the

J. C. Hung et al. (Eds.): FC 2021, LNEE 827, pp. 545–551, 2022.
https://doi.org/10.1007/978-981-16-8052-6_67

concept of innovative education, therefore, innovative education and quality improvement exists an interdependent relationship. First of all, innovative education is the means to improve the national quality. In traditional education, education attaches importance to the transmission and study of knowledge, but neglects the development of wisdom and the formation of character to a certain extent. Because of such a series of problems, China began to pay attention to the direction of higher education reform, and put forward the idea of innovative education, education aimed at improving the overall quality of people. It follows that innovative education and liberal education are unified, that they are two sides of the same coin and that their goals and directions are the same [3, 4]. Only successful innovative education can improve the national quality and strengthen the nation. Second, liberal education is the home of innovation education. If the emphasis is placed on improving the professional ability of the educated, but the cultivation of innovative quality is neglected, education still can not cultivate high-quality talents. One of the important conditions for high-quality education is to have innovative ability, and not only reflected in the rich after-school life and artistic accomplishment and so on. Liberal Education is the goal, foothold and destination of innovation education [5].

2.2 Innovative Education and Cultivation of Students Learning Motivation

As the saying goes, "teach a man to fish, rather than teach a man to fish" the same goes for education. Therefore, the emphasis of education should not be on inculcating theoretical knowledge, but on teaching students to learn. Only when students master the learning methods, improve their learning ability, and learn to explore and seek knowledge, can they really exert their initiative, become a lifelong learner. In addition to laying the basic quality of students, one of the important tasks of innovative education is to cultivate students' learning ability and spirit of exploration. All of these cannot be separated from the cultivation of learning motivation, learning motivation is the process of students forming subconscious thinking, which will have a great impact on the future career development of college students, people who really have learning ability and innovative wisdom, must have a good learning habits and strong desire to learn, therefore, innovative education is another goal is to stimulate students learning motivation. First, develop an interest in learning. The cultivation of interest cannot be separated from the students' confidence in the object of study [6]. Only when they are confident about something can they have a sense of achievement, and then they have a strong interest in it. The study interest of college students is different from the hobby interest of primary and middle school students, the professional interest needs to be persistent, because this major will probably accompany with them for a lifetime. Therefore, professional interest should be established at a higher level, so that they truly understand the role and value of the major for them, so that students can find their own goal positioning. Secondly, cultivate the ability of creative learning. Education should encourage students to question authority, not to accept knowledge passively, to explore and explore, not to stick to the beaten path, to discover problems, to solve them, and to develop the ability to think independently [7].

2.3   Creative Education and Students Critical Thinking

Critical thinking is a kind of thinking ability and attitude, which shows a person's thinking level and character. Cultivating students' critical thinking has been paid more and more attention by higher education and has become one of the core goals of the new teaching reform. In the 1990s, the United Nations Educational, Scientific and Cultural Organization published the World Declaration on Higher Education for the twenty-one St Century: Ideas and action. One of the most important contents is to take "cultivating critical and independent attitude" as one of the missions of higher education, which shows that critical thinking has already become the consensus of higher education all over the world. Innovation Education should be devoted to training students' ability of independent thinking and dialectical thinking, and guide students to carry out critical thinking [8]. First of all, cultivate students' ability of abstract thinking. Traditional cramming teaching is not conducive to students thinking more, they just passively accept the knowledge indoctrinated by teachers. Under this teaching model, teachers are the main body of the teaching process, and students understand and master knowledge according to the teacher's teaching contents, teachers have absolute authority over knowledge. This idea leads to students' blind obedience and passivity to classroom teaching, which is not conducive to the formation of the ability of independent thinking. Therefore, only students as the main body of teaching, through innovative education to cultivate their abstract thinking ability, can achieve the goal of students' critical thinking. Second, critical thinking is included in the curriculum. Only if the curriculum system contains the content of critical thinking, teachers and students will attach great importance to it, so as to ensure the deep learning and implementation of the curriculum from the system. Finally, cultivate students' critical thinking through practical activities. By finding and solving problems in practical activities, students learn to think deeply and question deeply, which can improve their thinking ability and logicality, thus improving the quality of work and life, critical thinking can effectively help students to analyze and orientate problems from different angles, to evaluate things more objectively and to discover the deep causes of problems, so as to solve the essential problems. Rational thinking and behavior choice [9].

3   On the Implementation Strategy of Innovative Education in Universities

3.1   Combining Subjectivity with Individuality to Improve Students' Quality in an All-Round Way

The cultivation of students' comprehensive quality cannot be separated from teachers' teaching process, teachers' purposeful curriculum setting and arrangement, and students-oriented teaching plan and teaching design, fully reflect the importance and leading role of students in the teaching process. It is the duty and mission of higher education to devote to the all-round development of students and the improvement of their comprehensive qualities. First of all, the importance of teaching experience, highlighting the subjectivity of students. With the coming of Information Age, education should emphasize students' learning experience and feeling, and students' initiative and dominance in the teaching process. The learning experience directly determines the learning effect, therefore, when teachers design classroom teaching, they should plan the classroom effectively according to students' age characteristics, professional characteristics and background knowledge. From another point of view, the comprehensive quality of college students includes psychological quality, thinking ability and humanistic quality. Specifically includes physical and mental health, General Education and basic accumulation, aesthetic ability and moral quality, professional level and social skills. The improvement of these qualities cannot be achieved overnight. It needs educators to permeate gradually according to the characteristics of students in the teaching process to achieve the effect of moistening things quietly. Secondly, cultivate students' personality development. Education should take into account not only the overall characteristics of students, such as their major and age characteristics, but also the individual differences of students. Each student is an independent and distinctive existence, the understanding of the problem, the desire to know and the interest hobby have the very big difference. Therefore, the teaching process of teachers should also take into account the individual characteristics of students [10]. The liberal education aims to develop students' abilities and uniqueness in all aspects, and it is their responsibility to foster their innovative thinking and spirit, both to make them good citizens and to be creative enough to make their due contributions to society, in order to achieve a good teaching effect, it is necessary to complete a specific historical mission and take into account students' general and individual characteristics. Therefore, the form of classroom teaching should be reconstructed according to different students, through the rich form of teaching means to stimulate students' learning initiative, really achieve the effect of innovative education requirements.

3.2   The Combination of Inquiry and Criticism Can Stimulate Students' Learning Motivation

College education requires students to have a strong logical thinking ability and abstract thinking, specialization is a prominent feature of college courses, and strong theoretical is also a feature of college courses. Therefore, learning to learn is the most basic learning quality that college students need to have. However, these qualities can be realized only through the guidance and inspiration of teachers. College teaching should encourage students to explore learning and critical thinking. First, inspire positive thinking. In the process of teaching design, teachers should arrange some difficult tasks for students, but the students can accomplish them effectively through hard work. If the task design is too difficult, the students will not achieve their goals no matter how hard they try, they will lack the initiative to explore and give up to solve the problem. Therefore, the arrangement of teaching contents and tasks is an art. Teachers should take a step-by-step approach to stimulate students' learning enthusiasm and achieve good results. Second, stimulate curiosity about knowledge. College learning content is no longer limited to basic knowledge and test-oriented skills. College courses cover many areas of specialization and general education that students have never covered in high school, and college life is very different from their previous campus life, therefore, everything is full of freshness for students, and students have a great curiosity [11]. As teachers, we should try to keep students in this state of mind through curriculum design and guidance, students are encouraged to explore the unknown, taught to respond critically to challenges, and motivated to learn. Finally, reinforce the learning motivation. Learning motivation is based on students' recognition and confidence in their field of study. The stronger the professional trust, the more clearly students will plan their future, thus generating strong learning motivation. Teachers should make full use of instructional design to promote students to form a positive cycle, in the teaching practice, through the actual operation or business simulation inspection to enable students to find their own potential, resulting in maximum self-realization desire, let them realize the value of hard study and struggle, so as to form self-motivation, stable learning motivation, achieve the desired results.

3.3   Combining Innovative Talents with Innovative Achievements to Cultivate Excellent Teachers

The cultivation of innovative talents cannot be separated from excellent teachers. Colleges and universities should strengthen the introduction and cultivation of teachers. The success of innovation education in universities depends on the agreement between managers and teachers, teachers and teachers, teachers and students, as well as their innovation ability and cooperation spirit. First of all, pay attention to the introduction of talent. The introduction of talents is the most direct mode of talent gathering. Colleges and universities can attract in-service teachers or graduates with advanced degrees from famous domestic and foreign universities to join the teaching team. They can also hire internationally and domestically renowned scholars as visiting professors, on the one hand, these high-quality talents can cultivate students' innovative spirit through teaching, and on the other hand, they will bring fresh breath to the teaching team in the process of cooperation with the teachers of this school, thus affects the teacher troop overall quality promotion. Secondly, secondly, self-cultivation. There are several channels for self-cultivation [12]. The first is that schools organize teachers to conduct seminars, select academic and teaching leaders to lead the teaching team, through regular exchange of experience and learning from each other to constantly improve the level of business and innovation. The second is to conduct academic exchanges. That is, to organize relevant teachers and other professional teachers to exchange and observe learning, learn from each other, and constantly improve the strength of teachers. The third is to arrange for teachers to conduct domestic and international visits and exchanges. These teachers, as visiting scholars, have broadened their academic horizons and expanded their professional depth through work and study in different universities, especially in famous universities abroad, strengthened the theoretical basis, increased the strength of innovation. Finally, strengthen social cooperation. The ways of cooperation with society are: First, to strengthen cooperation with enterprises. By sending teachers to enterprises and science and technology development companies for study and cooperation, teachers can not only master practical innovation, but also understand the true meaning of theory from practical operation, thus tamping innovation ability. The second is to invite enterprises and entrepreneurs with greater social influence to give students' reports to teachers and students in colleges and universities. Through these activities, students' innovative awareness will be stimulated, students' innovative motivation will be strengthened, and students will be encouraged to innovate and start their own businesses, to take innovation to a new level [13].

4   Conclusion

With the development of information technology and the progress of the Times, Innovation Education is called for. Therefore, we must grasp the value connotation and specific function of innovation education comprehensively and accurately from the height of history, through different approaches and strategies, the relationship between liberal education education and traditional education, Innovation Education and basic education education should be well dealt with, the correct direction of education development should be set, and the well-being of human beings brought about by innovation education should be put into practice.

References

1. Xuan, X., Duan, W.: Under the background of supply-side reform, the talent ecological training mode of innovative entrepreneurship education in application-oriented colleges and universities. Educ. Occup. (15) (2019)
2. Han, G.: Innovative entrepreneurship education in pilot universities of applied transformation in Jilin Province: realistic basis, main problems and reform path. Vocat. Tech. Educ. (36) (2018)
3. Fei, Z., Chen, M.: Research on teaching reform of innovative entrepreneurship education in applied universities under the background of integration of industry and education. Chin. Adult Educ. (18) (2018)
4. Mettam, G.R.: How to prepare an electronic version of your article. In: Jones, B.S., Smith, R. Z. (eds.) Introduction to the Electronic Age, pp. 281–304. E-Publishing Inc., New York (1999)
5. Bowie, L., Wang, R.: Curban innovation, innovative entrepreneurship education and college students' entrepreneurial behavior - based on Yangtze river delta economic zone data. Discuss. Mod. Econ. (09) (2018)
6. Chen, Y., Zhang, S., Zhang, W., Yang, D., Zhang, H., Kim, C.: Construction of hybrid teaching mode of MOOC + rain class – taking teaching of nursing anatomy as an example. The study of anatomy. A study on the influence of (04) (2019)
7. Zhang, Y., Xu, S., Xun, C., Niu, C.: Micro-teaching mode on college students' learning motivation. Outside School Educ. China (30) (2018)
8. Lu, L., Gao, .: Constructing a new mode of intelligence teaching in colleges and universities under the guidance of information technology. Electron. Test. (03) (2021)
9. Li, Y., Haiyan: Methods and strategies for innovative higher education based on modern educational technology. Agric. Netw. Inf. (05) (2018)
10. Xing, G.: Research on modern educational technology and innovative education in universities. Mod. Educ. (03) (2018)

11. Yang, J., Sun, H., Li, S., Fu, Z., Gu, X.: Innovation education laboratory and social maker space in American universities. Mod. Educ. Technol. (05) (2015)
12. Jiang, C.: Use of distance education resources. Sci. Consult. Educ. Sci. Res. (12) (2014)
13. Chow, H.P.: The importance of modern educational technology in educational teaching. Enterp. Sci. Technol. Dev. (16) (2008)

On Comprehensive English Classroom Model Based on PBL Theory

Yan Tang[(⊠)]

College of Foreign Languages and Literature, Wuhan Donghu University,
Wuhan, Hubei, China

Abstract. Success of foreign language acquisition depends on a variety of factors; the factors of language input and language output possess a very important position in foreign language learning. Positive and efficient input promote output whereas effective and accurate output further promote the input, the two promote each other. College English teachers should design a "learning-oriented" synthetic classroom teaching model based on PBL theory to promote students' classroom language input and output as well as motivate deep learning, in order to motivate students to build knowledge initiatively, and the college English classroom should become a language input and output platform for the students, and ultimately improve the quality of the acquisition of EFL students.

Keywords: PBL theory · Language input output · Deep learning ·
"Learning-oriented" synthetic classroom teaching model

1 Introduction

Project-driven learning, or project-based learning, referred to as PBL in this topic, is a student-centered education method and principle that requires students to pass a series of individual or cooperative tasks, with the help of others, including teachers and learning partners. The learners are required to use necessary learning materials to solve real-world problems based on acquired knowledge and skills. PBL has been widely used in the field of second language and foreign language education abroad for more than 20 years. Its influence on language learning is mainly reflected in the promotion of students' learning motivation, interest, confidence and learning autonomy. Students' ability to solve problems, cultivation of critical thinking skills will be proved and language learning skills will be cultivated. Therefore, PBL can create student-centered language learning environments that cultivate autonomous learning capabilities [1]. Although foreign researchers have some research on PBL learning model, the research of PBL learning model in China is still in the preliminary stage of development, and it still needs to be further deepened and discussed. In the process of constructing a college English mixed teaching, the task-based approach is conducive to improving students' initiative in English learning, cultivating students' ability to solve English problems, helping students have a comprehensive understanding of their own English ability, and improving students' comprehensive ability to use English [2]. Only by further in-depth discussion and research can the PBL learning model be effectively used to stimulate students' enthusiasm for learning.

J. C. Hung et al. (Eds.): FC 2021, LNEE 827, pp. 552–559, 2022.
https://doi.org/10.1007/978-981-16-8052-6_68

2 Conception of a Comprehensive English Classroom Teaching Model Based on PBL Theory

2.1 Language Input in a Comprehensive English Classroom

Based on PBL theory, language input needs to be strengthened. Krashen emphasized that the only way to acquire language is to "understand the input information" or "get comprehensible input". Therefore, "input" requires a large amount of target language corpus with moderate difficulty in order to make learners succeed. Without the input and accumulation of rich and natural target language and related knowledge, that is, if the language input is insufficient, the language knowledge accumulated by the foreign language learner will be ineffective, and the corresponding output cannot be achieved. The strategy to achieve the balance and sustainable development of college English ecological classroom is to improve the curriculum setup of school-based characteristics and develop the personalized textbook system [3]. Therefore, in college English classrooms, teachers should take into account the actual level of the English majors, take into account the specific learning environment and learning characteristics of the English majors, and fully consider the original knowledge structure and cultural level of the target audience. It is necessary to teach students in accordance with their aptitude, and choose appropriate teaching methods based on PBL theory. The target language corpus and related background knowledge provide a scientific platform for students' intelligible input.

2.2 Language Output in a Ccomprehensive English Classroom

Based on PBL theory, language output should be based on input, and input that deviates from output requirements will inevitably become passive and inefficient. Language output plays an important role in the process of language learning. Language output can help students obtain more appropriate input, stimulate autonomy, verify hypotheses, develop discourse skills, and form learners' personal style. It can be seen that language output is a very important link, which enables English majors to decode existing knowledge based on understanding learning and transform it into an ability to use language knowledge autonomously. Language output enables students to learn more actively and critically instead of mechanically and passively storing and accepting knowledge.

2.3 Research on "Learning-Oriented" Comprehensive Classroom Teaching Model: The Balance Between Language Input and Output

At present, the traditional teaching mode of "teaching-oriented" is commonly used in comprehensive English classes. "Teaching-oriented" emphasizes the central position of teachers, emphasizes the teaching of theoretical knowledge, and ignores individual differences among students. Teachers cannot boldly use teaching methods and means that embody new teaching concepts, and students cannot achieve a balance between input and output of language in the classroom in traditional classes. Students accept the theoretical knowledge taught by teachers passively. Such input is inefficient and is not conducive to the mastery and improvement of students' language ability. Constructing a harmonious and co-existing English classroom ecology, we should break the traditional class [4]. The American psychologist Kelly believes that due to different cultural backgrounds and talented environments, each person has a unique psychological structure. Any learning is a process in which learners absorb new knowledge in an existing knowledge system. The theory emphasizes that learners are cognitive subjects and active builders of knowledge meaning. According to this theory, teachers should change the traditional "teaching-oriented" teaching mode in college English classrooms, fully realizing that teachers are guides, not knowledge indoctrinators, and the classroom should be a "learning-oriented" classroom, and students are the main body of learning, teachers should focus on the initiative of students according to the actual level of students, allow students to actively participate in classroom practice activities, improve students' independent learning ability, and reduce dependence on teachers. Students should be allowed to switch between language input and output in a real language application environment to form their own learning style and improve their language ability.

3   Exploration of Practical Approaches to Comprehensive English Classroom Teaching Model Based on PBLTheory

3.1   Design of Comprehensive English Classroom Teaching Model Based on PBL Theory

In view of the new teaching environment and teaching mode, college English smart classroom needs to adopt different evaluation methods [5]. Stoller proposed that the design of PBL must follow the following principles: content-oriented; to produce a final product; to encourage learners' autonomy and responsibility in the project; the difficulty of the project tasks should be within the scope of the students' ability but with certainly challenging, which can arouse students' curiosity and interest; the project should last for a period of time; to encourage the training of learning skills in the project; and require students to learn cooperatively and independently. Furthermore, teachers are different from traditional classrooms in the project, and students have different roles and responsibilities; the formative and summative assessment must be carried out for students [6]. When designing related learning models, teachers should take into account the students' specific learning environment and learning characteristics, and fully consider the original knowledge structure and cultural level of the subject, and design an English learning model that utilizes various learning resources, especially network resources. For example: Role-Play Activity design and result feedback, cocurricula activity design and result feedback, classroom observation and interview design and result feedback, etc. A multi-level design model with multiple perspectives, centering on the core purpose of improving students' comprehensive English quality, through empirical research and qualitative and quantitative analysis of data, it proves how the learning model driven by the project will learn from students' learning attitude, learning ability and learning environment. Effectively cultivate the ability of autonomous learning in all aspects and improve the efficiency of English learning. The "learning-oriented" comprehensive classroom teaching model takes students as the core, and students are the active builders of knowledge theory, emphasizing the roles and responsibilities of teachers different from traditional classrooms, and teachers should be facilitators and guides. When designing the relevant "learning-oriented" comprehensive teaching model, teachers should take into account the specific learning environment and learning characteristics of EFL students, and fully consider the original knowledge structure and cultural level of the teaching object, and design and utilize various learning resources. It is a network resource, such as: Role-Play Activity design and result feedback, cocurricula activity design and result feedback, classroom observation and interview design and result feedback, etc., to guide students to consciously and actively carry out language input and output, so that students can use relevant knowledge flexibly in practice, and promote students' language input and output to achieve a balance. The "summary and reflection" stage is also an essential and important link. Through the summary and reflection of the previous classroom and the evaluation and feedback of the students' learning effects, teachers can adjust the classroom teaching mode in a timely manner, optimize the combination of resources, and build a "learning-oriented" comprehensive classroom that is more suitable for students' knowledge structure level, so that students can fully give full play to their autonomy, actively participate in the classroom, balance language input and output, and improve the efficiency of English learning.

3.2 Implementation and Application of the "Learning-Oriented" Comprehensive English Classroom Teaching Model Based on PBL Theory

Because of the limitation of theoretical understanding, the traditional teaching mode of "teaching-oriented" is still widely used in current college English teaching, and very few teachers use teaching methods and teaching methods that embody new teaching concepts. From the two aspects of teaching effect and students' proficiency in using English, the traditional teaching method of "teaching-oriented" can no longer adapt to the development of society. There are huge differences in the knowledge and skills learned by students of different ability levels. Alnofaie proposed that "the cultivation of critical thinking ability should be integrated with basic foreign language courses" [7]. In the EFL environment, it is difficult for college students to balance language input and output due to individual differences and different learning foundations, and they need to be trained under the guidance of teachers. Based on the PBL theory, through the design and application of the "learning-oriented" classroom teaching model, and through case studies, the relevant data and qualitative and quantitative analysis can be obtained, which can greatly improve the autonomous learning ability of EFL learners. It is important to investigate the quality of the motivational regulation outcomes in classrooms and it is imperative to examine how students' motivational regulation affects their engagement [8]. When English teachers implement the "learning-oriented" comprehensive classroom teaching model, they should first combine the characteristics of curriculum learning and the foundation of the students, encourage various classroom projects or classroom activities to cultivate students' English learning skills, and encourage students to exercise learning autonomy. College English should keep students' critical thinking development as the focus of its present reform [9]. The main role of the teacher in the comprehensive English classroom is to give full play to the role of facilitator and guide. They should strengthen their responsibility as an evaluator and strengthen the guidance and evaluation of students. At the same time, before the implementation of the project, the training and training of students' learning strategies should be strengthened, focusing on the balance of English knowledge input and output theory, and reducing individual differences among students, so that the design and application of the "learning-oriented" comprehensive classroom teaching model can achieve the best effect, for language reflects not only situational opinions, but also long-term opinions tied to social identity [10].

3.3   Feedback Evaluation of the "Learning-Oriented" Comprehensive English Classroom Teaching Model Based on PBL Theory

In the process of implementing the "learning-oriented" comprehensive classroom teaching model, teachers need to summarize and evaluate the teaching effect in a timely manner, and continuously improve teaching methods. Multi-level feedback and first-line teaching evaluation can help teachers adjust their own teaching design plans, and ultimately design a scientific and reasonable English classroom teaching model suitable for students. The author conducted a practical exploration of the "study-oriented" comprehensive English classroom teaching model among the English Classes of Wuhan Donghu University, and implemented the "study-oriented" comprehensive classroom teaching model (referred to as the experimental classroom) in English class 6, and the traditional teaching method (traditional classroom for short) is still used in the English class 3, and the teaching methods are implemented and compared through a semester. Through classroom observation methods, student interviews, and assignment completion statistics, it is found that the experimental class and the traditional class have large differences in learning effects and English skills improvement. The specific performance is as follows (Table 1):

Table 1. Class participation in different learning modes (in percentage terms)

Class	Week (90min/Two periods)														
	1	2	3	4	5	6	7	8	9	10	11	12	13	14	15
Experimental class	35.48	32.25	38.37	35.48	32.25	38.70	41.93	38.70	35.48	32.25	38.70	32.25	41.93	35.48	45.16
Traditional class	16.12	19.35	16.12	12.90	19.35	12.90	19.35	22.58	16.12	12.90	16.12	19.35	16.12	22.58	16.12
Points difference	19.32	12.90	22.25	22.58	12.90	25.80	22.58	16.12	19.36	16.13	22.58	12.90	25.81	12.90	29.04

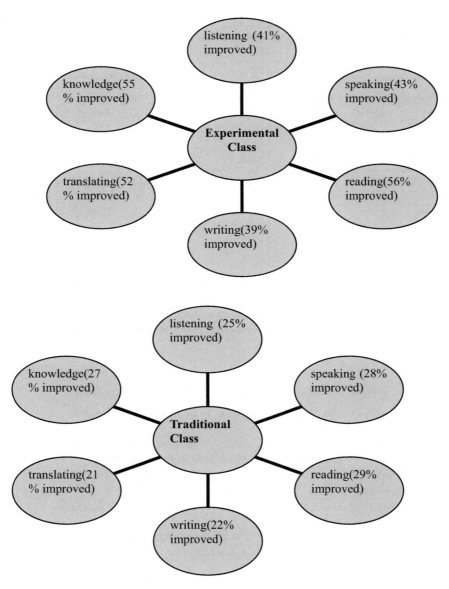

Chart 1. PBL and English skill (based on assignment completion)

Assignments include listening exercises, speaking exercises, reading comprehensione exercises, writing and translation. The student's assignments are rated from A to D. The grade from B to A indicates that the student has improved their English skill. Chart 1 shows that the English skill of students from the experimental class are improved compared with the students from traditional class.

Meantime, the final examination scores as shown in Table 2 and Table 3, which further demonstrates superiority of the "learning-oriented" teaching model.

Table 2. Analysis of test scores (in percentage terms)

Analysis of test scores (Grade 18 English class 6: 31 students) (Experimental class)							
Score	90–100 (Excellent)	80–89 (Good)	70–79 (Qualified)	60–69 (Pass)	<60 (Fail)	Average score	The standard deviation
Number	3	17	9	1	1	81.23	8.98
Proportion	9.68%	54.84%	29.03%	3.23%	3.23%		
Analysis of test scores (Grade 18 English class 3: 31 students) (Traditional class)							
Score	90–100 (Excellent)	80–89 (Good)	70–79 (Qualified)	60–69 (Pass)	< 60 (Fail)	Average score	The standard deviation
Number	2	11	12	5	1	77.97	8.14
Proportion	6.45%	35.48%	38.71%	16.13%	3.23%		

Table 2 shows the quantification of various indicators of the test paper scores of the two classes. Through Table 2, we can clearly see that the students in the experimental classroom have better knowledge absorption and mastery and higher grades, while the students in the traditional classroom are not very motivated to learn, and the test scores are relatively low.

Table 3. Analysis of examination papers (in percentage terms)

Analysis of examination papers (Grade 18 English class 6:31 students) (Experimental class)					
Score	90–100 (Excellent)	80–89 (Good)	70–79 (Qualified)	60–69 (Pass)	<60 (Fail) Attendance rate 98%
Number	10	20	1	0	0
Proportion	32.26%	64.52%	3.23%	0%	0%
Analysis of examination papers (Grade 18 English class 3:31 students) (Traditional class)					
Score	90–100 (Excellent)	80–89 (Good)	70–79 (Qualified)	60–69 (Pass)	<60 (Fail) Attendance rate 92%
Number	6	20	5	0	0
Proportion	19.35%	64.52%	16.13%	0%	0%

Table 3 shows the quantification of various indicators of the overall test scores of the two classrooms. In the overall test scores, 70% are on paper and 30% are day-to-day performance scores. The usual grades are composed of five indicators: class attendance rate, class performance, usual tests, class work and extracurricular group activities. From Table 3, we can clearly see that the attendance rate, learning enthusiasm, and the number of high scores in the experimental classrooms that implement the "learning-oriented" comprehensive classroom teaching model are far higher than those of traditional classrooms.

4 Conclusion

Foreign language teaching should be a process of promotion, not a process of forcibly instilling. There are individual differences among language learners, so teachers should adopt flexible teaching strategies in teaching. The "learning-oriented" teaching models based on PBL takes students as the core of learning, and emphasizes the cultivation of students' language input and output capabilities. Teachers strengthen language input in the classroom and provide suitable language materials for students' language output. The flexible use of teaching methods can create a platform for students' language output, and ultimately improve students' learning ability.

References

1. Dochy, F., Segers, M., den Bossche, P.V., Gijbels, D.: Effects of problem-based learning: a meta-analysis. Learn. Instr. **13**, 533–568 (2003)
2. Ren, L.: The application of task-based teaching to the construction of college English mixed teaching. Theory Pract. Educ. **39**(18), 59–60 (2019). (in Chinese)
3. Zhou, Y.: A probe into the construction of college english ecological classroom of local universities in the context of "internet plus." Theory Pract. Educ. **39**(36), 51–54 (2016). (in Chinese)
4. Ye, L., Zhang, G., Yao, Y.: A study of college English's flipped classroom in the era of "internet plus". TEFLE (175), 3–8 (2017). (in Chinese)
5. Zhu, Y., Chen, L.: Constructing the evaluation index system for college English smart classroom teaching. TEFLE (4), 94–111 (2020). (in Chinese)
6. Stoller, F.L.: Content-based instruction: perspectives on curriculum planning. Annu. Rev. Appl. Linguist. **24**, 261–283 (2004)
7. Alnofaie, H.: A framework for implementing critical thinking as a language pedagogy in EFL preparatory programmes. Think. Skills Creat. **10**, 154–158 (2013)
8. Ren, Q.: The effects of motivational regulation on student engagement in blended college English classroom teaching environment. TEFLE **197**(1), 44–60 (2021). (in Chinese)
9. Lin, X., He, L.: On the cultivation of critical thinking ability in college English class. J. Xi'an Int. Stud. Univ. **25**(1), 61–66 (2017). (in Chinese)
10. Byram, K.A.: Using the concept of perspective to integrate cultural, communicative, and form-focused language instruction. Foreign Lang. Ann. **44**(3), 525–543 (2011)

Analysis of the Dynamic Influence of the Train Operation Vibration on the Subway Tunnel Structure Based on the Artificial Intelligence Monitoring System

Chen Yang[✉] and Xingxing Liu

School of Civil and Architectural Engineering, Nanchang Institute of Technology, Nanchang 330000, Jiangxi, China

Abstract. With the rapid development of high-speed railways and urban subways, the scale of construction, construction technology and the follow-up operation of mechanical tracks have reached new heights. At the same time, many technical problems have been brought about in the design, construction and subsequent operation of the tunnel, such as the dynamic stability of the tunnel structure under the action of high-speed train vibration. The increase in the speed of high-speed trains further aggravates the impact force between the wheels and the rails, and the energy generated by vibration cannot be quickly dissipated, which leads to greater damage and intensifies the dynamic response of the tunnel structure. However, the urban subway has a ladder-like shape, a complete structure of complex network lines and tunnel stations, as well as the function of high-frequency trains, which makes the tunnel structure in a complex dynamic response for a long time. In this paper, the artificial intelligence-based monitoring system is used to study the dynamic influence of the train running vibration on the subway tunnel structure. First, the literature research method is used to summarize the dynamic response of the subway tunnel structure to vibration and the dynamic effect of vibration on the subway tunnel structure. The simulation experiment is used to test the dynamic influence of train operation vibration on the subway tunnel structure. The experimental data is analyzed and summarized, and corresponding measures are proposed. The experimental results are as follows: First: Under the influence of vibration, the displacement of the tunnel changes from 0.0045 to 0.0038 and then to 0.0015. The speed will continue to drop again, and the beginning will be most affected. The front section of the tunnel can be reinforced, and vibration-proof structures can be added. Second, after increasing the vibration frequency, all the data have increased, but the trend is the same, the displacement is 10 times the original, and the speed has also increased by 10 times.

Keywords: Monitoring system · Train vibration · Subway tunnel · Structural dynamics

J. C. Hung et al. (Eds.): FC 2021, LNEE 827, pp. 560–567, 2022.
https://doi.org/10.1007/978-981-16-8052-6_69

1   Introductions

From the 1990s to today, the vibration caused by the train passing through the tunnel has attracted widespread attention in the technical field. Designers from different countries have studied the dynamic response of investment structure under train dynamic load from different angles [1, 2]. Research methods include theoretical analysis and field measurement, model testing and software simulation [3, 4]. The research direction focuses on the response of the structure itself, the impact of dynamic loads on the environment, methods of vibration reduction and isolation, and the dynamic response of the soil around the tunnel [5, 6]. The results of previous studies have shown that the long-term effects of dynamic train loads are the main cause of tunnel diseases. In recent years, the rapid development of track technology has led to new forms of train vibration and a new round of attention, such as the acceleration of existing lines, the large-scale construction of high-speed railways, and the emergence of urban tunnels [7, 8].

Many scholars have studied the dynamic effects of train operation vibration on subway tunnel structure based on artificial intelligence monitoring system, and have achieved good results. For example, V, Lugin, EL have introduced numerical models to study subway vibration. The three-dimensional elastic dynamic response of the transition between the left and right parallel tunnels to the upper and lower parallel tunnels under load, and the deformation characteristics of the tunnel, the vibration acceleration [9]. Cai H B relies on the Nanpu Bridge Scheme Interval Tunnel Project, carries out a three-dimensional elastic finite element numerical simulation based on the measured vibration acceleration data on site, and carries out an elastic-plastic analysis of a typical section [10].

This paper studies the dynamic influence of the train operation vibration on the subway tunnel structure based on the artificial intelligence monitoring system. First, through the literature research method, the dynamic response of the subway tunnel structure to the vibration and the dynamic effect of the vibration on the subway tunnel structure are summarized, and then the simulation experiment is used to test the dynamic influence of train operation vibration on the subway tunnel structure. The experimental data is analyzed and summarized, and corresponding measures are proposed.

2   Research on the Dynamic Influence of the Monitoring System's Train Running Vibration on the Subway Tunnel Structure

2.1   Research Method

(1) Literature research

Reading books and articles on domestic and foreign related literature on the dynamic influence of the train operation vibration on the subway tunnel structure based on artificial intelligence monitoring system, the advantage is that you can understand the development process of the research object from the source, and understand the development status of the research object. And provide a clear and structured theoretical basis for in-depth thesis development.

(2) Quantitative analysis

Qualitative analysis is related to quantitative analysis. Quantitative analysis refers to the analysis of mathematical hypothesis determination, data collection, analysis and testing.

Qualitative analysis refers to the process of conducting research through research and bibliographic analysis based on subjective understanding and qualitative analysis.

2.2   The Dynamic Response of Subway Tunnel Structure to Vibration

(1) The response of the structure under the influence of dynamic load is mainly measured on the basis of dynamic displacement, speed, acceleration rate and other indicators. At the same time, certain additional dynamic stress and strain will also be generated inside the structure. When their strength exceeds a limit value of the material or component, the structure will be damaged with different strengths. In the environment of high-speed train operation, the highest value of the main tension in the tunnel structure is generally 0.6 mpa.

2.3   Dynamic Influence of Vibration on Subway Tunnel Structure

According to the cumulative fatigue damage of the underground train, during the operation of the underground train, between the slab columns in the sensitive part of the tunnel structure cross each other, the lower slab is reinforced concrete, and the inversion and other auxiliary structures need to be reworked. The load stress and compression stress caused by the action of steel bars, concrete and other steel bars alternately change, cracks open and close repeatedly, and the direction of welding stress between steel bars and concrete also alternates repeatedly, which greatly promotes the increase and destruction of residual deformation. With continuous accumulation, the structural cracks will continue to expand. Because if the highway tunnel structure is in a humid environment, the corrosiveness and the adverse effect of the environment on the durability of the structure must be fully considered. Therefore, when a macroscopic crack appears in the building structure, this corrosion is mainly groundwater. The corrosion of steel bars will produce a lot of volume and expansion after being corroded, which will aggravate the concrete cracks of the building. A large number of foreign experiments have proved that the volume of steel bars after corrosion is about six times the volume of steel bars before corrosion. Corrosion of steel bars will further increase fatigue cracks.

2.4   Various Vibration Isolation Measures for Vibration Phenomena

(1) Lead core rubber bearing

The elasticity of rubber and the plasticity of the lead core absorb seismic energy. By using the horizontal flexibility of the rubber support to form a shock-proof layer, the purpose of reducing damage to the upper structure can be achieved. The rubber lead bearing has a strong load-bearing capacity and can also provide a certain degree of strength to help fix the bracket.

(2) Other types of vibration isolators

The vibration isolator is an elastic element that attaches the device to the base to reduce and eliminate the vibration force transmitted from the device to the base and the vibration transmitted from the base to the equipment.

(3) Floating slab track structure

A concrete floor with a certain mass and rigidity is floated on a steel spring vibration isolator, 30 mm or 40 mm away from the upper surface of the floor mat. This is a mass-spring-vibration isolation system. This system is widely used for vibration control and noise reduction in subways. Its advantage is: good vibration isolation effect, which can reduce 25–40 dB vibration. The service life exceeds 29 years. At the same time, it has spatial elasticity and small horizontal displacement. It is very convenient to inspect or replace the track without disassembling the track, and will not affect the operation of the subway, and can be carried out by increasing or reducing the thickness of the steel plate.

(4) Other vibration isolation control methods

   1) Vibration-proof material is added between the guide rail and each layer.

   2) Increase the buried depth of the tunnel and increase the wall thickness of the tunnel.

2.5 Artificial Intelligence Monitoring System Algorithm

(1) Determine the value of the total efficiency coefficient of the monitoring section

Calculate the total efficiency coefficient of each sensor monitoring data in the early warning index system as

$$E_t = \sum_{i=1}^{s} w_{ti} e_{ti} \tag{1}$$

Where: E is the total efficiency coefficient value of the monitored section at time t (t = 1, 2, m); wti is the normalized weight of the i-th sensor at time t (I = 1, 2, ..., s).

(2) Calculate the efficiency coefficient value of each sensor

For the underwater shield tunnel, the water level is too large or too small to be detrimental to its stability, so the monitoring value Xu of a certain sensor i is an interval variable, and its efficiency coefficient value is calculated by Eq. (2).

$$e_{ti} = \frac{100 - 40\left(\frac{x_{ti} - x_{\mu mini}}{x_{mini} - x_{\mu mini}}\right)}{60} \quad 100 - 40\left(\frac{x_{ti} - x_{\mu mixi}}{x_{mini} - x_{\mu mixi}}\right) \tag{2}$$

Where: eti is the efficiency coefficient value of the i-th sensor at time t: Xi is the monitoring data of the i-th sensor at time 1; Xmaxi is the maximum value of the monitoring data interval of the i-th sensor: Xmini is the monitoring data interval of the i-th sensor the minimum value; Xxmini is the upper limit of the monitoring data of the i-th sensor is not allowed: Xmixi is the lower limit of the monitoring data of the i-th sensor.

3   Experiments on the Dynamic Influence of the Train Operation Vibration on the Subway Tunnel Structure Based on the Artificial Intelligence Monitoring System

3.1   Experimental Platform Construction

In order to minimize the impact of environmental interference on the experiment, the platform is built on the foundation pier of the laboratory (the foundation pier is connected to the base plate). Correct the monitoring system and accelerometer measurement points on the foundation pier in the same way. Use an electric drill to drive at the same distance from the two experimental measurement points. Taking into account the different running time of the train, the excitation duration is 10 s and 25 s respectively, and the amplitude of the vibration is recorded and analyzed by an accelerometer. When the amount of vibration meets the characteristics of vibration interference in train operation, the response output of the monitoring system collected by the data acquisition system is analyzed.

3.2   Vibration Interference Impact Test

In order to measure the impact of vibration interference on the actual measurement of the monitoring system, the resolution of the sensor selected for the monitoring system in this experiment is 0.02 μm. First, check the sensitivity of the sensor at a certain point in the monitoring system. The test results: the impact of vibration on the output of the system is about 52 mV, the sensitivity of the sensor is 47 mV/μm, and the corresponding displacement is 1.0 μV. The result is within the allowable error range of the system measurement.

3.3   Experimental Design

By building a small subway tunnel in the laboratory, using an artificial intelligence monitoring system to monitor the dynamics of the subway tunnel structure, using an electric drill to vibrate at different points to simulate train vibration, and increasing the frequency of the electric drill to obtain different vibration frequencies. Carry out two experiments with different frequencies (20–80 Hz, 0–20 Hz) respectively, by monitoring the section acceleration, velocity, and displacement response parameters at the beginning, end, and end of the subway tunnel.

4   Analysis of Experimental Results

(1) Through the experiment of the dynamic influence of the vibration of the simulated frequency of 0–20 Hz on the subway tunnel structure, the experimental results are shown in Table 1:

Table 1. Vibration of 0–20 Hz on the dynamic influence of subway tunnel structure

	Beginning	Middle	End
Displacement	0.0045	0.0038	O.0015
Acceleration	O.52	0.34	0.08

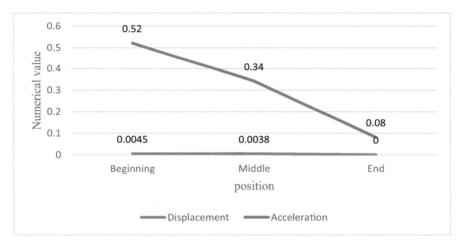

Fig. 1. Vibration of 0–20 Hz on the dynamic influence of subway tunnel structure

It can be seen from Fig. 1 that under the influence of vibration in the tunnel, the displacement changes from 0.0045 to 0.0038 and then to 0. 0015. The speed will continue to drop again, and the beginning will be the most affected. The front section of the tunnel can be reinforced, and vibration-proof structures can be added.

(2) Through the experiment of the dynamic influence of the vibration of the simulated frequency of 20–80 Hz on the subway tunnel structure, the experimental results are shown in Table 2.

Table 2. Vibration of 20–80 Hz on the dynamic influence of subway tunnel structure

	Beginning	Middle	End
Displacement	0.030	0.025	0.023
Acceleration	2.40	2.27	2.07

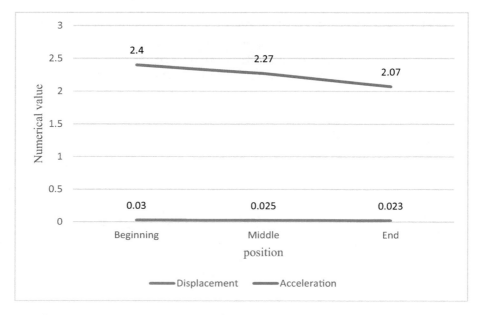

Fig. 2. Vibration of 20–80 Hz on the dynamic influence of subway tunnel structure

It can be seen from Fig. 2 that after increasing the vibration frequency, all the data have increased, but the trend is the same, the displacement is 10 times the original, and the speed has also increased by 10 times.

5   Conclusions

The study of tunnel vibration response is currently a hot topic. The initial travel speed is low, and the vibration caused by the train has little effect on the structure of the tunnel. The first batch of studies on subway vibration at home and abroad mainly focused on environmental impact. In recent years, as more and more cities are building underground railways, the tunnel structure model has also changed. The adverse effects of these changes on the tunnel structure are increasing rapidly. Then there were diseases such as water leakage and softening of the grassroots, which led to further deterioration of the tunnel structural system and seriously affected the safety of the railway company. In this paper, a small subway tunnel is built in the laboratory, and the artificial intelligence monitoring system is used to monitor the dynamics of the subway tunnel structure. The results are obtained. First: the displacement of the tunnel is changed from 0.0045 under the influence of vibration. To 0.0038 and then to 0.0015. The speed will continue to drop again, and the beginning will be most affected. The front section of the tunnel can be reinforced, and vibration-proof structures can be added. Second, after increasing the vibration frequency, all the data have increased, but the trend is the same, the displacement is 10 times the original, and the speed has also increased by 10 times.

References

1. Ding, Z., Li, D.: Dynamic response analysis on vibration of ground and track system induced by metro operation. Eng. Comput. **36**(3), 958–970 (2019)
2. Shi, W., Miao, L., Luo, J., et al.: The influence of the track parameters on vibration characteristics of subway tunnel. Shock Vibr. **2018**(PT.10), 2506909.1–2506909.12 (2018)
3. Mitroshin, V.A., Mondrus, V.L.: Analysis of the impact of shallow subway train traffic on urban development. In: IOP Conference Series: Materials Science and Engineering, vol. 1015, no. 1, p. 012019 (5pp) (2021)
4. Yang, J., Lan, K., Zhu, S., et al.: Dynamic analysis on stiffness enhancement measures of slab end for discontinuous floating-slab track. Comput. Sci. Eng. **PP**(3), 1 (2019)
5. Ren, D., Shao, S., Cao, X., et al.: Vibration characteristics of subway tunnel structure in viscous soil medium. Adv. Mater. Sci. Eng. **2021**(10), 1–12 (2021)
6. Wei, K., Wang, P., Yang, F., et al.: The effect of the frequency-dependent stiffness of rail pad on the environment vibrations induced by subway train running in tunnel. Proc. Inst. Mech. Eng. Part F J. Rail Rapid Transit **230**(3), 171–178 (2016)
7. Sun, C., Gao, L., Hou, B., et al.: Analysis of vibration characteristics and influence parameters of buildings adjacent to subway line. J. Beijing Jiaotong Univ. **41**(4), 23–30 and 39 (2017)
8. Yang, M., Li, H., Li, N., et al.: Effect of subway excavation with different support pressures on existing utility tunnel in Xi'an loess. Adv. Civ. Eng. **2020**(2), 1–14 (2020)
9. Lugin, I.V., Alferova, E.L., et al.: Integrated performance analysis of ventilation schemes for double-line subway tunnel. In: IOP Conference Series: Earth and Environmental Science, vol. 262, no. 1, pp. 12043–12043 (2019)
10. Cai, H.B., Li, M.K., Li, X.F., et al.: Analytical solution of three-dimensional heaving displacement of ground surface during tunnel freezing construction based on stochastic medium theory. Adv. Civ. Eng. **2021**(4), 1–16 (2021)

Emotional Consumption of Micro-blog Fans Based on Big Data Analysis

Zhujun Dai[✉]

Jilin Engineering Normal University, Changchun, Jilin, China

Abstract. With the Combined Influence of social media, big data power and the Digital Economy, Fan circle girls are expressing their support for celebrities through ritual data labor and collective emotional consumption on Weibo. Fan sentiment is quantified by data and becomes a source of traffic and capital. Traffic and data not only dominate fandom culture and fan economy, they also change fan behavior and community dynamics, and fandom girls become data workers, to some extent, it is a worldwide competition for data to become a form of capital.

Keywords: Big data · Traffic · Data · Fan culture

The Fan Culture, which follows the stars in the way of "making data", such as ranking, controlling and evaluating, has received wide attention from the media and academic circles [1]. In the digital age, "fans" have gone from scattered individuals to a large number of groups with formal organizational structures and rules. The Fan community has the basic characteristics of emotional connection, the main characteristics of active participation and the organizational characteristics of having a common goal. Fans come together with a clear purpose, driven by a love of and admiration for their idols. They share information and produce content within the community and actively participate in the promotion and support activities of their idols. In the digital age, huge crowds of fans have gathered to form a community. They show their support for the stars through ritualistic data labor and collective emotional consumption, and fans generate large data traffic through organized promotion of star related topics, forming a high level of discussion and exposure across the web, star effect on the promotion of considerable commercial value. Demonstrated a strong organizational capacity, productivity and action. Weibo is the social media of Fan circles and the main connecting field of fan communities. Weibo has more than 500 million monthly active users. As "the leader of the fan economy" [2]. With the development of big data and ALGO-RITHMIC technology, Weibo has gradually built a set of systems specifically designed for fans, a special kind of data production and traffic shaping rules. The number of fans, the number of reading, the number of user-generated content, and so on, has become a "hot topic" "hot search" standard, but also become the flow of visual reference. With the combination of big data and the Digital Economy, traffic has become a measure of fan loyalty, Star popularity and business value.

J. C. Hung et al. (Eds.): FC 2021, LNEE 827, pp. 568–571, 2022.
https://doi.org/10.1007/978-981-16-8052-6_70

1 Traffic Behind the Ring of Data Labor

Traffic refers to the level of user activity on a network platform. It is regarded as an objective description of a star's personal value at the market level, as well as a cultural evaluation of public perception [3]. In 2016, "traffic" began to become a high-frequency word in the entertainment industry. In the era of big data, star traffic not only represents fame and popularity, but also is directly related to commercial interests. The star effect brings the attention economy, the advertisement resources and the movie and TV play resources will also incline to the attention degree high, the fan many flow star, but the fan strength direct influence star's attract the gold ability. Those fans, high popularity, wide impact, commercial value of the stars often become "flow star". In 2019, China's new idol, Wang Yi Bo, gained a huge following due to his super-high looks and talent. The number of fans reached 38.18 million, holding more than 30 endorsements, China's top entertainment traffic.

Social Media, technology capital, entertainment capital, online content platforms, brands, marketing companies and other stakeholders are all working together to extract money from fans and create data, together, they contribute to the generation and operation of the flow economy. Traffic is both a data set, a Means of production, a catalyst and glue for the fan economy, a star power, and a driving force for emotional consumption.

The term "Data Labor" refers to the work that fans do to increase the popularity of their stars' traffic, which can be calculated and quantified, by collectively posting, reposting, liking, and so on under the guidance of social media [4]. Data is both the Labor of the fans and the Means of production of the fans' economy, as well as the traffic power that all the stakeholders chase and control. In the digital age, fans are an early and enthusiastic adopter of new media technologies to expand their reach and connectivity. One of the reasons that fan girl is different from the average fan, or fan, is "her obsession with data" [5]. According to Ayman, in 2018, 41.6% of fans in the official fan club's role was data, and 84.1% of fans participated in data group activities. Almost every celebrity fan club has a dedicated section called the Foshan data group, or fan support groups that set up separate data stations outside the terminal. On Weibo, Fan girl fandom has become a daily routine of posting and doing data. Compared with the behavior of fans who need corresponding technical skills such as writing propaganda copy independently, editing pictures and video clips, the data labor is easy to operate, reproducible and low in technical content, as a form of labor that maximizes the mobilization of fans. Girls in the fan circle often deride themselves as "data workers", turning their emotions into data labor by constantly posting, liking and reprinting content on Weibo. Fans have gone to great lengths to generate buzz and online buzz for Meisei in order to give their favorite idols more resources. Fans are pushing the data methodically, with new shows around idols, endorsements, new songs, etc. The Fan data group posts the Weibo content that needs to be retweeted and commented on every day, unifies the topic hot words, concentrates the time to publish, the massive retweet forms the huge flow. As far as platforms and capital are concerned, the quantity of content produced by fans is far more important than quality, attracting more users to work on data long term, spontaneously, in groups, free of charge, and steadily, constantly producing more data, to create a flood of flow. Weibo is precisely the "factory" that stimulates and promotes the data consciousness and labor of its fans. Weibo can constantly adjust the rules to stimulate and generate data awareness among its followers, creating new demand for lists and spurring new data production with the help of new algorithms.

2   The Collective Emotional Consumption Behind the Data

In Modern Society, emotional needs are more and more satisfied in the form of emotional consumption [6]. In the process of generating traffic, fans, out of their affection for idols, voluntarily become "data workers" and "leeks", participate in the capital game, and justify the capital logic of justify.

Fan Labor, as a free labor based on emotion, has also received a lot of attention. Whether it is providing text on an ongoing basis, or actively participating in idol related activities such as idol concerts, birthday parties, fans need to spend a lot of energy and material resources, the vast majority of fan activities are paid for free, for the media style (film, Television Works, etc.), marketing companies and so bring the corresponding value [7].

On social media, emotions become information that can be "digitized" and transformed into a source of traffic and capital. Fans experience mixed emotions in the process. On the one hand, fans love their idols and work for them. The practice of fan behavior is, in a sense, a kind of free labor. This kind of emotional labor brings fans self-satisfaction and self-enjoyment. Fans, on the other hand, try to find the perfect image of themselves in their idols to make up for their desire. When the source text cannot satisfy or satisfy the fans, the fans struggle with it through re-creation, trying to express to others the unrealized possibilities in the original work. So that they can willingly devote their attention and creativity to their loved ones and experience a sense of satisfaction and fulfillment in the process.

As one of the most active and dynamic groups of media consumers, fans can bring into play a huge potential "fan industry" and form a new "fan economy". In the study of Fan Economy, Zhang Qiang, a scholar from Taiwan, holds that "emotional capital is intangible capital such as consumers' liking, familiarity, loyalty and perception of brands and idols, as well as consumers' association with brands and idols" [8]. We can say that the basic logic of fan economy is emotional economy, fans use the power of the group to combine celebrity topic marketing with deductive products. By using this emotional connection, data labor maximizes the revenue opportunities for capital and attracts the attention of the public and the media, the connection between data labor and fan sentiment is undeniable.

In the social, economic, cultural, and technological changes influenced by big data, traffic and data not only dominate fandom culture and fan economy, but also change fan behavior and community mechanism, at the same time, it reflects the change of production and consumption in the digital age, and even further affects the cultural production and political ecology. "The impact of big data on society as a law of technology" [9]. Girls in the rice bowl become data workers, to some extent, it is a worldwide competition for data to become a form of capital.

References

1. Pearson, R.: Fandom in digital era. Popular Commun. **8**(1), 84–95 (2010)
2. 2018 white paper for Weibo fans. http://sina.aiman.cn/first-pc.html
3. Tong, Q.: Fandom girls' battleground for traffic: data labor, affective consumption and neoliberalism. J. Guangzhou Univ. (Soc. Sci. Edn.) **5** (2020)
4. Kucklich, J.: Precarious playbour: modders and the digital game industry. Fibrecult. J. (5) (2005). http://five.Fibreculturejournal.org/fcj-025-precarious-playbour-modders-and-the-digital-games-indus-try/
5. Whiteman, N.: The de/stabilization of identity in online fan communities. Converg. Int. J. Res. New Media Technol. **15**(4), 391 (2009)
6. Wang, N.: Affective consumption and the affective industry: a study on the sociology of consumption. J. Natl. Sun Yat-sen Univ. **40**(6), 109–113 (2000)
7. Kosnik, A.D.: Interrogating free fan labor. Spreadable Media (2013). https://spreadablemedia.org/essays/kosnik/index.html#. Xviq7y QRXDs
8. Zhang, Q.: Fan Power, p. 95. University Press (2010)
9. Ji, D.: Big data in China: from myth to political economy. Educ. Media Res. **3**, 42–47 (2018)

Incomplete Big Data Filling Algorithm Based on Deep Learning

Lujun Chen[✉]

Sichuan Technology and Business College,
No. 8 Juyuan Section of Tianfu Avenue, Dujiangyan, Sichuan, China

Abstract. In recent years, deep learning has made remarkable achievements in many fields, including incomplete big data filling algorithms. The incomplete filling algorithm based on deep learning is one of the most popular research topics in the current big data algorithm field. The traditional data filling algorithm fills all data sets with the corresponding data content, which increases the amount of calculation without considering the correlation between the data, and most algorithms cannot directly extract the missing data attributes for the iterative completion algorithm. Most of them have slow convergence speed and insufficient accuracy. This article discusses the incomplete data filling algorithm based on deep learning. This paper first uses deep learning methods to extract features from incomplete big data, uses kernel functions to calculate missing values based on non-parametric methods, and proposes a local distance strategy to initialize missing values, thereby reducing the number of repetitive fillings. The filling algorithm proposed in this paper is used to supplement and correct incomplete or incorrect data to ensure the completeness and accuracy of big data and maximize its value. According to the application results, it shows that the big data filling algorithm proposed in this paper can achieve a filling accuracy of about 85%, the filling effect is good, and it can meet the needs of further data analysis.

Keywords: Deep learning · Incomplete data · Big data · Filling algorithm

1 Introduction

With the development and application of data mining and artificial intelligence technology, a large amount of data must be processed [1, 2]. The increase in data provides a good prospect for the development of data applications in China, but it also brings many problems [3, 4]. Due to various reasons, there are a lot of incomplete data in big data, which directly affects further data analysis and extraction, so that the data cannot maximize its value [5, 6]. Incomplete data filling algorithms based on deep learning have become an increasingly important field of research [7, 8].

When studying the filling algorithm of incomplete big data based on deep learning, many scholars have conducted research on this and achieved good results. For example, Zhu D proposed a filling algorithm based on k nearest neighbor (kNNI), which uses k similar records to fill in missing values. This method first selects item k by calculating the Euclidean distance of different data, and then uses the average of k similar items to complete [9]. Deng W and others proposed an incomplete data integration algorithm based on AP grouping, using a new similarity measurement method

J. C. Hung et al. (Eds.): FC 2021, LNEE 827, pp. 572–579, 2022.
https://doi.org/10.1007/978-981-16-8052-6_71

to measure the similarity between incomplete data, and then directly grouping the incomplete data, and then using The same clustering [10].

This paper first proposes an incomplete big data filling algorithm based on deep learning, which uses deep learning with noise reduction function to directly extract features with specific robustness from incomplete data. Then use these grouping functions, use the common event table and part-time strategy to vote on the data in each class, and finally convert the scores into weights for weighting. This paper adjusts and optimizes the incomplete big data algorithm, uses the local distance strategy to make up for the missing values, improves the convergence speed of the algorithm, and applies it to data filling and standardization.

2 Research on Incomplete Big Data Filling Algorithm Based on Deep Learning

2.1 Feature Extraction Algorithm Based on Incomplete Data of Deep Belief Network

(1) Data preprocessing

First, the data is scored and standardized. Since Boltzmann's continuous con-straining machine has requirements for the range of data values (that is, the continuous value between [0, 1]), the data must be normalized. The attribute vector is expressed as $(a_{i1}\ a_{i2}\ a_{i3}...\ a_{im})^T$. Calculate the expectation and variance of the attribute on all X sampling points.

(2) Deep belief network for noise reduction

In the process of auto-encoder learning features, in order to obtain more robust features, the sample is added with random noise qD according to a certain distri-bution, that is, the corresponding input node does not deactivate the neuron, and then the sample is changed from x to, then Will be used to reconstruct the output y, and minimize the cost function L of the initial input and output y to obtain the model parameters. In order to make full use of complete data information and missing data, k-means are used to process attributes. Then, when you complete the records in a particular class, use all the records in that class to estimate missing values. While reducing calculations, related items also ensure the accuracy of completion.

2.2 The Weighting Method of Correlation Matrix and Partial Distance

(1) Co-occurrence matrix method

Since most of the attribute values of the data are continuous values, it is now necessary to divide them into a finite discrete space to represent the data again. Discretely transform the matrix X by means of data discretization. For propor-tional discretes and fixed frequency discretes, they are more concerned about the number of records in each category. If the number of records in a certain category is, then the next category may also contain the value of the previous category, or even Maybe the first category also contains the same value.

(2) part distance strategy

Define the sample data as $X' = \{X'(a_1), X'(a_2), \ldots (a_m), \}$, and there is a complete attribute, and the remaining attributes are complete. The correlation between two incomplete data can be measured in Euclidean space. Then calculate $X' = \{X'_1, X'_1, \ldots, X'\}$ the partial distance of the data to be filled, and normalize it into a correlation score $V_2 = \{V_1^2, V_2^2, \ldots, V_t^2\}$.

2.3   Research on Incomplete Big Data Weighted Sequence Filling Method Based on Deep Learning

First, use the attribute mean strategy to pre-populate the incomplete data set, and then use the co-occurrence matrix method to find the highest co-occurrence attribute data set of the attributes to be filled $X' = \{X_1, X_1, \ldots, X_i\}$ and follow the missing rate from High to low sort order, that is, fill in the low completeness first, and then continue to fill other items with the filling result. Then use the co-occurrence matrix and the correlation score obtained by the partial distance strategy to comprehensively evaluate the correlation between the data to be filled and other data $V = \{V_1, V_2, V_3, \ldots, V_t\}$. The specific method is shown in formula (1)

$$V = V_1 \times \lambda + V_2 \times (1 - \lambda) \tag{1}$$

For the selection of parameter, this paper believes that the correlation between attributes and the correlation between data tuples account for the same proportion, and $\lambda = 0.5$. In this way, we can get $X' = \{X_1, X_1, \ldots, X_i\}$, the tuple of which is the correlation set of the data record to be filled, and then convert it into the correlation weight $W = \{w_1, w_1, \ldots w_i\}$ through the following formula.

$$\begin{cases} \frac{W_{t-i+1}}{w_i} = \frac{V_{t-i+1}}{v_i} \\ \sum_{j=i}^{t} w_j = 1 \\ V_i \in V \\ i = 1, 2, 3, \ldots, t \end{cases} \tag{2}$$

Then according to formula (2), we can find out the set $X' = \{X'_1, X'_2, \ldots, X'_t\}$ of the value of the attribute position to be filled corresponding to the relevant attribute in $Y(a_i)$

$$X'(a_j) \rightarrow Y(a_i) = \{y_1(a_i), y_2(a_i), y_3(a_i), \ldots y_t(a_i)\} \tag{3}$$

$y_i(a_i)$ indicates the value of the relevant attribute data containing the attribute to be filled in the position of the attribute to be filled, and then use the weight obtained before to sum it,

$$\begin{cases} \sum_{i=i}^{t} w_i y_i(a_j), y(a_i) \in Y(a) \\ \sum_{j=i}^{t} w_j = 1 \end{cases} \tag{4}$$

Then continue to use formula (4) to fill in other attribute values of the current data, and fill in the data with high missing rate in order after completion.

3 Research on Incomplete Big Data Filling Algorithm Based on Deep Learning

3.1 Experimental Environment

In this research, multiple data sets and different data completion algorithms are used in experiments and comparative experiments, and the algorithm based on incomplete data in this article based on deep learning is compared with the comparison algorithm. The test uses Windows764bit operating system, and the processor is Intel(R) Core(TM) i52450CPU@ 2.50 GHz. The algorithm is implemented using Matlab programming language and Java programming language.

3.2 Experimental Procedure

The first step is to establish an inverted index on the original data, use the target record to be modified for full-text search to obtain Top-k, and then convert the record into a numeric vector. The second step is to score the relevance of the records through the partial distance strategy, and then use the longest field among the multiple records with the highest score to replace the fields to be filled or directly fill the fields to be processed. The third step is to sample 5000 pieces of data from the sample to analyze the accuracy.

3.3 Data Sources

The data used in this article is US import and export data. There are 133 fields, about 3.9 million records, and the size is about 6 GB. The application environment is an enterprise-class server with 64 GB of memory and Windows 10 operating system. Approximately 23% of the received data is invalid data. After preprocessing, such data will be deleted, and then the remaining 77% of the data will be cleared. At the same time, due to the particularity of the data, RMSE cannot directly measure the filling result, so the actual result of the application is manually determined by a technician with relevant professional knowledge.

4 Incomplete Data Filling Algorithm Based on Deep Learning

4.1 Verification Analysis of Feature Extraction Quality

Because the algorithm in this paper is to extract the features of incomplete big data through the filling algorithm based on deep learning, perform clustering, and then fill in each class separately. Therefore, it is first necessary to verify that the features extracted by the deep learning-based filling algorithm are sufficiently robust, so that the effect of the increase in the missing rate on the clustering is as little as possible. For different types of data, the features extracted by the filling algorithm proposed in this paper and the original data are extracted using k-means, and the results are shown in Table 1.

Table 1. The clustering accuracy of different data in different missing ratio

Missing Ratio	Abalone (dDBN)	Abalone (dAE)	Abalone
0	80.5	83	79
5	79	79.5	77
10	78	73	75
15	77	76	72
20	76	71	67
25	74	67	63
30	69	60	51

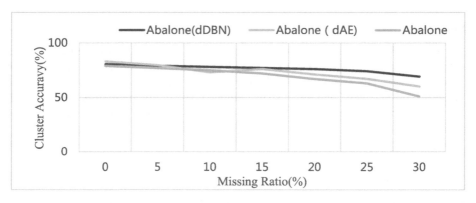

Fig. 1. The clustering accuracy of different data in different missing ratio

It can be observed from Fig. 1 that the use of the incomplete big data filling algorithm based on deep learning proposed in this paper can greatly enhance the robustness of the extracted features, making the result better than directly using the incomplete data set for calculation. And observe the experiment on the Iris data set, the effect of direct clustering drops sharply when the missing rate increases. The reason is that the Iris data has only 4 attributes, and the high missing rate has a greater impact on the data, which in turn leads to a decrease in clustering accuracy. At the same time, when the missing rate increases, the clustering accuracy slope of the incomplete big data filling algorithm based on deep learning proposed in this paper changes slowly, that is, the decline is slower, while the experimental group that directly uses k-means is missing. The rate of increase has changed drastically. It also proves that the features extracted by the filling algorithm are robust to a certain extent.

4.2   Verification Analysis of Filling Accuracy

This research uses two different types of data filling methods: k-nearest neighbor filling algorithm (kNNI) and expectation maximization (EMI), and on the data set, the three algorithms including this research are measured under different missing rates. For the processing of the complete data set, this article adopts completely random deletion, and

Table 2. Filling result

Original Value	0.3	0.2	5.7	1.1	5.1	6.3
Filled Value	0.3069	0.3078	0.3425	1.1088	5.6825	6.7532
Error	0.0093	0.1059	0.1421	0.0049	0.5623	0.5665

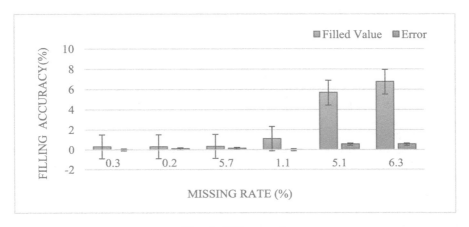

Fig. 2. Filling result

the experiment is conducted under six kinds of deletion rates. The filling accuracy is measured by the standard deviation:

It can be seen from Table 2 that the algorithm proposed in this paper is better than k nearest neighbor filling and expectation maximization filling methods in performance. At the same time, in order to observe the filling results more intuitively, the filling results of the algorithm in the Iris dataset with a 3% missing rate are shown in Fig. 2. In addition, in terms of execution time, the algorithm in this paper is higher than the k-nearest neighbor algorithm, but compared to the iterative filling algorithm of expectation maximization, the execution time of the algorithm in this paper is better than this when the iterative convergence is unstable. At the same time, it can be seen from the comparative experiments on different data sets that the incomplete big data filling algorithm based on deep learning proposed in this paper increases with the increase of the missing rate, the slope of the RMSE change is slow, it has a certain degree of robustness.

5 Conclusions

This paper also tested the filling accuracy under different missing rates under these data sets with kNNI and EMI. Incomplete big data filling algorithm using deep learning. In this study, incomplete data is directly processed to extract robust features; these features are then used to cluster and fill in each class; for the filling strategy, a co-occurrence matrix (obtained using a discretizer) and partial distance are used. To score the related tuples, and finally convert the scores into weights, and fill them by weighting. And apply it to the US import and export trade data, through the adaptation of string input data and the improvement of response speed. In the end, under the analysis and verification of professionals, the accuracy of the algorithm is as high as 85%, which can handle various missing situations, and obtain a relatively ideal filling result.

In recent years, people have paid more and more attention to filling incomplete big data, and various methods have emerged one after another. The simplest way to deal with this problem in the early days is to "ignore", that is, do not perform any processing on data containing missing values, or Delete directly. Therefore, in order to avoid this large-scale loss of data, researchers have invested a lot of energy to study the filling of missing values. With the continuous growth of data, deep learning plays an increasingly important role in the filling and analysis of incomplete big data. Especially the outstanding performance in feature extraction has been widely valued by scholars at home and abroad.

This research proposes an incomplete big data filling algorithm based on deep learning. The original data is first analyzed before application. The incomplete big data filling algorithm is first analyzed, and the algorithm proposed in this article is applied to real data. The text is specific to the specific data. The algorithm is adapted and optimized. This research not only has theoretical research significance but also has certain practical application value. In quantitative sampling, through experiments and professional analysis, the algorithm proposed in this paper can improve the filling accuracy while ensuring the accuracy rate, meeting the requirements of academic research and industrial application standards, and has certain theoretical and application value.

Acknowledgments. Our thanks to all of people who have contributed to this paper. My paper is supported by the project of the Sichuan Provincial Department of Education's Fund for Educational Research and Reform Projects (SCJG20A004-4).

References

1. Zeng, F., Hu, S., Xiao, K.: Research on partial fingerprint recognition algorithm based on deep learning. Neural Comput. Appl. **31**(9), 4789–4798 (2018). https://doi.org/10.1007/s00521-018-3609-8
2. Zhao, R., et al.: Deep learning and its applications to machine health monitoring. Mech. Syst. Signal Process. **115**, 213–237 (2019)
3. Han, K., Wang, Q.: Research on O2O platform and promotion algorithm of sports venues based on deep learning technique. Int. J. Inf. Technol. Web Eng. (IJITWE) **13**(3), 73–84 (2018)
4. Liu, J., Choi, W.H., Liu, J.: Personalized movie recommendation method based on deep learning. Math. Probl. Eng. **2021**(6), 1–12 (2021)
5. Shi, Y., Wu, X., Fomel, S.: Interactively tracking seismic geobodies with a deep learning flood-filling network. Geophysics **86**(1), 1–19 (2020)
6. Lu, L.: Design of visual communication based on deep learning approaches. Soft. Comput. **24**(11), 7861–7872 (2019). https://doi.org/10.1007/s00500-019-03954-z

7. Zhang, C., et al.: DEM void filling based on context attention generation model. Int. J. Geo-Inf. **9**(12), 734 (2020)

8. Liu, J., et al.: A quasi-human strategy-based improved basin filling algorithm for the orthogonal rectangular packing problem with mass balance constraint. Comput. Ind. Eng. **107**(MAY), 196–210 (2017)

9. Zhu, D.: Max–min bin packing algorithm and its application in nano-particles filling. Chaos Solitons Fractals Interdiscip. J. Nonlinear Sci. Nonequilib. Complex Phenom. **89**, 83–90 (2016)

10. Deng, W., et al.: A missing power data filling method based on improved random forest algorithm. Chin. J. Electr. Eng. **5**(4), 33–39 (2019)

Teaching Design Optimization and Application Research of College Students' Education Evaluation Data Analysis Based on Big Data

Ninghai Liu and Yanchun Ruan[✉]

College of Cuisine Science and Technology, Jiangsu College Tourism, Yangzhou 225100, Jiangsu, China

Abstract. With the influence of big data (BD) on the field of education (EDU), the scale of EDUal BD is becoming larger and larger. The explicit data of traditional EDU is not enough to meet the present EDUal situation. All kinds of static data are mined and invisible data appear. At the same time, Internet EDU has produced massive data, and the field of EDU has formally stepped into the era of BD, and the development prospect of EDU integrated with information technology is broad. The characteristics of the data analysis of college EDU evaluation are that it can comprehensively and systematically combine, process and analyze the information data of students and teachers, and then make a scientific and reasonable value judgment on the development potential and progress of students on this basis. The function is to promote students' self-growth in an all-round way. Through the analysis results of BD, we can accurately understand the development space and changes of students. Teachers can find the disadvantages of data processing in time and deal with them in time, which has good pertinence for teaching. This paper mainly uses the literature research method, the questionnaire survey and the interview type survey and so on, has carried on the study to the literature, elaborated the BD EDU, the EDU appraisal and so on basic concept in detail; the questionnaire survey object is a certain university 120 teachers and students, altogether distributed 120 questionnaires, the questionnaire altogether distributed 120 questionnaires, recovered 110 questionnaires, the recovery rate is as high as 91.1 Using the form of interviews to judge the use value of EDUal evaluation data, the results show that in the BD environment, educators' demand for EDUal evaluation data analysis is gradually rising. At the same time, the advantages and disadvantages of EDUal evaluation are found. In the information age, higher EDU has undergone unprecedented changes. In the future, we look forward to the innovative influence of BD in the process of higher EDU development, and promote the development of higher EDU to individualized, popularized and shared higher EDU.

Keywords: BD EDU · EDU evaluation · Data analysis · Higher EDU

1 Introduction

Under the action of BD, the analysis of EDUal evaluation data is accurate and can be analyzed according to the students' learning results and the evaluation data of learning process. It provides "one-to-one" guiding significance for students at different levels.

J. C. Hung et al. (Eds.): FC 2021, LNEE 827, pp. 580–586, 2022.
https://doi.org/10.1007/978-981-16-8052-6_72

The more thoroughly teachers are familiar with students, the more they can choose the most suitable learning content in teaching, formulate the most accurate learning goals, make the most objective learning evaluation and the most timely learning feedback, and give students individualized guidance [1].

Some foreign scholars believe that "can make students pay attention to and think about the most critical views and ideas of the topic, or the results of teaching can stimulate and enhance the interest in understanding, such classroom teaching can be considered the most effective [2, 3]." The key to realize teaching optimization is that teachers choose and organize the best scheme of effective teaching. Once teachers master the systematic method of organizing optimal teaching, teachers, especially experienced teachers, will be on the basis of the original teaching. At this time, teachers can get the maximum teaching effect in the specified time.

EDUal evaluation data is a kind of data with deep mining value that can be easily obtained. According to the subject knowledge model, the subject test is compiled to obtain effective student evaluation data, and the knowledge structure and understanding level of learners are analyzed. Instead of focusing on quantitative representation of academic performance, ranking and other information [4]. The deep mining of learner evaluation data can not only be used as the basis of resource push, but also provide teachers with more effective evaluation information that can reflect learners' current learning state. Teachers can organize teaching more effectively, analyze learning situation and adjust teaching strategies with the support of objective data, so as to give full play to the potential value of student evaluation data to classroom teaching.

2 Proposed Method

2.1 EDUal BD Research Methodology

The research method of EDUal BD is to apply BD to EDU. We can analyze and summarize all kinds of data to get useful information. EDUal researchers can also predict the development of EDU by using the analysis method of EDUal BD, so that researchers can understand the object of EDU more clearly and carry out more targeted EDU. According to the level of EDUal BD, some experts can divide the EDUal BD into national EDU BD, regional EDU BD, school EDU BD, individual EDU BD [5]. As the source of data in EDUal BD research, five levels of EDUal BD are of great significance to the construction of EDUal BDbase, and also to the use of EDUal BD research methods. EDUal BD research methods do not negate other EDUal scientific research methods, but will use the data collected by other methods for centralized processing. A lot of information will be collected using qualitative research methods such as EDUal survey, EDUal observation, EDUal experiment and EDUal ethnography. EDUal BD research methods are more about transforming these data into data that can be stored and processed. In order to make a more scientific analysis of these data, perhaps we can also use these data to predict the development of research problems [6, 7]. EDUal BD research methods also greatly expand the channels for collecting data in the field of EDUal research. Researchers can collect data through the Internet or other intelligent devices. In this way, we can use all kinds of data to carry out comprehensive research on EDU.

2.2   EDUal Evaluation

EDUal evaluation refers to the process of scientifically measuring and judging all kinds of EDUal activities, EDUal process and EDUal results by using certain techniques and methods under the guidance of certain EDUal values and according to the established EDUal objectives. The EDUal evaluation data is the data fact obtained for the EDUal effect or the development of the students in all aspects. The EDUal evaluation is the process of judging the value according to these data. In a certain period of time using conventional software tools to capture the processing of large data sets. Academic data collected in universities are mainly around the overall situation of the evaluation exercises and the investigation of the topics [8]. In the process of teaching, the application of EDUal evaluation data analysis software or tools, the regular collection and analysis of these data is the key to give full play to the value of the evaluation data. Further guide learners to adjust the pace of learning, constantly improve and optimize the teaching design process.

2.3   BD

BD is a kind of information asset with large volume, many kinds and fast speed. It cannot be collected and managed by ordinary processing tools, but must rely on virtualization technology such as cloud computing beyond a certain time and space boundary. In the era of BD, the most concerned is to be able to process a large amount of data to get the potential value behind the data and expectations for the future. The core and key of this project is to analyze and process data information, and to use cloud computing and other virtual technologies to ensure the real value of data. Besides relying on special processing tools and systems, when people carry out data collection, management, analysis and other steps, especially the staff of colleges and universities, they should take into account the wide range of information sources, implement specific system norms to protect privacy and attach importance to the training of specialized talents [9]. Whether in the BD era or other fields of information technology, Bayesian formulas can answer questions:

The Bayesian theorem formula is as follows:

$$P(B_i|A) = \frac{P(B_i)P(A|B_i)}{\sum_{j=1}^{n} P(B_j)P(A|B_j)} \tag{1}$$

where $P(A|B)$ is the possibility of A occurrence in the event of a B occurrence. A1, …An is a complete event group, that is:

$$\bigcup_{i=1}^{n} A_i = \Omega, A_i A_j = \emptyset, \ P(A_1^2) > 0 \tag{2}$$

When there are more than two variables, the Bayesian theorem still holds, for example:

$$P(A|B,C) = \frac{P(A)P(B|A)P(C|A,B)}{P(B)P(C|B)} \tag{3}$$

3   Experiments

3.1   Experimental Subjects

Based on the background of BD, this paper expounds the application status of data analysis of EDUal evaluation in colleges and universities. It is found that the analysis of EDUal evaluation data plays a guiding role in teachers' teaching, can feedback students' learning results in time, evaluate students' learning effectively and reflect accurately, which plays a decisive role in excavating students' development of learning ability and effectively arranging and adjusting the later teaching mode.

3.2   Experimental Methods

This paper mainly uses the literature research method to study the relevant literature, expounds the definition and development status of BD EDU, the definition of BD and the basic concept of EDU evaluation based on BD background, and studies the whole paper based on this concept. Using the questionnaire method, 120 teachers and students in a university are used as the survey object, the feedback degree of the questionnaire is good, the data is collected, integrated and analyzed, and it is found that the analysis of EDUal evaluation data can feedback the students' learning results to a certain extent, and the satisfaction degree of EDUal evaluation data analysis is good; the interview method is also used to further understand the advantages of EDUal evaluation data analysis for teachers, as well as to find out the problems in the process of teachers' application, and to analyze the future EDUal evaluation data for teaching activities.

4   Discussion

4.1   Investigation and Analysis on the Application Effect of College EDU Evaluation Under BD

In order to test the application effect of data analysis based on BD, whether it is accepted by teachers and students, whether there is satisfactory classroom effect there is satisfactory classroom effect and whether it can reflect good teaching success, the paper teachers and students were distributed, and 110 questionnaires were collected.

Through the investigation of the basic information of teachers, we know that most of them are teachers with rich knowledge and teaching experience. The data clearly show the distribution of teachers' academic qualifications. There are 25 undergraduate teachers, accounting for half of the total, and 14 teachers and 11 junior college teachers. On the whole, teachers have high comprehensive level, strong teaching ability and high information literacy. As shown in Fig. 1:

4.2   Learning Feedback from Data Analysis

Under the background of BD, the most important thing is the validity of data analysis and the improvement of students' ability. According to the following figure, all teachers think that the application of this model has fully aroused the enthusiasm of

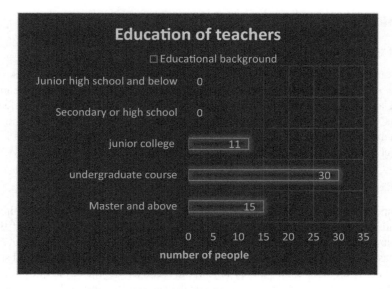

Fig. 1. Distribution of EDUal qualifications

learning, 50 teachers think that it is helpful to understand the students' acquisition and understanding of new knowledge, 47 teachers think that the application of this model has improved the students' sense and ability of teamwork, and 45 teachers think that teaching activities under this model can help students consolidate and internalize the learned knowledge, in addition, 36 teachers said that the students' ability of thinking development has also been greatly improved. Therefore, the introduction of EDUal evaluation data analysis into teaching has had a great impact on teaching and improved students' ability in many aspects. For the development of students to provide a supporting environment. Students have a positive attitude to the evaluation and feedback provided, some students think that they are the most helpful in the review stage," can let me understand their weaknesses, better into the review." "Can make my thinking clearer, timely leak filling and the use of thinking methods exercise." Some students also said that the feedback provided by teachers will basically look at the content, and then draw out what they think is worse, need to continue to review and consolidate the part. Some students have a good habit of taking notes. Through the targeted content provided by teachers, these students show that they have more understanding of their own cognition, which can make up for what they think they have mastered but have not mastered, which is a good spur (Fig. 2).

4.3   Satisfaction Survey Analysis

Whether the evaluation model can be better popularized and applied depends on the teacher's satisfaction with the application of the model, and puts forward whether the teacher is satisfied with the data analysis model of EDU evaluation in BD environment. The survey results are shown in Table 1. It can be seen from the diagram that 18% of

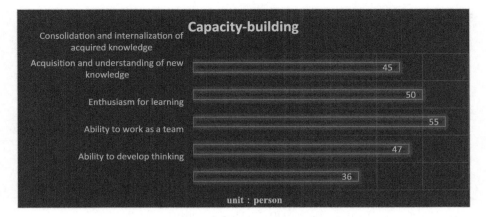

Fig. 2. Learning feedback from data analysis

Table 1. Model satisfaction questionnaire

	Very pleased	General satisfaction	Satisfied	Discontent
Number (persons)	34	58	22	8
Percentage (%)	28%	48%	18%	6%

teachers are satisfied with the development of the model, 28% are very satisfied with the application of the model, 48% are generally satisfied, only 6% are not satisfied with the model.

5 Conclusions

With the development of technology and the advent of the BD era, some information that is omitted or difficult to collect in teaching will be collected and processed in time by various means, which will become useful information that can assist teachers in teaching. EDUal system developers also hope to design a system to assist teachers in collecting, analyzing and using data. How to apply mathematical statistics, data mining and other technical means, in-depth and effective analysis and mining, find the law, find problems, not only form some common teaching strategies, but also put forward some personalized learning suggestions, so that the data for teaching services, it is very meaningful. This kind of purposeful and conscious teaching data collection and analysis can help us to find the right proximal development area and carry out accurate teaching, thus improving the effectiveness of the whole school teaching.

References

1. Lv, C.: Application and optimization of lifelong sports model in university physical EDU based on BD analysis. Bol. Tecnico Techn. Bull. **55**(14), 339–345 (2017)
2. Wu, D.F., Mao, H.: Research on optimization of pooling system and its application in drug supply chain based on BD analysis. Int. J. Telemed. Appl. **2017**, 1503298 (2017). (2017-02-15)
3. Zhai, S., Duan, J., Ai, X.: Research on the third party logistics system and economic performance optimization based on BD analysis. Bol. Tecnico Techn. Bull. **55**(11), 301–308 (2017)
4. Ding, L., He, X., Tang, Y.: Integration of Chinese traditional cultural elements in modern graphic design teaching based on BD analysis. J. Phys. Conf. Ser. **1533**(4), 042095 (2020). (7pp)
5. Xin, L., et al.: Common cancer genetic analysis methods and application study based on TCGA database. Hereditas (Beijing) **41**(3), 234–242 (2019)
6. Hua, R., et al.: Optimization of the dynamic measure in spillover effect based on knowledge graph. Int. J. Comput. Syst. Sci. Eng. **34**(4), 215–223 (2019)
7. Sun, W., Gao, Y.: The design of university physical EDU management framework based on edge computing and data analysis. Wirel. Commun. Mob. Comput. **2021**(2), 1–8 (2021)
8. Zhao, J., Zhu, C., Huang, Y.: Network consumption demand analysis and structure optimization based on BD. J. Phys. Conf. Ser. **1800**(1), 012013 (2021). (9pp)
9. Min, L., Wei, Z., Wen-Jing, L.: Evaluation of e-commerce listed companies' operating performance based on faceted data mining analysis. IPPTA Q. J. Indian Pulp Paper Techn. Assoc. **30**(6), 768–774 (2018)

Thought and Measures of Reforming About Database Courses in Universities

Tao Geng[1(✉)] and Aiping Wang[2]

[1] Department of Electronic and Information Engineering, BoZhou University,
Bozhou, Anhui, China
[2] Computer Institute, AnHui University, Hefei, Anhui, China

Abstract. Along with the wide application and the rapid development of the database technology, the teaching of the database principle will focus on developing the students' innovative abilities and their application skills On the basis of a comprehensive grasp of the concepts and principles, combined with the actual applications, the students will understand the development trend, and unceasingly enhance the level of the database knowledge, which has great benefits for training the comprehensive talents. In this paper, by analyzing the problems existing in the database courses, the author of this paper put forward the ideas and measures for the reform of the relevant curriculums.

Keywords: Database course · Teaching reform · Curriculum structure

1 System and Structure of Database Course

The database principle, as the basic subject, undertakes the theoretical education of the students, and is the teaching of the basic concepts, technologies and methods, which can lay the donation for the specific application of the students. The theoretical teaching is also the top priority, and therefore, it should be arranged as the compulsory course. The application of the database system can let the students know the mores specific implementations of the abstract concepts through the use of tools, and the students' practical abilities can also be improved [1]. The curriculum design of the database principles is to improve the abilities of the students to solve the practical problems through the basic knowledge and tools and is also the key of the structure of the curriculum system, Based on the mastering of the basic knowledge, the students can integrate the contents of other disciplines to solve the specific problems, to achieve the teaching purpose of the practical application.

Through the introduction of the above systems and structures of the course and the arrangement of the curriculum, it can be seen that the teaching contents of the database course need further optimization, and the contents of the optimization mainly include: the clarification of the teaching important points, the structure of the contents, merging with other courses but no repeat, and the maximization of the teaching goals. In the teaching arrangement of the database principle, the priority is the learning of the principles, so the teaching of the abstract contents also becomes one of the focuses of the optimization of the course. Through the studies, we found that the curriculum arrangement with the combination of the practice and the theory is a kind of the

J. C. Hung et al. (Eds.): FC 2021, LNEE 827, pp. 587–593, 2022.
https://doi.org/10.1007/978-981-16-8052-6_73

courses. Through the studies, we found that the curriculum arrangement with the combination of the practice and the theory is a kind of the mode with the better acceptance effect of the students. For example, on the basis of the introduced basic principles of the database, combined with the SQL language and the programming languages, with the concrete cases, carry on the design and ideas. In the introduction of the database security, we combine with the current forefront technologies and the successful cases, so that the students can integrate the abstract concepts into the specific cases. The structures of the curriculum systems should also be improved.

2   Reform of Teaching Contents of Database Course

According to the needs of the development of the IT industries and enterprises and the requirements of the knowledge, abilities and qualities needed to complete the actual work and tasks of the professional posts, choose the teaching contents [2]. This is the process of the selection of the teaching contents. First, carry out the investigations of the industries and enterprises, the analysis of the jobs, and the classification of the duties, and clarify the abilities and structures of the posts. By analyzing the abilities and the teaching contents, eventually establish the curriculum goals, and clarify the selection of the contents of the curriculum.

The learning contents of this course are divided into three stages, to realize the teaching goals using the progressive learning tasks: first, the stage of the basic knowledge of the database systems; second, the phase of the project leering; third, the stage of the project practice [7].

The main teaching contents in the stage of the basic knowledge of the database system include the basic concepts of the database system, the basic concepts of the database mode, the relational mode, and the theory of the normalization of the database.Through the study of this stage, the students can master the basic contents of the database system and the connotation of the field, to lay a theoretical basis for the later learning of the actual operational skills.

In the learning phase of the project, run the real project "The Company Management System" as the instance through the whole process. According to the structure of "The Company Management System" and the development process, the curriculum contents can be divided into the four learning situations: the project preparation, the creation of the database, the data sheets, the application and development, the database security management and maintenance [4]. The stage of the project practice is the practice of the entire development process from the analysis of the requirements to the final integration of the system with the "system of the selection of the students' courses" as an example. The students through the development of the full project can get familiar with the basic principles of the design and construction of the database, master the methods of the analysis and design of the database system, and improve the students' actual programming abilities, to provide the necessary skills for the future engagement in the development of the information system.

3 Analysis of Problems Existing in Reforming of Database Course

3.1 Teaching Mode is Single and Teaching Ideas Lag Behind

The college teachers often adopt the three-step teaching mode of the principle introduction, the problem illustration, and the curriculum practice. In the process, due to a lack of the contact and practice with the actual situations, the students cannot combine the knowledge of the database principle with the design of the actual system, but just passively accept the contents of the course. The teaching ideas and patterns of the database principle are similar, and we also carry out the teaching by using the concept introduction, and the serial practice. The basic concepts of the database principles, such as the database models, the concept of the database principle, and the logical mode, are relatively abstract for the students, with the strong theoretical nature, so that the students feel difficult to understand them in the learning process, and their learning enthusiasm will reduce gradually. Form the long time observation, we found that the effect of the practice of the students in the practice phase after the explanation of the theoretical concepts is not good, which is also caused by the not solid mastering of the theoretical learning, which directly led to the serious disconnection between the theory and the practice, so that the teaching qualities and effects gradually decline.

3.2 Teaching Management is Lack and Updating of Teaching Materials Lags Behind

Because of the continuous introduction of the young teachers, to exercise their teaching abilities, every year arrange the young teachers into such professional foundation courses as the database principles. The constant changes of the professional teachers led to the constant changes of the teaching cases and the teaching methods. In this process, the ire is not a unified and scientific management mode, which leads to the decline of the quality of the teaching. For the textbook used in colleges and universities, generally there will be no changes for a long time, and the publishing cycle of the same set of textbooks is long, and the updating of the teaching materials is slagging behind seriously, which will not be able to combine with the constantly changing new technologies in reality.

3.3 Theory Disconnects with Practice, Curriculum Assessment is Old

The weak ability to accept the theoretical knowledge led to a serious lack of the practical applications of knowledge of the students, so that the students in the development of the programs cannot well combine the knowledge of the database principle [10]. In the evaluation of the database principle, we have been using the method of the closed-book exams, and the questions tend to give priority to the theoretical questions, with little practice design, which can't check out the student's comprehensive levels and the students' practice ability of the theoretical application.

4 Implementation Methods of Teaching Reform of Database Course

4.1 Choose Quality Teaching Materials and Teaching Cases

Many higher vocational colleges prepare and perfect the syllabus for the curriculum teaching in strict accordance with the requirements of the personal training of the higher vocational colleges by the national Ministry of Education, but there are shortcomings in the selection of the teaching materials, and the selected teaching materials are intermingled. It is suggested that the school with such problems in the selection of the teaching materials should pay attention to: The teaching materials should select the universal textbooks for the higher vocational education, and the database management system of the selection of the teaching materials shall be in line with the markets, and should be the mainstream language, such as the SQL Server database management system. The instance of the database of the teaching materials should best be what the students are very familiar with. In addition, we should also pay attention to the collocation of the chapters in the textbooks and the cohesion of the contents [6].

4.2 Reasonable Arrangement of Leading Courses and Subsequent Courses

The database teaching should be arranged in the second semester of the freshmen's year or the first semester of the sophomore, and there are a few more important prerequisite courses of this curriculum. First, it is "An introduction to computers", and second, the student must master the programming language, such as the C language > At the same time, in view of the importance of the application of the database in the enterprise, we can offer some corresponding courses combined with the database in the subsequent lessons, thereby increasing the practical application of the course.

4.3 Reasonable Arrangement of Weight of Contents of Chapters

The higher vocational college students are facing the difficulties of the tight schedules, the burdensome tasks, and the higher demands and so on. So in the teaching process, we must teach clearly and thoroughly the contents that must be mastered, and reasonably arrange the weight of the knowledge, with the proper time schedule. Combined with my experience, the proportion of the database over vies, the E-R diagram, the relation database theory, the SQL language, the database development technologies and various other parts and the corresponding teaching periods are shown in the following table:

4.4 Let Students Know the Functions of the Course and the Practice

Through the application of the database in the real situations, the students can fully realize the importance of it, straighten the status of the course in their learning, and change the passive learning into the acting learning. Because the database is the course with the close integration of the theory and the practice, the traditional appraisal system depend rarely on the scores of the practice, so that the students from their psychologies don't attach importance to the practice. Therefore, in the teaching process, we should pay attention to the organic combination of the theory and the practice [3].

5 Application and Practice of Teaching Reform of the Database Course

5.1 Effect of Teaching Reform

In the reform of the teaching, through the improvement and construction of the teaching materials, we can improve the quality of the teaching. The college computer profession requires the teaching mode with the combination of the practice and the theory with the teachers as the main body and the students in the center [8, 9]. Through the reform of the practice courses, we can improve the innovative abilities of the students. We have increased the 10% of the past practice classed to 50% of the practice courses, and divide the unit practice and the group practice and other different methods in the practice course, and the final practice reports and results should be treated as part of the assessment. The ways of the reform directly promoted the students' ability of the autonomous learning and the interests in the studies and exploration.

Through the reform of the examinations, we can cultivate and improve the students' ability of the self-study, and the traditional end-of-term closed-book exams often cannot comprehensive assess the learning results of the students, so we have divided the performances of the database principle into several aspects: the normal classroom tests account for 15%, and the tests of the unit practice account for 20%, and the curriculum design accounts for 20%, and the end-of-term closed-book exams account for 45%. Through this test method, we can more comprehensively understand the accepting effects of the students, and can see the situations of their actual applications. In the peacetime classroom tests, we can carry out the random tests, mainly with the argument and the calculation problems. This way can test the students' learning situations of this period of time, which also can have the effect of checking on the working attendance, and discover the places needing to be improved in the process of teaching the course. The unit practice is mainly the machine operation of the students, and we can rate according to the situations of the experiments and the experimental reports, to promote the students' self-study abilities [5]. The curriculum design should take the teams as the unit, and the teachers give the design topics, and the students carry out the discussion an design in groups, In the process, the students can systematically understand the knowledge systems and contents of the database principle.

5.2 Application of Teaching Practice

The learning of the database principle is mainly aimed at the practical application, and we can carry out the further penetration of the curriculum contents with the tools of the programming language and the database system. SQLServer, as a kind of database management system easy to learn, is extremely convenient in the development of the small information systems. We can choose this kind of tool for the curriculum design to train the basic ways, skills and abilities of the students to process the data, and cultivate this system development ability of the students. Through the database management system, we can cultivate the thinking of the students to handle the data transference process. For the processing of the large amounts of the data, we need the constant practice, in order to achieve the effect of integrating the theory with the practice. For example, in the development of a system for the library information management, by the SQLServer development database system, the students can make the concepts gradually merged in the specific cases, and this application also makes the understandings of the students of the abstract concepts deep ended and they will no longer feel boring.

6   Conclusion

The teaching of the database principle is an important professional basic curriculum of the computer science. IN the process of the theoretical educations, we should focus on the practice education, which is the place that we should continuously explore and reform. How to let the students understand and learn this course to a great extent is also the most concerned problem of the teachers. Through the reform of the curriculum contents, the reform of the practice teaching, and the form of the learning methods, the students can improve their ability of the autonomous learning, combine the theory with the practice, and constantly improve the students' learning enthusiasms and their innovative abilities. We have achieved certain success in the process of therefrom, and the quality of the teaching has been greatly improved, which also prompted the teaching workers to explore more in the process of the reform, an constantly improve the quality of the teaching.

Acknowledgements. This work was supported by:

1. Quality Engineering Project of Anhui Education Department "Research on Database course Teaching Model based on Superstar Learning System" (Project ID: 2019 jyxm0542).

2. Quality Engineering Project of Anhui Education Department "Basic Operation Virtual Simulation Experiment Teaching Project of SQLServer" (Project ID: 2020 xfxm43).

3. Quality Engineering Project of Bozhou University "Online and Offline Hybrid and Social Practice Course of Database)" (Project ID: 2020 hhkc03).

References

1. Wang, S.: The practice of project-based teaching reform of fieldbus technology course guided by practical engineering application. Int. J. Soc. Sci. Educ. Res. **4**(4), 145 (2021)
2. Guan, P., Zheng, Z.: Research on animation majors' competition and cultivation of creative talents. In: 6th International Conference on Arts, Design and Contemporary Education (ICADCE 2020), vol. 515, pp. 348–353 (2020)
3. Cui, T., Daixin, F.: Teaching reform practice of international trade practice course based on OBE education concept. In: 2020 International Conference on Humanities, Arts, and Social Sciences (HASS 2020), pp. 146–150 (2020)

4. Zhang, Y.: Research on the course system of design in Guangdong universities based on regional cultural characteristics taking Guangdong university of finance and economics as an example. In: 6th International Conference on Arts, Design and Contemporary Education (ICADCE 2020), vol. 515, pp. 146–150 (2020)
5. Xia, G.L.: The application and research of corpus method in college English teaching reform. In: 2019 International Conference on Humanities, Management Engineering and Education Technology (HMEET 2019), pp. 372–381 (2019)
6. Liang, B.: Research on the cultivation of the intercultural communication ability of college students under the vision of globalization. In: Proceedings of the 37th Annual Conference of Liaoning Translation Society and the 8th International Academic Forum on Language, Literature and Translation in Northeast Asia, pp. 395–400 (2019)
7. Zhao, Y.: Exploration of the SPOC-based blended teaching model: case study of business English course. In: North-East Asia International Forum on Linguistics, Literature and Teaching, pp. 841–847 (2020)
8. Lin, X.: Research on the reform of physiology teaching mode based on CBL teaching method. In: 2020 3rd International Academic Conference on Wisdom Education and Artificial Intelligence Development, vol. 1, pp. 52–54 (2020)
9. Zhang, J.: Research on the cultural and tourism cooperation between Gansu province and Russian-speaking countries along the belt and road exploration of local Russian language talents cultivation. In: 2020 International Conference on Language, Communication and Culture Studies, (ICLCCS2020), pp. 174–180 (2020)
10. Ren, C.: Research on the health care service model of community sports under the active aging. In: 2020 International Symposium on Education, Culture and Social Sciences, (ECSS 2020), pp. 106–112 (2020)

Market Data Analysis and Forecast Based on Big Data Technology

Jie Liu[1($\boxtimes$)] and Feixiong Li[2]

[1] Chengdu University, Chengdu 610106, Sichuan, China
liuxin@uestc.edu.cn
[2] China University S&T Achievements Transformation Center (South China),
Huizhou 516000, Guangdong, China

Abstract. With the in-depth development of modern Internet technology, we have gradually entered a period of data prosperity. The information technology and big data technology that the Internet has also advanced have provided us with updated vitality and meaning, as well as new methods of analyzing and predicting data. This article takes stock market data as the main research center, analyzes market big data and looks for hypotheses based on Internet and big data technology. This article mainly studies stock data analysis, forecasting and mining technology based on big data and stock market-related theories. In-depth study of the stock market, discussion of the use of data mining in searching a single market, and the use of multiple mining methods, provide a market analysis and forecasting model. This paper analyzes and predicts the development trend of our country's securities market through data mining and other technologies. It uses a complex time series based on the highest density algorithm to estimate the value of stocks. The market conducts group analysis, and uses Lasso-based Logistic regression model to warn the financial status of listed companies and conduct empirical analysis and comparison. Research shows that data mining technology based on big data is meaningful for shareholders, stocks and listed companies; the complex time model based on the highest density shows almost all the true value of stock prices with 95% confidence. In the test data, the ST estimation accuracy of the early warning model proposed in this paper is 61% higher than that of the single Logistic regression model.

Keywords: Big data technology · Market data · Data analysis · Data prediction

1 Introduction

In various markets, it is easier than ever to collect and obtain a large amount of and complex information [1]. This article mainly focuses on our country's stock market as the main research object, discusses how to make full use of big data and related technologies to analyze the large amount of data that may be generated in our country's stock market in this era of big data, and provide us with in-depth research based on big data. Data technology provides assistance in analyzing and forecasting large amounts of data in other stock markets [2, 3]. The analysis of the stock market is an important link in the development of our country's securities investment. It means that investors can comprehensively evaluate the value and trend of their stocks through various

J. C. Hung et al. (Eds.): FC 2021, LNEE 827, pp. 594–603, 2022.
https://doi.org/10.1007/978-981-16-8052-6_74

scientific and effective means and methods, and judge their trends and changes in a timely manner. The behavior of the trend can effectively and accurately analyze the stock market conditions, which will help guide investors to make more accurate investment decisions in a timely manner and increase investment returns and rates of return [4, 5]. At present, many scholars have conducted research on new technologies for market data analysis and prediction, such as: Francisco Falcão-Reis and Francisco Falcão-Reis. The proposed method combines the two most commonly used forecasting methods within the company to solve players based on participant risks. Behavior adjustment: internal data analysis and external (or departmental) data analysis. By balancing these two components, the proposed model can dynamically adapt to the market environment, using the expected price of competitors as a reference, and market prices through artificial neural networks prediction [6].

Waseem Ahmad and Ajit Narayanan proposed a new time series data analysis method inspired by the natural immune system process. First, use segmentation to subdivide the entire time series data into sub-sequences, and secondly, use artificial immune system algorithms to segment the data. Perform analysis and clustering, and finally, the clustering information is used to build a model for prediction and prediction [7].

This article briefly introduces data mining technology based on big data, systematically introduces grouping and classification algorithms in data mining technology, and the ability to apply this technology to stock market research. In this article, the highest density algorithm is used to group the data, and a new complex time prediction model is used to predict the closing price of the stock exchange index. Aiming at the problem that the K-mean algorithm tends to drop to a local extreme during the search process and is sensitive to the initial grouping center, this paper uses the advantages of the artificial fish school algorithm to obtain the best solution in the world. Based on the characteristics of large scale of financial data of listed companies and large surpluses, this paper introduces an early warning model based on Lasso method and Logistic regression to judge whether the financial status of listed companies is in crisis.

2 Research on Market Data Analysis and Forecast Based on Big Data Technology

2.1 Data Mining Technology

(1) Related concepts

Data mining is a practice other than automatically searching large amounts of stored data to discover patterns and analyze simple trends [8]. It uses complex mathematical algorithms to segment data and evaluate the possibility of future events, also known as Knowledge Discovery in Data (KDD) [9, 10]. According to actual work needs, commonly used data mining methods include grouping and sorting.

(2) Cluster analysis

Cluster analysis is a convenient method to identify homogeneous groups (clusters) of objects [11]. Objects have many common characteristics in a particular cluster, and very different objects belong to multiple clusters [12]. The methods of cluster analysis are: the level-based method, and the density-based method.

1) Layer-based approach

The hierarchical grouping method is to obtain the grouping result by decomposing or aggregating a given data set. Usually, the number of user-defined categories or the distance between two adjacent categories is used as the criterion for terminating the algorithm. There are two types of hierarchical grouping methods: centralized method and division method. The class spacing is a key factor in the hierarchical grouping algorithm. The most common methods for measuring the distance between classes are: centripetal distance, average distance, nearest and farthest distance.

The centroid distance represents the distance between the centroids in the two classes and the distance between the classes. The formula is:

$$D_{mean}(C_i, C_j) = |m_i - m_j| \tag{1}$$

The farthest distance represents the distance between the farthest data points in the two classes represents the distance between classes, and the formula is:

$$D_{max}(C_i, C_j) = \max_{x \in C_i, y \in C_j} |x - y| \tag{2}$$

Using different methods to measure the gap between two classes will result in different hierarchical grouping methods. The application of the hierarchical grouping algorithm is relatively simple. It can adapt to changes in the spatial scale during the separation or aggregation process, and may encounter multi-scale grouping problems, but it is difficult to use a single hierarchical grouping method to determine its termination conditions and perform separation and aggregation functions, and the method has poor scalability. Therefore, when solving practical problems, the hierarchical grouping method is usually used in combination with other methods to improve the efficiency and quality of grouping.

2) Density-based method

The clustering algorithm based on hierarchical structure uses distance to describe the similarity between data. This type of algorithm is usually only applicable to spherical groups, and usually has limitations for non-spherical groups. In order to solve this defect, a clustering algorithm is proposed to replace the similarity between data and density. The density-based algorithm starts from the data distribution density, filters low-density areas, connects dense sample areas, finds clusters of any shape, and can effectively process a single point.

(3) Classification method

The classifier can use the model to analyze existing data, or it can predict data. The prediction is based on the known historical data and automatically obtains an extended description of the known data to achieve the result of the data prediction. Typical classification methods include decision trees, Bayesian classification, logistic regression and so on. Among them, QUEST is a binary decision tree classification algorithm, which represents a fast, fair and efficient statistical tree, mainly used to select industry variables and breakpoints, and uses different strategies for these two problems.

2.2 The Applicability of Data Mining Technology to the Research of Stock Market

The stock market generates a lot of trading information every day, and the financial information of listed companies also contains a lot of information. The reason for using data mining technology in the processing and analysis of stock data is based on the following reasons: A large amount of data reveals the financial and operational status of the enterprise and helps investors make more effective decisions.

2.3 Fuzzy Time Series Model Based on Density Peak Algorithm

(1) Related concepts of fuzzy time series
   1) U is the universe of discourse, divide it into n subintervals, define A as the fuzzy set on the domain U, denoted as:

$$A = \frac{f_A(u_1)}{u_1} + \frac{f_A(u_2)}{u_2} + \cdots + \frac{f_A(u_n)}{u_n} \tag{3}$$

   Where fA is the membership function defined on A, fA:U $\to$ [0,1], fA(ui) is the membership degree of ui to fuzzy set A, and fA(ui) $\in$ [0, 1], $0 \le i \le n$.
   2) Suppose F(t) is determined by F(t-1),...,F(t- m), satisfying

$$F(t) = (F(t-1) \times \cdots \times F(t-m))^{\circ} R(t, t-m) \tag{4}$$

   Among them, $m \ge 1$ is the Cartesian product, which represents the composite operation, which is called the definition between F(t−1), F(t−2),...,F(t−m) and F(t) Fuzzy relationship.
(2) Introduction to density peak algorithm
   The peak density algorithm is an algorithm based on density clustering. It maximizes the distance between the cluster centers by selecting points that are far away and densely distributed as the cluster centers. The algorithm is based on the following two assumptions: First, the distance between the class center point and other points with large local density values is larger; second, the local density value of the class center points is larger than the surrounding data points.

2.4 Financial Warning for Listed Companies

Under normal circumstances, the higher the financial information of a listed company, the greater the early warning effect. However, due to the influence of many factors, too many financial indicators will lead to a mix of variables. This paper proposes a logistic regression model based on the dragline method to provide financial early warning for listed companies. The method first uses the Lasso method to select variables for high-dimensional data to reduce the size of the data and eliminate the correlation between the variables, and then uses the Logistic regression method to provide early warning of the financial status of listed companies.

(1) Logistic regression model based on Lasso method
Logistic regression model is a non-linear probability model. It has no restrictions on the types of independent variables, does not require data to ensure the integrity of variance and normality, and has the advantage of being able to explain the coefficients well. Logistic regression analysis transfers the problem of solving the relationship between the independent variable and the dependent variable to determining the possibility of the event. The model can not only predict whether listed companies will be affected by the financial crisis, but also intuitively reflect the possibility of the company's financial crisis and provide the reality of the company's operations.

3 Market Data Analysis and Forecast Research Experiment Based on Big Data Technology

3.1 Subjects

This article selects various data generated in our country's stock market from 2017 to 2019 as the research experiment objects, and provides research data for various verification and comparison experiments.

3.2 Experimental Method

(1) Comparative analysis method
This paper compares the predicted value and the true value of the proposed stock prediction model, and finds that their error is within a reasonable range and the error range is extremely small.
(2) Test experiment
This paper selects two functions with different characteristics to test the artificial fish school algorithm and the kernel function artificial fish school algorithm, verifying that the kernel function artificial fish school algorithm can break the limitation of local extreme value, the convergence speed is fast, and the convergence process is relatively stable.

3.3 Basic Methods of Preprocessing and Analysis of Functional Data

(1) Smoothness of data
The original data collected in actual research are all discrete data, and due to external reasons, the observed original data will contain noise components, so the original data needs to be smoothed before analysis, the most commonly used smoothing in research It is smooth and interpolation method. The choice of the two methods lies in the accuracy of the observed original data. The data observed in

actual research will basically have observation errors. Therefore, the smoothing method is generally used.

Expansion of the original data through the basis is a smoothing processing technique often used. The core of this method is to use a linear combination of K basis functions φk, $(k = 1,...,K)$ to represent the function $x(t)$, the formula is as follows:

$$\hat{x}(t) = \sum_{k=1}^{K} C_k \varphi_k(t) \tag{5}$$

The following bases are commonly used in research: Fourier bases, polynomial bases, and B-spline bases.

(2) Rough punishment method

To use the collected observations to fit the function curve, generally two aspects need to be considered: one is to ensure that the residual sum of squares is minimized, so that the fitted function curve can well represent the various original sample data characteristics; second, when the obtained fitting curve has large local fluctuations, in order to preserve the smoothness of the curve, a higher degree of fit is not required. On this basis, a rough punishment method is introduced, which can meet the requirements of the above two aspects at the same time. The rough penalty method guarantees the perfect fit of the basis function to the original data and the smoothness of the fitted function curve, and also avoids their limitations. In this paper, the rough penalty method uses the spline smoothing method.

(3) Cross validation

The cross-validation method is mainly used to solve the smoothing parameter λ involved in the rough penalty method. The basic idea of the method is to group experiments, one group is used as the training set of the experiment, and the other group is used as the test set of the experiment. At the beginning, model training will be established on the data of the training set, and then the results of the training will be verified through the data of the validation set.

(4) Rough punishment method

To use the collected observations to fit the function curve, generally two aspects need to be considered: one is to ensure that the residual sum of squares is minimized, so that the fitted function curve can well represent the various original sample data characteristics; Second, when the obtained fitting curve has large local fluctuations, in order to preserve the smoothness of the curve, a higher degree of fit is not required. On this basis, a rough punishment method is introduced, which can meet the requirements of the above two aspects at the same time. The rough penalty method guarantees the perfect fit of the basis function to the original data and the smoothness of the fitted function curve, and also avoids their limitations. In this paper, the rough penalty method uses the spline smoothing method.

(5) Cross validation

The cross-validation method is mainly used to solve the smoothing parameter λ involved in the rough penalty method. The basic idea of the method is to group experiments, one group is used as the training set of the experiment, and the other group is used as the test set of the experiment. At the beginning, model training will be established on the data of the training set, and then the results of the training will be verified through the data of the validation set.

4   Market Data Analysis and Forecasting Research Experimental Analysis Based on Big Data Technology

4.1   Validation Analysis of Stock Forecasting Model

This article selects the closing price of the Shanghai Stock Exchange Index from January 4, 2019 to March 25, 2018 as sample data to predict stock prices, and applies the prediction model proposed in this article. In order to obtain more intuitive prediction results, the real value Compared with the predicted value, the results are shown in Table 1:

Table 1.   Forecast result table of fuzzy time series model based on density peak

	0	5	10	15	20	25	30	35	40	45	50
Actual	3321	3041	2916	2759	2754	2767	2878	2741	2891	2888	3021
Predict	3321	3031	2934	2750	2746	2773	2881	2706	2874	2901	3046
Predict-low	3115	3001	2894	2645	2711	2697	2815	2674	2796	2834	2989
Predict-high	3431	3101	3078	2910	2806	2884	2965	2904	2941	3097	3131

It can be seen from Fig. 1 that the prediction model proposed in this paper predicts the stock price, and its predicted value is not much different from the true value, and the true value is almost all distributed within the 95% confidence interval, indicating that the predicted result and the true value It is very close and the prediction effect is better.

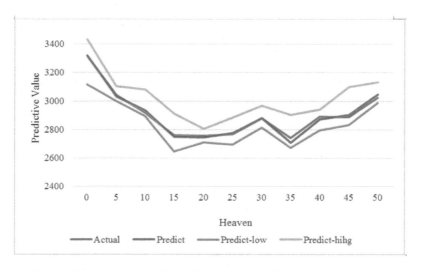

Fig. 1.   Forecast results of fuzzy time series model based on density peak

4.2 Experimental Analysis of Financial Early Warning of Listed Companies

This paper compares the accuracy of the regression logic model with the prediction results of the model proposed in this document through verification to test the effectiveness of the proposed model in predicting financial conditions. The comparison results are shown in Table 2:

Table 2. Data comparison result

	Training Data			Test Data		
	Total Correct Rate	Non-ST Prediction Accuracy Rate	ST Prediction Accuracy	Total Correct Rate	Non-ST Prediction Accuracy Rate	ST Prediction Accuracy
The Model Proposed in this Article	90.8	97.1	74.9	94.7	99.3	82.1
Logistic Regression Model	89.1	95.9	71.1	71.5	87.1	19.5

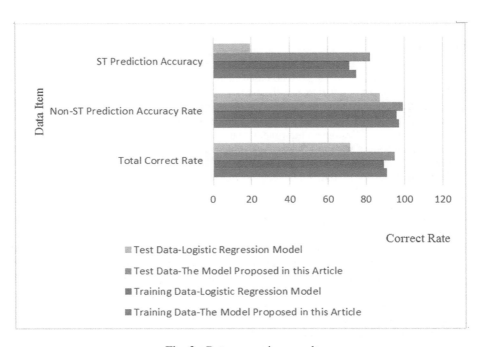

Fig. 2. Data comparison result

It can be seen from Fig. 2 that compared with the Logistic regression model, in the test data, the ST prediction accuracy rate is 61% higher than that of the Logistic regression model. It can be seen that the financial early-warning model for listed companies proposed in this article can better analyze the financial status of the company, and has a strong predictive ability. It can promptly discover the problems of the company and deal with it in a timely manner.

5   Conclusions

This article focuses on stock trading based on stock trading information and financial information of listed companies, combining computer science and technology, and solving problems related to big data association rule mining, data analysis and forecasting. This paper takes stock price information analysis and prediction as the research object, and proposes a series of complex time models based on the density peak algorithm in the big data environment, and uses them as stock price prediction models. Based on the stock price data provided, the effect of this model is studied. This paper takes the analysis and prediction of the financial information of listed companies as the research object, and proposes a logistic regression model based on the drago method. This model is used for the financial early warning of listed companies, and has been experimentally verified to prove that the model can provide effective results for listed companies.

References

1. Lin, H.Y., Yang, S.Y.: A smart cloud-based energy data mining agent using big data analysis technology. Smart Sci. **7**(JUN.), 1–9 (2019)
2. He, Z., Cai, Z., Yu, J.: Latent-data privacy preserving with customized data utility for social network data. IEEE Trans. Veh. Technol. **PP**(99), 1 (2017)
3. Li, W., Zhou, C.: Customer churn prediction in telecom using big data analytics. In: IOP Conference Series: Materials Science and Engineering, vol. 768, no. 5, p. 052070 (2020). (6pp)
4. Hye-Sun, K., et al.: A study on big data new technology trends and market prospects. Adv. Sci. Lett. **22**(11), 3563–3566 (2016)
5. Louhichi, K., Jacquet, F., Butault, J.P.: Estimating input allocation from heterogeneous data sources: a comparison of alternative estimation approaches. Agric. Econ. Rev. **13**(2), 83–102 (2017)
6. Pinto, T., Falcao-Reis, F.: Strategic participation in competitive electricity markets: Internal versus sectorial data analysis. Int. J. Electr. Power Energy Syst. **108**(JUN.), 432–444 (2019)
7. Ahmad, W., Narayanan, A.: Time series data analysis using artificial immune system. Intell. Decis. Technol. **12**(2), 119–135 (2018)
8. Fawzy, D., Moussa, S., Badr, N.: Trio-V wind analyzer: a generic integral system for wind farm suitability design and power prediction using big data analytics. J. Energy Resour. Technol. **140**(5), 051202.1–051202.13 (2018)
9. Ranco, G., et al.: Coupling news sentiment with web browsing data improves prediction of intra-day price dynamics. Plos One **11**(1), e0146576 (2016)

10. Li, W., Xing, L., Fu, S., et al.: Input data analysis for the thermal rating prediction of the overhead conductor. In: IOP Conference Series: Earth and Environmental Science, 2019, vol. 371, no. 5, p. 052008 (2019). (6pp)
11. Kamaludin, K., Sundarasen, S., Ibrahim, I.: Covid-19, dow jones and equity market movement in ASEAN-5 countries: evidence from wavelet analyses. Heliyon **7**(1), e05851 (2021)
12. Liu, K., Nakata, K., Li, W., et al.: An abductive process of developing interactive data visualization: a case study of market attractiveness analysis. In: Liu, K., Nakata, K., Li, W., Baranauskas, C. (eds.) Digitalisation, Innovation, and Transformation. ICISO 2018. IFIP Advances in Information and Communication Technology, vol. 527. Springer, Cham (2018). https://doi.org/10.1007/978-3-319-94541-5_29

Identification and Localization of Potato Bud Eye Based on Binocular Vision Technology

Yigao Wang, Jiwan Han, Chunqing Cao, and Fuzhong Li[✉]

Shanxi Agricultural University, Jinzhong 030801, Shanxi, China

Abstract. The solution to three dimensional coordinate of potato bud eye is of great significance to the improvement of seed potato cutting machine. In this study, checkerboard camera calibration method developed by Zhang Zhengyou was used to calibrate the binocular camera. On the basis of stereo correction, the potato bud eye was identified by image recognition technology and its pixel coordinate information was obtained. Then, the three-dimensional coordinate of the potato bud eye was calculated by binocular distance measurement method. The results showed that when the distance between the camera and the potato was 20 cm and the ambient light intensity was (351 lx–725 lx), the eye recognition accuracy was the highest. The absolute error and relative error of 3D coordinate measurement of potato bud eye are less than 3 mm and 5%. This study has a certain reference value for the future location of the bud eye and the improvement of the seed potato cutting machine.

Keywords: Potato · Binocular vision · 3D positioning · Identification of bud eye

1 Introduction

Potato grows mainly by asexual reproduction, the cost of sowing a whole potato is very high, so the potato is cut into pieces so that each piece of seed potato has one or more eyes, so as for germination. At present, the cutting of seed potato is still mainly done manually, and the operation is complicated. With the continuous development of computer vision technology, the recognition of potato bud eye can be realized, which lays a foundation for the automatic cutting of seed potato and accelerates the modernization and mechanization of seed potato cutting. At present, most of the researches are mainly related to the recognition of potato eye. Tian Haitao et al. [1] used machine vision technology to segment and recognize the eye region by Euclidean distance and dynamic threshold in color space, and the recognition rate of the eye reached 96%. Lu Zhaoqin et al. [2] proposed a method of potato image eye recognition based on Gabor feature. After collecting potato images, image filtering was carried out and morphological image processing was performed.

Zhang Jinmin et al. [3] proposed a recognition method for potato bud eyes based on local binary mode and support vector machine. Firstly, the local binary mode was used to extract the features, and then the support vector machine was trained to recognize the bud eyes. The empirical identification rate reached 97.33%. Although good results were achieved in the recognition of potato bud eye, there are few researches on the

J. C. Hung et al. (Eds.): FC 2021, LNEE 827, pp. 604–611, 2022.
https://doi.org/10.1007/978-981-16-8052-6_75

localization of potato bud eye, and the location information of potato bud eye cannot be obtained on the basis of the recognition of potato bud eye. Li Mingdong [4] et al. carried out the ranging experiment of binocular stereo matching, and the distance information of the object by using the parallax of the picture was obtained, as a result, the ideal experimental results were displayed. Cong Zhiwen et al. [5] designed a tomato picking system based on binomial vision, which was used to figure out the spatial coordinates of tomatoes and planned the shortest path through ant colony algorithm, laying a good foundation for realizing automatic crop picking. Wei Chun [6] et al. designed a set of fruit recognition and positioning system based on binocular vision technology and machine learning, which provides a reference for subsequent crop identification and positioning.

The position information of the bud eye is of great significance to improve the existing automatic cutting machine of seed potato. The cutting machine can make a decision to cut tuber according to the position information of the bud eye and improve the utilization rate of seed potato. In this paper, firstly, Zhang Zhengyou camera calibration method [7] was used to calibrate the binocular camera and to collect experimental data. After that, stereo correction was carried out on the binocular image, and then the three-dimensional coordinate of potato bud eye was obtained by combining the image recognition technology with binocular ranging technology.

2 Research Methods

2.1 Subjects and Materials

The potato variety used in this experiment is Jinshu 16 potato. According to the standard *NYT1066–2006- Potato Grade Specification of the Ministry of Agriculture of China*, a total of 30 potato samples of different quality were randomly selected, and the potatoes with incomplete surface, dry rot and damage were screened out. Among them, there were 10 major classes (>300 g), 10 middle classes (100–300 g) and 10 small classes (<100 g).

2.2 Data Collection Method

Under the condition of sufficient and uniform ambient light, the potato is fixed on the rotary table in a vertical way, and then the binocular camera is fixed on the side of the potato horizontally to take pictures. The resolution of the binocular camera used in this experiment is 2560 × 960, and the monocular resolution of the image after segmentation is 1280 × 960. The specific operation of the acquisition is to focus the camera to ensure that the eyes can be clearly captured by both the left and right eye lenses under close range conditions. In practical tests, three pairs of binocular photographs of each potato (one at a 120° interval) were taken to collect all the eyes on the potato's side.

2.3   Camera Calibration and Stereo Correction

Zhang Zhengyou's checkerboard calibration method is adopted in this paper, which is endowed with characteristics of simplicity and accuracy. Through camera calibration, the internal and external parameters, translation vector and rotation matrix of the camera can be obtained, so as to establish the relationship between the two-dimensional coordinates of the image and the three-dimensional coordinates of the space [8]. Here, the MATLAB calibration toolbox is used for calibration [9], and the steps are as follows:

(1) Print a checkerboard with a grid number of 6 × 8 with each side length of 25 mm, which can be fixed on the horizontal surface.
(2) First, the binocular camera was used to shoot the checkerboard from different angles for a total of 20 times. After importing the photos into MATLAB, the left and right cameras were calibrated respectively. Then, the calibration parameters of the left and right cameras were imported into the binocular calibration toolbox for stereo calibration of the binocular camera, and the calibration parameters of the binocular camera were obtained.
(3) Since the camera is prone to distortion when collecting images. Therefore, in this section, binocular calibration parameters are used for stereo correction after calibration to eliminate distortion.

2.4   Recognition of Potato Bud Eyes

2.4.1   Model Deployment

With EasyDL image recognition, the potato bud eye image recognition model can be quickly created and deployed to complete the bud eye recognition. First, potato photos under different growing environments and light conditions were taken to form the eye image data set. Then, the model was deployed locally after tagging and training the eye on the EasyDL platform.

2.4.2   Influence of Light Intensity and Distance on Potato Bud Eye Recognition Rate

In the process of image acquisition, brightness and distance will affect the clarity of the eye in the photo, which will affect the eye recognition. Therefore, under the condition that the distance between potato and camera is respectively 20 cm, 30 cm and 40 cm, and the light intensity is respectively weak (90–350) lx, medium (351–725) lx and strong (726–725) lx, the relationship between the light intensity and distance and the eye recognition rate is studied in this section.

2.4.3   Acquisition of Pixel Coordinates of Potato Bud Eyes

The potato eye detection box contains the position information of the eye, and the pixel coordinate information of the eye detection box in the two figures is obtained respectively. The pixel coordinate system is shown in Fig. 2, with the horizontal axis of U and the vertical axis of V. Where, point A is the vertex in the upper left corner of the detection box, and (U, V) is the pixel coordinate of point A; $(U + \frac{W}{2}, V + \frac{H}{2})$ is the pixel coordinate of center point B of the detection box. The first group of images is taken as an example for eye recognition. The coordinate information of the eye detection box is shown in Table 1, where A_L and A_R respectively represent the pixel coordinates of point A in the left and right eye images, and H and W respectively represent the height and width of the detection box (Fig. 1).

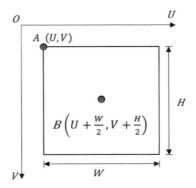

Fig. 1. Pixel coordinate system

Table 1. Pixel coordinate information of the eye detection box

No	A_L (pixel)	H_L (pixel)	W_L (pixel)	A_R (pixel)	H_R (pixel)	W_R(pixel)
1	(788,223)	44	69	(466,219)	46	72
2	(866,310)	44	64	(546,312)	40	61
3	(766,470)	40	97	(431,472)	36	100
4	(747,274)	23	65	(412,269)	31	60

The size and position of the detection frame generated by eye recognition will change due to different perspectives. However, in the two three-dimensional correction pictures, the error of the ordinate value of each center point is less than 5 pixels. Therefore, the initial coordinate information in Table 1 is calculated, and the center point B of each detection frame is taken to represent the position of the eye. When solving the three-dimensional coordinates of the eye through binocular ranging, it is necessary to calculate according to the pixel coordinates of the center point of each eye in the left and right images. Since the image has been stereoscopic corrected, the average of the vertical coordinates of point B in the binocular image is taken as the vertical coordinates of the bud eye, as shown in Table 2.

Table 2. Processing of the ordinate mean value of point B

No	U_L (pixel)	U_R (pixel)	Vavg (pixel)
1	822.5	504	243.5
2	898	576.5	332
3	814.5	481	490
4	779.5	442	285

2.5 Calculation of 3D Coordinates of Potato Bud Eyes by Binocular Ranging

The principle of binocular distance measurement is shown in Fig. 2. P is the eye on the potato surface, O_L and O_R are the optical centers of the left eye camera and the right eye camera respectively, the imaging points of point P on the camera sensor are P and P', f is the pixel focal length, B is the baseline distance between the two cameras, and f = 896.41565, B = 6.612053 were demonstrated after camera calibration. Z is the distance from point P to the camera.

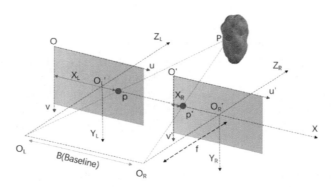

Fig. 2. Schematic diagram of binocular ranging

In order to facilitate the verification of the results and the follow-up research, the left eye camera is used as the origin of the world coordinate system here, so the origin of the pixel coordinate system needs to be converted from O, O' to OL' and OR' respectively. In this test, the resolution of monocular image is 1280 × 960. Assuming that the coordinate of a point is (U, V) in the original pixel coordinate system, and the converted coordinate is (X, Y). The conversion equation is as follows:

$$\begin{cases} X_L = U_L - \frac{1280}{2} = U_L - 640 & (1) \\ X_R = U_L - \frac{1280}{2} = U_R - 640 & (2) \\ Y = V_{avg} - \frac{960}{2} = V_{avg} - 480 & (3) \end{cases}$$

Presuming that the three-dimensional coordinate of potato bud eye be (x, y, z), and the following three-dimensional coordinate calculation equation is obtained from the principle of similar triangles:

$$
\begin{cases}
x = \frac{X_L \cdot b}{X_L - X_R} & (4) \\
y = \frac{Y \cdot b}{X_L - X_R} & (5) \\
z = \frac{f \cdot b}{X_L - X_R} & (6)
\end{cases}
$$

The 3D coordinates of the bud eye are calculated by Eqs. (4), (5) and (6) after the coordinates of point B was converted by Eqs. (1), (2) and (3), and it was compared with the actual measured values by hand. The measurement results of the first group of pictures are shown in Table 3.

Table 3. Measurement results of the first group of pictures

No	X value (cm)	X value (cm)	Y value (cm)	Y value (cm)	Z value (cm)	Z value (cm)
1	3.79	3.90	−4.91	−4.90	18.61	18.50
2	5.31	5.50	−3.04	−3.00	18.44	18.40
3	3.46	3.50	0.20	0.20	17.77	17.70
4	2.73	2.70	−3.82	−3.70	17.56	17.40

2.6 Evaluation of Positioning Accuracy of Potato Bud Eye

In the field of measurement, absolute error is usually used to reflect the range of error variation, and relative error is used to reflect the reliability of measurement methods or results. Here, the positioning accuracy is evaluated by the relative error and absolute error of the calculation results, and the detailed data are shown in Table 4.

Table 4. Absolute error and relative error

No	Absolute error of x (cm)	Relative error of x (%)	Absolute error of y (cm)	Relative error of y (%)	Absolute error of z (cm)	Relative error of z (%)
1	0.11	2.82	0.01	0.20	0.11	0.59
2	0.19	3.45	0.04	1.33	0.04	0.22
3	0.04	1.14	0.00	0.00	0.07	0.40
4	0.03	1.11	0.12	3.24	0.16	0.92

3 Test Results and Analysis

In the process of stereo correction of the acquired binocular images using binocular calibration parameters, the color images will be converted to grayscale images for calculation. After stereoscopic correction, eye detection was performed on the result image through the eye recognition model, and the detection results (partial) were shown in Fig. 3.

(a)Left eye based image bud detection   (b)Right eye based image bud detection

Fig. 3. Bud detection of binocular image

According to the test and analysis of the camera used in this experiment, it was found that the eye recognition rate was the highest under the condition that the distance between the camera and the potato was 20 cm and the light level was medium light (351 lx–725 lx).

Due to the influence of camera calibration accuracy, measurement tools and manual measurement methods, there are inevitable errors in the 3D coordinates of the bud eye measured by this method. However, the positioning accuracy of the potato bud eye was analyzed by relative error and absolute error (see Table 4), and it was obtained that the absolute error was controlled within 3 mm and the relative error was controlled within 5%. It shows that this method enjoys high reliability and can be applied to the automatic cutting machine of seed potato [10], which can also be applied to measurement in other fields.

4   Conclusions

In this paper, the potato is taken as the research object, in order to realize the recognition and location of the potato eye. In order to realize the recognition and location of potato eye, a scheme based on binocular vision and image recognition technology was designed. The experimental results showed that the recognition rate of the bud eye was the highest when the distance between the camera and the potato was 20 cm and the light level was medium light (351 lx–725 lx). Moreover, the absolute error of positioning results was less than 3 mm, and the relative error was less than 5%. In view of the irregular shape of potato bud eyes and the fact that it is feasible for each seed potato with bud eyes when it is cut. As a result, this measurement method isendowed with certain accuracy and reliability and can be applied to production activities. It is of certain reference value for the follow-up research on the identification and localization of potato bud eye and lays a foundation for the improvement of the automatic cutting machine for seed potato.

References

1. Tian, H.T.: Research on automatic seed cutting method of potato based on machine vision. J. Shandong Agric. Univ. (Nat. Sci. Edit.) (2017)
2. Lu, Z., Xieteng, Q., Wanzhi, Z., Liu, Z., Zheng, W., Mu, G.: Gabor feature based potato image eye recognition. J. Agric. Mech. Res. 201 **43**(02), 203–207
3. Zhang, J., Yang, T.: Identification of potato bud eye based on LBP and SVM. J. Shandong Agric. Univ. (Nat. Sci. Edit.) 202 **51**(04), 744–748
4. Li, M., Lu, B., Jin, C.: Research on binocular ranging based on binocular vision stereo matching technology. J. Langfang Norm. Univ. (Nat. Sci. Edit.) **19**(02), 18–20 (2019)
5. Cong, Z., Wang, H., Gao, M., Li, J., Wang, Z.: Tomato image processing and picking trajectory planning based on binocular vision. Mach. Tool Hydraul. **48**(23), 112–118 (2020)
6. Wei, C., Li, M., Long, J.: Research on fruit recognition and localization of picking robot based on binocular vision and machine learning. J. Agric. Mech. Res. 201 **43**(11), 239–242
7. Zhang, Z.: A flexible new technique for camera calibration. IEEE Trans. Pattern Anal. Mach. Intell. **22**(11), 1330–1334 (2000)
8. Liu, Y., Xu, X., Wang, C., Zhang, T.: Research on the method of camera calibration system. Intell. Comput. Appl. **9**(03), 133–136+141 (2019)
9. Chengze, L., Dancheng, M., Binghua, Y., Siyuan, X.: Camera calibration and object localization in images based on MATLAB. Electron. World **03**, 79–81 (2021)
10. Zuochang, X., Subo, T., Siyao, L., Xuewei, B., Zuli, Z.: Design of seed potato automatic cutting machine based on machine vision. J. Agric. Mech. Res. **38**(10), 69–73 (2016)

Application of Artificial Intelligence Technology in Digital Image Processing

Yongming Pan[(⊠)]

Department of Digital Media and Design, Neusoft Institute Guangdong, Foshan, Guangdong, China
panyongming@nuit.edu.cn

Abstract. In the era of rapid development of social science and technology, artificial intelligence technology has gradually matured. At present, it has become one of the most concerned technologies in the world, bringing opportunities and challenges to the development of society. Based on artificial intelligence technology, the development of all walks of life has made certain progress. At the same time, artificial intelligence technology is also applied in all aspects of life, especially in digital image processing, which provides great convenience. Digital image processing is the use of computers to process images, and then realize more valuable applications with the support of artificial intelligence technology. The purpose of this article is to study the application of artificial intelligence technology in digital image processing. Under normal circumstances, the original images are collected from various disturbing objective environments, so there are many uncertain factors. In addition, the imaging tools and storage systems will also bring unnecessary noise, which will reduce the quality of the pictures. Therefore, this article uses artificial intelligence technology to digitally process the image. Based on artificial intelligence technology, this paper mainly studies the application of digital image processing in fingerprint recognition, face recognition and gesture recognition. Recognition mainly includes three parts, first is preprocessing, second is feature extraction, and last is distance judgment. The preprocessing stage is mainly to remove the noise of the image, and then use the different numerical ranges of the foreground and background in the color model space to extract the detailed features, and finally compare the distances for recognition. The experimental research results show that: the most prominent is the face recognition technology of the digital image processing system B, accounting for 45.08%, followed by the digital image processing system C, the most prominent is the gesture recognition technology, accounting for 39.27%, and fingerprint recognition the technology is more than 30% on average. It shows that the application of artificial intelligence technology in digital image processing needs to be improved.

Keywords: Artificial intelligence · Digital image · Image processing · Application

1   Introduction

In the era of rapid development of social science and technology, artificial intelligence technology has gradually matured [1, 2]. At present, it has become one of the most concerned technologies in the world, bringing opportunities and challenges to the development of society [3, 4]. Based on artificial intelligence technology, the development of all walks of life has made certain progress [5, 6]. At the same time, artificial intelligence technology is also applied in all aspects of life, especially in digital image processing, which provides great convenience [7, 8]. Digital image processing is the use of computers to process images, and then with the support of artificial intelligence technology, more valuable applications can be realized [9, 10].

In the research on the application of artificial intelligence technology in digital image processing, many domestic and foreign scholars have studied it and achieved good results. Kalafi E Y pointed out that artificial intelligence is an emerging edge subject after the emergence of electronic computers. It studies how to make robots have the ability to obtain information and process information through the simulation of human intelligence, in order to complete some tasks that can only be done by humans [4]. Barzegar R focused on the impact of the development of science and technology represented by artificial intelligence on the theory of social development dynamics [11].

Under normal circumstances, the original images are collected from various disturbing objective environments, so there are many uncertain factors. In addition, the imaging tools and storage systems will also bring unnecessary noise, which will reduce the quality of the pictures. Therefore, this article uses artificial intelligence technology to digitally process the image. Based on artificial intelligence technology, this paper mainly studies the application of digital image processing in fingerprint recognition, face recognition and gesture recognition. Recognition mainly includes three parts, first is preprocessing, second is feature extraction, and last is distance judgment. The preprocessing stage is mainly to remove the noise of the image, and then use the different numerical ranges of the foreground and background in the color model space to extract the detailed features, and finally compare the distances for recognition.

2   Research on the Application of Artificial Intelligence Technology in Digital Image Processing

2.1   The Nature of Artificial Intelligence

(1) New tools for human practice

Artificial intelligence is a tool made by mankind for the purpose of better transforming the objective world. The research of artificial intelligence is to improve people's ability and efficiency in various tasks of transforming nature and governing society through intelligent machines. Human beings have invented various tools to expand their capabilities in practical activities due to the fragility of their own organisms and the limited power. The axe is actually an extension of human arm strength, the bow and arrow are the overcoming of human speed defects, and intelligent machines can help the human brain to deal with some tedious and repetitive tasks. Therefore, artificial intelligence, as a high-tech technology developed, is a new tool to help improve people's ability to transform the objective world.

(2) Human's simulation of its own intelligence

The intelligent simulation of humans by artificial intelligence can only stay on functional simulation. There is a big difference between artificial intelligence and human intelligence. First of all, from the perspective of internal mechanism, human intelligence is to process and store information through the neural network of the human brain, and the basis of artificial intelligence is integrated circuits. Secondly, from the perspective of the nature of intelligence, human intelligence is active, active, and unrestricted, but artificial intelligence is limited by human intelligence. Human intelligence is obtained through participation in social practice and can be continuously updated and accumulated. Human intelligence is driven by people's own goals, while artificial intelligence is driven by external objects, that is, human goals. Moreover, from the perspective of the way of intelligence transmission, human intelligence is inherited and accumulated through human ears and eyes or through education, but artificial intelligence is a kind of mechanical filling. Human wisdom itself is a dynamic process that can be constantly updated and supplemented, and it is in the eternal process of inquiry. Wisdom comes from the reflection on life experience, it is complex and multidimensional. Artificial intelligence technology absorbs a lot of human intelligence, but artificial intelligence technology does not have the ability to weigh the pros and cons, and there is no moral and ethical concept of human beings. Artificial intelligence technology is only the processing and utilization of human knowledge, and artificial intelligence does not possess human wisdom.

2.2   Image Quality Evaluation

(1) Subjective quality evaluation

Improve image display equipment, processing equipment, and data acquisition equipment. We look at the image, and determine the quality of the image mainly from two points. One is to get the required information in it, and to get the visual satisfaction.

(2) Objective quality evaluation

Overall difference compared to the reference image and target image, because the image of pure mathematics statistics is peak signal-to-noise ratio and the characteristics of the mean square error, so that cannot correctly reflect the difference in the local, so there will be fewer big differences and appear more small difference is processed by the same situation, human visual characteristics and practical evaluation result is not consistent, This is the biggest defect of traditional objective quality evaluation method.

2.3   Status Quo of Application of Intelligent Identification Technology

(1)   Fingerprint recognition application
Attendance in enterprises still dominates. As the foundation of today's business management, attendance has also become an important means of evaluating the level of business management. Employee attendance equipment using fingerprint recognition can fundamentally solve the related problems in various attendance methods in the past. Providing fingerprint recognition can not only enhance the validity and authenticity of attendance data, but also fully improve the management capabilities of the enterprise, Improve management efficiency. It is precisely because fingerprint recognition time attendance equipment is currently the most simple and convenient to use, the most authentic and reliable biometric device for processing data, and the product price is relatively cheap, so many small and medium-sized enterprises in China will choose this technical product as their corporate time attendance equipment.
Providing security for smart communities has become a popularization hotspot. As people pay more and more attention to property safety, more and more people will choose smart communities. When designing and selecting a community, it also pays more attention to the feelings of community users. The design usually integrates functions, performance, cost and other aspects. Therefore, many smart communities usually choose products with high technology content, relatively more mature technology, higher product safety and reliability, and relatively best cost-effective products. Therefore, fingerprint recognition technology is the most mature technology of this kind, the product performance is the most stable, and the use cost is relatively low, so this technology is particularly suitable for use in smart communities.

(2)   Face recognition application
The face recognition system has simple equipment requirements, easy-to-use operating procedures, wide application range, strong applicability, supports multi-point simultaneous collection and comparison, and has low requirements on the collection environment. It can determine whether the user's identity is legal in a very short time. Various subjects involved in face recognition have also achieved a lot of results, such as biometric analysis, image processing algorithms, pattern recognition algorithms, etc., which have played a strong role in promoting the research of face recognition technology, and electronic hardware and computer. The development of software technology has created a better software and hardware application environment for it, and human face recognition technology has truly entered the field of practical application. Therefore, the derivative results of this technology are likely to promote the development of other recognition fields and artificial intelligence fields to a certain extent. Therefore, face recognition technology has very important theoretical research value.

(3)   Gesture recognition application
The human hand is the most flexible part of the human body. First of all, it can be combined with the arm to complete some simple movements. At the same time, the

different positions and movements of the fingers can express a variety of different meanings. If these meanings can be correctly understood by the computer, then its application prospects it is indescribable.

2.4   Population and Sample

Suppose a finite population contains N population units, and the k-th population unit is identified by k, where the finite population can be expressed as:

$$U = \{1, \cdots, k, \cdots N\} \tag{1}$$

Let y denote a research variable, which is the value of the research variable of the k-th overall unit in the population. Usually before the investigation, the value is unknown. And what we need to estimate is the total value of the research variable y

$$t = \Sigma_u y_k \tag{2}$$

Or study the overall mean of variable y

$$\bar{y}_u = t/N = \Sigma_u y_k/N \tag{3}$$

3   Experimental Research on the Application of Artificial Intelligence Technology in Digital Image Processing

3.1   Experimental Subjects and Methods

This experiment takes the digital image processing system as the research object, and conducts research on its practical application. Perform experimental research on image conversion and analyze how accurate the conversion is. Then conduct experimental research on intelligent recognition and analyze its most prominent recognition technology, use actual investigations to conduct tests.

3.2   Data Collection

Investigation task decomposition makes a detailed breakdown of the total amount of investigation activities, assigns investigation tasks to investigators or investigation teams, and assigns tasks equally by default. The survey task is broken down into specific task assignments, such as how many answers each investigator or survey team needs to collect, the time range for collecting the answer sheets, and the location of the survey. The survey tasks can also be manually assigned. If the system automatically divides them into an inappropriate way, you can modify the assigned tasks.

4 Experimental Research and Analysis of the Application of Artificial Intelligence Technology in Digital Image Processing

4.1 Analysis of the Accuracy of Digital Image Type Conversion

This experiment is based on a digital image processing system and analyzes its image conversion accuracy (unit percentage). The image types include true color images, indexed images, grayscale images and binary images are mainly used for experimental research. The experimental results are shown in Table 1:

Table 1. Image type conversion accuracy

	True color image	Index image	Grayscale image	Binary image
Digital Image Processing System A	36.01	29.53	43.08	31.26
Digital Image Processing System B	41.27	36.92	25.19	37.79
Digital Image Processing System C	22.72	33.55	31.73	30.95

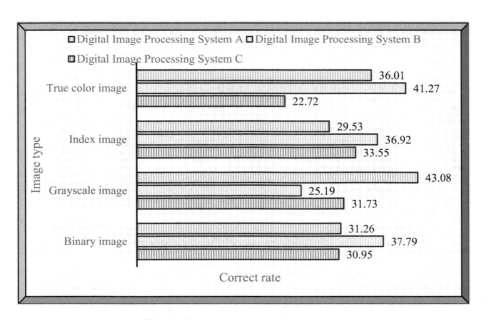

Fig. 1. Image type conversion accuracy

As shown in Fig. 1, the conversion rate of digital image processing system B is the most accurate in true color images, index images, and binary images, which are higher than the accuracy of other systems, which are 41.27%, 36.92%, and 37.79%, grayscale.

The image has the highest accuracy rate in the digital image processing system A, with 43.08%. In summary, the main application should refer to the digital processing system B, which is excellent in all aspects, followed by the digital image processing system A.

4.2   Intelligent Recognition and Analysis of Digital Images

This experiment focuses on the digital image processing system, and studies its fingerprint recognition, face recognition, and gesture recognition, and analyzes where its applications are most prominent. The experimental results are shown in Table 2:

Table 2. Proportion of intelligent image recognition

	Fingerprint recognition	Face recognition	Gesture recognition
Digital Image Processing System A	31.79	31.23	36.12
Digital Image Processing System B	33.14	45.08	24.61
Digital Image Processing System C	35.07	23.69	39.27

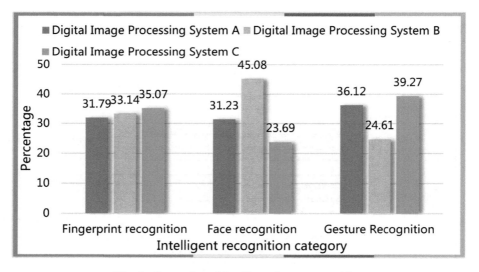

Fig. 2. Proportion of intelligent image recognition

As shown in Fig. 2, the most prominent is the face recognition technology of the digital image processing system B, accounting for 45.08%, followed by the digital image processing system C, the most prominent is the gesture recognition technology, accounting for 39.27%, fingerprints the recognition technology is more than 30% on average. It shows that the application of artificial intelligence technology in digital image processing needs to be improved.

5 Conclusion

Artificial intelligence will play an increasingly important role in future social development. Artificial intelligence provides a powerful tool support for the improvement of social productivity and the accumulation of material wealth. Mankind will finally usher in the era of intelligent machines, and society will become an intelligent society. A completely intelligent society is characterized by an intelligent economic society in which intelligence-intensive industries dominate. At that time, smart machines enter factories, offices, homes and other places, and all aspects of life are infiltrated with smart factors, and more and more smart factors are infiltrated into people's lives. The driving force of the economy has changed from being mainly driven by material forces to being driven by intelligence, and the driving effect of intelligence has become more obvious. Due to the intelligentization of the production process and management process in the intelligent age, it is possible that, as Marx had foreseen, workers just stand by the side of the production process to assist or monitor. Science and technology represented by artificial intelligence have played a huge role in the process of mankind's acquisition of freedom and all-round development. However, artificial intelligence technology also hides a series of social risks. This requires the rational use of high-tech technologies such as artificial intelligence and vigilance against their negative effects in the overall development of human freedom.

References

1. Hu, S.: Research on data acquisition algorithms based on image processing and artificial intelligence. Int. J. Pattern Recognit. Artif. Intell. **34**(06), 1–13 (2020)
2. Tan, X., Konietzky, H., Chen, W.: Numerical simulation of heterogeneous rock using discrete element model based on digital image processing. Rock Mech. Rock Eng. **49**(12), 4957–4964 (2016). https://doi.org/10.1007/s00603-016-1030-0
3. Hosseininia, S., Kamani, M.H., Rani, S.: Quantitative determination of Sunset Yellow concentration in soft drinks via digital image processing. J. Food Meas. Charact. **11**(3), 1–6 (2017)
4. Kalafi, E.Y., Tan, W.B., Town, C.: Automated identification of monogeneans using digital image processing and k-nearest neighbour approaches. BMC Bioinform. **17**(19), 259–266 (2016)
5. Sundararajan, D.: Image enhancement in the spatial domain. In: Digital Image Processing, pp. 23–64. Springer, Singapore (2017). https://doi.org/10.1007/978-981-10-6113-4_2
6. Zhang, S., et al.: The characteristic analysis of temperature sensor based on a fabricated microstructure fiber by digital image processing technique and finite element method. J. Phys. D: Appl. Phys. **53**(17), 175108 (2020)
7. Yao, J., Xiao, X., Liu, Y.: Camera-based measurement for transverse vibrations of moving catenaries in mine hoists using digital image processing techniques. Meas. Sci. Technol. **27**(3), 035003 (2016)
8. Silvia, B., et al.: Quantitative assessment of mouse mammary gland morphology using automated digital image processing and TEB detection. Endocrinology (4), 1709–1716 (2016)

9. Barbedo, J.G.A.: A novel algorithm for semi-automatic segmentation of plant leaf disease symptoms using digital image processing. Trop. Plant Pathol. **41**(4), 210–224 (2016)

10. Villibor, G.P., Santos, F.L., Queiroz, D.M.D., et al.: Determination of modal properties of the coffee fruit-stem system using high speed digital video and digital image processing. Acta Sci. Technol. **38**(1), 41 (2016)

11. Barzegar, R., Adamowski, J., Moghaddam, A.A.: Application of wavelet-artificial intelligence hybrid models for water quality prediction: a case study in Aji-Chay River Iran. Stoch. Environ. Res. Risk Assess. **30**(7), 1797–1819 (2016)

Safe Face Recognition Based on Partial Ranking

Yu Jin[✉], Hang Xiong, and Jialiang Yang

Wuhan University of Technology, Wuhan 430070, Hubei, China
jinyu929@whut.edu.cn

Abstract. In privacy protection, the protection of facial features is a focus of many companies and governments. The leakage of facial feature data will often bring great losses to the parties. There have also been many incidents of face data leakage recently. We propose a face feature protection method using partial ordering, the method first performs the exclusive OR operation on the face data and the applicationspecific character string; Then divides the execution result into several blocks and divides these blocks into several groups, and blocks in each group will be sorted by their decimal value; Finally the initial block is converted into a sorted value for system to storage, and the initial data and intermediate data are deleted. By this method we can simultaneously meet the irreversibility, revocability and irrelevance of the biometric template proposed by the international standard ISO/IEC 24745, and the method can achieve good recognition performance. We also discuss the Unlinkability, Irreversibility and Revocability of the method in the end.

Keywords: Security algorithm · Face template protect · Ordering ranking

1 Introduction

Biometric recognition technology has developed very rapidly in recent years, especially after the introduction of deep learning technology in traditional biometric recognition, it is possible to extract biometrics and perform recognition with a very high recognition rate. Many applications also introduce biometric recognition, such as Apple's FaceID [1], Microsoft's Windows Hello, and Alipay can use face recognition to authenticate payments on the mobile phone. Biometric recognition is to distinguish individual organisms from their biological characteristics. The biological characteristics studied by biometric recognition technology include fingerprint, face feature, iris, palm, print, retina, voice, physical characteristics, personal habits, etc. The corresponding recognition technology includes face recognition, fingerprint recognition, palmprint recognition, iris recognition, and retina recognition, Voice recognition, body shape recognition, keyboard stroke recognition, signature recognition, etc. Most of the biometric information of the organism cannot be changed. If the biometric information is leaked, it is generally difficult to modify the biometric information to invalidate the leaked information. Therefore, it is significant to protect the biometric data [2].

Facial recognition is a very important field in biometric recognition [3], and it is also a rare biometric technology that is very attractive and can inspire people's enthusiasm. As artificial Intelligence enters the field of biometrics, facial recognition technology has also developed rapidly. Facial recognition is the process of identifying or verifying a person's identity through the face. It captures, analyzes and compares patterns based on the details of a person's face. Generally, it is divided into three steps, namely face detection, face capture and face matching. Facial information is considered to be the most natural of all biological characteristics, compared to fingerprint and iris characteristics.

At present, facial biometric recognition is generally the preferred biometric benchmark. Because it is easy to deploy and realize, the end user does not need to perform physical interaction, and the process of face detection and face matching is also very fast. At present, the main problems of face recognition still focus on the protection of characteristic information and privacy. According to ISO/IEC 24745 [4], the biometric data protection method need to meet the three basic security requirements of "irreversibility [5, 6], revocability and nonconnectivity". Irreversibility requires that the initial biometric data be restored from the tem plate used for identification is different; revocation means that if the secure biometric template in the server or other persistent storage is leaked, the administrator can easily revoke the leaked template, and New biometric templates can be released for identification; nonconnectivity means that security biometric templates in different applications cannot be cross matched, which means that the attacker cannot judge whether these templates come from same source of bio metrics or not.

Currently, many face feature encryption methods are insecure. In our article, we put for ward a face feature protection method that is based on partial ranking to address the problem that the current face feature security cannot be fully guaranteed. By using this method to en crypt the face data, the template data obtained It can simultaneously meet the irreversibility, revocability and irrelevance of the biometric template proposed by the international standard ISO/IEC 24745, and use this method for encryption to obtain good recognition performance.

2   Related Work

This section, we will reviewed the related research on face template protection and introduced the Partial ranking algorithm proposed in this article.

At present, many face feature protection methods have been proposed. These methods mainly include hash transformation [7], unlinkable and irreversible transformation [8], as well as the recently proposed change encryption using machine learning and deep learning methods. There are also some methods specifically studied for cancelability [9].

In the article A hybrid approach for face template protection [10], FC Feng proposed a face protection method that mixes random projection, class distribution preserving transformation and hash function, which can achieve diversity and revocability, which is better than unobstructed random multiple The spatial quantitative biohashing algorithm, but this method has no clear experiment to prove its irreversibility.

In the article Unlinkable and irreversible biometric template protection based on bloom filters [11], Matrta proposed a biometric template protection method that uses unlinkable and irreversible changes. This paper proposes a general framework for the evaluation of the unlink ability of biometric template protection schemes, and an improved, unlinkable and irreversible system based on Bloom filters. This method maintains the performance of an unprotected bio metric system and has certain protection against attacks.

In the article SecureFace: Face Template Protection [12], Mai proposed a method using random CNN to generate a face biometric template given an input face image and a userspecific key. The use of a userspecific key brings security to the security template, thereby enhancing the security of the template, and by not storing the key, but storing a security sketch to generate the key, the security is further improved. At the end of this article, the method verifies the irreversibility and cancelability [10], while guaranteeing part of the performance, but the method does not guarantee irrelevance.

In addition to the above methods, related methods of deep learning [13] have also been applied to feature extraction and template protection. In the last 5 years, many papers [14, 15] using related methods have been published. In addition, many new dedicated methods have been proposed [11].

3   Approach

We propose a face feature protection method using partial ordering, which can simultaneously meet the irreversibility, revocability and irrelevance of the biometric template proposed by the international standard ISO/IEC 24745, and the method can achieve good recognition performance.

This method first performs the exclusive OR operation on the face data and the application specific character string; then divides the execution result into multiple blocks and divides these blocks into several groups, and blocks in each group will be sorted through to their decimal value; finally The original block is converted into a sorted value for storage, and the original data and intermediate data are deleted.

3.1   Details of Method

The process of the local sorting method used in this article is introduced as follows.

First, we have a face data sequence. In order to simplify the representation of the algorithm, we assume that the face data sequence we have now is the m bit sequence x, and the i bit of x is all It can be represented by xi.

Step 1: Perform a bitwise exclusive OR operation between the face data and the string specified by the application.

Perform an exclusive OR operation on the face data x with any m bits and a specific string.

p with the same m bits to get the string t;

$$t_i = x_i \oplus p_i$$

Step 2: Divide the execution result into multiple blocks.

Divide the t string into n blocks, each block includes b bits, you can know that n*b = m, m is the length of the original face data; the converted length is n u sequence, and $u_i = u_{i,1} \ldots u_{i,b}$, for any bit in u, $u_{i,j} = t_{j+(i-1)*b}$, j = 1,...., b.

Step 3: Divide these blocks into several groups.

Divide u = $u_1 \ldots u_n$ into g groups to get $U_i = U_1 \ldots U_g$, where $U_i = \{u_{1+(i-1)\times d,\ldots}, u_{i\times d}\}$, that is, each U contains d blocks in u, each group includes d bits, and n = g * d.

Step 4: Sorted these blocks in each group through to their decimal value For i = 1... g, calculate decimal value of $\{u_{(i-1)*d+1},\ldots, u_{i*d}\}$ v_1, v^d

$$v_j = \sum_{k=1}^{b} u_d \times (i-1) + j, k \times 2^{b-k}$$

Sort $u_{(i-1)*d+1},\ldots, u_{i*d}$ based on their decimal value to get the new sort value $r_{1+(i-1)\times d},\ldots,r_{i\times d}$ (Just like the previous $u_{1+(i-1)\times d},\ldots, u_{i\times d}$):

Step 5: Store the converted sort value of the original block $r_{1+(i-1)\times d},\ldots,r_{i\times d}$ as a template, and store the rest of the data x, t, u, U, v are deleted.

In this way, we get a sorted sequence $r_{1+(i-1)\times d},\ldots,r_{i\times d}$ if the current two values vi and vj Is equal, then we will compare them by their indexes, namely i and j. Among the above methods, the initial face data is well protected since we just store the after ordering value, and the original face data is deleted by us. We would explain in the security analysis that it is difficult to infer the original face data from the ordering sorted value.

The analysis shows that when the input data size is m, we can know that since the calculation process is mainly linear and limited operations, the time complexity of the algorithm is O(m).

According to ISO/IEC 24745, the biometric data protection method should meet the three basic security requirements of "irreversibility, revocability and nonconnectivity". Irreversibility requires that the initial biometric data be restored from the template used for identification is different; revocation means that if the secure biometric template in the server or other persistent storage is leaked, the administrator can easily revoke the leaked template, and New biometric templates can be released for identification; nonconnectivity means that safe biometric tem plates in different system cannot be crossmatched, which means that an attacker cannot judge whether the templates come from the same source of biometrics.

If we want to perform authentication and identification, we can calculate the distance/difference between any two template sequences such as the sequence $r_{1+(i-1)\times d},\ldots,r_{i\times d}$ and $r'_{1+(i-1)\times d},\ldots,r'_{i\times d}$ (converted from any two face data sequences) is as follows:

$$\text{Dis}\left(r, r'\right) = \sum_{i=1}^{n} |r_i - r'_i|$$

This algorithm can simultaneously meet the irreversibility, revocability and irrelevance of the biometric template proposed by the international standard ISO/IEC 24745, and the method can obtain good recognition performance.

3.2   Face Template Protection

The goal of this method is to better protect private information such as facial features. In this section, we will give an example process of the application of this method, and explain how this method realizes the revocability of the biometric template.

Figure 1 shows the architecture of this process. To put it simply, we obtain the face data through the face recognition device and send it to the corresponding application. In the applica tion, we apply our method to encrypt the face data. We first generate a P data in the application, and then Send P data and the acquired face data as parameters to our method. The specific process of the method has been explained in the previous chapter, so I won't describe it here. Through our method, we can get a V data, this is the data we can use for storage, we can store this data in the database, and store the corresponding P data at the same time [16]. If it is desired to match the existing face data through V data, then the face data must be processed through the corresponding P data to perform correct matching. If there is a leak of V data, we only need to reset P data and reencrypt the face data to obtain new V data and perform face matching. We will discuss the security of the final V data in the next chapter.

It should be noted here that in order to ensure security, for each V data, P data is randomly generated and requires a certain degree of complexity.

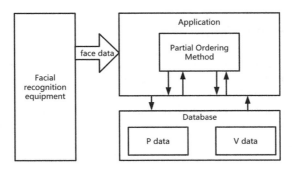

Fig. 1.   Application of our method in the system

3.3   Safety Analysis

In this section, we will expound the unlinkability, irreversibility and revocability of the face feature protection method based on local sorting.

The method used in this section mainly refers to the proof of the safety of local ranking in the article Iris Template Protection Based on Local Ranking [10].

For any given template, there are a lot of original data that can be mapped, so the attacker has no way to reverse the original face data through the encrypted data.

1. Unlinkability

   The encrypted data used for identification needs to meet the unlinkability, so as to ensure that the attacker cannot determine whether the two data from different applications correspond to the same user.

   In this method, since the p cited in the data processing is randomly generated, the distance between the finally calculated results will also make the results unlinkable.

2. Irreversibility

   Irreversibility requires that the attacker cannot recover the initial data from the template data which is used for identification.

   For any given template, the amount of facial feature data that could be mapped to the current template obviously very large.

   Therefore, for an attacker, even if the template data for matching is obtained by some means, the original face data cannot be obtained in reverse.

3. Revocability

   Revocability [17] requires that when the template data used for identification is leaked, the method used to protect the data should be able to revoke the leaked data, make it invalid, and be able to reconstruct new template data that can be used for identification.

   In the method we provide, we can easily achieve revocability by using p, which is refer to P data, corresponding to a specific application, by deregistering p, and regenerating template authentication data by generating a new p.

   Since the p data has been changed, the old template data will not be used for authentication.

4   Conclusions and Future Work

We mainly introduces the research background of this article, the current research situation at home and abroad, and puts forward the research content of this article based on the current situation. Then, we introduces the method proposed in this article, introduces the realization principle of the local sorting algorithm, the basic framework and corresponding steps of this method, and analyzes the safety of the method in this article, and analyze. Then the recognition performance of the method is demonstrated through experiments and its effectiveness is verified. Finally, the method was compared with the existing methods, and the overall effect was evaluated.

In future work, we will attempt to design more efficient and safe p data management methods, and improve current methods to support the protection of other biological characteristics. We also learned about a method [18] using negative databases to process encrypted data, and we are considering introducing it into our method in the future.

References

1. Bud, A.: Facing the future: the impact of apple faceID. Biom. Technol. Today **2018**(1), 5–7 (2018)
2. Wang, Y., Hatzinakos, D.: Face recognition with enhanced privacy protection. In: 2009 IEEE International Conference on Acoustics, Speech and Signal Processing, pp. 885–888. IEEE (2009)
3. Zhao, D., Luo, W., Liu, R., Yue, L.: Negative iris recognition. IEEE Trans. Dependable Secur. Comput. **15**(1), 112–125 (2018). https://doi.org/10.1109/TDSC.2015.2507133
4. Information TechnologySecurity TechniquesBiometricInformation Protection. ISO/IEC JTC1 SC27 IS 24745 (2011). http://www.iso.org/iso/cataloguedetail?csnumber=52946
5. Nagar, A., Jain, A.K.: On the security of non-invertible fingerprint template transforms. In: 2009 First IEEE International Workshop on Information Forensics and Security (WIFS), pp. 81–85. IEEE (2009)
6. Jin, Z., Lim, M.H., Teoh, A.B.J., et al.: A non-invertible randomized graph-based hamming embedding for generating cancelable fingerprint template. Pattern Recognit. Lett. **42**, 137–147 (2014)
7. Furukawa, M., Muraki, Y., Fujiyoshi, M., Kiya, H.: A secure face recognition scheme using noisy images based on kernel sparse representation. In: 2013 Asia-Pacific Signal and Information Processing Association Annual Summit and Conference, pp. 1–4. Kaohsiung, Taiwan (2013). https://doi.org/10.1109/APSIPA.2013.6694155
8. Sardar, A., Umer, S., Pero, C., et al.: A novel cancelable face hashing technique based on non-invertible transformation with encryption and decryption template. IEEE Access **8**, 105263–105277 (2020)
9. Teoh, A.B.J., Yuang, C.T.: Cancelable biometrics realization with multispace random projections. IEEE Trans. Syst. Man Cybern. Part B (Cybern.) **37**(5), 1096–1106 (2007)
10. Feng, Y.C., Yuen, P.C., Jain, A.K.: A hybrid approach for face template protection. In: Proceedings of the SPIE 6944, Biometric Technology for Human Identification V, p. 694408, 17 March 2008. https://doi.org/10.1117/12.778652
11. Mai, G., Cao, K., Lan, X., Yuen, P.C.: SecureFace: face template protection. IEEE Trans. Inf. Forensics Secur. **16**, 262–277 (2021). https://doi.org/10.1109/TIFS.2020.3009590
12. Jindal, A.K., Chalamala, S., Jami, S.K.: Face template protection using deep convolutional neural network. In: 2018 IEEE/CVF Conference on Computer Vision and Pattern Recognition Workshops (CVPRW), pp. 575–5758. Salt Lake City, UT (2018). https://doi.org/10.1109/CVPRW.2018.00087
13. Sun, W., Song, Y., Chen, C.: An orthogonal facial feature learning method based on convolutional-deconvolutional network. J. Shenzhen Univ. Sci. Eng. **37**(5), 474–481 (2020). https://doi.org/10.3724/SP.J.1249.2020.05474
14. Chen, L., Zhao, G., Zhou, J., et al.: Face template protection using deep LDPC codes learning. IET Biom. **8**(3), 190–197 (2018)
15. Pandey, R.K., Zhou, Y., Kota, B.U., Govindaraju, V.: Deep secure encoding for face template protection. In: 2016 IEEE Conference on Computer Vision and Pattern Recognition Workshops (CVPRW), pp. 77–83. Las Vegas, NV, USA (2016). https://doi.org/10.1109/CVPRW.2016.17
16. Rathgeb, C., Busch, C.: Multi-biometric template protection: issues and challenges. New Trends Dev. Biom. 173–190 (2012)

17. Patel, V.M., Ratha, N.K., Chellappa, R.: Cancelable biometrics: a review. IEEE Signal Process. Mag. **32**(5), 54–65 (2015). https://doi.org/10.1109/MSP.2015.2434151
18. Zhao, D., Luo, W., Liu, R., Yue, L.: Negative iris recognition. IEEE Trans. Dependable and Secur. Comput. **15**(1), 112–125 (2018). https://doi.org/10.1109/TDSC.2015.2507133

Design of Control System for Optical Cable Sheath Production Line Based on AI Algorithm

Xiaojin Mo and Fei Zhou[✉]

Department of Electrical and Electronic Information,
Wuhan Institute of Shipbuilding Technology, Wuhan, Hubei, China
Feizhou@mail.wspc.edu.cn

Abstract. At present, the production and demand of optical cables in my country has ranked third in the world. People have also put forward higher and higher requirements for various optical cable performance indicators. The fiber optic cable jacket production line is the last process in the fiber optic cable manufacturing industry. The quality of the outer shell will have a vital impact on the mechanical, thermal and chemical properties of the optical cable. In this paper, the design of the control system for the optical cable sheath production line based on the AI algorithm, firstly uses the literature research method to describe the composition of the optical cable sheath production line and the design content of the cable sheath diameter control system, and summarize the shortcomings of the production system. Aiming at the shortcomings of traditional PID control algorithms in parameter optimization, the ant colony algorithm in the AI algorithm is proposed to solve this problem, and then based on the analysis of system requirements, the optical cable sheath production line control system is designed, and the overall system architecture is proposed. Select PLC as the control unit to enhance the coordination of control functions and subsystems: use industrial computers instead of touch screens as the host, and form a host monitoring system with configuration software to improve monitoring and management of the entire production line.

Keywords: AI algorithm · Optical cable sheath · Control system · System design

1 Introduction

Special equipment dedicated to the manufacture of optical cables is an important basis for updating and developing new varieties of optical cables, improving the quality of optical cables, improving the efficiency of optical cable production lines and improving production technology [1, 2]. Now, the large-scale use of optical communication and optical information has also promoted the continuous and rapid development of optical cable manufacturing technology [3, 4]. In all of China's national economic projects, the optical cable industry occupies a vital position. At present, the use of optical cables has been inseparable from human life and economic development [5, 6]. Therefore, the in-depth research and exploration of optical cable manufacturing technology not only play a strong role in enhancing the competitiveness of my country's optical cable industry [7, 8].

© The Author(s), under exclusive license to Springer Nature Singapore Pte Ltd. 2022
J. C. Hung et al. (Eds.): FC 2021, LNEE 827, pp. 629–637, 2022.
https://doi.org/10.1007/978-981-16-8052-6_78

In the research on the control system design of the optical cable sheath production line based on the AI algorithm, many companies have conducted research on it and achieved good results. For example, the optical cable housing production line developed by the Austrian Rosendahl (Nasir M) company is suitable for external insulation diameters of Processing and manufacturing of 1.05–5.20 mm optical cables. The maximum production line speed of the production line can reach 110 m per minute, and the maximum barrel capacity can reach 5.11 km per barrel [9]. Another example: The fiber optic cable jacket production line developed by Finland NEXTROM also uses production control technology similar to Austria's ROSENDAHL. The maximum line speed of the 0FC40 production line can reach 300 m per minute, and the maximum barrel volume can also reach 4.8 km. The YORK optical cable production line in the United States (Zhu B W) adopts fieldbus-based control technology, which can monitor production and operation status in real time, and is equipped with a remote service subsystem, automated management subsystem production line and optical cable, quality monitoring subsystem [10].

This article is based on the design of the optical cable sheathing production line control system based on the AI algorithm. Firstly, using the literature research method, the composition of the optical cable sheathing production line and the cable sheath diameter control system are described, and the shortcomings of the production system are summarized. According to the analysis of system requirements, the control system of the optical cable sheath production line is designed, and the overall system architecture is proposed to select PLC as the control unit to enhance the coordination of control functions and subsystems: use industrial computers instead of touch screens as the host, and form together with configuration software Host monitoring system to improve monitoring and management of the entire production line.

2   Research on AI Algorithm and Optical Cable Sheath Production Line Control System

2.1   The Composition of the Optical Cable Sheath Production Line

(1) Pay-off mechanism

The main function of the pay-off unit is to send the cable by controlling the speed of the cable tray and the cable voltage. When loading and unloading the rack, this unit controls the speed of the cable roller, the speed of the optical cable wheel and the voltage of the optical cable. In order to ensure that the linear speed of the optical cable is constant, the speed of the cable roll must be adjusted in real time, so a set of speed control system is required. In addition, the mechanical mechanism of the device also includes a control panel for controlling and adjusting the loading and unloading of the tray, the rotation stop on site and the speed of the cable tray.

(2) Extrusion unit

The function of the extruder unit is to extrude a certain amount of colloid as needed. The main requirements include that the amount of colloid should increase or decrease linearly with the change of production speed, and the cable diameter should not have too much error. In addition, each extruder heating zone needs to be temperature controlled, and the hopper must continue to be heated.

(3) Traction mechanism

The traction mechanism is used to provide power to move the core wire during the entire production process. In actual production, the productivity is usually determined by adjusting the adhesion rate. The bonding speed can also be adjusted to reduce the deviation between the cable diameter and the nominal value. The traction mechanism can not only provide the cable traction forward, but also can be used in conjunction with the repayment mechanism to control the voltage stability during the cable transmission process.

(4) Testing equipment

The testing equipment of the sheath production line includes a diameter pressure gauge and a spark machine. The controller can compare the real-time cable diameter measured by the diameter gauge with the specified value to obtain the wire diameter deviation, and then check whether the product cable diameter is stable within the allowable range. Spark detection equipment is used to detect the high voltage resistance of the cable behind the housing. The insulation performance of the cable is controlled by applying high frequency and high voltage kV to the cable environment, and the specific applied voltage can be based on specific specifications and specifications of the cable. The required standard stipulates that if there is a defect in the cable casing when passing through the spark machine, the cable will break at this time, and the spark detection equipment will generate an alarm fault point and automatically shut down the production line and cut off the power supply.

(5) Take-up mechanism

The main task of the cable receiving mechanism is to wind the cable into a coil as required. The production line that adopts the double-axis rewinding method for rewinding has high rewinding speed, and the two rewinding shafts work alternately. There is no need to stop when changing the cylinder, so it has a high production efficiency.

2.2   Design Content of Cable Sheath Wire Diameter Control System

(1) The realization of the start and stop of the production line and the inching and linkage control functions of the extruder;

(2) All-digital DC speed regulation across the entire production line;

(3) Pneumatic fully automatic control of the opening and closing of the crawler tractor;

(4) Temperature control and display of the extruder, set the required temperature on the operating interface of the host computer;

(5) The tension of each sub-part is constant and controllable, and the matching of the full line speed is realized:

(6) Over-limit alarm of the water level of the cooling water tank to keep the water temperature constant;

(7) Real-time monitoring and feedback of fiber optic cable diameter, sparks, and bulge;

(8) Motor start and stop control of the circulating water pump of the water tank;.

(9) Automatic cable routing control of the take-up part of the optical cable;

(10) The whole production line's global production process, real-time curve, historical curve, dynamic real-time display of important parameters, and alarm settings.

2.3 Problems in the Cable Sheath Wire Diameter Control System

(1) For wire diameter control, some related cable diameter control systems still use the wire diameter tester as the controller, and check the cable diameter together with the diameter tester. The cable diameter tester compares the real-time cable diameter data from the diameter gauge with the specified nominal cable diameter in real time to obtain the deviation, obtains the control value according to the calculated calculation method, and outputs it to the ruler to check the cable diameter. Adjust the size of the adhesion speed.

(2) In order to control the temperature, most of them are executed with a temperature controller. Each temperature control point corresponds to a temperature control instrument. The temperature controller obtains the deviation by comparing the temperature information measured by the thermocouple with the specified temperature. At the same time, the control value obtained after PID calculation is used to drive the heating or cooling unit to achieve temperature control.

(3) In order to wind and arrange the cable, the current sheathed cable diameter control system usually uses a manual adjustment method to wind the cable into a coil. The absorption mechanism is controlled by a simple relay circuit that intelligently performs basic action control. Most of the work must be done manually according to the production situation, which will waste a lot of manpower.

2.4 AI Algorithm

(1) Although the research of various intelligent control algorithms by experts and researchers in the control field has become more and more mature in recent years, due to their application in actual production control, they cannot be successful due to the following reasons Separated from the laboratory stage. Various complications. At present, the most widely used in the process control field is still the simple PID control algorithm. The PID control rules are shown in the following formula:

$$u(t) = K_p \left[e(t) + \frac{1}{T_i} \int e(t) + T_d \frac{d}{dt} e(t) \right] \tag{1}$$

In the formula: Kp — proportional coefficient; Ti — integral time constant; Td — differential time constant.

(2) PID parameter optimization method based on ant colony algorithm

1) Problem description

Assume that the significant figures of the three parameters Kp, Ti and Td are all 5 digits. According to experience, among the 5 significant digits of Kp, 2 digits before the decimal point and 3 digits after the decimal point; among the 5 significant digits in Ti and Td, 1 digit before the decimal point and 4 digits after the decimal point. So a set of parameters (Kp, Ti, Td) corresponds to a 15-bit number sequence one-to-one.

2) The objective function is established. Reasonable selection of the objective function can ensure that the system has good performance. The main performance indicators of the system include stability, accuracy and rapidity. In order to obtain a better system control effect, we also consider the system's control amount, control error and rise time as constraints. Therefore, the following performance index function (objective function) is used:

$$J = \sum_{i=1}^{\infty} \left(w_1 |e(i)| + w_3 u^2(i) + w_3 t_u \right) \tag{2}$$

3) Construction of the path. This step is the main step in the entire ant colony algorithm. First, assume that each artificial ant is at the initial point 0; then, each ant on (x_{i-1}, y') crawls to the next node (x_i, Y_j) until each ant reaches $(15, y'')$, completing a cycle. In this process, we assume that the time for each ant to climb from the previous node to the next node is equal, which is equal to the previous two nodes. The distance of the points is irrelevant. In addition, each ant on (x_{i-1}, y') enters the next node (x_i, y_j) according to the random ratio rule as follows:

$$P(x_j, y_j, t) = \frac{\tau^\alpha(x_j, y_j, t) \eta^\beta(x_j, y_j, t)}{\sum_{J=0}^{9} \tau^\alpha(x_j, y_j, t) \eta^\beta(x_j, y_j, t)} \tag{3}$$

4) Update of pheromone. All ants start from the initial point, after 15 time units, climb to the end and complete a cycle, the amount of information on the node (x_i, y_j) will change.

3 Design of Control System for Optical Cable Sheath Production Line Based on AI Algorithm

3.1 Analysis of the Functional Requirements of the Control System

Cable sheath the main function of the cable diameter control system is to perform automatic sheathing of the cable sheath. The users of the control system are on-site operators and technical administrators, and their operating requirements are different. Therefore, it is necessary to consider various factors, thoroughly analyze the functional requirements of the housing cable diameter control system, and use it as the starting point of the system design. The functional requirements of the sheath wire diameter control system are shown in Fig. 1.

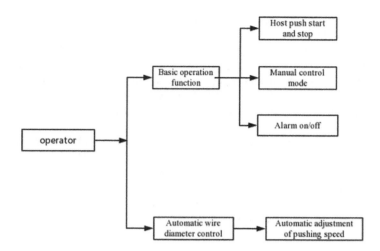

Fig. 1. Function requirements of sheath wire diameter control system

3.2   The Overall Architecture Design of the Control System IPC+ PLC+ PROFINET

Traditional process control systems use touch screens as HMIs to perform basic production control in the operating space. At the monitoring level, the computer performs important display of historical data curves, display of dynamic cable diameter curves, recording of production information, adjustment of process parameters, and management of alarm information to achieve monitoring functions. This method is not suitable for the case production line, because the entire production line requires multiple on-site operators, and in addition to the controller, the main operator must also understand the production information in real time. It analyzes the shortcomings of the current sheathed cable diameter control system from three aspects of the control unit, host monitoring system and field communication network, and proposes corresponding improvement plans. Specifically, select PLC as the control unit to enhance the coordination of control functions and subsystems: use industrial computers instead of touch screens as the host, and form a host monitoring system with configuration software to improve monitoring and management of the entire production line; introduce PROFINET Into the shell cable diameter control system to create a more developed and stable communication network. Briefly, it refers to the "concentrated management and decentralized control thought" in the distributed industrial control system. The IPC+ PLC+ PROFINET overall architecture of the entire sheath production line is shown in Fig. 2.

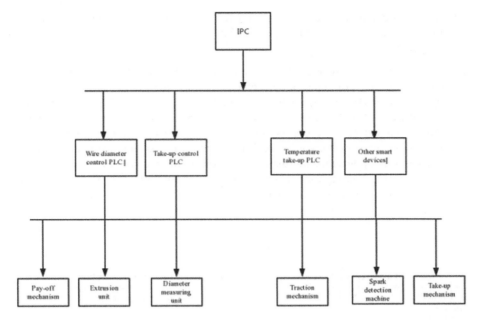

Fig. 2. Overall architecture diagram

3.3 System Software Development Platform

(1) The PLC project development platform of the control system is Siemens TIA Portal software. Based on the concept of fully integrated automation, Portal integrates STEP7 and WINCC. The configuration and programming of the upper computer and the lower computer can be in one software. Complete and easy to complete control system design Portal also has online simulation functions, so you can download and debug programs without a real PLC. It also supports online diagnosis. When the PLC fails, you can use the online diagnosis function to find the problem, which is very convenient and fast.

(2) This article uses SIMATIC WINCC (Windows Control Center)-Windows Control Center as the development software of the host computer monitoring system.

4   Control System Hardware Design

According to the production process of the optical cable sheath and the actual control requirements of the production equipment, and on the basis of full investigation, we choose the PLC of Germany Siemens TIA Portal as the core part of the control system of the optical cable sheath production line. The entire control system of the production line is controlled by the industry. Composed of computer and electric control cabinet. The industrial computer is a series of products of Taiwan Delta Electronics Industry Co., Ltd. The industrial computer and PLC communicate through InovanceMD500.

The industrial computer is used for parameter setting and monitoring screen display. The PLC realizes the input and output of switch quantity and the input and output of analog quantity.

(1) By connecting abundant expansion cards, Inovance MD500 series converters can support multiple communication methods. Among them, the MD38TX1 communication expansion card specially developed by Innovation Technology Co., Ltd. is used to carry MD500 series converters that support 485 communication functions. The circuit design of the expansion card adopts an isolation scheme, and all parameters and performance indicators conform to international standards. Through this expansion card, the user can control the speed of the inverter through the host or PLC through serial communication, or configure the inverter as needed. The function description of MD38TX1 expansion card terminal is shown in Table 1:

Table 1. MD38TX1 expansion card terminal function description

Terminal identification	Terminal name	Function description
485+	485 communication signal positive	485 communication terminal input, isolated input
485−	485 communication signal negative	485 communication terminal input, isolated input
CGND	485 communication signal reference ground	The power supply is an isolated power supply

5   Conclusions

The optical cable sleeve production line is the core of the optical cable industry. The production line is a necessary process in the process of producing optical cables. On the other hand, the quality of the cable jacket plays a vital role in the overall performance of the cable. The optical cable production line control system combines many modern control technologies, including not only detection and detection technology at the executive level, fieldbus technology at the communication level, computer control technology at the monitoring level, but also DC drive technology motors. Various technologies are coordinated with each other and work together in every link of the entire production line.

References

1. Jiang, X., et al.: Design of simulation verification platform for ship sub assembly digital production line control system. J. Phys. Conf. Ser. **1650**(3), 032167 (2020). (6pp)
2. Lu, W., Yang, H., Yan, J., et al.: Design of control system for on-line ultrasonic testing device of nuclear power hollow flange bolt based on LabVIEW. ATW **64**(2), 98–102 (2019)
3. Jin, Z., Meng, L.: The design of pneumatic manipulator for optical shaft production line. J. Phys. Conf. Ser. **1633**(1), 012036 (2020). (5pp)
4. Pangaribuan, T., Sihombing, F.: Design of a digital controller for linear plant based on model following control system. In: IOP Conference Series: Materials Science and Engineering, 2018, vol. 420, no. 1, p. 012053 (2018). (5pp)
5. Qiu, M., Wang, X., Bian, H.: Design and implementation of the control system for the traction motor of electric vehicles based on AUTOSAR. Qiche Gongcheng Autom. Eng. **40**(6), 659–665 (2018)
6. Weibin, L., et al.: Design and implementation of digital control system based on PAC architecture for large-capacity pulse power supply. IEEE Trans. Plasma Sci. **47**(12), 5339–5344 (2019)
7. Wang, C., Zhang, B., Ma, X.: Design of microcomputer anti-misoperation alarm system for substation based on intelligent agent. J. Phys. Conf. Ser. **1846**(1), 012006 (2021). (6pp)
8. Tahtawi, A., Somantri, Y., Haritman, E.: Design and implementation of PID control-based FSM algorithm on line following robot. JTERA (Jurnal Teknologi Rekayasa) **1**(1), 23 (2017)
9. Nasir, M.A.: Design of polarimetric-based optical current sensor for electric power system application. In: IOP Conference Series Materials Science and Engineering, 2021, vol. 1041, no. 1, p. 012016 (2021)
10. Zhu, B.W., Wang, X.L.: Design of diffractive optical elements based on firefly algorithm. J. Optoelectron.·Laser **28**(8), 817–823 (2017)

Design and Implementation of the Financial Big Data Visualization Analysis Platform Under the Internet Background

Meilian Ge[✉]

Department of Economics and Management, Taishan University, Taian 271000, Shandong, China

Abstract. In the Internet era, financial data is expanding rapidly. In order to be able to process these data quickly, a data visualization platform came into being. Faced with today's urgent needs for fast and accurate data analysis in all walks of life, data visualization technology can meet it. It can display data intuitively and reduce the difficulty of analysis. Naturally, we cannot lag behind in the financial field. The purpose of this article is to study the design and implementation of a financial big data visualization analysis platform under the background of the Internet. This paper combines the design and implementation of data visualization analysis platforms at home and abroad, studies and learns typical visualization algorithms, and masters the preprocessing technology of data acquisition. The purpose of this article is to design and implement a flexible and intelligent financial big data visualization platform, using its own visual perception to select the key points in the visualization analysis. This intuitive way of expression reduces the difficulty of data analysis and improves the speed of data analysis. First, the overall design of the platform is introduced, and then several tools used in the development process are introduced and analyzed. This paper introduces visualization technology in the process of designing and implementing the financial big data visualization analysis platform, and develops it by using related data calling technology. At the same time, in order to ensure the reliability and stability of the system, a test experiment was carried out on the enterprise visual data analysis system. The experimental results show that: there is a parallel coordinate view for the choice of platform design algorithm, and its use frequency is the lowest, only 76%, and it is mostly used for high-dimensionality. Geometric visualization may not be suitable for financial big data analysis. The contrast stack flow view, which is used to compare and analyze the current situation and development trend of the two types of data, is more popular, with a frequency of use as high as 89%. In today's such a developed Internet background, with the rapid advancement of information technology, mass data can only be understood through visual analysis to understand the law of change and the hidden value in order to play the greatest role.

Keywords: Internet · Financial Big Data · Visualization Platform · Design and Implementation

J. C. Hung et al. (Eds.): FC 2021, LNEE 827, pp. 638–645, 2022.
https://doi.org/10.1007/978-981-16-8052-6_79

1   Introduction

In the Internet era, financial data is expanding rapidly. In order to be able to process these data quickly, cloud storage came into being [1, 2]. It can quickly obtain effective information from massive data, and meets the urgent needs of today's era [3]. At the same time, with the development of science and technology, data visualization technology was born, which can display data intuitively [4, 5]. In recent years, many effective visualization tools have been developed in all walks of life. Naturally, the financial field cannot lag behind [6, 7].

In the research on the design and implementation of the financial big data visualization analysis platform under the Internet background, many domestic and foreign scholars have studied it and achieved good results. Liu et al. proposed the first large-scale integrated semantic zoom technology Time series data visualization system-LiveRAC [8]. The source of data analyzed by the system is several attribute values of thousands of network devices. The more prominent feature is that the system has been able to compare multi-granular detail information from any combination of attributes of any device in parallel [9]. Al Mamun has improved the system and implemented an interactive electronic medical record visualization tool, which makes up for the defects of traditional query language and table display that are not intuitive, and it is difficult to find hidden patterns in the data set [10].

This paper combines the design and implementation of data visualization analysis platforms at home and abroad, studies and learns typical visualization algorithms, and masters the preprocessing technology of data acquisition. The purpose of this article is to design and implement a flexible and intelligent financial big data visualization platform, using its own visual perception to select the key points in the visualization analysis. This intuitive way of expression reduces the difficulty of data analysis and improves the speed of data analysis. First, the overall design of the platform is introduced, and then several tools used in the development process are introduced and analyzed.

2   Research on the Design and Implementation of Financial Big Data Visualization Platform Under the Internet Background

2.1   Visualization Platform Design Requirements

(1) Navigation tree display time
    After the GUI client is started or when the model is changed, the directory/file in the model will be displayed in a tree structure. The time for displaying the navigation directory usually requires less than 2 s.

Client real-time response time

In the process of interacting with the user, the response time of events such as mouse clicks to expand or collapse the directory in the navigation tree, keyboard input search conditions, etc. generally requires less than 1 s.

File opening time

Double-click the file in the navigation tree, and the file content (data table/picture/video playback) will be displayed in a new tab page. Generally, the response time should be less than 1 s.

Curve drawing time

The time for the user to select specific data drawing or mouse zoom and other operations to redraw the curve will vary depending on the amount of data. The time is required to be within the time range that the user can tolerate, usually within 5 s.

Navigation tree update delay time

Users can add/delete directories/files in the model through the navigation tree, and the synchronization update time of the navigation tree should be less than 2 s.

2.2   Visualization Tools

(1) HighCharts

It is a chart library frequently used by developers. HighCharts has the following outstanding advantages: good compatibility, HighCharts can be compatible with most browsers currently in use, such as Safari, IE and Firefox, etc.; independent of language constraints, HighCharts can be used in most web development and supports current Various mainstream languages, such as ASP, PHP, JAVA, .NET, etc.; eye-catching information prompts, in the image drawn with HighCharts, you can set implicit prompt information on some data according to actual needs. When the data point is triggered by the mouse, the written information will be displayed, which is convenient for users to understand the image, and is more conducive to their grasp of the information behind the image; external data, HighCharts can realize the dynamic loading of server data.

(2) amCharts

It is a visualization tool that mainly provides icon components. Currently supports basic commonly used graphics such as histograms, bar charts, area charts, and radars. It has the advantage of being platform-independent, so that the results of data analysis can be applied across platforms, and it also has the advantage of unlimited tags.

(3) Anychart

This tool is based on Flash technology and can provide users with basic charts and graphs. At present, it is widely used in business data analysis occasions. This tool has the advantage of being able to cross browsers, facilitates the development work of developers, and improves the work efficiency of users.

2.3   Implementation of the Visualization Platform

(1) Building an indicator model

Before constructing the index model, the data must be preprocessed. The original data is generally larger in size and contains more information, but at the same time, the problem is that it is difficult to deal with. Therefore, the original data must be cleaned up, and useless data, dirty data, and invalid data are processed to obtain usable and cleaner data. When analyzing data, it is necessary to pre-determine the characteristics. Some characteristics are obtained through statistical calculation and information entropy. After the data is processed, the index library is obtained, and the index model can be constructed through the index library.

(2) Select visualization algorithm

After the indicator model is constructed, it is necessary to select a suitable visualization algorithm to analyze the data. Each visual expression method has its own specific characteristics and adaptation scenarios. Therefore, the visualization algorithm should be selected by observing the indicator model. Through the visualization of the data, we can know the hidden knowledge and the anomalies in the data.

(3) Comparison of visualization algorithms

After the visualization algorithm is selected, perform experimental analysis on the visualization algorithm, select the correct and effective visualization algorithm that presents the information contained in the data, and discover the internal connection between the data at a glance from the selected visualization algorithm. Through comparative experiments on visualization algorithms, the visual analysis method used in this article is composed of contrast stacked flow graphs, parallel coordinate graphs, and thermal views.

(4) Visual analysis of data

Before realizing the analysis and display of the data, we must first customize the data we need, in addition to the secondary processing of the data model, and finally select the content we want to display from the data model library, and perform summary display. A global display based on operational data can be realized. Graphical information is displayed through the data table cascade selection method.

2.4   Implementation of the Main Functional Modules of the Platform

(1) System management module

This module mainly includes three sub-modules. The functions of each sub-module are described in detail below. The first one is user management: manage the registration and login information of all users. The second is menu management: manage the detailed information of all menus of the system, including number, menu name, menu address, order, and superior menu. The third is permission management: different users are assigned different permission levels. Basic operations on user information can be realized.

(2) Data statistical analysis module

This module mainly analyzes the data of the database. By categorizing each attribute of the data in the database, it is displayed in a visual form. Classify different data so that the administrator can clearly see the proportion of the number of users of different protocols.

(3) Data visualization display module
Managers can view data in real time through this module, have an intuitive review of the results, and then discover problems and improve their level.

2.5   Visualization Algorithm

(1) K-means algorithm
The basic idea of the K-means algorithm is to divide the selected n data objects into k clusters according to certain criteria. Through the clustering process, the distance within the cluster is as small as possible, and the distance between clusters is as large as possible. In the process of clustering, the criterion for judging when to end is until the standard measure function starts to converge. In the actual operation of the algorithm, the mean square error is generally selected as the standard measurement function, which is defined as follows:

$$E = \sum_{i=1}^{k} \sum_{p \in c_i} |p - m_i|^2 \tag{1}$$

Where E is defined as the sum of the mean square deviations of all objects; p represents a point in the space of the object; is the mean value of the cluster (p and are multi-dimensional).

(2) Improvement of K-means algorithm
In view of the shortcomings of the K-means algorithm, many researchers have put forward their own improvement methods after studying these problems in depth. The clustering stability is improved, running time is saved, and time complexity is reduced. The following is the research on the improved algorithm. The individual contour coefficient is a branch concept contained therein, and the expression of the contour coefficient S is as follows:

$$S = \frac{1}{n} \sum_{i=1}^{n} s_i \tag{2}$$

3   Experimental Research on the Design and Implementation of Financial Big Data Visualization Platform Under the Background of the Internet

3.1   Experimental Subjects and Methods

This article takes the data visualization algorithm and the functional requirements of the data visualization platform design as the experimental objects, through the investigation of relevant personnel in the financial field, and analyzes and summarizes the recovered data.

3.2 Data Collection

In the context of the Internet, more storage devices and faster I/O mechanisms are needed to access data. In order to achieve more convenient and rapid access to data in big data applications, academia and industry continue to propose new technologies and methods to shorten the gap between high-performance CPU and low-speed I/O. Secondly, although solid-state drives achieve efficient random I/O and alleviate storage difficulties, they cannot achieve efficient sequential I/O at the same time. High-efficiency and low-cost storage devices are still in the continuous development process.

4 Experimental Research and Analysis on the Design and Implementation of the Financial Big Data Visualization Platform under the Internet Background

4.1 Comparison of Data Visualization Algorithms

The purpose of this experiment is to select an optimal algorithm suitable for the design and implementation of a financial big data visualization platform. Data visualization algorithms have their own characteristics and the most suitable places for analysis. Through surveys of relevant talents in the financial field, the use frequency of visualization algorithms such as parallel coordinate view, contrast stack flow view, thermal view, and bubble view are analyzed respectively. The experimental results are shown in Table 1:

Table 1. Frequency of use of visualization algorithms

	Relevant talents in the financial field
Parallel coordinate view	76%
Contrast against stacked flow view	89%
Thermal view	90%
Bubble view	80%

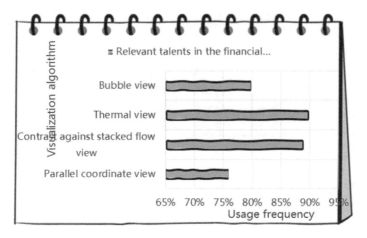

Fig. 1. Frequency of use of visualization algorithms

As shown in Fig. 1: The use frequency of parallel coordinate views is the lowest, only 76%. It is mostly used for the visualization of high-dimensional geometry, and may not be suitable for financial big data analysis. The contrast stack flow view, which is used to compare and analyze the status quo and development trend of the two types of data, is more popular, and is used as high as 89%.

4.2   Functional Requirements Analysis

The purpose of this experiment is to understand the functional requirements of the design and implementation of the financial big data analysis platform, taking some data visualization platforms as the research object, analyzing the key points of their needs, and giving help in the design of this platform. The experimental results are shown in Table 2:

Table 2. Platform design functional requirements

	High performance	Timeliness	Ease of use	Automation	Maintainability
Data visualization platform A	36%	35%	28%	44%	30%
Data visualization platform B	30%	25%	33%	19%	29%
Data visualization platform C	34%	40%	39%	37%	41%

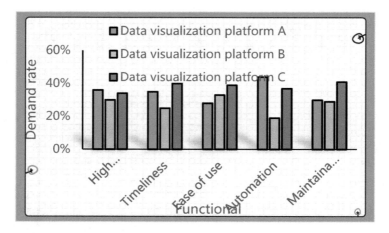

Fig. 2. Platform design functional requirements

As shown in Fig. 2, the most automated data visualization platform A accounts for 44%, and the most average functionality is data visualization platform C, with an average demand accounting for about 38%. In conclusion, the design of this platform should be close to platform C, or even better.

5 Conclusions

In the era of data explosion, how to obtain effective information from massive data has become a major challenge now. Data visualization technology allows people to make full use of their own visual perception and select key points in visual analysis. This intuitive way of expression reduces the difficulty of data analysis and improves the speed of data analysis. Therefore, visualization technology has received more and more attention. This paper studies the design and research of the financial big data visualization analysis platform under the background of the Internet. The purpose is to provide a more convenient and effective platform for the work of financial industry practitioners through the visualization and analysis of financial data. Therefore, in the future, we can study the application of data mining technology to the realization of visualization to assist users in analyzing and predicting the knowledge in the data, and to conduct a comprehensive and in-depth analysis and exploration of the data.

References

1. Větrovsk, T., Petr, B., Daniel, M.: SEED 2: a user-friendly platform for amplicon high-throughput sequencing data analyses. Bioinformatics (13), 13 (2018). https://doi.org/10.10 93/bioinformatics/bty071
2. Patterson, E., et al.: Dataflow representation of data analyses: toward a platform for collaborative data science. IBM J. Res. Dev. **61**(6), 9:1–9:13 (2017)
3. Andres, V., Alan, M., Ari, L.: Wasabi: an integrated platform for evolutionary sequence analysis and data visualization. Mol. Biol. Evol. **4**, 1126–1130 (2016)
4. Giuseppe, D.D., et al.: THU0608 visualization and analysis platform in the swedish rheumatology register for real-time data feedback. Ann. Rheum. Dis. **75**(Suppl 2), 412.1–412 (2016)
5. Borrill, P., Ramirezgonzalez, R., Uauy, C.: expVIP: a customizable RNA-seq data analysis and visualization platform. Plant Physiol. **170**(4), 2172 (2016)
6. Zhu, Q., Fisher, S.A., Dueck, H., et al.: PIVOT: platform for interactive analysis and visualization of transcriptomics data. BMC Bioinformatics **19**(1), 6 (2018)
7. Long, Z., Saddik, A.R, NeedFull, A.E: A Tweet analysis platform to study human needs during the COVID-19 pandemic in New York State. IEEE Access **PP**(99), 1 (2020)
8. Liu, J., Gao, Y., Shan, G., Chi, X.: VASEM: visual analytics system for electron microscopy data bank. J. Vis. **22**(6), 1145–1159 (2019). https://doi.org/10.1007/s12650-019-00597-y
9. Li, W., Chen, G., Liao X.: Countermeasures of Chinese traditional commercial banks to meet the challenges of internet finance based on big data analysis–evidence from ICBC. J. Phys. Conf. Ser. **1648**(3), 032066 (2020). (8pp)
10. Al Mamun, M.A., et al.: Theoretical model and implementation of a real time intelligent bin status monitoring system using rule based decision algorithms. Expert Syst. Appl. **48** (Apr.15), 76–88 (2016)

Analysis of Transportation Risk of International Crude Oil Trade Based on RK Neural Network Technology

Xiangqin Ni[✉]

Chengdu College of University of Electronic Science and Technology of China, Chengdu 611731, Sichuan, China

Abstract. Data is everywhere, analysis is everywhere, and it is the vision of various foreign trade companies to maximize profits through data mining and analysis. Use mathematical analysis methods to find valuable information from the chaotic mass data, use different algorithms to combine business processes, and create automated data calculation modules and friendly link modules to form an automated operating system. The rapid economic growth and the acceleration of the process of industrialization have led to the growth of our country's energy demand. The proportion of crude oil resources in our country's energy structure has steadily increased, and our country's crude oil imports have increased rapidly. This article takes the transportation risk of international crude oil trade as the research theme. First, it introduces the status quo of our country's international crude oil trade, and then uses the RK neural network technology in big data to design a transportation risk system model for international crude oil trade, and finally adopts the model. Corresponding data analysis was made on the correlation of our country's international crude oil trade routes and pirate attack factors, and corresponding transportation improvement measures were derived based on the results of the analysis.

Keywords: Big data · International crude oil trade · RK Neural Network Technology · Risk analysis

1 Introduction

Since the introduction of big data, the research of big data by scientific researchers has increased day by day, and the application system of big data has become more and more mature, bringing convenience to various industries and people's lives in society. Big data has gradually spread to the public's consciousness, allowing people to feel the power of technology and enjoy the fruits of technology. In the mid-1990s, our country changed from a net exporter of crude oil to a net importer of crude oil. For more than ten years, the national economy has grown at an annual rate of more than 8%, and our country's demand for petroleum resources has become increasingly serious [1]. In the 21st century, the growth rate of our country's crude oil consumption has almost doubled. At the same time, our country's crude oil output increased by only 26 million tons, an increase of only 16%. However, the gap between crude oil supply and demand has rapidly increased by 155 million tons. However, oil is currently the most important energy source in the world, and its alternative varieties have rapidly expanded to 215 million tons. However, oil is currently the most important energy source in the world. Its alternative varieties have not completely replaced oil. Therefore, our country must import crude oil through international crude oil trade and transport crude oil to the country through various transportation methods. Therefore, crude oil transportation is also an important part of ensuring energy security [2]. However, as far as crude oil imports are concerned, our country mainly transports imported crude oil by sea, and there are many unpredictable factors in the transportation process. Nowadays, there are few researches on big data in the risk of international crude oil trade transportation, but the power of big data technology is very huge. It is very necessary to use big data technology to serve the transportation risk analysis system of crude oil trade [3].

Chinese scholar Zhu Kongchao and others believe that the security of crude oil supply occupies an important position in China's national security. The crude oil resource status of the supplier country, transportation risks, and political risks are the main obstacles restricting the security of supply in most countries. To prevent risks, China needs to persist in multilateral development. "Energy diplomacy", expansion of strategic oil reserves and other policy policies [4]. Zheng Guofu believes that as China's economy continues to rise, energy demand is becoming increasingly strong, crude oil is known as the "industrial blood", and there is still a huge increase in the quantity of China's crude oil import trade. High dependence on foreign sources, unreasonable source structure, and major producing areas. Political turmoil, significant risks along the transportation route, and high import unit prices [5]. Chang Chang believes that with the rapid development of the national economy, people's living standards have greatly improved, and their consumption ability in various fields has greatly increased, and people have begun to pursue a higher quality of life. Industrial production capacity has also been rapidly improved. As one of the important pillar industries of the national economy such as transportation, electronic information and national defense, the petrochemical industry has received much attention, and petrochemical products have a relatively high proportion of the entire trade market. However, due to the international environment and the domestic economic development status and the overall environment of the industry, there are certain risks in the domestic trade of petrochemical products. This requires the national government and relevant petrochemical companies to timely adjust their strategic decisions and formulate corresponding risk early warning mechanisms to avoid risks in time and lay a solid foundation for the healthy and stable development of the petrochemical industry [6].

In the past, international crude oil trade research mainly used methods such as supply and demand and economic growth. This article takes the risk of developing international crude oil trade as a pioneer and develops a unique method to analyze our country's crude oil trade [7]. This article first studies the current situation of our country's crude oil trade, then uses big data to design a transportation risk system model, and finally uses the designed risk system model to analyze the transportation risks of certain aspects of China's crude oil import trade, and make the results of the analysis. Formed tables and images, conducted a vivid and objective analysis of transportation risks, reached relevant conclusions, and put forward relevant policy recommendations based on the conclusions.

2   The Status Quo of Transportation Risk Analysis of International Crude Oil Trade Based on Big Data

2.1   The Status Quo of China's Crude Oil Trade

In the context of high oil prices in recent years, global crude oil trade flows can be divided into three types: strategic orientation, countries choose trading partners, regions and trade volumes according to their long-term development strategies; it is estimated that if the transportation cost is in line with the country's industrial economic development. According to the requirements of the plan, the trade mode is chosen; it is geographically oriented, and when expanding trading partners, more attention will be paid to countries with narrow geographic locations or geographic advantages as potential actual trading partners [8, 9]. By 2020, our country's crude oil imports will continue to grow for 10 years, and imported crude oil will exceed 250 million tons. The reason for the continued increase in crude oil imports is mainly due to the reduction in our country's crude oil production and the commissioning of new refineries. In terms of crude oil imports, our country's crude oil imports from the Middle East and Africa accounted for 76% of total imports in 2020, and the share of imports from the Middle East declined slightly, but still accounted for 39%. Crude oil trade mainly includes spot contracts, futures, stocks and long-term contracts. our country's participation in international crude oil trade is mainly spot spot, which accounts for more than half of the spot transaction volume, with less participation in other ways [10].

2.2   The Supply and Demand Gap and Forecast of Crude Oil

Foreign scholars have analyzed the economic growth of the United States in the past ten years through empirical analysis, and found that there is a strong correlation between energy use and GDP, and the ratio of energy to GDP has changed a lot. Directly or indirectly affected by changes in energy structure [11]. Foreign researchers used the distribution correlation test to find that there is an integrated relationship between the energy consumption and economic growth of seven countries on the African continent, and based on the error, the energy consumption of the four countries has a long-term positive impact on economic growth. The causality of the correction model the relationship test also found two-way causality. He studied the relationship between the economies of the United States, Europe and other countries and international oil prices, and found an asymmetric co-integration relationship, and found that rising oil prices slowed down the economy more strongly than falling oil prices [12]. Regarding the forecast of our country's crude oil demand, domestic researchers have established relevant models and combined with recent statistical data to predict our country's crude oil imports. By 2025, our country's crude oil imports are expected to reach 42.637 million tons. From the perspective of our country's crude oil demand structure, the impact of the three major industries on crude oil demand is analyzed, and crude oil consumption is predicted. By 2025, our country's crude oil consumption will reach 612 million tons.

2.3   Crude Oil Import Trade and Transportation

Due to the rapid growth of our country's crude oil imports, domestic crude oil consumption is increasingly dependent on the international crude oil trade market, and the research on crude oil transportation has also attracted the attention of scholars. In view of the current situation of our country's import of crude oil from the sea, the transportation routes and the conditions for China to import crude oil, the development plan of our country's state-owned shipping companies has put forward new proposals. In terms of crude oil pipeline transportation, the reasons for the delay in the construction of the China-Myanmar crude oil pipeline were analyzed from the perspective of stimulating the economic development of the pipeline to the west and the diversification of crude oil transportation methods. Regarding the safety and risk of oil transportation, study the current situation of crude oil in the United States and ask questions, establish a mathematical model for oil spill risk analysis, design relevant methods that are helpful for decision-making and research, and propose relevant measures to reduce risks.

3 Design of Transportation Risk System of International Crude Oil Trade Based on Big Data

3.1 Construction of Risk Index System for Transportation System

The international crude oil trade and transportation system is a complex dynamic system. The transfer process involves many risks. Establishing a scientific, reasonable, comprehensive and feasible risk indicator system is very important for assessing the risks of the transmission system. In order to comply with the actual situation of the international crude oil trade and transportation system, certain principles must be followed when establishing a risk indicator system. The principle of system optimization, the number of indicators and the structure of the structure, not only must understand the potential risk factors of the international crude oil trading system in many aspects and depth, but also must consider the interdependence of various aspects, and also seek to optimize the overall risk indicators system. Scientific principles are the most basic requirement for constructing any evaluation system. That is to say, the risk indicators of the international crude oil trade and transportation system must be combined with theory and practice, but not only must meet the objective elements of the evaluation object, but also meet the needs of theoretical standards. Only by adhering to scientific principles can we guarantee the practicability and value of the risk indicator system for promoting the international crude oil trading system; the practicability principle refers to simplifying the selection of indicators for evaluation objects and simplifying evaluation methods. It is easy to implement, and the rating index data is easy to obtain, and the entire operation process should be standardized. Otherwise, it will be difficult to continue the evaluation or make the work difficult; the principle of comparability includes horizontal comparison and vertical comparison. Level comparison is a comparison between different objects, which can be compared with the weight distribution method. Vertical comparison is the comparison of the same object in different time periods, and the index is variable.

3.2 Risk Assessment Model of International Crude Oil Trade Transportation System Based on RK Neural Network

Transfer risk assessment methods usually use traditional methods, such as analytic hierarchy process and fuzzy overall assessment, and the calculations of these methods are very complicated. As an intelligent evaluation method, the RK model has the advantages of anti-interference, high efficiency and accurate evaluation. At present, the model has been applied in many fields, but has not been applied to the risk assessment of international trade and crude oil transportation. The RK neural network is a multi-level feedforward neural network model trained in the error propagation algorithm, and its structure is similar to a multi-step perceptron. Because of its simple structure, strong plasticity, strong versatility and good functionality, it has been widely used. However, it also has disadvantages, such as slow convergence, over-adaptation, and difficulty in obtaining the best local solution. The RK algorithm belongs to the minimum variance algorithm. The basic idea is to input the training sample data into the neural network, and use the error between the actual network output data and the target output data to correct the network weights so that the actual network output is as close to the Output target as possible, so the target output and the actual output are between The mean square error of tends to be minimized. Use the particle output of the output layer and the output of the hidden layer to correct the weights of the hidden layer and the output layer to perform the backward propagation process:

$$w_{ho}^{n+1}(k) = w_{ho}^{n}(k) + \alpha \delta_0(k) ho(k) \tag{1}$$

The hidden layer and the vector derivative of the input sample are used to correct the weights of the input layer and the hidden layer. The reverse forwarding process is as follows:

$$w_{ih}^{n+1}(k) = w_{ih}^{n}(k) + \beta \delta_h(k) x_i(k) \tag{2}$$

3.3   Determination of the Number of Neurons in Each Layer

The input level is a bridge connecting external information and the RK neural network. The number of neurons in the input layer is determined by the dimension of the input vector. It will include the normal quality of transportation personnel, the psychological quality of transportation personnel, the professional quality of transportation personnel, and the professional ethics of transportation personnel and transportation personnel. Seventeen sub-indices including personnel quality, transportation safety, transportation equipment maintenance, road factors, natural environment and surrounding traffic environment are used as input nodes of the neural network model. Therefore, the number of input nodes of the RK neural network evaluation model is 17. The number of hidden layer neurons is usually obtained through experiments. If the number of neurons in the hidden layer is too small, the evaluation model lacks the ability to receive information. Generally, an empirical formula is used to determine the optimal number of neurons in the lattice:

$$H = 0.51 + \sqrt{0.43IO + 0.12I^2 + 2.54O + 0.77I + 0.35} \tag{3}$$

The adjustment of the number of neurons in the output layer of the RK neural network is relatively flexible and can be determined according to the actual situation of the research object. Because the expected result of the evaluation model is the risk evaluation result of the open-pit mining route, the evaluation result of each route is represented by a risk level, and each risk level is coded as a corresponding target result.

4 Analysis of Risk Results of China's Crude Oil Import Trade and Transportation Based on Big Data

4.1 Relevance and Ranking of China's Crude Oil Import Trade Routes

Table 1. Relevance and ranking of trade routes

Transportation route	Degree of relevance	Sequence
Middle East routes	0.9193	1
South American routes	0.4895	4
North Africa routes	0.6372	2
East African routes	0.6210	3

Fig. 1. Relevance of trade routes

According to Table 1 and Fig. 1, it can be known that the routes of East Africa and North Africa have roughly the same degree of correlation. Since they are located in Africa, their relative voltage curves are roughly the same. The lag in political and economic development has led to a generally high port risk factor; the distance is average, so the relative route risk factor is significantly lower than that of the South American route. Close to the risk level of the Middle East route; the most important risk factor for the routes in the African region is the risk of pirate attacks, while the risk of pirate attacks on the North African routes has increased sharply. In view of the fierce competition for oil resources in the Middle East, countries have strengthened their resource strategies. The cost of acquiring oil in the Middle East will inevitably increase. Although the risk of shipping to Africa is higher than that of the Middle East, it is not very obvious.

4.2 China's Crude Oil Import Trade Has Been Affected by Pirate Attacks

Table 2. Statistics of pirate attacks from 2016 to 2020

Year	Pirate attacks
2016	236
2017	185
2018	194
2019	218
2020	203

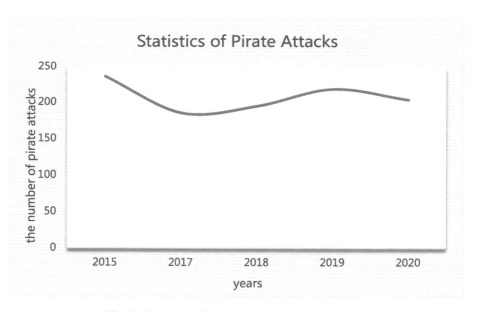

Fig. 2. Statistics of pirate attacks from 2016 to 2020

According to Table 2 and Fig. 2, in the past five years, China's crude oil import trade has been attacked by pirates on average more than 200 times. It can be seen that the risk of piracy has become the main factor affecting the risk of foreign oil transportation. Compared with other modes of transportation, the cost of transporting oil abroad is relatively low, but at the same time, the risk is the highest. In some geo-economic and politically unstable areas, piracy is common, seriously threatening the safety of oil tankers. The Somali Peninsula, the Gulf of Aden, the Strait of Malacca and other nearby waters have become one of the most uncontrolled areas by pirates. If the risk of piracy can be reduced reasonably and effectively, then the African region will be a strong growth point for oil imports.

5   Conclusions

Crude oil resources continue to play a leading role in the current international energy strategy. Safe and sufficient crude oil resources are a prerequisite for economic growth. As our country's economy continues to grow, its dependence on crude oil imports continues to increase. The risk assessment of China's international crude oil trade and transportation has become more and more important. This article first conducts an in-depth study of the current status and forecast of China's international crude oil trade, then uses the RK neural network in big data technology to design an international crude oil trade transportation risk system model, and finally uses the model to correlate with China's crude oil import trade routes. Conducted a risk analysis with the pirate attacks in the past five years, and proposed corresponding trade improvement measures based on the results of the analysis.

References

1. Jiaxin, L.: Research on risk analysis and avoidance measures of foreign trade enterprises—taking Fuyao glass group as an example. Mod. Bus. Trade Ind. **41**(19), 28–29 (2020)
2. Jinqiu, H.: Risks and countermeasures of China's container manufacturing and export under the new situation of international trade. Future Dev. **43**(01), 70–74 (2019)
3. Zhang, Y.: Analysis and countermeasures of Cargo rights issues in international trade ocean logistics transportation. Railw. Purch. Logist. **15**(164(05)), 29–32 (2020)
4. Kongchao, Z., Shuwen, N., Yuan, Z., Xin, Q.: Quantitative assessment of the supply security of the country of origin of China's crude oil imports. J. Nat. Resour. **35**(11), 63–78 (2020)
5. Guofu, Z.: The status quo, problems and improvement of China's crude oil import trade development–Taking the data from 2001 to 2018 as an example. Foreign Econ. Relat. Trade Pract. (005), 72–74 (2019)
6. Chang, C.: Domestic trade risk analysis and preventive measures of petrochemical products. China Pet. Chem. Stand. Qual. **039**(017), 130–131 (2019)
7. Wei, W.: Transnational logistics and transportation project risks and management countermeasures. Enterp. Reform Manag. **369**(04), 30–31 (2020)
8. Wang, Y.: A safety net for trade activities. China Foreign Exch. **406**(16), 18–19 (2020)
9. Chen, B.: Thinking on the writing of the header of the ocean bill of lading triggered by a case. Foreign Econ. Relat. Trade **316**(10), 24–26+55 (2020)

10. Yang, H.: Research on Taxation Risk Management of International Freight Forwarders Co., Ltd. Public Investment Guide, vol. 325, no. 05, pp. 139–141 (2019)
11. Liu, H.Q., Yu, J.: Sino-Russian economic and trade cooperation in the changing world. Eurasian Econ. **238**(01), 7–25+131+133 (2019)
12. Liang, J.: The design of container information management system based on blockchain. Containerization **31**(347(07)), 7–10 (2020)

Hybrid Storage System Planning for Power Quality Improvement in Power Distribution System with Solar Photovoltaic Sources

Sichao Chen[1(✉)], Bin Yu[2], Liguo Weng[2], Guohua Zhou[1], and Rongjie Han[2]

[1] StateGird Zhejiang Hangzhou Xiaoshan Power Supply Co., Ltd., Hangzhou, Zhejiang, China
[2] Zhejiang Zhongxin Power Engineering Construction Co., Ltd., Hangzhou, Zhejiang, China

Abstract. The distributed photovoltaic sources have demonstrated the benefits of clean and environmental protection, flexible construction, nearby utilization and small impact on the power grid. It is considered of paramount importance to optimize the energy provision and efficiency as well as the emission reduction to realize the low carbon energy provision. In this paper, an algorithmic solution for sitting and sizing of hybrid storage systems in power distribution systems has been developed and presented. The proposed solution is evaluated through the IEEE-33 test network, and the numerical result clearly confirmed the effectiveness of the proposed solution in appropriately allocating the energy storage systems for improving the power quality in active distribution systems with distributed solar energy resources.

Keywords: Hybrid storage system · Power distribution systems · Power quality · Photovoltaic power generation

1 Introduction

New green energy has the characteristics of green environmental protection, which can effectively avoid the disadvantages of traditional energy combustion, but there are still many problems. In addition to nuclear energy, wind energy and light energy cannot provide stable energy output like coal, and it is difficult to control, so it is easy to affect the operation of the power grid once the renewable resources are connected with the grid. Even in island operation mode, it cannot provide stable power to the load. With the development of power electronics technology and modern control theory, microgrid as effective access to distributed generation and organization management has become an effective way to solve the above defects.

As widely used energy storage equipment, the battery has a large energy density, which can provide more energy. However, it should be noted that the power density of the battery is generally small, and the flexibility of power charging and discharging are not satisfactory in practice. This indicates that, under the condition of sudden load variations, the battery cannot release or absorb the target power as quickly as expected, and the dynamic performance of the system is poor. On the contrary, the super-capacitor has a smaller energy density but higher power density, which can discharge power energy in a very short time. The composite energy storage system composed of a battery and supercapacitor improves the overall performance and enables fast response and large stable power supply.

With the rapid and massive deployment of distributed photovoltaic power sources in the medium and low voltage power distribution networks, the traditional distribution network has been transformed into an active distribution network [1–4]. The battery has been developed and adopted for years and are considered can meet the energy density requirements. However, due to the limited electrochemical reaction rate, the battery power density is often considered small, and hence cannot absorb or release the power as expected. Under the condition of load demand variations, the batteries cannot meet the dynamic requirements of the system. During the charging and discharging of the supercapacitor, the internal changes are physical, which has the characteristics of high power density. It can provide large power in a short time and provide a buffer for other equipment, but the energy density is relatively low. Therefore, the supercapacitors and batteries have been considered to work in a complementary manner. These two types of energy storage units can be connected to form a composite energy storage system. This can provide the benefits of both types of energy storage units to obtain better performance. On the one hand, the massive connection of distributed photovoltaic sources to the distribution network, there will be large power instantaneous fluctuation in the distribution network. The supercapacitor provides short-time power grid energy support; On the one hand, for the long-time power fluctuation of the distribution network, the battery provides the power grid energy support. On the one hand, the energy storage system can provide reactive power compensation and improve power quality when there are power quality problems or reactive power shortages in the distribution network area. Especially after the popularization and application of the new generation of the intelligent distribution automation system, the access of low-voltage distribution network terminal devices has gradually become a trend. This work takes this as the starting point to develop the integrated device of composite energy storage and reactive power compensation, aiming to address the aforementioned technical challenge induced by the massive penetration of high power density distributed generation.

It is widely agreed that the massive connection and access to distributed photovoltaic power can directly bring a great impact on the power quality of the power supply (e.g., [5–10]). In practice, the transformer capacity in the distribution network area is limited, and the ability to resist the interference of large load impact and the ability of instantaneous large power fluctuation is weak, especially in the power grid under the condition of high power density distributed photovoltaic access, it is difficult to meet the impact of large instantaneous power fluctuation of distribution network. In this paper, the main technical contributions are as follows:

(1) This paper proposes an optimization model with multiple objectives for location planning of energy storage system.
(2) The corresponding algorithm is developed to implement the location of energy storage systems and ensure that the system power fluctuation is in the optimal range. Taking the node voltage as power quality index, the power quality is exploited and evaluated through quantitative analysis.

2   Hybrid Storage Systems and Configuration

The typical composite energy storage for the power distribution system with photovoltaic power generation sources is shown in Fig. 1. The photovoltaic module plays the role of energy conversion to convert the solar energy into electric energy. The unidirectional DC-DC converter can boost photovoltaic power generation unit and can work in the maximum power point tracking mode to maximize resource utilization. The bidirectional DC-DC converter can control the charging/discharging actions of the energy storage system. It can convert direct current into alternating current, and the transformer can convert the alternating current into suitable voltage to supply power for the load. Also, the system is connected to the power grid through a common connection point. In this section, the independent operation mode of composite energy storage photovoltaic power generation system is considered.

The power quality degradation induced by the massive penetration of distributed photovoltaic sources mainly includes voltage overrun and harmonic pollution. The sitting and sizing of the composite energy storage systems need to fully consider the economy of the composite energy storage system and the effect of improving power quality.

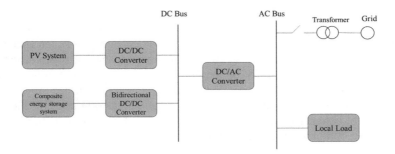

Fig. 1. Composite energy storage system in the photovoltaic power generation system

3   Optimal Planning of Hybrid Storage in Distribution Systems

3.1   Objective Function

3.1.1   Node Voltage Deviation

The high photovoltaic output will lead to the voltage limit of some nodes. Therefore, this paper selects the sum of node voltage deviation as one of the objective functions of location and capacity determination of composite energy storage system as follows:

$$f_U = \sum_{i=1}^{N} \sum_{j=1}^{T} |U_{ij} - U_N| \tag{1}$$

Here, N and T are the number of system nodes detection time; U_{ij} is the voltage of i node at time j, and U_N is the rated system voltage.

3.1.2 Total Harmonic Distortion Rate of Current

The photovoltaic sources require a large number of power electronic inverters connected to the distribution network. As a result, the total harmonic distortion rate of the current is considered that can be formulated as follows:

$$f_{THD_1} = \frac{\sqrt{\sum_{h=2}^{M} I_h^2}}{I_1} \times 100\% \tag{2}$$

Where I_h and I_1 are the root mean square of the hth harmonic current and the fundamental current.

3.1.3 Cost of Composite Energy Storage System

The cost and the power quality management need to be considered in the composite energy storage system. The total cost is formulated as (3)

$$f_P = c_1 \frac{P_c}{\eta_1} + c_2 \frac{P_b}{\eta_2} + c_m P_{sum} \tag{3}$$

Where, P_c and P_b is the capacity of supercapacitor and battery to be configured in the system; η_1 and η_2 represent the energy conversion efficiency of supercapacitor and battery, respectively; c_1 and c_2 are the price of the unit capacity of supercapacitor and battery; P_{sum} is the total capacity of the composite energy storage system, and c_m represents the daily maintenance cost of the composite energy storage system.

Therefore, the node voltage deviation, total harmonic distortion rate of the current and total cost of composite energy storage are considered in the formulation of the the multi-objective optimization model, that can be expressed in (4).

$$\begin{cases} \min F = \alpha f_U + \beta f_{THD_I} + \gamma f_P \\ \alpha + \beta + \gamma = 1 \end{cases} \tag{4}$$

Here, α, β and γ are the normalized weight coefficients.

3.2 Constraints

3.2.1 Node Voltage

The voltage of each node in the system should be maintained within a certain range

$$V_{\min} \leq V_i \leq V_{\max} \tag{5}$$

Where, V_i is the voltage of the i node, and $V_{\min}$ and $V_{\max}$ are the lower limit and upper limit of the system voltage, which are determined by the system's requirements for voltage overrun.

3.2.2   Power Balance

Power balance should be maintained in the system at any time, that is:

$$P_S = \sum_{i=1}^{N} P_{Li} - \sum_{j=1}^{L} P_{DG_j} - \sum_{k=1}^{M} P_{storek} \qquad (6)$$

Among them, P_S is the total power of input distribution network, P_{Li} is the load power of ith node; P_{DG_j} is the distributed photovoltaic output of the jth node; P_{storek} is the capacity of the kth composite energy storage system, and is positive when the energy storage is discharged.

3.2.3   Composite Energy Storage Power Constraint

There are upper and lower limits for the power provided by the hybrid energy storage system

$$P_{oc\_\min} \leq P_{oc\_i} \leq P_{oc\_\max} \qquad (7)$$

Among them, $P_{oc\_\min}$ and $P_{oc\_\max}$ are the lower and upper limits of the output power of composite energy storage systems.

4   Experiments and Numerical Results

This work adopted the IEEE-33 bus test network for the performance assessment and analysis. In the test network, the total active and reactive loads are 3715.0 kW and 2300.0 kvar, respectively. The system voltage reference value is 12.66 kV. The installation location of distributed photovoltaic sources is shown in Fig. 2. It is assumed that all the nodes are connected with the distributed photovoltaic sources with a capacity of 300 kW.

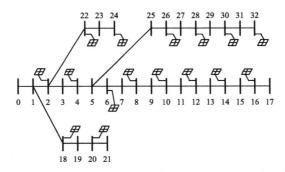

Fig. 2. Experimental IEEE-33 node distribution system

In the simulation of this paper, in principle, the composite energy storage system is allowed to access 1–32 nodes. On the one hand, the composite energy storage system composed of the supercapacitor and lead-acid battery uses the rapid response ability of the supercapacitor to meet the high-power fluctuation, on the other hand, it uses the relatively low cost of the lead-acid battery to stabilize the system fluctuation and control harmonics. Considering the node voltage deviation, total harmonic distortion rate of the current and total cost of composite energy storage system, the node sensitivity analysis method is applied to select four node-set including {25, 30, 13, 20} The maximum installed power is 600 kW.

In this study, for {25, 30, 13, 20} and the other four nodes, the improved particle swarm optimization algorithm is implemented to adopted to obtained the composite energy storage installation location and capacity. The numerical results are presented in Table 1.

Table 1. Installation position and capacity of composite energy storage device

Node	Rated power/kW
25	570
30	554

Further, the power quality control indexes of the distribution network before and after the installation of composite energy storage devices in the test network are calculated, as shown in Table 2.

Table 2. Power quality control index of distribution network

Power quality measures	Before storage installation	After storage installation
Node voltage deviation/pu	1.1570	0.9203
Total harmonic distortion rate of current/%	0.6815	0.2516

5 Conclusions

This paper proposed an algorithmic solution for sitting and sizing of hybrid storage systems in power distribution systems. Through the study, it can be concluded that the composite energy storage cannot only effectively solve the voltage overrun caused by distributed photovoltaic, but also suppress harmonics, which can meet the demand of high power density distributed photovoltaic access to the distribution networks.

Acknowledgments. This work was supported by the research of operation control technology for multifunctional energy storage systems.

References

1. Paatero, J.V., Lund, P.D.: Effects of large-scale photovoltaic power integration on electricity distribution networks. Renew. Energy **32**(2), 216–234 (2007)
2. Brekken, T.K.A., Yokochi, A., Jouanne, A.V., et al.: Optimal energy storage sizing and control for wind power applications. IEEE Trans. Sustain. Energy **2**(1), 69–77 (2011)
3. Teleke, S., Baran, M.E., Bhattacharya, S., et al.: Optimal control of battery energy storage for wind farm dispatching. IEEE Trans. Energy Convers. **25**(3), 787–794 (2010)
4. Demirören, A., Zeynelgil, H.L., Sengör, S.N.: The application of neural network controller to power system with SMES for transient stability enhancement. Int. Trans. Electr. Energy Syst. **16**(6), 629–646 (2006)
5. Kanchev, H., Lu, D., Colas, F., Lazarov, V., Francois, B.: Energy management and operational planning of a microgrid with a PV-based active generator for smart grid applications. IEEE Trans. Ind. Electron. **58**(10), 4583–4592 (2011)
6. Mohamed, F.A., Koivo, H.N.: System modelling and online optimal management of microgrid using mesh adaptive direct search. Int. J. Electr. Power Energy Syst. **32**(5), 398–407 (2010)
7. Nunna, H.K., Doolla, S.: Energy management in microgrids using demand response and distributed storage–a multiagent approach. IEEE Trans. Power Del. **28**(2), 939–947 (2013)
8. Jiang, Q., Xue, M., Geng, G.: Energy management of microgrid in grid-connected and stand-alone modes. IEEE Trans. Power Syst. **28**(3), 3380–3389 (2013)
9. Mahmoodi, M., Shamsi, P., Fahimi, B.: Economic dispatch of a hybrid microgrid with distributed energy storage. IEEE Trans. Smart Grid **6**(6), 2607–2614 (2015)
10. Vasilj, J., et al.: Day-ahead scheduling and real-time economic MPC of CHP unit in microgrid with smart buildings. IEEE Trans. Smart Grid **10**(2), 1992–2001 (2019)

Monthly Power Consumption Forecast of the Whole Society Based on Mixed Data Sampling Model

Shanshan Wu[1(✉)], Xiang Wang[1], and Hengyue Hou[2]

[1] State Grid Energy Research Institute, Beijing, China
[2] School of Statistics, Renmin University of China, Beijing, China

Abstract. Based on the daily power generation data, this paper constructs a combined model to forecast the monthly power consumption of the whole society. For the linear part of power consumption, mixed data sampling (MIDAS) model is used for prediction, while for the non-linear part, the autoregressive moving average (ARMA) model is used to forecast, and the results of the two parts are added together to get the final results. The research shows that the daily power generation has a high accuracy in forecasting the monthly power consumption of the whole society, and it can also ensure the time advance, which is of great significance for guaranteeing the power supply and studying the economic trend.

Keywords: Mixed data · Combined model · Power consumption forecast

1 Introduction

Power is closely related to economy. Power supply is an important support for economic development, and the trend of power consumption is also regarded as a barometer and thermometer of economic operation (Tan Xiandong, 2017 [1]). At present, the COVID-19 epidemic prevention and control situation in China continues to improve. In order to monitor the economic recovery situation in a more timely and accurate manner, meet the demand for electricity, and ensure the stable operation of the economy and society, it is urgent to grasp the power demand dynamics in time. The power consumption data of the whole society published by the government has a certain time lag. Based on the daily power generation data, this paper combines the mixed data sampling model and the time series model to realize accurate and stable prediction of the monthly power consumption of the whole society in China.

2 Literature Review

Wang Wensheng et al. took the annual data of GNP, the added value of the primary, secondary and tertiary industries and the total population of Sichuan Province as the independent variables, the annual power consumption of Sichuan Province as the

J. C. Hung et al. (Eds.): FC 2021, LNEE 827, pp. 662–668, 2022.
https://doi.org/10.1007/978-981-16-8052-6_82

dependent variable. The data from 1978 to 1993 were used as the training set, while the data from 1994 to 1998 were used as the test set. They constructed the partial least square regression model to fit and predict, and the result was more reliable than the least square regression model [2]. Zhang Xuan established ARIMA product seasonal model based on the monthly data of power consumption of the whole society from 2007 to 2012, compared the fitting effects of four different models, and selected the best model to predict the power consumption in the next six months [3]. According to the power consumption data of 5360 users in Shanghai from 2005 to 2014, Zhao Teng et al. used a top-down clustering algorithm to divide users by different electricity consumption modes. They searched 15 factors with the strongest correlation with power consumption by using mutual information theory, and introduced them into random forest model as input values to realize the prediction of power consumption [4]. Hu Yishuang et al. took into account the correlation between electricity and economy, built a production function based on the data of power consumption and output value of the three industries in Zhejiang Province from 2010 to 2016, and predicted the power consumption of the whole society in Zhejiang Province from 2018 to 2020 by combining industrial power consumption and residential power consumption [5]. According to the monthly power consumption of the whole society published by the national energy administration, Liu Zhi built a multi-layer neural network with the data from 2010 to 2017 as the training set and the data from 2018 as the test set, and introduced the early stop algorithm to improve the model. He used the data in the first half of each year to predict the monthly power consumption in the second half of each year, and controlled the deviation rate of the test set within 3.5%. The prediction result was relatively stable [6]. Cao Min et al. proposed a combined prediction model based on the power consumption data of 5 enterprises from 2016 to 2018. They used ARMA model to predict the linear part of the data and subtracted the predicted results of the linear part from the historical electricity consumption data to get the residual series. For the residual series, they introduced load, planned power consumption, temperature and other factors and used support vector machine model for prediction. The prediction results of linear and non-linear parts were added together to get the final power consumption, and the prediction effect has a high accuracy [7].

Generally speaking, the existing methods for power consumption prediction mainly include regression model, time series model and machine learning model. Among them, due to the limitation of sample size and other reasons, the prediction accuracy of traditional regression model is low. While for ARIMA model, support vector machine, random forest, neural network and other methods, as a black box tool, although they can improve the prediction accuracy to some extent, they can not specifically describe the deep explanation relationship between independent variables and dependent variables. Therefore, in order to better predict the power consumption of the whole society, this paper introduces the daily power generation data with more information, and uses the mixed data sampling model to forecast the monthly power consumption of the whole society, which is conducive to improving the explanatory ability and prediction accuracy of the model. At the same time, due to the time advance of the daily power generation index, we can monitor the trend of power consumption in a more timely manner.

3 Theoretical Model

3.1 MIDAS Model

In traditional econometric models, when dealing with data of different frequencies, interpolation is generally used to unify the mixed data into data of the same frequency, and then the macroeconomic model will be applied to the processed data. However, the processing method of interpolation will lead to the loss of information in the original data. The use of high-frequency data can improve the effectiveness of model estimation and the accuracy of prediction.

In this paper, Mixed Data Sampling (MIDAS) model is used to predict the monthly power consumption data of the whole society in China based on the daily power generation data. MIDAS model can be expressed as follows:

$$y_t = \alpha + \beta W(\theta, L) x_t^{(m)} + \varepsilon_t \tag{1}$$

Here, y_t represents the power consumption data of the whole society in the t month in this paper, $x_t^{(m)}$ is the daily power generation data with high frequency, α is the constant term, β is the substitute parameter, ε_t is a random error term and $\varepsilon_t \sim N(0, \sigma^2)$. $W(\theta, L)$ represents polynomial weight, which can be expressed as follows:

$$W(\theta, L) = \sum_{i=0}^{qm} w_i(\theta) L^{\frac{i}{m}} \tag{2}$$

Among them q is the lag order, m is the frequency of high frequency data, $L^{\frac{i}{m}}$ is the high frequency lag operator, $L^{\frac{i}{m}} x_t^{(m)} = x_{t-i/m}^{(m)}$, so the original form can be written as follows:

$$y_t = \alpha + \beta(w_0 x_t^{(m)} + w_1 x_{t-1/m}^{(m)} + \ldots + w_m x_{t-1}^{(m)} + \ldots + w_{2m} x_{t-2}^{(m)}$$

$$+ \ldots + w_{qm} x_{t-q}^{(m)}) + \varepsilon_t \tag{3}$$

Note that the unconstrained model is U-MIDAS, which has no constraints on the estimation parameters of the lag polynomial, so there is no weight function in it, and the model can be written as follows:

$$y_t = \alpha + \sum_{i=0}^{qm} \beta_i x_{t-i/m}^{(m)} + \varepsilon_t \tag{4}$$

3.2 ARMA Model

The model with the following structure is called autoregressive moving average (ARMA) model, abbreviated as ARMA (p, q):

$$
\begin{cases}
x_t = \varphi_0 + \varphi_1 x_{t-1} + \ldots + \varphi_p x_{t-p} + \varepsilon_t - \theta_1 \varepsilon_{t-1} - \ldots - \theta_q \varepsilon_{t-q} \\
\varphi_p \neq 0, \theta_q \neq 0 \\
E(\varepsilon_t) = 0, Var(\varepsilon_t) = \sigma_\varepsilon^2, E(\varepsilon_t \varepsilon_s) = 0, s \neq t \\
E(x_s \varepsilon_t) = 0, \forall s < t
\end{cases}
\tag{5}
$$

There are three restrictions in the model:

Condition 1: $\varphi_p \neq 0, \theta_q \neq 0$, which guarantees that the highest order of autoregressive is p and the highest order of moving average is q.
Condition 2: $E(\varepsilon_t) = 0, Var(\varepsilon_t) = \sigma_\varepsilon^2, E(\varepsilon_t \varepsilon_s) = 0, s \neq t$, which requires the random error sequence $\{\varepsilon_t\}$ is a white noise sequence with zero mean.
Condition 3: $E(x_s \varepsilon_t) = 0, \forall s < t$, which ensures that the current interference value is independent of the historical sequence value.

4 Results

The independent variable used in this paper is the daily data of the national full caliber power generation, which is the sum of capacity of all generating units in the country, including not only the units under the control of dispatching (based on the generator terminal meter), but also the non-regulated self-provided units (based on the meter or submitted data), and the low-voltage distributed generation.

The dependent variable used in this paper is the monthly data of power consumption of the whole society, and its trend from 2017 to 2020 is shown in the figure below (Fig. 1):

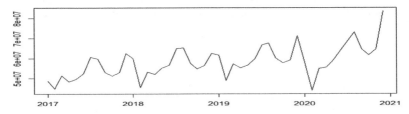

Fig. 1. The trend of power consumption in the whole society

As can be seen from the figure above, the power consumption of the whole society has a similar trend of change every year, and the seasonal effect is significant. The power consumption is high in summer and winter, and the overall trend is gradually on the rise.

In order to forecast the power consumption of the whole society, this paper adopts a combined model, which divides the power consumption into two parts: linear and non-linear. Specifically, considering the difference between the generation on weekdays and weekends, we calculate the smooth value for the power generation on Saturday and Sunday of each week according to the trend from Monday to Friday, and make the difference between the actual value and the smooth value to get the residual value, which represents the non-linear part. For the monthly power consumption, the linear part of power consumption can be obtained by subtracting the sum of the residual values in that month.

First of all, for the linear part of power consumption, MIDAS model is used to fit the data. The training set uses the data from January 2017 to December 2020, and the test set uses the data from January to March 2021. At the same time, in order to compare the prediction effect, we use the first 10, 15, 20, 25 days of daily power generation each month as the independent variables respectively.

Secondly, for the non-linear part of power consumption, ADF test shows that the series is stable, and white noise test shows that the series is not white noise, which meets the establishment conditions of ARMA model. The ACF figure and PACF figure of the series are as follows (Fig. 2):

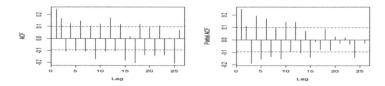

Fig. 2. The ACF figure and PACF figure of the series

Both of two figures show tailing, then try to fit ARMA (1,2) model, and predict the non-linear part from January to March 2021. The fitting and prediction results are shown in the figure below. The black part is the original value of the sequence, the red part represents the fitting result, and the blue part represents the prediction result (Fig. 3).

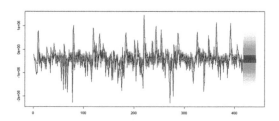

Fig. 3. The prediction results of non-linear part

Finally, the prediction results of the linear part and the non-linear part are added together to get the final results, which are sorted and obtained in the following Table 1:

Table 1. The prediction results of power consumption of the whole society

The range of daily data	January 2021		February 2021		March 2021	
	Predicted value	Deviation ratio	Predicted value	Deviation ratio	Predicted value	Deviation ratio
First 10 days	74488612	0.0170	62109139	0.1799	66818206	0.0077
First 15 days	74687406	0.0197	57517526	0.0927	68188686	0.0283
first 20 days	76271274	0.0413	61410739	0.1667	65979991	0.0050
First 25 days	78992172	0.0785	56834699	0.0797	63691367	0.0395

5   Conclusion

Based on the mixed frequency data, this paper proposes a combined model to forecast the monthly power consumption of the whole society. For the linear part of power consumption, we use MIDAS model to fit and forecast, and for the non-linear part of power consumption, we adopt ARMA(1,2) model. The final forecast result can be obtained by adding the two parts.

From the perspective of the prediction effect, the deviation ratio is controlled within 10% for the prediction in January and March 2021, and within 20% for the prediction in February 2021. Considering the first 10 days, the first 15 days, the first 20 days and the first 25 days of the daily power generation data, the prediction deviation ratio of the first 15 days and the first 25 days is small and relatively stable. Taking the time advance into account, using the power generation data of the first 15 days to predict the monthly power consumption can guarantee the prediction accuracy and time advance at the same time.

The significance of this paper is to make full use of the information contained in the daily data, and to forecast the power consumption data of the current month in advance on the basis of ensuring the prediction effect. On the one hand, it is helpful to grasp the trend of electricity demand change in time and take corresponding measures in advance to ensure the power supply; on the other hand, it gives full play to the value of big data through data mining, which provides a key basis for the power data to support the research and judgment of economic trend and support macro decision-making.

Acknowledgments. This work was financially supported by the State Grid Science & Technology Project No. SGNY0000GXJS2100065 (Research on bidirectional data value mining technology and typical application of "power economy" in SGCC).

References

1. Xiandong, T., Shan Baoguo, W., Shanshan, S.B.: Study on the difference between China's total electricity consumption and economic growth under the new normal. Electr. Power **12**, 1–5 (2017)
2. Wensheng, W., Jing, D., Yulong, Z., Xiaoming, Z.: Study on the long term prediction of annual electricity consumption using partial least square regressive model. Proc. CSEE **10**, 17–21 (2003)
3. Xuan, Z.: The application of ARIMA product seasonal model in the forecast of power consumption in the whole society. Mod. Econ. Inf. **13**, 234–235 (2012). (in Chinese)
4. Teng, Z., Lintong, W., Yan, Z., Shiming, T.: Relation factor identification of electricity consumption behavior of users and electricity demand forecasting based on mutual information and random forests. Proc. CSEE **36**(03), 604–614 (2016)
5. Yishuang, H., Yi, D.: Algorithm of societal electricity consumption forecasting based on society-electricity-economy production function. Distrib. Energy **3**(05), 16–21 (2018)
6. Zhi, L.: Monthly total power consumption forecast of the whole society based on big data and deep learning. Power Syst. Big Data **22**(08), 28–34 (2019)
7. Cao Min, J., Jian, B.Z., Juntao, L.: Forecasting method of power consumption of electricity selling users based on ARMA-SVM combined model. Energy Environ. **01**, 49–51 (2021). (in Chinese)

Artificial Intelligence (AI) Innovation Enhance Children's Reading Engagement and Attainment

Ziyou Wang[1(✉)] and Xiaobo Fang[2]

[1] South China Normal University, Guangzhou, Guangdong, China
wangziyou88@eimail.cn
[2] Guangzhou Institute of Educational Research, Guangzhou, Guangdong, China

Abstract. The positive relationship between learners' engagements and their overall reading performance has been well documented. Governments, educators and global societies have worked hard on discovering ways to improve young people's participation in reading, in order to create a literate and well-read society for young people to live in. Owing to the widespread concern at the perceived low engagement in reading after schools amongst young people in Guangzhou, the local educational executives, recently, has conducted an AI-assisted reading project—so-called "Smart Reading"—to tackle this issue. The project has been enacted within 110 pilot Guangzhou primary schools for one year. This study investigates timely on how and whether this Smart Reading project improves learners' reading engagement. In this paper, a comparative response of participant learners from pilot schools versus non-pilot schools towards questionnaires about engagement in reading activities is presented. Results indicate that learners from pilot schools exhibit higher engagement and attainment in reading than their counterpart peers from non-pilot schools. The results can offer implications for future AI-enhanced learning.

Keywords: AI-Powered learning · AI enhanced · Reading engagement

1 Introduction

1.1 Current Issues in Reading Amongst Young People

It is well documented that learners' engagement and motivation towards learning has direct influence to their achievement in study, and the influence is no less concerning literacy competence [1]. A wealth quantity of studies have indicated that the close relationship between reading engagement and reading achievements [2, 3]. Globally, governments, educators, parents are working on ways to promote young people's participation in reading in and out of schools, intending to improve literacy attainments. In country like China, this work is confronted by particular challenges. It is reported that there has been a crisis in Chinese education that reading engagement amongst young people after schools are decreasing. However, there are limited research studies on how AI and teachers' collaboration can manipulate to increase student reading engagement. In the latest research that has been done in Guangzhou with random

© The Author(s), under exclusive license to Springer Nature Singapore Pte Ltd. 2022
J. C. Hung et al. (Eds.): FC 2021, LNEE 827, pp. 669–675, 2022.
https://doi.org/10.1007/978-981-16-8052-6_83

chosen 19168 teenagers, nearly half (48%) young learners, including primary and secondary years, spend less than 30 min on reading per day after schools, which are deemed not met the expected reading time (at least 30 min) advocated by Chinese Government.

Two main factors could lead to this phenomenon. First, it is patience wasting and time taking to select appropriate reading materials from millions of public book resources. The appropriated reading materials here means materials that suitable for young readers to read according to their age, gender, reading competence, and reading interests. Second, it is difficult to provide supports and guilds to learners' individual reading wherever they might need assistance. Hence they may run away from reading. Mayer [4] suggests in his study that deep learning could only occur when learners are engaged with relevant and appropriate materials. Schunk and DiBenedetto's self-efficacy theory [5] also indicates to us that effective learning only happens if the learners are confident and comfortable with the materials they are encountering with. In this sense, without appropriate materials, learners have no clue to organise information and relate prior knowledge, hence demotivated in continuous learning.

The Government and educators have long recognised the need to improve young people's engagement in reading after class and the difficulty of doing so in the face of the limited tunnels to stimulate young people's motivation in reading. In order to tackle this problem, Guangzhou educational government has advocated implementing AI-powered applications — which is called Smart Reading Platform into schools to help to develop learners' motivation and engagement in reading. Though embedding technology to assist teaching and learning are voiced in educational academic areas, there is still a widely held acknowledgement on the positive side of the potential for technology to enhance learning in robust and meaningful ways [6, 7]. The application platform is designed to bridge the gap between what learners have read after school and how teachers may support in their way. To tackle the current problems. This platform has been piloted with 110 schools in Guangzhou for nearly two years. It is timely to investigate the feedback of the project. We will quantitatively analyse and evaluate the comparison results between pilot and non-piloted students.

This paper will go to first introduce how the Smart Reading Platform functions to support young people's reading. Then the article will display the research data on the effectiveness of Smart Reading in pilot schools. Before drawing a conclusion, the article will provide an implementation on how AI innovation may better serve its job in future education.

1.2   Smart Reading Platform

Smart Reading Platform (https://portal.zhydgz.cn/#/home) is designed to develop young people's reading engagement in two aspects. The first is to accelerate the effectiveness of young people in searching for appropriate and relevant reading materials, using intelligent personalisation improves student engagement in reading; The second is to create young people personal reading profile to make their reading trace visible for them and teachers. The third is to provide teachers actionable data which offer teachers insights to give young people timely support, targeted interventions.

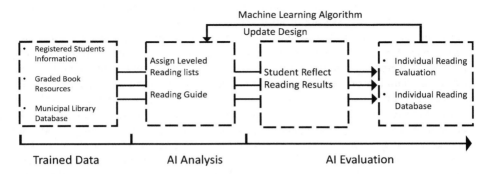

Fig. 1. The process of Smart Reading Platform

Smart Reading Platform (Fig. 1) is designed for young people from primary and secondary years. The platform adopts AI-powered Algorithm to provide optimal reading lists and reading guilds to children scientifically, mathematically, which involve the consideration of their age, gender, literacy abilities, and reading interests. This saves learners time to judge the creditability of reading materials such as authoring, intended message, and textual characteristics.

1.3 How Does AI-Powered Smart Reading Platform Assists Reading

As discussed above, one of the urgent issues for Chinese young people is the low engagement in reading after class. Without securing the reading engagement, learners would easily fall behind the curriculum. The Smart Reading Platform is an application that designed to tackle this issue and support teaching and learning in two aspects. The first one is to help teachers and students to improve the efficacy in looking for appropriate books resources. It is said that reading effectively necessitates critical evaluation and selection of materials [8]. There is usually an overwhelming amount of reading resources, online and offline. Deciding which ones are appropriate for individual young students to read can be difficult. The Smart Reading Platform utilises cognitive neuroscience and data analytics to create personalised reading list and reading plans for each registered learners. In this way, both learners and teachers reduced workloads in searching and evaluating reading resources. With an AI-powered algorithm on the existing data, the Smart Reading Platform automatically provide young people booklists, which are most likely to match their reading competence, interesting topic, and cognitive measurement. Also, the platform automatically upgrades and sequences the reading list according to students' updated reading data. The platform adopts an Item-Based Collaborative Filtering algorithm for making book recommendations. This algorithm first analyses the matrix of books to identify relationships between different books, and then use these relationships to indirectly compute recommendations for learner end users. This Smart Reading Platform also provides teachers tunnel to access to relevant reading resources and in turn to reduce their time on planning, grading, and managing students' homework reading.

The second one is to create a reading database profile for each registered student in terms of their ID and reading history, reading active index, and general reading evaluation. The platform tracks student progress and identifies knowledge gaps and offers personal reading recommendations and feedback. The Smart Reading Platform adopts AI-powered algorithms to track how young students interact with reading texts, such as the duration of time they spend on reading over certain topics. The platform automatically makes an evaluation on learners' reading comprehension levels by learning their reading records and matching the results with updated reading materials that are more sequenced relevant, and purposeful. These data-driven reading records, to some degree, may offer teachers with insightful and statistical information about students reading performance. Teachers could adopt these information to facilitate their classroom teaching [9]. The data also serves to help schools to collect students' reading data and increase students' reading engagement. The database also provides teachers overviews and insights into a student's reading situation and designates areas in which the student needs supports. Teachers then can provide targeted support to help students build on their literacy skills.

It is noteworthy to argue here that the Smart Reading Platform does not replace the role of teachers in analysing students' reading competence and sequence teaching strategies to meet learners' needs. The function of this AI technology in education not only detects different patterns of reading materials at scale, but also provides data-driven evaluations and suggestions to students for improving the quality of students' reading. Namely, the Smart Reading Platform creates teachers' integrative learners' reading and learning experience with collaborative data from various factors, and identifies individual students' reading patterns and makes decisions that are supportive to all learners.

Smart Reading Platform now has been trialled in 110 pilot schools in Guangzhou for over one year. In order to explore the effectiveness of this project in improving learners' engagement in reading, we conducted to comparative research between young people from AI assisted pilot schools and non-pilot schools in terms of their reading engagement and awareness in using reading strategies.

2   Methodology

2.1   Research Questions

The broad research goal of this study is to exam the effectiveness of AI-assisted platform in developing young people's participation in reading after schools by addressing the following two research questions:

- Research Question 1 (RQ1): Is there more population of young people from AI-supported pilot schools who actively engage reading after schools than those of young people from non-pilot schools?

- Research Question 2 (RQ2): Is there more population of young people from AI-supported pilot schools who find it more interesting in reading than their counterpart peers from non-pilot schools?

2.2 Method

Questionnaires are adopted to collect data of students. Questionnaire surveys have a long tradition in educational motivation and engagement research and have the merit of gathering representative data allowing statistical analysis, but would ideally be complemented by a qualitative study, which was not possible in this case.

2.3 Participant

Participant students are from two types of primary schools—pilot schools (where Smart Reading Platform implemented) and non-pilot schools. The online questionnaire was accessed by the entire sample population of 12 pilot schools (n = 16896) non-pilot schools (n = 16626), and 12902 responses from pilot schools and 11754 response from counterpart students from the non-pilot school were collected, representing the response rate of 76% of pilot schools and 70% of non-pilot schools. The sample included a slightly higher number of students from pilot schools than from non-pilot schools.

3   Result

3.1   RQ1: Reading Engagement

Drawing from the data (Fig. 2), we can see the overall population of young people from pilot schools who actively engage in reading after schools is more than those of young people from non-pilot schools. The survey results show that the number of pilot schools with an average daily investment of more than 30 min per day on reading (meet the national reading time requirement advocated by the government) has reached 45.94%, which is nearly 10% higher than those of population proportion collected from non-pilot schools (34.45%). The chi-square test results showed that the difference in young people's engagement in reading between pilot schools and non-pilot schools is significant (p < 0.0001). This result indicates that the implementation of Smart Reading Platform has a positive impact on improving students' reading engagement.

3.2   RQ2: Interesting in Reading

The data (Fig. 2) also results show that more students from pilot schools prefer to read extracurricular books than those of from non-pilot schools, with the proportion 33.68% versus 23.98%. revealed that there was a statistically significant difference between students from pilot schools and students from non-pilot schools in terms of interests in reading (p < 0.0001). In this sense, we can be indicated that the Smart Reading Platform can effectively promote students' interest in reading.

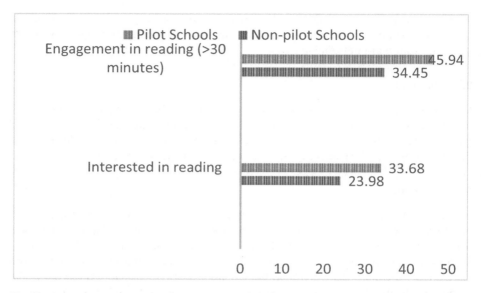

Fig. 2. A comparative population proportion between pilot schools and non-pilot schools in terms of reading engagement and interest

4   Discussion

Drawing from the result above, students from pilot schools appear to be more aware of the importance of reading, and tend to invest more time on reading after with the assistance of the Smart Reading Platform. The Smart Reading Platform automatically provide appropriate and relevant reading resource to students. This not only saves their time on looking for books in the ocean of reading resources, but also purposefully intrigue students interest to stick to the recommended books and continue to read them. However, we should argue here that the significant role of the student–teacher-parent relationship in supporting students' reading development and engagement should not be ignored. The important roles that teachers and parents play in recommending students' books and analysing students' reading should not be replaced by AI technology. The result of this study, however, may offer insight to researchers, teachers, and governments on the possibility that AI technology may improve students' reading engagement in and out of classrooms.

5   Limitation

This thesis is a small-scale study that is contextually limited to the participants in limited schools. The findings of this study are not generalisable to the positive impact of AI-powered technology in developing reading engagement to students of all ages level. However, the insights from the findings of this study contribute towards a more informed consideration of implications for future research on embedding AI technology in and out of classrooms.

6   Conclusion

In this study, AI-powered technology works in its to absorb and retain information that could work in favour of engaging students in learning rather than against it. The sound performance of the Smart Reading Platform in Guangzhou primary schools in improving students' reading engagement opens up possibilities to adopt AI in education to fill the gap between teaching and learning in and out of schools. The future for AI-enhanced learning in education is promising. This study offers us a possible vision for customising curriculum for every student's needs in the future.

Acknowledgments. Guangzhou Education Science Planning 2020 Annual Project, Number: 202012712.

References

1. Adesope, O.O., Rud, A.G. (Eds.): Contemporary Technologies in Education: Maximizing Student Engagement, Motivation, and Learning. Springer International Publishing, Cham (2019). https://doi.org/10.1007/978-3-319-89680-9
2. Taboada, A., Tonks, S., Wigfield, A., Guthrie, J.: Effects of motivational and cognitive variables on reading comprehension. Read. Writ. **22**, 85–106 (2009)
3. Wang, J.H.Y., Guthrie, J.T.: Modeling the effects of intrinsic motivation, extrinsic motivation, amount of reading, and past reading achievement on text comprehension between US and Chinese students. Read. Res. Q. **39**(2), 162–186 (2004)
4. Mayer, R.E.: Cognitive Theory of Multimedia Learning. In: Mayer, R.E. (ed.) The Cambridge handbook of multimedia learning, 2nd edn., pp. 43–71. Cambridge University Press, New York (2014)
5. Schunk, D.H., DiBenedetto, M.K.: Self-efficacy theory in education. In: Wentzel, K.R., Miele, D.B. (eds.) Handbook of motivation at school, 2nd edn., pp. 34–54. Routledge, New York (2016)
6. Herrington, J., Oliver, R.: An instructional design framework for authentic learning environments. Educ. Tech. Res. Dev. **48**(3), 23–48 (2000). https://doi.org/10.1007/BF02319856
7. Tamim, R.M., Bernard, R.M., Borokhovski, E., Abrami, P.C., Schmid, R.F.: What forty years of research says about the impact of technology on learning: a second-order meta-analysis and validation study. Rev. Educ. Res. **81**(1), 4–28 (2011). https://doi.org/10.3102/0034654310393361
8. Rouet, J.-F., Ros, C., Goumi, A., Macedo-Rouet, M., Dinet, J.: The influence of surface and deep cues on primary and secondary school students' assessment of relevance in web menus. Learn. Instr. **21**(2), 205–219 (2011)
9. McBroom, J., Yacef, K., Koprinska, I., Curran, J.R.: A Data-Driven Method for Helping Teachers Improve Feedback in Computer Programming Automated Tutors. In: Penstein Rosé, C., et al. (eds.) AIED 2018. LNCS (LNAI), vol. 10947, pp. 324–337. Springer, Cham (2018). https://doi.org/10.1007/978-3-319-93843-1_24

Analysis and Research on the Techniques of Digital Painting Creation

He Feng[✉]

Dalian Neusoft Information Institute, Dalian 116000, Liaoning, China
fenghe@neusoft.edu.cn

Abstract. With the development of digital art and the needs of business, digital painting has become modular, standardized and flow-oriented in the process of creation. At present, there are mainly two drawing processes and techniques in digital painting creation. To improve the learning efficiency of digital painting and master the techniques of digital painting creation that meet people's own needs, this paper analyses and studies the two main creative methods.

Keywords: Digital painting · Creative techniques · Drawing process

1 Drawing Method of Line Draft

The creation technique of line draft is a relatively traditional creation process. The advantage of line draft drawing is that it is easy to adjust and modify in the early stage of design consideration, and there is a lot of room for change [1]. The disadvantage is that the time cost of drawing high-standard line manuscripts is also high. The conventional step of drawing a line draft is the sketch stage, in which the general composition and edge structure are drawn, and the details are not involved so much. After that, we enter the line drawing stage, which needs to consider many details, such as: the structure, light, shadow and material of the target need to be matched in the line draft stage. One of the key points of this stage is that the lines drawn are as single and closed as possible [3]. This can provide great convenience for subsequent drawing. Line draft drawing on the complexity of the drawing target corresponds to the corresponding number of times to clean the line draft, the general experience of cleaning the line draft will be two or three times. After the completion of the line draft, we will draw and match the fixed color of the drawing target. Only the color matching is considered in this part, and the color of the drawing is flat painting. After the completion of solid color rendering, we enter the shading relationship rendering stage. There are several software layers commonly used in this stage, which are positive overlay layer, overlay layer and so on. The first problem to be solved in the stage of drawing the relationship between light and shade is to make three-dimensional shape of the line draft. According to the drawing requirements and creation goals, we should grasp the degree of three-dimensional. For example: in the two-dimensional drawing style, the number of shading levels should not be too many and should be flat [4]. In the realistic rendering style, it is required to open the relationship between light and shade as far as possible and enrich the hierarchical relationship. Therefore, the specific number and degree of

J. C. Hung et al. (Eds.): FC 2021, LNEE 827, pp. 676–679, 2022.
https://doi.org/10.1007/978-981-16-8052-6_84

drawing layers should be adjusted according to the creative requirements [7]. Then we need to draw the light and shadow relationship. According to the number of light sources to draw, conventional painting will appear one or two light sources, which are the main light source and the auxiliary light source [9]. There are also differences between cold light and warm light in color. The two light sources usually make the target more solid and full. The last part is the detail drawing of the material. In this link, we need to paint the material and the details until the picture meets the creation expectation or the commercial requirement (Fig. 1).

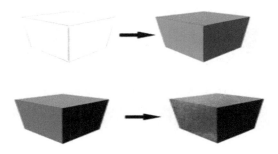

Fig. 1. Sketch of line drawing

2 Drawing Method of Ambient Occlusion

"AO" means "ambient light shading". It uses the layer mixing function of Photoshop and other software to mix different material layers and light and shadow layers together to form a complete image [2]. The advantage of AO is that each layer is treated separately during the drawing process, which means we don't have to consider shape, light and shadow, material and color at the same time as traditional painting. The traditional AO drawing method is relatively complex, requiring some corresponding layers, and each layer should correspond to the blending mode of PhotoShop layer. For example, the occluded shadow area corresponds to the overlay layer, the light-shade relationship corresponds to the overlay layer, the inherent color corresponds to the normal layer, the material expression corresponds to the normal layer, and the light-shadow relationship corresponds to the overlay or soft light or color filter layer (depending on the drawing target). In this paper, the method of AO is simplified and easy to understand [6]. The normal drawing process chooses silhouette to start the draft. First, we are supposed to determine the outer outline of the drawing target and draw the correct sketch relationship. When drawing a sketch relationship, we only need to pay attention to shaping the correct black-and-white relationship. This link can greatly help beginners to learn, in the process of starting digital painting learning, there are often many difficulties, such as: the inadaptability of the conversion from easel painting to digital painting, unskilled software operation and so on. To some extent, the drawing way of AO is equivalent to dividing the complex problem into multiple links, which makes the mastery of each link relatively simple. After the sketch relationship drawing link, we will use layer mode to match the fixed color. The commonly used layer modes are generally "color", "overlay" and other layer modes [10]. The color of the drawing target can be given without changing the big relationship of the sketch. Then we could enter the stage of drawing detail material, which is the same as the drawing method of the line draft and belongs to the final in-depth drawing stage (Fig. 2).

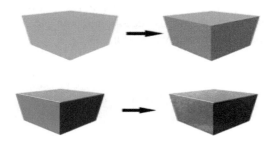

Fig. 2. Sketch of white model drafting

3 Summary

Strictly speaking, the creative process of digital painting is free and highly malleable. At present, the main purpose of the two main drawing processes is applied to commercial services, and the current painting process is formed by repeated summary and refinement in the practice of commercial services. Currently, many universities have also set up courses related to digital painting, so the conclusions based on creative techniques for different groups of people are as follows: for those who have just come into contact with digital painting, it is suggested to do more decomposition exercises in the initial part of the painting processes, that is, the practice of line drafts and the drawing of white models [5]. The reason is that the drawing exercises of line drafts can be completed by using fragmentation time, which greatly improves the efficiency of the exercises. In addition, the shape, structure, perspective and other problems of the drawing target can be accomplished through the practice of the line draft. In addition, in the initial stage of practicing the line draft, it is recommended to turn off the pressure sensing function of the digital board and simply use the line draft to solve the key points mentioned above. The reason why the drawing method of AO is suitable for novice practice is that most creators' method of digital painting is converted from easel painting to digital painting, and the creators are very familiar with sketch. In addition, they can concentrate on solving the problems encountered in painting one by one at this stage and simplify the complex problems. Experienced digital painting creators should pay more attention to the in-depth drawing of the two painting processes [8]. If it is said that beginners need to repeatedly practice the ability of observation, memory and practice, then creators with certain creative experience should practice the thinking ability and imagination ability of painting more. When you are proficient in the both drawing processes, you can find efficient creative techniques and painting processes according to your needs. The two have the same goal.

References

1. Zibiao, L.: The value and development of digital painting. Fine Arts Grand View **11**, 207 (2010)
2. Xianbing, C.: Creative experience of digital painting and traditional painting. J. Hunan Univ. Sci. Technol. **1**, 214–216 (2009)
3. Xiao, Q.M.: On the relationship between traditional painting and computer painting. J. Taiyuan City Polytech. **4**, 177–178 (2013)
4. XueGuo, Y.: Unlocking CG Thinking: The Artistic Enlightenment of Digital Painting. Electronic Industry Press, Beijing (2018)
5. Lu, L.: Analysis and research on the creative characteristics of digital painting and traditional painting. Beauty Times (middle) (7), 23–24 (2017)
6. Zhou, Q.: Application of traditional painting techniques in digital painting. Art Educ. Res. (9), 88–90 (2018)
7. Jun, Z.: Research status and development trend of digital illustration art. Mod. Commun. (1), 70–73 (2013)
8. Linyun, F.: Review of domestic research on digital painting. Beauty Times (middle) **11**, 29–33 (2018)
9. Junhong, C.: Research on the current situation and development trend of Chinese digital illustration. Guangdong University of technology, Guangzhou (2018)
10. Chen, L.: The development trend of illustration art in the digital era. Art Educ. Res. **12**, 115 (2012)

The Integration of Network Application Platform to Promote the Process of Regional Education Informatization in the Era of Big Data

Wenli Zhong[✉]

School of Yunnan Technology and Business University,
Kunming, Yunnan, China

Abstract. With the continuous promotion, development, popularization and improvement of mobile Internet technology, people's daily life, work and study will be gradually free from the constraints of time, space and region. The network has developed into an indispensable part of people's daily life and study in modern society. In the field of education, with the development of educational informatization, higher requirements are put forward for the construction of educational service platform and educational database. The purpose of this paper is to study how to use the big data era to integrate the campus network application platform to promote the process of regional education information construction. This paper mainly uses the interview method, content analysis method and practice summary method and other exploratory research methods to study the key influencing factors of informatization promoting the development of regional education modernization, and the specific practices and typical experience of informatization promoting the construction of education modernization in the city, and summarizes the practical experience of informatization promoting education modernization in the city. The research results show that the development process of education informatization in our city has been developing year by year since 1978, and the relevant data of regional informatization is typical. In addition, the four generations project can be carried out on a pilot basis to continuously explore and optimize the development of urban education informatization.

Keywords: In the era of big data · Network application platform · Regional education informatization · Education resource database

1 Introduction

The application of modern information technology plays an important role in the process of education informatization and modernization [1, 2]. From the initial traditional technology to the later audio-visual media technology, and now to the computer communication technology marked by the Internet, education has been undergoing great changes because of the emergence of these new media technologies [3, 4]. Now we are in the information age, with the rapid development of information technology marked by computer technology and communication technology. The emergence of

J. C. Hung et al. (Eds.): FC 2021, LNEE 827, pp. 680–687, 2022.
https://doi.org/10.1007/978-981-16-8052-6_85

Internet, big data, cloud computing, artificial intelligence and other advanced technologies not only affects the development track of human society, but also deeply affects the educational concept, educational system, educational means, educational content and people's learning methods in the field of education And the transformation of advanced thinking ability [5, 6].

Many scholars at home and abroad have given their own views on the research of integrating network application platform to promote regional education informatization under big data [7, 8]. For example, Oide t has made research on the development planning of regional informatization. Firstly, it analyzes the important aspects that can realize the development of regional education informatization. Secondly, it formulates the corresponding principles. Finally, it puts forward the basic model, that is, the development model of regional education informatization with three-level Co Construction of government, school and enterprise and multi-agent participation and sharing [9]. Qun and Chen summed up two successful cases of promoting the development of regional education informatization, one is Beijing Daxing education application store, the other is Prince Charles and his prince micro class, and drew a new road for enterprises to participate in promoting the development of education informatization [10].

According to the sociological research methods, this paper first designs and prepares the research scheme before conducting the research. Then, in the specific implementation process, it uses the research method of interview as the main method and assisted by a variety of information means to collect the relevant information on the promotion of education modernization by education informatization in our city through multi-channel and multi-dimensional. Finally, after the research is completed, Combined with the collected literature data, this paper makes a scientific and detailed analysis of the research [11, 12].

2   The Renewal Service of Educational Resources Network

2.1   Resource Modeling of Educational Resource Network Update Service

The construction of parameters in the resource model is not as complicated as the fields in the database. We only need to select the attributes that can accurately reflect the resources and play an important role in the model matching calculation. When building a resource model, we need to consider the following factors: first, the ownership of resources, the same kind of resources in different grades may have keyword crossing situation; Second, the subject of resources; the third is to describe the weights of various characteristics of resources.

2.2   Personalized Recommendation Method Based on Data Mining

2.2.1   Content Based Approach

At present, there are many methods to calculate the similarity, usually Minkowski distance method and cosine distance method.

1) Minkowski distance method

Let n be the dimension of the feature, I and j be natural numbers, and I (x, y) be the distance between them. The Minkowski distance formula of X and Y is as follows:

$$R(x, y) = \left[\sum\nolimits_{i=1}^{n} |x_i - y_i| \right]^{\frac{1}{j}} \tag{1}$$

If J = 1, the formula becomes absolute distance; if J = 2, the above formula becomes Euclidean distance.

2) Cosine distance

The distance formula of cosine is as follows:

$$R(x, y) = \frac{\sum_{i=1}^{n} x_i y_i}{\sqrt{\sum_{i=1}^{n} x_i^2 \sum_{i=1}^{n} y_i^2}} \tag{2}$$

The main advantage of this recommendation method is that it is simple and effective. However, the success or failure of push is often related to the correctness of user personalized feature selection. If the selection of features is not accurate, the final result is likely to have no direct relationship with users at all. In addition, only one information base can be used to find the resources that users already have interest in, which will not enable them to find a new content of interest through an information base. Sometimes, it will be relatively single.

2.3   Updating Service Technology of Education Resource Network

2.3.1   Construction of Educational Resource Database Updating Service Model

In the whole model, the collection of user information is the foundation of updating service. It adopts the method of combining active and passive collection to obtain the basic information and behavior information of users. According to the collected user information, it is easy to build an accurate user interest model library. By analyzing the situation of resources, we can build resource model base. The accurate establishment of the two model bases can provide the necessary data support for further matching and filtering. Therefore, the construction of user interest model base and resource model base is an important part of the whole service. When the number of users is small, that is to say, in the initial stage of the resource database, because the amount of data collected is too small and incomplete, only the user interest model library and resource model library are filtered based on content, and the processed content is generated into recommendation data, then select the most valuable data to push to the resource manager. When the number of users reaches a certain level, with the continuous improvement and continuous operation of the update service system, more and more data will be collected. In this way, it is too thin to filter the content only once, which will lead to too much data pushed to the manager, and eventually lead to too much pressure on the manager, therefore, the collaborative filtering method is integrated, that is, a large number of user interest models are clustered to analyze the overall user interest categories, and then the reasonable and high-quality resources are pushed through the collaborative filtering method with the resource model library. In this way, not only the pushed data has a strong basis, but also the focus is prominent and the organization is clear, at the same time, it can also solve the problem of excessive workload of managers.

3 Integrate the Network Application Platform to Promote the Research of Regional Education Informatization

The development of education informatization in China presents obvious regional characteristics, and the development level of education informatization varies greatly between regions.

3.1 Research Objects

In order to facilitate and cost as low as possible, taking the education situation of major schools in the city as the research object, this paper studies the process of promoting regional education informatization by integrating network application platform in Colleges and Universities under the background of big data.

3.2 Research Methods

The topic of this paper is based on the practical exploration of promoting the modernization of education by regional education informatization, mainly through the interview method and literature research method, to investigate the main methods of promoting the modernization of education by informatization in Suzhou, and then refine and summarize the successful experience of promoting the modernization of education by informatization in Suzhou, and make further exploration combined with the basic national conditions of our country. In addition to face-to-face interview, this paper also uses some common information technology means to assist the research, such as e-mail, telephone, wechat as a supplement to face-to-face interview research.

3.3 Basic Principles of Research

As the main research method, interview method follows the following principles in the specific implementation process: the principle of target gathering, the principle of seeking truth from facts, the principle of open reflection.

3.4 Design of Investigation Scheme

3.4.1 Determination of Research Objects

This research chooses the city as a typical research object, takes a series of practical explorations to promote the modernization of education carried out in recent years as the main research content, deeply excavates the deep connotation behind these explorations, and finally analyzes and summarizes the typical experience of the city's modernization of education, so the interviewees selected for this research activity must be able to directly participate in the city's education Education is the main personnel of information construction. After preliminary consideration, the director of the Municipal Education Bureau will be the main candidate, because the education work of a city is mainly carried out under the leadership of the local education bureau. Considering from all aspects, director Gu is a good candidate. The Municipal Education Bureau is in charge of the education development of the whole city. The scope is too large, and the work is more inclined to the macro decision-making and leadership. Therefore, in order to make the interview more targeted, we finally choose the director of the municipal audio-visual education museum as the main interview object. At the same time, we are ready to interview other main staff in the audio-visual education museum.

3.4.2   Selection of Research Methods

This research mainly uses the method of interview and literature analysis. The interview method mainly uses unstructured interview method. The specific interview methods are direct interview method and indirect interview method. The literature analysis method is mainly used in the preparation stage before the survey and the data analysis stage after the survey.

3.4.3   Development of Research Outline

The purpose of this survey interview is to deeply explore the main experience of the city in promoting the development of education modernization by education informatization. Specifically, a series of relatively successful education informatization projects carried out in the city are taken as the breakthrough point to investigate the specific practices and advance exploration of the city in promoting education modernization by informatization. The whole interview outline is mainly divided into three parts: overall planning, specific issues and in-depth discussion.

4   Survey Results and Analysis

4.1   Feasibility Analysis of Research Object

Looking at the statistical yearbook, we can see that since the reform and opening up, the economic development structure of the city has shown exponential growth since around 2005. This period is the period of rapid development of the Internet in China. The city resolutely seizes the opportunity, vigorously develops the tertiary industries such as commerce, finance, transportation and transportation, and services. At the same time, it has the courage to carry out system reform and truly implement the "science and technology" policy It is the first productive force, and the economic development is growing. The GDP of the city is shown in Table 1.

Table 1. GDP of the city

Particular year	1978	1993	2008	2010	2016	2020
Gross GDP	32.45	205.21	4256.43	7942.6	9003.78	11345.34

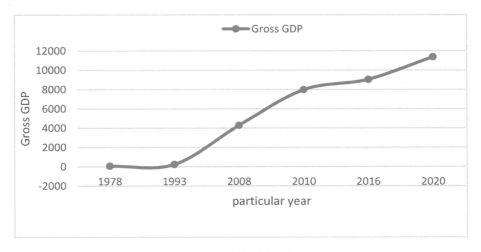

Fig. 1. GDP of the city

As can be seen from Fig. 1, the city's economy has been growing year after year. The development of economy will inevitably lead to great progress of society and education. Therefore, the Research Report on the process of regional education informatization in this city is completely feasible.

4.2 The Development Process of Educational Informatization

Referring to the development process of educational informatization in foreign developed countries, combined with China's specific national conditions and local policies, the development of educational informatization in our city can be roughly divided into the following four stages: the initial stage from 1992 to 1997; the infrastructure stage from 1998 to 2005; the application promotion stage from 2006 to 2012; and the application integration from 2013 to now stage. With the national reform and opening up, China's economic policy has been comprehensively adjusted. Comrade Deng Xiaoping put forward the educational development goal of "Three Orientations" in 1983, pointing out the way forward for the development and construction of China's educational modernization. The city actively carries out the practical exploration and application of education modernization, and actively applies information technology to education and teaching. As shown in Table 2, it is the development process of education informatization in this city.

4.3 Suggestions on Education Informatization

In order to realize the modernization of education, it is actually to realize the modernization of people. Suzhou Education always attaches importance to the cultivation of talents, strives to change the management concept, and increases the investment in education. In addition, the development of educational informatization needs "four generations in one hall", that is, one generation of research, one generation of pilot, one generation of promotion and one generation of popularization, so as to continuously explore and optimize the development road of educational informatization in Suzhou. Because many projects of educational informatization construction need to be carried out simultaneously, so in a period of time, there should be not only a certain project in the promotion generation, but also some new exploration generation or pilot generation. As shown in Table 2, it is the part of "four generations in one house" project.

Table 2. "Four generations together" project.

Project	Research generation	Pilot generation	Promote a generation	Popularize a generation
Education e-card	2009	2010	2011	2012
Future classroom	2012	2013	2014	2015
Smart campus	2014	2015	2016	2017
Gigabit broadband on campus	2016	2017	2018	2019

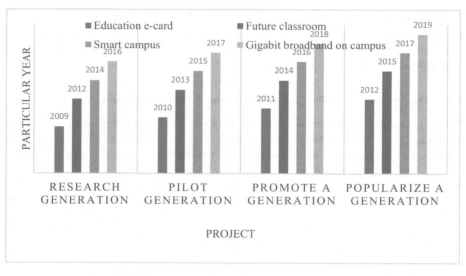

Fig. 2. "Four generations together" project.

As shown in Fig. 2, the "four generations in one" project is implemented simultaneously. When a project is being promoted, another project may have begun to study. When the current generation is popularized, the latter can also be promoted on a pilot basis. In this way, all construction projects are seamlessly connected, there will be no so-called empty window period or congestion period, and all work will be carried out all the time in an orderly way.

5   Conclusions

In view of the current situation of promoting the modernization of regional education by information technology in China, Suzhou city is selected as a typical representative. Through the methods of interview, literature research and content analysis, this paper studies the current situation and advance exploration of promoting the modernization of education by information technology in Suzhou City, aiming to provide help for the optimization and development of comprehensively promoting the modernization of education by information technology in other regions. The specific research includes the research on the key influencing factors of promoting regional education modernization by informatization, and the research and experience summary of typical projects of promoting education modernization by informatization in this city.

References

1. Wang, M.: Problems performance and path optimization of rural education informatization construction. Adv. Soc. Sci. **10**(3), 510–515 (2021)
2. Chen, X.: Analysis on the development of higher vocational education information. Tak. J. Liaoning Teach. Coll. (Nat. Sci. Edit.) **021**(002), 45–49 (2019). 83
3. Feng, Y.C., Ren, Y.Q.: Education focusing wisdom in education information 2.0 era: simultaneous elimination of the three-layer gap and the intergenerational transmission of poverty: the interpretation of education informatization 2.0 action plan (3)%. J. Distance Educ. **036** (004), 20–26 (2018)
4. Gj, A., Dw, A., Dy, A., et al.: Role and object domain-based access control model for graduate education information system. Procedia Comput. Sci. **176**, 1241–1250 (2020)
5. Zaslavsky, A.A.: Prospects for the use of blockchain algorithms to ensure security in the management of the educational organization. RUDN J. Inform. Educ. **15**(1), 101–106 (2018)
6. Kou, S.: Application of education informatization promoting educational equity in remote areas of China. Int. J. Inf. Educ. Technol. **10**(8), 608–613 (2020)
7. Devi, S., Roy, S.: Physiological measurement platform using wireless network with android application. Inform. Med. Unlocked **7**, 1–13 (2017)
8. Cao, D., et al.: Construction of red culture heritage in digital application platform. Bol. Tecnico Techn. Bull. **55**(12), 399–406 (2017)
9. Oide, T., Abe, T., Suganuma, T.: COSAP: contract-oriented sensor-based application platform. IEEE Access **1** (2017)
10. Qun, C.R., et al.: Strategies for informatization construction of continuing education management in higher vocational colleges: taking pharmacology courses as an example. Asian Agric. Res. **12**(04), 66–68 (2020)
11. Fanfan, Z., Wenna, L., Chengcheng, Y., et al.: The application of network platform in the formative application in the teaching of surgery. China Contin. Med. Educ. **010**(026), 41–43 (2018)
12. Azeez, N., Mustafa, A.M.: Use of viber platform as a social media application in TEFLat erbil polytechnic university. Int. J. Psychosoc. Rehabil. **23**(2), 622–633 (2019)

A Proposed Method for Improving Picking Efficiency in Multiple Distribution Zones

Jing Zhang[✉]

School of Mathematics, Physics and Information Science,
Zhejiang Ocean University, Zhoushan 316022, Zhejiang, China
j.zhang.gr@zjou.edu.cn

Abstract. Internal zone layout and picking are extremely important in the operation of a distribution warehouse. It is the most time-consuming and costly task in a distribution warehouse. Therefore, it is very important to study the efficiency and optimization of picking in the warehouse. In this paper, we propose an asynchronous parallel setting method for shelf zone layouts in logistics warehouses and build a simulation to show the efficiency improvement effect of the proposed method on real data.

Keywords: Total picking · Method · Parallel · Simulation

1 Introduction

1.1 Background

Picking is labor intensive and the most expensive aspect of distribution warehouse operations. Previous studies for improving the efficiency of various picking methods have been conducted [1]. In particular, many studies on efficiency improvement of the total picking method of putting together orders for multiple shipping destinations have considered this as an order batching problem (OBP). In this method, a zone picking method is assigned to each divided zone.

Total picking is a method in which a picker picks multiple customer orders at once. It is necessary to sort items for each customer order, generally in a pick-and-sort or sort-while-picking procedure [2]. On the other hand, methods in which only one customer order is picked at a time are called single picking methods.

In this research, completing customer orders in total picking is called "batching," and batched customer orders are called "batch orders." In sort-while-picking methods, the picker sorts items for each customer while picking. In this case, there is a sorting area separate from the picking area. Generally, material handling equipment such as a "put to light" (PTL) method or other automated methods are used in the sorting area. In this research, post-pick sorting is used on the premise of utilizing sorting equipment.

Recently, with increased demand for online shopping, picking work volumes can exceed processing capacities assumed at the time of warehouse design. Large online shopping distribution warehouses in particular assume conventional total picking. In methods where one picker covers the whole area, movement distances become inefficient. A combination of total picking and zone picking would be more efficient, but

J. C. Hung et al. (Eds.): FC 2021, LNEE 827, pp. 688–694, 2022.
https://doi.org/10.1007/978-981-16-8052-6_86

this approach is limited in existing methods. The proposed method addresses this limitation.

This research considers a warehouse consisting of picking, temporary storage, and sorting areas. We define "asynchronous pick-up total picking" as a method for batch order picking that is not affected by other zones in each zone. We verify that the proposed method improves processing capacities in distribution warehouses. We also propose an algorithm for improving comprehensive processing capacity, including placement and sorting.

1.2 Purpose of Research

Order batch processing efficiency in total picking has been solved as an OBP. OBPs are used to group customer orders based on the arrangement of items in a warehouse and a given picking equipment capacity (batching). In OBPs, the solution is searched for using the movement distance or time necessary to collect all items as an evaluation function. The solution is a set of batch orders, to which all customer orders are assigned. As a constraint, many studies set an upper limit on the amount of items that one picker can carry at one time.

Solutions in OBPs are divided into constructive solutions and meta-heuristics solutions. The former are algorithms that do not reassign customer orders once assigned to a batch order, first proposed by Eslayed et al. [3] in the 1980s. The latter are algorithms that search for improved solutions by incorporating a re-assignment step for customer orders. Metaheuristics have been applied to OBP since around 2005 [4–10].

The authors are unaware of any studies that consider the efficiency of sorting devices in OBP. Functions in these studies are simplified to travel time, and it is not possible to comprehensively evaluate processing capacity capturing warehouse operations.

Based on present conditions and previous research, this research verifies processing capacity improvements resulting from introducing asynchronous parallel processing in a multi-zone layout in the total picking of a distribution warehouse.

To that end, we perform the following:

Definition of an asynchronous and parallel processing total picking method.

Design and construction of a picking list output method.

Because efficiency problems related to the picking method considered in this research have not been studied, we define this as a new batching problem. We build a method that outputs an efficient solution to the defined problem. As its internal algorithm, we construct algorithms for batch order creation, picking list creation, and picking list ordering. A picking list is a list that indicates the items to be picked by the picker.

2 Assumptions of the Proposed Method

There are two primary differences from problem definitions in existing OBPs. First, it is necessary to batch customer orders in two ways, with batch order processing by sorting devices and with picking lists for performing picking together. Next, it is necessary to consider the order of outputting picking lists to improve the operation rate of sorting devices.

We uniquely set an evaluation function that evaluates the processing capacity of the asynchronous parallel processing-type total picking method that is the subject of this research. This function is defined as the time required from the start of picking the first order to the end of sorting all orders under a certain input resource, as shown in Eq. (1). TI is a function that evaluates processing power in the target method. $T_{work-times}$ is the time taken from the start of picking the first order to the end of sorting all orders. $C_{resource}$ is a cost conversion for the amount of equipment and personnel introduced.

$$TI = T_{work-times}/C_{resource} \tag{1}$$

We consider three constraints:

① As in existing OBPs, there is an upper limit on the number of items that the picker can pick in a single picking.
② As in existing OBPs, the storage unit price remains the same.
③ Unlike existing OBPs, there is an upper limit on storage volume in the temporary storage area.
④ Unlike existing OBPs, there is an upper limit on the number of customer orders that the sorting device can sort at one time.
⑤ As in existing OBPs, The hourly cost of sorting remains the same.

The picking area is divided into multiple zones, and items are placed in each zone. A group of multiple products is called a category, and each zone comprises multiple categories. Picking bases are located in one corner of each zone. Each location has a starting point for a conveyor that leads to the temporary storage location.

3   Proposal of Picking List Output Method

3.1   Method Overview

Figure 1 shows the system flow from receiving a customer order to evaluating processing capacity. The proposed system reads customer orders according to this flow, and outputs picking lists in order. With this system, it is possible to quantitatively evaluate the processing capacity of the targeted picking method.

First, read customer order data ordered at regular intervals. When the read number becomes a constant, the picking list is output as an input value. Next, create a batch order based on the read customer order. This determines the order in which the sorting device sorts. After that, divide the batch order into zones. Finally, create a picking list based on the divided batch order. The picking list is output by further batching the divided batch orders. This process is based on the premise that the volume of batch orders assigned to each zone is less than the volume that can be performed in one picking.

The internal algorithm of this method organizes the processing capacity improvement. To improve the processing capacity in the range defined by the evaluation

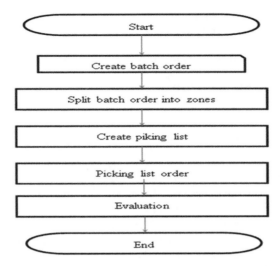

Fig. 1. Asynchronous parallel processing-type total picking method flow chart

function, it is necessary to make the sorting and picking operations more efficient. We prepared four approaches toward such efficiency improvements:

- Organize customer orders up to the upper limit of sorter capacity
- Reduce downtime of the sorting device
- Reduce movement distances for each picking
- Reduce the number of pickings

3.2   Batch Order Creation Algorithm

The goal for batch order creation is twofold: preparing customer orders up to the upper limit of sorter capacity, and reducing the travel distance for one picking. To reduce the travel distance for one picking in category-based placement, we apply Seed Algorithm.

Seed Algorithm continuously creates batches in two phases. The first phase assigns the first customer order to be newly batched. The second phase keeps assigning as many unassigned customer orders as possible. This creates one batch order. Seed Algorithm batches all customer orders by continuing to make assignments to unassigned orders. Allocation is based on seed selection rules that determine the first order and on order combination rules that determine additional orders.

The efficiency of a Seed Algorithm is expressed as the total time required for picking. The time taken for picking can be divided into time spent inside and outside of a category. It is assumed that the required time in the category is proportional to the number of product types, and that the required time outside the category is determined by the round-trip time between the base and the category farthest from the base.

A joint rule is set by giving priority to the category with the fewest additional categories in the batch, and by selecting the category with the shortest movement between additional bases and categories. The following rules for X and Y can be considered as policies for selection rules. In the case study, four rules combining X and Y (X1Y1, X1Y2, X2Y1, X2Y2) are considered.

① Prioritize clusters with many categories (X1), or prioritize clusters with few categories (X2)
② Prioritize clusters where the farthest category is closest to the base (Y1), or prioritize distant categories (Y2)

3.3   Picking Simulation

Divide the created batch order into each zone and create a picking list based on them. The goal is to reduce the number of pickings. It is assumed that the batch order divided by zone does not exceed the upper limit that can be carried by a picker at one time. The reason is that "The total number of items in each zone is 1 picker as a constraint condition the reason is that it is set to be "Picking Capacity" or less, and depending on the zone, a batch order by zone is created, in which the total number of talents is sufficiently lower than Picking Capacity. Therefore, in this algorithm, we combine as many batch orders as possible, divided into zones such that this constraint is satisfied.

By using an algorithm in the order of outputting the created picking list, we aim to shorten the non-operating time of the sorting device. The condition for operating the sorting device is a completed batch order existing in the temporary storage area. That is, in the case where there is a non-operating sorting device, a divided batch order from some zones has arrived at a temporary storage location but an order from other zones has not arrived, or immediately after the start of picking work. We first avoid picking list output timings for divided batch orders that are extremely separated between zones Then, we aim to complete batch orders as quickly as possible after the start of picking operations, and propose a picking list ordering algorithm.

We create a time-series picking simulation every minute, allowing us to evaluate the warehouse processing capacity from the viewpoint of both picking and sorting. The Parameters used in the simulation are shown in Table 1.

Table 1. Simulation parameters

Number of zones
Number of aisles
Number of pickers
Number of sorting machines
Number of orders processed by sorting equipment

4   Method Comparison Evaluation

We verified and compared the proposed method using data from an Internet mail order company that uses the total picking method. and has established picking, temporary storage, and sorting areas in a warehouse. Simulation evaluations were performed under the following conditions:

① A picker can handle up to 6 units in one picking.
② The sorting device can sort up to 20 customer order items at once.
③ The temporary storage area capacity is assumed to be sufficiently large.
④ There are 5 sorting devices and 20 pickers.
⑤ The farthest category requires one minute for round-trip travel from the base.

Efficient picking list output was performed for 1,228 customer orders accumulated over an average nighttime, and we verified that the processing capacity improved as compared with the current total picking method.

The simulation results in Fig. 2 show the superiority of the proposed method.

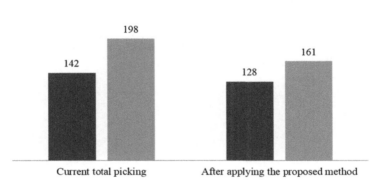

Fig. 2. Simulation results of the proposed method

5   Conclusion

Using an asynchronous parallel processing method for multiple zones, we proposed a reduced order batch method that improves final processing time over existing total picking methods and improves processing capacity in distribution warehouses. We demonstrated the proposed method's effectiveness using actual data.

References

1. De Koster, R., Le-Duc, T., Roodbergen, K.J.: Design and control of warehouse order picking: a literature review. Eur. J. Oper. Res. **182**(2), 481–501 (2007)
2. Parikh, P.J., Meller, R.D.: Selecting between batch and zone order picking strategies in a distribution center. Transp. Res. Part E: Logist. Transp. Rev. **44**(5), 696–719 (2008)
3. Elsayed, E.A., Unal, O.I.: Order batching algorithms and travel-time estimation for automated storage/retrieval methods. Int. J. Prod. Res. **27**(7), 1097–1114 (1989)
4. Hsu, C.M., Chen, K.Y., Chen, M.C.: Batching orders in warehouses by minimizing travel distance with genetic algorithms. Comput. Ind. **56**(2), 169–178 (2005)
5. Hsieh, L.F., Huang, Y.C.: New batch construction heuristics to optimise the performance of order picking systems. Int. J. Prod. Econ. **131**(2), 618–630 (2011)
6. Tsai, C.Y., Liou, J.J., Huang, T.M.: Using a multiple-GA method to solve the batch picking problem: considering travel distance and order due time. Int. J. Prod. Res. **46**(22), 6533–6555 (2008)
7. Albareda-Sambola, M., Alonso-Ayuso, A., Molina, E., De Blas, C.S.: Variable neighborhood search for order batching in a warehouse. Asia-Pac. J. Oper. Res. **26**(05), 655–683 (2009)
8. Žulj, I., Kramer, S., Schneider, M.: A hybrid of adaptive large neighborhood search and tabu search for the order-batching problem. Eur. J. Oper. Res. **264**(2), 653–664 (2018)
9. Won, J., Olafsson*, S.: Joint order batching and order picking in warehouse operations. Int. J. Prod. Res. **43**(7), 1427–1442 (2005)
10. Zhao, Z., Yang, P.: Improving order-picking performance by optimizing order batching in multiple-cross-aisle warehouse systems: a case study from e-commerce in China. In: Industrial Engineering and Applications (ICIEA), 2017 4th International Conference on, pp. 158–162. IEEE (April 2017)

Data and Analysis of Transportation Human Resources for in Guizhou Province Under the Background of Big Data

Yong Liu[1(✉)] and Chenggang Li[1,2]

[1] School of Big Data Application and Economics, Guizhou University of
Finance and Economics, Guiyang 550025, Guizhou, People's Republic of China
[2] Collaborative Innovation Center for Poverty Reduction and Development in
Western China, Guizhou University of Finance and Economics,
Guiyang 550025, Guizhou, People's Republic of China

Abstract. Human resource is an important factor for transportation development. From the actual situation of transportation human resources in Guizhou
Province, this paper mainly analyzes the current situation and current problems
of transportation human resources in Guizhou Province by combining the data
of the last ten years, draws some basic conclusions, and gives policy
suggestions.

Keywords: Guizhou province · Human resources · Transportation industry

1 Introduction

The report of the 19th Party Congress for the first time clearly put forward the
development strategy of building a "strong transportation country". In February 2021,
the Central Committee of the Communist Party of China (CPC) and the State Council
issued the "Outline of National Comprehensive Three-dimensional Transportation
Network Planning", which requires all regions and departments to implement it seriously with the actual situation. Guizhou Province is located in the southwest region of
China, and the terrain factor is the main factor restricting the development of transportation in Guizhou, and the human resource allocation structure of transportation
industry is also the secondary factor affecting the development of transportation in
Guizhou Province. The so-called transportation human resources refer to the staff
engaged in various transportation industries, mainly in railroad, highway, waterway,
aviation, urban public transportation, pipeline transportation, and other transportation
industries, covering various fields such as infrastructure, transportation services, and
maintenance management [1]. To understand the overall development trend and stage
characteristics of human resources in the transportation industry in Guizhou province,
and to compare the development of human resources in the transportation industry in
the country and neighboring provinces, it is beneficial for the government to formulate
corresponding policies and the layout of transportation human resources in accordance
with the local actual situation, and to respond positively to the development strategy of
"strong transportation country". Strengthening the construction of human resources in

J. C. Hung et al. (Eds.): FC 2021, LNEE 827, pp. 695–702, 2022.
https://doi.org/10.1007/978-981-16-8052-6_87

the transportation industry, enhancing the capacity of talent protection and improving the working environment of talents are of great significance to accelerate the development of modern transportation industry in Guizhou Province, speed up the implementation of innovation-driven and open-driven strategy, build a comprehensive three-dimensional transportation network system in the province, and promote the economic and social development of Guizhou Province. This paper then combines the data of recent years to make a brief exploration of the human resources of the transportation industry in Guizhou Province from the current situation of development and the existing problems.

2   Current Status of Transportation Human Resources

2.1   Total Human Resources Continue to Run at a High Level, the Individual Transport Industry has a Downward Trend

As shown in Fig. 1, the number of employees in the transportation, warehousing and postal industry in Guizhou Province increased overall, but the growth trend was relatively slow. By the end of 2019, the number of transportation workers in Guizhou Province is 131,662, an increase of 11,662, or 9.72%, compared with 120,000 in 1995. The total amount of human resources continues to operate at a high level. The high level operation of the total amount of human resources is not only reflected in the number of employees, but also in the number of legal entities in the industry. In 2015, the number of legal entities in the transportation, warehousing and postal industry in Guizhou Province was 4,392. By 2019, the number of legal entities in the industry was 8,270, 1.88 times that of 2015.

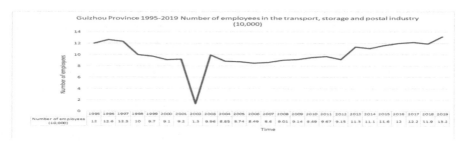

Fig. 1. Number of employees in the transportation, storage and postal industry in Guizhou Province, 1995–2019. Data source: Guotaian database

The number of employees in the major transportation industry in Guizhou Province from 1995 to 2019 is shown in Fig. 2, from which it can be seen that the total number of employees in the major transportation industry in Guizhou Province has been growing in the last two decades. In particular, the number of employees in the road transportation industry has grown most rapidly in the past decade, and the fluctuation is also the largest, from 32,000 in 1995 to 59,640 in 2019, compared with 1995, the number of employees in the road transportation industry increased by 86.38%, which transferred a large number of labor force population for the towns and rural areas. The railroad transportation industry has a large workforce and has maintained a steady development, with no significant fluctuations in the number of employees. Compared with 1995, the number of people employed in the railroad transportation industry has even decreased slightly, from 39,000 in 1995 to 33,450 in 2019, a decrease of 14.23%. The reason for the decline may be the emergence of intelligent ticketing systems and network ticketing in the context of the development of the big data industry, which requires the release of a large number of employees, and also reflects from the side that the railroad transport industry in Guizhou has developed more slowly than the road transport industry investment. At the same time, the human resources demand for water transportation industry in Guizhou Province is on a declining trend, from 3,000 in 1995 to 193 in 2019. Guizhou, located in the southwest inland, although very rich in water resources, but due to the influence of topographic factors, many of its advantages can not be brought into play, compared with the amount of funds for the construction of roads and railroads, the amount is very small, for water transport industry manpower funding investment is minimal. Water transportation is particularly important to the construction of a comprehensive three-dimensional transportation network system and economic and social development in Guizhou Province, which is also a short board that needs to be filled. The air transportation industry, on the other hand, has maintained a steady growth momentum, and as of 2019, the number of people employed in aviation in Guizhou province was 15,109, 15.11 times that of 1995. In particular, Guiyang City relies on the industrial structure layout and location advantages of "one airport and two districts" (Guiyang Longdongbao International Airport, Gui'an New District and Shuanglong Airport Economic Zone), which promotes the mutual development and deep integration of local economy and society and civil aviation industry [2]. As of 2019, the number of employees in the postal industry is 17,942, which is 1,058 less than the 19,000 in 1995, and from the line graph we also find that the scale of human resources in the postal industry has not fluctuated significantly and has been in a relatively stable state of development.

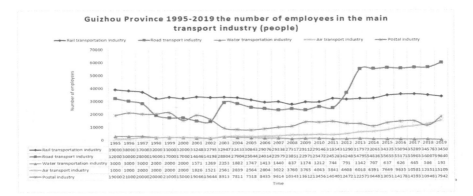

Fig. 2. Number of employees in major transportation industries in Guizhou Province, 1995–2019. Data source: Guotaian database

2.2   The Industry Environment has Improved Significantly

First, the province's transportation system has made a series of innovative reforms in the human resources work mechanism, strengthening communication and contact between the top and bottom, forming an efficient and reasonable work pattern and improving the operational efficiency of the industry; second, it has strengthened the investment in human resources funding and education and training for the transportation industry in Guizhou Province, improving the service level and soft power of the industry; third, it has improved the wages and benefits of the employees year by year. The living environment has been continuously improved to stimulate the vitality of in-service workers. The changes in the average wages of employees in the transportation, storage and postal industry in Guizhou Province from 2000 to 2019 are shown in Fig. 3.

As can be seen from Fig. 3, as of 2019, the average salary of employees in the transportation, storage and postal industry in Guizhou province is 91,525 yuan, an increase of 81,730 yuan compared with 9795 yuan in 2000. The number of employees in the transportation, storage and postal industry in Guizhou province is 131,662 in 2019, compared with 91,000 in 2000, and when converted into total salary calculation, the salary funding has increased from 891 million in 2000 to 12.050 billion in 2019, an increase of about 12.52 times in funding investment during the period.

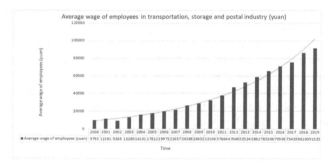

Fig. 3. Average wages of employees in the transportation, storage and postal industry in Guizhou Province, 2000–2019. Data source: Guotaian database

2.3   Uneven Regional Development of Human Resources

The total human resources in the transportation industry in Guizhou province show a general growth trend, but the regional distribution is indeed extremely unbalanced, and this unbalance is mainly reflected in the number of non-private sector employees in the transportation, storage and postal industry. Considering the availability of data, the number of non-private sector employees in transportation, storage and postal industry in six cities and states of Guizhou Province is listed in this paper, as shown in Table 1.

From Table 1 we can see that the total number of non-private sector employment in the transportation, storage and postal industry in Guizhou Province in 2019 is still relatively large, but the total human resources in the transportation industry shows a trend of regional imbalance, mainly concentrated in two regions, Guiyang and Zunyi, and the total human resources in other regions are relatively small. This imbalance in the total number of talents then leads to a serious shortage of professional and technical personnel, skilled personnel and highly educated personnel in some fields and units, especially in grassroots units in ethnic minority areas and backward rural areas, such as ship inspection and maintenance, turbines, ship driving, highway and waterway maintenance machinery operation, modern logistics, business management and other talents are particularly lacking [3].

Table 1. Number of non-private sector employees in the transportation, storage and postal industry in Guizhou Province 2019. Data source: Statistical Yearbook of each city and state

Region	Guiyang	Liupanshui	Zunyi	Anshun	Qiannan	Qiandongnan
Indicators	Transportation, storage and postal industry non-private sector employees (people)					
2019	89867	4745	11035	3353	4842	5980

2.4   The Overall Quality of Talent Continues to Improve

By the end of 2020, the overall size of the number of workers with higher education among the management talents of the system of Guizhou Provincial Department of Transportation and restructured enterprises, municipalities (states), counties (districts and cities) and their affiliated units has expanded, the number of workers with graduate degrees and bachelor's degrees has relatively improved compared to the previous ones, and professional and skilled talents have not only increased in quantity but also improved in structure.

3   Existing Problems

3.1   Serious Lack of Human Resources

There is a serious shortage of talents within the transportation enterprises in Guizhou Province. Firstly, there is a shortage of high-quality practical talents and management talents, especially the leading talents in major construction projects and safe transportation, and there is a shortage of senior engineers and senior managers in the transportation industry, which makes it difficult to carry out a reasonable allocation and management of hardware resources and human resources within the transportation enterprises, and thus unable to achieve reasonable work expectations; secondly, there is a shortage of basic employees. The situation of "more work and less manpower", unable to complete the stage tasks on time, ultimately affecting the image of the enterprise and the overall operational efficiency. At the same time, the province's transportation industry with a graduate degree in talent is significantly less, more is less than college education, low-level talent is sufficient, the level of education needs to be further enhanced, but also to increase the investment in staff training funds.

3.2   Internal Structure of Employees is Not Reasonable

Firstly, the age level of employees within the transportation enterprises in Guizhou Province is unreasonable, and there is an age discontinuity of employees within the industry. The majority of employees are of middle and senior age, and there are few young employees, especially the leading cadres tend to be old, the grassroots employees are old and lack of vitality, and there are few young reserve cadres; secondly, the design of staff positions does not take into account the actual situation, and the unclear job responsibilities lead to unclear distribution of staff tasks and unclear boundaries, and conflicts often occur between departments; thirdly, the quality and ability of employees vary, and internal management lacks coordination and unity, which often leads to inefficiency and failure to complete work on time.

3.3   The Internal System Mechanism of the Industry is Not Sound Enough

The internal system mechanism of the industry is not sound enough mainly in the following aspects. First of all, the talent assessment and evaluation and incentive system is not perfect. There is no separate assessment by post, most units have the phenomenon of "sharing food from the same big pot", lack of corresponding incentives and restraint mechanisms, and do not focus on stimulating the enthusiasm, initiative and creativity of employees; secondly, Guizhou Province is not enough training for young employees, the lack of career planning, ignoring staff development. Training is superficial and relatively single, without a complete and comprehensive training plan, lack of integrated planning and effective allocation of resources for employees.

4   Conclusions and Policy Recommendations

By studying the development of human resources in transportation in Guizhou Province and analyzing the data of previous years, this paper draws the following conclusions. First, the total human resources in Guizhou Province continues to run at a high level, with a declining trend in individual transportation industries; second, the industry environment has improved significantly, and the treatment of employees continues to improve, which stimulates the passion and vitality of employees; third, the development of total human resources shows a regional imbalance, with a serious lack of talents in minority and backward areas, which affects the overall development of the transportation industry in Guizhou Province; fourth, the overall quality of talent continues to improve accompanied by a serious shortage of talent resources, but there is also a problem of unreasonable internal structure of the staff and the lack of sound internal system mechanisms in the industry.

In response to the above analysis and related findings, this paper presents the following policy recommendations:

Continuously increase investment in waterway transportation, introduce human resources in the waterway transportation industry and make up for the shortcomings of waterway transportation. Water transportation has the advantages of low cost, high capacity, low pollution and high versatility compared with road transportation, railroad transportation and other transportation [4], and has a broad development potential. Guizhou Province is relatively rich in water resources, but due to topographic factors, water transportation in Guizhou Province started late compared with other strong waterway transport provinces, and there are still certain gaps, which are mainly the gaps in waterway transportation infrastructure and human resources. The gap of waterway transportation infrastructure is mainly the aging of ships, poor navigation conditions, small ship tonnage, etc. To improve the navigation infrastructure conditions in Guizhou Province, open up new water transport areas and speed up the development of shipping, comprehensively coordinate the planning and construction of coastal, adjacent river regional railroads, highways and port connections, enhance the radiation-driven role of ports, and gradually realize the seamless connection of water transport and other modes of transport [5]. The gap of talent resources is mainly the total number of talents is low, and there is a serious shortage of technical talents in the shipping industry; in the future, we should continuously improve the talent introduction system, vigorously introduce technical talents, and encourage the construction and cultivation of the professional water transportation talent team.

Increase investment in infrastructure of transportation industry in Anshun, Liupanshui and other areas as well as rural and backward areas to encourage talents to go to the grassroots. From Table 1 we can see that the current center of gravity of human resources in the transportation, storage and postal industry in Guizhou province is concentrated in the Guiyang and Zunyi regions, which has led to an increasingly serious polarization of talent resources in regional distribution. In order to balance the regional development of talent resources, the construction projects and policies of the transportation industry are more inclined to non-Guiyang and Zunyi areas, and professional and technical talents, skilled talents and highly educated talents are encouraged to go to the grassroots units in minority areas and backward rural areas.

Improve the talent introduction system, internal structure of employees and internal system mechanism. Firstly, Guizhou Province currently has a serious shortage of talents in the transportation industry, which requires continuous improvement of the talent introduction system in Guizhou Province, introducing various talents in the transportation industry through various channels and ways to make up for the serious shortage of talents in Guizhou Province; secondly, further improving the situation of unreasonable internal structure of employees, introducing young talents, training and reserving young cadres, and strengthening the investment and training of talents; thirdly, introducing scientific human resource management methods and models, adhere to the staff-centered, do a good job of staff career development planning, unify the staff job planning, the intrinsic needs of employees and corporate development goals, and strive to seek the unity of strategic planning of human resources and corporate development goals.

References

1. Zuping, L., Chen, Y., Shunxuan, L., Yu J.: Analysis of human resources situation in the transportation industry and its development trend. J. Beijing Traffic Manag. Cadre Coll. **2005**(01), 7–12 (2005)
2. Tian, P.: An analysis of the advantages of air transportation development in Guizhou and its development and utilization–take changlong airlines co. Bus. News **2020**(13), 11–12 (2020)
3. Guizhou Provincial Department of Transport. Guizhou transportation talent 13th Five-Year Plan development plan. http://jt.guizhou.gov.cn/xxgk/zdgk/ghjh_16064/zxgh/201609/t20160908_27168592.html,2016-05-31
4. Wensheng, Z.: The current situation and development suggestions of human resource management in Anhui water transportation industry. Ship Sea Eng. **43**(01), 58–61 (2014)
5. Wei, L., Jia, S.: Thinking about accelerating the development of shipping in Guizhou. China Water Transp. (Second Half Mon.) **13**(06), 3 (2013)

Application of Partial Least Squares Method Based on Big Data Analysis Technology in Sensor Error Compensation

Xiaoli Wang, Fang Wang, and Kui Su[✉]

Mudanjiang Medical University, Mudanjiang 157011, Heilongjiang, China

Abstract. In practice, uncertainties such as humidity and temperature cause unavoidable random errors in the sensor data. In order to reduce the error, a quick and easy way to deal with it is to use the least squares method to correct the data by linear regression. The linear compensation method can solve many sensor measurement errors for random phenomena such as noise, but usually the actual function of the measurement data itself is nonlinear, and using a linear function to simulate nonlinear measurement data often results in a lack of accuracy. For this reason, this paper designs a way to increase the order of independent variables and variable coefficients to improve the compensation accuracy, and because the increased variable coefficients may lead to problems such as multiple correlations, a single dependent variable partial least squares method is used instead to build the compensation model.

Keywords: Sensor data · Least square · Data compensation

1 Introduction

Due to the influence of the sensor's property or humidity, temperature, noise, etc., random errors will inevitably occur in the measurement of the sensor [1, 2]. Nowadays, sensor data compensation methods are mainly divided into hardware compensation and soft compensation (digital compensation) [3, 4]. The former method is to improve the sensor process, the accuracy, or optimize the measurement circuit and soft compensation by combining their respective advantages to achieve the sensor measurement data compensation [5]. In the latter way, intelligent algorithms, including numerical analysis or neural network learning, are used to regress the collected data [6]. Hardware compensation can also be integrated to improve the purpose of accurate measurement [7].

Regression compensation based on least square method has been widely used in soft compensation methods of various sensors due to its wide application range, simple modeling, convenient operation and other characteristics [8–10]. According to Gauss-Markov theorem, when the sensor input and output functions are linear and the measurement error is completely random and normally distributed, the compensation function obtained by Least Squares is the unbiased estimation of the actual function. However, it is difficult to guarantee that the actual function must be linear in the real state, and the measurement error may not be completely normal distribution.

J. C. Hung et al. (Eds.): FC 2021, LNEE 827, pp. 703–709, 2022.
https://doi.org/10.1007/978-981-16-8052-6_88

In order to solve the error when the actual function may be non-linear, this paper discusses the way to increase the order of independent variable and variable coefficient to improve the compensation accuracy. Theoretically, as long as n is large enough, a polynomial of order n can represent any free function. As long as the order of the independent variable is increased enough, more accuracy can be improved in the data compensation. However, the increase of order leads to the increase of coefficient terms, which may cause multiple correlation problems when the number of samples is insufficient. In order to solve this problem, partial least-squares is used to establish the solution model to eliminate the irrelevant terms and reduce the error.

2   Linear Compensation Method for Sensor Data Error Based on Least Squares

Assume that the sensor has input value matrix $X(X_{n \times p})$ and output value matrix $Y(Y_{n \times 1})$, and the real function relationship can be expressed as $Y = f(X)$, where n is the number of input and output measured values, and p is the total number of input variables.

In the actual measurement, the error matrix caused by random factors such as ambient humidity and temperature is denoted as $\varepsilon(\varepsilon_{n \times 1})$, thus $Y = f(X) + \varepsilon$.

When the relation is approximately linear with respect to the parameter and approximates the standard normal distribution, there is an approximate compensation function relation $Y_{n \times 1} = X_{n \times (p+1)} B_{(p+1) \times 1} + \varepsilon_{n \times 1}$.

In this case, the coefficient matrix $B_{(p+1) \times 1} = \begin{bmatrix} \beta_0 & \beta_1 & \cdots & \beta_p \end{bmatrix}^T$, $(\beta_0, \beta_1, \beta_2, \cdots, \beta_p)$ of compensation function f is the corresponding linear coefficient of f.

At this point, the input value matrix X can be expressed as

$$X_{n \times (p+1)} = \begin{bmatrix} 1 & x_{11} & x_{12} & \cdots & x_{1p} \\ 1 & x_{21} & x_{22} & \cdots & x_{2p} \\ \vdots & \vdots & \vdots & \ddots & \vdots \\ 1 & x_{n1} & x_{n2} & \cdots & x_{np} \end{bmatrix}$$

Let $B_{(p+1) \times 1}$ be evaluated as $\hat{B}$, and the sum of squares of least squares Y and estimated $X\hat{B}$ should be minimized, that is $\left\| Y - X\hat{B} \right\|^2 \to min$, and $\left(Y - X\hat{B} \right)^T \left(Y - X\hat{B} \right) \to min$. Expand the left side of the preceding expression and take the partial derivative of $\hat{B}$ to get:

$$\hat{B} = \left(X^T X \right)^{-1} X^T Y \tag{1}$$

3   Problems in Realistic Modeling and a Way to Improve

In the ideal state, when n groups of sensor observation data are given, the regression coefficient $\hat{B}$ of the linear compensation model can be obtained from Eq. 1, and then the corrected output $\hat{Y}$ can be obtained according to the regression equation. The linear compensation model established by least squares needs to meet the following basic assumptions: 1. There is no significant linear correlation between the input variables. 2. The actual function is a linear function. 3. The error is independent and random.

In practical measurement, only the data sets of each input and output can be observed, but the specific model function relationship is not clear. In other words, we do not know whether the function is linear or nonlinear, so we can only assume that the function relation is linear through estimation.

According to Weierstrass approximation theorem, any continuous function in a closed interval can be approximated by m-degree polynomials, as long as m is large enough.

In this paper, the linear compensation model based on the least square method is improved by using the basic idea of the theorem above, and the compensation accuracy is improved by increasing the order of independent variables and the coefficient of variables.

Now, it is assumed that the input-output relationship of the sensor is $y = f(X_p)$. The linear compensation model established is $y = \beta_0 + \beta_1 x_1 + \beta_2 x_2 + \cdots + \beta_p x_p + \varepsilon$. We can increase the order of the independent variable by the approximation theorem, when the order is increased to 2

$$y = \beta_0 + \beta_{01} x_1 + \cdots + \beta_{0p} x_p + \beta_{11} x_1^2 + \cdots + \beta_{pp} x_p^2 + \beta_{12} x_1 x_2 + \\ \cdots \beta_{(p-1)p} x_{(p-1)} x_p + \varepsilon \tag{2}$$

It can also be expressed by a substitution of variables: $y = \beta_0 + \beta_1' x_1' + \beta_2' x_2' + \cdots \beta_m' x_m' + \varepsilon$, it's essentially the same thing as

$$y = \beta_0 + \beta_1 x_1 + \beta_2 x_2 + \cdots + \beta_p x_p + \varepsilon \tag{3}$$

When the order m is large enough, and even the sum of the coefficients after the order is increased approximates to the observed number n, the n-element linear equation can be solved directly, and the compensation function passes every observed value point.

However, due to the addition of too many variable coefficients, the random error is treated as a real function, so the model established in this way cannot truly reflect the actual function, and there is no error compensation for the output value of the function. At the same time, the increasing order brings up serious linear correlation problems. Since the added additional terms do not always truly reflect the actual function, this paper uses least squares to eliminate the irrelevant terms to solve the problem caused by the linear correlation and the sum of the coefficients may be greater than or equal to the data items.

4   Least Squares Solve Sensor Data Error Compensation

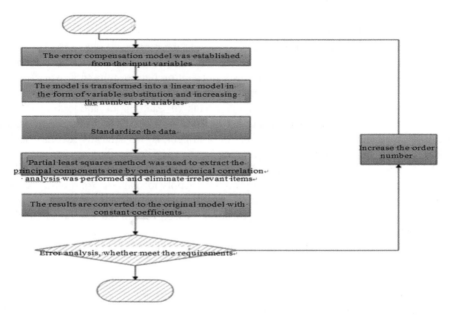

Fig. 1. Flow chart of sensor data error compensation method based on partial least squares

4.1   The Basic Idea of Solving the Problem

Firstly, the data compensation model is established through least squares and the corresponding coefficients are solved through Eq. (1). The error accuracy of the data error compensation model obtained is verified. If the accuracy is satisfied, stop, otherwise increase the order. Variable substitution is carried out at the same time as increasing the order, and all items in the form of Eq. (2) are changed into the form of Eq. (3).

After the transformation of Eq. (3), the data are standardized and substituted into the standardized data. Least squares is applied to carry out principal component analysis and canonical correlation analysis, and a regression model is established to eliminate the irrelevant items. If the accuracy is within a reasonable range, the appropriate term will be retained as the independent variable, and the corresponding coefficient will be obtained and replaced to the original model form. If the error is not enough, the order will be further improved. The flow chart of the algorithm is shown in Fig. 1.

4.2   Data Standardization

Data centralization means that the coordinate center of the transformation is shifted to the bary-center of the sample point set through translation transformation, which makes the calculation simple and does not change the correlation between variables. The compression of data is to make the variance of each variable equal to 1, which eliminates the dimensional effect of variables and makes each variable have the same dimension. Standardization of data means centralized and compressed data at the same time.

Assume that has matrix $X = (x_{ij})_{n \times p}$ for each of these elements $x_{ij}^* = (x_{ij} - \overline{x_j})/s_j$ among which, $i = 1, 2, \cdots, n$, $j = 1, 2, \cdots, p$, $\overline{x_j}$ *is the mean of* x_j, *and* s_j *is the variance of* x_j. Input matrix X and output matrix Y can be changed into $E_0 = (x_{ij}^*)_{n \times p}$ and $F_0 = (y_i^*)_{n \times 1}$ through data normalization transformation.

4.3   Partial Least Squares Solving Method

When the input matrix X and the output matrix Y are transformed into E_0 and F_0 through data standardization, the first component t_1 and u_1 are extracted from E_0 and F_0. t_1 is a linear combination of $x_1, x_2, \cdots x_p$, and since it's a single output, u_1 is equal to $u_1 = F_0$.

These t_1 and u_1 should carry more specificity in their respective data and maximize the correlation between t_1 and u_1. In the formula, t_1 and u_1 have the largest co-variance, and t_1 and u_1 have the largest variance respectively.

Let's say we have $t_1 = E_0 w_1$, and have $w_1 = E_0^T F_0 / \|E_0^T F_0\|$ according to the Lagrangian simplification. So we can set up a regression equation for E_0 to t_1, $p_1 = E_0^T t_1 / \|t_1\|^2$,

$$E_0 = E_1 + t_1 p_1^t. \tag{4}$$

According to Eq. (4), E_1 and F_1 can be obtained, and then E_1 and F_1 are used to replace E_0 and F_0 for substituting, until the crossover meets the conditions (Fig. 2).

5   Experiment Analysis

In order to facilitate the experiment and present intuitive two-dimensional coordinate observation effect, a single set of input and output data of the voltage sensor is taken for testing. When the voltage gradually increases, the average observed values of the sensor input voltage and output are shown in Table 1.

The difference between using least squares and using improved least squares with increased order can be intuitively felt in Fig. When using least squares for linear compensation, the fitting curve becomes a straight line. When increasing the order to 2, the fitting curve can better reflect the real point and the fitting effect is better.

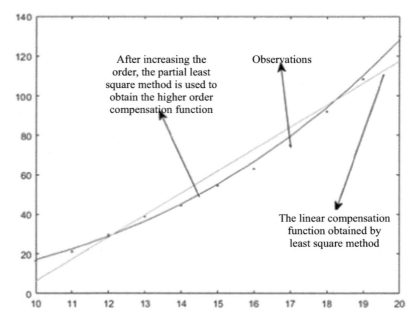

Fig. 2. Comparison of least squares and modified partial least squares

Table 1. Average observed values of sensor input voltage and output when voltage is gradually increased

10	11	12	13	14	15	16	17	18	19	20
21.569	26.742	26.742	33.74	47.72	57.7	68.97	81.91	90.019	111.92	127.63

6 Conclusion

In order to visualize the experimental data with single input and single output, the algorithm is also suitable for data compensation with multiple input and single output. It can be seen from the experiment that appropriately increasing the order can better fit the compensation data. In order to solve the problem of linear correlation and too few data samples caused by increasing order, partial least squares method is an effective way to replace least squares. The results show that the improved method based on least squares is effective for sensor error compensation.

References

1. Volponi, A.J.: Sensor error compensation in engine performance diagnostics. In: Turbo Expo: Power for Land, Sea, and Air, vol. 78873, p. V005T15A008. American Society of Mechanical Engineers (1994)

2. Yi, B., Chu, B.C.B., Chiang, K.S.: Temperature compensation for a fiber-Bragg-grating-based magnetostrictive sensor. Microw. Opt. Technol. Lett. **36**(3), 211–213 (2003)
3. Zhang, H., Hong, Y., Qiu, J.: An off-policy least square algorithms with eligibility trace based on importance reweighting. Clust. Comput. **20**(4), 3475–3487 (2017)
4. Kiani, M.: Extensions to the modified Gram-Schmidt strategy and its application in the steepest ascent method. J. Stat. Comput. Simul. **80**(4), 389–400 (2010)
5. Veena, P.V., et al.: Least square based image denoising using wavelet filters. Indian J. Sci. Technol. **9**(30) (2016)
6. Mohebbi, M., Nourijelyani, K., Zeraati, H.: A simulation study on robust alternatives of least squares regression. J. Appl. Sci. **7**(22), 3469 (2007)
7. You, K., Song, S., Qiu, L.: Randomized incremental least squares for distributed estimation over sensor networks. IFAC Proc. Vol. **47**(3), 7424–7429 (2014)
8. Dickow, A., Feiertag, G.: A framework for calibration of barometric MEMS pressure sensors. Procedia Eng. **87**, 1350–1353 (2014)
9. Srivatsa, S., et al.: Application of least square denoising to improve ADMM based hyperspectral image classification. Procedia Comput. Sci. **93**, 416–423 (2016)
10. Hanlon, P., Lorenz, W. A., Strenski, D.: Least-squares fit of genomic data by sums of epistatic effects. J. Parallel Distrib. Comput. **63**(7), 683–691 (2003)

A Study on the Exterior-Interior Relationship of Pansystems

Weiwen He[(⊠)]

School of Information Engineering, Guangzhou Nanyang Polytechnic,
Guangzhou 510000, Guangdong, China

Abstract. The paper extends the exterior-interior relations in the meaning of pansystems and puts forward these relationships in computer science. The exterior-interior relationship here means that a pair of categories those are used to illustrate the relationship in the objective reality which is the target of both the cognition activity and the practice activity. It can help us to have a better understanding of both the objects and the relationship among these objects. In computer science, many relationships among the research objects can be concluded in the relationship of the exterior-interior relations. It would make sense to provide theoretical foundation in the cognition of the complex computer systems.

Keywords: Pansystems · Exterior-interior relationship · Computer systems · Algebraic-Systems

1 Introduction

With mathematical foundation of classical set theory, pansystems theory is formally put forward by Chinese scholar Xuemou Wu in 1976 [1, 2]. This theory deals with generalized systems, relations and transformations. There are many subsequent studies of it, e.g. [3, 4].

Studies of pansystems theory have been taken place in science, technology, literature and arts, and in various human activities [5, 6]. There are many artificial or virtual subjects, such as coordinate systems, reference systems, topology spaces, knowledge databases, theoretical frameworks, windows, finders, axiomatic systems, rough set systems, various panboxes, various manmade concentration-decentration-observo control tools, general manmade eight-counter tools, etc. [7]. There are also some studies in computer science based on pansystems, e.g. [8].

In this paper, we put forward the exterior-interior relations and illustrate the relations by two aspects: one is the relation between the algebraic system and the product algebra, and the other is the relations in the developing of Management Information Service.

2 "Coordinate System"-"Product Algebra" Model is Generalized Exterior-Interior Relationship

2.1 The Pansystems View of Coordinate Axes and the Algebraic System

The algebraic system can be denoted as R=<R, ■>, ■ denotes all the operations of arithmetic defined on R.

The x axes illustrated in Fig. 1 has an origin O.

In the meaning of geometry, O denotes the original point of the x axes. Its visual meaning is that O is certain position in space. When the coordinate axes are given algebra sense and put into a one-to-one correspondence with real number, some operations on the field of real numbers can be denoted as the relative position among points on the coordinate axis. For example, 2 < 3 can be denoted as the relative position that 2 is in front of 3 on the coordinate. In a sense, coordinate axes can be treated as the algebraic system R=<R, ■>. 0 in R can be seemed as O on X axes. We can see that 0 and O are the same things, and we also can see that 0 and O are different. Why?

0 and O are the same thing; it means that there is corresponding relationship between 0 and O. 0 and O are different; it means that 0 is a real number, but O is a certain position on the X axes.

In the meaning of exterior-interior relationship, the coordinate axes are exterior, and the algebraic system is interior. Geometric representation is intuitively, image, but one-sided. Algebraic representation is nature, precise and comprehensive.

2.2 The Pansystems View of Coordinate System and Product Algebra

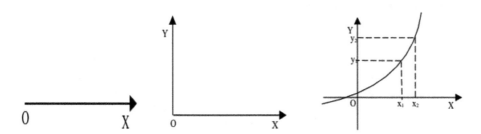

Fig. 1. The X axes   **Fig. 2.** The coordinate system   **Fig. 3.** The increasing function

Coordinate system is also geometrical expression of product algebra as show in Fig. 2.

The X axes and the Y axes are both the algebraic system R=< R, ■>, and the origin O of both X axes and Y axes is also 0 in real number R. Function and geometric graph in the coordinate system are one-to-one correspondence. Function is algebraic representation, and graph is geometric representation. They are the same thing in a sense, because people see that the graph is rising, they will say that the function is increasing.

According to exterior-interior relationship in pansystems [9, 10], coordinate system is exterior, because geometric representation is intuitively, image, but one-sided. Product algebra is interior, because algebraic representation is nature, precise and comprehensive.

In coordinate system, people see that the graph is rising, as show in Fig. 3. In product algebra, x$\hat{\mathrm{I}}$ R, if x1 < x2 and F(x1) < F(x2), the function F(x) is increasing.

Graph is geometric representation, but it isn't product algebra. We are using the grid of the coordinate to observe the graph. Let the grid be small to an infinitesimal, the coordinate system can be the algebraic system.

In the view of exterior-interior relationship, it is exterior that graph of a function is ascending, and it is interior that the function is increasing. In the view of subject-object relationship, the result, geometrical graph is ascending, means that coordinate system observes the graph. The result, function is increasing, means that product algebra

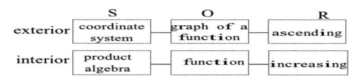

Fig. 4. The mode of S.O → R in computer science

observes the function. We can see that, the integrated relationships consisting of both exterior-interior relationship and the subject-object relations as show in Fig. 4.

2.3 The Exterior-Interior Relations Between "Coordinate System" and "Product Algebra" is Relative

From the discussion above, we can get a notion that the relationship between the coordinate system and the product algebra is the exterior-interior relationship. Function and geometric graph in the coordinate system are one-to-one correspondence. They are the same thing in a sense, because people see that the graph is rising, they will say that the function is increasing.

We can treat the coordinate system and the product algebra as a whole and consider them to be the "interior". The exterior-interior relationship is a pair of categories. What is the corresponding "interior"? The answer to this question is the correspondence tables between two quantities such as the hash table between keywords and the physical storage address, the mapping table between the plain text and the cipher text, and so on.

Therefore, the coordinate system is the "exterior" in the exterior-interior relationship between the coordinate system and the product algebra. But it is also part of the "interior" in the exterior-interior relationship, when we treat the coordinate system and the product algebra as the same thing. And the "exterior" is the corresponding table.

2.4   The Changes of "Interior" Will Be Reflected in the "Exterior"

In the exterior-interior relationship of the coordinate system and the product algebra, coordinate system is exterior, and product algebra is interior.

If a function in the product algebra, F(x) is increasing in a certain definitional domain; we can see that the corresponding graph is ascending in the coordinate system. If the function in the product algebra, F(x) is decreasing in another definitional domain, we can see that the corresponding graph is descending in the coordinate system. So the change of the function can be reflected in the graph.

Considering the coordinate system and product algebra as a whole, the "interior" is both the coordinate system and product algebra. The "exterior" is the corresponding table.

In the exterior-interior relationship between the hash table and the hash function, if the hash function is increasing, or the graph of the hash function is ascending, the hash value in the hash table is getting bigger and bigger. If the hash function is decreasing, or the graph of the hash function is descending, the hash value in the hash table is getting smaller and smaller.

2.5   In Some Cases, the "Exterior" Can Affect the "Interior"

In the exterior-interior relationship of the coordinate system and the product algebra, coordinate system is exterior, and product algebra is interior.

If a graph is ascending in a certain range, we can judge that the corresponding function is increasing in the corresponding definitional domain. If a graph is descending in a certain range, we can judge that the corresponding function is decreasing in the corresponding definitional domain. So we can get some recognition only through the graph. The judgment of the function usually is made through the corresponding graph for that the "exterior" is intuitively and image.

Considering the coordinate system and product algebra as a whole, the "interior" is both the coordinate system and product algebra. The "exterior" is the corresponding table. In the exterior-interior relationship between the hash table and the hash function, if the hash value in the hash table is getting bigger an bigger, we can get a rough and quick judge that the hash function may be increasing, or the graph of the hash function may be ascending,. If the hash value in the hash table is getting smaller and smaller, the hash function may be decreasing, or the graph of the hash function may be descending.

3   The Relationship in Computer Programming Between the Exterior and Interior

With the development of technology, computer programming language is also undergoing changes. Yet whatever changes have taken place in programming languages, a number of programming ideas, operational processes still play a strong role. The paper below discusses the exterior–interior relationship in computer programming through the developing processes of management information system.

3.1   The Exterior and Interior Relations of Programming is Relative

In applications of computer programming, research and development of Management Information Systems (MIS) has great market potential, for example, ERP enterprise management software. This kind of systems can be abstracted into "Database - Code - Interface" model, which fully shows in the relationship between exterior and interior in programming.

Database (DB) is the real place to store the data. It is the foundation of the system, that is, the "interior" of the system. Code can access the database through a number of features in the data (such as the SqlCommand class, SqlDataAdapter class in.NET). That is, relative to the database, code layer is the "exterior".

If you want to query all the information related to 25 years old in table Student, and get the results displayed in the form with the dataGridView control (using Visual Studio 2005 environment, C# language and SQL Server 2005 database). The SQL is "select * from Student where age = 25".

The processed data, (usually through SQL statement processed), are stored in the DataSet class or the DataReader class. In this case, the query results are from the database, but they are saved in memory in the form of a copy. It needs further operation through codes.

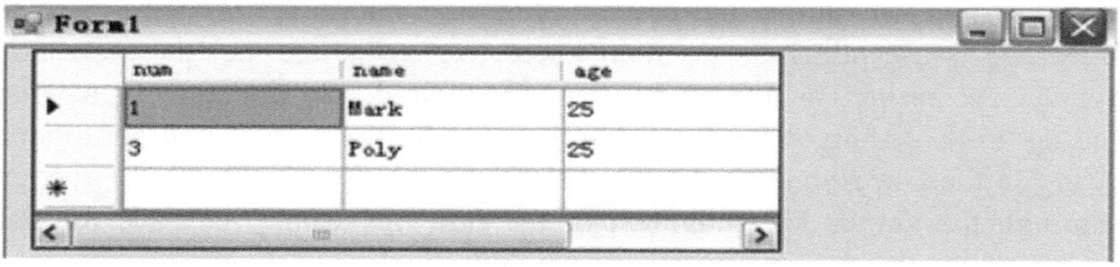

Fig. 5. The program results

When the processed data are seen by common users, the result in the code layer becomes "interior" and can be displayed some controls, such as the TextBox control, ListBox control. This case uses the dataGridView control. The processing results are shown in Fig. 5 as follows:

At this time, relative to the results in code layer, the data displayed on dataGridView interface becomes the "exterior". It is the result of code processing.

Through this case we can see, the exterior and interior relations of programming is relative. In a different frame of reference, the "exterior" can be converted into "interior".

Thus, we can see from the above case that the relationship between exterior and interior in programming is relative. With different reference, the "exterior" can be converted into "interior".

3.2   The Changes in "Interior" Will Be Reflected on the "Exterior"

Since the exterior and interior in programming are closely interwoven, the change of "interior" will inevitably affect the "exterior". For example, the change of the name Poly to Mary in the third line in table Student.

At this time, the interior of the system—the database—has changed, which will be reflected on the "exterior" level. After code execution, the copy of query result saved in the DataSet class will change accordingly.

Relative to the interface, the code layer becomes the "interior", and the contents of the interface will change as well.

This above case shows that when the database, as the "interior", has changed, the code layer and the interface, as "exterior", will change as well.

What changes will it cause to the "interior and exterior" if the code layer changes

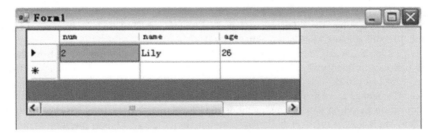

Fig. 6. The implementation of the results

only?

We are still using the database information of Fig. 1. Change the SQL to "select * from Student where age = 26" to get all relevant information to 26 years old. The relative "interior" –the code layer—has changed, and it will be directly reflected on the "exterior", that is, the interface. The implementation of the results shown in Fig. 6:

3.3   In Some Cases, the "Exterior" Can Affect the "Interior"

In the Management Information System, the interface, as the "exterior", can affect the data stored in the database. For example, when the data is inserted, changed or deleted in the interface, the processing results can be saved into the database through the code layer.

For example when we want to add a record from the interface, we click the "Insert" button. Interface operate can affect the data in the database through the code.

4   Conclusions

Based on pansystems theory, we extended the exterior-interior in the field of computer science. Other relationships in computer science can also be concluded to the extended exterior-interior relationships such as rough sets theory and artificial intelligence. The major advantage of this paper is that we used the analysis methods of pansystems to get some cognition in computer science. It would make sense to have more comprehension in both technology and philosophical meaning.

References

1. Xuemou, W.: The Pansystems View of the World, Beijing, pp. 1–316 (1990)
2. Li, Y., Li, Z., Qu, J., Wang, X.: The sorites paradox from the view of pansystems. In: Proceedings of the 2007 International Conference on Foundations of Computer Science, pp. 219–223. CSREA Press (2007)
3. Xuemou, W.: Pansystems Theory A Historical Record, pp. 289–362. University of Science and Technology of China Press, Hefei (2005)
4. Li, Y., Liu, Y., Wang, X.: An initial comparison of fuzzy sets and rough sets from the view of pansystems theory. In: 2005 IEEE International Conference on Granular Computing, pp. 175–179 (2005)
5. Wu, C., Wang, L.: An improved limited and variable precision rough set model for rule acquisition based on pansystems methodology. Kybernetes **37**(9/10), 1264–1271 (2013)
6. Chen, M., Yu, L., Xu, Y., et al.: Pansystems variation theory and its application to computer science. Kybernetes **38**(10), 1787–1793 (2009)
7. Yi, L., You, W., Li, C., Wu, X.: Pansystems ecology, management and knowledge rediscovery. Kybernetes **38**(1/2) (2009)
8. Lin, Y., Chuan-bin, D., Peng, C.: Broad-spectrum analysis and pansystems methodology. Kybernetes **40**(5/6), 824–830 (2011)
9. Smith, M., Xuemou, W.: Pansystems 0**-logoi of philosophy, mathematics, technology, systems and society. Kybernetes **40**(1/2), 213–250 (2011)
10. Fu, Q., Gong, F., Jiang, Q., et al.: Risk assessment of the city water resources system based on pansystems observation-control model of periphery. Nat. Hazards **71**(3), 1899–1912 (2014)

Exploration of the Innovative Path of National Cultural Symbol Communication in the Big Data Era

Yan Zeng[✉]

College of Cultural Tourism and International Exchange,
Yunnan Open University, Kunming, Yunnan, China

Abstract. This paper uses data mining technology to establish a mathematical model of ethnic cultural symbols, and uses the Cholisky decomposition method to study the path of innovation in the propagation of ethnic cultural symbols. This paper compares the national culture that uses big data public opinion and cultural projects to promote the national culture and the national culture that does not use big data technology to carry out general publicity, and analyzes the speed of the spread of the two and the people's love for this culture to explore the spread of national cultural symbols New path. Through the experimental analysis of this article, it is found that the national cultural symbols spread through using big data will be a simpler way to be accepted by the public, spread faster, and have a wider path, its popularity reaches 70%. Through experiments, it is concluded that ethnic cultural symbols use a modern scientific method such as data mining to study more in-depth ethnic culture, dig out relevant and more interesting ethnic cultures, and then disseminate them, greatly increasing the public's acceptance of ethnic cultural symbols.

Keywords: Big data · National culture · Cultural communication

1 Introduction

Today with the vigorous development of computer science and technology, big data has exerted a great influence on people's life. Today everyone is a participant and beneficiary of big data. This is largely due to the rise of computers, the internet and technology capable of capturing data from the world people live in. Big data is a manifestation of fast-growing Internet technology [1–3]. In today, the construction of various information systems, the Internet, local area networks, databases, other software and hardware has also been gradually upgraded and improved with the development of national culture. Video, photos, news and other data information showed explosive growth. China has excellent national cultural and historical heritage and regional unique folk cultural properties [4]. Under the conditions of modern information technology, cultural visual databases of different ethnic groups with functions such as collection, integration, preservation, and opening are also being established [5]. For example, Yunnan University established a human imaging base, used modern communication technology to protect the cultural heritage of ethnic minorities, and achieved good results. These attempts will greatly enrich the types of cultural transmission resources in the country.

Through experimental analysis, the researchers need to innovate a new path for the spread of national cultural symbols from the following three points. The first is to establish the understanding of big data, to implement the collection and processing and application of information and data, and to provide the possibility of achieving innovative changes in the content and methods of popularizing national culture; the second is to use big data technology to strengthen and optimize the national Human thinking and psychological analysis; the third is to use big data technology to realize the direction and service of popularizing national culture.

2 Research Methodology

2.1 Application of Data Mining Technology

At first the process of data mining chooses the required data samples coming from the database, and then according to the specific requirements for the next step in the sample data it sorts data. In accordance with the principles and standards of data mining, the researchers will adjust the pre-processed data, and then for the adjusted data in the sample, the researchers will use statistics and probability light box light model to bring these data into the analysis, and finally evaluate the data for finding gaps. The data that comes out of this way is through data mining technology [6].

Cholisky decomposition method needs to calculate the coefficient by mathematical model. From the data sample the researchers randomly select model parameters y and m self-parameters $X_{m-1},.....X_0$, sorting and analyzing the relationship between these parameters which can be expressed by the formula (1):

$$y = a_0x_0 + a_1x_1 + \ldots + a_{m-1}x_{m-1} + a_m \tag{1}$$

Put these parameters into the formula, and then do a linear analysis, where $a_0, a_1, \ldots, a_{m-1}, a_m$ are equation coefficients, which are fixed values.

According to the mathematical linear law, the researchers need to find the minimum value of

$$q = \sum_{i=0}^{n-1} [y_i - (a_0x_0 + a_1x_1 + \ldots + a_{m-1}x_{m-1} + a_m)]^2 \tag{2}$$

and the parameter coefficient $a_0, a_1, \ldots, a_{m-1}, a_m$ needs to adapt to the following linear system of equations:

$$(CC^T) \begin{pmatrix} a_0 \\ a_1 \\ a_2 \\ \ldots \\ a_{m-1} \\ a_m \end{pmatrix} = C = \begin{pmatrix} y_0 \\ y_1 \\ y_2 \\ \ldots \\ y_{n-1} \\ y_{n-1} \end{pmatrix} \tag{3}$$

2.2 New Ways to Spread National Culture

The digitization of traditional culture has two parts: resource collection and processing. Resource collection usually uses digital cameras or high-resolution scanners to statically collect text documents, and dynamically collect opera, folk music, and dance audio and videos through photo and video technology. Resource processing usually uses multimedia information processing technology or image creation software to classify audio, video, and text, and save various forms of storage [7]. After the resources are digitized, a related graphic image database has been established, which can be easily obtained and extracted, providing a large amount of digital resource can support for the research and utilization of traditional literature and art. Digital protection model has obvious advantages over traditional protection model. Digitization can break geographical restrictions and realize resource sharing in different places. In addition, the digitization of art resources can increase the amount of data. The larger the database, the more comprehensive the data, and the higher the reliability of the analysis [8].

Establish a big data platform for the traditional culture of ethnic minorities. Big data requires a big quantity of data, and the expansion of the data volume depends on the number of reads, so the increase of the data volume requires the public database. A large amount of research data is used in the internal resources of various research institutions, universities and cultural enterprises. Due to the realization of resource sharing, many academic researches cannot conduct interaction and comprehensive analysis between databases, and the data utilization rate is low, and the advantages of database establishment cannot be realized [9]. In view of such problems, China has proposed a big data action plan to promote the open sharing of data resources in various industries, regions, and fields. In order to dispose of the segmentation of current data resources, the Ministry of Culture has launched the National Cultural Information Resource Sharing Project, established the National Folk Culture Protection Committee, and launched national traditional cultural information sharing. The traditional culture of the nation constitutes the data application platform [10].

3 Experiment

3.1 Sample Data

In order to better understand the research on the transmission of ethnic cultural symbols with the background of big data technology, this experiment selected a certain ethnic minority cultural symbols as the object of research and investigation, and conducted experiments in the English Department of Jiaotong University. There are two groups (the experimental group and the control group).The experimental group uses big data mining technology for dissemination, and the control group uses general publicity. Analyze the transmission speed of the two and the degree of acceptance and love of the English majors to explore new paths for the spread of national cultural symbols. By issuing questionnaires to students, to investigate their thoughts, satisfaction, opinions, etc., the researchers can get the influence of big data technology on national culture transmission.

3.2 Research Design

Through the analysis of the relevant literature on the influence of big data technology on the development of national culture communication the researchers collected the theoretical and applied research on the current situation and development prospect of this research, which laid a foundation for composeing of this paper. Questionnaire survey method is used in this study. A total of 70 questionnaires are distributed and 70 valid questionnaires were collected. The researchers performed data analysis and word processing on the questionnaires. In order to avoid incomplete or insufficiently specific and in-depth information in the questionnaire survey, the literature research method is also used on the basis of the questionnaire survey to make the research more comprehensive and accurate. The following data analysis organically combined the results of questionnaire and literature research to analyze and discussed the current application of big data technology to the transmission of ethnic culture.

4 Discussion

4.1 Experimental Research Report Statistics

The innovation of the communication of national cultural symbols should follow the path of sustainability, health and value, and keep meaningful data and information. Through the use of big data, in-depth analysis of the real needs of the audience, carefully judge the situation of the audience. Through big data technology and information technology, the correct demand information can be obtained from the huge data on the Internet, and the diversified and networked data characteristics can be objectively analyzed. Therefore, the audience can choose to collect and re-read the beneficial national cultural symbols that meet their needs. Of course, in the process of using this technology, it is necessary to effectively distinguish the reliability of data information and strengthen the function of information integration. Table 1 shows the scope of national cultural symbols transmitted by big data technology and without the use of such technology. It can be seen from the table that the spread scope of using the technology is larger than that of ordinary means.

Table 1. Comparison of the scope of national cultural transmission under the two methods

	Place	Number of people	the way
Based on big data technology	The entire campus and surrounding schools	More than 20,000	The internet
Ordinary spread	Two or three classes	About a thousand people	Oral communication

4.2   Survey of Students' Acceptance of Ethnic Culture

This paper analyzes the impact of big data technology on the spread of national culture through a questionnaire.The resource database can effectively improve the people's cognition level of national cultural characteristic resources, and the resource database can be better sharing the national cultural resources and maintain the diversity of national culture. The Internet has increased the popularity of national cultures, maintained the diversity of national cultures, realized the sharing of resources on a global scale, and enriched the meaning of national cultures in the world. As shown in Fig. 1, it is the degree to which English students like the spread of national culture. It can be found that 70% of students consider that ethnic culture is very interesting and they can learn more from it. Only 5% of students think that ethnic culture is very uninteresting. It can be seen that the public's acceptance of national culture is quite high.

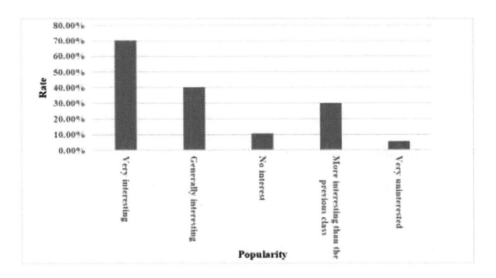

Fig. 1. Students' views on national culture

Figure 2 shows the degree of national culture exploration under big data technology. After big data processing, ethnic cultural symbols become more meaningful and easier for the audience to accept and understand.The combination of modern technology and national culture can make national culture more vigorous and full of vitality. Analysis of the data in Fig. 2 can be seen that useful information from data mining accounts for 44% and the hidden information accounts for 13%, indicating that most of the information mined by data mining is valuable and can actively promote the spread of ethnic cultural symbols. The information excavated from the national culture can be combined with the development of modern culture, which will mutually promote the development of culture and make the Chinese national culture to be better inherited. Figure 2 shows that useful information extracted from data mining accounts for 44% and the hidden information accounts for 13%. This indicates that most of the information extracted from data mining is valuable.

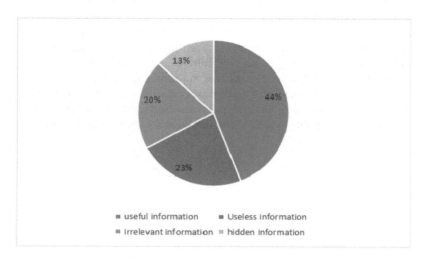

Fig. 2. National cultural information after big data processing

5   Conclusions

In order to promote Chinese culture, promote the innovation of cultural content and communication methods from the starting point of the high age, liberate and develop cultural productivity, use advanced technology to innovate cultural production methods, cultivate new cultural formats, and accelerate the construction of extensive cultural exchanges for high-speed transmission. In order to popularize national culture, if the researchers use big data technology, the symbol of national ethnic culture becomes more easier to be accepted by the public, and the faster popularization can take a wider road. Through the investigation of this article, it can be seen that the focus of national cultural symbol communication innovation lies in the form and content of communication. The innovative path of national cultural symbol communication includes the change of thinking mode, innovative analysis mode, innovative conclusion direction, relying on traditional paper media, using big data techniques, and rational use of communication tools. Therefore, continue to strengthen the advanced design of database construction, improve the level of various types of database construction, establish and improve the information sharing mechanism, continue to form a long-term mechanism for maintaining the database platform, and make the popularization of national culture more efficient and convenient.

Acknowledgements. This work was supported by Yunnan Provincial Education Department Scientific Research Fund Project: NO. 2020J1385.

References

1. Sang, H., Kyoung, K., Eun, C., et al.: System framework for cardiovascular disease prediction based on big data technology. Symmetry **9**(12), 293 (2017)
2. Alvarezdionisi, L.E.: Envisioning skills for adopting, managing, and implementing big data technology in the 21st century. Int. J. Inf. Technol. Comput. **9**(1), 18–25 (2017)
3. Wei-Chih, C., Wen-Hui, C., Sheng-Yuan, Y.: A big data and time series analysis technology-based multi-agent system for smart tourism. Appl. Sci. **8**(6), 947(2017). (in Chinese)
4. Gabriel, L.P., Rodrigues, A.A., Macedo, M., et al.: Electrospun polyurethane membranes for tissue engineering applications. Mat. Eng. C Mat. Biol. Appl. **72**(3), 113–117 (2017)
5. Tacconelli, E., et al.: Analysis of the challenges in implementing duidelines to prevent the spread of multidrug-resistant gram-negatives in Europe. BMJ Open **9**(5), e027683 (2019)
6. Salehan, M., Kim, D.J., Lee, J.N.: Are there any relationships between technology and cultural values? A country-level trend study of the association between information communication technology and cultural values. Inf. Manag. **55**(6), 725–745 (2018)
7. Prim, A.L., Filho, L.S., Zamur, G.A.C., et al.: The Relationship between national culture dimensions and degree of innovation. Int. J. Innov. Manag. **21**(1), 1323–1339 (2017)
8. Yongfeng, Q., Chao, Z., Lingwei, W.: Studies on the belt and road initiative and the China—Mongolia—Russia culture industry corridor. Contemp. Soc. Essence **17**(1), 43–59 (2017). (in Chinese)
9. Inanlou, Z., Ahn, J.Y.: The effect of organizational culture on employee commitment: a mediating role of human resource development in Korean firms. J. Appl. Bus. Res. **33**(1), 87–94 (2017)
10. Tomaselli, K.G., Jun, Z., Armida, D.L.G.: Culture, communication and cross-media arts studies: transnational cinema scholarship perspectives. Transnatl. Cine. **8**(2), 160–163 (2017)

Intelligent City Data Acquisition System Based on Artificial Neural Network BP Algorithm

Xiao Tao[(✉)]

FiberHome Telecommunication Technologies Co., Ltd., Wuhan 430205, Hubei, China
taoxiao@fiberhome.com

Abstract. Smart city is an advanced form of urban modernization and information technology, which is based on information and communication technology facilities, takes promoting social development, strengthening social management and enriching social life as its core tasks, and is mainly characterized by being more optimized, greener, more beneficial to the people and more sophisticated. This paper mainly studies the design of smart city data acquisition system based on artificial neural network BP algorithm. In this paper, the empirical formula and trial method are used to determine the number of hidden layers, the number of neural nodes in the hidden layer and the learning rate. Through analysis and comparison, the TANH function is finally determined as the activation function to complete the BP neural network model design. NET, VBA, Visual LISP, C++ are used as the main development languages to develop "Smart City Spatial Data Acquisition Subsystem", "Construction Land Data Processing Subsystem", "Public Facilities Survey Subsystem", "Data Transformation Subsystem" and so on.

Keywords: Artificial neural network · BP algorithm · Smart city · Data acquisition

1 Introduction

Nowadays, with the rapid development of science and technology, the development of artificial intelligence follows closely. Artificial intelligence plays a pivotal role not only in academia, but also in industry [1]. One of the research hotspots in the field of artificial intelligence is smart city management. Smart city management is to use data mining, video image algorithm, scheduling algorithm and deep learning algorithm to solve urban management problems with maximum efficiency, minimize labor force and make the most reasonable use of information resources. In recent years, more and more cities adopt smart city management system to manage urban operation more conveniently and quickly [2, 3]. Not only has it been successfully implemented in some first-tier cities, but most second-tier cities have gradually become trial points for smart cities. These cities have successively achieved smart city management with excellent technology and innovation strength. Driven by the concept of smart city, great changes have taken place in urban life. No matter in transportation, health, education,

J. C. Hung et al. (Eds.): FC 2021, LNEE 827, pp. 724–731, 2022.
https://doi.org/10.1007/978-981-16-8052-6_91

environment or other fields, a new ecology that is more intelligent, green and efficient has been bred [4, 5]. Internationally, governments around the world have recognized the value of information interconnection in smart cities, started to adopt the concept of smart cities and implement big data application, with the purpose of further narrowing the distance between government work and citizens' lives and improving citizens' living standards and quality through the realization of open data sharing.

The basic idea of a smart city is to manage various social services such as medical treatment, transportation and people's livelihood by means of network information according to the development and information construction of the local city, so as to achieve the purpose of building a smart city [6]. Foreign literatures on the construction of smart cities are mostly about the realization of smart management and operation of cities. Foreign countries have long started to build a new model of smart city. For example, in the first half of 2006, researchers in the European information industry built the first smart city by using network technology [7]. Subsequently, Singapore also joined the construction of smart city in the second half of the year, while the establishment of smart city in China started relatively late. Shanghai became the first city in China to realize the concept of smart city in 2010 [8].

Based on the actual needs of "smart city" construction for basic geographic information data, this paper establishes unified "smart city" basic geographic information basic topographic map data standards and current land use data standards, standardizing the survey items, attribute data and other contents of "smart city" basic spatial geographic information data.

2   Data Acquisition Based on BP Neural Network

2.1   BP Neural Network Model

(1) Determination method of the number of hidden layers

Although increasing the number of hidden layers can reduce network errors and improve network prediction accuracy, it will also make the network more complicated, thus increasing the training time of the whole network, and the phenomenon of "overfitting" will occur [9]. Moreover, the research in this paper adopts a high-performance GPU deep learning environment, whose processing capacity can make up for the extra overhead of the training network when using multiple hidden layers. Higher training accuracy is obviously more important, so this model adopts the design of multiple hidden layers. By using the trial and error method in the experiment, the model is finally determined to adopt two hidden layer structure.

(2) The method to determine the number of hidden layer nodes

In the process of designing BP neural network model, the number of nodes in the input layer and the output layer is unique, but the number of nodes in the hidden layer can only be determined by relevant calculation, but the existing empirical formula can only be used:

$$h = \sqrt{m+n} + a \tag{1}$$

In the training process in order to be able to avoid the phenomenon of "fitting", make sure that the network reached the highest prediction performance and generalization ability of network hidden layer node number of the most basic principle is: on the premise of ensure that can satisfy the precision requirement, make the network structure is compact, namely make the hidden layer nodes at least [10]. It is found that the number of nodes in the hidden layer is not only related to the number of nodes in the input and output layers, but also related to the complexity of the problem to be solved, the type of conversion function and the characteristics of the sample data.

(3) Determine the learning rate

The convergence rate of BP neural network algorithm is also affected by the learning rate. If the preset value of network learning rate is too small, it will take more time to train the sample for many times before finding the minimum value. If the preset value of network learning rate is too large, the training results will be oscillated, resulting in the model's failure to converge [11, 12]. Therefore, the minimum error of the model cannot be obtained, and the network training cannot be completed. The learning rate is usually set at 0.001–1. In this paper, the training model is started from the minimum 0.001, and then the learning rate is gradually improved in each iteration process, and the optimal learning rate is finally estimated to be 0.01.

(4) Determine the activation function

The relationship between output and input of TANH function can keep the non-linear monotone rising and monotone falling, which is consistent with the gradient solution of BP network. It has good fault tolerance and is bounded. Moreover, it is asymptotic to 0 and 1, which is consistent with the law of human brain neuron saturation. Compared with Sigmoid function, the saturation period is delayed. Therefore, Tanh function is selected as the activation function in the model in this paper:

$$f(x) = \tanh(x) = \frac{2}{1 + e^{-2x}} - 1 \tag{2}$$

2.2   Data Acquisition Platform

With the help of a mature, open, advanced, easy expansion and reusability of the multi-layer architecture, from meet "wisdom city" basic geographic information data processing, storage, distribution, management and multidimensional, tenses and forms of data analysis, use AutoCAD 2010 or 2012 platform, develop "wisdom city" space data acquisition platform.

The development language is C++, .NET, Lisp, VB language, etc., nested ActiveX controls, ObjectARX application, ArcGIS and other technologies, and based on SQL

Server2010 or Oraclellg database and Atuocad2010 or 2012. Develop "Smart City Spatial Data Acquisition Subsystem", "Construction Land Data Processing Subsystem" and "Public Facilities Survey Subsystem", compile data conversion procedures, and implement the production of basic geographic information data and conversion among different data formats for the construction of smart cities. The functions of each subsystem are as follows:

(1) Spatial data acquisition subsystem

Unified linearity, symbol, color and coding are adopted to develop spatial data acquisition subsystem, which can meet the collection, editing, updating, statistics and quality check functions of basic topographic maps of L: 500, 1:1000 and 1:5000. At the same time, combined with the daily production needs, the field measurement data to achieve automatic spread point mapping, attribute automatic generation, batch data quality inspection, common data format reading, data exchange output.

(2) Data processing subsystem

Realize the mapping, extraction, classification, statistics and quality inspection functions of 1:1000 current urban construction land and 1:5000 current urban and rural land.

(3) Subsystem of public facilities survey

According to the data standards, the subsystem of public public facilities survey is developed to realize the marking and statistical functions of point-like and line-like facilities of public public facilities, and to provide a map of public public facilities for the layout and planning of public public facilities.

(4) Digital conversion subsystem

FME is the complete spatial ETL solution. It is based on the new data transformation concept "semantic transformation" proposed by OpenGis organization. By providing the function of reconstructing data in the transformation process, it realizes the conversion between different spatial data formats. Through the development of data conversion subsystem and nested FME technology, the data conversion functions of the 1-day version of basic topographic map and the new version of basic topographic map (meeting the basic geographic information data standards of smart city space), GIS data and DWG data are realized.

(5) Database design

The database design mainly includes the database of the spatial data acquisition platform, which is used to support the basic data drawing and editing of the platform, quality inspection, data support of codes, lines and symbols during data conversion, which is the embodiment of the map surface and attribute metadatabase of ground object elements. The platform needs to meet the requirements of "Smart City" project for basic topographic map, land use status map and building attribute information. As there are many types of ground feature data produced on this platform, data editors need to frequently query and modify the ground feature attribute information.

The involved databases include SQL2010 database, AutoCAD database and ACCESS database. Among them, SQL2010 database mainly stores data,

AutoCAD database stores linear library, symbol library and text library, access database stores code table, linear table, symbol table, text annotation table and data conversion corresponding table.

3   Testing of Spatial Data Acquisition System

3.1   Platform Use Environment

According to the operating environment of the platform and the development trend and performance-price ratio of the computer, the hardware environment used by the platform is as follows:

Computer model: IBM-PC compatible. Such as: HP, ThinkPad, Lenovo, Dell and other original installed or good performance group installed, 4 GB memory, 8 GB is recommended; 2 GB of available disk space (for installation); 1920X1080 resolution true color display; 2 G video memory or more independent video card; Intel Pentium or AMD dual-core processor (1.6 GHz or higher).

Database server: Dell R720 server, 1 TB hard disk.

3.2   Platform Function Test

Functional testing of the spatial data acquisition platform is divided into two steps:

(1) Testing of core modules. In the spatial data acquisition subsystem, the ground feature element rendering module and the quality inspection module are the core modules. Using the completed mapping data of a residential area, the relevant functions of the ground feature element rendering module are used to draw symbols, linearly and graphs, and test whether the symbols, linearly and graphs are correct and whether the corresponding functions are realized. Using the data drawn by the cell, test whether the function of the quality check module is realized and whether it is consistent with the data standard.

(2) Test other major functions. On the basis of the core module test, the basic topographic map data of 6 km^2 is used to test whether the functions of the spatial data acquisition subsystem, such as data editing module, attribute data auxiliary processing module, data management module, data format conversion module and data conversion subsystem, are realized. The data of 6 km^2 of current urban construction land and public utilities are used to draw and process the data of construction land and public utilities, and test whether the corresponding functions are realized.

3.3   Platform Performance Testing

Performance test uses two sets of data with different performance and two sets of different sizes to test the operating index and operating efficiency of the system.

4   Platform Test Results

4.1   Function Testing

Table 1. List of residual defect issues

Function	Data collection	Construction land	Facilities of the census	Data conversion	Quality inspection
Quantity	41	34	15	9	12

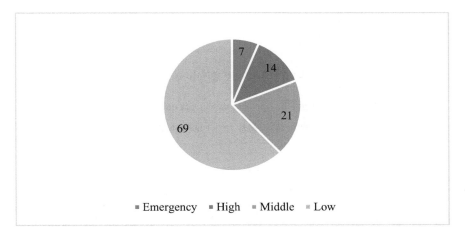

■ Emergency   ■ High   ■ Middle   ■ Low

Fig. 1. Summary chart of defect classification

As shown in Fig. 1 and Table 1, the test results show that the main residual defects of the spatial data acquisition subsystem include 41 problems that the size of individual symbols does not meet the standard requirements and the link of dynamic library is not good. The main residual defects of the subsystem of construction land include 34 problems in total, such as individual stratification error, surface accuracy not meeting the requirements, area statistical location not meeting the requirements, etc. The main residual defects of the subsystem of public utilities survey include 15 problems, such as the disunity of individual point attributes and surface attributes, the insufficient reservation of extended field length, etc. The main defects of the data conversion subsystem include 9 problems, such as the incomplete correspondence between old and new codes and the inability to convert some blocks. At the same time, 12 problems were found in the quality inspection module, such as the omission of individual attributes, the failure of wrong ground objects to level, and the slow running speed of inspection. From the point of view of the severity of defects, the emergency, high, medium and low defects were 7, 14, 21 and 69, respectively. In view of the above defects, according to the data standards and functional requirements, we have carried out thorough modifications one by one, until the platform is stable and reliable and put into the mass production of the project.

4.2   Performance Testing

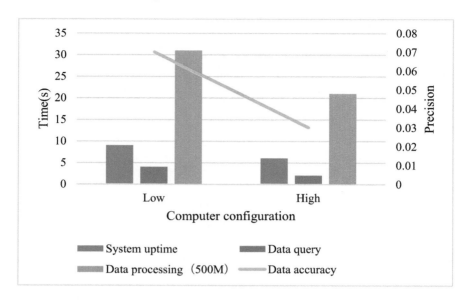

Fig. 2. Performance test results

As shown in Fig. 2, the above tests show that the system platform requires fast operation, fast response, reasonable organization of all kinds of data, smooth data query, update and drawing, the operation habits follow the habits of surveying and mapping workers, and the system runs stably in the generation of large quantities of data. At the same time, the system can be developed in the future to expand the underground space data acquisition and processing, municipal facilities acquisition and processing and other functions, with scalability.

5   Conclusions

In the construction process of a smart city, it is in an environment with large data throughput and multiple and complex information departments. Therefore, the importance of information collaboration and data collection is self-evident. Based on the actual demand, the spatial data acquisition platform is designed, and the spatial data acquisition subsystem, construction land acquisition and processing subsystem, public utilities survey and processing subsystem and data conversion subsystem are developed. This platform can automatically input, process and output data that cannot be represented by general topographic maps, such as housing ownership, area, height, land use nature, traffic and water conservancy information, etc., thus meeting the requirements of diversification of spatial data and information of smart cities.

References

1. Zhen, C.: Empirical research on the wisdom port-city coupling system. J. Comput. Theor. Nanosci. **13**(3), 2014–2020 (2016)
2. Dell, K.J.: Wisdom and folly in the city: exploring urban contexts in the book of proverbs. Scott. J. Theol. **69**(4), 389–401 (2016)
3. Kassens-Noor, E.: Failure to adjust: Boston's bid for the 2024 Olympics and the difficulties of learning Olympic wisdom. Environ. Plan. **51**(8), 1684–1702 (2019)
4. Healey, K.: Information is not wisdom, convergence is not integrity: proverbs for an era of digital humanism. Explor. Media Ecol. **15**(3), 355–372 (2016)
5. Temudo, M.P., Cabral, A., Talhinhas, P.: Petro-landscapes: urban expansion and energy consumption in Mbanza Kongo City. North. Angola. Hum. Ecol. **47**(4), 565–575 (2019)
6. Rizqi, A., Wulandari, L.D., Utami, S.: Architectural style of riverside settlements in Banjarmasin City. Local Wisdom Jurnal Ilmiah Kajian Kearifan Lokal **11**(2), 121–131 (2019)
7. Gu, B., Sun, S., Xiao, B., et al.: Analysis of wisdom economic electricity load based on vector autoregressive model. C e Ca **42**(6), 2407–2412 (2017)
8. Liu, Y.: Political parties' wisdom and strength for global economic governance keynote speech at the CPC in dialogue with the world 2016. Contemp. World **04**, 10–13 (2016)
9. Tang, S., Yu, F.: Construction and verification of retinal vessel segmentation algorithm for color fundus image under BP neural network model. J. Supercomput. **77**(4), 3870–3884 (2020). https://doi.org/10.1007/s11227-020-03422-8
10. Zhang, Y.-G., et al.: Application of an enhanced BP neural network model with water cycle algorithm on landslide prediction. Stoch. Env. Res. Risk Assess. **35**(6), 1273–1291 (2020). https://doi.org/10.1007/s00477-020-01920-y
11. Yuan, S., Wang, G., Chen, J., et al.: Assessing the forecasting of comprehensive loss incurred by typhoons: A Combined PCA and BP neural network model. J. Artif. Intell. **1**(2), 69–88 (2019)
12. Liang, Y.J., Ren, C., Wang, H.Y., et al.: Research on soil moisture inversion method based on GA-BP neural network model. Int. J. Remote Sens. **40**(5–6), 2087–2103 (2019)

Application of Meta-cognitive Strategy in Autonomous Learning Ability of Online College English

Qian Zhou[✉]

School of Foreign Language Studies, Anhui Sanlian University,
Hefei, Anhui, China

Abstract. This study aims to prove that meta-cognitive strategies have a positive effect on the cultivation of students' autonomous learning ability. In order to cater to the reform of College English online teaching and improve college students' English autonomous learning ability, this study conducted a one-semester meta-cognitive strategy training for 90 non English Majors in the experimental group. The results show that meta-cognitive strategy training can improve students' College English performance. It is effective in cultivating learners' planning and self-monitoring ability, but it has little effect in using learning strategies, extracurricular activities, self-assessment and learning reflection. The research shows that the development of college students' autonomous learning ability is unbalanced.

Keywords: Meta-cognitive strategy training · Autonomous learning ability · College English

1 Introduction

As a new model, online English teaching is becoming the mainstream trend of second language acquisition teaching reform for college students [1]. Colleges and universities actively adopt the online teaching model and encourage students to actively use the online platform and online materials to conduct personalized and autonomous English learning [2]. However, college students' weak ability of autonomous English learning has become one of the obstacles in the online teaching environment. Most students in application-oriented colleges and universities have weak learning ability, especially meta-cognitive ability. Most students are accustomed to passive learning and lack of autonomous learning ability. Nowadays, with the wide application of Internet teaching platform in colleges and universities, it is necessary to cultivate English autonomous learning ability of students in application-oriented colleges and universities [3].

2   Meta-cognitive Strategy

The research of this paper is based on meta-cognitive strategy theory and autonomous learning theory. O'Mally&Chamot and Oxford separate meta-cognition from second language acquisition and argue that meta-cognitive strategies are of cognitive processes [4]. It is the adjustment of cognitive process through planning, monitoring and evaluation, which plays an important role in improving the learning effect. Wen Qiufang believes that on the basis of meta-cognitive strategies, cognitive processes need to be self-managed, and management strategies are divided into several steps: goal determination, plan making, strategy selection, self-monitoring, self-evaluation and self-adjustment [5].

Autonomous learning is a new learning concept developed on the basis of humanistic psychology. Holec believes that autonomous learning means that learners can be responsible for their own learning, and determine learning goals, learning plans, learning strategies and evaluate learning effects in the learning process according to their own needs and existing knowledge [6]. Dickinson points out that autonomy is not only a learning attitude, but also an independent learning ability, which is referred to as autonomous learning ability [7]. Wenden emphasized that meta-cognitive strategies are the key to developing autonomous learning ability [8]. In the context of the popularization of online teaching, is meta-cognitive strategy applicable to the fostering of autonomous SLA (second language acquisition) learning ability in application-oriented universities? This paper will focus on this aspect.

3   Research Design

3.1   Research Object

This study chose two parallel classes of freshmen majoring in traffic engineering of Anhui Sanlian University as the research object, a total of 91 students, and they were divided into two groups. One group was the experimental group, with 46 people. The other group was a control group of 45 people. The English scores of the two groups were similar and there was no significant difference in average scores. Students of the two classes all use New Horizon College English Book 1 as the textbook on the Xuexitong APP and are taught by the same teacher with the same teaching method and same periods (2 h per week of traditional offline teaching, 2 h online teaching). Students in the experimental group were additionally given meta-cognitive strategy training.

3.2   Meta-cognitive Strategy Training Procedure

The training of students in the experimental group is mainly carried out by five steps of meta-cognitive strategy.

(1) Cultivate students' learning plan
     Students in the experimental group set learning goals and made learning plans.
(2) Cultivate students' ability to use learning strategies

Students in the experimental group were guided to actively use the learned learning strategies, demonstrate the learning strategies, provide contextualized learning strategies and discuss the effectiveness of the learning materials.

(3) Cultivate students' ability of self-monitoring.

Students use Xuexitong APP to monitor their own learning data statistics, and regularly study videos and read aloud on the learning APP.

(4) Cultivate students' ability to study after school

Urge students to participate in extracurricular activities, such as English Corner, English Competition for College Students, English Competition for FLTRP, etc., and judge their advantages and disadvantages.

(5) Cultivate students' self-assessment and evaluation

Guide students to do self-evaluation and mutual evaluation regularly, and reflect on the problems of study. According to the learning data and feedback, students are required to summarize the learning experience, adjust learning goals, plans and strategies.

The five steps are all student-centered. Students can use various learning websites or learning APP. The strategies given by teachers are used for practice and self-monitoring by students in this process. It is necessary for teachers to explain students' learning strategies in detail. Students receive clear guidance from the teacher on how and when to use these strategies. Learning strategy guidance is realized through various forms of online teaching means, such as Xuexitong and other various English learning APPs (baicizhan, daily English listening, Tencent classroom), etc. Teachers use APP to monitor students' learning by assigning homework and asking questions in online class; the way to assign homework is to give tasks through the learning platform. At the beginning of the experiment, the teacher asked experimental group to make learning plans and strategies individually, evaluate the implementation effect of last week's plan every week, and adjust next week's plan and strategy according to the effect. The experiment lasted for a semester during which, the electronic data were exported through the learning platform once a week to obtain relevant information and materials.

3.3 Research Tools

Questionnaire survey and test were used as data collection tools in this study. The questionnaire is designed around students' autonomous learning ability, and Likert five-level scoring system is adopted. The students were asked to choose the option that was closest to their actual situation on a scale of five. (1 = completely or almost not my case 2 = mostly not my case 3 = half my case 4 = mostly my case 5 = completely or almost my case). The researchers also conducted irregular personal interviews with the students. The method of the test is mainly the results of College English Test at the end of the first semester, in order to evaluate the application degree of Meta-cognitive strategies.

3.4   Data Analysis

In this study, SPSS was used to analyze the data. Before the final exam, autonomous English ability of the experimental group and the control group in the same period of time was investigated and analyzed. The analysis was divided into two steps: 1. The final test results of the experimental group and the control group were compared and analyzed, and independent sample t test was conducted based on the results (independent samples t-test, Table 1); 2. The self-learning ability of the experimental group and the control group was compared and analyzed, and the t-test was carried out on the results (Table 2).

Table 1. Descriptive analysis and t-test results of test group and control group's final grade

Group name	Number of students	Average	Standard deviation	t value	P value (two-tailed tests)
Test group	46	87.15	6.54	2.135	0.020
Control group	45	83.21	8.65		

Table 2. Descriptive analysis and t-test results of test group and control group's self-learning ability

Category	Test group		Control group		Independent sample t-test	
	Average	Standard deviation	Average	Standard deviation	t value	P value (two-tailed tests)
Plan making	17.32	3.65	15.86	3.89	2.136	0.037*
Strategy application	17.77	2.67	17.69	3.72	0.056	0.867
Self monitoring	33.54	2.79	31.32	4.81	2.463	0.015*
Learning after school	10.98	4.56	11.64	2.58	−1.348	0.213
Self assessment	14.42	3.17	14.07	2.96	0.492	0.537
Total	94.03	10.79	90.58	13.45	1.489	0.174

4   Results and Analysis

4.1   Influence of Meta-cognitive Strategy Training on English Performance

As can be seen from Table 1, the average final score of students in the experimental group was higher than that in the control group, and the independent sample t-test results showed that there was a big gap between the two groups in the final exam. (t = 2.135, P = 0.020 < 0.05) It can be concluded that meta-cognitive strategy training is effective in improving English performance.

4.2   Influence of Meta-cognitive Strategy Training on Autonomous Learning Ability

Table 2 shows the autonomous learning ability (average = 94.03, standard deviation = 10.79) of the experimental group was higher than that of the control group (average = 90.58, standard deviation = 13.45) on the whole. It is reflected that under the meta-cognitive strategy training, the students' autonomous learning ability has been improved to a certain extent. However, the results of independent sample t-test showed that there was no significant difference in autonomous learning ability between the two groups (t = 1.489, P = 0.174 > 0.05). Secondly, from the five aspects of autonomy, the results of independent sample t-test show that there are significant learning ability differences in planning and self-monitoring between the experimental group and the control group (P = 0.037 < 0.05; P = 0.015 < 0.05). This shows that after meta-cognitive strategy training, the planning and self-monitoring ability of the experimental group is better than that of the control group. In other words, the experimental group is better than the control group in setting goals, making plans, planning learning and self-monitoring. However, there was no significant difference in learning strategy use, self-assessment and reflective learning between the experimental group and the control group. The data also show that meta-cognitive strategy training has quite slight effect on the development of strategy application, learning activities after school and self-assessment abilities (P = 0.867 > 0.05; P = 0.213 > 0.05; P = 0.537 > 0.05).

4.3   Feedback from Individual Interview

Interviews with some experimental group students show that after receiving meta-cognitive strategy training, many students have fundamentally improved their previous aimless English learning. Most of the participants were able to set learning goals according to their actual situation and the guidance of the teacher. Most of the goals were exam-oriented, such as passing CET-4 and CET-6. Second, the abilities, such as listening, speaking, reading and writing were improved a lot. Most of the participants said that they spent more time learning English than before, finished learning hours per week according to the plan, and made self-assessment and self-reflection on the learning effect. According to the feedback, if the learning plan is not completed, the reasons are mainly as follows: 1. The plan is strict. For example, it is difficult to memorize and recite vocabulary within the specified time. This frustration will make them adjust their learning plans. 2. The plan is affected by other factors. For example, the school's experimental training classes occupy much time of evening self-study or may be influenced by other extracurricular activities. After being disturbed by other factors, students often make adjustment and add extra class hours in the second week's learning plan. The above facts indicate that meta-cognitive strategy training has a positive effect on students' awareness of autonomous learning, plan making and self-monitoring ability.

In terms of the use of learning strategies, the tested students expressed that they did not quite understand meta-cognitive strategies before. Under the guidance of researchers, the awareness of learning strategies has been strengthened, but in the short term, there was still a long way to go to achieve the proficiency of learning strategies.

In terms of extra-curricular learning ability, most of the tested students said that they only cared to the intake of textbook content and the level of test scores as they did in high school and therefore had little awareness about extra-curricular activities and extra-curricular reading materials. The reasons may be as follows: 1. Lack of self-confidence. Some students said they knew that extracurricular activities such as English Corner and English Competition for College Students could improve their English ability, but they were afraid to speak and gave up since they thought their English level was not qualified enough. 2. Too many professional courses. The students said the university offered too many professional courses which were much more important than other auxiliary courses in their minds, so they had no time to do other extracurricular activities. In terms of self-monitoring and self-reflection, the students in the test reported that they had attempted to do reflection in high school, but the result was still not good. Now they can not only find the learning problems, but also make appropriate adjustments. At the same time, they generally believe that the learning result can be better if there is a teacher to guide. The feedback above indicates that although meta-cognitive strategy training has certain influence on cultivating learners' usage of learning strategies, participation in extracurricular activities and self-assessment, the influence is still quite limited. This is also consistent with the statistical results of the questionnaire survey.

In the later stage of meta-cognitive strategy training, the author found that a small number of students did not accept this strategy psychologically, and they did not apply this strategy to train their autonomous learning ability. There are two main reasons. First, the attention paid to the second language acquisition is not enough. Some of the college students hold that College English is one of the common required courses and the most important courses in college are the professional courses. Therefore, second language acquisition should not occupy too much study time. 2. The influence of exam-oriented education. It is a consensus among non-English college students that as long as tests like CET-4 and CET-6 are passed, no additional training is required. These factors indicate that after a series of meta-cognitive strategy training, these students are still unable to get rid of their inherent thinking and learning habits and still lack of strong learning motivation. It is difficult to improve their autonomous learning ability only though a period of a semester. Murayama proposed that the key of strategy training lies in the learning goals existed in the learner's mind [9]. If students cannot face up to the importance of active learning, no strategy can foster their learning autonomy [10].

5   Conclusion

In this study, meta-cognitive strategies were used to cultivate the English autonomous learning ability of non-English major freshmen. The result indicates: 1. Meta-cognitive strategy training can improve academic performance and self-study ability, but it has no significant effect on learning awareness and autonomy learning. Meta-cognitive strategy training has positive effects on learners' planning and self-monitoring, but it has limitations on learners' abilities such as using learning strategies, choosing extra-curricular materials and making self-assessment. Although strategy training helps learners to master the methods and skills of management, it does not necessarily enable learners to form the ability to manage the cognitive process and grasp the elements of autonomy learning. And this kind of control ability is exactly the key to meta-cognitive strategy training. In view of the effectiveness and limitations of meta-cognitive strategy training in improving autonomous learning ability, the author believes that it is necessary to conduct meta-cognitive strategy training for college students in the online teaching environment. Training is not only conducive to the improvement of English performance, but also to the development of autonomous learning ability. Secondly, as for college students with weak learning awareness, meta-cognitive strategy cultivation can be appropriately extended to one academic year. Finally, the research shows that it is unrealistic to improve the autonomous learning ability of college students only by cultivating one specific strategy such as meta-cognitive strategy. The teaching platform of colleges and universities should actively adopt diversified teaching materials and mixed teaching modes to foster students' inner motives of autonomous learning and improve their autonomous learning awareness and interest.

Acknowledgements. This work was supported by Anhui Provincial Humanities & Social Sciences Project (Code: 2019zyrc115) and Anhui Sanlian Research Project "Business English" (Code: 18zlgc008).

References

1. Ho, W., Tai, K.: Doing expertise multilingually and multimodally in online English teaching videos. System **94**(3), 102340 (2020)
2. Saunders, P.: Teaching academic and professional English online. System **38**(3), 508–510 (2010)
3. Yvc, A., Ts, B.: Profile of second language learners' metacognitive awareness and academic motivation for successful listening: a latent class analysis. Learn. Individ. Differ. **70**, 62–75 (2019)
4. Gumartifa, A., Agustiani, I., Aditiya, F.: Differences of language learning strategies in English as a second language classroom of social collage students. J. Engl. Educ. Study (JEES) **3**(2), 119–128 (2020)
5. Wen, Q.F.: The use of cognitive contrastive analysis in foreign language teaching. Foreign Lang. Learn. Theory Pract. **01**, 1–5 (2014). (in Chinese)
6. Muhammad, M.: Promoting students autonomy through online learning media in EFL Class. Int. J. High. Educ. **9**(4), 320 (2020)
7. Jamrus, M., Razali, A.B.: Using self-assessment as a tool for English language learning. Engl. Lang. Teach. **12**(11), 64 (2019)
8. Komljanec, K., Ebalj, L.: Encouraging learner autonomy development in distance teaching in primary school. J. Foreign Lang. **12**(1), 277–294 (2020)
9. Cheng, L., Wong, P., Chi, Y.L.: Learner autonomy in music performance practices. Br. J. Music Educ. **37**(3), 234–246 (2020)
10. Klegeris, A.: Mixed-mode instruction using active learning in small teams improves generic problem-solving skills of university students. J. Furth. High. Educ. **3**, 1–15 (2020)

The Application of Virtual Reality Technology in Art Design

Ya Li[✉]

Yunnan Communications Vocational and Technical College,
Kunming, Yunnan, China

Abstract. With the advent of the 21st century, with the continuous progress and development of science and technology, information technology has also developed rapidly, and the most popular vocabulary in the near future is Virtual Reality. Because it can achieve an immersive virtual experience, bring users a virtual environment integrating vision, hearing, and touch, and realize behaviors that cannot be achieved in the real environment, so it has been widely used in various fields, mainly including games, medical care, Military architecture and design, etc. Virtual reality is becoming closer to our lives, and people are becoming more aware of the huge development potential of virtual reality. How to improve the simulation effect of virtual reality and bring users a more realistic feeling and experience has become a research hotspot nowadays. In art design, the combination of virtual reality technology (V R T) and art design will be the main trend in the future, including traditional art protection, art design environment analysis, and virtual art roaming. The purpose of this article is to study the application of V R T in art design. This article summarizes the basic theory of V R T and extends the core technology of V R T. And through the analysis of the current state of art design in our country. To discuss the application of V R T in art design. Experimental research shows that compared with traditional art design, the application of art design using V R T is more extensive and its feasibility is higher.

Keywords: Virtual reality technology · Art design · Visual elements · Applied research

1 Introduction

In recent years, the use of V R T has brought great achievements to people [1, 2]. In all aspects of our country, such as the use of V R T in the national defense and military to simulate scenes that cannot be achieved in reality, this not only saves the funds of the troops, but also enables the sergeants to be fully trained, which can bring huge benefits [3, 4]. Also in the medical field, using V R T can simulate many difficult operations for teaching, experimentation, etc. It can also provide necessary auxiliary operation functions during the operation to deal with more difficult surgical risks [5, 6].

The United States is a country with early research on V R T. NASA's AMES laboratory has engineered the data glove to make it more usable, and complete the real-time simulation of space station manipulation at the Johnson Space Center [7, 8].

J. C. Hung et al. (Eds.): FC 2021, LNEE 827, pp. 739–746, 2022.
https://doi.org/10.1007/978-981-16-8052-6_93

Now NASA has established a virtual reality training system for aviation and satellite maintenance, a space station virtual reality training system, etc., and has established a virtual reality education system that can be used nationwide [9, 10]. Compared with developed countries, the research level of V R T in our country still has a big gap, but it has attracted the attention of relevant government departments and scientific research departments, mainly focusing on the construction of software and hardware [11, 12, 13].

This article analyzes the application of V R T in art design by analyzing the characteristics of V R T and the application of reality, combined with its broad application prospects. Finally, compare the advantages of traditional methods to analyze and discuss the specific problems and solutions in the application.

2   The Application of Virtual Reality Technology in Art Design

2.1   Virtual Reality Technology

1) Immersive virtual reality system
    The immersive virtual reality system isolates the user from the real environment through a VR helmet or other equipment, and uses a hand-controlled input device to collect the user's actions, sounds, etc., so that the user interacts and communicates with the virtual environment, making it appear as if it is in the virtual environment. Or use holographic projection to create a virtual reality environment for users.
2) Augmented reality virtual reality system
    The augmented reality virtual reality system not only uses V R T to establish a virtual environment, but also combines the real environment with the virtual environment, and the information that cannot be felt or felt inconvenient in the augmented reality environment.

2.2   Features of Virtual Reality Art Design

(1) Reality
     Virtual reality is digital and virtual in form, but it needs to reflect the information of the real world. Users need to get a sense of reality when browsing it, which determines that there must be a mapping relationship between it and the real world, that is, the various attributes and characteristics of the virtual reality world are visualized. The inextricable connection with reality is reality, which is what distinguishes it from other art forms.
(2) Transcendence
     The virtual world originates from the real world, and can surpass the real world at the same time. Compared with the real world and traditional art design works, the transcendence of virtual reality art design works makes it uniquely attractive.

2.3   Analysis of Visual Elements of Virtual Reality Art Design

The characteristic of the visual elements of the virtual world is that it separates the elements that are mixed in the traditional three-dimensional art form-form, material, light, etc., which makes the production steps of the visual elements of virtual reality art design have their own characteristics.

(1) The three-dimensional geometric description is converted into a two-dimensional view, on a computer screen or a large projection screen, or a helmet-mounted display
(2) What is displayed on the screen is a two-dimensional view one by one. Use the perspective of the scene to transform it.
(3) In order to obtain realistic visual effects, all objects in the scene should be light-distributed and colored according to the observer's position, distance and angle to make them have a real texture.

2.4   Problems in Virtual Reality Art Design and Solutions

The advantages of VR technology in art design applications are undoubted, but at the same time it also faces many problems. Not only are there problems in the application of VR technology in art design, but there are also some potential side effects of continuous technological progress on humans and society, which require certain thinking. Only by thinking from multiple dimensions can we understand the essence of the problem.

(1) There is a problem
   1) Decline of designer's professional ability
      In the design review stage of VR technology, technology replaces the designer's brain's ability to control space. If you rely on VR to review space for a long time, it will directly lead to a decline in its spatial thinking ability. VR scrutiny requires the use of tools and equipment. If the tools are missing, if the designer does not have good spatial thinking ability, how to undertake the task of design work.
   2) Blindly dazzling design works
      In the use of some VR technologies, there is a phenomenon that the form is greater than the connotation, which weakens the user's attention to the design content. Even some works blindly focus on the research of the technology because they want to highlight the effect of the VR technology, instead of spending energy and time, improve the artistic beauty and depth of thought of design works.
   3) Poor experience of design works
      In the production process, the contradiction between the current network transmission speed and the size of its generated files is also an important issue facing VR in the design. If you want to produce better results, you need to make the model finely. However, even with the advanced network technology and the good transmission speed, the computer still suffers from freezes and unsmooth operation. In order to avoid this problem, the amount of data is generally reduced, which will cause the distortion of the model and affect the final imaging effect.

(2) Solutions

The reason for the poor production effect and experience effect is due to technical constraints. For VR, how to continuously innovate related technologies and equipment is a major bottleneck in front of the industry. However, the problems of VR technology are related to the actual level of productivity. The current level of technology is limited, and people cannot be completely immersed in the VR experience, and the speed of the processor cannot simulate things 100% truly, and it has not reached the full sense of virtual reality. However, this constraint is a temporary reality constraint, which is constantly changing and evolving. Its restriction on technology is like the restriction on the development of science by the level of knowledge. Its restriction on technology is not in principle, and will eventually be broken.

2.5 Research on VR 3d Stereo Algorithm

B-spline (Basic spline) can overcome the shortcomings of not having local properties due to the overall representation, and solve the connection problem when describing the copied shape. The curve equation proposed is:

$$p(u) = \sum_{i=0}^{n} d_i N_{i,k}(u) \tag{1}$$

Inverse calculation of the control vertices of cubic B-spline interpolation curve should meet the interpolation conditions:

$$p(u_{i+3}) = \sum_{j=1}^{i+3} d_j N_j \, u \in [u_{i+3}, u_{i+4}] \subset [u_3, u_{n+3}] \tag{2}$$

3 Experimental Research on the Application of Virtual Reality Technology in Art Design

3.1 Subjects

(1) In order to make the results of this experiment more scientific and effective, this experiment will investigate the art design majors of colleges and universities in a certain place. In order to investigate the feasibility of V R T in art design, this experiment will use traditional. The performance of art design and art design using V R T is compared to analyze the research results of this article.

(2) In order to further analyze the application of V R T in art design, this experiment also analyzes the functions of virtual reality art design applications. This article uses the form of online questionnaire to question relevant scholars. The survey, in order to make the experimental data more accurate and scientific, the gender ratio of the survey subjects is equal, and the relevant working years are all more than three years. Have their own opinions on the advantages and disadvantages of art design.

3.2 Research Methods

(1) Questionnaire survey method
   This paper sets up targeted questionnaires and adopts a semi-closed form to investigate the survey subjects. The purpose is to promote the correct filling of the questionnaires by relevant personnel and to ensure the accuracy of the experimental data.

(2) Field research method
   This article analyzes and sorts out the data collected from the interviews by going deep into the school's art design majors and related enterprises and companies, and conducting face-to-face interviews with relevant personnel. These data provide a reliable reference for the final research results of this article.

(3) Depth analysis method
   This article analyzes the essence and connotation of things in depth through the appearance of things. This topic focuses on environmental art design under VR technology, but the problem cannot be analyzed in a simple way. You should see some potential negative effects of VR technology.

(4) AHP
   Use relevant software to conduct statistics and analysis on the final research results.

4 Experimental Analysis of the Application of V R T in Art Design

4.1 Comparative Analysis of Art Design

In order to analyze the application of V R T in art design, this experiment compares the application of V R T in art design with traditional art design. To explore the feasibility of V R T. The data obtained is shown in Table 1.

Table 1. Comparative analysis of art design

	Intuitiveness	Collaboration	Sensitivity	Others
Virtual Reality	87.4%	76.7%	67.4%	70.5%
Traditional	42.0%	46.8%	52.5%	49.7%

It can be seen from Fig. 1 that compared to traditional art design, art design using V R T has improved intuition and collaboration in all aspects, especially in terms of intuition. Based on V R T, it is better than traditional art design. As high as 40%, this reflects the feasibility of V R T in the application of art design.

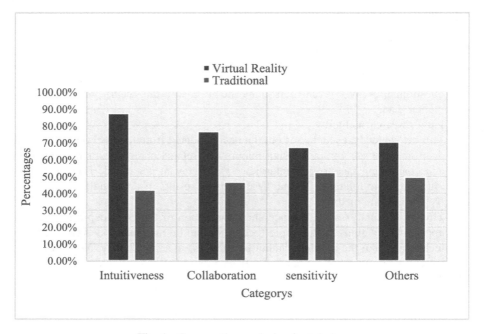

Fig. 1. Comparative analysis of art design

4.2   Performance Analysis of Virtual Reality Technology in Art Design

In order to further study the application of V R T in art design, this experiment adopted a questionnaire survey method to survey relevant experts, and adopted a ten-point scoring system, and the degree of satisfaction gradually increased from 1 to 10. The results obtained are shown in Table 2.

Table 2. Performance analysis of VR technology in art design

	Reality	Transcendence	Integration	Immateriality
Man	7	8	7	6
Woman	6	9	5	6

It can be seen from Fig. 2 that most of the relevant experts agree with the application of V R T in art design, especially the recognition of the transcendence of art design is very high, which fully demonstrates the good function of V R T in art design use.

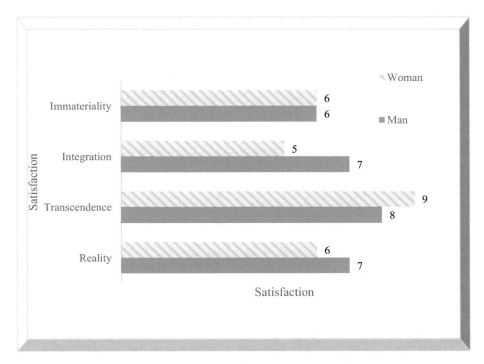

Fig. 2. Performance analysis of VR technology in art design

5   Conclusion

Although it is foreseeable that VR can cause a revolution in artistic design and artistic expression in the future, the V R T at this stage can only be regarded as a means to fill in the gaps in traditional design. Realizing a complete virtual reality viewing process requires not only the so-called 360° panoramic visual experience, but also an all-round interactive experience. Judging from the current development of V R T, visual effects are only the beginning stage. Assists such as handles and seats require huge and heavy wired equipment, which leads to a poor viewing experience. The immersive sense of interaction required by art design needs to be further improved in hardware. With the rapid development of computer hardware and software technology and people's gradual realization of its role, virtual technology has begun to be used in many fields, and it reflects the broad application prospects of virtual technology. As a kind of extremely novel art form, virtual art has strong interactivity and can connect works and audiences closer than ever before.

References

1. Kim, K., Lee, M.: Flocking in interpretation with visual art design principles. Wirel. Pers. Commun. **93**(1), 211–222 (2016). https://doi.org/10.1007/s11277-016-3925-1
2. Zhang, T. Analysis on the development trend of environmental art design model from the perspective of cultural heritage and classical reconstruction. Int. Technol. Manag. (3), 82–83 (2017)
3. Zhu, W.: Study of creative thinking in digital media art design education. Creat. Educ. **11**(2), 77–85 (2020)
4. Cao, X.: Three-dimensional image art design based on dynamic image detection and genetic algorithm. J. Intell. Fuzzy Syst. **5**, 1–12 (2020)
5. Yu, C.: Climate environment of coastline and urban visual communication art design from the perspective of GIS. Arabian J. Geosci. **14**(4), 1–16 (2021). https://doi.org/10.1007/s12517-021-06692-5
6. Yamato, T.P., et al.: Virtual reality for stroke rehabilitation. Phys. Ther. **10**, 10 (2016)
7. Bastug, E., Bennis, M., Medard, M., et al.: Toward interconnected virtual reality: opportunities, challenges, and enablers. IEEE Commun. Mag. **55**(6), 110–117 (2017)
8. Serino, M., Cordrey, K., Mclaughlin, L., et al.: Pokémon Go and augmented virtual reality games: a cautionary commentary for parents and pediatricians. Curr. Opin. Pediatr. **28**(5), 673 (2016)
9. Gou, F., Chen, H., Li, M.C., et al.: Submillisecond-response liquid crystal for high-resolution virtual reality displays. Opt. Express **25**(7), 7984 (2017)
10. Riva, G., Bacchetta, M., Baruffi, M., et al.: Virtual reality environment for body image modification: a multidimensional therapy for the treatment of body image in obesity and related pathologies. Cyberpsychol. Behav. **3**(3), 421–431 (2016)
11. Lenoir, J., et al.: Workshop on virtual reality interaction and physical simulation (2005)
12. Ganovelli, F., Mendoza, C. (Eds.): Interactive physically-based simulation of catheter and guidewire. J. Prev. Med. Inf. **61**(13), 2132−2141 (2017)
13. Hilfert, T., König, M.: Low-cost virtual reality environment for engineering and construction. Vis. Eng. **4**(1), 2 (2016)

Combination of the Cultivation of Translation Competence and Information Technology

Hong Zhou[✉]

School of Foreign Languages, Anhui Sanlian University, Hefei, Anhui, China

Abstract. From the perspective of the combination of translation teaching and information technology, this paper takes the "Translation Workshop" which is a translation practice course of undergraduate Translation major as an example, and optimizes the teaching process of this course according to PACTE Translation Competence Model. It discusses how to combine information technology with translation workshop teaching in three stages before, during and after translation activities. Then, taking the translation workshop course of our university as an illustration, it focuses on the teaching process of the specific course. It is believed that the translation teaching assisted by corpus technology can directly compare translation skills with data, and can also be effectively applied to the evaluation of translation quality, which plays a significant role in promoting the overall improvement of students' translation competence and translators' professional accomplishment.

Keywords: Translation workshop · Information technology · Translation teaching

1 Introduction

Translation workshop is a practical course offered to undergraduate translation majors in colleges and universities. It has strong applicability and practicality. In the practice of translation teaching, the PACTE model of translation competence plays a guiding role, and corpus technology can maximize the efficiency of translation operation and the reliability of translation evaluation. The combination of the two aims at cultivating students' strong translation and application ability and making them qualified applied translators.

1.1 Translation Teaching Assisted by Computer Information Technology

Translation teaching has been in a marginalized position for a long time, lacking effective methodology guidance, many well-conceived experimental means and handy research tools. Traditional translation teaching is mostly based on teachers' experiential, perceptive and mentor-apprentice teaching methods, and has some shortcomings in translation of textbooks, classroom teaching and students' independent learning. In 1997, the first academic conference on "Corpus Use and Learning to Translate, CULT," was held in Bertinoro, Italy. The second and third CULT academic conferences were held in 2000 and 2004 respectively. The translation text generated by using a specialized monolingual corpus as an auxiliary tool is significantly better than the translation text supplemented by

J. C. Hung et al. (Eds.): FC 2021, LNEE 827, pp. 747–754, 2022.
https://doi.org/10.1007/978-981-16-8052-6_94

traditional means (dictionary) in terms of understanding topics, choosing terms and expressing idioms. Corpus plays an important role in the field of translation evaluation. On the one hand, teachers can verify their intuitions about language expression with the help of corpus; on the other hand, corpus can provide strong evidence for the judgment of the quality of translation. According to Bowker, ideally, the goal of all translation teaching should be let students be independent from the teacher so that they can continue to learn and practice translation after they leave the classroom. In order to achieve this independence, students must receive constructive feedback from the teacher, including understanding their translation mistakes and the reasons for them. [1] This means that the translator must be able to point out specific problems rather than make judgments based solely on intuition or habit. The corpus-based approach to evaluate translation quality can minimize subjective factors as much as possible, and evaluators can also obtain a large number of real corpus texts as a benchmark, which can help to evaluate the quality of translation. Translation teaching assisted by corpus technology can truly realize "discovery learning" and "data-driven learning". Bernardini once suggested that translation teaching should be supplemented by parallel corpus retrieval, so that students majoring in translation can form a kind of translation consciousness, reflection and strain. Using a corpus, students can actively enter the learning process by collecting texts, evaluating corpus, extracting terms, establishing cross-language equivalence between bilingual texts, etc. Bower believes that the best way to make translation learners aware of translation is to expose them to the translation. [2] Bilingual parallel corpora or comparable corpora can not only enable learners to directly observe the rules of the language system of the source language, but also enable learners to analyze the characteristics of the translated text itself, so as to establish perceptual understanding of translation conversion. Wang Kefei pointed out that bilingual parallel corpora can present abundant examples of word-to-word translation, structure-to-structure translation and sentence-to-text translation, which has a broad application prospect and development value in translation teaching. [3] He maintains that students can use the bilingual parallel corpus to independently summarize and summarize translation skills through observation of a large number of translation examples, timely evaluate or reflect on their own translation behaviors, and master translation strategies and skills [4].

1.2   PACTE Translation Competence Model

PACTE (Procés d'Adquisició de la Competència Traductora Avaluació) refers to translation acquisition process and evaluation group, which is organized by translators and translation teachers from the Autonomous University of Barcelona. The group proposed the translation competence model (see Fig. 1) which includes the following sub-competencies: bilingual sub-competence, extra-linguistic sub-competence, strategic sub-competence, instrumental sub-competence, translation knowledge sub-competence and psycho-physiological components. Translation competence is a kind of professional knowledge, which consists of a variety of interrelated sub-competences. The model has a guiding role in the teaching design of translation workshop and the evaluation of translation ability for undergraduate students majoring in translation. With the assistance of information technology, the translation instruction based on these competences can greatly improve the teaching effect.

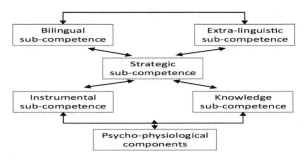

Fig. 1. PACTE translation competence mode [5]

1.3 About Translation Workshop

According to the definition proposed by Wolfson, "a translation workshop is extracurricular program and a voluntary encounter; It is a place for bilateral communication between the coordinator and the participants. In a workshop one does not inculcate knowledge in the way a traditional course does; rather, ways of reading, interpreting, and communicating are discussed freely". [6] As an organizational form different from traditional teaching, translation workshops are widely used in the daily translation teaching of BTI in Chinese colleges and universities. The workshop provides an opportunity for participants to learn from each other and discuss translation issues. The form of the workshop is free from the limitations of traditional courses, and it is also more creative, which let the students get rid of blind obedience and attachment, and then cultivate the habit of thinking diligently and translating independently. The workshop provides students with a large number of high-intensity translation training platforms, so that students can constantly improve their translation ability and translator competence through the ways of "learning translation through translation", "learning translation through cooperation" and "learning translation through discussion". And through the interactive learning environment inside and outside the class, the workshop let students feel, understand and grasp the true meaning of translation, independent translation activities, and implementation of translation projects. It also lays a foundation for the students to undertake different translation tasks in the future. By placing students at the center of teaching and focusing on the practical problems that cognitive subjects (students) encounter in the process of translation, translation workshops not only help participants master translation skills and absorb knowledge from the outside world, but also cultivate participants' attitudes about acquiring texts, i.e. the way texts are handled. This approach can be used for professional translation in the future [6]. An important part of the workshop teaching is to translate the first draft through group cooperation, compare the different translations from multiple angles and multiple levels, and finally modify the translations till the team members are satisfied with.

2 Information Technology Assisted Translation Workshop

Computer information technology can be effectively combined with translation teaching in the three stages before, during and after translation activities in workshops to assist teaching activities at different stages. Before translation, we should focus on the cultivation of translation technical ability to make adequate preparation for the effective implementation of translation work; in the process of translation, human-computer interaction professional translation ability training should be carried out with translation projects as the highlight; after the translation is done, text editing and translation quality evaluation with the assistance of computer information technology are carried out.

2.1 Early-Period

In the "workshop" teaching mode assisted by information technology, the teacher plays the role of leader, promoter and supervisor. Before the translation activities are carried out, the teacher should determine the specific content of the course, choose the translation texts and assign translation tasks. According to the content of the lecture, assign a bibliography and design appropriate translation tasks and requirements. Students are required to be familiar with the operation of web-aided translation and related translation software at the pre-translation stage, including: computer editing, translation of corpus and bilingual parallel text retrieval and web search skills. Before the translation, each group receives the translation task, previews and grasps relevant theoretical knowledge, collects relevant background knowledge and professional terms, and makes full preparations for the follow-up translation work.

2.2 Mid-period

In the process of student translation, the "translation workshop" focuses on "translators (students) and practice, and assists translation practice with translation theories and skills [7]". In the process of translation, students should carry out translation practice based on the data collected from the bilingual parallel text corpus in the early stage, and make full use of the teaching platform of the translation laboratory to search for similar sentence structure and word collocation information. The group members will initially translate the translation, discuss the difficulties and doubts, share the translation, check the translation with each other, and then summarize the translation in the group to form the best version that the group members think, and finally proofread the version and submit it. Throughout the process, team members acted as "translators," "reviewers," and "proofreaders." In this process, teachers play the role of "leader" and "person in charge", leading students to participate in translation practice, coordinating various relationships and supervising the development of students' translation activities.

2.3 Late-Period

After the students complete the translation practice, the other groups will evaluate the translation quality of the best translation selected by each group. Based on the PACTE competence assessment model, teachers evaluate students in each group according to their classroom performance and translation quality, and the specific grading rules can be differentiated according to the characteristics of different schools. Therefore, in the whole process, teachers should give professional guidance and judgment in place, guide students to discuss and ask questions independently, and summarize the problems in students' translation. Through this interaction and feedback process, students can find out the problems in translation in time and learn from the translation experience and lessons of other students. Teachers can also learn about students' translation in a timely manner and adjust relevant teaching methods and teaching designs accordingly, so as to further improve the quality of translation teaching.

3   An Illustration of Corpus-Based Translation Practice

Due to the diversity and openness of translation, the interpretation of any translation phenomenon or translation problem cannot be unitary or one-way, but multi-dimensional and multi-perspective. When different translators think about problems from different perspectives, they will have different understandings and interpretations. Even if the same translator translates the same work in different periods, there will be different translations due to different perspectives. Not to mention, different translators will have different translations in different periods or the same period. Gentzler once commented on the variety of translation, "Despite all the education and proper training in the right methodologies, research has shown that if one gives two workshop translators the same text, what evolves are two different translations" [8].

For this reason, the translation workshop in our university adopts a student-centered teaching process (as shown in Fig. 2), which can be roughly divided into six steps: a. text selection for the translation task;b. randomly group and assign translation tasks;c. independent translation by each student;d. reading translation versions in each group;e. group presentation;f. the refinement and induction of translation theories and skills. In the first and second steps, teachers take the lead, while students take the initiative in the remaining four steps.

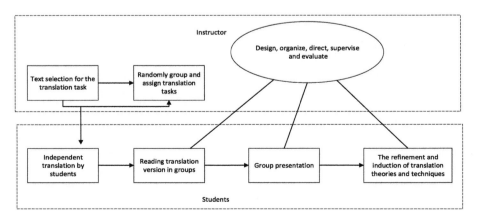

Fig. 2. Operation process of translation workshop teaching

As mentioned above, an important link in the workshop teaching is to present the first translation draft through group cooperation, compare the different translations from multiple angles and layers, and then get a satisfactory translation version after repeated modification. Among these steps, multi-angle and multi-level comparison can be completed with the help of corpus technology. For the purpose of deepen the development of undergraduate translation major in our university and optimize the BTI cultivation conditions, we collected translation corpora from the practice work of "translation workshop" and established a dedicated English-Chinese bilingual parallel corpus which comprises two sub-databases for the teaching of translation major. One of the sub-database is English-Chinese parallel corpus for classic works, and the other is English-Chinese parallel corpus for students' works.

When we were doing translation exercises involving English passive voice structure, the students searched and observed the English passive voice structure and its Chinese translation forms in the classic works sub-database and the students' works sub-database respectively. The Chinese translation forms are divided into ten types according to the common characteristics of translation, and the frequency of different translation forms in the two sub-databases is counted respectively, and then the data is compared. The purpose is to find out whether the translation tendency of English passive voice in students' translations is consistent with the tendency reflected in classical translations, and what are the translation forms with significant differences (see Table 1 for data comparison). After finding out the translation forms with obvious differences, we can search the corpus again, observe the specific corpus one by one, explore the translation skills, and reflect on our students' translation exercises.

Table 1. Comparative statistics on the Chinese versions of English "be-passive"

Translation Form	Frequency (students' work)	Frequency (classic translation work)	Types
Passive structure	14.73%	8.94%	Syntactic and lexical passivity
Active structure	30.78%	39.10%	General active sentences; Replacement of logical relation
Patient theme	20.23%	30.46%	……shi……de;……you……
grammaticalization of verb-object construction	5.08%	3.86%	(dui)……jin xing/yu yi/zuo……
Disposal construction	6.83%	2.57%	…… ba ……
subjective-object structure	1.20%	0.75%	…… shi/ling ……
Attributive phrase	10.65%	3.17%	Center verb "de" +Patient nouns
Nominal structure	6.09%	4.38%	nominalization
Analogous grammaticalization	1.21%	1.67%	…… de yi / huo de ……
Ellipsis	3.20%	5.10%	Ellipsis
Total	100%	100%	

In the subsequent translation quality assessment, both teachers and students can use corpus retrieval data for reference to judge the quality of translation. [9] Corpus - based translation quality assessment has unique advantages. Firstly, the samples are selected and naturally occurring texts, which are easy to be retrieved and analyzed by computer. Secondly, analyzing the real corpus in corpus is empirical and objective in nature. Thirdly, if the corpus technology is combined with data statistics method, teachers and students can intuitively understand the characteristics of language use. [10] On this basis, teachers explain and let learners master translation methods, which is difficult to do with traditional methods.

4   Conclusions

In the undergraduate practice course of translation major, "Translation Workshop" is a course that can comprehensively improve students' translation ability. All kinds of computer information technology can be combined with teaching in different forms in the early, middle, and late translation practice periods. This course lays basis on the PACTE model of translation competence, focuses on the interaction between teachers and students, and through the effective combination of computer information technology, especially corpus technology which has played an incomparable role in traditional translation teaching methods, does well in cultivating translators' professional qualities and improving their translation skills.

Acknowledgements. This work was supported by Domestic Visiting Research Project for Outstanding Young Talents of Colleges and Universities of Anhui Province Education Department (gxgnfx2019070), by Research Fund Project of Anhui Sanlian University "Construction of Special Corpus for BTI Teaching" (2020060201), and by Anhui Provincial Quality Engineering Project "English and Chinese Translation" (2020kfkc201).

References

1. Bowker, L.: A corpus-based approach to evaluating student translations. Translator **2**, 183–210 (2000)
2. Bowker, L.: Using specialized monolingual native-language corpora as a translation resource: a pilot study. Meta: Transl. J. **43**(4), 631–651 (1998)
3. Kefei, W.: The use of parallel corpora in translator training. Technol. Enhanc. Foreign Lang. Educ. **100**(12), 27–32 (2004). (in Chinese)
4. Hongwu, Q., Kefei, W.: Parallel corpus in translation teaching: theory and application. Chin. Transl. J. **5**, 49–52 (2007). (in Chinese)
5. PACTE. Investigating translation competence: conceptual and methodological issues. Meta **50**(2), 609–619 (2005)
6. Wolfson, L.: The contact between text, mind, and one's own word in a translation workshop. Transl. J. (2005). http://accurapid.com/journal/34workshop,htm. Accessed 26 Apr 2021
7. Liangqiu, L.: Exploration on the teaching mode of translation workshop. Shanghai Transl. **4**, 48–51 (2014). (in Chinese)

8. Gentzler, E.: Contemporary Translation Theories, 2nd edn. Multilingual Matters, Clevedon (2001)
9. Guangrong, D., Shangjun, Z.: The application and research of corpus in translation quality assessment. Foreign Lang. Educ. **42**(2), 92–96 (2021). (in Chinese)
10. Bowker, L.: Towards a methodology for a corpus-based approach to translation evaluation. Meta: Transl. J. (2), 345–364 (2001)

Computer Search for Non-isomorphic Extremal Trees of Harmonic Index with Given Degree Sequence

Haiyan Wu[✉] and Zhuoyu Peng

Zhixing College of Hubei University, Wuhan, Hubei, China

Abstract. With the wide application of computer technology in chemical research, the research on the topological index of molecule and its relationship with the invariant of graph is of great value in academic research. Harmonic index, as an important topological index of chemical molecules, is closely related to the chemical and physical properties of molecules. This thesis mainly discusses the harmonic index of trees with given degree sequence, and focuses on the key problem of "how to search all non- isomorphic extremal trees for the harmonic index". In this paper, we propose a method of equivalent transformation of the local switching of graphs, and delete the isomorphic extremal trees by means of the longest path degree sequence, so as to search all the non-isomorphic extremal graphs corresponding to a given degree sequence.

Keywords: Harmonic index · Degree sequence · Extremal trees

1 Introduction

With the wide application of computer technology in chemical research, the research of molecular topological index theory reflects the important value of graph theory application, and the research of this problem has great significance and application background for the purposeful synthesis of molecules or drugs [1, 2]. A topological index, the harmonic index, is defined in reference [3], where $E(G)$ is the edge set of graph G and $d(u)$ is the degree of vertex u.

$$H(G) = \sum_{uv \in E(G)} \frac{2}{d(u) + d(v)} \tag{1}$$

The harmony index is also closely related to the physical and chemical properties of chemical molecules, and the search for molecules with certain chemical or physical properties is the central problem of combinatorial chemistry [4]. The solution of these problems has a strong application background, so more and more scholars pay attention to it, and have achieved a lot of results [5–10]. Literature [10] points out that greedy tree is a maximum tree of harmonic index based on degree sequence, but it may not be the only maximum tree. This paper will do further research on it.

This research focuses on the key scientific problem of "how to construct extremum tree and the algorithm to determine non-isomorphism tree". On the one hand, based on

© The Author(s), under exclusive license to Springer Nature Singapore Pte Ltd. 2022
J. C. Hung et al. (Eds.): FC 2021, LNEE 827, pp. 755–759, 2022.
https://doi.org/10.1007/978-981-16-8052-6_95

the given degree sequence and the properties of the extreme graph, a method of preserving local torsion is proposed. On the other hand, by means of the longest path degree sequence, the isomorphic extremum tree is deleted, and all non isomorphic extreme graphs are searched.

2   Structural Characteristics Analysis of Extremal Trees

Given a sequence of nonnegative integers $\pi = (d_1, d_2, \cdots, d_n)$, that is $d_1 \geq d_2 \geq \cdots \geq d_n$. Let T_π be the set of trees with degree sequence π. For convenience, we will call a Max-tree if it maximizes the harmonic index among all the trees belong to T_π.

Proposition 1. [10] A greedy tree T^* is an optimal Max-tree with degree sequence for the harmonic index, but it may not be the only maximum tree.

Definition 1. Suppose that G is a tree and G' is a tree obtained by the local switching edges st and uv with $G' = G - uv - st + su + tv$, where $d(s) = d(v) = d(u)$. It is easy to see that local switching preserves degrees and $H(G') = H(G)$. We call G' is obtained by equivalent transformation of the local switching.

By using the connection relationship of edges in greedy tree and the method of equivalent transformation of the local switching of graphs, all the maximum trees with the same harmonic index as greedy tree are searched.

3   Isomorphism of Trees with the Same Degree Sequence

Definition 2. Select a maximal path so that the sum of degrees of each vertex of the maximal path is the largest. We record the degrees of all non leaf points of the selected maximal path in order to get the longest path degree sequence. As shown in Fig. 1, $v_2 v_1 v_3$ is a longest path of T^*, so the longest path degree sequence of T^* is (4, 4, 4).

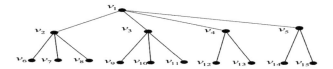

Fig. 1. T^* with degree sequence $\pi = (4, 4, 4, 3, 3, 1, 1, 1, 1, 1, 1, 1, 1, 1, 1)$

We know that trees with different longest path degree sequence or inverse sequence are definitely not isomorphic, but trees with the same longest path degree sequence may or may not be isomorphic. Therefore, we need to get the secondary sequence set of the longest path degree sequence. The degree sequences of G_1, G_2 and G_3 in Fig. 2 are all $(3, 3, 2, 1, 1, 1, 1)$, the longest path degree sequences are $(2, 3, 3)$, $(3, 3, 2)$ and $(3, 2, 3)$, hence G_1 and G_3 are not isomorphic, however, the longest path degree sequences or inverse sequences of G_1 and G_2 are the same, and the secondary sequence sets of the longest path degree sequences of G_1 and G_2 are $\{(1), (1), (1,1)\}$ and $\{(1,1), (1), (1)\}$, so G_1 is isomorphic to G_2.

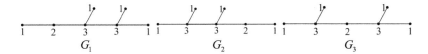

Fig. 2. G_1, G_2, G_3 with degree sequence $\pi = (3,3,2,1,1,1,1)$

4 Search for All Non-isomorphic Extremal Trees

Denote Γ_π the set of maximum trees obtained. A tree selected from Γ_π each time is recorded as T'. The algorithm process is as follows:

Step 1: Select a tree T' from Γ_π to judge whether there are two non-adjacent edges st and uv in $E(T')$, such that $d(s) = d(v) = d(u)$. If not, perform the second step. If there are, stop.

Step 2: The local switching edges of st and uv is implemented to change from T' to T'', that is, $T'' = T' - uv - st + su + tv$.

Step 3: Judge T'' and T' are isomorphic or non-isomorphic. If isomorphic, directly delete, and jump to the fourth step; If T'' and T' are not isomorphic, judge whether they are isomorphic to a tree in Γ_π. If so, delete T'' and execute step 4. Otherwise, set $\Gamma_\pi = \Gamma_\pi \cup \{T''\}$ and execute step 4.

Step 4: Repeat steps 1–3 for the other satisfying edges st and uv in step 1 until there are no satisfying edges st and uv in $E(T')$ and go to step 5.

Step 5: Repeat steps 1–4 for each new addition T'' in Γ_π until no T'' appear.

5 Example Demonstration of Searching

For example, consider the degree sequence $\pi = (4,4,4,3,3,1,\cdots,1)$ with $n = 15$. First, select the only tree T^* from Γ_π as T' and find v_1v_2 and v_3v_9 that satisfy the condition in step 1. Then $T'' = T' - v_1v_2 - v_3v_9 + v_1v_9 + v_2v_3$. In order to distinguish, we record the new T'' as T_1 shown in Fig. 3. The longest path degree sequence of T' and T'' is $(4,4,4)$ with $(4,4,4,3)$, hence T' and T'' are not isomorphic, thus $\Gamma_\pi = \Gamma_\pi \cup \{T_1\}$. Then for the other edges st and uv in T', we repeat steps 1–3 and get the tree T'' which are all isomorphic to T', so we go to step 5 and repeat steps 1–4 for the new addition T_1.

The operation of T_1 is briefly described below. $T_2 = T_1 - v_1v_3 - v_2v_6 + v_1v_2 + v_3v_6$. T_2 is isomorphic to T_1. Then we delete T_2, and continue to choose other edges in T_1. $T_3 = T_1 - v_1v_4 - v_2v_3 + v_1v_2 + v_3v_4$. T_3 and T_1 are not isomorphic, moreover T_3 is not isomorphic to any one of Γ_π, thus $\Gamma_\pi = \Gamma_\pi \cup \{T_3\}$. While there are no satisfying edges, we go to step 5, and repeat steps 1–4 for the new addition T_3. Continue with the algorithm and we'll get T_4, T_5, T_6, T_7 and T_8 in Fig. 3. After isomorphism judgment, the set of all the non-isomorphism trees is $\Gamma_\pi = \{T^*, T_1, T_3, T_5\}$.

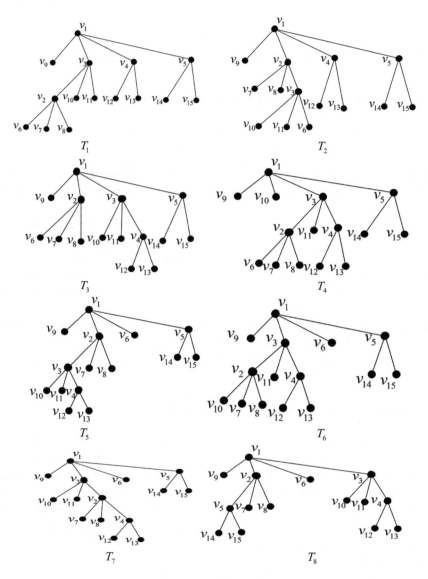

Fig. 3. $T_1, T_2, T_3, \cdots, T_8$.

Acknowledgements. This work was supported by the guiding project of scientific research plan of Hubei Provincial Department of Education (No. B2020330 Computer Search for Non-Isomorphic Extremal Trees of Harmonic index with Given Degree Sequence), and by the teaching reform research project of Zhixing College of Hubei University (No. XJY202019 Research and Practice on the training mode of Applied Mathematics talents in Independent Colleges Based on Mathematical Experiment).

References

1. Ahmadi, M.B., Sadeghimehr, M.: Atom bond connectivity index of an infinite class NS1 of dendrimer nanostars. Optoelectron. Adv. Mat. **4**, 1040–1042 (2010)
2. Clark, L.H., Gutman, I.: The exponent in the general randic index. J. Math. Chem. **43**, 3–44 (2008)
3. Fajtlowicz, S.: On conjectures of Graffitiff I. Congr. Numer. **60**, 187–197 (1987)
4. Golaman, D., Istrail, S., Lancia, G., Piccolboni, A., Walenz, B.: Algorithmic strategies in combinatorial chemistry. In: Proceedings of the 11th ACM-SIAM Symposium on Discrete Algorithms, pp. 275–284 (2000)
5. Zhong, L.: The harmonic index for graphs. Appl. Math. Lett. **25**, 561–566 (2012)
6. Zhong, L.: The harmonic index on unicyclic graphs. Ars Comb. **104**, 261–269 (2012)
7. Zhong, L., Xu, K.: The harmonic index for bicyclic graphs. Utilitas Math **90**, 23–32 (2013)
8. Lv, J., Li, J.: On the harmonic index and the matching number of a tree. Ars Comb. **116**, 407–416 (2014)
9. Zhang, G.J., Chen, Y.H.: Functions on adjacent vertex degrees of graphs with prescribed degree sequence. MATCH. Commun. Math. Land Comput. Chem. **80**, 129–139 (2018)
10. Wang, H.: Functions on adjacent vertex degrees of trees with given degree sequence. Central Eur. J. Math. **12**(11), 1656–1663 (2014). https://doi.org/10.2478/s11533-014-0439-5

Network Security Issues in Universities and Exploration of Defense System

Weixiong Wang[✉], Gui Wang, Xiaodong He, Zhen Guo, Ziqin Lin, and Zhitao Wang

Airport Management College, Guangzhou Civil Aviation College, Guangzhou 510403, Guangdong, China

Abstract. Campus network is an important infrastructure of colleges and universities, and it plays an important role in university teaching, research, management and external communication. As the demand for campus networks continues to grow and network users continue to grow, some security problems have emerged one by one, which sometimes seriously hinders the normal operation of campus networks. The main purpose of this article is to study the research and protection system of network security in colleges and universities. Based on the current situation of school network security, this paper analyzes the hidden dangers of network security, and establishes a security protection system that is compatible with the reality of the school campus network. Based on the actual situation of the school campus network, this article focuses on the current implementation of some technologies, such as secure password verification technology, firewall technology and intrusion detection technology, and conducts detailed research on the problems to be solved and verifies the results. This article focuses on firewall technology, intrusion detection, and encryption and decryption technologies to build a platform for managing network security system operations. Research experiments show that Sophos's detection rate of Trojan horse viruses reached 77.4%, and Av's detection rate of the virus reached 73.4%. After the implementation of the network security strategy, hacker attacks can be avoided scientifically and efficiently.

Keywords: Network security · Defense system · Security issues · College network

1 Introduction

With the continuous development of network technology, colleges and universities have begun to widely use campus networks for teaching, office and information exchange. Due to the interconnection and openness of the network and the gaps and defects of the network system itself, campus networks are facing threats and attacks from both inside and outside the network [1, 2]. At present, although many schools are equipped with network security products, they have not played a real security protection role, causing heavy losses [3, 4]. Designing advanced security solutions based on the specific current situation of campus networks, choosing reasonable network security products, and constructing an effective campus network security defense system are the main problems that need to be solved urgently [5, 6].

J. C. Hung et al. (Eds.): FC 2021, LNEE 827, pp. 760–767, 2022.
https://doi.org/10.1007/978-981-16-8052-6_96

Currently, many scholars have conducted research on network security issues and defense systems in colleges and universities. For example, Chih-Che Suna and Adam Hahn and others have conducted the latest overview of the most relevant network security research in the power system, reviewed the research to demonstrate network security risks, and constructed solutions to enhance grid security [7]. This provides a reference direction for studying network security in colleges and universities. Y.Danyk and S.Vdovenko, in the context of introducing the possibility and urgency of the experience of Ukraine's cyber security and cyber defense system, analyzed the general principles of establishing cyber security and cyber defense systems in advanced countries in the world, and analyzed the cyber security and urgency of Ukraine. The conditions, status quo and existing problems of the formation of cyber defense systems [8]. It provides theoretical support for the study of network security issues and defense systems in colleges and universities.

This article mainly discusses the safety of campus network. The authentication encryption technology, firewall technology and intrusion detection technology are discussed in detail. The principles of these technologies are first analyzed, and then these technologies are flexibly applied to the school's network security design. This article uses the campus network of our school as the background to analyze the security threats faced by the campus network, and then analyzes the security requirements of the campus network of Lanzhou Petrochemical Institute from various aspects such as campus network and external network, campus network internal network and campus network security management. And put forward the campus network topology deployment plan, the external network adopts firewall technology and network intrusion prevention system, the deployment and configuration of firewall and the deployment and configuration of network intrusion prevention system, the division of VLAN plan in the network and so on.

2 Research on Network Security Issues and Defense System in Colleges and Universities

2.1 Definition of Cyber Security

The field of network security technology is a complex and interdisciplinary research field, which requires the accumulation of long-term basic knowledge and the latest development of many majors in mathematics, physics, communications, and computers [9, 10]. The basic purpose of network security is to ensure that there is no change or omission when information is transmitted to the desired computer through the network to the target position, so that the authenticity, integrity and usability of the information are protected.

2.2 Network Security Technology in Colleges and Universities

(1) Password encryption and decryption technology

    1) Symmetric encryption algorithm
       Symmetric encryption system is also called private key encryption method [11]. The private key cryptographic system management system is a traditional

cryptographic system management system [12]. The most representative contemporary designers are des, aes, idea, rc5, etc. Their certainty and certainty are based on complex mathematical calculations.

If M is the cost of all private key plaintext information, C is the cost of all ciphertext information, and K is all keys, then the cryptographic system of the private key is composed of such a set of function pairs:

$$Ek:M \to C \qquad\qquad (1)$$

$$Dk:C \to M, \quad k \in K \qquad\qquad (2)$$

Here, all $m \in M$ and $k \in K$ have $Dk(Ek(m)) = M$. When using this system, both ends of the connection must have access to the preset $k \in K$, and they can obtain each other's keys through a face-to-face connection or a trusted third party. Obviously, the properties of the decoding system should be: n and m are easy to implement, and after removing the third party of secret system information other than the key selection method, it is still impossible to obtain M (or k) according to C.

  2) Asymmetric encryption algorithm

  This encryption method uses a pair of "public key" and "private key" to encrypt and decrypt information. The key pair is processed by the recipient of the information, the "public key" is used as part of the encrypted information, the key is sent to the sender, and the recipient has the "private key" required for decryption. Since encryption keys are public keys, the distribution and management of keys is very simple, and digital signatures can be easily implemented, so they are more suitable for the needs of e-commerce applications.

(2) Firewall technology

  All information transmitted between the internal network and the external network must be controlled by the firewall, and only authorized data can be passed. The firewall data packet filtering technology is to filter data packets in the network layer, and its filtering is based on the filtering logic specified in the system, which is called an access control list. The proxy firewall is the most secure firewall. Use proxy technology to participate in the entire TCP communication process, so it is transmitted internally; after the firewall has processed the data packet, it seems to come from the external network, and the firewall card is suitable for concealing the internal network architecture.

(3) Antivirus technology

  There are many prevention and control technologies for computer viruses. Commonly used technologies include: signature scanning technology, virtual execution technology, real-time file monitoring technology, antivirus technology, virus immunity technology, etc.

2.3   College Campus Network Security Faces Problems

(1) Network layer security issues
Network platform security includes network topology, network routing status and network environment. Network attacks mainly include: unauthorized access to network application services; violation of the confidentiality of information sharing; virus transmission and penetration into the network; network denial of service attacks; network monitoring; network fraud; vulnerability attacks, etc.

(2) System-level security issues
System security refers to the reliability of the entire network operating system and network hardware platform. Currently, there is no completely secure operating system to choose from, and the system itself has many security vulnerabilities, which will lead to major network security risks. Although no operating system is completely secure, if security is configured, it is difficult for any intruder to successfully invade the network.

(3) Application layer security issues
Application security refers to the security level of the application software of the host system. Although the introduction of application system software brings convenience to users, it also brings new threats to the network. This is because most online application system software does not have a good security design, and network server programs often use super user privileges, thus causing security problems in some applications.

2.4   Campus Network Security Defense System

(1) Firewall deployment
The firewall is deployed between the intranet and the extranet of the campus network to effectively separate the campus network from the extranet so as to enhance the security of the intranet. Configure network address translation in the firewall, configure the IP address pool in the external network port of the firewall as exchange addresses, and use the access control list set during the configuration process, which shows the conversion of computer addresses. It can realize the address and make full use of the limited number of legal IP address resources to protect the campus intranet private IP.

(2) Deployment of intrusion prevention system
Research and analysis of various network behaviors that cross the border, and follow the pre-established strategy information, so that illegal and abnormal behaviors are controlled and prevented. This design method can filter some attacks by the firewall first, and then use the intrusion prevention system to analyze data packets for the attacks that the firewall cannot handle, filter illegal data packets, and block them from entering the campus network.

(3) The division of VLAN in campus network
The virtual local area network in the campus network of this school is set as a static VLAN. The purpose of using VLAN is to protect the key network segments in the campus network. While making full use of network resources and reducing

overhead, it resolves conflicts in the broadcast domain and improves the performance of the system as much as possible. The physical location of each functional department in the university is relatively fixed, so the VLAN division of the campus network of Lanzhou Petrochemical Institute is divided according to the switch port.

3 Research on Network Security Issues in Colleges and Universities and Experimental Research on Defense Systems

3.1 Intrusion Detection Technology Test

(1) Test environment

The test environment of the system is the intranet of the school campus.

In order to facilitate the security management of these hosts, set the host IP as 192.168.111.102 and the external IP of the gateway as the default IP.

(2) Filter function test

This article uses hierarchical filtering detection technology, mainly to detect the ability to block IP addresses, URLs, email addresses, email subjects, WEB scripts, web page titles, etc.

(3) Virus detection

In order to test the virus detection performance of this subsystem, the virus sample database used in this test is the benchmark virus database in the ASTS#6 sample of Antiy Labs. The benchmark sample library used in this article completely covers PE viruses, script viruses, worms, and Trojan horses. Network content security analysis provides two virus scanning engines, Sophos and Av. Therefore, in this test experiment, the ability of Sophos and Av to detect virus samples is tested at the same time. The test index is its detection rate. In the benchmark virus database used in this article, there are a total of 15,617 samples, 2231 for each type of virus.

3.2 Analysis of the Effect of Defense System Deployment

(1) Analysis object

This paper mainly analyzes the effect of the defense system proposed in this paper after it is deployed on the school's network, and analyzes the performance of its core router CPU usage and message delay, data flow, multi-port service balance system deployment effect, and defense against ARP attacks.

(2) Analysis method

This analysis mainly uses a comparative analysis method, which compares the changes in some indicators of the school's network before and after the deployment of the defense system.

4 Research on Network Security Issues in Colleges and Universities and Experimental Research and Analysis of Defense System

4.1 Intrusion Detection Technology Test-Virus Detection

The main threats to the system at present come from PE viruses, script viruses, worms and Trojan horses. The viruses used in the test can be divided into several categories. Specific virus attack categories and test results are shown in Table 1.

Table 1. Security test results

	Guided area mirror	Dos document type	Macro virus	Trojan horse	Script virus	NE/PE	Other
Sophos	75.6%	69.3%	74.1%	77.4%	72.6%	71.9%	75.7%
Av	72.8%	74.9%	65.4%	73.4%	70.4%	68.1%	69.3%

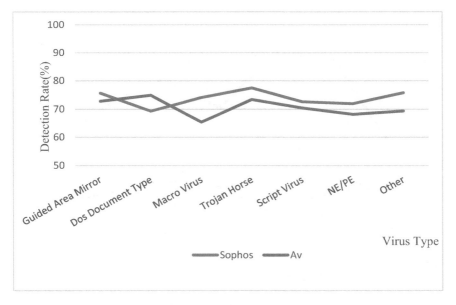

Fig. 1. Security test results

It can be seen from Fig. 1 that both Sophos and Av virus scanning engines have good detection capabilities for network viruses. Among them, Sophos has a detection rate of 77.4% for Trojan horse viruses, and Av has a detection rate of 73.4 for this virus. %. In general, Sophos's detection rate is about 3% higher than Av's, but Av's detection rate for Dos file-type viruses is higher than Sophos. The above results show that after the implementation of the network security strategy, hacker attacks can be avoided scientifically and efficiently, making the school's services more reliable and more efficient.

4.2   Analysis of the Deployment Effect of the Defense System-Traffic Test

Data traffic is a common indicator among many data on the network, but it is a barometer of network security. Table 2 shows the data sampling on the router exit.

Table 2. Analysis of router exit data

	Mon	Tue	Wed	Thu	Fri	Sat	Sun
Before	75M-	1M-	5M-	26M-	10M-	73M-	34M-
	186M	200M	173M	196M	153M	200M	197M
After	30M-	29M-	1M-	32M-	40M-	25M-	18M-
	84M	177M	183M	121M	154M	124M	79M

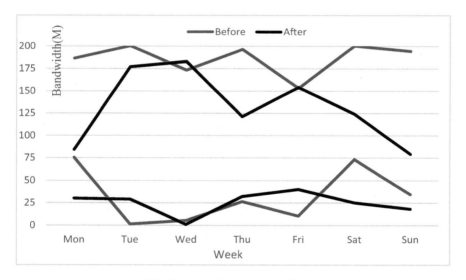

Fig. 2. Analysis of router exit data

It can be seen from Fig. 2 that before the adjustment of the security system, the bandwidth from 0M to 200M was basically occupied during peak hours. After the implementation of the security system, there was basically no peak of the export bandwidth limit, which only accounted for 1M to 183M. It shows that after the system is deployed, traffic will be reduced and the number of illegal accesses will be reduced.

5   Conclusions

Currently, the security of the campus network environment is always facing major challenges. Viruses, internal attacks, external intrusions, denial of service attacks, internal abuse, a large number of disasters and accidents always threaten the security of the normal operation of the network and network information. Therefore, we must attach great importance to the safety and environmental protection of college campus networks and strengthen management. When deploying campus network security protection, real-time detection of network operating conditions should be improved, potential security attacks should be discovered in time, security incidents should be prevented, network information and security vulnerabilities should be resolved, and the security of network detection should be ensured. Discover and repair network and system vulnerabilities in time to prevent problems and further improve the security of the campus network environment.

Acknowledgements. The project will be funded by the Special Fund for Science and Technology Innovation Strategy of Guangdong Province in 2020.

References

1. Cowan, G.: SSITH effort targets hardware cyber security issues. Jane's Int. Def. Rev. **50** (JUN), 25–25 (2017)
2. Akhmetov, L.: System of decision support in weakly- formalized problems of transport cybersecurity ensuring. J. Theor. Appl. Inf. Technol. **96**(8), 2184–2196 (2018)
3. Demchak, C.C.: Three futures for a post-western cybered world. Mil. Cyber Aff. **3**(1), 6 (2018)
4. Du, C., Zhao, S., Wang, W.: RRPOT: a record and replay based honeypot system. J. Phys. Conf. Ser. **1757**(1), 012183 (2021)
5. Michalski, D.: System approach to coherent cybersecurity strategy. Saf. Def. **5**(1), 31–36 (2019)
6. Chen, Z., Guo, Y., Bai, D., et al.: Research on cyber security defense and protection in power industry. J. Phys. Conf. Ser. **1769**(1), 012040 (7pp) (2021)
7. Sun, C.C., Hahn, A., Liu, C.C.: Cyber-physical system security of a power grid: state-of-the-art. Int. J. Electr. Power Energy Syst. **99**(JUL), 45–56 (2018)
8. Danyk, Y., Vdovenko, S.: Problems and prospects of ensuring a state cyber defense. Collection of scientific works of the Military Institute of Kyiv National Taras Shevchenko University, **2019**(66), 75–90
9. Buchler, N., Rajivan, P., Marusich, L.R., et al.: Sociometrics and observational assessment of teaming and leadership in a cyber security defense competition. Comput. Secur. **73** (MAR), 114–136 (2018)
10. Reagin, M.J., Gentry, M.V.: Enterprise cybersecurity: building a successful defense program. Front. Health Serv. Manage. **35**(1), 13–22 (2018)
11. Dawson, M.: National cybersecurity education: bridging defense to offense. Land Forces Acad. Rev. **25**(1), 68–75 (2020)
12. Shen, G., Wang, W., Mu, Q., et al.: Data-driven cybersecurity knowledge graph construction for industrial control system security. Wirel. Commun. Mob. Comput. **2020**(6), 1–13 (2020)

A Model of the Impacts of Online Product Review Features on Trust in Ecommerce

Lianzhuang Qu[✉], Yi Guan, and Xiaolin Yao

School of Information and Business Management, Dalian Neusoft University of Information, Dalian, Liaoning, China
qulianzhuang@neusoft.edu.cn

Abstract. The pandemic of COVID-19 makes consumers rely on e-commerce to an even higher extent. Since consumers utilize online product reviews (OPRs) for decision making, it is vital to study how various OPR features impact consumer behavior. This research proposes a model for understanding effects of OPR features on consumers' trust based on HSM. For data collection, we carry out lab experiments and use PLS-SEM analysis to test the model. The results indicate that OPR features influence trust via usefulness of OPRs and attitude toward website. We contribute to the OPR literature by applying the HSM model to the context of OPRs. We also examine the interactions between heuristic mode and systematic mode.

Keywords: Trust in OPRs · Attitude toward website · Usefulness of OPRs · OPR features

1 Introduction

In recent years, e-commerce has been increasingly pervasive in consumers' daily life. The pandemic of COVID-19 makes consumers rely on e-commerce to an even higher extent. A report by eMarketer [1] shows that retail ecommerce sales grew 27.6% for 2020, and the total reached $4.280 trillion. Therefore, it is essential to study how various aspects of ecommerce impact consumers. One prominent aspect that deserves attention is online product reviews (OPRs), whose importance has been acknowledged by both researchers and practitioners. The reason is that a major part of online consumers utilizes OPRs for decision making. According to a report by Diana Kaemingk [2], 93% of customers read online reviews before buying a product.

Given the importance of OPRs, it is critical to examine how OPRs affect online consumer behaviors. Such insights on effects of OPRs can help ecommerce companies to make the full use of OPRs. However, only after consumers develop trust in those OPRs, they are likely to reply on those OPRs for evaluating products. Put differently, trust in OPRs is a prerequirement for OPRs to generate influences on consumers. In fact, building trust in online environment has been an important topic receiving much attention from researchers. In addition, OPRs have distinctive features from offline product reviews. therefore, we focus on effects of OPRs' features on consumers' trust.

In this study, we take the perspective that online consumers engage in both an effortless mode and a more cognitively demanding mode when processing OPRs. We

J. C. Hung et al. (Eds.): FC 2021, LNEE 827, pp. 768–775, 2022.
https://doi.org/10.1007/978-981-16-8052-6_97

further shed some light on OPR features that drive consumers to develop their trust in OPRs. The lens that consumers engage in both an effortless mode and a more cognitively demanding mode allow us to develop a theoretical model for examining how OPR features influence consumers' trust in OPRs. Additionally, we investigate how involvement, the heterogeneity in consumer information processing motivation would moderate the propositions.

2   Theoretical Development

Compared with traditional product reviews, online product reviews have two major features, as highlighted in Fig. 1. The first feature is that OPR can accumulate votes from readers. Those votes indicate the extent to which other readers think the OPR is useful. The second one is managerial response, which are messages from sellers on OPRs. It provides evidence that sellers pay attention to customers' experience. In this paper, we study the influences of those two factors on online consumers' trust in OPRs. The theoretical lens we use is the heuristic-systematic model (HSM). Research model of this paper is shown in Fig. 2.

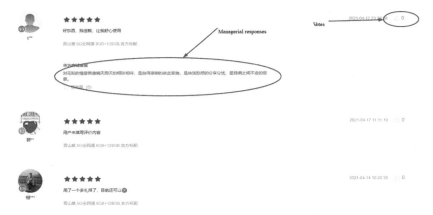

Fig. 1. Examples of OPRs from vmall.com

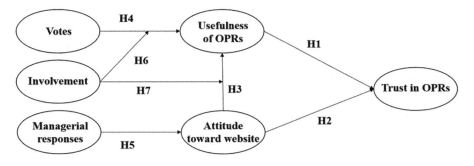

Fig. 2. Research model

Trust is an important concept for online environment, we focus on trust as a construct for this study. It is composed of several beliefs, such as competency, integrity and benevolence [3]. In this paper, we define trust in OPRs as the online consumers' beliefs that the OPRs have attributes that are beneficial to them. Such beliefs include competence, integrity and benevolence.

HSM states that in different situations, individuals will vary their effort in cognitively processing a particular piece of information contained in social interaction [4]. Specifically, HSM proposes that, when processing information, people may engage in either a relatively effortless heuristic mode or a more cognitively demanding systematic mode. When they are unable to or not highly motivated to process information, they make judgment via the heuristic processing mode. In this mode, decision cues are critical in influencing people's judgment. A potentially infinite number of these cues exists [5]. For instance, people often use decision cues pertaining to the source of information. In contrast, people with high motivation and capability tend to use systematic strategy to evaluate the information and achieve a judgment. In the systematic processing mode, people count on an assessment of the information distinguishing a stimulus, an inspection of one's own thoughts and beliefs about the information of the stimulus and an integration of the information to make a decision. When systematic process occurs, argument quality, defined as the strength or plausibility of persuasive argumentation [6], plays the major role.

For this research, we argue that when consumers are in the systematic mode, usefulness of OPRs impacts trust in OPRs. Usefulness of OPRs refers to the extent to which an individual perceives OPRs to be useful in performing shopping tasks. If consumers perceived OPRs as useful, they tend to think that OPRs are able to provide information to help them evaluate product. In other words, usefulness of OPRs enhance consumers' perception of OPR's competence. Therefore, we propose,

H1. Usefulness of OPRs positively impacts trust in OPRs.

Attitude toward a website, similar to website quality, indicating a predisposition to respond favorably or unfavorably to a website. Therefore, consumers with a positive attitude tend to like the website. If consumers are in a heuristic mode, they are likely to use attitude as a heuristic of the trustworthiness of the OPRs. The reason is that likeable websites have higher chance to act in consumers' interests. Therefore, we propose,

H2. Attitude toward a website positively impacts trust in OPRs.

Previous studies on halo effects show that people are inclined to base their evaluations of individual attributes on an overall evaluation. Attitude toward a website reflects consumers' general assessment resulting from their interactions with the website. Therefore, once online consumers from a positive attitude toward a website, they are likely to regard OPRs on this website as instrumental for their decision-making. Therefore, we propose,

H3. Attitude toward a website positively impacts usefulness of OPRs.

With the development of IT technology, websites allow online consumers to voice their opinions on the OPRs by helpfulness votes. After reading an OPR, consumers can decide whether this review provides insights on their evaluation of products and vote for its helpfulness. Therefore, OPRs that have accumulated a large number of votes indicate that numerous consumers perceive those OPRs to be useful in performing shopping tasks. Votes are an important cue for consumers to evaluate the OPR usefulness. Therefore, we propose.

H4. Votes positively impacts usefulness of OPRs.

As OPRs become an important outlet for consumers to voice their opinions, e-commerce companies pay great attention to OPRs. To enable timely response, companies make their feedbacks public by managerial responses on OPRs. Those responses include solutions to issues raised by consumers and comments on consumers' compliments. Thus, managerial responses provide evidence that companies care about consumers. It is natural that online consumers use those responses as a cue and develop a favorable evaluation of e-commerce companies and their website. Therefore, we propose.

H5. Managerial responses positively impact attitude toward a website.

Studies on HSM suggest that involvement is a prominent factor impacting consumers' processing modes. When individuals are highly involved in a situation, they are in a systematic mode and exert much effort in processing related information. In contrast, when the involvement is low, individuals are likely to engage in the heuristic mode, spending few time and energy and relying on cues for decision-making. Therefore, we propose,

H6 The effects of votes on usefulness of OPRs are greater when involvement is low than when it is high.
H7. The effects of attitude toward website on usefulness of OPRs are greater when involvement is low than when it is high.

3   Research Method

For data collection, we conducted a two (votes) by two (managerial responses) between subject experiment. Subjects were randomly assigned to a treatment group. The product we used for the experiment was a time management book. For one thing, customers often read online book reviews, therefore, using a book made the experiment realistic. For another, time management was a relevant issue to our subjects, undergraduates.

The experiment used a scenario which depicts the time management problems encountered by an undergraduate student. The student determines to fix those problems and decides to buy a time management book. The participants were asked to think from the perspectives of a friend and help evaluate the book by reading online product reviews on a website.

During the experiment, subjects were informed that the objective of the experiment was to learn searching and browsing behaviors of online consumers. To minimize demand effects, we did not mention the focus of this study. The experiment tasks included asking subjects to read reviews, adapted from real online reviews, on the time management book. After finishing the reviews, subject proceeded to a questionnaire for collecting information on demographic data, background information, control variables, dependent variables and manipulations.

Measurements for usefulness of OPR, attitude towards website, involvement, and trust in OPR are constructed by adapting reliable and valid scales from studies by Luo and colleagues [7], Shin and colleagues [8], Peng and colleagues [9], as well as Banerjee and colleagues [10]. Control variables include hours spent on reading reviews during last 30 days, gender, age, and years of reading online product reviews.

4   Analysis and Results

For the randomness and manipulation checks, we carried out ANOVA and Chi-square tests. Results show that subjects are randomly assigned to treatment groups and manipulations are successful.

We analyze the proposed research model with Smart-PLS. Individual item reliability, measured by outer loadings of individual items on related constructs, is satisfactory since all of them are greater than .70. Table 1 displays the results of individual item reliability test. Table 2 reports the composite reliability and Cronbach's alpha. All entries have value higher than .70, showing high internal consistency [11].

Table 1. Outer loadings of indicators

Construct	Indicator	Loading
1 INVO	Item 1	.830
	Item 2	.814
	Item 3	.833
	Item4	.836
	Item5	.852
2 TRU	Item 1	.921
	Item 2	.893
	Item 3	.901
	Item 4	.888
3 ATW	Item 1	.914
	Item 2	.901
	Item 3	.871
	Item 4	.908
4 UOPR	Item 1	.917
	Item 2	.927
	Item 3	.947

Note: INVO = involvement; TRU = trust in OPRs; ATW = attitude toward website; UOPR = usefulness of OPRs

Figure 3 presents the PLS-SEM testing results for the proposed model. The results show that votes have positive and significant effects on perceived usefulness of OPR (β = .334, p < .001). The effect size f2 is .135, a medium effect. The managerial responses significantly impact attitude toward website (β = .374, p < .001). It has an effect size f2 of .163, a medium effect. The results show that usefulness of OPR significantly influences trust in OPR (β = .486, p < .001). The effect size f2 is .263, a large effect. Attitude toward website has a significant effect on trust in OPR (β = .236, p < .001), with a close to medium effect size f2 of .077. Therefore, H1, H2, H4 and H 5 are supported. The study indicates that attitude toward website has influence on usefulness of OPR, almost significant at .05 level (β = .119, p = .08). This research shows that the moderating effects of involvement on the relationship between votes and usefulness as well as the relationship between attitude and usefulness are not significant. Therefore, H3, H6 and H7 are not supported.

Table 2. Internal consistencies

Construct	Cronbach's Alpha	Composite Reliability
1 INVO	.893	.919
2 TRU	.923	.945
3 ATW	.921	.944
4 UOPR	.923	.925

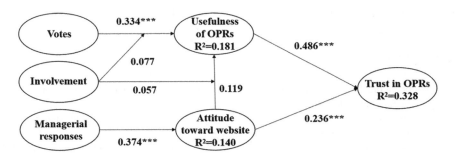

Fig. 3. PLS testing results for the research model

5 Conclusions

The findings of this research show that OPR features have impacts on online consumers' trust in OPRs. In particular, votes have significant impacts on perceived usefulness of OPRs. On websites, OPRs with votes enhance usefulness of OPRs. The data supports that managerial responses significantly affect consumers' attitude toward website. In other words, websites that provide managerial responses increase consumers' attitude toward the website. Our results illustrate that although effects of attitude toward a website on usefulness of OPRs is not significant, the effect size is very large. Another finding is that usefulness of OPRs have significant effects on trust in OPRs. The higher the perceived usefulness of OPRs, the more likely consumers trust OPRs. We also show that attitude toward website significantly influences trust, such that more favorable attitude leads to higher level of trust. However, the results show that impacts of involvement as a moderator on the relationship between votes and usefulness are not significant. Similarly, the moderating effects of involvement on the relationship between attitude toward website and usefulness are not significant.

This research contributes to the OPR literature in three ways. First, we construct a model for understanding effects of OPR features on consumers' trust in product reviews, taking the view that online consumers engage in both an effortless mode and a more cognitively demanding mode to process OPRs. Second, we apply the HSM model to the context of OPRs by using usefulness and attitude as the pillar predictors of trust. Third, we examine the interactions between heuristic mode and systematic mode. For practice, this research can help ecommerce companies understand the impacts of various features and develop their tactics to increase their sales.

Nevertheless, several limitations exist for this study. First of all, this research only tests the proposed model for B2C websites. To increase the generalizability of the findings, future research can test the model for C2C and B2B websites. Second, the subjects for data collection are college students. This might limit the applicability of the findings to other subjects. Future studies should be carried out to investigate the effects of OPR features on consumers with different backgrounds.

Acknowledgements. This work was supported by 2019-BS-014.

References

1. Cramer-Flood, E.: Global Ecommerce Update (2021). https://www.emarketer.com/content/global-ecommerce-update-2021. Accessed 27 Apr 2021
2. Kaemingk, D.: Online reviews statistics to know in 2021, (2020). https://www.qualtrics.com/blog/online-review-stats/. Accessed 28 Apr 2021
3. Leimeister, J.M., Ebner, W., Krcmar, H.: Design, implementation, and evaluation of trust-supporting components in virtual communities for patients. J. Manag. Inf. Syst. **21**(4), 101–135 (2005)
4. Chaiken, S., Liberman, A., Eagly, A.H.: Heuristic and Systematic Information Processing within and beyond the Persuasion Context. In: Uleman, J.S., Bargh, J.A. (eds.) Unintended Thought, pp. 212–252. Guilford Press, New York (1989)
5. Gergen, K.J.: Toward Transformation in Social Knowledge. Springer, New York (1982)
6. Eagly, A.H., Chaiken, S.: The Psychology of Attitudes. Fort Worth, Harcourt Brace Jovanovich College Publishers, T. X. (1993)
7. Luo, C., et al. The influence of eWOM and editor information on information usefulness in virtual community. In: Pacific Asia Conference on Information Systems (2014)
8. Shin, S.Y., et al.: Investigating moderating roles of goals, reviewer similarity, and self-disclosure on the effect of argument quality of online consumer reviews on attitude formation. Comput. Hum. Behav. **76**, 218–226 (2017)

9. Peng, L., et al.: Moderating effects of time pressure on the relationship between perceived value and purchase intention in social E-commerce sales promotion: considering the impact of product involvement. Inf. Manag. **56**(2), 317–328 (2019)
10. Banerjee, S., Chua, A.Y.K.: Trust in online hotel reviews across review polarity and hotel category. Comput. Hum. Behav. **90**, 265–275 (2019)
11. Nunnally, J.C., Bernstein, I.H.: Psychometric Theory. McGraw-Hill, New York (1994)

Redundant Data Transmission Technology Supporting Highly Reliable 5G Communication

Shuhua Mao$^{(\boxtimes)}$, Xiaoli Xie, Shenghui Dai, Bolu Lei, and Liya Li

Department of Information Engineering, Yangtze River College, East China University of Technology, Nanchang, Jiangxi, China
shhmao@ecit.cn

Abstract. This paper studies how to support 5G high reliability through redundant transmission in the user plane, and proposes a redundant transmission scheme between RAN and UPF. Redundant packets in this scheme are transmitted between UPF and RAN through two independent N3 interfaces. These channels are associated with a single PDU session, and the reliability of the service is enhanced through different transport layer paths, which can be activated on demand according to QoS traffic Redundant transmission.

Keywords: 5G · RAN · UPF · QoS traffic · Redundant transmission

1 Introduction

URLLC, as one of the three application scenarios of 5G mobile communication networks, is suitable for autonomous driving, Internet of Vehicles, smart home, augmented reality (Augmented Reality, AR), virtual reality (Virtual Reality, VR), industrial control, and other highly delay-sensitive types Widespread application of business is very critical. If the network delay is high, the normal operation of URLLC services will be affected, and there will be control errors. In view of this, 3GPP has defined the indicators of URLLC delay and reliability [1]. The uplink and downlink user plane delay target of the URLLC service is reduced to 0.5 ms. The reliability requirements of URLLC are: the reliability of transmitting 32-byte packets within 1 ms of the user plane delay is $1*10-5$. The eMBB service is the basic service in the future 5G network and will occupy a dominant position. It is characterized by a large amount of data and high requirements for transmission rate. In contrast, the data packets of URLLC services are small, and the requirements for delay and error rate are very strict. Therefore, on the premise of the existing e MBB service, how to design a transmission scheme that can meet the requirements of the URLLC based on the 5G unified system framework, especially the design of a more reliable user plane transmission scheme, [2, 3]. so as to ensure the service quality requirements of the URLLC service It is an urgent problem to be solved. It has important research significance.

© The Author(s), under exclusive license to Springer Nature Singapore Pte Ltd. 2022
J. C. Hung et al. (Eds.): FC 2021, LNEE 827, pp. 776–779, 2022.
https://doi.org/10.1007/978-981-16-8052-6_98

2 Description of Redundant Data Transmission Scheme

Assuming NG-RAN nodes, the reliability of UPF and CP NF is high enough to meet the reliability requirements of URLLC services served by these NFs. In order to ensure that two N3 interfaces can be transmitted through disjoint transmission layer paths, NG-RAN nodes, SMF or UPF will provide different routing information in the channel information, which will be mapped to disjoint transmissions according to the network deployment configuration Layer path. The RAN node and UPF support data packet replication and elimination functions. By modifying the GTP-U protocol to realize packet duplication and elimination. In the case of DL traffic, UPF copies packets from DN and assigns them the same GTP-U sequence number for redundant transmission. These data packets are transmitted to NG-RAN through N3 interface 1 and N3 interface 2 respectively. In order to eliminate duplicate packets, the NG-RAN forwards the packet received first from any channel to the UE, and discards the duplicate packet with the same GTP-U sequence number as the forwarded packet. In the case of UL services, NG-RAN duplicates packets and assigns them the same GTP-U sequence number, and UPF eliminates duplicate packets based on the GTP-U sequence number. One of the N3 interfaces is determined to be the main transmission path during the establishment of the GTP-U channel. For those QoS flows of the same PDU session that do not require redundant transmission, use this main N3 interface [4, 5].

In addition, two intermediate UPFs (I-UPF) can also be inserted between UPF and NG-RAN to extend the scheme to support redundant transmission based on a single NG-RAN mode and two N3 and N9 interfaces. UPF, RAN nodes and UPF shall support data packet replication and elimination functions [6, 7].

In Fig. 1, there are two N3 and N9 interfaces between NG-RAN and UPF for redundant transmission. The UPF, which connects to the DN and acts as a traffic distributor for DL traffic, copies the packets of the URLLC service from the DN and assigns them the same GTP-U serial number. These duplicate data packets pass through N9 interface 1 and N9 interface 2 respectively Transfer to I-UPF1 and I-UPF2. Each I-UPF forwards packets with the same GTP-U sequence number, which is received from UPF to NG-RAN through N3 interface 1 and N3 interface 2, respectively. NG-RAN eliminates duplicate packets based on the GTP-U sequence number. In the case of the UL service, the NG-RAN serving as the service distributor of the UL service duplicates the packet for the URLLC service of the UE, and the UPF eliminates the duplicate packet.

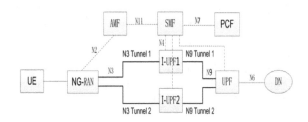

Fig. 1. Redundant transmission scheme diagram

3   Impact of Redundant Data Transmission Scheme

The PCF can determine whether redundant transmission needs to be activated for QoS flows based on its QoS requirements, [8, 9]. UE subscription and network deployment conditions. SMF can determine whether redundant transmission needs to be activated for QoS flows based on the local policy of DNN or S-NSSAI. In the case that SMF allocates CN channel information, it should provide CN channel information for the two interfaces of the redundant transmission path. Operators can control whether to activate redundant transmission through the PCC mechanism. There is no further impact on the existing control plane mechanism. [10] There is no impact on the air interface and UE. During the handover process, the NG RAN node should provide the CN with two N3 interface AN channel information, and the CN will also provide the NG RAN node with two N3 interface CN channel information [11].

In the home route roaming scenario, the NG RAN node is connected to the anchor UPF in the HPLMN through the intermediate UPF node in the VPLMN, There are two disjoint transmission paths between the I-UPF and the PSA UPF supporting the aforementioned HR redundant transmission request. To support redundant transmission with two N3 and N9 interfaces. V-SMF selects two intermediate UPFs and sets up redundant transmission channels on N3 and N9 to prevent it from receiving two CN channel information from the URLLC QoS flow of H-SMF. As shown in Fig. 2.

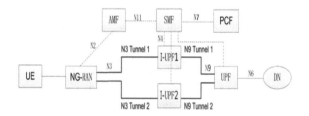

Fig. 2. Diagram of redundant transmission schemes in roaming scenarios

4   Conclusions

This solution provides a highly reliable transmission mechanism by performing redundant transmission between Anchor UPF and RAN nodes via discontinuous CN channels. This solution is used to set up redundant paths on N3 when the transport layer cannot meet the reliability requirements. Redundant transmission can be activated on demand based on QoS traffic.

Acknowledgements. This work was supported by Science and Technology Research Project of Jiangxi Provincial Department of Education in 2019: Research on User Plane Redundant Transmission Technology for Highly Reliable 5G URLLC.

References

1. Abreu, R., et al.: A blinder transmission scheme for rltra-reliable and low latency communications. In: 2018 IEEE 87thVehicular Technology Conference, pp. 1–5. IEEE, Porto (2018)
2. Singh, B., et al.: Interference coordination in rltra-reliable and low latency communication networks. In: 2018 European Conference on Networks and Communications, pp. 251–255. IEEE, Ljubljana, Slovenia (2018)
3. Jehad, M., et al.: OFDM-subcarrier index selection for enhancing security and reliability of 5G URLLC services. IEEE Access **5**, 25863–25875 (2017)
4. Jacobsen, T., et al.: System level analysis of uplink grant-free transmission for URLLC. In: 2017 IEEE Globecom Work-shops, pp. 1–6. IEEE, Singapore (2017)
5. Pocovi, G., Pedersen, K.I., Mogensen, P.: Joint link adaptation and scheduling for 5G ultra-reliable low-latency communications. IEEE Access **6**, 28912–28922 (2018)
6. Alfadhli, Y., et al.: Latency performance analysis of low layers function split for URLLC applications in 5G networks. Comput. Netw. **162** (2019)
7. Electronics - Electronics and Communications; Researchers from Technical University Report Recent Findings in Electronics and Communications (Mission Reliability for URLLC in Wireless Networks). Computers, Networks Communications (2018)
8. Chochliouros, I.P., et al.: Enhanced mobile broadband as enabler for 5G: actions from the framework of the 5G-DRIVE project. In: MacIntyre, J., Maglogiannis, I., Iliadis, L., Pimenidis, E. (eds.) Artificial Intelligence Applications and Innovations. AIAI 2019. IFIP Advances in Information and Communication Technology, vol. 560. Springer, Cham (2019). https://doi.org/10.1007/978-3-030-19909-8_3
9. Mendiboure, L., Chalouf, M.A., Krief, F.: Towards a 5G vehicular architecture. In: Hilt, B., Berbineau, M., Vinel, A., Jonsson, M., Pirovano, A. (eds.) Communication Technologies for Vehicles. Nets4Cars/Nets4Trains/Nets4Aircraft 2019. LNCS, vol. 11461. Springer, Cham (2019). https://doi.org/10.1007/978-3-030-25529-9_1
10. Albonda, H.D.R., Pérez-Romero, J.: Reinforcement learning-based radio access network slicing for a 5G system with support for cellular V2X. In: Kliks, A. et al. (eds.) Cognitive Radio-Oriented Wireless Networks. CrownCom 2019. LNICS, Social Informatics and Telecommunications Engineering, vol. 291. Springer, Cham (2019). https://doi.org/10.1007/978-3-030-25748-4_20
11. Nguyen, V.-D., Duong, T.Q., Vien, Q.-T.: Correction to: editorial: emerging techniques and applications for 5G networks and beyond. Mob. Netw. Appl. **25**, 1987 (2020)

Research and Innovation of Interior Design Teaching Method Based on Artificial Intelligence Technology "Promoting Teaching with Competition"

Hang Cao[✉] and Hui Li

School of Arts and Humanities, Yunnan Technology and Business University, Kunming, Yunnan, China

Abstract. In recent years, the active attempts and breakthrough progress of artificial intelligence in interior design teaching have been amazing. The rapid development and progress of contemporary scientific and technological means have brought great challenges and impacts to interior design teaching, which has brought material and technical changes to the development of interior design teaching, and it also has an impact on interior design from the perspective of teaching modes and ideological concepts. Innovation in teaching plays an important role in promoting development that cannot be underestimated. The purpose of this article is to study the teaching method of interior design based on artificial intelligence technology, "promoting teaching with competition". This article describes the value of artificial intelligence technology applied to interior design education, which is explained from the macro and micro perspectives. This paper studies the contradiction between the demand for talents and higher education in the teaching model of interior design in colleges and universities under the economic development of my country, and explores the methods and application scenarios of the teaching model of interior design in actual teaching. It is found that there are problems in the current teaching mode, such as the traditional teaching mode, the backward curriculum, and the weak teaching staff. Through analysis and research, the main hope is to use the game-promoting teaching model to re-formulate the overall curriculum goals of the interior design major, re-integrate the curriculum content, and build a double-qualified teacher team to make the teaching system more complete. Experimental research shows that more than 75% of students are satisfied with artificial intelligence technology combined with interior design teaching methods, indicating that the use of artificial intelligence interactive teaching enhances the appeal and influence of the teaching situation, and can satisfy students' curiosity and knowledge of information.

Keywords: Artificial intelligence · Use competition to promote teaching · Interior design teaching · Innovative teaching

J. C. Hung et al. (Eds.): FC 2021, LNEE 827, pp. 780–789, 2022.
https://doi.org/10.1007/978-981-16-8052-6_99

1   Introduction

The research on the role of artificial intelligence technology in contemporary interior design teaching is the strong development trend and trend that has been reflected in the world's interior design education in recent decades [1, 2]. Whether it is a developed country that has mature enough teaching theory or a developing country that is still actively exploring the road of education [3, 4]. They strive to explore a new kind of interior design education and teaching development space and ideas, and gradually abandon backward and unscientific educational concepts, so as to meet and adapt to the requirements and training of interior design talents in today's society [5, 6].

In the research on the teaching method of interior design based on artificial intelligence technology, many scholars have studied it and achieved good results. For example, Lancet T believes that education should unify "doing" and "learning" into Only together can we understand this thing, so as to achieve the effect of exercise. In the organization of teaching content, we should take activity theory as a framework to integrate professional skills practical knowledge and professional technical theoretical knowledge [7]. Ma L believes that starting from the perspective of knowledge theory and artificial intelligence, learning from practical knowledge theory, contextual learning theory, cognitive apprentice theory, etc., analyze the nature of interior design professional education in college education, and the importance of practical teaching in secondary vocational education Sex, provide a theoretical basis [8].

This article is mainly used to study the feasibility and effectiveness of the teaching model of promoting learning through competition. According to the theoretical and practical needs of the county, the teaching method of the teaching model of promoting learning by competition is proposed and practiced in the actual classroom. What problems may be encountered in practice by simulating the learning-by-game-promoting curriculum model, and what specific implementation measures can be developed. This article discusses and analyzes the multimedia interactive teaching concepts, teaching methods, and digital teaching characteristics of interior design creative thinking courses by expounding the application value and significance of creative thinking in interior design creativity and design performance.

2   Method of Interior Design Teaching Based on Artificial Intelligence Technology "Promoting Teaching with Competition"

2.1   Value of Artificial Intelligence Technology in Interior Design Education

(1) Integrate and optimize the level of teachers

For the field of interior design education, on the one hand, it directly optimizes the overall teacher qualifications in the market, because some teachers who are not "teachers" in the true sense will be replaced by machine teaching. On the other hand, teachers can use artificial intelligence technology to learn and charge more quickly, quickly and efficiently, thereby further improving their own teaching level and standard. This is a virtuous circle, and those who are only satisfied with the status quo, are not active Teachers who make self -improvement will be gradually eliminated by the market [9].

(2) Improve students' design standards in general

The development of artificial intelligence technology will break geographical restrictions and deadlocks, and can provide interior design learners with excellent interior design resources and fair learning opportunities.

(3) Improve the teaching quality and efficiency of teachers

Using the big data analysis of artificial intelligence technology, we can accurately understand the students' learning background, learning process and their own advantages and disadvantages in the shortest time, as well as how students behave and respond in the learning process [10, 11]. In this way, the teacher can quickly enter the role of the teacher and arrange the teaching content reasonably.

(4) Improve student learning efficiency

Interior design learning is happy and interesting, but it is also boring. To master a skill, it requires a lot of technical hard training and mental training. Not every learner can persist to the end, not every All learners can do well. But in the process of studying this article, artificial intelligence technology can light up the study of this article in the dark and seek comfort in loneliness.

2.2   Vocational Schools Promote the Training of Professional Talents Through Competition

(1) Integrate teaching content and integrate the competition into the teaching content system

1) Pertinence of teaching content

Under the current college education, professional practical knowledge has been ignored in this article. The reason is that many times the task of teachers is to teach academic or theoretical knowledge, and this knowledge is not combined with practical situations and becomes empty and general. In the teaching model of promoting learning by competition, the relevant practical knowledge and practical ability corresponding to the curriculum should be paid attention to in the formulation of teaching content, and the competition should be integrated into the design of the curriculum content. For example, the "Basics of Interior Design" course should focus on the students' Among the eight technical practical abilities, the ability to observe, think, imagine, innovate, and express, aiming at these abilities, a skill competition is incorporated into the teaching content, like a hand -painted skill competition [12]. Then use the competition requirements to measure the level of students' ability to meet t he target requirements in the "Graphic Creativity" course.

2) Applicability of teaching content

The era of big data has arrived. The era of information explosion has also spawned many corresponding job demands, such as web designers, interface designers and other new professional positions. At the same time, interface designers are also one of the most sought-after talents in my country's information industry. The popularity of the interface design industry can also be reflected in the competition items of the skill competition. Many provinces and cities have also added interactive interface (UI) design vocational skills competition items in the skill competition, and selected outstanding talents through the competition. The content of college education courses for interior design majors should also closely follow the competition items of the skill competition, close to the needs of the society, and formulate course content suitable for meeting the social needs of the industry.

3) Organization and arrangement of teaching content

Organizing project-based teaching content according to the model of promoting learning by competition is a prerequisite for arranging course content. Take the skill competition as a task to connect professional skills knowledge courses and professional theoretical knowledge courses. The knowledge structure is organized in the order of practice first and theory. The content of each course is composed of several units. According to the tasks of the skills competition, professional skills courses and professional theory courses are integrated into the same course, and then practice to theory The content of the course is carried out in the order of the order. But this is not a simple way to organize course content from practice to theory, but to establish an interactive relationship between the two.

The arrangement of the development sequence of teaching content includes two major categories: general cultural courses and professional courses, which can be carried out at the same time in the entire teaching model of promoting learning by competition because they are not deductive but parallel. The order of the development of teaching content should mainly consider the internal teaching content of professional courses. Because the teaching model of promoting learning by competition breaks the pyramidal course classification of professional ability from coarse to fine, and uses the competition project as the core course classification method, it can better integrate professional theoretical knowledge and professional practical knowledge. Can prepare comprehensive practical courses.

(2) Choose suitable competitions and carry out the reform of the model of "promoting

The school leaders have been strongly supporting students to participate in the skill competition, focusing on the teaching reform and development that promotes learning, and finally chose the best method from the many skill competitions, and choosing the skills competition suitable for student development and practical ability. Integration with the classroom. For example, the college has insisted on participating in the national college digital art design competition for many years, and participating in other provincial skill competitions as a supplement.

(3) The competition is integrated into the reform of the teaching method of the curriculum

Vocational colleges can effectively promote the reform of college teaching mode by participating in the vocational skill competition and analyzing the key points in the competition. During the competition, students use the knowledge they have learned to consolidate in the competition, improve their professional skills, and exercise their willpower and professionalism. Colleges and teachers can communicate with each other and improve each other when participating in the skill competition, which also provides a greater platform for exchanges and cooperation between schools and enterprises.

1) Targeted, choose the right skill competition

Participating in the skills competition greatly mobilizes students' enthusiasm for learning, and at the same time, teachers can also enhance their awareness of learning through the training of the skills competition, and can also enhance the teachers' practical teaching ability when guiding students to participate in the skills competition. However, as all walks of life have seen the benefits of the skill competition, more and more competitions have sprung up. Nowadays, there are so many names for the skill competition, which makes people overwhelmed. The scale of the competition is also uneven, which adds to the burden of learning, especially the development of teaching work. Some competitions even appear to be commercially profitable, the fees are too high, the proportion of awards is large, and the award-winning works do not reach the required level, and so on. At this time, it is the key to have a purpose and choice to select suitable students to participate in the skill competition.

2) Clarify teaching objectives to improve teaching efficiency

Technical knowledge is not simply a deduction of scientific knowledge, but has its own unique structure and nature system, independent of scientific knowledge. In addition, the interior design major has higher creative requirements, so the interior design major of the secondary vocational education must have its own unique curriculum theory and practice mode, which determines the uniqueness of the vocational education curriculum of the interior design major. Furthermore, college education needs to choose a practice-oriented teaching model, not a learning-oriented model. The teaching model that takes practice as the purpose and regards the promotion of learning with competition as the logical core of the education curriculum is to infiltrate the promotion of learning with competition into all aspects of the interior design professional curriculum of secondary vocational education, and reconstruct the theory and curriculum of interior design professional vocational education.

(4) Specific training methods for the ability to promote learning through competition. In the teaching model of using competition to promote learning, the main task of the design teacher of ability project training is to use the existing competition resources to coordinate the learning environment. Teachers should meet the following criteria when selecting and designing competition teaching items:

1) The skill competition is used to learn specific teaching capabilities, to cultivate the corresponding one or several technical practice abilities of students, with a specific practical value or clear task description;

2) Able to combine professional theoretical knowledge and professional practical knowledge to complete the task of this skill competition;

3) Able to meet the requirements for participating in the relevant skill competition;

4) Provide opportunities for students to plan their own projects, organize their own school behaviors within the specified time frame, and cultivate creativity;
5) Customer service and deal with the problems and difficulties in the training;
6) At the end of the training, teachers and students jointly evaluate the training results and training learning methods.

2.3 Regularized Kernel Regression Learning Algorithm Based on Sampling Survey

Regularization algorithms are an important research topic in learning theory. Regularized kernel regression algorithms include regularized least squares algorithm, coefficient regularization algorithm, spectral algorithm, etc. The least squares algorithm is mainly used to learn the regression function, defining the expected risk functional of the function f: $X \rightarrow Y$:

$$\varepsilon(f) = \mathbb{R}(f(x) - y)^2 = \int_Z (f(x) - y)^2 d\rho \tag{1}$$

The regression function is defined as $f_\rho(x) = \mathbb{R}(yx) = \int_Y y d\rho(yx)$, which describes how the output value y depends on the input value x, because ρ is unknown but exists objectively. The learning algorithm is Estimate or approximate f_ρ by learning random sample points.

Since $\varepsilon(f) = \mathbb{R}(f(x) - f_\rho(x))^2 + \varepsilon(f_\rho)$ minimizes the expected risk functional $\varepsilon(f)$. The empirical risk error corresponding to the expected risk functional is:

$$\varepsilon_z(f) = \frac{1}{m} \sum_{i=1}^{m} (f(x_i) - y_i)^2 \tag{2}$$

The function that makes the empirical risk functional is denoted as.

3 Experimental Research on Interior Design Teaching Based on Artificial Intelligence Technology "Promoting Teaching with Competition"

3.1 Main Purpose of the Assessment

After the introduction of an interactive artificial intelligence teaching platform, it is necessary to conduct surveys and evaluations on the "series of artificial intelligence interactive teaching processes" among the teaching objects. Through data analysis of the evaluation results, the value of interactive applications in this teaching mode is tested. The comprehensive evaluation of digital teaching effects provides quantitative standards, thereby effectively promoting the in-depth research of digital teaching and the gradual improvement of teaching ideas.

3.2   Main Content of the Evaluation

Based on the effectiveness certification of the interactive artificial intelligence teaching platform for creative thinking in interior design, according to the main purpose of the survey and evaluation, I chose two methods: staged teaching effect evaluation and satisfaction survey questionnaire, targeting at the core of artificial intelligence interactive technology The teaching content design, teaching technology model, teaching method application, exercise training methods, teaching evaluation information feedback, etc. have been investigated, and the research results of Israel have been digitally quantified, and the development and application of the digital teaching platform is proved by scientific evidence that the development and application of the digital teaching platform is integrated with the curriculum.

3.3   Results of the Evaluation and Questionnaire Survey

This article chooses the three classes taught in this course as the main objects of the survey and evaluation. As the three classes of the two departments are Visual Communication Design and Production, Architectural Design and Digital Media Art, I will use 3 classes According to the method of inter-professional interaction, it is divided into two classes: traditional teaching class and artificial intelligence interactive teaching class. In the traditional teaching class, I used traditional PPT courseware to teach, and in the digital interactive teaching class, I used artificial intelligence interactive teaching software to teach. After teaching for a period of time, I evaluated the teaching effect of 122 students in 3 classes.

4   Experimental Research and Analysis of Interior Design Teaching Based on Artificial Intelligence Technology "Promoting Teaching with Competition"

4.1   Class Evaluation Analysis

Through the analysis of the data of the test results, a large difference is revealed. Classes that use artificial intelligence interactive teaching software to teach are all in many aspects such as the understanding and application of knowledge, the flexibility of thinking, and the scope of creative imagination. It is ahead of the traditional teaching class, and the comparison results are shown in Table 1 (Fig. 1).

Table 1. Comparison of the scores of the assessment papers

Participate in the survey and evaluation of the class	Traditional teaching class	Artificial intelligence interactive teaching
Number of people	46	43
Multiple choice	26.4	27.4
Case study questions	20.7	25.6
Design creation topic	30.9	31.3

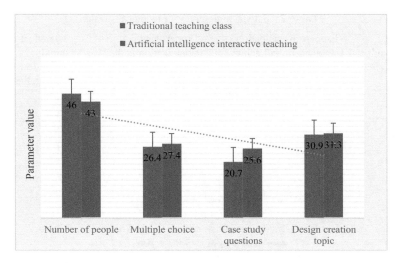

Fig. 1. Comparison of the scores of the assessment papers

The comparison of the scores of the above assessment papers shows that there is a big difference in the learning effect of the students under the two teaching methods. Because the artificial intelligence interactive teaching class adopts the teaching process of the artificial intelligence multimedia teaching platform, it makes the students' reception of the teaching content more visual and intuitive, and has a clearer understanding of the abstract concepts of things and the imaginative creative process in the past. Reflects the remarkable effect of this participatory interactive teaching based on digital technology.

4.2 Evaluation and Analysis of Teaching Effect

This article is to further understand students' specific attitudes towards the multimedia teaching platform and teaching mode of creative thinking in interior design, and whether they recognize the implementation effect of digital teaching, learning efficiency, teaching methods, and the conversion of thinking modes, and make an objective decision Evaluation, so as to provide a reference basis for standardizing digital teaching and perfecting the design of teaching ideas. The experimental results are shown in Table 2.

As shown in Fig. 2, more than 75% of students are satisfied with the combination of artificial intelligence technology and interior design teaching methods, and generally hold full recognition for the intelligent, diversified and flexible development trend of teaching brought by artificial intelligence interactive platforms Attitude. The use of artificial intelligence interactive teaching to enhance the appeal and influence of the teaching situation, so that the students' audiovisual perception has been hit unprecedentedly, and they have obtained a brand-new experience process in the senses. The traditional teaching methods are Obviously unable to satisfy the students' the curiosity and thirst for knowledge.

Table 2. Satisfaction survey of technical means

Teaching process and teaching effect	Increase participation enthusiasm and achieve good interaction	Proper teaching organization and rich teaching resources	The effect of thinking expansion is remarkable
70 or less	4	2	6
70–79	7	9	3
80–89	29	31	34
90–100	74	77	79

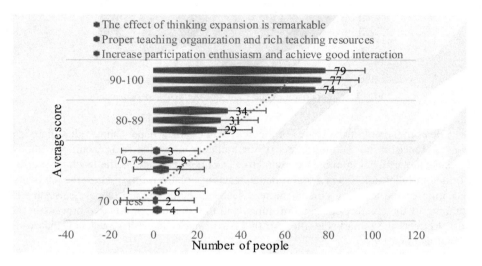

Fig. 2. Satisfaction survey of technical means

5   Conclusions

The use of the interactive artificial intelligence teaching platform for creative thinking in interior design based on the competition to promote teaching can improve students' enthusiasm for participating in learning more than traditional teaching methods. The higher simulation degree of simulated practice teaching can bring students immersively. The sense of experience has created conditions for expanding students' creative thinking fields and individualized learning methods, and the teaching effect and quality of teaching have also been significantly improved. The development and realization of an interactive artificial intelligence teaching platform for creative thinking in interior design, as well as the use of digital teaching methods, and a teaching demonstration based on the "competition to promote teaching" teaching method, leading teachers of the same profession to the "competition to promote teaching" teaching model Active exploration has been carried out, and a valuable and meaningful teaching demonstration has been made for the effective promotion of teaching reform.

Acknowledgements. This work was supported by Science Research Fund of Yunnan Education Department 2021.

References

1. Chao, Z.: An improved design and mode innovation of physical education teaching evaluation based on AI system. Revista de la Facultad de Ingenieria **32**(12), 610–616 (2017)
2. Triatmaja, S.: Designing a design thinking model in interior design teaching and learning. J. Urban Soc. Arts **7**(2), 53–64 (2020)
3. Li, M.: On the teaching of interior design for commercial buildings in the age of experience economy. J. Landsc. Res. **10**(04), 169–171 (2018)
4. Yang, J.: Teaching optimization of interior design based on three-dimensional computer-aided simulation. Comput. Aided Des. Appl. **18**(S4), 72–83 (2021)
5. Hegde, A.L., Bishop, N.: Simulation and reflective experience: an effective teaching strategy to sensitize interior design students to the visual needs of older adults. Art Des. Commun. High. Educ. **19**(1), 33–49 (2020)
6. Popenici, S.A.D., Kerr, S.: Exploring the impact of artificial intelligence on teaching and learning in higher education. Res. Pract. Technol. Enhanc. Learn. **12**(1), 1–13 (2017). https://doi.org/10.1186/s41039-017-0062-8
7. Lancet, T.: Artificial intelligence in health care: within touching distance. Lancet **390** (10114), 2739 (2018)
8. Ma, L.: Teaching quality monitoring system based on artificial intelligence. Agro Food Ind. Hi Tech **28**(1), 2002–2006 (2017)
9. Yu, D.D., Ding, M.R., Li, W.J.: Designing an artificial intelligence teaching service to assist university student in art and design to develop a personal learning experience. In: 3rd Eurasian Conference on Educational Innovation 2020 (ECEI 2020) (2020)
10. Medeiros, L., Junior, A.K., Moser, A.: An artificial intelligence teaching on artificial intelligence: experience report. Braz. J. Dev. **7**(1), 4734–4744 (2021)
11. Huang, J., Saleh, S., Liu, Y.: Promotion of artificial intelligence technology in the teaching of biochemistry. Int. J. Learn. High. Educ. **28**(1), 85–95 (2021)
12. Liu, S., Wang, J.: Ice and snow talent training based on construction and analysis of artificial intelligence education informatization teaching model. J. Intell. Fuzzy Syst. **40**(3), 1–11 (2020)

Human Object Recognition Technology Based on Internet of Things in Pedestrian Wandering Detection of Traffic Hub

Jiali Zhang[1(✉)], Haohua Qing[1], and Haichan Li[2]

[1] Guangzhou College of Technology and Business, Guangzhou 510850, Guangdong, China
155128373@sasu.edu.cn
[2] Software Engineering Institute of Guangzhou, Guangzhou 510990, Guangdong, China

Abstract. In this paper, human target recognition technology is used to detect pedestrian wandering state in traffic hub. In this paper, the current mainstream target recognition algorithm YOLOv4 is selected to lighten the model. By integrating multiple SSP modules to optimize the performance of the model, the lightweight Ghost convolution module combined with channel pruning limit compression method is used to compress and accelerate the model efficiently, so as to obtain a target recognition model with high precision, small volume and high efficiency, so it is suitable to be deployed on the embedded platform with limited resources, and the local features are collected by Cauchy model. The accuracy of this model is 92.5% and the time complexity is T(n). The local detection ability of the optimized Cauchy model is about 30% higher than that of the simple YOLO model. The frame rate of YOLOv4 used in this paper is 20–25 FPS, which can achieve smooth multi-target recognition in ordinary pedestrian detection.

Keywords: Human target recognition technology · Pedestrian wandering detection · Intelligent transportation · Internet of things image analysis

1 Introduction

Intelligent traffic target detection plays an important role in the field of video and image, which is closely related to face recognition, human behavior analysis, automatic driving and other technologies. However, due to the low resolution and fuzzy image, the general target detection algorithm has weak feature expression ability and insufficient feature extraction. In practical application, there are some problems such as missed detection, false detection and low positioning accuracy, which seriously affect the target detection accuracy. It is an important problem to be solved in the field of target detection. The traditional target detection method mainly uses manual extraction of physical signs, and then through the sliding window to detect.

On the research of target recognition algorithm, Zhao ZQ proposed the lightweight backbone network mobilenets, introduced the deep separable convolution module, reduced a lot of convolution operation and floating-point operation, but the network

J. C. Hung et al. (Eds.): FC 2021, LNEE 827, pp. 790–798, 2022.
https://doi.org/10.1007/978-981-16-8052-6_100

itself has few parameters, the ability of feature extraction is insufficient, and the recognition accuracy is low [1]. Based on vovnetv2 backbone network, DALAL n proposes a lightweight real-time target detection algorithm centermask lite, which optimizes the backbone network and alleviates the problem of channel information loss. On the coco data set, the mean average accuracy (map) reaches 40.7%, but it needs the computing power support similar to RTX 2080ti graphics card [2]. Luo J proposed the extensible target recognition model efficientdet, and proposed a variety of scale models based on neural network architecture search technology to adapt to different platforms. However, there are some problems such as insufficient accuracy of small-scale models and difficulty in training large-scale models [3]. Felzenszwalb P F proposed a new lightweight network GhostNet, which uses linear change to generate more feature graphs and mine more feature information from the original features at a very low cost. It is superior to the latest lightweight network in efficiency and accuracy [4]. Girshick r combines the current mainstream optimization techniques and more complex network architecture to design YOLOv4, which can carry out fast and accurate training and detection on the graphic processing unit (GPU). It is a highly practical target detection algorithm, but there are still a lot of model parameters and large model volume [5]. Most of the above researches rely on high-performance graphics card, even if it can run smoothly on a single graphics card, the overall performance of the algorithm is still poor. Therefore, this paper optimizes the Cauchy model to reduce the hardware requirements of the overall target detection.

In this paper, human target recognition technology is used to detect pedestrian wandering state in traffic hub. In this paper, the current mainstream target recognition algorithm YOLOv4 is selected to lighten the model. Multiple spp modules are integrated to optimize the performance of the model, and the lightweight ghost convolution module combined with channel pruning limit compression method is used to compress and accelerate the model efficiently, which is suitable for the embedded platform with limited resources. This paper mainly changes the feature fusion structure, that is, adds a new detection scale, which is more conducive to pedestrian wandering detection in traffic hub. However, the small target information in different detection scenarios is different, that is, the multi-scale fusion method is not suitable for transfer learning because of its poor transferability.

2   Target Recognition Algorithm and Pedestrian Target Statistics

2.1   Pedestrian Target Recognition and Detection Algorithm

The main research content of target recognition is to locate and recognize predefined special targets efficiently and accurately from massive images. The powerful representation and modeling ability of deep learning overcomes the influence of complex environment, illumination transformation and object deformation on recognition results to a large extent [6]. In various fields, traditional human activities will be replaced by intelligent and unmanned intelligent devices [7]. In addition, in real life, visible light images are mostly affected by light, occlusion or complex background, which affects the accuracy of target recognition. Moreover, people can not carry large devices to support the target detection algorithm based on deep learning. Only when the light-weight target detection model is implanted into the embedded device can it be closer to the actual life [8]. At present, most of the target detection algorithms need the acceleration support of GPU to train the deep learning algorithm in a short time to achieve real-time detection effect [9]. With the development of computer hardware, computer computing resources have been greatly improved, and the development of target recognition algorithm based on deep learning is also changing with each passing day [10]. The structure of darknet-53 is different from that of traditional CNN. It does not contain full connection layer and pooling layer. It is a full convolution network, which contains many 1×1 and 3×3 convolution layers, including 53 convolution layers. In road target detection, due to the distance between the vehicle and the camera, the size of the vehicle in the image is different, and the size of the last layer output feature image is only 13×13, which is 1/32 of the original image [11]. The deeper the network layer is, the more features of small objects are lost. In deep neural network, semantic information is often contained in deep network structure; The shallow network structure has a larger resolution and retains more location information [12]. YOLOv3 network consists of a large number of DBL structures, including convolution layer, batch standardization layer and activation function. This kind of algorithm adopts the way of direct regression to the target category, which improves the speed of target detection to a certain extent. Although the target detection algorithm has achieved good results in the traditional way and deep learning method, there are still the following problems: the research on small target detection is not mature, the resolution of small target is low, and the proportion of pixels is small. The small object itself has a fixed low resolution, and the effective information extracted in the process of target detection is very limited; In convolutional neural network, the deep receptive field is large, and the feature map is decreasing after many times of down sampling, which makes it more difficult to extract features, resulting in serious object miss detection, false detection and so on.

2.2 Statistical Modeling Algorithm of Pedestrian Target

In order to achieve the purpose of dynamic contour detection, this paper uses Cauchy model to deal with the statistical modeling of pedestrian objects. Cauchy distribution, maximum likelihood estimation and the calculation of the second kind of statistics are three application links. Because the imaging mechanism of the target pedestrian contour image in high resolution state can directly lead to the uneven distribution of the extracted region, the traditional Kinect statistical method can not well describe the coefficient condition of the pedestrian contour image, the Cauchy model with thick tail and sharp peak can model the original pedestrian contour image with high accuracy. The probability density function of Cauchy model is defined as follows:

$$C_i = \left(H(f_i) \bullet u^{f_i}\right)^x |\sigma_i \in i = 1, 2, \cdots, n \tag{1}$$

Where, h is the given coefficient, and C is the pedestrian position parameter based on Cauchy model:

$$Y = \begin{cases} \frac{n}{\Delta_{ikjl}} \sqrt{\sum_{s=1}^{n} \left(x_{ik}(\varepsilon) - x_{j1}(\varepsilon)\right)^2 \Delta(\varepsilon)} & \Delta_{ikjl} > 0; \\ x_{ji}(\varepsilon))^2 \Delta(\varepsilon) & \Delta_{ikjl} < 0 \end{cases} \tag{2}$$

The peak value can be obtained at X; Y represents the scale parameter of pedestrian contour based on Cauchy model. When the extraction permission x is 0, the extraction probability density function of Cauchy model is always symmetrical about the ordinate axis. The probability function distribution of Cauchy model with different values of y can directly show the distribution states P and R of the target pedestrian contour data:

$$P_{ij} = \frac{e^{b_{ij}}}{\sum_k e^{b_{ik}}} * X \tag{3}$$

$$R_{ssim} = P - \frac{\left(2\mu_x\mu_y + C_1\right)\left(2\sigma_{xy} + C_2\right)}{\left(\mu_x^2 + \mu_y^2 + C_1\right)\left(\sigma_x^2 + \sigma_y^2 + C_2\right)} \tag{4}$$

Cauchy model has obvious statistical characteristics of peak and thick tail, and the statistical model of high-resolution pedestrian contour image always shows impact characteristics. Therefore, Cauchy model is completely suitable for modeling the original pedestrian contour image coefficients. In the concept of statistics, moment estimation and similarity estimation are the two most classical numerical estimation methods. They estimate the location parameters of each extracted information in Cauchy model through the target pedestrian contour data samples. However, these classical numerical estimation methods can only be applied to some simple mathematical distribution models. For Cauchy model, the above two estimation strategies can not fully adapt to the Gaussian distribution behavior of data samples. Therefore, this model is different from the conventional probability processing method, and needs to follow a new maximum likelihood estimation rule. Maximum likelihood estimation (MLE) can determine the information density condition related to the extracted constant value for the traversal samples of the target pedestrian contour data. In the process of practical application, the maximum likelihood estimation method can identify the original probability model related to the target pedestrian contour data, and the occurrence probability tree of data samples will gradually change with the change of extraction information conditions. Generally speaking, the maximum likelihood method is a kind of application behavior that completely follows the data tree reconstruction rule. Let I denote the minimum value limiting condition associated with variable x under Cauchy distribution, and N denote the maximum value limiting condition:

$$\begin{cases} \text{Cauchy} = \prod_{i=s_1}^{s_l} \sigma_i^{v_i} = \prod_{i=s_1}^{s_l} H(v\|i)^{v_i} u^{v_i s_i} \\ NX = \sum_{i=s_1}^{s_l} v_i f_i \end{cases} \tag{5}$$

Where NX represents the value of the X variable based on the value n.

3 Experimental Design of Pedestrian Wandering Detection

3.1 Research Methods

In this paper, human target recognition technology is used to detect pedestrian wandering state in traffic hub. In this paper, the current mainstream target recognition algorithm YOLOv4 is selected to lighten the model. By integrating multiple spp modules to optimize the performance of the model, the lightweight ghost convolution module combined with channel pruning limit compression method is used to compress and accelerate the model efficiently, so as to obtain a target recognition model with high precision, small volume and high efficiency, so it is suitable to be deployed on the embedded platform with limited resources, and the local features are collected by Cauchy model.

3.2 Experimental Design

YOLOv4 target recognition algorithm is a real-time and efficient one-stage target recognition algorithm considering both detection efficiency and accuracy. Based on the original YOLO framework, this paper optimizes the backbone backbone network for feature extraction, neck network for feature fusion and head output of prediction head for classification and regression, and integrates the excellent algorithms and models of deep convolution neural network in recent years. Among them, the backbone part is based on the original darknet53 of yolv3, combined with cross stage partial (CSP) to design the cspparknet53 feature extraction network. The new backbone network can reduce the computational complexity and ensure the recognition effect of the network. When there are multiple pedestrian contour targets in the range of Cauchy model, the new backbone network can reduce the computational complexity, according to the principle of depth distance information in target recognition image, the original pedestrian target can be segmented in three-dimensional space fundamentally, that is, the target image to be processed in depth direction can be processed by adaptive distance stratification. Cauchy model is used to segment the extracted pedestrian contour image adaptively. It must be carried out under the premise of adaptive distance stratification and horizontal segmentation in 3D space. So far, the processing of various index parameters and the verification of various theoretical principles are completed. With the support of Cauchy model, pedestrian contour extraction and target detection are realized.

3.3 Results and Analysis

As shown in Fig. 1, the input layer, as the first image processing path of the target pedestrian contour data, can preliminarily extract individual feature parameter values under the promotion of convolution kernel organization according to the practical application requirements of Cauchy model, and finally derive a clear output feature image.

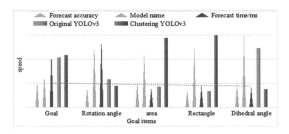

Fig. 1. Pooling unit for target detection

Each convolution layer structure must traverse the whole pedestrian contour image in a form similar to sliding window, and can convolute the small packets in each window separately. Generally, there is an obvious pooling unit between the convolution layer and the output layer, that is, the down sampling layer and the full connection layer.

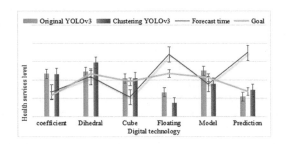

Fig. 2. Pedestrian contour extraction image after convolution processing

After convolution processing, the target pedestrian contour data can maintain the two-dimensional vector features for a long time, and the final output extracted image information can always maintain the contour application state under the coordination of Cauchy model. As shown in Fig. 2, the pedestrian contour extraction image after convolution processing can directly become the existence state of heat map. However, due to multiple pooling operations, part of the feature information is easy to be lost, and the size of the information not lost is relatively small, which is difficult to meet the final pedestrian target detection requirements, resulting in the continuous decline of the contour boundary clarity. Under the effect of convolution principle, convolution neural network can make the target pedestrian contour image maintain the original output state for a long time, but the overall extraction effect is poor, and it can not get ideal processing effect in the actual detection process, so it is necessary to add deconvolution process on the basis of convolution. Usually, the nodes of convolution and deconvolution are the same, but the order is opposite.

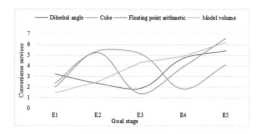

Fig. 3. Pedestrian wandering detection results in traffic hub

As shown in Fig. 2, e0-e2 is the target pedestrian contour extraction node of convolution processing, and e2-e0 is the target pedestrian contour extraction node of deconvolution processing. The accuracy of this model is 92.5% and the time complexity is t (n). The local detection ability of the optimized Cauchy model is about 30% higher than that of the simple YOLO model. The frame rate of YOLOv4 used in this paper is 20–25 FPS, which can achieve smooth multi-target recognition in ordinary pedestrian detection. Considering the practicability, in the same convolutional neural network, with the enhancement of Cauchy model, the number of target pedestrian contour data to be extracted will continue to increase, resulting in the disorder of individual extraction nodes. Because of the convolution and deconvolution, these displaced nodes can be fixed obviously, which can provide sufficient image data information for subsequent detection and processing. With the enhancement of the function of the full connection layer, the target pedestrian contour data can be freely converted between a node and a node, so as to realize the error free connection between the established neural nodes.

As shown in Table 1, Gabor feature extraction is very similar to the visual stimulus response behavior of neurons in the target visual system. In this case, the extraction host is good at capturing the frequency domain information and local space of the target pedestrian contour data, so it can maintain a relatively sensitive processing state to the edge information parameters of the target image. With the support of Cauchy model, Gabor features can provide both scale selection and direction selection characteristics of target pedestrian contour data, which has a strong practical value in visual information understanding.

Table 1. Gabor special conquest proposal target visual system

Item	Goal	Rotation angle	Area	Center of rectangle	Dihedral angle
Forecast accuracy	2.16	1.64	2.19	1.38	1.97
Model name	2.68	5.35	4.61	5.11	5.94
Forecast time/ms	4.62	5.89	1.61	1.78	1.64
Original YOLOv3	4.14	2.33	1.72	1.33	4.94
Clustering YOLOv3	4.34	1.8	5.78	5.99	1.52

4   Conclusions

In order to verify the value of human object recognition technology based on the Internet of things in pedestrian wandering detection of traffic hub, this paper selects the area with relatively suitable pedestrian flow as the experimental data extraction interval, and imports all image information into the core detection host, and uses the software analysis function to determine the change of pedestrian object contour detection accuracy in the given experimental time, the experimental group is equipped with pedestrian contour extraction and target detection algorithm based on Cauchy model, while the control group is equipped with kinect detection algorithm. The accuracy value of pedestrian target contour detection can reflect the real-time stable tracking ability of the control host to the monitored target. Generally, the higher the accuracy value is, the stronger the stable tracking ability of the control host is, and vice versa. In addition, with the development of deep learning and the update of hardware equipment, we can use the more advantageous YOLOv4 network to detect pedestrian wandering in traffic hub in the future, so as to further improve the performance of the algorithm.

Acknowledgements. This work was financially supported by School-Level Quality Engineering Construction Project in 2020 of Guangzhou College of Technology and Business (Teaching Reform of Big Data Course Based on Chaoxing Fanya Network Teaching Platform—Take "Data Analysis and Mining Actual Combat (Python)" as an example, No. ZL20201243).

References

1. Zhao, Z.Q., Zheng, P., Xu, S.T.: Object detection with deep learning: a review. IEEE Trans. Neural Netw. Learn. Syst. **30**(11), 3212–3232 (2019)
2. Dalal, N., Triggs, B.: Histograms of oriented gradients for human detection. In: Proceedings of the IEEE Conference on Computer Vision and Pattern Recognition (CVPR), vol. 9, no. 2, pp. 886–893. IEEE (2020)
3. Luo, J., Oubong, G.: A comparison of sift pca-sift and surf. Int. J. Image Process. (IJIP) **3**(4), 143–152 (2019)
4. Felzenszwalb, P.F., Girshick, R.B., Mcallester, D.: Object detection with discriminatively trained part-based models. IEEE Trans. Pattern Anal. Mach. Intell. **32**(9), 1627–1645 (2019)
5. Girshick, R., et al.: Rich feature hierarchies for accurate object detection and semantic segmentation. In: Proceedings of the IEEE Conference on Computer Vision and Pattern Recognition (CVPR), vol. 9, no. 14, pp. 580–587. IEEE (2019)
6. Girshick, R.: Fast r-cc. In: Proceedings of the 2015 IEEE Conference on Computer Vision and Pattern Recognition (CVPR), vol. 12, no. 8, pp. 1440–1448. IEEE (2019)
7. Ren, S., He, K., Girshick, R., et al.: Faster r-cnn: towards real-time object detection with region proposal networks. IEEE Trans. Pattern Anal. Mach. Intell. **39**(6), 1137–1149 (2016)
8. Liu, W., et al.: SSD: single shot multibox detector. In: Leibe, B., Matas, J., Sebe, N., Welling, M. (eds.) Computer Vision – ECCV 2016. ECCV 2016. LNCS, vol. 9905, pp. 21–37. Springer, Cham (2016). https://doi.org/10.1007/978-3-319-46448-0_2
9. Redmon, J.: You only look once: Unified, real-time object detection. In: Proceedings of the 2016 IEEE Conference on Computer Vision and Pattern Recognition (CVPR), vol. 9, no. 12, pp. 779–788. IEEE (2016)

10. Redmon, J.: YOLO9000: better, faster, stronger. In: Proceedings of the 2017 IEEE Conference on Computer Vision and Pattern Recognition (CVPR), vol. 2, no. 5, pp. 7263–7271. IEEE (2017)

11. Liu, Y., Liu, H.-Y.: Survey of research and application of small target detection based on deep learning. Acta Electron. **48**(3), 590–601 (2020)

12. Geiger, A.: Vision meets robotics: the kitti dataset. Int. J. Robot. Res. **32**(11), 1231–1237 (2019)

Fault Classification of Outage Transmission Lines Based on RBF-SVM and BP Neural Networks

Yingpei Liu[1], Xiangyu Wang[1(✉)], Feifei Zhang[2], Jianxiang Gao[2],
Yujing Su[2], Xuewei Zhang[2], and Haiping Liang[1]

[1] Department of Electrical Engineering, North China Electric Power University
(Baoding), Baoding 071003, Hebei, China
Wangxy616@ncepu.edu.com
[2] State Grid Hebei Electric Power Co., Ltd., Shijiazhuang, Hebei, China

Abstract. In this paper, the fault state of the outage line is judged based on the effective value of the single terminal voltage in the hot standby state of the outage line in the same tower and plays a guiding role in closing operation and troubleshooting of outage line in dispatching work. It has higher practical significance. This method inputs the single terminal voltage measurement value of outage line into radial basis kernel function support vector machine (RBF-SVM) and BP neural network to judge the fault state and fault type on outage line respectively. In order to verify the recognition accuracy of the proposed scheme, four different 500 kV double-circuit transmission line models with the same tower are established, and then simulated under different fault states. The results show that the accuracy of this method is 100% for the fault judgment on the hot standby line, and 99.4% for each fault type.

Keywords: Double circuit transmission line with tower · Induced voltage · RBF-SVM · BP neural networks · Failure recognition

1 Introduction

Double-circuit transmission line on the same tower is widely used because it can realize long-distance transmission and save land occupation area. Because double-circuit transmission lines share the same power corridor, the distance between each circuit line is close. When one of the circuits is out of operation due to fault or maintenance, the operating line will generate induced voltage on the outage line through electromagnetic field. For 500 kV double-circuit transmission lines, inductive voltage can even reach tens of kV [1].

At present, the research on inductive electricity focuses on the theoretical calculation of inductive electricity [2–4], the simulation calculation of inductive electricity [5–7], and the analysis of the influencing factors of inductive electricity [8].

Timely detection and troubleshooting of transmission lines is the key to ensure the reliability of power supply. Currently, a large number of documents are used to detect and judge the faults of operating line, and few documents identify the fault states of

J. C. Hung et al. (Eds.): FC 2021, LNEE 827, pp. 799–805, 2022.
https://doi.org/10.1007/978-981-16-8052-6_101

transmission lines in hot standby state. In order to prevent the closing to the grounding line, it is usually necessary to judge the state of the closing line. In practice, the fault state of the line is judged according to the simulation or the experience of the staff. For the same tower and double circuit transmission lines, when one is out of operation and the other is in normal operation, the inductive voltage will be generated. When the line is in different fault state, the effective value of induced voltage is different. The fault state of the outage line can be judged by measuring the effective value of the voltage of the outage line. Accurate judgment of line fault state can provide reference for scheduling operation and troubleshooting.

An intelligent ground state identification method based on the effective value of single terminal inductive voltage of outage transmission line is proposed in this paper. The single-terminal three-phase inductive voltage of outage transmission line is measured, and the measured value is standardized. The processed data is input RBF-SVM to determine whether there is a fault in the line, and if there is a fault in the line, the data is input BP neural network to judge the fault type. In order to verify the fault judgment results of this method, the simulation experiment is carried out based on four simulation models of the same tower double circuit transmission line with different parameters, and the simulation data are collected to train and test the fault identification algorithm.

2 Intelligent Identification Method of Grounding State

The ground state identification method proposed in this paper consists of two parts. First, the RBF-SVM is used to judge the fault state of the outage line, that is, to judge whether there is a fault on the outage line. If there is a short circuit fault, the sample is input BP neural network to determine the current fault type in 11 different types of faults.

2.1 Support Vector Machine

Support vector machine (SVM) is a powerful supervised machine learning algorithm, which can solve linear or nonlinear classification, regression, and even abnormal value detection tasks, and is widely used in binary classification problems [9].

For linearly separable data, SVM finally want to find a hyperplane to classify the data, which can be expressed as

$$\omega^T x + b = 0 \tag{1}$$

where both ω and x are n dimensional vectors.

The distance between each data point and the hyperplane is also called classification interval, the larger classification interval will obtain better classification effect, the calculation formula of classification interval d is

$$d = \frac{|\omega^T x + b|}{\|\omega\|} \tag{2}$$

Hence the linear SVM can transform the problem into

$$\begin{cases} \min \frac{\|\omega\|^2}{2} \\ s.t. \quad y_i[\omega^T x_i + b] \geq 1, i = 1, 2, \cdots, n \end{cases} \tag{3}$$

The problem satisfies the condition of Karush-Kuhn-Tucker (KKT). The sequence minimum optimization (SMO) algorithm can be used to solve the hyperplane by converting the problem into its dual problem.

For nonlinear separable data, the data can be mapped to a linearly separable high-dimensional feature space by introducing kernel functions. Kernel functions do not require specific calculations to map to high-dimensional sample points. A better classification effect is obtained without adding too much computation. When the sample is input into the linear SVM, the SVM cannot accurately distinguish whether the line fails, so this paper uses the SVM with the RBF kernel function to judge whether the outage line fails. The expression of RBF function is

$$\phi(x, l) = \exp(-\gamma \|x - l\|^2) \tag{4}$$

where x is input sample, l is the center of RBF function.

2.2 BP Neural Networks

Neural networks are composed of four parts: input X, input weight W, activation function $f(\cdot)$ and output y. For each neuron, the output is

$$y = f(W \cdot X + b) \tag{5}$$

where b is bias.

BP neural network is a backpropagation neural network. BP neural network is mainly composed of one input layer, several hidden layers and one output layer. And each layer is composed of several neurons. The number of neurons in the input layer is the same as the characteristic number of the input sample. The number of hidden layers and the number of neurons in each layer are selected according to the actual demand. And the number of output neurons is determined by the form of output.

BP neural network training is divided into two parts: forward propagation and back propagation. In the process of forward propagation, the output of each neuron in the input layer is used as the input of all neurons in the next layer, and the predicted value is obtained from this kind of push to the output layer. The predicted value is compared with the expected value. In the process of backpropagation, the error is propagated from the output layer to the input layer, the weights and biases of each layer are adjusted. Repeat the above process until the minimum loss function completes the training.

3   Simulation and Analysis

3.1   Data Generation

The ATP-EMPT electromagnetic transient simulation software is used to establish the models of four 500 kV double-circuit transmission lines with different parameters to obtain the measured value of single-terminal inductive voltage under different fault states of outage lines.

When the outage transmission line in the hot standby state without fault, the operating line voltage is set, the effective value of the power supply voltage is adjusted to 475 kV–525 kV, and the step size is 5 kV. To reflect the power flow of the running line, the power supply angle is set in the range of −20° and 20°, and the step size is 2°. A total of 924 simulation experiments were carried out on four lines and the effective value of induction voltage of outage lines was collected.

The 21 fault points are set for each line model, and 11 fault types are simulated for each fault point. Considering the different transition resistance when the fault occurs. A total of 10164 simulation experiments were carried out on the four lines, and the effective values of the induced voltage of the outage line were collected respectively.

3.2   Feature Extraction

Selecting appropriate sample features according to data can greatly improve the accuracy of classifier. In order to reflect the change of voltage before and after the fault, the average value of each phase fault-free voltage V_{avg} in each line model is taken respectively. Increasing the ratio of the measured value to the average value V/V_{avg} and the average value of the fault-free voltage V_{avg} are new features. The preprocessed input data is a combination of the outage line voltage measurement value and the added feature, so each input sample has 9 features.

3.3   Standardization

When the data ratio of the input machine learning algorithm is different, it often leads to the poor performance of the algorithm, so in order to improve the generalization ability of the algorithm, the feature scaling of the data is needed. In this paper, the data are preprocessed by standardized method. Each feature in the sample is standardized, that is, the difference between the feature and the average value μ of the feature in all samples divided by the variance σ of the feature. Through standardized processing, the data conform to the standard normal distribution, that is, the mean value is 0 and the standard deviation is 1. This method makes the results less affected by the outliers.

3.4   Model Training

SVM and BP neural networks are supervised learning algorithms, which need to be trained by inputting labeled samples into the model.

For RBF-SVM, two hyperparameters of penalty coefficient C and RBF kernel function parameter *gamma* need to be adjusted for RBF-SVM. The selection of appropriate parameters is helpful to improve the ability of the algorithm. When the final C is 0.01 and the *gamma* is 1, the model has good judgment effect.

BP neural network will judge the current fault type after RBF-SVM determined the fault of the line. As mentioned above, there are 11 kinds of short-circuit faults in outage lines, so they are marked with 0–10 and coded by one-hot encoding. Each sample consists of 9 features, so BP neural network input layer neurons are set to 9. The output of the neural network is 0–10 encoded by one-hot encoding, and the number of neurons in the output layer is 11. In this paper, the number of hidden layers is 1, and the number of neurons in hidden layer is 20. For the input layer and the hidden layer activation function is ReLU. It can overcome the problem of gradient disappearance during training and speed up the training. Sigmoid function is used as the activation function of the output layer to realize the classification. The neural network training uses the cross-entropy loss function. Adam is selected as the neural network training optimizer. Adam [10] is the most widely used optimizer at present, which combines the characteristics of many optimizers.

The SVM is trained first, the data collected by four simulation lines with different parameters are labeled, the data is scrambled and 80% is taken as the training set, and the remaining 20% is used as the test set. After standardized operation on the training set, the SVM is trained with the training set data. After training, the test set is standardized with the same parameters of the training set. Test set data is used to finally test the model. The training of BP neural network is similar to the SVM training process, so we will not repeat it.

3.5 Evaluation of Fault Identification Methods

The SVM model and BP neural network are evaluated by confusion matrix. The diagonal elements of the confusion matrix are the correct sample is judged, and the off-diagonal elements are the sample which are erroneous judgement. In the classification problem, we can see the result of the algorithm intuitively.

The model is investigated on the training set and the test set, respectively. The confusion matrix on the training set and test set is shown in Fig. 1.

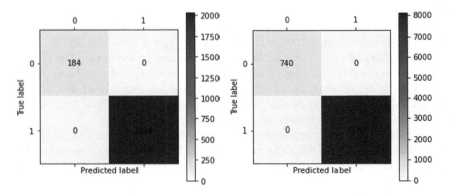

Fig. 1. The confusion matrix on the training set and test set of fault status identification

Since RBF-SVM have no misjudgment on both the training and test sets, the accuracy are 1.

Above all, on the training set and the test set, RBF-SVM can accurately judge whether there is a fault, showing a good classification ability. The method can make accurate judgment on whether the fault occurs.

Confusion matrix of BP neural network on training set and test set are shown in Fig. 2.

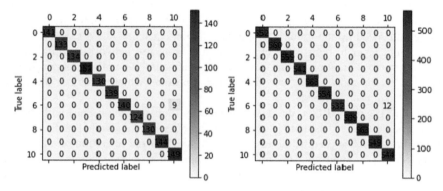

Fig. 2. The confusion matrix on the training set and test set of fault type judge

The figure shows that the BP neural network has a good classification effect on the training set. Only 12 samples are classified wrong, 12 ABG fault samples are judged ABCG fault, and the accuracy is 99.8%. There are only 9 sample classification errors in fault type judgment, and 9 ABG fault samples are judged as ABCG faults, and the accuracy is 99.4%.

4 Conclusion

In this paper, based on the actual requirements of the project, a fault state identification method for outage transmission lines using the measured value of single terminal inductive voltage of outage transmission lines as input is proposed. Using the characteristic that the effective value of the voltage measured by the outage line in different fault states shows different characteristics, it effectively solves the problem of judging whether there are artificial or natural factors grounding situation of the outage line before closing. Further reduce the probability of closing to the fault line. And make sure the safe and stable operation of the power system.

The method is divided into two parts: first, the RBF-SVM is used to judge whether the line has a fault. When the SVM judges the line has a fault, the measurement data is input BP neural network to judge the fault type in order to further judge the cause of the fault and even to troubleshoot the fault. Then, by the confusion matrix, accuracy, precision, recall rate, F1 value and other indicators to examine the proposed method. The results show that the method can accurately identify the fault situation on the outage line, and the accuracy of fault type identification is 99.4%, which has a better fault type identification effect.

Acknowledgments. This work was financially supported by the fund of State Grid Hebei Electric Power Co., Ltd. "Research on Intelligent Monitoring Technology of Grounding Safety Based on the Features of Induced Voltage Characteristics", 2020.

References

1. Yuwei, H., Yakun, L., Nannan, Y., et al.: Evaluation of induced voltage and current on low-voltage distribution lines under EHV/UHVAC and DC transmission lines. Power Syst. Technol. **39**(09), 2640–2646 (2015). (in Chinese)
2. Minchuan, L., Jinghui, L., Xiangyang, P., et al.: Calculation method of induced voltage and induced current for complex transmission network. High Voltage Eng. **42**(10), 3308–3314 (2016). (in Chinese)
3. Guolong, L., Jianzhengl, D.: Calculation and analysis of induction voltage between bus and line. Northeast Electr. Power Technol. **40**(12), 20–24 (2019). (in Chinese)
4. Zhu Jun, W., Guangning, C.X., et al.: Electromagnetic coupling calculation and analysis of lines Non-Parallelly erected entirely in one common AC/DC transmission Corridor. High Voltage Eng. **40**(06), 1724–1731 (2021). (in Chinese)
5. Yunlong, Z., Richeng, L., Ming, Z., et al.: Induced voltage and current analysis for four-circuit transmission line with different voltage classes on the same tower. Northeast Electr. Power Technol. **36**(01), 216–222 (2021). (in Chinese)
6. Ma, A., Xu, D., Wang, H., et al.: Induced voltage and induced charge of 0.38 kV lines operated parallel under 500 kV AC double-circuit transmission lines on same tower. High Voltage Eng. **41**(01), 306–312 (2015). (in Chinese)
7. Baoju, L., Hao, Z.: Calculation and analysis on induced voltage and current of 1 000 kV transmission line adopting structure of double circuit on the same tower. Power Syst. Technol. **35**(03), 14–19 (2011). (in Chinese)
8. Shen, Z., Xiaofeng, K., Taishan, H., et al.: Research on calculation of induced voltage and current on 110 kV crossing transmission lines under UHV AC transmission lines. Electr. Measur. Instrum. **54**(13), 30–35 (2017). (in Chinese)
9. Liu, Y., Xu, Z., Li, C.: Online semi-supervised support vector machine. Inf. Sci. (2018)
10. Kingma, D., Ba, J.A.: A method for stochastic optimization. Comput. Sci. (2014)

Application of Image Defogging Algorithm Based on Deep Learning

Changxiu Dai[(✉)]

South China Business College, Guangdong University of Foreign Studies,
Guangzhou 510545, Guangdong, China

Abstract. Affected by atmospheric particles such as fog, haze, sand and dust, the images taken outdoors will appear gray and white. The dark channel priori defogging algorithm relies too much on prior information, which makes the calculation of the transmission map inaccurate. In order to solve these problems, In this paper, the image defogging is improved based on deep learning. The algorithm first distinguishes whether the image is foggy based on the VGG16 model, and then obtains the depth map of the foggy image based on the convolutional neural network, three different scales of feature extraction, feature learning and feature fusion are designed to improve the accuracy of the depth map. Then the depth map is converted into a transmission map. Finally, a clear image is obtained according to the atmospheric scattering model. The defogging experiment shows that the algorithm in the article has a good defogging effect on outdoor fog images, can obtain better color visual fidelity, and has strong applicability.

Keywords: Dark channel prior · VGG16 · Convolutional neural network · Atmospheric scattering model

1 Introduction

As the environment changes, smog has become a common occurrence. Taking pictures in the haze weather, the image becomes gray and white, and the visual effect is poor, which severely limits the application of computer vision.

Image defogging algorithms are mainly divided into two categories: one is the method of image enhancement [1, 2], and the other is the method of image restoration [3–10]. Tarel [3] uses white balance to simplify the physical model, but this method produces halos. He [7] proposed the dark channel prior to estimated the global A and the preliminary transmission map, and finally used soft matting or guided filtering [8] to optimize it to obtain a clear image. However, the estimation of the transmission map will be inaccurate, resulting in unsatisfactory dehazing effect.

With the development of deep learning technology, Cai [11] proposed the use of convolutional neural networks [12, 13] to learn the characteristics of foggy images. However, the estimation of the propagation map in a specific scene will be inaccurate and the dehazing effect will not be ideal.

This paper improves the literature [11]. First, design three convolution kernels of different scales for parallel feature extraction,then feature learning, and finally deep

J. C. Hung et al. (Eds.): FC 2021, LNEE 827, pp. 806–811, 2022.
https://doi.org/10.1007/978-981-16-8052-6_102

fusion. The method in this paper allows the network to learn more feature information of the original image, so it can obtain a better dehazing effect.

2   Image Dehazing Based on Dark Channel Prior

The dark channel a priori is based on the degradation model of imaging in the foggy scene to obtain a clear image from the foggy image. The mathematical model:

$$I(x) = J(x)t(x) + A(1 - t(x)) \tag{1}$$

I(x): Original foggy image, J(x): Image after defogging, t(x): transmission image, A: The global atmospheric light. The dark channel a priori believes that there are always some pixels in a fog-free image that have a small value and close to 0 in a certain channel. Therefore, the above formula can be further transformed into the following mathematical model:

$$J(x) = \frac{I(x) - A}{t(x)} + A \tag{2}$$

3   The Algorithm of This Article

The defogging algorithm in this paper can be shown in Fig. 1.

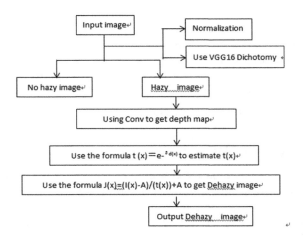

Fig. 1. The algorithm of this paper

3.1 Image Classification

The image is divided into a foggy image and a fogless image. This is a two-class classification problem. The VGG16 network does not set many hyperparameters. It is a simple network that focuses on building a convolutional layer. It is used in image classification and target detection tasks. Both have very good results. Therefore, this article uses VGG16 to classify the images. The steps for VGG16 to distinguish whether the image is foggy are shown in Fig. 2.

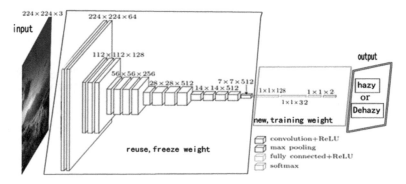

Fig. 2. VGG16 to dichotomy

3.2 Convolutional Neural Network Seeking Depth Map

The steps for obtaining the depth map based on the convolutional neural network are as follows.

(1) Input a foggy image and perform normalization processing;
(2) Multi-scale feature extraction of shallow features;
(3) Feature learning, deep features;
(4) Connection and fusion of shallow features and deep features;
(5) Non-linear regression processing for output results.

The solution process of the above steps is shown in Fig. 3.

3.3 Optimize the Depth Map and Estimate A

Estimate the depth map based on the convolutional neural network model, obtain the initial transmission map according to the relationship between the depth map and the transmission map, and then guide filtering to optimize it, smooth the details to obtain the optimal transmission map [14]. Select the first 1% of the low-brightness pixel portion of the transmission image as t0, and select the corresponding highest pixel value as A in the foggy image $I(x) = 0$. $x \in \{y|t(y) \leq t0\}$.

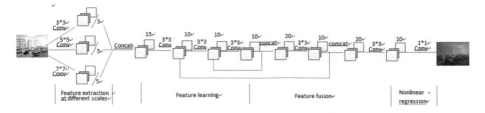

Fig. 3. Use conv to get the depth map.

3.4 Dehazing Reduction Based on Atmospheric Scattering Model

Knowing t(x) and A, a clear image can be obtained by formula 2.

4 Simulation

The experimental data in this paper is divided into two categories. The first category is derived from the foggy image RTTS dataset in natural scenes, and the second category is derived from the indoor ITS sub-dataset. The experimental environment is based on Python's Tenserflow the deep learning framework.

4.1 Dehazing Effect of RTTS Dataset

Two images are selected from the RTTS dataset for defogging using the algorithm in this article. It can be seen from the defogging effect diagram that the method in this article effectively removes the fog in the image without the sky distortion, and the overall brightness is moderate, and the visual effect is good. The defogging effect is shown in Fig. 4.

Original image   This algorithm   Original image   This algorithm

Fig. 4. RTTS dataset defogging

4.2 Comparison of the Algorithm in This Article with Other Algorithms

The comparison of defogging algorithms is shown in Fig. 5.

Reference [8] has a darker color and excessive defogging. Reference [11] has less distortion after defogging, but where the depth of field is far, the degree of defogging is low. The algorithm in this paper makes the color of the sky area more real and natural, and the overall contrast of the image is improved.

Original image   Reference [8]   Reference [11]   The algorithm

Fig. 5. The comparison of defogging algorithms

5 Conclusions

The algorithm in this paper is an improved image defogging algorithm based on deep learning. First, use VGG16 to distinguish whether the image is foggy, and then use 3 different scale convolution kernels to extract features, feature learning and feature fusion on the foggy image, so that the network can learn the mapping between the foggy image and its depth map more comprehensively Relationship, and then guide filtering to optimize the transmission map. The optimized transmission map has a better dehazing effect on the image edge, texture and color distortion.

References

1. Kim, J.Y., Kim, L.S., Hwang, S.H.: An advanced contrast enhancement using partially overlapped sub-block histogram equalization. IEEE Trans. Circ. Syst. Video Technol. 11(4), 475–484 (2001)
2. Jang, J.H., Bae, Y., Ra, J.B.: Contrast-enhanced fusion of multi- sensor images using subband- decomposed multiscale retinex. IEEE Trans. Image Process. 21(8), 3479–3490 (2012)
3. Tarel, J.P., Hautière, N.: Fast visibility restoration from a single color or gray level image. IEEE Int. Conf. Comput. Vis. 2201–2208 (2010)
4. Fattel, R.: Single image dehazing. ACM Trans. Graph. 27(3), 721–728 (2008)
5. Dong, X., Hu, X., Peng, S., et al.: Single color image dehazing using sparse priors. IEEE Int. Conf. Image Process. 3593–3596 (2010)
6. Zhu, Q., Mai, J., Shao, L.: A fast single image haze removal algorithm using color attenuation prior. IEEE Trans. Image Process. 24(11), 3522–3532 (2015)
7. He, K., Sun, J., Tang, X.: Single image haze removal using dark channel prior. IEEE Conf. Comput. Vis. Pattern Recogn. 1956–1963 (2009)
8. He, K., Sun, J., Tang, X.O., et al.: Guided image filtering. In: Proceedings of the 11th European Conference on Computer Vision, Heraklion, pp. 1–14 (2010)
9. Meng, G., Wang, Y., Duan, J., et al.: Efficient image dehazing with boundary constraint and contextual regularization. IEEE Int. Conf. Comput. Vis. 617–624 (2014)
10. Tang, K., Yang, J., Wang, J.: Investigating haze-relevant features in a learning framework for image dehazing. IEEE Conf. Comput. Vis. Pattern Recogn. 2995–3002 (2014)

11. Cai, B., Xu, X., Jia, K., et al.: DehazeNet: an end-to-end system for single image haze removal. IEEE Trans. Image Process. **25**(11), 5187–5198 (2016)
12. Lin, H.-Y., Lin, C.-J.: Using a hybrid of fuzzy theory and neural network filter for single image dehazing. Appl. Intell. **47**(4), 1099–1114 (2017). https://doi.org/10.1007/s10489-017-0942-z
13. Eigen, D., Krishnan, D., Fergus, R.: Restoring an image taken through a window covered with dirt or rain. IEEE Int. Conf. Comput. Vis. 633–640 (2013)
14. Wang, W., Yuan, X., Wu, X., et al.: Fast image dehazing method based on linear transformation. IEEE Trans. Multimedia **19**(6), 1142–1155 (2017)

Customer-Side Energy Management Controller Design Based on Edge Computing and Docker Technology

Jingyi Lin[1(✉)], Wen Li[1], Bin Yang[2], Sirui Zhang[1], and Yongxian Yi[3]

[1] Department of Electricity Power Consumption and Energy Efficiency, China
Electric Power Research Institute, Beijing, China
linjingyi@epri.sgcc.com.cn
[2] State Grid JiangSu Electric Power Co., Ltd., Nanjing, China
[3] State Grid JiangSu Marketing Service Center, Nanjing, China

Abstract. There is an increasing demand for efficient energy management at the customer side for low carbon energy provision and consumption. The study aims to focus on the energy consumption scenarios of commercial buildings, industrial enterprises and parks. Through in-depth analysis of common energy demand on the customer side, it breaks through key technologies such as demand response, multi-energy complementarity and energy efficiency improvement. This work builds a replicable and promotable energy consumption control system on the customer side, develops an energy controller supporting the ubiquitous access and edge optimization control of comprehensive energy on the customer side, e.g., commercial buildings. The proposed design can benefit the customer side energy management in practice.

Keywords: Hybrid storage system · Power distribution systems · Power quality · Photovoltaic power generation

1 Introduction

To further promote green, low-carbon and sustainable development of energy, the state actively promotes the revolution of energy consumption, deeply develops comprehensive energy services, suppresses unreasonable energy consumption and improves energy utilization efficiency. However, at present, there are many problems on the customer side (e.g., [1–3]). The comprehensive energy efficiency level is low, and the energy cost is high. Therefore, the comprehensive energy business needs to be further developed and improved.

At present, there are many kinds of energy consumption control systems on the customer side, mainly focusing on the functions of acquisition monitoring and energy consumption optimization, but there are still some problems: 1) there are many application scenarios on the customer side, and the common requirements such as data acquisition monitoring, demand response, energy efficiency improvement and security protection are not fully understood, and the multi-functional collaborative demand analysis is lacking; 2) The system is lack of interactive ability, high development cost,

J. C. Hung et al. (Eds.): FC 2021, LNEE 827, pp. 812–818, 2022.
https://doi.org/10.1007/978-981-16-8052-6_103

poor portability and reusability; 3) The coverage of electric, heat, cold, gas and other energy acquisition terminals is low, the communication access and protocol standards are not unified, and the edge computing ability is lack; and 4) The function of the system is mainly collection and monitoring, which is lack of strategic support, such as demand response and energy efficiency improvement. Therefore, it is urgent to carry out research on key technologies such as ubiquitous access, demand response, multi-energy complementarity and energy efficiency improvement on the customer side, build a replicable and scalable energy consumption control system on the customer side, support multi-scale demand response on the grid side, and promote the company's transformation from a power supplier to an integrated energy service provider, We will work together to build an open, shared and win-win ecosystem of comprehensive energy services (e.g., [4–11]).

In this paper, through the analysis on the key technologies of data acquisition and monitoring of energy consumption control system on the customer side, master the technical architecture of energy consumption data acquisition and monitoring under multiple scenarios such as commercial buildings, industrial enterprises and park services, the dynamic perception and online analysis technology of energy consumption information based on Internet of things and edge computing, and the application method of edge computing gateway in the power Internet of things, to realize the business sinking, energy-saving and energy-saving Reduce the underlying data processing from the cloud to the ground, solve the problem of low efficiency of resource utilization and time processing under the single cloud computing mode, and provide support for building a safe, intelligent, professional and integrated Internet of things management platform.

2 Edge Computing and Docker Technology with Internal System Isolation

2.1 Edge Computing

In the practical deployment and application of edge computing, there are many difficulties. To ensure low latency and reduce bandwidth resource consumption, there are still security and reliability problems to be solved. To solve these problems, many technologies are introduced into edge computing. The key technologies include heterogeneous computing (HC), data storage and virtualization. For example, the CPU (central processing unit) is good at system control, task decomposition and task scheduling; GPU (graphics processing unit) has powerful floating-point and vector computing capabilities and is good at the matrix and vector parallel computing. FPGA (field-programmable gate array) has the characteristics of hardware programmable and low delay. Therefore, the heterogeneous computing architecture can integrate all the independent processing units with computing resources on an edge computing platform to form a close collaborative whole to deal with different types of computing load, to improve the overall efficiency of data processing.

To meet the needs of data storage, edge computing chooses temporal data storage technology to solve the problem. The new generation of the temporal database can store complete historical data in chronological order, and support the functions of fast writing, persistence, distributed storage and aggregate query of temporal data.

This study adopts the microservice architecture of virtualization technology, divides the single application into a set of small services during software development, and the small services cooperate, and finally realizes the complete business function. Individual microservice modules are running in the process of mutual isolation. Microservice modules are isolated from each other and cooperate by simple and fast communication mechanisms.

2.2 Docker Technology

Energy controller is an important tool for the realization of energy consumption information sensing access for edge computing. In addition to protocol conversion, data acquisition, manageability and other common functions, it can also carry edge computing and intelligent services, including edge side data storage and stream data processing, intelligent reasoning and decision-making, collaborative application and other application services. Especially for massive energy information access, an energy controller is an important carrier of edge computing and intelligent services. As the link between the energy control and production management system of the customer side and the smart energy service platform, the energy controller collects a large number of static information and real-time dynamic information in the heterogeneous network of the data acquisition site, converts the communication protocol and transmits the data packet; On the other hand, it focuses on core services such as data processing, integration, analysis, decision-making and sharing.

The hardware function virtual (HFV) technology is introduced into the energy controller to separate the functions of the IoT terminal hardware equipment. The hardware equipment only performs the functions of information collection and control execution. The docker container technology is used to form a one-to-one corresponding HFV hardware virtualization module with the hardware equipment through the network. In the network access stage of terminal equipment, the terminal equipment will send the data packet containing its equipment information to the gateway in the form of a network access request through a unified communication protocol. After the lower network access layer analyzes and uploads the data, the virtualization layer realizes the data preprocessing. After the terminal equipment is allowed to enter the network, it can read the number information sent by the terminal equipment, Extract the container parameters that need to be configured, then pull the corresponding template image from the private cloud image warehouse, add the configuration parameters extracted in advance on the template image, complete the creation and operation of the container corresponding to the terminal device, deploy the HFV module in the container to process the corresponding business, and finally encapsulate the software function of the HFV module into an API interface, At the same time, it integrates the cloud synchronization module to synchronize the API interface formed by the HFV module directly to the cloud.

3   Customer-Side Energy Management Controller

As a miniaturized, low-cost, high intelligent edge data transmission and data processing equipment, energy controller can comprehensively improve the energy data transmission capacity of the customer side, sink the business to the edge equipment, realize data hierarchical processing, improve energy efficiency diagnosis, optimize the regulation effect, and reduce the network operation and maintenance cost.

3.1   Structural Design of the Management Controller

The core board interface module includes one WIFI-BT module, three RS-485 interfaces, one RS-232 interface and two DI/DO interfaces. The mainboard communication module includes a 4G/5G communication module. The interface module of the main board includes two 100M Ethernet interfaces and one PLC/Lora pluggable external interface. PLC/Lora can plug external interface to realize plug and play of PLC /Lora communication module that can determine to access PLC/Lora communication module according to field communication needs. The WIFI-BT module, RS-485 interface, RS-232 interface and PLC/Lora interface are used to communicate with the energy consumption system control module and parameter monitoring equipment. The data storage module has a built-in energy efficiency diagnosis model library and optimization control strategy library which support real-time updates. The structural design of the energy management controller is illustrated in Fig. 1.

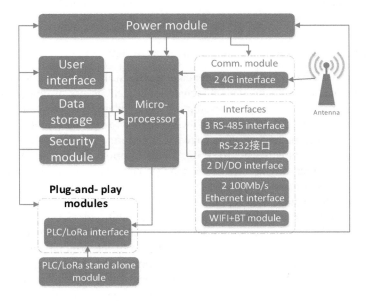

Fig. 1.   The structural design of the energy management controller

The energy controller mainly focuses on miniaturization and compact design, with the following advantages:

- The energy controller provides users with powerful edge computing performance, with cortex-a35 quad-core processor, 1.5 GHz main frequency (adjustable), 1 GB DDR3 memory and 8GB flash. It provides computing resources for edge node data optimization, real-time response, agile connection, intelligent application, security and privacy protection, and effectively shares the load of cloud computing resources.
- Rich and flexible transmission system, to widely apply to the power energy industry, the energy controller is compatible with a variety of wired/wireless mainstream communication interfaces, e.g., RS485, RS232 and WIFI.
- The energy controller software management is containerized. Based on the container, it forms code and component reuse and simplifies the maintenance of the application program. The ecological partners can automatically allocate resources on the cluster according to the requirements of APP, and scale on-demand to achieve high-level resource isolation and strong reliability.
- The energy controller provides a complete security protection scheme, including encryption technology to protect data transmission security, firewall function to protect network security and user Hierarchical Authorization Mechanism to ensure equipment management security.
- The energy controller adopts industrial standards from processing chip, memory chip, memory chip, communication module to power supply device. It can reach the industrial use level index in EMC 3, IP51 protection level and wide temperature characteristics. It is solid as a rock, meeting the harsh environment of industrial sites and durable.

3.2   Software Functionality of the Controller Design

According to the actual business and demand analysis, the main functions of energy controller application software system include: intelligent access of monitoring equipment and acquisition of energy consumption data; Intelligent processing of energy consumption data; Energy consumption data is uploaded to multiple remote service cloud platforms; For the remote device management platform, it supports remote control commands such as remote access to device status information, remote restart, remote update, remote server migration, remote management data reporting platform, etc.; Support the acquisition and transmission of sensor data, support the recording of system operation log information. The application system of energy controller can realize intelligent access to various types of monitoring equipment, and realize the acquisition of various types of energy consumption data. The intelligent processing of energy consumption data mainly includes the format processing of the original data collected by energy consumption equipment and the intelligent processing of the energy consumption data reported to the platform to avoid a large number of invalid data transmissions.

The energy controller can report energy consumption data to multiple remote service cloud platforms and can add, delete and modify multiple reporting platforms as needed. In the process of transmission, TCP, UDP, COAP and other transport protocols are supported. When docking with multiple remote server platforms, it can process and upload energy consumption data as needed, and support real-time data transmission, that is, reporting energy consumption data every minute, hourly average data transmission, daily average data transmission and monthly average data transmission. At the same time, for specific energy consumption data, it supports encrypted transmission to ensure the security of the energy consumption data transmission process.

The main control module is the main module of the system application, which is responsible for maintaining the normal operation of the whole system, as illustrated in Fig. 2. The main control module reads the configuration file to complete the initialization of the system and obtains the parameter information of each module of the system. The multi-thread method is used to realize the intelligent scheduling, dynamic configuration and timely maintenance of multiple subroutine modules. The log files and data files generated by the system operation are regularly deleted and backed up. The figure below shows the implementation process of the main control module.

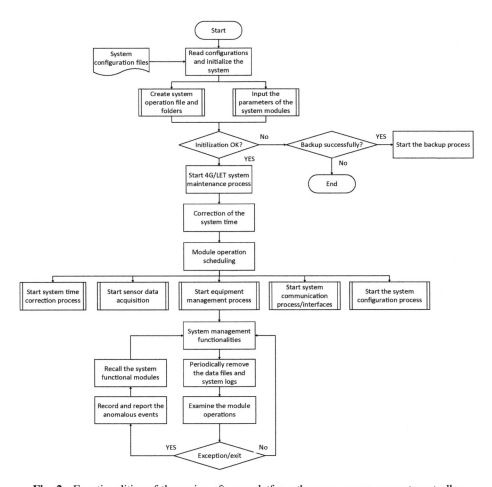

Fig. 2. Functionalities of the main software platform the energy management controller

4 Conclusions and Remarks

In this paper, based on the mobile edge computing system, an energy controller is developed, which can work with the cloud platform and energy efficiency terminal to realize energy efficiency diagnosis according to the monitoring data, and automatically implement the optimal control strategy, changing the working mode of the original energy management system that only monitors but not controls or manual controls. The edge computing and Docker technologies are introduced into the energy controller design, and the hardware virtualization module corresponding to the hardware device is formed through the network that ensures the independent implementation of various energy efficiency optimization and control strategies in the energy controller.

Acknowledgments. This work is supported by the technical project of the state grid corporation of China "Research and demonstration application of key technologies of customer-side energy consumption control system based on smart energy service platform" (5400-202018213A-0-0-00).

References

1. Ma, J., Chen, H.H., Song, L., Li, Y.: Residential load scheduling in smart grid: a cost efficiency perspective. IEEE Trans. Smart Grid **7**(2), 771–784 (2016)
2. Khan, F., Siddiqui, M.A.B., Rehman, A.U., Khan, J., Asad, M.T.S.A., Asad, A.: IoT based power monitoring system for smart grid applications. Int. Conf. Eng. Emerg. Technol. (ICEET) **2020**, 1–5 (2020)
3. Rahimi, F., Ipakchi, A.: Overview of demand response under the smart grid and market paradigms. Innovative Smart Grid Technol. (ISGT) **2010**, 1–7 (2010)
4. Raza, R., Hassan, N.U.L.: Quantifying the impact of customized feedback on user energy consumption behavior with low-cost IoT setup. Int. Conf. UK-China Emerg. Technol. (UCET) **2020**, 1–4 (2020)
5. Li, Y., Zeng, L., Chen, R., Wu, H.: Research and development of an on-line electric energy data acquisition and monitoring system, In: 2017 29th Chinese Control And Decision Conference (CCDC), pp. 6041–6045 (2017)
6. Lu, X., Wang, S., Li, W., Jiang, P., Zhang, C.: Development of a WSN based real time energy monitoring platform for industrial applications. In: 2015 IEEE 19th International Conference on Computer Supported Cooperative Work in Design (CSCWD), pp. 337–342 (2015)
7. Memari, P., Mohammadi, S.S., Ghaderi, S.F.: Cloud platform real-time measurement and verification procedure for energy efficiency of washing machines. Federated Conf. Comput. Sci. Inf. Syst. (FedCSIS) **2018**, 697–700 (2018)
8. Marinakis, V., Papadopoulou, A.G., Anastasopoulos, G., Doukas, H., Psarras, J.: Advanced ICT platform for real-time monitoring and infrastructure efficiency at the city level. In: 2015 6th International Conference on Information, Intelligence, Systems and Applications (IISA), pp. 1–5 (2015)
9. Kolar, H.R., et al.: The design and deployment of a real-time wide spectrum acoustic monitoring system for the ocean energy industry. MTS/IEEE OCEANS - Bergen **2013**, 1–4 (2013)
10. Demin, C., Xiaobo, L., Yang, G.: System design and development of distribution network planning quality indicator monitoring platform. In: 2017 IEEE Conference on Energy Internet and Energy System Integration (EI2), pp. 1–6 (2017)
11. Medina, J., Muller, N., Roytelman, I.: Demand response and distribution grid operations: opportunities and challenges. IEEE Trans. Smart Grid **1**(2), 193–198 (2010)

The Application of Statistical Analysis Method in the Comprehensive Evaluation of Enterprise Employees' Promotion

Songshan Zhang[✉]

School of Business, Zhujiang College, South China Agricultural University, Guangzhou 510900, Guangdong, China

Abstract. Nowadays, the competition of social enterprises is not only the competition of technology and capital, but also the competition of talents. If social enterprises want to survive for a long time, they must know how to choose talents, how to match talents with suitable positions, and how to keep talents in the enterprise. It is the vision of every company to train employees into the talents needed by the company and let employees and the company grow together. At the same time, statistics, especially statistical analysis methods, have become more and more common in the analysis of employee promotion. This phenomenon is worthy of attention and research. In view of this, this article studies the application of statistical analysis methods in the comprehensive evaluation of enterprise employee promotion. This article first analyzes the importance of corporate talents, conducts an in-depth study of statistical analysis methods, and then implements comprehensive evaluation of employee promotion with 165 employees of a domestic company with promotion space, and finally does a review of a certain employee of the company. Whether it conforms to the promotion principle and the analysis of the promotion comprehensive evaluation score.

Keywords: Statistical analysis method · Enterprise talent · Employee promotion · Comprehensive evaluation

1 Introduction

With the rapid development of the global economy, companies are facing increasingly fierce market competition. Technological reforms, employee retention, and the acquisition and maintenance of competitive advantages are all difficult problems that companies must consider in their future development [1]. Human resources are gradually being valued by enterprises. It is not only the primary resource of the enterprise, but also plays a vital role in the utilization of other resources in the development of the enterprise. In the final analysis, the competition between modern companies is human competition, the development and utilization of human resources. Therefore, many companies pay more and more attention to human resource management, and actively explore effective methods to improve the attraction and retention of company talents, so that they can maintain or even improve the company's core competitiveness. In today's enterprise human resource management, the promotion and evaluation system of

J. C. Hung et al. (Eds.): FC 2021, LNEE 827, pp. 819–827, 2022.
https://doi.org/10.1007/978-981-16-8052-6_104

employees is a very important task, which has a huge impact on the competitiveness of enterprises [2]. Talent is the core competitiveness of an enterprise, the source of motivation for the growth of an enterprise, and the resource for many companies to compete with each other. Brain drain will cause huge losses to the company, including obvious costs [3]. According to research, among the many reasons for resignation, obstacles to personal career development account for a large proportion, which shows that a healthy employee promotion system plays an important role in attracting and retaining talents.

Chinese scholar Zeng Guanghui believes that performance appraisal is mainly an overall assessment of the performance and efficiency of employees, and provides basic information about employee compensation and benefits, job promotion and job adjustments. In the company's daily operations and management, performance evaluation methods are essential to improve employees' work efficiency, economic efficiency and good operating conditions [4]. Fang Min believes that in my country's human resources management system, corresponding salary increases and promotion of outstanding employees are the usual ways to maintain employee business, but not all employees can have such opportunities, so this is necessary. An effective incentive mechanism can retain the company from other aspects, such as strengthening human environmental care and increasing employee benefits. In the maintenance and social welfare mechanism, the social security free market for employees is one of the key interests and an important part of the social management of human resources [5]. Huang Nan believes that the compensation system is the most effective tool for companies to retain and motivate employees. The impact of the compensation system on employee engagement can be attributed to the impact on job satisfaction, while low job satisfaction of employees directly leads to The occurrence of turnover. Evaluating the salary system from the perspective of employee turnover behavior and job satisfaction can reveal the shortcomings of the company's fixed salary performance rating system and promotion route design [6].

Nowadays, with the continuous improvement of employees' education, employees of today's companies are more interested in promotion than in salary. To some extent, whether the company's promotion system is correct determines whether employees leave or stay. Therefore, it is of great practical significance to explore appropriate methods for the overall evaluation of enterprise employee promotion to reduce employee turnover [7]. At present, the human resource management projects of many companies are divided into modules such as recruitment, training, performance evaluation, and salary. And it lacks a comprehensive view of human resource management and the integration of various modules. Through the combination of statistical analysis methods and the overall evaluation of employee promotion, employee training, performance appraisal, salary and career planning can be linked. Interaction and interdependence can promote human resource management, thereby increasing business potential. The application of statistical analysis methods in the comprehensive evaluation of enterprise employee promotion has guiding significance for improving the company's human resource management.

2   The Application of Statistical Analysis Method in the Comprehensive Evaluation of Enterprise Employees' Promotion

2.1   The Status Quo of Statistical Analysis

Generally speaking, the key content of data statistical analysis is the collected data and objective reality. Through statistical methods and data analysis, the internal situation of each factor can be explained, and the fundamentals and rules of the matter can be understood to complete the task [8]. To conduct research conceptually, statistical analysis methods are the main content of statistical analysis methods, and quantitative analysis is the common feature of all statistical analysis methods. From the study of the use process, the data statistical analysis method finds the fundamental attributes of things through quantitative analysis, obtains the quantity situation of things and things, and understands the original laws of things [9]. In summary, we can know that the core content of statistical analysis is data, and the fundamental of statistical analysis is data. Quantitative analysis shows that all conclusions can be accurately expressed with data. The effective collection, processing and utilization of data is the core of statistical analysis, which can help people understand the original laws of society. The method of collecting, sorting, and analyzing statistical data is called statistical analysis, and statistical analysis can be used to give corresponding answers to existing problems. Because the content of statistical data is different, the effects of different statistical analysis methods are also different. To use statistical analysis methods, we must maintain certain principles and always adhere to the fundamental concept of speaking through data. At each stage, we must clearly know what the content of the data is desired, clearly know where the data comes from, and carefully integrate the collected data.

2.2   The Selection Basis of Statistical Analysis Method

First, according to the research object, specify the type of design, research factors and number of layers. Confirm the attributes of the corresponding data, including whether it is normally distributed and the size of the sample; accurately analyze the attributes, counts, and levels of statistical data, and calculate accurate statistics through the corresponding content of statistical methods measure value; also through the objective facts of professional technology and design content, using statistics and management knowledge, accurately select data statistical analysis methods at any time. Secondly, according to the current status of the concept of design methods, through data statistical analysis and practical collation, a landmark and accurate design method has been produced, which can provide practical guidance for work [10]. Investigate the potential needs of users and create a standard user preference model. Finally, based on the statistical analysis of the above data, a standardized and simplified interface interaction method was constructed and converted into software engineering, thus forming a "public data" system, so that everyone can use the system for design and development. If the data can be used by as many people as possible, all the information in the system can be open to everyone, and you can easily work on the interface. At the same time, this information can achieve interdisciplinary, interdisciplinary business, and interdisciplinary business. Mode integration, manual design can create higher quality, practical and most advanced products for users and the world at a faster and faster speed [11].

2.3   The Significance and Purpose of the Research

An excellent promotion system can mobilize the enthusiasm and enthusiasm of employees, improve their personal qualities and skills, and form a benign competition mechanism for the company [12]. As a company facing severe competition and challenges from the outside world, it must find a scientific, effective and appropriate analysis method to evaluate the promotion of Chinese employees in order to organically integrate the company's personal growth and development goals. Promoting the sustainable development of enterprises in competition has become an important issue that urgently needs to be solved in the development and management of enterprise human resources. The importance of studying the application of statistical analysis methods in the overall evaluation of enterprise employee promotion lies in: promoting enterprises to better adapt to the fierce competition for talents. Through the combination of statistical analysis methods and the comprehensive evaluation system for promotion of business employees, you can better understand the advantages and problems of the business system. If the monitoring can be improved, it will greatly improve the management efficiency of business personnel and save business. Human resource cost: In order to promote the better development of business employees, the comprehensive evaluation system of the promotion of business employees through statistical analysis methods can also enable employees to better understand the purpose, importance and leadership of the system design. In this way, he can have a clearer understanding of his position, value and personal growth, and help employees achieve better business growth.

3   The Method Process of the Application Research of Statistical Analysis Method in the Comprehensive Evaluation of Enterprise Employee Promotion

3.1   The Role of Employee Promotion

In order for employees and the company to grow together, employees must have a clear understanding of their career development direction. If the relationship between a company and its employees is merely hiring and hiring, then employees will always be "machines", their capabilities will not develop as they should, and the company will not truly prosper. If the company regards the relationship between the two as a combination of mutual trust and mutual benefit, and provides a corresponding development platform for employees at every step, then it will greatly encourage employees to work hard to achieve their goals and continue to achieve them. Use your talents. When they stand out from the enterprise, it is time to get the most benefit from the enterprise. Many outstanding employees are ambitious, strong, full of vigor, willing to learn and work hard, and strive to equip themselves with talents commensurate with their honor and success. For managers, to use these talents for me, they must clearly provide them with a platform for career development so that they can fully release their potential and seek a sense of accomplishment in their careers. Through professional management, you can create professional development standards, provide employees with the knowledge and skills standards to be achieved, and then use a wide range of assessments to test employees' performance over the past period of time and increase professional knowledge and skills. It takes a certain amount of time to accumulate. Before completing the purpose of job promotion, employees still have to meet the company's standards and requirements at work, work correctly and improve their work ability. Only in this way can they lay a solid foundation for higher-level promotion.

3.2 Principles of Employee Promotion

There are many principles for employee promotion, and loyalty to the company is the primary principle for the company to determine whether an employee is promoted. If employees are loyal enough to the company and gain the full trust of the company, then employees have more opportunities for promotion and the cycle is relatively short; conversely, if employees are not loyal to the company and always want to change jobs, then employees have fewer opportunities for promotion. The promotion cycle is relatively long, and promotion is even impossible. Attitude to work, only those employees who have a serious work attitude and devote themselves to their work can get promotion opportunities. On the contrary, if they are not strong in their sense of responsibility and work sloppyly, it is difficult to get promotion opportunities. Democracy and principle of employees' work and whether they can be down-to-earth, truth-seeking and pragmatic in their work. Employees can not only adhere to principles, but also fully promote democracy, have a good work style, and it is also one of the principles of whether employees can be promoted. After being promoted, employees need to be responsible for more difficult tasks than ever before, and they need stronger working ability. If employees can't even do the tasks in their current jobs effectively, most companies will not promote this kind of employees. Even if an employee who does not match their work abilities is promoted, it is just a waste of resources and a job. The spirit of cooperation, whether you can integrate with colleagues, and establish good cooperation and personal connections is also one of the principles to be considered for employee promotion. After being promoted, employees become leaders. Not only do they need to be responsible for difficult tasks, but they also need to manage a department to ensure that the team's work in this department can be completed well. Without a good spirit of cooperation, it is difficult to lead employees to complete their tasks smoothly and with high quality. When judging whether an employee's promotion complies with the principles, the statistical analysis formula used in the single-principle evaluation is:

$$Y = \frac{1}{2}E[Z(x) - Z(x+h)]^2 \tag{1}$$

3.3 Application of Statistical Analysis Method to Comprehensive Evaluation of Enterprise Employees' Promotion

This article takes a domestic company as the research object. The company has 165 employees with room for promotion. The standardization of data related to promotion comprehensive evaluation mainly depends on whether the data source is disclosed. In terms of data sources, the promotion comprehensive evaluation scores of 165 employees are all calculated by statistical analysis. Among the 165 comprehensive evaluation score samples with data sources indicated, there are 34 from primary data only, 115 from secondary data, and 16 from both. First-hand data refers to the data obtained by the company from within the company, conducting employee surveys, or other databases generated, while second-hand data refers to the data collected by other companies. From the data, it can be reflected that the comprehensive evaluation score data of enterprise employee promotion researched in this paper rely more on external data in terms of data sources. The data sources are relatively rich and not limited to a single channel. The problem is that the proportion of the company's own comprehensive evaluation scores for employee promotion to obtain first-hand data is not too high. The reliability and persuasiveness of the comprehensive evaluation score for employee promotion lies not only in the final data presentation results, but also in the rigorous and standardized data statistical analysis and processing process, so the source of the data must be explained. The first-hand data of the company's comprehensive evaluation score for employee promotion is lower than the second-hand data, indicating that there are still some irregularities in the process of calculating the comprehensive evaluation score for enterprise employee promotion using statistical analysis. The statistical analysis formulas used in the calculation of the comprehensive evaluation of enterprise employee promotion are:

$$Z^*(x) = \sum_{i=1}^{n} \lambda_i Z(x_i) \tag{2}$$

$$\sigma_E^2 = E[Z(x) - Z^*(x)]^2 \tag{3}$$

4 Analysis of the Results of the Evaluation Example of the Application Research of the Statistical Analysis Method in the Comprehensive Evaluation of Enterprise Employee Promotion

4.1 Analysis of Whether an Employee of the Surveyed Company Meets the Promotion Principles

As shown in Table 1 and Figure 1, Table 1 records the total number of votes for each principle of the surveyed employee, and Figure 2 shows the score of each principle calculated according to the statistical analysis method. The result shows that the employee's loyalty score is 22 points, the work attitude score is 24 points, the work style score is 15 points, the work ability score is 24 points, and the cooperative spirit score is 23 points. The employee's scores in all principles have reached the qualification requirements, and can be assessed for comprehensive evaluation of promotion.

Table 1. The employee's principle score evaluation situation

Principle	Good	Better	General	Poor	Very poor
Loyalty	3	2	2	0	0
Work attitude	4	2	1	0	0
Work style	2	3	1	1	0
Work ability	4	2	1	0	0
Team spirit	4	1	2	0	0

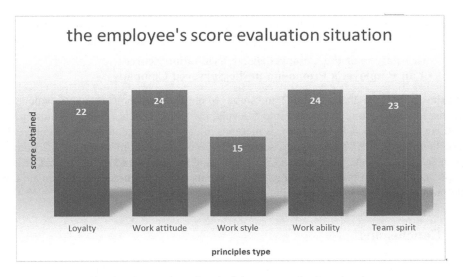

Fig. 1. The employee's principle score evaluation situation

Table 2. Evaluation of the comprehensive evaluation score of the promotion of the employee

	Excellent	Not bad	Medium	Poor	Very poor	Total score
Score	27	21.6	11.2	9	7.5	76.3

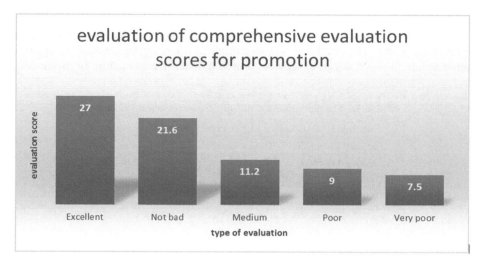

Fig. 2. Evaluation of the comprehensive evaluation score of the promotion of the employee

4.2   An Analysis of the Comprehensive Evaluation Scores of an Employee's Promotion in the Surveyed Company

As shown in Table 2 and Figure 2, the surveyed employees' comprehensive evaluation scores for promotion are divided into 5 levels, namely excellent, good, medium, poor and poor. Through statistical analysis methods, it can be calculated that the employee's excellent score is 27 points, good score is 21.6 points, medium score is 11.2 points, poor score is 9 points, poor score is 7.5 points, and the total score is 76.3 points. If it is stipulated that the total score is more than 70 points can be promoted, the employee can be promoted to a first-level position.

5   Conclusions

Reasonable and objective determination of employee promotion opportunities is necessary to mobilize employees' enthusiasm, initiative and creativity, enhance organizational cohesion, centrifugal force, and effective combat, and to successfully achieve organizational goals. This article uses statistical analysis methods to study the promotion opportunities of employees and considers the main factors that affect the promotion of employees. An intuitive and clear quantitative value can be obtained, which provides a reference for the correct selection of talents.

References

1. Liu, C.: Research on enterprise personnel promotion based on Peter's principle. Guangxi Qual. Supervision Herald **236**(08), 120–121 (2020)

2. Xiaoli, S., Na, W., Maolin, Y.: The relationship between enterprise employees' career adaptability, career satisfaction and job performance. Chin. Ment. Health J. **33**(001), 77–79 (2019)
3. Li, Y., Cheng, W.: Strategies for the improvement of employee stock ownership plans in private enterprises: the enlightenment of Shanxi's stock ownership system. South. Financ. **531**(11), 21–29 (2020)
4. Zeng, G.: Building a performance appraisal system for all employees to achieve a win-win situation for both the company and employees. Chinese Foreign Corp. Cult. **605**(04), 74–76 (2020)
5. Fang, M.: The status and role of social security in human resource management. Chinese Market **1054**(27), 98–99 (2020)
6. Huang, N.: Evaluation of Starbucks salary system from the perspective of employee turnover behavior and job satisfaction. China Bus. J. **820**(21), 124–126 (2020)
7. Jiang, T., Wu, X., Xu, S.: Investigation and analysis of employee satisfaction in a tertiary hospital. Hosp. Manage. Forum, **37**, 280(02), 61–64 (2020)
8. Xiaoqiang, S.: Design of human resources specialty sequence and rank system in commercial banks. Hum. Resour. Develop. **389**(02), 97–98 (2019)
9. Zhao, J., Ding, J.: Research on the elimination of job burnout of hotel staff from a dual perspective. Tourism Overview **323**(14), 13–17 (2020)
10. Wu, Q., Pang, B.: Problems and countermeasures in the management of non-staff hired staff in scientific research institutes. Chin. Pers. Sci. **36**(12), 21–26 (2020)
11. Zhang, S.: Research on the turnover tendency factors of Z bank marketing staff. Econ. Outlook Bohai Sea **311**(08), 145–146 (2020)
12. Wang, Z.: Analysis of the problems and countermeasures in the performance appraisal of middle-level managers in the company. Employ. Secur. **235**(17), 21–23 (2019)

Mechanism Modeling of SCR Flue Gas Denitration Reaction System

Yan Pan[1(✉)], Feng Yan[1], Jingya Yang[1], Xiangji Zeng[1], Xue Li[2], and Xiao Qi[3]

[1] Changsha Nonferrous Metallurgy Design and Research Institute Company Limited, Changsha, Hunan, China
[2] School of Avionics Maintenance, Changsha Aeronautical Vocational and Technical College, Changsha, Hunan, China
[3] Energy and Electricity Research Center, Jinan University, Zhuhai, China

Abstract. To eliminate the emission of nitrogen oxide created by coal combustion power plant becomes a new mission in front of the electric producing industry and nonferrous metal smelting to achieve rigorous environmental demand. After the present situation of SCR flue gas denitration automatic control system is analysed, the simplified dynamical transfer function model of SCR flue gas denitration reaction system is proposed by adopting mechanism modeling. The relationship between intrinsic and extrinsic parameters of SCR flue gas denitration reaction system is studied. To validate the exactness of the model, experimental data from a power plant is used. The given model can use for design of SCR flue gas denitration automatic control system or parameter tuning of controller, and has great practical value.

Keywords: SCR reactor · Flue gas denitration · Thermal power generation units · Dynamic modeling · Automatic control

1 Introduction

To eliminate the emission of nitrogen oxide (NO_x) created by coal combustion power plant becomes a new mission in front of the electric producing industry to achieve rigorous environmental demand. Many countries have already taken their responsibility by imposing rules and laws on power plant. For example, the United States has published a huge regulation system to aim at different combustion situations, the newly Chinese industry enforcing standards mandates a reduction of nitrogen oxide (NO_x) of 100 mg/m^3, and 50 mg/m^3 is suggested, and so on.

Among the denitration technologies, the selective catalytic reduction (SCR) method has been widely used since it was invented twenty years ago. The technique itself is constantly supplemented and completed during this period, and is accepted by professionals. But there still are the problems of power plant SCR denitration automatic system to be dealt with. Usually the operation quality is optimal only at the condition which the running parameters are the same as parameters that are choose when designing. If the operational states such as sorts of coal, load command or some other key factors is changed, the result could be worse. Also, the dynamic characteristics can be not stay at the best along if the disturbances are occurs. The problems often cause

J. C. Hung et al. (Eds.): FC 2021, LNEE 827, pp. 828–838, 2022.
https://doi.org/10.1007/978-981-16-8052-6_105

the automatic control system to break out and SCR denitration automatic system of many power plants is still in the manual state. As a result, the concentration of nitrogen oxide (NO_x) at outlet duct or the ammonia slip rate may exceed the upper limit value.

The effect of SCR denitration control schemes is mainly depended on designers who comprehend operational mechanism of controlled objects and control requirements. To gain a full understanding of controlling system, an accurate model of controlled objects is established in the first place. The achievements of SCR denitration system modeling can be divided into three categories, the first kind of research mainly focus on the model designing [1–5] of SCR reactor by using numerical simulation methods and fluid simulation software; the second kind provides input-output models [6–10] based on experimental investigation method by using field data; the third kind presents object model by studying the operational mechanism of SCR denitration system. The amount of researches in third area is much less than the other two areas mentioned above, and most mechanism models [11–16] is given in the form of time domain to meet the requirements of simulation training.

In this paper, the SCR flue gas denitration reaction system is simplified from an automatic control point of view, and the transfer function model that can be used for control system design and control quality study is founded by using mechanism modeling method.

2 SCR Flue Gas Denitration Reaction System

Power plant SCR flue gas denitration reaction system can be described as complicated system that is affected by chemical reaction process on the surface of catalyzer and flow distribution of flue gas. In chemical reactions process, liquid ammonia or urea could be used as deoxidizer and a variety of catalysts have been developed. Although the characteristics of different schemes are not equal, each of them has the same basic principles. The SCR flue gas denitration system is consisted of three major subsystems including denitration reaction system, ammonia storage-supply system and ammonia injection system. The mechanism control-oriented model of SCR denitration reaction system is founded. The simplified process diagram of SCR flue gas denitration reaction system is as follows.

2.1 Mechanism of Flue Gas Denitration Reaction

In the SCR denitration reactor, nitrogen oxide (NO_x) in the flue gas reacts with injected ammonia (NH_3) to form nitrogen (N_2) and water (H_2O) under the help of catalysts $[V_2O_5 - WO_3(MoO_3)TiO_2]$. It is a redox exothermic reaction. Because the ninety-five percent of nitrogen oxide (NO_x) in the flue gas is nitric oxide NO, and the oxygen O_2 account for more than one percent of the flue gas, the main chemical equation with the temperature range of 290 °C to 430 °C can be described as follows (Fig. 1).

$$4NO + 4NH_3 + O_2 \xrightarrow{V_2O_5 - WO_3MoO_3TiO_2} 4N_2 + 6H_2O \tag{1}$$

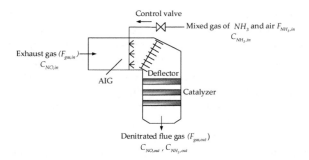

Fig. 1. Simplified diagram of SCR flue gas denitration reaction system

The reaction is accompanied by the evolution of energy. The experiment indicates that adding tungsten trioxide (WO_3) can improve the catalytic efficiency and is good for nitric oxide (NO) to be reduced, but the conversion rate for sulfur dioxide (SO_2) to be oxidized to sulfur trioxide (SO_3) is lifted too. The monoammonium sulfate (NH_4HSO_4)formed by adverse chemical reaction that sulfur trioxide (SO_3) reacts with ammonia (NH_3) has strong viscous properties. This feature poses a threat to stable operation of catalyst, and air preheater that the flue gas passes by may be blocked up. To protect air preheater from corrosion and maintain activity of catalyst, the amount of tungsten trioxide (WO_3) that is added to catalyst and the flow of ammonia (NH_3) should be controlled seriously.

2.2 Control Tasks of SCR Denitration System

There are three tasks should be accomplished simultaneously, the fundamental task is to ensure nitrogen oxide (NO_x) concentration at the boiler end part to be accept by mandatory standards, this is also the reason that the whole SCR denitration optimize system to be designed; the second control objective is to eliminate the ammonia (NH_3) slip; the third task is to avoid the blockage of air preheater. The last two tasks are caused by mechanism of the system, the whole system block diagram can be described as shown below (Fig. 2).

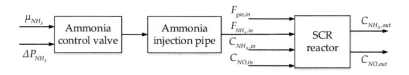

Fig. 2. Block diagram of SCR flue gas denitration reaction system

The block diagram gives a brief introduction of the SCR flue gas denitration reaction system. In the diagram, $C_{NO,out}$ stands for concentration of nitric oxide (NO) in the flue gas at the reactor outlet; $C_{NH_3,out}$ stand for concentration of ammonia (NH_3) in the same place, both variables are controlled variables. The SCR denitration control system accomplishes control tasks by changing valve opening of ammonia (NH_3)

control valves according to the real-time feedback concentration. Here, μ_{NH_3} is used as opening of ammonia (NH_3) control valve. The major disturbance can be summarized as follows: the pressure difference of the ammonia (NH_3) control valve, which is represented by ΔP_{NH_3}, the concentration of ammonia (NH_3) and nitric oxide (NO) in the reactor entrance, which are represented by $C_{NH_3,in}$ and $C_{NO,in}$ respectively, the flue gas flow in the reactor entrance, which is represented by $F_{gas,in}$.

3   Mechanism Modeling of SCR Denitration System

Denitration reactor is the core equipment of SCR denitration system. It is the place where mixed gas reacted chemically. On the surface of the catalysts in the reactor, the process of the ammonia (NH_3) adsorption and desorption as well as the chemical reaction between nitric oxide (NO) and adsorbed ammonia (NH_3) take place simultaneously, these add the complexity of modeling. To build the mechanism mathematical model based on mass balance and the chemical equilibrium of the reaction process, the dynamic characteristics and the static characteristics of the SCR denitration system are considered, design features of the reactor is taken account too.

3.1   The Mass Balance Equations of Ammonia (NH_3) on Catalyst Surface

According to adsorption and desorption process of ammonia on the surface of catalyst, the ammonia mass balance equations are as follows.

$$\frac{d\theta_{NH_3}}{dt} = k_a C_{NH_3}(1 - \theta_{NH_3}) - k_d \theta_{NH_3} \tag{2}$$

$$k_a = k_a^0 e^{-\frac{E_a}{RT}} \tag{3}$$

$$k_d = k_d^0 e^{-\frac{E_d\left(1 - \alpha\theta_{NH_3}\right)}{RT}} \tag{4}$$

Where, k_a = kinetic constant for ammonia (NH_3) adsorption [m^3/(mol · s)]; k_a^0 = pre-exponential factor for ammonia (NH_3) adsorption [m^3/(mol · s)]; k_d = kinetic constant for ammonia (NH_3) desorption [1/s]; k_d^0 = pre-exponential factor for ammonia (NH_3) desorption [1/s]; C_{NH_3} = gas-phase ammonia (NH_3) concentration [mol/m^3]; θ_{NH_3} = ammonia (NH_3) surface coverage; α = ammonia (NH_3)/nitric oxide (NO) molar feed ratio; E_a = activation energy for ammonia (NH_3) adsorption [kJ/ mol]; E_d = activation energy for ammonia (NH_3) desorption [kJ/mol]; R = ideal gas constant, R = 8.314[J/mol · K]; T = temperature in the reactor [K]. Formula (2) is the mass balance equation of ammonia on catalyst surface. It is linearized at the rated operating point. The result is formula (5).

$$\frac{d\theta_{NH_3}}{dt} = k_a(1 - K_1)C_{NH_3} - (k_a K_2 + k_d)\theta_{NH_3} \tag{5}$$

Where, both K_1 and K_2 are the linearized coefficients. $K_1 = \theta_{NH_3}^0$ stands for the coverage rate for ammonia (NH_3) on the surface of catalyst at the rated operating point; $K_2 = C_{NH_3}^0$ stands for ammonia (NH_3) molality of the flue gas at the rated operating point. The block diagram corresponded with formula (5) is as follows.

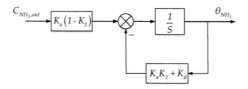

Fig. 3. Block diagram of NH_3 mass balance on the surface of catalysts

3.2 The Mass Balance Equations of Nitric Oxide (NO) in SCR Denitration Reactor

The investigation indicates that if $V_2O_5-WO_3(MoO_3)TiO_2$ is used as catalyst, the selective catalytic reduction reaction between nitric oxide (NO) and ammonia (NH_3) works on the basis of Eley-Ridea mechanism. The mass balance equations and the interaction equations are as follows.

$$\frac{dC_{NO}}{dt} = \frac{1}{V}\left(F_{gas,in}C_{NO,in} - F_{gas,out}C_{NO,out}\right) - r_{NO}\Omega_{NH_3} \tag{6}$$

$$r_{NO} = k_{NO}C_{NO}\theta_{NH_3} \tag{7}$$

$$k_{NO} = k_{NO}^0 e^{(-E_{NO}/RT)} \tag{8}$$

Where, C_{NO} = molality of nitric oxide (NO) in the reactor [mol/m³]; $C_{NO,in}$ = molality for nitric oxide (NO) in the reactor entrance [mol/m³]; $C_{NO,out}$ = molality for nitric oxide (NO) at the reactor outlet [mol/m³]; $F_{gas,in}$ = volume flow of flue gas in the reactor entrance [m³/s]; $F_{gas,out}$ = volume flow of flue gas at the reactor outlet [m³/s]; r_{NO} = rate of nitric oxide (NO) consumption [1/s]; Ω_{NH_3} = catalyst ammonia (NH_3) adsorption capacity [mol/m³]; V = volume of reactor [m³]; k_{NO} = kinetic constant for the $deNO_x$ reaction rate constant [m³/(mol · s)]; k_{NO}^0 = pre-exponential factor for the $deNO_x$ reaction rate constant [m³/(mol · s)]; E_{NO} = activation energy for the $deNO_x$ reaction [kJ/ mol].

Formula (6) is the mass balance equation of nitric oxide (NO), formula (7) and formula (8) stand for the reaction rate function of reaction between the nitric oxide (NO) and the ammonia (NH_3) that is adsorbed on the catalyst surface. Substituted formula (7) and formula (8) into formula (6), and then it is linearized at the rated operating point. The result is as follows.

$$\frac{dC_{NO,out}}{dt} = \frac{1}{V}(K_5 - K_7)F_{gas} + \frac{1}{V}K_6 C_{NO,in} - \left(\frac{1}{V}K_8 + k_{NO}\Omega_{NH_3}K_3\right)C_{NO,out} - k_{NO}K_4\theta_{NH_3}\Omega_{NH_3}$$

$$(9)$$

Where, $C_{NO} = C_{NO,out}$; $F_{gas} = F_{gas,in} = F_{gas,out}$; $K_3 \sim K_8$ are the linearized coefficients, $K_3 = \theta^0_{NH_3}$ stands for ammonia (NH_3) coverage rate on the surface of catalyst at the rated operating point, $K_4 = C^0_{NO,out}$ stands for nitric oxide (NO) molality of flue gas at the reactor outlet at the rated operating point. $K_5 = C^0_{NO,in}$ stands for nitric oxide (NO) molality of flue gas in the reactor entrance at the rated operating point. $K_6 = F^0_{gas,in}$ stands for volume flow of flue gas in the reactor entrance at the rated operating point. $K_7 = C^0_{NO,out}$ stands for nitric oxide (NO) molality of flue gas at the reactor outlet at the rated operating point. $K_8 = F^0_{gas,out}$ stands for volume flow of flue gas at the reactor outlet at the rated operating point. The block diagram correspond with formula (9) is as shown Fig. 4.

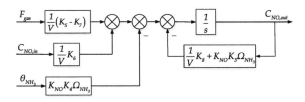

Fig. 4. Block diagram of NO mass balance in the reactor

3.3 The Mass Balance Equations of Ammonia (NH_3) in SCR Denitration Reactor

Similar to the mass balance equations of nitric oxide (NO) in the reactor, the ammonia (NH_3) mass balance equations in SCR denitration reactor are as follows.

$$\frac{dC_{NH_3}}{dt} = \frac{1}{V}\left(F_{NH_3,in}C_{NH_3,in} - F_{gas,out}C_{NH_3,out}\right) - r_{OX}\Omega_{NH_3} \tag{10}$$

$$r_{OX} = k_{OX}\theta_{NH_3} \tag{11}$$

$$k_{OX} = k^0_{OX}e^{-\frac{E_{OX}}{RT}} \tag{12}$$

Where, C_{NH_3} = molarity of ammonia (NH_3) in the reactor [mol/m^3]; $F_{NH_3,in}$ = volume flux for ammonia (NH_3) in the reactor entrance [m^3/s]; $C_{NH_3,in}$ = molarity of ammonia (NH_3) in the reactor entrance [mol/m^3]; $C_{NH_3,out}$ = molarity of ammonia (NH_3) at the reactor outlet [mol/m^3]; r_{OX} = rate of ammonia oxidation [1/s]; k_{OX} = rate constant for ammonia oxidation [m^3/(mol · s)]; k^0_{OX} = pre-exponential factor for ammonia oxidation [m^3/(mol · s)]; E_{OX} = activation energy for ammonia oxidation [kJ/mol].

Formula (10) is the mass balance equation of ammonia (NH_3), formula (11) and (12) are equations that describe the ammonia oxidation rate of the reaction. Substitute formula (11) and (12) into formula (10), and then it is linearized at the rated operating point. The formula (13) can be obtained as follows.

$$\frac{dC_{NH_3,out}}{dt} = \frac{1}{V}K_9 F_{NH_3,in} - \frac{1}{V}K_{11}F_{gas} + \frac{1}{V}K_{10}C_{NH_3,in} - \frac{1}{V}K_{12}C_{NH_3,out} - k_{OX}\theta_{NH_3}\Omega_{NH_3}$$

(13)

Where, $K_9 = C^0_{NH_3}$ stands for the ammonia molarity of flue gas in the reactor entrance at rated operating point; $K_{10} = F^0_{NH_3,in}$ stands for the volume flux of ammonia in the reactor entrance at rated operating point; $K_{11} = C^0_{NH_3,out}$ stands for the ammonia molarity of flue gas at the reactor outlet at rated operating point; $K_{12} = F^0_{gas,out}$ stands for the volume flux of flue gas at the reactor outlet at rated operating point. The block diagram correspond with formula (13) is as shown in Fig. 5.

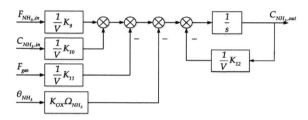

Fig. 5. Block diagram of NH_3 mass balance in the reactor

3.4 The Equations of Ammonia Injection Control Valves and Pipelines

The control valves have the same characteristics as throttles that the cross section is variable. Flow characteristics of the control valves are not only relevant to valve opening, but also be bound up with the pressure difference of the valve, corresponding mathematical expression can be expressed as follows.

$$F_{NH_3} = f\left(\mu_{NH_3}, \Delta P_{NH_3}\right)$$

(14)

Where, F_{NH_3} = the volume flow of ammonia that flowed through the control valve [m^3/s]; μ_{NH_3} = opening of ammonia injection control valve [%]; ΔP_{NH_3} = the pressure difference of the ammonia injection control valve [Pa].

The functional relationship of control valve is determined by internal mechanical structure of valve, usually the relation can be described as nonlinear. For the convenience of study, it is linearized at the rated operating point. The results are as follows.

$$F_{NH_3} = K_{13}\mu_{NH_3} + K_{14}\Delta P_{NH_3} \tag{15}$$

$$K_{13} = \frac{\delta f}{\delta \mu_{NH_3}}\bigg|_{f0} \tag{16}$$

$$K_{14} = \frac{\delta f}{\delta \Delta P_{NH_3}}\bigg|_{f0} \tag{17}$$

Where, both K_{13} and K_{14} are the linearized coefficients.

To make the model have more pertinence, the throttle loss of ammonia pipelines is usually neglected and the characteristics of the pipeline can be regarded as pure time delay element. The corresponding expression is as follows.

$$F_{NH_3,in} = F_{NH_3}(t - \tau) \tag{18}$$

Where, $F_{NH_3,in}$ = the volume flow of ammonia that flow into the reactor under the condition that the pure delay element is considered [m³/s]; F_{NH_3} = the volume flow of flue gas in the ammonia injection control valve entrance [m³/s]; τ = the pure delay time of the ammonia injection pipeline [s].

Formula (15) and (18) are used equations to describe the ammonia injection control valve and the pipelines, the corresponding block diagram is as shown in Fig. 6.

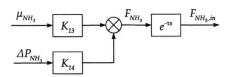

Fig. 6. Block diagram of ammonia control valves and pipelines

3.5 Combining and Analyzing

According to Figs. 3, 4, 5 and 6, the simplified block diagram of SCR denitration reaction system is as shown in Fig. 7.

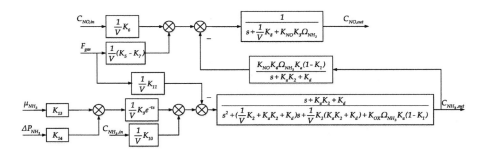

Fig. 7. Block diagram of SCR flue gas denitration reaction system

The model provided above gives a detailed description for dynamic relationship between input-output variables. As is shown in Fig. 7, there are five variables used as input for the SCR denitration reaction system: the opening of ammonia injection control valve (μ_{NH_3}); the concentration of injection ammonia in the reactor entrance ($C_{NH_3,in}$); the pressure difference of the ammonia injection control valve (ΔP_{NH_3}); the flue gas flow (F_{gas}) and the nitric oxide (NO) concentration of flue gas in the reactor entrance ($C_{NO,in}$). According to the control tasks, there are two output variables: the nitric oxide (NO) concentration of flue gas at the SCR denitration reactor outlet ($C_{NO,out}$); the ammonia concentration of flue gas at the SCR denitration reactor outlet ($C_{NH_3,out}$). The pressure difference of the ammonia injection valve (ΔP_{NH_3}) depends on the diluted ammonia pressure in the control valve entrance when the control valve is maintained a constant opening. Because the pressure of diluted ammonia is remain unchanged while the unit is in good operation (The diluted ammonia pressure is maintained constant by ammonia supply system), the pressure difference (ΔP_{NH_3}) can be regarded as constant too. The concentration of ammonia in the reactor entrance ($C_{NH_3,in}$) depends on the mechanism of ammonia-air mixer system. Because it is maintained constant while the system operates normally, the concentration of ammonia in the reactor entrance ($C_{NH_3,in}$) is regarded as a constant while it is involved in computation. Above all, the transfer function models that describe the dynamic characteristics of SCR denitration reaction system are shown as follows.

$$\frac{C_{NO,out}}{C_{NO,in}} = \frac{K_6}{V} \times \frac{1}{s + \frac{1}{V}K_8 + K_{NO}K_3\Omega_{NH_3}} \tag{19}$$

$$\frac{C_{NO,out}}{F_{gas}} = \frac{1}{V}\left[(K_5 - K_7) + \frac{1}{s^2 + \left(\frac{K_2}{V} + K_aK_2 + K_d\right)s + \frac{K_2}{V}(K_2K_a + K_d) + K_{OX}\Omega_{NH_3}K_a(1 - K_1)}\right]$$
$$\times \frac{1}{s + \frac{K_8}{V} + K_{NO}K_3\Omega_{NH_3}} \tag{20}$$

$$\frac{C_{NO,out}}{\mu_{NH_3}} = -\frac{K_9K_{13}e^{-\tau s}}{V} \times \frac{1}{s + \frac{K_8}{V} + K_{NO}K_3\Omega_{NH_3}}$$
$$\times \frac{K_{NO}K_4\Omega_{NH_3}K_a(1 - K_1)}{s^2 + \left(\frac{K_2}{V} + K_aK_2 + K_d\right)s + \frac{K_2}{V}(K_2K_a + K_d) + K_{OX}\Omega_{NH_3}K_a(1 - K_1)} \tag{21}$$

$$\frac{C_{NH_3,out}}{F_{gas}} = -\frac{K_{11}}{V}$$
$$\times \frac{s + K_aK_2 + K_d}{s^2 + \left(\frac{K_2}{V} + K_aK_2 + K_d\right)s + \frac{K_2}{V}(K_2K_a + K_d) + K_{OX}\Omega_{NH_3}K_a(1 - K_1)} \tag{22}$$

$$\frac{C_{NH_3,out}}{\mu_{NH_3}} = \frac{K_9K_{13}e^{-\tau s}}{V} \frac{s + K_aK_2 + K_d}{s^2 + \left(\frac{K_2}{V} + K_aK_2 + K_d\right)s + \frac{K_2}{V}(K_2K_a + K_d) + K_{OX}\Omega_{NH_3}K_a(1 - K_1)} \tag{23}$$

4 Conclusions

The mathematical model of power plant SCR denitration reaction system is provided by the mechanism modeling method. It gives a detailed description of the relationship between intrinsic and extrinsic parameters and possesses strong generality. So long as the different groups of field data are put into the model, the model aimed at some condition can be received. The model helps to design the controller and tune the controlling parameter. All of these have proven that the model possess the practical significance.

Acknowledgements. This work was supported by National Key R & D Program of China (2019YFB1704705).

References

1. Liu, H.Q., Guo, T.T., Yang, Y.P., Lu, G.J.: Optimization and numerical simulation of the flow characteristics in SCR system. Energy Procedia, **17**, 801– 812 (2012). International Conference on Future Electrical Power and Energy Systems
2. Ogidiama, O.V., Shamim, T.: Performance analysis of industrial selective catalytic reduction (SCR) systems. Energy Procedia, **61**, 2154– 2157 (2014). The 6th International Conference on Applied Energy–ICAE2014
3. Ogidiama, O.V., Shamim, T.: Investigation of dual layered SCR systems for NOx control. Energy Procedia. **75**, 2345–2350 (2015). The 7th International Conference on Applied Energy–ICAE2015
4. Dong, J.X.: Experimental research and mathematical modeling of SCR DeNO$_X$ for fossil-fired power plant. Doctoral Dissertation North China Electr. Power Univ. 4 (2007)
5. Chen, L.F.: Simulation optimization of flow field and reaetants mixing on denitration reactors. Doctoral Dissertation Shandong Univ. 5 (2011)
6. Tan, P., Xia, J., Zhang, C., Fang, Q.Y., Chen, G.: Modeling and optimization of NO$_X$ emission in a coal-fired power plant using advanced machine learning methods. Energy Procedia. **61**, 377–380 (2014). The 6th International Conference on Applied Energy–ICAE2014
7. Forzatti, P., Nova, I., Tronconi, E., Kustov, A., Thøgersen, J.R.: Effect of operating variables on the enhanced SCR reaction over a commercial V$_2$O$_5$–WO$_3$/TiO$_2$ catalyst for stationary applications. Catal. Today, 184, 153–159 (2012)
8. Liu, J.Z., Qin, T.M., Yang, T.T., Lv, Y.: Variable selection method based on partial mutual information and its application in power plant SCR system modeling. Proc. CSEE. **36**(9), 2438–2443 (2014)
9. Qin, T.M., Liu, J.Z., Yang, T.T., Zhang, W.: SCR Denitration system modeling and operation optimization simulation for thermal power plant. Proc. CSEE. **36**(10), 2699–2703 (2015)
10. Hou, Y.T., Xue, J.Z., Wang, L., Wang, Z.: Recursive instrumental variable estimation algorithm for ammonia flow modeling SCR denitration reactors. Therm. Power Gener. **44**(11), 75–80 (2016)
11. Schaub, G., Unruh, D., Wang, J., Turek, T.: Kinetic analysis of selective catalytic NOx reduction (SCR) in a catalytic filter. Chem. Eng. Process. **42**, 365–371 (2003)

12. Colombo, M., Nova, I., Tronconi, E., Schmeier, V., Weibel, M.: Mathematical modelling of cold start effects over zeolite SCR catalysts for exhaust gas aftertreatment. Catal. Today **231**, 99–104 (2014)

13. Chen, C.T., Tan, W.L.: Mathematical modeling, optimal design and control of an SCR reactor for NOx removal. J. Taiwan Inst. Chem. Eng. **43**, 409–419 (2012)

14. Li, X.J., Li, Y.N., Cui, R.: Modeling and simulation analysis of the mechanism of denitrification reactor in 600 MW Plant. Instrum. Technol. **11**, 26–31 (2016)

15. Zhou, X.Q., Yang, C.: Mathematical model of selective catalytic reaction of NO_x in reactor and its smiulation. J. Chongqing Univ. (Nat. Sci. Ed.). **30**(6), 39–43 (2014)

16. Zhang, Z.C.: Modeling and optimal control of SCR denitration system in thermal power plant. Master Dissertation North China Electr. Power Univ. 3 (2015)

The Legal Concept of E-Commerce—Protection of the Rights of Participants in Relations

Peiyuan Lin[✉]

Peoples' Friendship University of Russia, Moscow 10100, Russia
1042208004@rudn.ru

Abstract. E-commerce is an integral element of the modern economy. The rapid development and change of e-commerce forces not only business and society, but also the state to adapt to such changes. In the last few years, the choice of optimal rules for regulating e-commerce has already become not only a scientific, but also a clearly expressed practical problem. At the same time, this problem is far from a final solution, since, as e-commerce evolves, it raises more and more complex questions. In the context of the study, an attempt is made to analyze the dynamics of the development of the legal foundations of electronic commerce and to develop directions for their improvement, taking into account the peculiarities of the development of electronic commerce.

Keywords: E-commerce · Legal basis of E-commerce · Legal basis of E-commerce · Inconsistency of legislative norms · Unfair contractual conditions · Protection of the rights of participants in relations

1 Introduction

Today, few people remember that the first transaction in the field of e-commerce was illegal - in 1972 (long before the advent of eBay or Amazon, students at the Massachusetts Institute of Technology conducted the first ever online transaction, using an account in Arpanet (the forerunner of the Internet) to buy marijuana from students from Stanford University [20]. Since then, almost fifty years have passed, and the importance and significance of e-commerce for the modern world is difficult to assess – from an illegal transaction, it has become the basis of the market, and the procedure for implementing e-commerce has become outlined by the relevant legal framework-the legislation of various countries on the procedure for implementing e-commerce. By the end of 2021, it is projected that this volume will be about $ 5 trillion. At the same time, it should be noted that the legislation in the field of e-commerce is probably one of the most unstable. This is due to the constant changes taking place in the field of e-commerce, to which the legislation is forced to adapt. In the last few years, the choice of an approach to the optimal legal regulation of e-commerce has become a global scientific and practical problem. And this problem, unfortunately, has not yet been solved for a number of reasons. The first reason why it is impossible to solve the problem of optimizing legislation in the field of e-commerce is the constant development of information technologies, which give rise to new and new rules and issues. The second reason is the lack of clear international rules in the field of e-commerce and the

J. C. Hung et al. (Eds.): FC 2021, LNEE 827, pp. 839–845, 2022.
https://doi.org/10.1007/978-981-16-8052-6_106

inconsistency of national legislation in the field of e-commerce with each other [1–5]. And if the first reason can be somehow eliminated by bringing into line with the current realities of national regulatory legal acts regulating e-commerce, then the second reason is more difficult to eliminate. On the one hand, e-commerce blurs the boundaries, making it possible to buy in different countries of the world, on the other hand, each country to which the trading entity belongs has its own rules that must be followed, but there is no single global digital market. For example, in the countries of the European Union (hereinafter referred to as the EU), a digital single market strategy is being implemented, which is based on the corresponding e-commerce structure outlined by the legal framework – European legislation in the field of e-commerce is aimed at preventing unfair discrimination against consumers and businesses that access content or buy goods and services online. In the limited scope of this study, unfortunately, it is not possible to consider all the problems identified, so we will focus only on those issues related to the optimization of the legal framework of e-commerce, taking into account historical retrospect and necessity, and ensuring legal certainty for businesses and citizens.

2   Materials and Methods

When writing the article, a comprehensive approach to the problems of research was implemented, since an extensive list of issues related to the evolution of the legal foundations of e-commerce and the prospects for the development of such foundations were considered. To identify the features of the development of the legal foundations of e-commerce, both general research methods and special legal methods were used. Among the general methods used in the study are the method of systematic, quantitative and qualitative analysis, synthesis, as well as the method of formal-logical method and theoretical generalization. For the purpose of cognition of the external and internal forms of legal phenomena, special legal methods are used: the formal legal and dogmatic methods used for cognition. The combined use of general and formal legal methods in conducting the study allowed us to achieve the validity of the scientific conclusions. E-commerce over the past fifty years has become not just a banal purchase and sale of goods, works or services via the Internet, but something in which citizens of many countries of the world participate on a daily basis, since e-commerce can take many forms and include various transactions, including payment for the services of government agencies. As for the legal framework of e-commerce, taking into account historical retrospect, the most important fact here is the fact that the content of the rules on e-commerce in a particular period of historical development corresponded to the level and features of the development of such commerce, as well as the level of theoretical understanding of the e-commerce process itself. Therefore, in order to more fully and clearly approach the issue of the prospects for the development of legislation on e-commerce, we focus on the content of this concept. The first specialist who drew attention to such a phenomenon as e-commerce was the American economist David Kozier, who considered e-commerce as e-commerce, the structure of which is based on traditional trade, and the use of electronic networks gives such trade its inherent flexibility and borderlessness [8]. E-commerce as an Internet trade is also considered by other Western economists, for example, D. Eymor, who points out that e-commerce involves the sale of goods, in which the demand for goods is carried out through the Internet [11]. Russian experts use a slightly different approach to the category of e-commerce, pointing out that e-commerce is an integral part of e-business, and e-commerce, in turn, is only a special case of e-commerce [6, 9, 12], which is "entrepreneurial activity for the implementation of commercial operations using electronic means of data exchange" [7]. In turn, S. V. Pirogov defines e-commerce as "the technology of performing commercial operations and managing production processes using electronic means of data exchange" [10]. All the above approaches to the content of e-commerce agree on one thing-e-commerce is somehow connected with the trade in goods, works, and services using the Internet, respectively, and the prospects for the development of legislation on e – commerce cannot be considered in isolation from the main component of e-commerce-the Internet. Despite the fact that at that time transactions were de facto carried out with the use of technologies, de jure, such payments were not regulated by anything, except for separate rules established by the market participants themselves-banks and large users [13]. In 1982, the Boston Computer Exchange was founded, becoming the world's first e-commerce company. On this exchange, buyers and sellers, using online databases, could buy and sell computers, spare parts and other consumables for computer equipment. Despite the interest of business in e-commerce and the creation of the Boston Computer Exchange, e-commerce in general, both for the state and for the entire civil society in the 80s of the last century, continued to be a small niche service, framed in the legal framework only by local acts regulating the features of the Boston Computer Exchange [17–19]. In 1994, The Rolling Stones broadcast the Voodoo Lounge tour on M-Bone, charging those who joined the broadcast, and Pizza Hut began accepting pizza orders on the network. The removal of all restrictions and legislative easing in terms of allowing e-commerce in the US and EU countries led to the fact that by 1995 more than 120,000 domain names were registered, which increased to more than 2 million by 1998. This situation has led to the fact that an increasing number of business entities began to look for ways to earn money on the Internet, including through e-commerce. Around this time, modern Internet giants-Amazon and eBay-begin to appear. And if the first, Amazon, was originally created in order to become a convenient platform for buying individual goods-books, then the second, eBay, aimed to simplify the e-commerce process and was the world's first trading platform that allowed people to sell and buy goods from each other using the Internet [14]. By the end of the 90s of the last century, when it became more or less clear that there was a new element of the economy, business began to actively develop services and applications for e-commerce stores. At the end of 1999, Cofinity, founded in 1998, and Elon Musk's company. x.com. merged and formed PayPal, which has now grown into one of the largest online payment systems in the world. Today, there are other online payment systems, but no other system managed to displace PayPal from the first place. Thus, it can be stated that in the 90s of the last century, an unprecedented development of e-commerce was observed in Western countries. This led to a kind of "gold rush" in the field of e-commerce-investors invested huge amounts of money in Internet companies, and received unprecedented profits, users were satisfied with the convenience of purchasing goods via the Internet. The only problem was the lack of a proper legal framework for e-commerce at that time, which led to a large amount of speculation in this market, and as a result, its collapse in 2000. So, investor speculation pushed the value of web companies to unsustainable levels, and in 2000, the dot-com bubble burst. In many ways, this unstable situation and the collapse of the market contributed to the lack of legislative regulation of e-commerce at that time, due to the fact that at the state level, for example, in the United States, before the dot-com bubble burst, they simply did not think about how serious the consequences of improper legal regulation of e-commerce could be, taking into account the dynamics of its development. Today we have come to the conclusion that the e-commerce market has changed significantly, along with the changes in the e-commerce market, the attitude towards it has also changed, both on the part of buyers and on the part of states. Some people or businesses prefer to make online purchases for convenience, others because of the lowest competitive price offered by e-commerce platforms. Businesses and citizens using e-commerce are affected by various e-commerce trends[1] (Fig. 1).

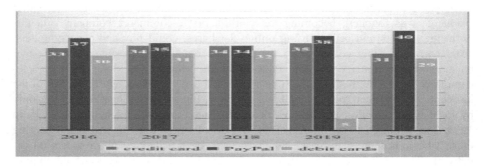

Fig. 1. Dynamics of the use of various means of payment in e-commerce 2016–2020

As can be seen from the data presented in Fig. 1, PayPal continues to be the preferred payment method in e-commerce over the past few years, with the traditional credit card taking second place, followed by debit cards. The range of Internet-connected devices available for e-commerce purchases is diverse, but nevertheless, statistics allow us to identify the main devices from which purchases are made, preferred by users[2] (Fig. 2).

Now we focus on how the growth of digital buyers around the world is projected[3] (Fig. 3).

As can be seen from the data presented in Fig. 3, more than 2.14 billion people worldwide are expected to buy goods and services online in 2021, compared to 1.7

[1] Electronic resource: Access mode: https://www.statista.com/statistics/251666/number-of-digital-buyers-worldwide/ (accessed 02.04.2021).

[2] Electronic resource: Access mode: https://www.statista.com/statistics/251666/number-of-digital-buyers-worldwide/ (accessed 02.04.2021).

[3] The statistics from the report Number of digital buyers worldwide from 2014 to 2021.
https://www.statista.com/statistics/251666/number-of-digital-buyers-worldwide/ (accessed 02.04.2021).

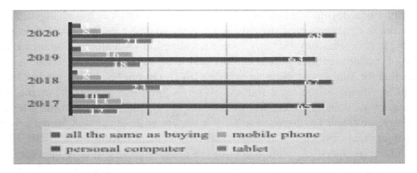

Fig. 2. Dynamics of user preferences in terms of using the type of devices for making purchases

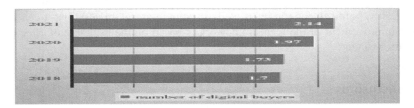

Fig. 3. Forecast of the number of digital buyers worldwide from 2018 to 2021 (billion a person)

billion shoppers who made purchases in 2018. The most distinctive characteristic of the contracts offered to users is that they are essentially accession contracts. The essential features of such contracts are that they cannot be considered the result of equal negotiations and that one of its parties must simply adhere to the terms proposed by the other party. In e-commerce, the contractual intention of one of the parties – the buyer-is only a more or less voluntary submission to the conditions dictated by the other party-the platform or the seller. And when the platform or the seller draws up a draft contract, the inconsistent terms may be unilateral. Due to an irrational process in e-commerce contracts, which is assumed to be equally enjoyed by both parties, it turns out to be directed in one direction – the platform or the seller. In other words, there is a kind of unilateral autonomy. Allowing the choice of legal provisions resulting from unilateral autonomy exacerbates the already unfair distribution of opportunities and reduces the freedom of the parties to the contract. All this leads to a situation in which the buyer in the process of electronic commerce, joining the contract, has no choice but to adhere to the conditions to which it may otherwise not agree, and at the same time will already be bound by conditions about which it may not know anything at all. The logic is that anyone who refrains from reading the contract and, in conscious ignorance of its terms, voluntarily agrees to it, will not be exempt from the consequences of the concluded transaction. The next question, which indirectly follows from the above question, is the question of the method of establishing and executing contractual relations. This is due to the fact that the use of artificial intelligence in e-commerce is constantly expanding. Equally important are the social consequences of the rapid technological progress of the last two decades-the content of information society services has changed. Despite the fact that e-commerce is constantly improving, adapting to changes in science and technology, it is necessary that any future act aimed at regulating e-commerce in the context of modern realities should comply with existing legal schemes. This is due to the fact that the subjects of the e-commerce market will not be able to quickly adapt to a complete change in the legal framework for e-commerce [15, 16]. For the implementation of e-commerce rules have determined the existing order, according to which a legislative structure has been created for a society without internal borders, without any geographical obstacles. This was especially evident during the COVID-19 pandemic. The importance of e-commerce for the entire market will only increase, as the opportunities offered by the offline world will continue to be limited for a long period of time, which is difficult to foresee at the moment. With regard to consumer protection, the legal framework for e-commerce should be developed in such a way that it can be considered as a global standard for online spaces, based on a balance between human rights, political, economic, as well as social interests. In addition, it is necessary that the improvement of the legal framework is based on the principle of effective and rapid enforcement.

3 Conclusions

Summing up, it should be noted that the improvement of the legal framework of e-commerce, aimed at solving the existing market problems in this direction, is seen as very relevant. This is due to several reasons. First, the global changes taking place in the field of e-commerce. Secondly, the need to protect the rights of the weakest participants in the relationship from unfair contractual conditions. At the same time, the modernization of the legal framework should not occur spontaneously and globally, that is, a rapid and complete change in legislation. The main thing is to create such a framework that will be adapted on the one hand to the changes in the market, and on the other hand, will eliminate the failures of the e-commerce market. Based on this, we can identify the prerequisites for improving the legal framework of e-commerce: First, in order to achieve the goals of effective regulation, it is necessary that all changes comply with existing legal schemes. Secondly, given the fact that e-commerce is mostly focused not so much on the relationship between the business, but on the relationship between the business and the person, it is necessary that the legal framework is focused on the person. Third, there must be a balance between human rights and political, economic and social interests.

References

1. Alekseev, P.V.: Problems of digital trade development in the EAEU member states. **2**, 26–32 (2020)
2. Barsova, U.V., Sinenko, O.A., Sitchikhina, V.E.: The practice of state regulation of e-commerce in the countries of the Asia-Pacific region. Education. Right. **1**(46), 111–116 (2019)

3. Demirchyan, V.V.: Problemy pravovogo regulirovaniya mezhdunarodnoy e-commerce [Problems of legal regulation of international e-commerce], **11**, 128–130 (2016)
4. Dolya, A.A., Odintsova, T.M.: Internet-trade: questions of the current legislation and the Internet marketing complex. Actual Issues Acc. Manage. Inf. Econ. **1**, 261–265 (2019)
5. Ivanova, N.I., Vlezkova, V.I.: Cross-border electronic trade: world experience and Russian practice. Probl. Enterp. Dev. Theor. Pract. 1–2, 175–178 (2019)
6. Klimchenya, L.S.: E-commerce: Textbook. Mn. Vysh. shk. 191 (2004)
7. Kobelev, O.A.: E-commerce: Textbook/Edited by Pirogov, S.V. (ed.) 3rd ed., reprint. and add. M.: Publishing and trading Corporation "Dashkov, K.", p. 684 (2008)
8. Kozie, D.: E-commerce/Trans. from English. M.: Publishing and trading house "Russian Edition", p. 2 (1999)
9. Kozinets, N.V.: Evolution of legal regulation of cross-border e-commerce in the US legislation. Actual Probl. Russ. Law, **9**, 195–202 (2015)
10. Pimenova, O.V.: Problems of customs administration in cross-border electronic trade and the role of interaction between customs and authorized operators in their solution, **12**(139), 150–153 (2019). ISSN 2073–4506
11. Pirogov, S.V.: E-commerce: Textbook/Edited by Pirogov, S.V.M.: Publishing House "Social Relations", Publishing House "Perspektiva", p. 428 (2003)
12. Slepchenko, E.A.: The state and prospects of development of legal regulation of electronic commerce in Russia, **6**, 193–195 (2019)
13. Strelets, I.A., Chebanov, S.V.: Digitalization of world trade: scales, forms, consequences. World Econ. Int. Relat. **64**(1), 15–25 (2020)
14. Tuzhilova-Ordanskaya, E.V.: Grazhdansko-pravovoe regulirovanie distantsionnogo kontrakta [Civil-legal regulation of a remote contract]. Vestnik Volgogradskogo gosudarstvennogo universiteta. Ser.5: Yurisprudentsiya, **1**(1b), 244–24b (2012)
15. Chebotareva, A.A.: Electronic commerce: strategic directions in the development of the state and problems of legal regulation. Bull. Acad. Law Manage. **2**(55), 51–58 (2019). Number 2074–9201
16. Shaidullina, V.: International legal aspects of conducting electronic commerce. Ekonomika. Taxes. Right, **4**, 157–166 (2020)
17. Eymor, D.: Electronic business. Evolution and/or revolution. Williams, Moscow, p. 320 (2001)
18. Yurasov, A.V.: Fundamentals of e-commerce. Moscow, 2007, p. 38, Schneider G. E-commerce. Boston (2008)
19. Gary, P.: E-Commerce, Ninth Edition, Schneider, Ph. D., CPA. Printed in the United States of America, p. 4 (2011)
20. Bartlett, J.: The Dark Web: Inside the Digital Underworld Paperback. Melville House, p. 320 (2016)

The New Trend of Media Development in the Context of the Chinese Dream-Media Convergence

Bokun Zhu[⊠]

Peoples' Friendship University of Russia, Moscow 10100, Russia

Abstract. Driven by new media, technologies, media convergence has gradually become an important development direction for the reform of the media industry. In 2012, the imagination of the Chinese Dream was put forward. Under this framework, Chinese media has evolved from traditional media as the main battlefield in the past to being gradually influenced by new media, and now it has developed to the gradual integration of traditional media and new media. Under the influence of new media, how traditional media and new media will integrate and develop in the future is an important topic of this article. This article analyzes the problems and challenges in the operation mode, media literacy, and supervision system faced by the integration of traditional media and new media, explores the development trend of integration of traditional media and new media, and provides reference for the development of media integration. More importantly, in this article, the author calculated the lifestyle and media usage based on the Chinese population. And used the model to measure and predict the trend of new media usage.

Keywords: Media convergence · New media · Traditional media · New media model

1 Introduction

With the development of network digital technology in the era of new media, new media has gradually become the main way for people to obtain information resources due to its efficient transmission speed and the advantage of information resources on the sea. The emergence of new media has changed the channels and ways of information transmission, making the interactivity of social information transmission gradually enhanced. In the development of TV, newspapers and other traditional media, they also gradually adapt to the changes of the new media environment through reform, giving birth to the integration of traditional media and new media. Since the proposal of the Chinese Dream in 2012, Chinese media has undergone major changes in all aspects. New media has developed rapidly in China, and traditional media are constantly being combined with new media. Make full use of new technologies and applications, innovate new media, communication methods, and accelerate the integration of new media and traditional media. Traditional media and new media have not only realized cooperation, but also to achieve the true integration of "There was a bit of me in you and a bit of you in me".

© The Author(s), under exclusive license to Springer Nature Singapore Pte Ltd. 2022
J. C. Hung et al. (Eds.): FC 2021, LNEE 827, pp. 846–852, 2022.
https://doi.org/10.1007/978-981-16-8052-6_107

The topic center of this paper is "media convergence", and "media convergence" mainly refers to the convergence of traditional media and new media. Since 2013, media convergence is still in the stage of rapid development. There are many research literatures on media convergence, and the main research directions are as follows: 1) The concept of media convergence [1, 2]; 2) The development of media convergence [3, 4]. Although the research literature on this topic has made a preliminary exploration of media convergence, there is still a lack of comprehensive and systematic information integration and analysis. On the basis of previous studies, this paper focuses on the analysis of the characteristics of integration of traditional media and new media in China, and takes the digital TV media "Mango TV" founded by Hunan Satellite TV as an example to elaborate the concept of media integration in a more specific way. In China, traditional media once occupied the position of official mainstream media, and TV media even covered the whole China. Most ordinary people relied on the official information of TV media and trusted traditional media very much. Since the rapid development of new media in China, including the global popularity of TikTok, a new media platform developed by Chinese companies, more and more people have changed their access to information from reading newspapers, listening to the radio and watching TV to surfing the Internet. Therefore, the integration of traditional media and new media in China is an inevitable result. The significance of the research and analysis of the fusion characteristics of traditional media and new media in China lies in making research examples for the fusion of media in other countries, and providing more mature experience for other media through the analysis of specific examples.

2 Materials and Methods

In China, new media under the prompting of rapid development, China's traditional media and new media integration to media from the media fusion depth fusion, from the merger of media cooperation media fusion, from digital to present to all media, from the traditional media " +" to "three micro end" is gradually realize the objectives of the reform of all media. Now the basic characteristics of Chinese media convergence mainly show three basic forms: content convergence, network convergence and terminal convergence. From the perspective of content convergence, media convergence mainly forms multi-level and multi-type media products through the use of cross-media and cross-platform in the production of media content. From the perspective of network convergence, it is the development trend of the convergence of telecommunications network, Internet and radio network, aiming at creating an integrated network transmission information system. From the point of view of terminal fusion, it is mainly the three-screen fusion of TV screen, computer screen and mobile phone screen. In this trend of content, network and terminal integration, the integration process of traditional media and new media will continue to accelerate.

At present, media convergence has the following characteristics:

1) *Technical* leadership

Under the background of continuous development of digital technology, WeChat technology and multimedia technology, many new forms of business in media industry have been born. For example, the development of IP technology enables the integration of wired network and wireless network. In the process of traditional media and new media, technology is taken as the guide to drive the change of relevant media industry, so as to promote the innovation and development of the media industry. Hunan Satellite TV used to broadcast mainly on TV, but now the technology is fully mature, thus creating Mango TV [5, 6].

2) *Media Content*

In the context of the continuous development of media convergence, the types of content resources produced and produced by media also show a trend of diversified development. In the era of great information technology, the use of a variety of means for communication is the only way for the integration of traditional media and new media. For example, China's People's Daily has changed from paper to electronic newspaper, mobile newspaper and other new media products.

3) *Make your* structure *systematic*

Media convergence is a systematic process that continues to deepen and expand. Whether it is technology integration, content integration or channel integration, the development dimensions of the integration all constitute a multi-dimensional development system.

Through the above analysis of media convergence, it can be concluded that the integration development of traditional media and new media is still facing many problems and challenges [7–10].

1) *The big* difference *in operation modes*

The operation mode of traditional media is mostly a combination of traditional enterprise organization and enterprise management mode, and the production process is relatively standardized and rigid. An article needs "three schools and three audits", which also has great credibility. However, new media makes everyone a disseminator of information, which makes it difficult to control low authenticity of information. This is a big problem facing media convergence.

2) *Media* literacy *is uneven*

Media literacy refers to the professionalism of media workers. The average knowledge level of traditional media practitioners is high, and traditional media enterprises also pay great attention to the cultivation of professional quality of personnel. However, now the entry threshold of new media practitioners is low, and the content and forms are diverse. It is precisely because of the great difference in the professional level of communicators that the quality of news and information has a great level of difference.

3) *Internal* supervision *is not perfect*

In the current media products formed by the integration of traditional media and new media, due to a lack of corresponding internal and external regulatory system, the quality of media convergence products are different to some extent. From the inside of the media, the information of traditional media needs to go through multiple processing and verification before it can be broadcast, but in the new media environment, information products are more based on the operation of individuals and related teams, and internal supervision and check link is weak. Then media convergence products in the regulatory system will have problems.

3 Conclusions

In the development of traditional media and new media, there are several ways to solve the problems mentioned above.

1) Build an integrated information platform to optimize and organize information
To strengthen the smooth integration of traditional media and new media, it is necessary to build a relatively perfect information platform in the integration process, so as to promote the development process of media convergence in the platform construction. First of all, the massive information resources are optimized and sorted based on the technology application of big data and cloud computing. For example, WeChat H5 products realize the high-quality transmission of information resources through new media formats.

2) Strengthen public opinion guidance to ensure the credibility of the media
With the emergence and growth of new media products, traditional media and new media can complement each other's advantages and better promote the development process of media convergence. In the process of content integration, traditional media should change the former lofty official language, produce more interesting content to guide the direction of public opinion on the basis of ensuring public credibility, and give full play to the influence of opinion leaders in traditional media and new media environment.

3) The internal and external supervision system and strengthening media control
Strengthening the supervision of information environment is the development trend of media convergence. Especially in the face of rumors and malicious marketing in the network environment generated by the network group gathering behavior, it is more necessary to achieve public opinion guidance and network space purification through the supervision of media and related organizations. On the one hand, comprehensively improve the media literacy of employees. On the other hand, it is necessary to build a sound media supervision system on the basis of media convergence.

4) The author uses the model to measure and predict the trend of new media usage.

To answer the third research question, seven logistic regression analyses were performed to assess the relationship between demographics, lifestyle factors, the media use, and the adoption of these media technologies and services. Logistic regression has been selected over discriminant analysis its adaptation provides the best fit for analysis of the dichotomous outcome (dependent or response) variables as well as interval or ratio variables.28 The pseudo R2 value measures the improvement of the loglikelihood obtained through the introduction of independent variables in a stepwise manner. The results are shown in Table 1. The overall fitness statistics show that the models successfully predicted from 3.1 to 16.1% of the total variance in the probability of

Table 1. Predicting use of new media technologies by demographics, lifestyles, and media use in urban China (Logistic regression models)

Predictors	Karaoke bars	Video stores	Personal computer	VCR	Cable TV	Pagers	Cellular phone
Block1:Demographics							
Age	0.07^{**}	-0.05^{**}	-0.07^{**}	-0.02^{**}	0.01^{**}	-0.02^{**}	-0.00
Gender(Male)	0.32^{*}	0.50^{**}	0.66^{**}	-0.08	-0.25^{**}	0.41^{**}	0.20
Income	0.39^{**}	0.25^{**}	0.07	0.14^{**}	0.19^{**}	$0.08\#$	-0.09
Education	0.08	0.03	0.58^{**}	0.13^{**}	0.08	0.21^{**}	-0.10
Disposable income	0.22^{**}	0.01	0.33^{**}	0.04	0.04	$0.10\#$	0.58^{**}
Pseudo $R^2(\%)^b$	14.3^{**}	6.8^{**}	14.7^{**}	3.1^{**}	2.2^{**}	5.2^{**}	3.7^{**}
Block 2:Lifestyles							
Sophisticated and fashionable	0.03^{*}	0.06^{**}	0.01	$0.02\#$	-0.00	$0.02\#$	0.03
Life expansionists	0.03	-0.02	0.09^{**}	-0.00	0.01	0.03	0.06
Pleasure and enjoyable life	-0.03	-0.02	-0.01	0.04^{*}	0.06^{**}	$-0.03\#$	-0.08
Preference for foreign product	0.02	0.02	0.01	$0.05\#$	0.04	0.04	-0.03
Credibility of mass media	-0.04	-0.09^{**}	-0.06	-0.04	-0.03	-0.01	0.06
Change in pseudo $R^2(\%)$	1.3	1.1	0.6	0.5	0.5	0.5	0.8
Change in $x^2(df = 5)^c$	12.27^{*}	19.69^{**}	11.56^*	14.17^{*}	14.12^{**}	13.66^{*}	5.43
Block 3:Media Use							
Newspaper reading	0.02^{*}	0.01	0.01	0.02	$-0.01\#$	0.02^{**}	0.01
Magazine reading	-0.01	-0.02	0.04^{*}	0.01	-0.02	0.02	-0.03
Primetime TV watching	0.04^{*}	0.02	-0.01	$-0.03\#$	-0.00	0.00	-0.05
Change in pseudo $R^2(\%)$	0.4	0.3	0.5	0.2	0.1	0.5	0.7
Change in $X^2(df = 3)$	9.18^{*}	5.96	6.02	5.03	3.35	14.06^{**}	3.52
Final pseudo $R^2(\%)$	15.4	8.2	16.1	3.8	3.1	6.2	5.2

Notes: aFigures are standardized regression coefficients for variables entered. Dependent variables were measured by asking respondents to self-report if they currently own the new media technology at home. Data were coded with 1"yes and 0"no. bA pseudo R2 is expressed in percent of variance accounted for by the corresponding block. Pseudo R2 is defined as 1!(L1/L0), where L1 stands for the log likelihood of the theoretical model and L0 stands for the null model with only the constant in the equation. Pseudo R2 value measures the improvement of the log likelihood obtained through the introduction of independent variables in a stepwise manner. cX2 is used to test the statistical significance of the model's coefficients. It plays the role of the F test in a regression. The degrees of freedom used to test the null hypothesis are equal to the number of variables added to the constant. dp(0.1; * p(0.05; ** p(0.01; *** p(0.001; N"2,020

respondents adopting a new media technology in each of the seven adoption questions. First, when a total of five demographic variables were entered as the first block in the logistic regression equations to predict respondents' media technologies adoption patterns, this study found that educated young males with more disposable income seemed to be more likely to own personal computers, pagers, visit karaoke bars and video stores. Female respondents who were young and with a higher socioeconomic status tended to own VCRs and subscribe to cable TV. Not surprisingly, the amount of disposable income appeared to be the single strongest predicting factor for the adoption of cellular phones. Demographics explained a 2.2—14.7% variance in the adoption of these technologies. Second, lifestyle variables were entered next into the seven logistic regression equations. With the exception of cellular phones, all other new media technologies adoption patterns in urban China were explained by one or more lifestyle variables. Such findings validate the A.I.O. and COFREMCA concepts which suggest that lifestyles play an important role in consumption patterns. Cellular phone adoption was explained only by a demographic variable expendable income (Beta"0.58). This may be rationalized by the novelty, high cost, and the low adoption rate (3.6%) of the technology in urban China. The block of lifestyle variables contributed from 0.5% to 1.3% variance in explaining adoption behavior, after controlling for five demographic variables. Although small in explanatory power (change in pseudo R2), the X2 statistics showed the significant improvement in seven models' coefficients. Specifically, being seen as 'sophisticated and fashionable' appeared to be an important lifestyle, explaining why individuals enjoyed visiting karaoke bars and video stores, and to a less extent owning VCRs and pagers (p(0.1).

Through the analysis of the characteristics of the media convergence at the present stage, it is found that there are still problems in the process of media convergence, such as operation mode, professional quality of employees and supervision system. For these problems, the author describes the possible ways and forms of improvement as a whole. Therefore, it can be concluded that in the new era background, media convergence is an inevitable trend, but also the common development and prosperity of traditional media and new media. Radio, TV, the newspaper and other traditional media must actively embrace the Internet, relying on the Internet as a carrier, further play to the traditional advantages, in order to find a real way out of survival and development. Of course, there are still quite significant problems in the process of media convergence. Some problems, such as changes in the form and mode of transmission of content, are relatively easy to realize quickly. However, for example, it is a much longer process to cultivate the quality of employees and to establish and improve the supervision system.

References

1. Callahan, W.A.: China's "Asia Dream" the belt road initiative and the new regional order. Asian J. Comp. Polit. 1(3), 226–243 (2016)
2. Lu, G.: Theoretical connotation and practical significance. important discourse on media convergence. Educ. Media Res. Soc. Sci. 1, 6–9 (2021)
3. Guzman, E.O.: Interstory: a study of reader participation and networked narrative in media convergence (2013)
4. Peters, M.A.: The Chinese Dream: Educating the Future: An Educational Philosophy and Theory Chinese Educational Philosophy Reader, vol. VII. Routledge (2019)

5. Lin, J.: Analyze the integration development path of traditional media and new media under big data. Media Forum, Inf. Technol. 28–29 (2021)
6. Liu, P. (2020, April). Investigation on the Development Trend and Characteristics of Media Convergence Based on Big Data Analysis. In Journal of Physics: Conference Series (Vol. 1533, No. 2, p. 022130). IOP Publishing
7. Lan, P.: Trend of media convergence: from an aspect of digital news papers and digital magazines. J. Int. Commun. **7** (2006)
8. Yang, D.T.: The Pursuit of the Chinese Dream in America: Chinese Undergraduate Students at American Universities. Rowman & Littlefield (2015)
9. Yang, G., Calhoun, C.: Media, civil society, and the rise of a green public sphere in China. China Inf. **21**(2), 211–236 (2007)
10. Zhang, Y.: From media convergence to 'internet plus'. New Media Trans. Soc. Life China, 19–56 (2017)

Time Interval on Language Comprehension—Simulation Experiment Using Celeron 333 Microcomputer

Xue Chen[1(⊠)] and Shuaian Zu[2]

[1] Pushikin State Russian Language Institute, 101000 Moscow, Russia
[2] Financial University Under the Government of the Russian Federation, 101000 Moscow, Russia

Abstract. In the fields of cognitive linguistics and cognitive psychology, there are still many controversies about the discussion of information processing, the characteristics of the time dimension of time information in the construction of context models, the extraction of time information, and the updating of time context models (Project "The study of time Conceptualization in Russian idioms" supported by Chinese Academy of Management Sciences. The number of project: (KJCX9700). http://www.zhongguanyuan.org.cn). This article is mainly based on Anderson's scenarios account [1] and Zwaan's strong iconicity assumption as the theoretical basis, with the help of Celeron 333 microcomputer to conduct simulation experiments to try to explore the influence of time sequence and distance on text understanding.

Keywords: The concept of time · Cognitive linguistic · Celeron 333 microcomputer · Scenarios account · Strong iconicity assumption

1 Introduction

Time, the basic way in which everything exists in the world, is the most basic feature of human existence and communication. In our daily life, we often contact time information and can use it skillfully. For example, in Chinese idioms, there are the following words: (1) Time flies, the sun and the moon are like shuttles; (2) Time is the water in the sponge, it is still possible to squeeze it; (3) The young and strong don't work hard, when he is getting old, he will have a hard time etc. [2–6].

1.1 Basic Understanding of Time Concept

Many scholars have made profound philosophical speculations on time: Kant believes that time and space are two characteristics and classifications of human cognition. For time, people cannot be truly aware of it, and more are based on intuition. An internal processing process. Hedegger believes that time is not a known objective existence in the world, nor is it a subjective conception of human understanding of the world, but the boundaries of events that have meaning and value to individuals [7]. McTaggart has carried out the types of time representations. Divided, time has two chronological sequences: "first-last" sequence and "past-present-future" sequence.

© The Author(s), under exclusive license to Springer Nature Singapore Pte Ltd. 2022
J. C. Hung et al. (Eds.): FC 2021, LNEE 827, pp. 853–859, 2022.
https://doi.org/10.1007/978-981-16-8052-6_108

Among them, the "first-last" sequence characterizes time, state and process with temporal continuity, while the "past-present-future" sequence needs to use the specific time angle of "now" as a reference to characterize time Make a distinction. A famous saying by Heraclitus: "Man cannot step into the same river twice" [8–12], this is a dialectical argument that time is like other things in the world, in constant change and development.

In the fields of cognitive linguistics and cognitive psychology, many scholars have studied the continuity of time from the perspective of individuals' subjective perception of time. The research in this area originated from the experiment of human perception and the analysis of text from the perspective of the combination of cognitive linguistics and linguistic culture.Researchers try to find the smallest unit of human time perception in this way. Specifically, it is the shortest time interval perceivable by the various receptors of the individual and the idioms of different nationalities in the world. The concept of time in literary works Analyze the differences in expressions to find the minimum time span and explore people's experience of the time attribute of "this moment".Among them, Poeppel (1988) found through a series of experiments that the shortest time range that people perceive in perceptual experience is 30–40 ms. Some researchers have further divided the different sensory channels and found that people are auditory. The shortest time that can be perceived is about 10 ms, it is about 25 ms in touch, and 100–200 ms in vision.

Individuals' subjective judgments of duration often have a clear contrast with the duration of objective time. This is because individuals' subjective judgments of duration are easily affected by some factors. For example: (1) The physical properties of the stimulus. Weak, familiar, and simple stimuli will have shorter time awareness, and vice versa. (2) The content and nature of time. The more time that occurs in a period of time, the more interesting the event will be. The individual The shorter the time perception and estimation will be. (3) Individual emotional state. In a happy, joyful emotional state, the shorter the perception and estimation of time, on the contrary, the longer. (4) Different nationalities Cultures have different understandings of the nature of the concept of time.

1.2   Differences in Time Structure Between Colonies

There are also differences between colonies in the structure of the concept of time. One of the most obvious influencing factors is social culture. Generally speaking, in all human cultures in the world, there are concepts about ancestors or ancestors, and based on this As a connection between the present and the past, the resulting concept of time appears to be different between different social cultures. For example, in China, Japan, Russia and Egypt, people will connect the present with a chain linear structure passed down from generation to generation Unlike the past, people in this cultural background also have a linear concept of time; in Greek and Indian cultures, there is no such obvious chain linear concept of time. Therefore, there are also many cognitive psychologists discussing the problem of time coding in the process of language or text reading from the perspective of language. The reasons are: First of all, regardless of the social and cultural background, people in the social environment will understand and understand through language. To express time, this process also expresses the continuity of time in the form of language. Secondly, almost all human languages have developed a variety of language components about time, such as: in English and Russian, through the form of verbs (Past tense, present tense, future tense) to express time, especially in Russian, there is a special kind of vocabulary that expresses the concept of time, which is the Adjective verbs. In Chinese, people use time adverbials to express time.

At the same time, people often use some time vocabulary or grammar to understand the nature of the event (past, present, future), the duration of the event (instant, short time, medium length time, long time) and frequency of event occurrence (Often, occasionally, never) express, and use this information to convey the sequence of events either explicitly or implicitly.

2 Experiments on the Influence of Time Sequence Relationship on Language and Text Comprehension Through Celeron 333 Microcomputer

The time information of the language allows readers to calibrate the time in the context model described in the article. A context model is a mental model that uses existing information to describe the situation, and is used to integrate the knowledge from the article and the reader's long-term memory. (Long) Time memory is denoted by LTM below). A considerable amount of existing data indicates that the understanding of the article involves the construction of the context model, and some studies also show that the context model affects the organization of information and its extraction from LTM.

The research on the time information in the context model mainly focuses on two issues: (1) How does the change of time prompt readers to update the current model or switch to a new model; (2) How does the organizational aspect of time affect the integration of information into a single Situational model.

The results show that the items arranged in the order of time (such as after heating-warm) are faster and more accurate than the items arranged in the reverse order of time (such as before heating-cold). Kintsch discusses In view of the influence of knowledge on the construction of text-based context models, he believes that the connections between the various elements of the text are usually non-directional.

Another study by van der Meer found that in high-frequency events, there is no difference in people's reaction time for very short presentation times, whether it is not related to the order of the text, or whether it is a sequential or reverse sequential text.; For medium-length presentation time, the reaction time of sequential text is significantly shorter than that of reverse sequential text and the reaction time not related to the text.

2.1 Use Celeron 333 Microcomputer to Simulate the Influence of Different Time Intervals on Language Text—Take Chinese Text with Time Conjunctions as an Example

In this part, the author will explore whether the experimental participants all use time information to construct a situational model with time conjunctions, and whether they all prioritize the processing of future-oriented events, and on this basis, further explore the impact of different time intervals on language The impact of understanding, that time interval is more suitable for readers to judge, understand, remember, whether to use time information to judge or semantics itself, or both.

The purpose of this experiment is mainly to verify whether the items arranged in the forward order are arranged in the reverse order and the text processing is faster and more accurate, and on this basis, the time correlation variable is added to explore whether the text processing that is not to be controlled It is faster and more accurate than processing related items.

The expectation of this experiment is: because texts that are not related to time are not affected by time factors, readers should process such texts in a shorter time and more accurate. Since pro-chronological and inverse-chronological are related in time, time correlation and time-sequence are combined into one variable, namely time-sequential correlation, which includes three levels of pro-chronological, inverse-chronological, and uncorrelated.

Test Subject: 130 senior high school students in Shanxian County's fifth high school.

Experimental Text Material: The selected corpus is composed of 30 sequences of four events (A, B, C and D, such as waiting, boarding, flying and disembarking). From the first event A or the last event of the sequence D starts, and combined with the second event in the sequence, this forms a sequential text (A-B: such as waiting for a plane-boarding) and a reverse sequential text (D-C: such as getting off a plane-taking a plane). Use the following four standards to represent the seven equivalent levels to test the experimental materials. The results show: (1) Time correlation: $X2$ (1, $N = 30$) = 9.677, P > 0.893; (2) Typical text Property: $X2$ (1, N = 30) = 7.8, P > 0.997; (3) Time duration: $X2$ (1, N = 30) = 6.0, P = 2.00; (4) Time distance of the event: $X2$ (1, N = 30) = 4.0, P = 2.00. The $X2$ values of these four standards are far less than the value that reaches the significance level, indicating that the experimental materials are homogeneous.The experimental materials consist of three texts, each of which contains 10 exercises and 30 test items. Each text is composed of previous information and goals. Here, the previous information contains two components: the temporal conjunction "before" or " After" and an event (for example, waiting for an airplane); the target is an event (for example: getting on an airplane). The time sequence between the previous information and the target or the sequence of events in reality is the same (for example: in an airplane After getting off the plane), these texts are called sequential texts; Or it is the opposite of the sequence of events, and these items are called reverse-chronological texts (e.g., waiting for the plane before getting on the plane); or they are not related to the chronological sequence, and these texts are called texts that are not related to the time factor. Before getting on the plane—germinate). These items all contain 7 words. The previous information consists of 3 words of time conjunctions and 2 words of events. The target consists of 2 words of events. The number and composition are strictly controlled.

Experimental Procedure: The experiment was carried out on a Celeron 333 micro-computer. At the beginning of each experiment, a red gaze point "-" appeared in the center of the display, and the participants were asked to always stare at this red dot during the experiment, and then it was presented The previous information is presented at an interval of 30 ms or 50 ms or 100 ms, and then the target is presented. The experiment participants are required to accurately determine whether the target time accurately describes the context characterized by the previous information, that is, the target event and the target event as quickly as possible. Whether the event described in the previous information is relevant. If relevant, press the E button; if not, press the O button. After the participant presses the button, the next text is presented after an interval of 1500 ms.

Experimental design: This experiment uses a 3*3 completely within-subject design. The first independent variable is the time sequence relationship, including three levels of pro-chronological, inverse-chronological, and uncorrelated with time factors; the second independent variable is The time interval includes three levels of 200 ms, 500 ms and 1000 ms. The dependent variables are the response time to the target word and the error rate. 1/3 of the text is presented with 200 ms of SOA (text composition method 1), 1/3 of the The project is presented in 500 ms of SOA (text composition method 2), and 1/3 of the text is presented in 1000 ms of SOA (text composition method 3). Each text composition arrangement includes 1/4 of the sequential text, 1/4 inverse temporal texts and 1/2 of the texts that are not related to the time factor. Therefore, all experimental participants have to complete the reaction time test of the nine conditional texts. In the three text composition methods, each text arrangement method The order in which the text is presented is randomly arranged.

Experimental material samples (Table 1).

Table 1. The experimental sample

Chronological relationship	Previous information	The goal
Reverse timing	Before boarding the plane	Waiting
Irrelevant	Before boarding the plane	Germination
Sequential	After flying	Get off a plane
Irrelevant	After flying	Know

3 Conclusions

In order to ensure the reliability of the data, the author deleted the extreme data beyond 3 standard deviations of the average reaction time. Therefore, the experimental data of 3 experimental participants were deleted here, and the valid data of 122 experimental participants were retained. The average reaction time and error rate of this experiment under various conditions are shown in Table 2.

Table 2. Average response time (ms) and error rate under various conditions

Time	Sequential		Reverse timing		Irrelevant	
Time interval	Reaction time	Error rate	Reaction time	Error rate	Reaction time	Error rate
200	1081	18.18	922	13.23	856	10.45
500	994	14.24	939	6.36	896	13.18
1000	896	2.73	780	8.41	773	3.86

Repeated measurement analysis of variance was performed on the reaction time and error rate in Table 2, and it was found that:

(1) In terms of reaction time, the main effect of time sequence correlation is significant, $F(2, 21) = 31.849$, $p < 0.001$. The results of multiple comparisons show that readers have faster processing time for sequential text than reverse sequential text. And there is a significant difference, $P < 0.001$. This shows that the sequence has a positive impact on the cognition of the correlation between daily events, and the participants of the experiment give priority to processing future-oriented events. (2)The main effect of the time interval is significant, $F(2, 21) = 12.709$, $P < 0.001$. The multiple comparison results found that the experimental participants reacted faster under 1000 ms of SOA than under 200 ms of SOA. An extremely significant difference, $P < 0.01$; and the processing effect of text under 500 ms SOA and 200 ms SOA is not good enough and insufficient. (3) In terms of error rate, the main effect of time sequence correlation is not obvious, $F(2, 21) = 1.562$, $P > 0.05$. But from the average in Table 2, it can be seen that the processing accuracy of irrelevant text is better than The processing accuracy of the related text is high, and the processing accuracy of the reader's text in the sequence is also higher than the text processing in the reverse sequence, but the two have not reached the level of significant difference. (4)The interaction between time sequence correlation and time interval is very significant. $F(4, 21) = 5.576$, $P < 0.001$. The simple effect test results found that under 200 ms SOA, the sequential text and the irrelevant text are more inverse than the inverse time sequence. The accuracy of text reader processing is higher, and has reached the level of significant difference, $P < 0.05$; under the condition of 500 ms of SOA, the error rate of sequential text is lower than that of reverse sequential text, and has reached the level of significant difference, $P < 0.05$; Under the condition of SOA of 1000 ms, the error rate of irrelevant text is lower than that of the sequential sequence, and reaches a significant difference level, $P < 0.05$. This experiment proves that people use time information to construct situational modeling and tend to prioritize events with future orientation.

Acknowledgements. First and foremost, I would like to show my deepest gratitude to our supervisor, Dr. Vladimir I Karasik. He is respectable, responsible and resourceful scholar. Without his enlightening instruction, impressive kindness and patience, I could not have completed my thesis. Finally, I want extend my thanks to the China Academy of Management Science for providing support and assistance to my thesis research. Project "The study of time Conceptualization in Russian idioms" supported by Chinese Academy of Management Sciences. The number of project :(KJCX9700). http://www.zhongguanyuan.org.cn.

References

1. Anderson, A., Garrod, S.C., Sanford, A.J.: The accessibility of pronominal antecedents as a function of episode shifts in narrative text. Q. J. Exp. Psychol. **35**(3), 427–440 (1983)
2. Bestgen, Y., Vonk, W.: Temporal adverbials as segmentation markers in discourse comprehension. J. Mem. Lang. **42**(1), 74–87 (2000)
3. Bestgen, Y., Costermans, J.: Temporal markers of narrative structure: studies in production. In: Processing Interclausal Relationships. Studies in the Production and Comprehension of Text, pp. 201–218 (1997)
4. Brent, M.R., Cartwright, T.A.: Distributional regularity and Phonotactic constraints are useful for segmentation. Cognition **61**(1–2), 93–125 (1996)
5. Freyd, J.J.: Dynamic mental representations. Psychol. Rev. **94**(4), 427 (1987)
6. Gernsbacher, M.A.: Attenuating interference during comprehension: the role of suppression (1997)
7. Heidegger, M.: Only a God can save us: der Spiegel interview with Martin Hedegger (1966). The Heidegger Controversy: A Critical Reader, pp. 91–116 (1993)
8. Heraclitus: The cosmic fragments. Cambridge University Press (1954)
9. Karasik, V.I., Gillespie, D.: Discourse personality types. Procedia Soc. Behav. Sci. **154**, 23–29 (2014)
10. Karasik, V.I.: Linguistic conceptualization of pride in Russian and English cultures. Russ. Lang. Stud. **16**(2), 157–171 (2018)
11. Karasik, V.I.: Language circle: personality, concepts. Discourse, pp. 21–27 (2002)
12. Landeros-Dugourd, E.: Quasi-Experimental Study: DCog and Travel Autonomy for Young Adults with Cognitive Disabilities. Capella University (2011)

Multi-level Wavelet Packet Prediction for Image Super-Resolution

Yunlong Ding[✉]

School of Mathematical Sciences, Beihang University, Beijing, China
dingyunlong@buaa.edu.cn

Abstract. Convolutional neural network (CNN) has made great progress on image super-resolution (SR) tasks. However, most SR models work on spatial domain and try to reconstruct context pixels. In the paper, we present a multi-level wavelet packet CNN (MWPCNN), which enhances the performance of CNN in SR tasks through residual-learning, in which residual blocks and wavelet residuals are used in the middle layer and last layer to predict wavelet coefficients. With the modified Deep Wavelet Super-Resolution (DWSR) architecture, wavelet packet transform is introduced to mine the advantages of wavelet packet transform, which makes full use of knowledge embedded in different levels from the wavelet spectrum. A key benefit of such an introduction is that it has an active influence on the image reconstruction performance, especially for reconstructing more detailed image content. Experiments show that MWPCNN is effective for SR tasks and produces better visual results than state-of-the-art methods.

Keywords: Super-resolution · Wavelet packet transform · Residual-learning

1 Introduction

Recovering a high-resolution image from a low-resolution image is denoted as Single Image Super Resolution (SISR). SISR is widely used for various tasks ranging from security and medical imaging to image generation. Early methods are mostly example-based [1–3] or interpolation-based [4]. Currently, convolutional neural network (CNN) has achieved remarkable progress on SISR problem because of its powerful generalization capabilities. Traditionally, applying CNN to resolve SISR problem by designing deeper network [10] or using more complex architecture [7]. However, such networks and techniques are limited by heavy computation costs. Additionally, network input is heavily dependent on the spatial domain, which lacks consideration of image structural information.

To solve the input data problem, many methods take wavelet domain into consideration and combine wavelet transform with CNN for processing SISR problem. Among them, Bae et al. [5] propose a novel residual learning algorithm, in which the input and label are wavelet sub-bands. Similarly, Guo et al. [6] utilize residual-learning to restore missing wavelet sub-bands in the low-resolution images. However, both of them only utilize 1-level wavelet packet decomposition, which leads to a question: How to reconstruct more detailed image content? Inspired by DWSR [6] and MWCNN

© The Author(s), under exclusive license to Springer Nature Singapore Pte Ltd. 2022
J. C. Hung et al. (Eds.): FC 2021, LNEE 827, pp. 860–866, 2022.
https://doi.org/10.1007/978-981-16-8052-6_109

[7], we recognize that taking the four sub-bands generated by the 1-level wavelet decomposition as input can improve the image reconstruction performance, whether taking the sub-bands generated by the 2-level wavelet decomposition as input can improve the restoration performance of every sub-band from the 1-level wavelet decomposition. Moreover, different levels include different image information, ignoring the knowledge embedded in different levels from the wavelet spectrum will have an adverse influence on the image reconstruction performance, especially for reconstructing more detailed image content.

To reconstruct more detailed image content, multi-level wavelet packet transform (WPT) is embedded in the CNN learning process. WPT scatters the original image information to different levels. If the input utilizes only 1-level WPT, which may be harmful to the reconstruction performance. Therefore, a multi-level WPT is introduced to supplement information loss of 1-level wavelet decomposition as network input. Considering that residual net [8] has shown strong ability in reducing training time, we use residual-learning to speed up the training process. In addition, wavelet coefficients decompose the image into sub-bands which provide structural information which benefits CNN learning.

In summary, the main contributions of this paper lie in three respects: (1) we combine multi-level WPT with CNN to enhance performance (2) Promising ability to predict image more detailed information. (3) We provide the State-of-the-art PSNR and SSIM results on SISR tasks.

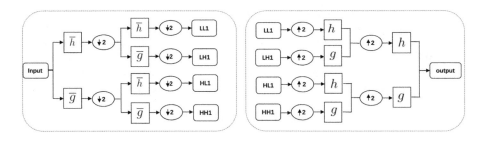

Fig. 1. The procedure of 1-level 2dWPT decomposition and reconstruction

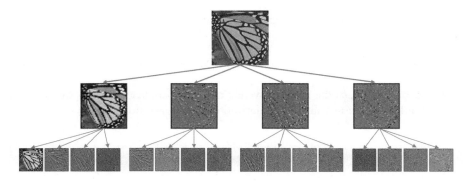

Fig. 2. The procedure of 2-level 2dWPT decomposition.

2 Method

We first review the multi-level wavelet packet transform, then introduce the MWPCNN architecture that can take advantages of good time-frequency localization properties from the multi-level wavelet packet transform.

2.1 2D Wavelet Packet Transformation (2dWPT)

The 1-level wavelet packet transform decomposes the image into discrete wavelet representation. As shown in Fig. 1, if Input represents image, the left part shows the procedure of 1-level 2dWPT decomposition, where h and g represents the low pass and high pass filter. The LL1, LH1, HL1 and HH1 are sub-bands containing average, vertical, horizontal and diagonal detailed information, respectively. After obtaining four sub-bands, we can reconstruct output by the 2d inverse WPT (2dIWPT) in right part of Fig. 1.

2.2 Training Procedure

The sub-bands at different levels reveal variances of the image signal on different levels. The decomposition provides good information for image restoration, which is difficult to learn from an end-to-end neural network. When we take wavelet sub-bands image as network input, SR can be regarded as a problem of restoring the image details given the input LR image. However, most of networks based wavelet consider only 1-level wavelet decomposition, ignoring the knowledge embedded in different levels may be harmful to the image restoration performance due to information loss caused by the 1-level wavelet decomposition. As shown in Fig. 2, the second row represents the 1-level 2dWPT results, the third row represents the 2-level 2dWPT results. To supplement information loss that 1-level wavelet decomposition as network input, we consider the 2-level 2dWPT with haar wavelet to produce wavelet sub-bands:

$$2dWPT\{LR, Level = 2\} \tag{1}$$

The label images use the same way to obtain the following:

$$2dWPT\{HR, Level = 2\} \tag{2}$$

As the input and output images are largely similar, the optimized goal (residual) is denoted as:

$$GR = 2dWPT\{HR, Level = 2\} - 2dWPT\{LR, Level = 2\} \tag{3}$$

We denote by (W, b) the network weight and bias parameters, $f(\cdot)$ is the network output. The loss function with L2 regularization is then given by

$$Loss = \frac{1}{2}||GR - f(\cdot)||_2^2 + \lambda||W||_2^2 \tag{4}$$

2.3   Network Structure

The full network structure is illustrated in Fig. 3. In training process, there are 64 filters of size $16 \times 3 \times 3$ in the first layer and 16 filters of size $64 \times 3 \times 3$ in the last layer. In the middle hidden layers, each layer is composed of convolution kernel with $64 \times 3 \times 3$ filters and rectified linear unit operations. In the end, we adopt the Adam optimizer [14] to predict every sub-band differences (residuals) by minimizing the loss function.

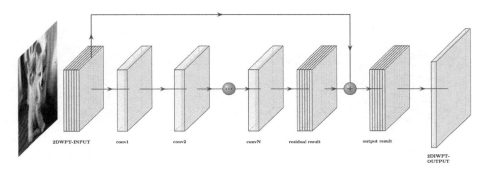

Fig. 3. Deep wavelet packet CNN (DWPCNN) network structure.

2.4   Generating Final Result Image

To generate final HR image, the input image is decomposed by 2dWPT, which produces the wavelet sub-bands as Eq. (1). Then the trained network takes the wavelet sub-bands as input to generate the corresponding residual. Finally, wavelet sub-bands and residual generate corresponding HR wavelet sub-bands (FRSB):

$$FRSB = f(\cdot) + 2dWPT\{LR, Level = 2\} \tag{5}$$

The 2dIWPT generates the final results (FR) is denoted as:

$$FR = 2dIWPT\{FRSB, Level = 2\} \tag{6}$$

3   Experimental Evaluation

3.1   Experimental Setting

To train DWPCNN, the NTIRE [13] 800 training images are selected in the training phase. The HR images are down-sample by the factor x4. To get the LR images, We use bicubic interpolation with the factor x4. The HR and LR images are cropped to the corresponding 160×160 sub-images with no overlapping. During the training process, we use the Adam optimizer [14] to updates (W, b). The learning rate is set to 0.01 and decays 50% every 40 epochs. To prevent overfitting, the weight regulator is set to

1×10^{-3}. Moreover, following [9], both training and testing phases of DWPCNN only utilize the luminance channel information. All the experiments are conducted in the TensorFlow package with Python 3.6 interaction interface. We use one NVIDIA Tesla K20c for both the training and testing [11, 12].

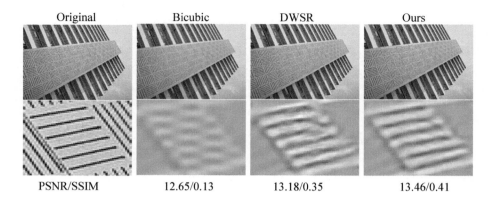

Fig. 4. SR results in the red region with upscaling factor x4

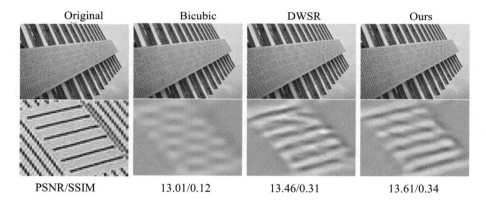

Fig. 5. SR results in the blue region with upscaling factor x4

3.2   Quantitative Evaluation

The Effectiveness of 2dWPT: To demonstrate the effectiveness of 2dWPT, we replace the four wavelet sub-bands in DWSR [6] with the wavelet sub-bands generated by Eq. (2), other settings remain the same in training phase. In testing phase, we especially choose image "img092" from Urban100 because it contains from coarse to detailed angle of image content. Figure 4 shows SR results in red region from the test image with scale factor x4. Compared with DWSR, our results provide higher PSNR/SSIM values and more pleasure results. Moreover, for more detailed text content in blue region from the test image, Fig. 5 shows the corresponding SR results, we also get better results. These results from coarse to detail angle show that multi-level 2dWPT can reconstruct more detailed image content.

Comparison with State-of-the-Art: In this section, DWPCNN is compared with VDSR [10], DWSR [6] and test on three datasets: Set5, Set14, and Urban100. Table 1 shows the different methods PSNR/SSIM results on three datasets. The best result is shown in red. From the numerical results of the experiment, DWPCNN perform excellently both PSNR and SSIM. Compared with VDSR, DWPCNN achieves about 1dB by PSNR on Set14. On Urban100, DWPCNN outperforms VDSR by about 0.8 dB. Figure 6 shows the visual comparisons of the different methods on the images "img024" from Urban100 with the scale factor x4. Compared to other methods, our DWPCNN shows the visual comparisons, especially in terms of the detailed textures, which is owing to time-frequency localization characteristics of the multi-level wavelet packet decomposition.

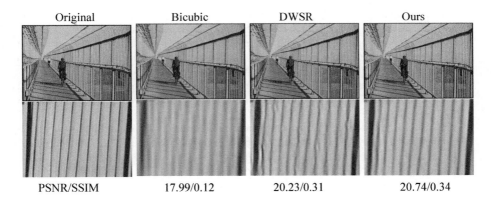

Original	Bicubic	DWSR	Ours
PSNR/SSIM	17.99/0.12	20.23/0.31	20.74/0.34

Fig. 6. SR results in blue region with upscaling factor x4

Table 1. Average PSNR/SSIM results of competing methods on three datasets.

Dataset	Scale	Bicubic	VDSR	DWSR	Ours
Set5	x4	28.42/0.8104	31.35/0.8838	31.39/0.8833	31.43/0.8842
Set14	x4	26.00/0.7027	28.01/0.7674	28.04/0.7669	28.92/0.7742
Urban100	x4	23.14/0.6577	24.52/0.7221	25.26/0.7548	25.28/0.7570

4   Conclusions

This paper proposes a multi-level wavelet packet CNN (DWPCNN) architecture for SISR, which uses wavelet packet transform to reconstruct the missing details. Compared with most state-of-the-art methods, DWPCNN has achieved excellent numerical and visual effects, which owes to the good time-frequency information provided by the wavelet packet transform. What is worth exploring in future work is how to choose the optimal wavelet basis and the number of wavelet packet decomposition layers for the SISR tasks.

References

1. Chang, H., Yeung, D.-Y., Xiong, Y.: Super-resolution through neighbor embedding. In: Proceedings of the 2004 IEEE Computer Society Conference on Computer Vision and Pattern Recognition 1, pp. I–I (2004)
2. Freedman, G., Fattal, R.: Image and video Upscaling from local self-examples. ACM Trans. Graph. 30(2), 1–11 (2011)
3. Glasner, D., Bagon, S., Irani, M.: Super-resolution from a single image. In: 2009 IEEE 12th International Conference on Computer Vision, pp. 349–356 (2009)
4. Lei, Z., Xiaolin, W.: An edge-guided image interpolation algorithm via directional filtering and data fusion. IEEE Trans. Image Process. 15(8), 2226–2238 (2006)
5. Bae, W., Yoo, J., Jong, C.Y.: Beyond deep residual learning for image restoration: persistent homology-guided manifold simplification. In: Proceedings of the IEEE Conference on Computer Vision and Pattern Recognition Workshops, pp. 145–153 (2017)
6. Guo, T., et al.: Deep wavelet prediction for image super-resolution. In: Proceedings of the IEEE Conference on Computer Vision and Pattern Recognition Workshops, pp. 104–113 (2017)
7. Liu, P., et al.: Multi-level Wavelet-CNN for image restoration. In: Proceedings of the IEEE Conference on Computer Vision and Pattern Recognition Workshops, pp. 773–782 (2018)
8. He, K., et al.: Deep residual learning for image recognition. In: Proceedings of the IEEE Conference on Computer Vision and Pattern Recognition, pp. 770–778 (2016)
9. Dong, C., et al.: Image super-resolution using deep convolutional networks. IEEE Trans. Pattern Anal. Mach. Intell. 38(2), 295–307 (2015)
10. Kim, J., Lee, J.K., Lee, K.M.: Accurate image super-resolution using very deep convolutional networks. In: Proceedings of the IEEE conference on computer vision and pattern recognition, pp. 1646–1654 (2016)
11. Lim, B., et al.: Enhanced deep residual networks for single image super-resolution. In: Proceedings of the IEEE Conference on Computer Vision and Pattern Recognition Workshops, pp. 136–144 (2017)
12. Zhang, Y., et al.: Residual dense network for image super-resolution. In: Proceedings of the IEEE Conference on Computer Vision and Pattern Recognition, pp. 2472–2481 (2018)
13. Timofte, R., et al.: Ntire 2017 challenge on single image super-resolution: methods and results. In: Proceedings of the IEEE conference on computer vision and pattern recognition workshops, pp. 114–125 (2017)
14. Kingma, D.P., Ba, J.: Adam: a method for stochastic optimization. *arXiv preprint* arXiv: 1412.6980 (2014)

An Improved DBN Method for Text Classification

Yuanyuan Shi[1], Jingqing Jiang[2], Xiaojing Fan[3], Jie Lian[2],
Zhili Pei[2(✉)], and Mingyang Jiang[2(✉)]

[1] College of Mathematics and Physics, Inner Mongolia University for
Nationalities, Tongliao, Inner Mongolia, China
[2] College of Computer Science and Technology, Inner Mongolia University for
Nationalities, Tongliao, Inner Mongolia, China
[3] College of Engineering, Inner Mongolia University for Nationalities, Tongliao,
Inner Mongolia, China

Abstract. Traditional Deep Belief Networks (DBN) is prone to the problems of too long training time and falling into local extreme values when dealing with text classification. Therefore, the DBN is improved by introducing adaptive learning rate and additional momentum items, and is invested in text classification tasks: Propose a Deep Belief Network (LMDBN) based on an adaptive learning rate-additional momentum term. The algorithm can make the convergence of the network tend to the global minimum, and ensure that the convergence process is gentler and more stable, and at the same time accelerate the convergence process of the network. Experimental results show that the LMDBN network is better than the DBN in terms of classification accuracy and convergence time. Compared with several other traditional classification models, the LMDBN network also shows good classification performance.

Keywords: Deep Belief Network · Text categorization · Adaptive learning rate · Additional momentum term

1 Introduction

The DBN [1, 2] has outstanding feature learning advantages, especially for data with a large amount of and high complexity. Although the DBN network is widely used in the field of text classification, it also has some disadvantages in theoretical algorithms and solving practical problems. The biggest problem is that the DBN network takes a long time in the training process and the network delays in reaching convergence [3]. In order to overcome the disadvantages of the DBN network [4], this paper proposes an improved DBN—LMDBN by explaining the advantages of the adaptive learning rate and the benefits of the relationship between the additional momentum term and the gradient on the network convergence. Experiments show that the network has good performance.

J. C. Hung et al. (Eds.): FC 2021, LNEE 827, pp. 867–873, 2022.
https://doi.org/10.1007/978-981-16-8052-6_110

2   Classification Model of LM-DBN

The DBN network is a probability generation model that can perform deep information mining and deep feature extraction for data [5]. The LM-DBN network adopts a layer-by-layer bottom-up unsupervised training method in the pre-training stage, and uses the output of the last Restricted Boltzmann Machine (RBM) [6] as the output layer of the network; and then uses the actual output of the output layer to find the network error. Supervised optimization of the network through BP algorithm, the output layer of the network and each RBM are trained and tuned, so that the error of the network is within an acceptable threshold and converges towards the global minimum [7].

2.1   Pre-training Process of LM-DBN

The composition of the LM-DBN network is the same as the DBN network, and the training method of the LM-DBN network in the pre-training phase is the same as that of the DBN network. The network is trained through RBM. The RBM [8] is a two-layer random neural network composed of a visual layer v and a hidden layer h. The connection between the two layers of neurons is characterized by no connection within the layer and full connection between the layers. The training process of RBM is clearly explained in the literature [9].

2.2   Tuning Process of LM-DBN

2.2.1   Adaptive Learning Rate Adjustment Method

In order to overcome the disadvantage of the fixed learning rate [10], in the training process of the deep belief network, after the pre-training of the RBM, an adaptive learning rate is proposed in optimization the stage of BP algorithm. The idea is that the adaptive learning rate can carry out adaptively learned according to the error of this generation and the previous generations, and the step length is adjusted. If the error of the network increases after one generation selection, the step length of the network learning should be reduced, that is, the learning rate should be reduced; The decrease of the error indicates that the step size of network learning should be increased, that is, the learning rate should be increased. Therefore, the adaptive learning formula proposed in this paper is as follows,

$$A\{i\} = \begin{cases} 0.4A\{i-1\}, ER\{i\} > 1.05ER\{i-1\} \\ 0.7A\{i-1\}, ER\{i-1\} < ER\{i\} < 1.05ER\{i-1\} \\ 0.9A\{i-1\}, 0.5ER\{i-1\} < ER\{i\} < ER\{i-1\} \\ 1.25A\{i-1\}, ER\{i\} < 0.5ER\{i-1\} \end{cases} \tag{1}$$

Among them, i represents the number of iterations, $A\{i\}$ represents the learning rate of the i th iteration, and $ER\{i\}$ represents the root mean square error of the i th iteration.

2.2.2 The Adjustment Method of the Additional Momentum Term to the Gradient

In the fine-tuning stage of the DBN [11], an additional momentum term is added to the gradient adjustment of the BP neural network, so that the gradient of the previous iteration affects the gradient of this iteration, so that the network reaches the state of convergence faster and more smoothly. The formula between the additional momentum term and the gradient is as follows,

$$DW\{i\} = 1/2(1 - \eta)DW\{i\} + 1/2\eta DW\{i - 1\} \tag{2}$$

Among them, η is the momentum factor, $\eta \in [0, 1)$, $DW\{i\}$ represents the gradient of the i th iteration.

Formula (2) indicates that the gradient of the i th iteration is not only related to this iteration, but also adds the memory of the previous iteration, which is affected by the gradient of the $i - 1$ th iteration. If the direction of the gradient of the i th iteration is the same as that of the $i - 1$ th iteration, the gradient of the i th iteration is increased to increase the update weight of the i th iteration and speed up the convergence process of the network. Otherwise, reduce the gradient to reduce the weight of the update. Slow down the convergence speed of the network.

2.2.3 Weight Update

In order to make the DBN have a better classification effect, in the optimization stage of the DBN, that is, the back propagation process of the BP neural network, the adaptive learning rate can be used to make the network find the appropriate learning rate with each iteration of training, and speed up the convergence speed of the network; the additional momentum term can make the network learn towards the minimum value of the objective function, and make the network train smoothly to obtain the global convergence value. Therefore, the DBN based on adaptive learning rate-additional momentum term is proposed. The weight formula is updated as follows,

$$W\{i + 1\} = W\{i\} + A\{i\}(1/2 * (1 - \eta)DW\{i\} + 1/2 * \eta DW\{i - 1\}) \tag{3}$$

Among them, $W\{i\}$ represents the weight of the i th iteration.

This method can solve the problem that the convergence speed of traditional deep belief networks is too slow. In network training, the learning rate is automatically adjusted by the error changes after the network iteration. This process will also be affected by the gradient of the previous iteration. When it has a positive effect in correcting the weights and ensures the stability of network training.

3 Experiment

3.1 Data Preprocessing

The experiment is completed in the environment of Matlab2015b, the computer is Win10 64-bit operating system, and the computer configuration is Intel(R) Core (TM) i7–6700 CPU @3.40 GHz triple core.

The experiment uses the ModApte version data set of the Reuters-21578 corpus, which includes a total of 8293 documents with 65 categories and 18933 participles. Only 7285 documents of the top 10 categories are selected for the experiment. First, feature selection is performed on the data set, and the first 3000 feature words of the data are selected as the final feature of the data; secondly, the data after the feature selection is normalized, and the mapminmax algorithm is selected as normalization function. The training set and the test set are divided into approximately 70% and 30% of the total data volume of this data set, and 5228 train set and 2057 test set are obtained.

3.2   Experimental Results and Analysis

In the case that the DBN network and the LMDBN network have the same network structure and network parameters, this section compares the DBN and the LMDBN in terms of the RMSE of the training set and the network loss value. The specific results are shown in Table 1.

Table 1. Comparison of the RMSE of the training set and the loss value of DBN and LMDBN

Algorithm	Root mean square error of training set	Network loss value
DBN	0.061743	0.08435
LMDBN	0.054380	0.06217

It can be observed from Table 1 that the performance of the LMDBN network is higher than that of the DBN network, which indicates that the LMDBN network has more sufficient training for data, higher fitting degree for the data, and smaller network loss.

In the experiment, when the DBN network and the LMDBN network have determined the completely consistent network structure and network parameters, Table 2 compares the performance of the DBN algorithm and the LMDBN algorithm in the classification error rate, the number of iterations during convergence and the running time.

Table 2. Classification error rate and running convergence time statistics of DBN and LMDBN

Algorithm	Error rate (%)	Number of iterations at convergence	Convergence time (s)
DBN	8.70	7831	4305.3773
LMDBN	6.27	6887	3581.4375

As can be seen from Table 2 above, when the two networks converge separately, the error rate of the LMDBN algorithm is only 6.27%, and the error rate of the DBN algorithm when converging is 8.70%, so the classification error rate of LMDBN algorithm is lower than that of the DBN algorithm. At the same time, the number of iterations when the LMDBN network converges is 6887, which is significantly less than that of iterations of the DBN network during convergence of 7831. The convergence time of the LMDBN network is only 3581.4375s. Compared with the convergence time of the DBN network of 4305.3773s, it has been significantly improved. While the LMDBN algorithm not only speeds up the convergence speed, but also improves the classification accuracy of the network.

Figure 1 shows the network convergence process comparison of the DBN and the LMDBN.

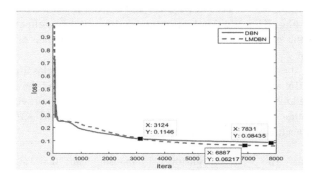

Fig. 1. Comparison of network convergence between DBN algorithm and LMDBN algorithm

As can be seen from the figure above, compared with DBN network, LMDBN network has better effect in the iterative process. When the number of iterations reaches 6887, the network can reach convergence. By comparing the convergence of the two networks, it can be found that when the number of iterations reaches 3124, the two networks reach the same network loss. When the number of iterations is greater than 3124, the advantages of LMDBN network have been highlighted. Compared with the DBN network, the loss per unit time decreases more, and the convergence speed of the network is also significantly faster, which indicates that the network has been effectively learning and updating in the learning process, which greatly improves the operation efficiency of the network, and makes the classification results more accurate and the classification effect more obvious.

At the same time, in order to further verify the classification performance of the LMDBN algorithm, the experiment compare the LMDBN algorithm with the classical CNN algorithm, SAE algorithm, SVM algorithm, and BP algorithm in terms of classification accuracy and network running time while the data is consistent.

Table 3. Classification accuracy of several different algorithms

Algorithm	Accuracy (%)	Running time (s)
CNN	77.78	11242.9210
SAE	88.33	4234.8064
SVM	88.82	3595.2098
BP	89.26	3908.9925
DBN	91.30	4305.3773
LMDBN	93.73	3581.4375

It can be seen from Table 3 that in terms of classification accuracy, the classification results of the CNN network, SAE network, SVM algorithm, and BP network are not as high as the classification accuracy achieved by the DBN network and LMDBN network, indicating that the DBN network has trained data more fully by learning more abstract information in data, and has better performance in text classification tasks. And the performance of the LMDBN network is also significantly improved than the DBN network, indicating that the adaptive learning rate of the LMDBN network will be adjust itself according to the network training, and adjust the learning rate most suitable for network training. The additonal momentum term controls effect on the smooth convergence of the network and realizes the superior performance of the network for text classification.

In summary, the LMDBN algorithm greatly improves the classification results of the traditional DBN algorithm because of its high robustness, and the running time has a significantly reduced, so that the network has been further improved. For the same data set, when other networks are used for classification, the LMDBN algorithm also achieves a high accuracy rate and has certain advantages.

Acknowledgements. This work was supported by Industry Innovation Talent Team of Inner Mongolia Grassland Talent Engineering (2017), Science and Technology Projects of Inner Mongolia Autonomous Region (2020GG0190), Research Program of Science and Technology at Universities of Inner Mongolia Autonomous Region (NJZY20112), Program for Young Talents of Science and Technology in Universities of Inner Mongolia Autonomous Region (NJYT-19-B18), Natural Science Foundation of Inner Mongolia Autonomous Region of China (2019MS08036), Inner Mongolia University for Nationalities doctoral research start fund project (BS543).

References

1. Hinton, G.E., Salakhutdinov, R.R.: Reducing the dimensionality of data with neural networks. Science **313**(5786), 504–507 (2006)
2. Geoffrey, E.: Hinton, Simon Osindero and Yee-Whye Teh: a fast learning algorithm for deep belief nets. Neural Comput. **18**(7), 1527–1554 (2006)
3. Junfei, Q., Gongming, W., Xiaoli, L.: Design and application of deep belief network based on adaptive learning rate. J. Autom. **43**(08), 1339–1349 (2017). (in Chinese)
4. Gongming, W., QingShan, J., Junfei, Q.: A sparse deep belief network with efficient fuzzy learning framework. Neural Netw. Official J. Int. Neural Netw. Soc. **121**, 430–440 (2020)
5. Yi, X., Beibei, L., Wei, S.: Research on improved deep belief network classification algorithm. Comput. Sci. Explor. **13**(04), 596–607 (2019). (in Chinese)
6. Fischer, A., Igel, C.: An introduction to restricted boltzmann machines. In: Alvarez, L., Mejail, M., Gomez, L., Jacobo, J. (eds.) CIARP 2012. LNCS, vol. 7441, pp. 14–36. Springer, Heidelberg (2012). https://doi.org/10.1007/978-3-642-33275-3_2
7. Li, L., Zhou, Z., Chen, D.: Offline handwritten Chinese character recognition based on the integration of DBN and CNN. J. Harbin Univ. Sci. Technol. **25**(03), 137–143 (2020). (in Chinese)
8. Hinton, G.E.: A practical guide to training restricted Boltzmann machines. In: Montavon, G., Orr, G.B., Müller, K.-R. (eds.) Neural Networks: Tricks of the Trade. LNCS, vol. 7700, pp. 599–619. Springer, Heidelberg (2012). https://doi.org/10.1007/978-3-642-35289-8_32

9. Cai, L., Cai, X.: Application of DBN in Chinese text classification. Comput. Eng. Des. **39** (09), 2974–2978+2991 (2018). (in Chinese)
10. Shi, J., Wang, D., Shang, F.: Research progress of stochastic gradient descent algorithm. J. Autom. 1–17[-04–03] (2021). (in Chinese)
11. Sun, Y.: Design of Doherty Power Amplifier Based on BP Algorithm Based on Adaptive Learning Rate Additional Momentum Term. East China Jiaotong University (2021). (in Chinese)

Fuzzy Adaptive PID Control of Waste Heat Power Generation

Jingya Yang$^{(\boxtimes)}$, Feng Yan, Yan Pan, and Xiangji Zeng

Changsha Nonferrous Metallurgy Design and Research Institute Company Limited, Changsha, Hunan, China

Abstract. In the cogeneration control system based on oxygen bottom blowing process, the highest temperature point in the whole system is the superheated vapor temperature at the outlet of superheater, and it is one of the parameters that represent the operating state of the system. The temperature of superheated steam affects the safe and economic operation of the whole system and the service life of the equipment. This paper analyzes the temperature characteristics of superheated steam and establishes the mathematical model of superheated vapor temperature control system. According to the characteristics of large control delay and inertia, a fuzzy adaptive PID controller is designed. A variable proportion factor fuzzy adaptive PID controller was proposed, and the SIMULINK tool was used to simulate the control system under different states.

Keywords: Oxygen bottom blowing process · Waste heat power generation · Superheated vapor temperature · Fuzzy adaptive PID

1 Introduction

Oxygen enriched bottom blowing smelting process is a copper smelting technology with independent intellectual property rights developed in China. The flue gas and heat generated in the smelting process are large, and the heat source is stable. If the afterheat of flue gas generated in oxygen enriched bottom blowing furnace can be comprehensively utilized or generated, and the electric energy generated by waste heat power generation can be used in the smelting process, It not only recovers a large amount of waste heat produced in the production process, but also reduces environmental pollution and improves economic benefits.

Boiler is one of the important equipment in waste heat power generation system. In the boiler thermal system, The temperature of superheated vapour affects the economic and safe operation of the whole system and the service life of the equipment [1]. Therefore, the temperature of superheated vapor at the outlet is too high, too low or the temperature change rate is too large, which has a great potential safety hazard for the safe operation of the system. It is essential to keep t the temperature of superheated vapor within a certain range.

Based on the study of process and control requirements, the mathematical model of superheated vapor temperature control system is established, A fuzzy adaptive PID

J. C. Hung et al. (Eds.): FC 2021, LNEE 827, pp. 874–881, 2022.
https://doi.org/10.1007/978-981-16-8052-6_111

controller with variable scale factor is proposed, and the simulation of superheated vapor temperature control system in different states is carried out by using SIMULINK simulation tool.

2 System Structure

The main process of waste heat power generation is: using steam with a certain temperature to drive the turbine to realize the conversion of heat energy to mechanical energy, and using the generator set to convert mechanical energy into electrical energy [2]. In the application of steam power generation by waste heat boiler, the method of superheated steam power generation is often used, that is, the saturated steam generated by waste heat boiler is superheated and then sent to superheated steam turbine generator unit for power generation. The superheated steam power generation process is as follows (Fig. 1):

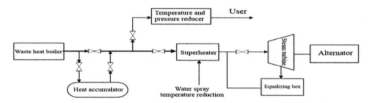

Fig. 1. Flowsheet of cogeneration

In the control system of cogeneration, the superheater inlet temperature, steam flow rate and desuperheating water volume are the factors that affect the temperature of superheated vapor. In addition, because the structure of superheater is complex, long pipeline and large heating area, when disturbance exists, the system will have the characteristics of large inertia and propagation delay. Therefore, it is difficult to achieve the stability control of superheated vapor temperature.

3 Fuzzy Adaptive PID Control

3.1 Fuzzy Adaptive PID Controller for Superheated Vapor Temperature

The temperature of superheated vapor control system has the characteristics of large inertia and propagation delay, and there are many disturbance factors in the system. The parameters and characteristics of the controlled object under different disturbances are also different. Because the conventional PID control needs accurate mathematical model, it is difficult to achieve stable control.

When the deviation and rate of change between the expected value and actual value are used as the input of the controller, the PID controller can be expressed as follows:

$$u(t) = K_P e(t) + K_I \int e(t)dt + K_D \frac{de(t)}{dt} \tag{1}$$

The expected output variable of the controller is $u(t)$; The deviation between the expected value and the actual value is $e(t)$(input quantity); The magnification factor is K_P; The integral time coefficient is K_I; The differential time coefficient is K_D.

The fuzzy controller takes superheated vapor temperature deviation e and rate of change ec as multi-input variables, and the proportional amplification coefficient K_P, integral time coefficient K_I and differential time coefficient K_D as output variables (Fig. 2).

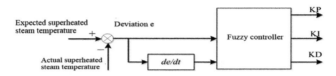

Fig. 2. Two input three output fuzzy controller

3.2   Fuzzy Adaptive PID Controller with Variable Scale Factor for Superheated Vapor Temperature

In the fuzzy controller, it mainly includes quantization factor, fuzzy rule, scale factor, membership function and control variable universe, In theoretical research, many scholars use the methods of variable universe [3–5], variable fuzzy rules [6], variable membership function [7], variable factor [8–10] and scale factor to adjust the relevant links of fuzzy controller online to meet the relevant performance indicators of the system. Through analysis, the above improved methods need to regulate the quantification and proportion factor online. Since the quantification factor of superheated vapor temperature control system has been determined, this paper improves the rapidity of the system by adjusting the proportion of fuzzy controller in real time.

In the fuzzy controller, there is a positive correlation between the scale factor and the input variables, when $|e|$ and $|ec|$ are larger, the calculated value of proportion factor K_u is larger; When $|e|$ and $|ec|$ are small, the calculated value of proportion factor K_u is small. In this paper, the scale factor can be calculated by the following formula:

$$K_u = K \cdot K_{u0} \tag{2}$$

In the above formula, it is obtained through expert experience. Therefore, the fuzzy controller is mainly used to regulate the coefficient K to achieve the online adjustment of the proportion factor K_u. There are two common methods of variable scale factor:

(1) An auxiliary fuzzy controller is established to control the scale factor coefficient;
(2) Using expert knowledge, the algorithm between scale factor and input variable is designed.

Method 2 algorithm is difficult to determine accurately and very complex, so method 1 is used to control the scale factor. The block diagram of Fuzzy adaptive PID controller with variable scale factor for superheated vapor temperature control system is shown in the figure below (Fig. 3).

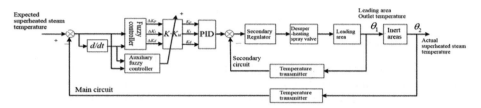

Fig. 3. Fuzzy adaptive PID controller with variable scale factor for superheated vapor temperature

In the control system, the regulator in the main circuit is composed of fuzzy controller and auxiliary fuzzy controller in parallel. The fuzzy controller adjusts the three parameters of PID control, while the auxiliary fuzzy controller adjusts the scale factor in the main controller. The parallel connection of the main controller and the auxiliary fuzzy controller can improve rapidity and dynamic characteristic of the system and reduce the overshoot of the system [11].

In the auxiliary fuzzy controller, the corresponding fuzzy subset is divided into seven levels, the fuzzy universe is $[0, 4][0, 4]$, and triangle and trapezoid membership functions are used, as shown in the figure below (Fig. 4).

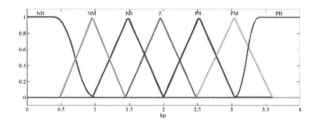

Fig. 4. Membership function of K

The control regulation of auxiliary fuzzy controller can be described as follows (Table 1).

Table 1. The regulation table of K

K		e						
		NB	NM	NS	Z	PS	PM	PB
ec	NB	PB	PB	PM	NS	Z	NS	NB
	NM	PB	PB	PM	Z	NS	NS	NS
	NS	PB	PM	PM	Z	NM	NM	NS
	Z	PM	PS	NM	NB	NM	PS	PM
	PS	NS	NM	NM	Z	PM	PM	PB
	PM	NS	NS	NS	Z	PB	PB	PB
	PB	NB	NS	Z	NS	PB	PB	PB

3.3   Simulation Result

Through the mathematical modeling of the superheater, we can know that the superheated vapor temperature control system consists of the main circuit and the auxiliary circuit. The auxiliary circuit mainly adjusts the superheated vapor temperature roughly, which can eliminate all kinds of interference; The main circuit mainly realizes the accurate regulation of superheated vapor temperature.

The transfer function of the superheated vapor temperature control system is given by reference [12]:

$$W(s) = W_1(s)W_2(s) = \frac{9}{(1+15s)^2(1+25s)^2} \tag{3}$$

The transfer function of the leading zone is $W_1(s) = \frac{8}{(1+15s)^2}$; Inert zone is $W_2(s) = \frac{1.125}{(1+25s)^2}$.

(1) Simulation results with step signal (Fig. 5).

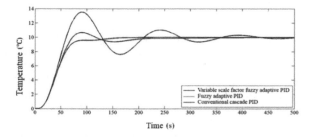

Fig. 5. Simulation results under step signal

(2) Simulation results under desuperheating water disturbance. When t = 300 s, the disturbance of desuperheating water is added (Fig. 6).

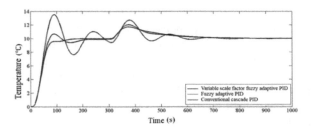

Fig. 6. Simulation results under desuperheating water disturbance

(3) Simulation results under vapor flow disturbance.When t = 300 s, the steam flow disturbance is added (Fig. 7).

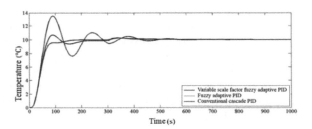

Fig. 7. Simulation results under vapor flow disturbance

(4) Simulation results under flue gas heat disturbance.When t = 300 s, the flue gas heat disturbance is added (Fig. 8).

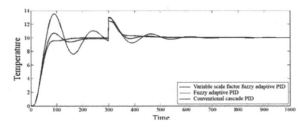

Fig. 8. Simulation results under steam flow disturbance

From the above analysis, when the flue gas heat disturbance is added, compared with the routine PID controller, the response curve of fuzzy adaptive and variable scale factor fuzzy adaptive PID controller fluctuates less, and the time to return to steady state is shorter, and the adjustment time is shorter. According to the control results of the three control methods, it can be make a conclusion that the fuzzy adaptive PID controller has regulation ability to the heat disturbance of flue gas, and can overcome the disturbance on site, which meets the needs of production.

To sum up, the control effect of the variable scale factor fuzzy adaptive PID controller designed under the action of step, steam flow disturbance and desuperheating water disturbance is ideal state, and has better anti interference capability and robustness.

4 Conclusions

In the cogeneration control system, the temperature of superheated vapor is an important standard to measure the quality of steam and an important parameter in the cogeneration control system. Therefore, the superheated vapor temperature control is of great significance in the whole operation process of the unit. However, the temperature of superheated vapor has the characteristics of large delay, time-varying and nonlinear, so it is hard to achieve the ideal control result. Therefore, this paper designs a variable scale factor fuzzy adaptive PID control system to realize the optimal control of superheated vapor temperature. Through the simulation test, It is proved that the control system improves the dynamic characteristics of the system to a certain extent and achieves good control effect.

Acknowledgements. This work was supported by National Key R & D Program of China (2019YFB1704700).

References

1. Yu, M., Wang, Y., Shao, S.: Application of critical proportioning method based on Matlab in PID parameters auto-tuning numerical simulation in engineering. J. Liaoning Teachers Coll. (Nat. Sci. Ed.), **20**(2), 6 (2018). (in Chinese)
2. Bersini, H., Bontempi, G.: Now comes the time to defuzzify neuro-fuzzy models. Fuzzy Sets Syst. **30**(7), 53–58 (1997)
3. Tong, S., Li, H.X., Wei, W.: Comments on direct adaptive fuzzy-neural control with state observer and supervisory controller for unknown nonlinear dynamical systems. IEEE Trans. Fuzzy Syst. **11**(5), 703–705 (2003)
4. Punjani, A., Abbeel, P.: Deep learning Helicopter dynamics models. In: Proceedings of the 2015 IEEE International Conference on Robotics and Automation, pp. 3223–3230. IEEE, Seattle, WA (2015)
5. Zhang, Z.: Optimization of furnace temperature PID control parameters based on frequency domain characteristics. Metall. Ind. Autom. **41**(1), 62 (2017). (in Chinese)
6. Luo, D.: Self-integration practice of automatic system revamping project of Baosteel 1 580 mm hot rolling mill. Metall. Ind. Autom. **39**(6), 7 (2015). (in Chinese)

7. Tang, H.: Development and application of the PLC monitoring system for high pressure blower. Metall. Ind. Autom. **42**(3), 68 (2018). (in Chinese)

8. Jiang, J., Li, Y., Guo, H.: Kivcet lead smelting process practice. World Nonferrous Met. (18), 7 (2018). (in Chinese)

9. Jing, X., Xie, B., Zhou, R.: Development and application of off-line simulation mathematical model for KIVCET furnace. Metall. Ind. Autom. **41**(1), 49 (2017). (in Chinese)

10. Zadeh, L.A.: Fuzzy sets, information and control. Inf. Control **8**(3), 338–353 (1965)

11. Mamdani, E.H.: Application of fuzzy algorithms for control of simple dynamic plant. Proc. Inst. Electr. Eng. **121**(121), 1585–1588 (1974)

12. Zadeh, L.A.: A fuzzy-algorithmic approach to the definition of complex or imprecise concepts. Int. J. Man Mach. Stud. **8**(3), 249–291 (1976)

Comparative Analysis of Structured Pruning and Unstructured Pruning

Zhengwu Yang[1(✉)] and Han Zhang[2]

[1] Fan Gongxiu Honors College, Beijing University of Technology,
Beijing, China
yzw@emails.bjut.edu.cn
[2] Faculty of Information Technology, Beijing University of Technology,
Beijing, China

Abstract. In recent years, neural networks have achieved great success promising superior in various fields of artificial intelligence, including image recognition, target detection, natural language processing, and data analysis. However, the training and inference of neural networks require a large number of parameters and calculations, which hinders the deployment of embedded devices with limited resources. Therefore, using neural network pruning to reduce memory overhead and increase inference speed has become a hot topic. In this paper, the pruning algorithm is analyzed and concluded from two aspects: structured pruning and unstructured pruning. Moreover, the effects of structured pruning and unstructured pruning algorithms after running different data sets in different neural network models have been comparatively analyzed. Furthermore, we have proposed certain suggestions to the future development of neural network pruning based on the experimental results.

Keywords: Neural network · Structured pruning · Unstructured pruning · Comparative analysis

1 Introduction

The neural network is a research hotspot hot issue in the field of artificial intelligence. It has made remarkable achievements in natural language processing, computer vision, speech recognition and other fields. In recent years, with the rapid development of GPU computing power, the network scale of neural networks has become larger and larger, and data processing capabilities have increased day by day. Many deep neural network models have been developed, for example CNN, VGGNet, AlexNet, and ResNet, and etc. [1–3]. The renewed wave of artificial intelligence has made machines surpass human accuracy in many tasks. For example, in the ILSVRC competition, the classification accuracy of the deep neural network model is close to 98% (human accuracy is around 95%). The accuracy performance of the neural network model is mainly enhanced by the specific deep structure of the model.

The improvement of the accuracy and the more complete functions of neural networks come from the continuous complexity and deepening of its structure. As shown in Table 1, only an 8-layer AlexNet requires 61 million network parameters and

J. C. Hung et al. (Eds.): FC 2021, LNEE 827, pp. 882–889, 2022.
https://doi.org/10.1007/978-981-16-8052-6_112

729 million floating-point calculations, occupying about 233 MB of memory. VGG-16 network parameters reached 138 million, the number of floating-point calculations was 156 million, and about 553 MB of memory was required. Also, InceptionNet network parameters reached 23.2 MB, occupying 90 MB of storage space. The above-mentioned network model obtains excellent performance, while the large scale of parameters makes it difficult to deploy in embedded devices with limited computing resources such as mobile phones, smart watches, and on-board computers. This greatly limits the development space of neural networks and reduces the number of nerves. The application of the network. The further development and application of the neural network are limited, which can be attributed to the above descriptions.

Table 1. Parameter indexes of mainstream network model (MB)

Neural network	Parameters	Number of multiply-add operations	Storage
AlexNet	60	720	240
VGG-16	138	15300	>500
GoogleNet	6.8	1550	50
Inception-V3	23.2	5000	90

By sharp contrast, the pruning method of the neural network can solve the above problems effectively. To be specific, the tremendous parameters of the neural network can be cut by the pruning method such that the complexity and the inference speed of the neural networks can be strongly improved. For instance, certain corresponding models based on the pruning method could achieve superior performance in various tasks such as image classification and emotion analysis.

According to the influence of pruning on the structure of the neural network, pruning methods can be divided into two categories (unstructured pruning and structured pruning), as shown in Fig. 1. The structured pruning uses the channel as the granularity of the pruning. Since the calculation is performed by matrix operations in the GPU, the reduction of the channel reduces the calculated matrix and increases the speed of the operation. Therefore, the advantage of the structured pruning is that it prunes the latter's speed increase. Yet because the pruning process is too rough, the information is seriously lost, and the compression ratio cannot be particularly high. Unstructured pruning uses the weight between connected neurons as the granularity of the pruning. The cutting granularity of this method is small, which reducing the loss of information, so the compression ratio of the Neural network is higher.

The purpose of training is to minimize the training cost function to make the output of the neural network in the process of approaching the true value, which can be written as Eq. (1):

$$\min_{W} C(W, D) \tag{1}$$

We assume that C represents training cost function, D represents training data, W represents network weights, and $\hat{W}$ represents pruned network weights.

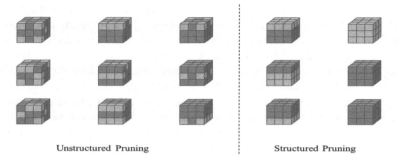

Unstructured Pruning   Structured Pruning

Fig. 1. Schematic diagram of structured pruning and unstructured pruning

The aim of pruning is to minimize the change of the pruned network weights to the training cost function, which can be represented in Eq. (2):

$$\min_{\hat{W}} \left| C(\hat{W}, D) - C(W, D) \right| \tag{2}$$

The pruning process is shown in Fig. 2. The specific pruning process of the neural network is to use importance factors to evaluate the importance of pruning objects and to delete them. Then, the parameters after pruning are re-trained and adjusted to determine whether the expected precision is achieved. The pruning process is repeated until the accuracy requirements are met if the accuracy loss is lower due to the pruning. It can be seen that the model pruning is an iterative process, pruning and training alternate, and therefore the process of pruning can also be called iterative pruning.

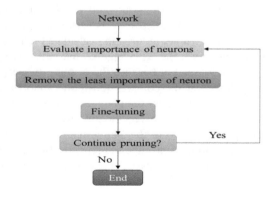

Fig. 2. Pruning flowchart

The main contributions of this work can be summarized as follows:

1. Novel and efficient pruning algorithms in recent years are introduced.
2. The performance of various neural network pruning methods are discussed and compared in detail.
3. The characteristics of neural network pruning methods are summarized.
4. This study would promote the research of pruning methods and provide guidance and assistance for further work.

The rest of this paper is organized as follows: in Sect. 2, the pruning methods of the neural network are introduced. In Sect. 3, different neural network pruning methods were carried out in experiments, and their characteristics were summarized and compared. Finally, Sect. 4 concludes this paper and the future research trends of neural network pruning methods are discussed.

2   Comparative Analysis of Pruning

2.1   Structured Pruning

Ye proposed a method to calculate the threshold through OBTD, and then combined with a random pruning algorithm to trim the eigenvalue gradient, which could reduce the calculation amount in the gradient descent stage and increase the calculation speed [4]. The Eagleeye method proposed by Li regards model pruning as a problem of searching for the optimal substructure in the search space [5]. Eagleeye used the adaptive batch normalization technique to prune the model. Lee et al. proposed a network pruning method based on connection sensitivity (SNIP) [6]. They employed the gradient of weight connection to measure the standard of saliency, which could be utilized for multiple sampling by mini-batch before in-depth model training. Based on this, the importance of connection with different weights can be determined, and therefore pruning masks from pruning targets is generated which can save time for iterative pruning. In [7], Frankle developed a method called The Lottery Ticket Hypothesis (TLTH) which could naturally reveal subnetworks, and the initialization of these subnetworks could effectively train them. The main aim of this method is to find subnetworks with the same test accuracy as the original network.

2.2   Unstructured Pruning

In this section, we introduced certain unstructured pruning methods in detail. For instance, Guo proposed a DMCP method to model channel pruning as a Markov process, where each Markov state represents the preservation of the corresponding channel during the pruning process and the transition between states represents the pruning process. The appropriate number of channels in each layer is implicitly selected through the Markov process after optimizing the transition probability [8]. Furthermore, Li proposed a pruning algorithm based on Artificial Bee Colony called

ABCPruner, which effectively found the best pruning structure. That is the number of channels in each layer is selected layer by layer and the search for the best pruning structure is formulated as an optimization problem [9]. To be specific, the ABC algorithm is integrated to solve the problem in an automatic manner. Similarly, Lin et al. proposed a pruning method called HRank, and they observed empirically that the rank of the feature map was robust to the input image [10]. They compact the convolutional neural network layer by layer, inputs a small number of images for each layer, calculates and sorts the rank of the feature map, and cuts off the convolutional kernel corresponding to the feature map with a small average rank in a certain proportion for each layer. After that, the remaining filter parameters are used as the initial value for fine tuning to get the pruned network. Ye et al. proposed a method called Optimal Threshold (OT), which works out the local thresholds layer by layer to best separate the important channels from the negligible ones [11]. He et al. proposed a method called Filter Pruning via Geometric Median (FPGM), which compresses the CNN model by pruning the redundant convolution kernel [12].

3   Experiments

3.1   Datasets

The MINIST dataset contains 60,000 training images of 28×28 pixels and 10,000 test images, which are grayscale images of handwritten digits. The Cifar10/100 dataset has a total of 60,000 32×32 pixel color images, which can be divided into 50,000 training images and 10,000 test images, and 10/100 classifications. The ImageNet dataset contains approximately 15 million full-size images and 22,000 categories. ILSVRC-2012 contains about 1.2 million images in 1,000 categories. The datasets utilized in the experiments is listed in Table 2.

Table 2. The datasets utilized in the experiments

Dataset	Sample size	Number of categories	Pixel
MINIST	60000	10	28*28
CIFAR-10	60000	10/100	32*32
ILSVRC-2012	1200000	1000	256*256
ImageNet	15000000	22000	256*256

3.2   Evaluation Metrics

ΔACC can be utilized to measure the accuracy variation of different models, which is obtained by subtracting the accuracy of the original model from the accuracy of the compressed model.

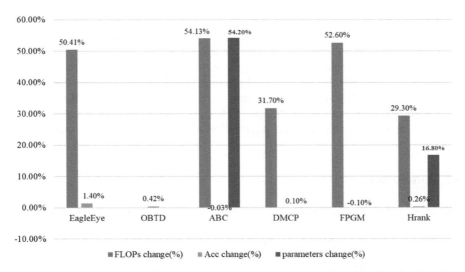

Fig. 3. The performance of different pruning methods tested on ResNet-56 using the CIFAR-10 dataset.

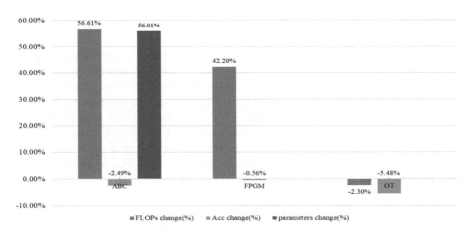

Fig. 4. The performance of different pruning methods tested on ResNet-50 using the ILSVRC-2012 dataset.

The parameter change is obtained by dividing the original model parameters and the pruning model parameters, which is used to measure the decrease of the parameter amount. Indicator FLOPs can be utilized to measure how much the computation is reduced. The performance of different pruning methods tested on two datasets (CIFAR-10 and ILSVRC-2012) is illustrated in Figs. 3 and 4.

From an analysis of the experimental results in Figs. 3 and 4, it can be observed that:

(1) The FLOPs change, and parameters change of ResNet-56 are higher than that of ResNet-50, which indicates that large-scale networks have higher parameter redundancy and are more suitable for compression.

(2) The parameters change of unstructured pruning is greater than that of structured pruning, indicating that unstructured pruning method achieved better performance.

(3) Compared with CIFAR-10, the pruning method performs worse on ILSVRC-2012 dataset, and the accuracy of the model dropped even more, which seems to show that pruning method has a great negative influence on the model of complex tasks.

4   Conclusion

This paper mainly focuses on the research background and basic concepts of the neural network pruning, and the tremendous experiments are conducted to analyze and compare the structured pruning method and the unstructured pruning method. Neural network pruning is one of the most used effective methods in model compression research methods. However, neural network pruning is targeted and can't be used in a multitasking environment. Because of the plasticity of the neural network itself, the clipped neural network can achieve the effect of network compression while ensuring certain precision. It is difficult to find an appropriate threshold value in a short time, which is mainly due to that the traditional clipping method requires repeated iteration to find the most suitable threshold value.

Network pruning can stably adjust the results of the network and increase the compression ratio when the accuracy loss is smaller. However, neural network pruning has certain risks, because it will not make the memory regularization which will hinder the further acceleration of the network, thus it is necessary to combine hardware and software together to solve this problem. Although the network generated by the structured pruning method has hardware-friendly characteristics, the loss of convolution kernels and channels will cause the loss of features and reduce the accuracy of the model. Therefore, the relationship between balance precision and overhead should be considered when designing the pruning algorithm. At present, the evaluation of neural network pruning methods is the evaluation of accuracy, running time, and model size, etc., and more practical and comprehensive model evaluation indexes should be used, for instance, operation amount, derivation time, data throughput, hardware energy consumption, etc. In the process of the neural network pruning, the same proportion is used for each layer, yet the degree of redundancy of parameters in different layers of the network is different. Continuing the same clipping proportion will lead to the over-pruning and the over-fitting of the model. The research direction in the future is to find the method of the appropriate threshold value and clipping proportion and make the pruning network have comprehensive performance. Furthermore, we note that the pruning technology of neural networks should be combined with the hardware architecture design to make it perform competitively on more devices.

References

1. Simonyan, K., Zisserman, A.: Very deep convolutional networks for large-scale image recognition. Comput. Sci. (2014)
2. Krizhevsky, A., Sutskever, I., Hinton, G.: ImageNet classification with deep convolutional neural networks. Adv. Neural Inf. Process. Syst. (2012)
3. He, K., Zhang, X., Ren, S., Sun, J.: Deep residual learning for image recognition. IEEE Conf. Comput. Vis. Pattern Recogn. (2016)
4. Ye, X., Dai, P., Luo, J., Guo, X., Chen, Y.: Accelerating CNN training by Pruning Activation Gradients (2020). (Arxiv No.: 1908.00173v3)
5. Li, B., Wu, B., Su, J.: EagleEye: fast sub-net evaluation for efficient neural network pruning. Comput. Vis. ECCV **12347**(37), 639–654 (2020)
6. Lee, N., Ajanthan, T., Torr, P.: SNIP: single-shot network pruning based on connection sensitivity (2018). (Arxiv No.: 1810.02340v2)
7. Frankle, J., Carbin, M.: The lottery ticket hypothesis: training pruned neural networks (2018)
8. Guo, S., Wang, Y., Li, Q., Yan, J.: DMCP: differentiable markov channel pruning for neural networks. In: 2020 IEEE/CVF Conference on Computer Vision and Pattern Recognition (CVPR), pp. 1536–1544 (2020)
9. Lin Mingbao, Ji Rongrong, Zhang Yuxin, Zhang Baochang, Yongjian Wu, and Tian Yonghong. "Channel Pruning via Automatic Structure Search," *Twenty-Ninth International Joint Conference on Artificial Intelligence and Seventeenth Pacific Rim International Conference on Artificial Intelligence*, pp. 673–679
10. Lin, M., Ji, R., Wang, Y., et al.: HRank: filter pruning using high-rank feature map. In: 2020 IEEE/CVF Conference on Computer Vision and Pattern Recognition (CVPR) (2020)
11. Ye, Y., You, G., Fwu, J.-K., Zhu, X., Yang, Q., Zhu, Y.: Channel pruning via optimal thresholding (2020). (Arxiv No.: 2003.04566v5)
12. He, Y., Liu, P., Wang, Z., Hu, Z., Yang, Y.: Filter pruning via geometric median for deep convolutional neural networks acceleration. In: 2019 IEEE/CVF Conference on Computer Vision and Pattern Recognition (CVPR), pp. 4335–4344 (2019)

Countermeasures of Management Information Innovation and Development Under the Background of Big Data

Jin Chen[✉]

Wuhan University of Technology, Wuhan, Hubei, China

Abstract. With the advent of the era of big date (BD), management informatization has become an important step in management development and an important part of realizing management modernization. For management informatization in the context of BD, management positions and management objects are all innovating, and traditional management methods require continuous innovation and development. The purpose of this article is to study the countermeasures for the development of management informatization innovation under the background of BD. This paper adopts the method of questionnaire survey to investigate the current situation of the development of management informatization in H colleges in terms of infrastructure, resources and applications, and management informatization. Then through the statistics and analysis of the data, the main problems existing in the current university management informatization are summarized and the reasons are analyzed. Based on the actual situation, the countermeasures for the innovation and development of management informatization under the background of BD are proposed from five aspects. The survey data shows that universities have specialized informatization management organizations and those without specialized informatization management organizations account for 54.6% and 10.2% respectively, and 35.2% of the survey respondents choose not to be clear, which also explains informatization from the side.

Keywords: Big data · Management informationization · Innovation and development · University management informationization

1 Introduction

The change of the times has brought about the upgrading of various information technologies, and the scale of all walks of life is showing an increasing trend [1, 2]. Especially in the wave of BD, massive and highly complex data brings many foundations and challenges to management [3, 4]. Under such background and challenges, how to promote the double improvement of management quality and efficiency through efficient management information construction is particularly important [5, 6]. As we all know, information management is a complicated process that requires a lot of investment and a long period. Therefore, even after more than ten years of development, most universities in our country still face many problems and deficiencies in the

J. C. Hung et al. (Eds.): FC 2021, LNEE 827, pp. 890–898, 2022.
https://doi.org/10.1007/978-981-16-8052-6_113

construction of information technology management, and they need continuous adjustment and improvement.

In the research on the countermeasures for the development of management informatization in the context of BD, many scholars at home and abroad have conducted research on it, and achieved good research results. For example, Kai Wj analyzed this with the help of corresponding data. The simulation method further judges and analyzes the various factors that affect enterprise information management [7]; Wang Y defined the concept of enterprise information management in his research and pointed out that the actual operation of enterprise information management in China's environment The characteristics of [8]; Zhang Y proposed the use of BD technology to establish a basic data center library to integrate all system data in the research of university information management. That is, all data can be maintained uniformly and space can be saved. It can provide more decision-making support strategies for school leaders [9, 10].

This research firstly collects a large number of documents related to management informatization, campus construction, BD technology, cloud computing technology, network technology, etc., through theoretical analysis, expounds the theoretical basis of college management informatization, and clarifies the importance of college management informatization. Subsequently, this paper takes H colleges as an example, through investigation and research methods, and questionnaire surveys to investigate the status quo of its management informatization, and then through statistics and analysis of data, summarizes the existence of H colleges in management informatization. The main issues, and then analyze these issues, and use some excellent examples in the management informatization construction of some developed countries at home and abroad as a reference, sort out the management informatization construction of H universities and universities with the same problems in our country.

2   Countermeasures of Management Informatization Innovation and Development Under the Background of BD

2.1   Problems and Causes of BD Management Informationization

(1) Lack of understanding of management informationization.

At present, most colleges and universities in our country have certain limitations in their understanding and views on the connotation and development of management informatization, and they have not considered from a long-term perspective. Many university teachers and students, managers and even some leaders think that management informatization is the application of information technology in management. They simply call the application of information technology in management as management informatization, while ignoring the modern and scientific Management thoughts and systematic management concepts play a leading role in the construction of management informationization. In addition, as far as professional managers are concerned, they are still stuck in traditional management concepts and methods, they have not carried out a unified study of the management concepts of informatization, and lack correct understanding and understanding.

(2) Lack of systematic planning for management informationization

In the information age, "data" has become a vocabulary that everyone is talking about. With the continuous expansion of the scale of colleges and universities, the substantial increase in the number of teachers and students, and the increase in business departments, the amount of data has reached an astonishing amount. Although colleges and universities have also established office OA management systems, educational administration management systems, student management systems, etc., information is blocked and data is not communicated between the various systems. Whether it is a teacher or a student, whether it is exam registration, information inquiry or handling a certain business, it has to log in to a different management system and wait for approval. These management systems have not yet unified identity authentication. Teachers and students need to remember multiple different accounts and passwords, which not only brings great inconvenience to the daily life of teachers and students. And the additional workload of the management staff is increased, which greatly reduces the efficiency of management.

(3) Incomplete management information construction system

In the construction of informatization, the construction of an informatization management system is a guarantee, and the construction of an informatization management system also needs to be regulated and restricted by related systems to eliminate randomness [11, 12]. At present, the formulation of the management information system is more biased towards hardware and network, while ignoring the management of network process, network resource content and software management. The existing management system cannot scientifically and comprehensively cover all aspects of the management information system. Management work, which makes loopholes in information management.

(4) Faculty and staff are not strong in participating in information construction

Information technology is constantly being updated, and universities are constantly introducing new development policies. However, the replacement of personnel has always been unable to keep up with the pace of the times and the innovation of science and technology. This has led to the emergence of the management team of some universities at this stage. There has been a "generation gap" in terms of academic background, knowledge structure and technical capabilities and the requirements of management information construction. There are several aspects. First, some managers are affected by traditional management ideas and do not have a thorough understanding of modern information management ideas; second, they are not proficient in management science and information science; in addition, their own information awareness and ability to apply information technology are relatively weak, and the use of information management systems is even more difficult for them.

2.2   Countermeasures of Management Informatization Innovation and Development Under the Background of BD

(1) Improve the construction of the management information system in colleges and universities

First of all, from the perspective of information system technology, it is necessary to formulate unified and standardized information coding rules to ensure the standard and consistent processing of data and avoid subsequent statistical analysis of data due to form confusion. In addition, in the operation of the management system, it is necessary to form a support system for various teaching affairs, strictly limit and correct the use of all information, and ensure the fairness, transparency and standardization of the operation of the information system. At the same time, establish rules and regulations for handling various services to facilitate the monitoring of circulation information, so as to promote the standard, stable and sustainable development of the entire management information structure.

(2) Create a basic data sharing center

The core data sharing center is an important part of management informatization. Through a unified access interface, network technology, storage technology, cloud computing technology and other integration and implementation to create a platform system for data transmission and synchronization between independent systems. The data sharing center acts as a repository for the basic data of all business systems, and can provide data export for each business system, thereby sharing data among different business systems.

(3) Strengthen the organization and leadership of management informatization

A qualified CIO not only needs to have information system planning and reform leadership capabilities; it also needs to have a proactive work attitude, so as to upload and issue various informatization policies and implementation plans, opinions, etc.; as a CIO, he must have enough innovative consciousness of the company must adapt to the changes and development of the times, and actively promote work innovation, whether it is technological innovation or application innovation.

2.3   BD Connection Algorithm

A sample unit is randomly selected from the overall sample as a starting point, and the remaining sample units are selected in a certain interval order. The specific method is as follows: Let N be the number of overall units, n be the sample size, and the sampling distance is S = N/n. Number the overall sample from 1 to N consecutively, first randomly select a number s1 from 1 to S as the first unit of the sample, and then sample s1 + S, s2 + S,... until the number of samples reaches n Up to one.

The probability of occurrence of A event can be set to p, and the number of occurrences of A in n repeated trials is m. When n is sufficiently large, it is approximately as follows:

$$\frac{m - np}{np(1 - p)} \sim N(0, 1) \tag{1}$$

Taking the confidence interval as 100%, there are:

$$3 < \frac{m - np}{np(1 - p)} < 3 \tag{2}$$

And so:

$$np - 3np(1 - p) < m < np = 3np(1 - p)$$
$$1 - \frac{3np(1 - p)}{np} < \frac{m}{np} < \frac{3np(1 - p)}{np} \tag{3}$$

It can be seen that the larger the n, the smaller the error of m, and the higher the accuracy of sampling.

3 Experimental Research on the Development of Management Information Innovation Under the Background of BD

3.1 Research Methods

This article uses a questionnaire survey method, taking H colleges and universities as an example, taking students, teachers and administrators of the surveyed colleges and universities as the survey objects, the actual situation of college management informatization is actually investigated, and the corresponding problems in the current situation are given.

3.2 Investigation and Research Implementation

(1) Implementation of the questionnaire
    The questionnaire is mainly distributed online. Relying on the questionnaire star, the online questionnaire distribution platform, questionnaires are distributed to students, full-time teachers and administrators of H colleges and universities, and paper questionnaires are filled in in some cases. In the end, 304 questionnaires were returned, and the number of valid questionnaires was 300.
(2) Reliability analysis of the questionnaire
    This article uses the more commonly used Cronbach reliability coefficient method to test the reliability of the questionnaire. The reliability of the questionnaire is tested by SPSS20.0 statistical software. The reliability coefficient of the questionnaire is $P = 0.000$, which is less than 0.05, so the reliability is Good, indicating that the questionnaire has a good level of reliability.
(3) Analysis of the validity of the questionnaire
    This article uses SPSS statistical software to test the structural validity of the questionnaire. The data shows that KMO value = 0.809, which is greater than 0.80, indicating that the correlation between variables is strong and the structure

validity is good. The P value of Bartlett's sphericity test is 0.000, which is less than 0.05. Therefore, there is a correlation between variables and the data is distributed in a spherical shape. The validity of the questionnaire is good.

4 Data Analysis of Management Information Innovation and Development in the Context of BD

4.1 Commonly Used Information Management Systems in Colleges and Universities

At present, most of the functional departments of colleges and universities in our country have established management systems, such as office automation (OA) systems, teaching and educational information systems, student management information systems, library information systems, in addition to finance, personnel, logistics services, and file management. Such as various information systems, and even some colleges and universities have created decision support systems and so on. These information management systems have improved management efficiency to a certain extent.

The survey data shows that the commonly used management systems of the surveyed schools are teaching and educational administration information system accounting for 87.46%, library information system accounting for 82.16%, student management information system accounting for 82.14%; the other two more commonly used information systems are OASystems, financial information systems; nearly half of colleges and universities use logistics service information systems, personnel management information systems, and scientific research information systems; while equipment asset management information systems and decision-making information systems are only used by a very small number of colleges and universities. As shown in Table 1:

Table 1. Commonly used information management systems in colleges and universities

Serial number	Common information management system	Proportion (%)
1	office automation	65.43%
2	Library information system	82.16%
3	Teaching and educational information system	87.46%
4	Student management information system	82.14%
5	Research information system	46.00%
6	Decision support systems	14.56%
7	Financial information system	61.4%
8	Personnel management information system	46.68%
9	Logistics service information system	49.24%
10	Equipment asset management information system	26.48%
11	Archives management information system	45.52%

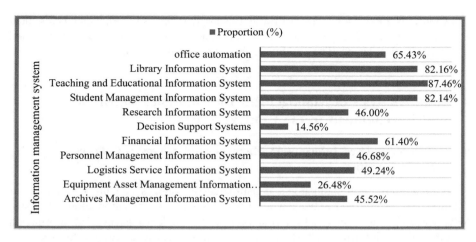

Fig. 1. Commonly used information management systems in colleges and universities

It can be seen from Fig. 1 that the surveyed colleges and universities have their own related information systems, but the construction of various systems is scattered and separate, and the use of them is extremely different. Some are used frequently, and some are basically not used. To a certain extent, it caused duplication and waste of resource construction.

4.2   Information Organization Management

The survey data shows that 54.6% of the surveyed universities have a special management agency responsible for the construction of informatization; only 10.2% of the survey respondents believe that their school does not have a special informatization organization, and 35.2% of the survey respondents It is not clear whether there is a specialized information organization. As shown in Table 2:

Table 2. University information management institutions

	Informationized organization and management agency	Informationized organization leadership
Yes	54.6%	52.1%
No	10.2%	18.9%
Unclear	35.2%	29%

As can be seen from Fig. 2, about whether there are information organization management institutions and information organization leaders, half of the colleges and universities have specialized information organization management institutions and information organization leaders, but one-third of them did not answer. This also illustrates the opaque system of university information management from the side.

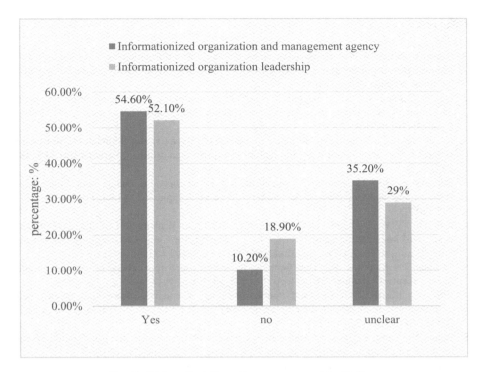

Fig. 2. University information management institutions

5   Conclusion

In the era of BD, advanced technologies such as computers and multimedia are important means to realize the informatization of university management. Carry out management informationization, close to the requirements of the development of the times. Nowadays, management relies more on information networks and online decision-making methods, which is a kind of support and an improvement. In the era of information management, under the influence of advanced information, there are higher requirements for human abilities. The comprehensive quality requirements of management personnel have also changed the personnel structure. Therefore, it is more urgent to achieve innovation to meet the requirements of the information age and take management to a higher level. This research has conducted certain research on the construction of university management informationization, and summarized its own thinking and the promotion strategy for the innovation and development of university management informationization, in order to provide a useful theory and time reference for management informationization.

References

1. Saatova, S.L.: Ways of effective management of information and communication technologies in innovative development of real sector of economy. Econ. Innovative Technol. **2018**(3), 27 (2018)
2. He, Y., Tsibizova, T.Y., Neusypin, K.A., et al.: Methodology of management of innovative development of complex socio-economic systems in the information society. ITM Web Conf. **35**(70), 06001 (2020)
3. Krasil'Nikova, L.: Digital technologies as a factor of innovative development of information support for an agricultural dairy production enterprise management. Agrarian Bull. **13**, 38–45 (2021)
4. Batkovskiy, M., Kravchuk, P., Sudakov, V.: Diversification management information system integrated structures of the military-industrial complex. Bull. Sci. Pract. **6**(1), 237–247 (2020)
5. Padalkin, V.Y., Dudchak, V.V., Prygunkov, A.M.: Implementation of the development strategy of integrated structures on the basis of enterprise's innovative activity assessment. Proc. Voronezh State Univ. Eng. Technol. **79**(4), 339–343 (2018)
6. Tatiana, P., Katerina, V.: Development of information imperatives of socio-economic development of Ukraine. Technol. Audit Prod. Reserves, **5**(43) (2018)
7. Kai, W.: Restricted factors and countermeasures of enterprise economic management informationization based on data interaction analysis. IPPTA: Q. J. Indian Pulp Paper Tech. Assoc. **30**(6), 227–233 (2018)
8. Wang, Y., Yang, Y.: Research on problems and countermeasures of Jiangsu agricultural and rural informatization development. Int. Core J. Eng. **5**(10), 226–232 (2019)
9. Zhang, Y., Zhou, X., Leng, P., et al.: Current development status of cucumber industry in Linyi City and countermeasures for improving quality and benefits. Asian Agric. Res. (5) (2020)
10. Liu, H.: Opportunities, challenges and countermeasures for the development of China's sports industry in the era of BD. J. Phys. Conf. Ser. 1237(2), 022012 (4pp) (2019)
11. Fu, C., Jiang, H., Chen, X.: BD intelligence for smart educational management systems. J. Intell. Fuzzy Syst. **40**(9), 1–10 (2020)
12. Lu, L., Zhou, J.: Research on mining of applied mathematics educational resources based on edge computing and data stream classification. Mob. Inf. Syst. **2021**(7), 1–8 (2021)

3D Model Retrieval Method Based on the Study of Shape Retrieval Method

Weiwen He[(✉)]

School of Information Engineering, Guangzhou Nanyang Polytechnic,
Guangzhou 510000, Guangdong, China

Abstract. Contour is an important feature of a three-dimensional model, and it represents the approximate shape of a model. Because the projection angles of three-dimensional model can be represented with a group of two-dimensional contours, so eigenvalues match on the two-dimensional plane is more stable and efficient. In this paper, use a simple and reliable method to extract the initial projection profile of a three-dimensional model, on this basis, through the noise processing, point of ordering and also other methods to obtain the contour line from the projection of three-dimensional model, and extracting the eigenvalue by two-dimensional Fourier function and saving into the eigenvalue database, and then use square variance method to match the eigenvalue. Finally, we compare the results about recall and precision rate of retrieval between our method and the other methods.

Keywords: Three-dimensional model · Contour line · Eigenvalues · Noise processing · Point of ordering

1   Introduction

With the number of three-dimensional models increasing on the Internet, the efficient retrieval method to take a large number of three-dimensional models is a hot research topic, including major research focused on the study of shape based retrieval method. Shape based three-dimensional model retrieval technology is mainly divided into two categories retrieval ideas: eigenvalue to extract direct effect on the three-dimensional model or approximate the first three-dimensional model represented with graphics on a two-dimensional plane, and then to extraction the characteristic value of two-dimensional graphics. Because the former need depends on the generate accuracy from two-dimensional to three-dimensional model, and also with algorithm defects, to reliability is not high due to the two-dimensional to three-dimensional method of automatic modeling, indirect extract three-dimensional model of the eigenvalue method is more conducive to the three-dimensional model retrieving the improvement of accuracy.

J. C. Hung et al. (Eds.): FC 2021, LNEE 827, pp. 899–906, 2022.
https://doi.org/10.1007/978-981-16-8052-6_114

2   Background

The three-dimensional model retrieval system of Princeton University [1–6] is one of the most complete experimental systems in the field of current three-dimensional model retrieval. The system main research directly extract the three-dimensional characteristics describe the character of the technology, respectively extract three-dimensional model library model characteristic values and Teddy system [7] to produce the three-dimensional characteristics of the model value, thereby matching the search result. This approach is based on Gaussian EDT which is a direct three-dimensional characteristic value extraction method, and the model representation of the value of such features intuitive and precise method, but because of the Teddy system itself inaccuracies caused by the user's query inconvenience, so the final efficiency of the search results is not high. And the other systems are also be established, such as Light filed by National Taiwan University [8], Retrieval of Konstanz University [9] and so on.

3   Methodology

Three-dimensional model retrieval method based on the outline of the main process: First, the three-dimensional model get a projection on a two-dimensional plane of any direction, and then de noising extracting contours calculate the eigenvalues of the data saved to the database side, last match with the eigenvalues of the hand drawn query graph the same or similar search results.

3.1   Projection

Complete three-dimensional model is formed by a polygonal mesh, in the original model file, polygon mesh data defined only by point and the face, when projected onto the entire three-dimensional model on the two-dimensional plane, the human eye actually only can see a one side surface of a three-dimensional model, that is, if the line of sight to see the direction is an acute angle sheet where the triangle faces of the model surface's angle of normal we will be able to see this face, but if the viewing direction and the surface normal to the sheet direction is an obtuse angle, we will cannot see the dough sheet, but only see the other side of the dough piece, to the whole model is the internal model. Therefore, according to this principle, the use of computing the angle between the normal direction and the viewing direction of each dough piece can be obtained, normal is defined only to the polygon, while the point of the normal line is not defined, so in this article gives another way to define the contour: when the angle between the normal and the line of sight direction of the triangular faces is an acute angle, this face is visible, that is called a forward surface, and vice versa other face is not visible to the user, referred to a backward surface, so we define the outline is the common edge between the forward surface and the backward surface. Therefore, the most direct method is to use as defined in the section of the contour, to detect the adjacent faces of the model's each edge, to identify the shared edges of the opposing faces, which also name the outline edge between the forward and backward faces. The formula is:

$$(p2 - p1) \times (p3 - p1) \cdot (E - p1) \tag{1}$$

Points p1p2p3 are three vertices of the model in any one face sheet, E is the viewpoint the location. In order to reduce the amount of calculation, from the departure of the data structure of vertices, lines and surfaces to improve this method, a data set has been defined:

class Model

{class Vertex

{double x,y,z;}; Class Line

{int beginN;

int endN;

int mnface;//The state flag of determine the angle between the normal and the viewing angle int mncheck;};//The state flag of determine whether the line is detected

class Face

{int p1,p2,p3; Line ml [3];};};

Designed two flags, one flag represent positive and negative angle and another represent whether detect or not. We only calculate the angle between the line segment normal and line of sight, and perspective of the definition of an acute angle to 1, and an obtuse angle to 0, and then push them to the flag. Further, since the face scan sequentially, edge is no longer undirected, but rather to the edge: only when the opposite state of the status flag of one edge which both forward and reverse direction of the normal viewing angle, it is the need to find contour edges from a two-dimensional projection. Through improvements in data structures, for edge traversal without double counting, therefore allows the calculation of the amount reduced by more than half, but the above algorithm to generate the contour of the actual effect is not completely satisfactory. For a simple model, in particular, the result of the model with smooth and less concave point is ideal, but for some models with more facets and more convex points, the contour is not satisfactory. Generally, projected operating model has the face of a small number of simple structures in other ways from the same type papers, and therefore most of the projection operation using the brute force method to generate. Because detection is not the outer contour line of the characteristic value extraction, so it needs to do further processing, such as removing the internal contour, the contour line sort, etc. Taken to abandon the practice of internal small contour is taken as to reduce the error to make up the multi viewing projection, as complete as possible retained the external contour information of the three-dimensional model.

3.2 Denoising Extracted Contour Edge

First, calculating the bounding box of the projection edge information from Sect. 3.1, to determine the max value about four direction of top, bottom, left, right around the bounding box, then traverse all the contour edge detected in Sect. 3.1 and extraction side of the midpoint of P, by four most midpoint of the previously calculated value of 4 connections Top P, Left P, P Bottom, P Right, which are composed of four test line, whether the intersection with the other to be detected silhouette edges, as long as any one test line segment don't blocked the contour,then that line is the outer contour (Fig. 1).

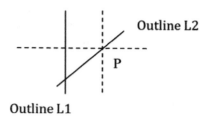

Fig. 1. Scanning line.

Since the projection denoising is based on the segment processing, rather than the point processing, so after denoising there will be some problems: based on the line segment as the unit of processing is performed in accordance with the order of the scanning lines, rather than in accordance with the order of the contour point, so the middle point of the sequence scanned the outline is disordered. Because the matching use a two-dimensional manner as a result of the decision finally, therefore, the two-dimensional point information obtained in the projection should also be processed in accordance with the algorithm of Fourier descriptors, particularly, this algorithm obtain an unordered set of points in Sect. 3.1, therefore, we need to sort of the point of order.

After projection denoising, to observation projection line segments we can found, despite these segments are disordered, but the points within each segment are ordered, the need to deal with these orderly internal segments combined into contour point sequence; and at the same time, both the end and start points of the line segment are correlation. According to this principle, we only need to calculate the order of those discrete of the line segment points obtained in Sect. 3.1, the specific algorithm: beginning from the first line segment L1 to scan the start and end points of all other line segments, and then respectively, to calculate the geometric distance from the point of L1 to the start and end points of the other discrete line segments, from the result of geometric distance to find out the shortest distance from L1 to L2, and then set the segment has been checked flag which will no longer be checked; to repeat operation L1 to L2, traverse completed until all segments. The final contour for the selected point sequence p1, p2, p3, and p4, p6, p5, as shown in Fig. 2, a black dotted line is added for the calculation, the red dashed line segments is finally added. Last point sequence is such as p1, p2, p3, p4, p6, p5.

As it relates to the search function, can't put just the first few steps of the calcu-lation into the array on the program, but the results of the retrieve data should be permanently saved. Therefore, it is necessary to create a separate database to store eigenvalues. The benefits of doing so allows the permanent preservation of data for user queries, as well as an important advantage that we can simplify the retrieval

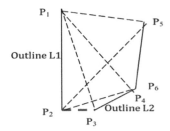

Fig. 2. Sorting point.

process. The database can only save the eigenvalues of the model and location of the files, rather than save the entire model data file, so that we can greatly reduce the amount of computation in the search process, in which the characteristic values in order to find a match using the location of model file to find the file directly to the user to find the model files.

According characteristic value extraction method of present system, to compare characteristic values similarity about the two-dimensional graphics using that restrict the characteristic values of three-dimensional models to the two-dimensional graphics. Therefore, restrict the contour lines of the projection with the same way of Fourier descriptors, calculated by Fourier descriptors to a contour of the series with a set of 10 values. Because the three-dimensional model is higher than the two-dimensional graphical in space dimension, so projection is not only one direction. Taking into account the habitual users, there are totally six of the direction of the three-dimensional orthonormal projection direction, including up and down, left and right, front and rear, that every coupe projection are similar. All above, the system project models with three directions: top view, front view and left view.

In addition to one of the measures to accelerate the speed of retrieval is to save the models features classified values after classification models. First, understanding classified models by are normalized to a category, and then we are established for each category of the table class, uncommon and bad normalization class models are grouped together into a category.

Calculated for the first time need to search the entire model library, computer processing time is very long, as a result, processing eigenvalue for almost 1800 models spend about 10 h.

Eventually store all the characteristics of values to a table, this table contains the model ID(id), the model name(name), the location of the model file(path) and three perspectives respectively model eigenvalues (top view, front view and left view).

Above all, the whole produce is that three-dimensional model data project into three groups of two-dimensional graphics data, next these data are converted into a one dimensional array through the Fourier descriptors, and the final match is in the one dimensional array.

3.3   Match

Because Feature extraction applies the method described by Fourier descriptor, the eigenvalue restrict in one dimensional space and the final values array numeric comparison, therefore, not related to the model itself panning, zooming, rotation invariant processing, due to the characteristic value is calculated by Fourier descriptors stored in the database and therefore the match happens in the numerical calculation areas. Values for the two groups were compared in similarity, the probability of referencing the variance concept, respectively into the two sets of values of the mean square deviation, to obtain the smaller the value of the discrete nature of the smaller, it means that the graphic of the user sketched is more similar to the model which searched in the model library.

$$S = \sqrt{\frac{\sum_{i=1}^{n}(x - x)^2}{n}} \tag{2}$$

Through the characteristics which mean square deviation can compare a set of the numerical discrete to measure the differences between the Fourier value extracting from hand painted results and Fourier value saved in the three-dimensional features library.

$$S^2 = \frac{1}{n}\sum_{m=1}^{n}(x_n - x_n')^2 \tag{3}$$

Since the viewing angle includes three angles about the front, the left and the top, therefore we can obtain the id list of the most similar model drawn by the user in the direction of the respective projection the graphics, and then, by contrast, first determine the priority higher the closer the top of the list id. Then, according to the rankings, respectively, we weighted discharged and painted graphics on the whole the most similar model sequence in the three views.

4   Experiments

Hardware configuration: core2 duo CPU p8600 with 2G memories; software configuration: Windows 7, Microsoft Visual Studio 2015 integrated development environment for C++ development platform, and rendering of three-dimensional model and draw using openGL package. The three-dimensional model of the experiment can read VRML, OFF, OBJ format which free download from the benchmark library of Princeton University [10], and the three-dimensional characteristics of database use SQL server 2010 to store.

Test for each search method precision/recall rate graph to represent the quality of search results, the accuracy the class search model number/total completion of the search model number; check the full rate in the search for the model number/the total number of class model. Calculated for each model match score, match the results obtained are ranked from best to worst, measured in the same class, the number of queries which model is closest to the ranking system top.

The results of the experimental are shown in Fig. 3.

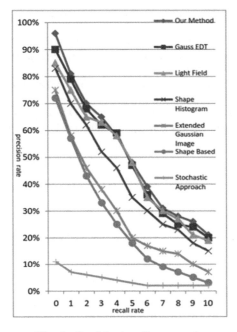

Fig. 3. Precision/recall rate graph.

5   Conclusion

Direct extraction of characteristic values of the three-dimensional models is the best method to express the models. Due to the lack of algorithmic support for the extraction value, this approach also requires a lot of improvements. Such as, we can find a better indexing method, because the method to the models based on design of the model library, and to retrieve search indexing technology will have a direct impact on the retrieval efficiency. On the other hand, we can retrieve the three-dimensional models more time in the details of the match in order to achieve improved retrieval accuracy. Also, we can design this system to get a variety of retrieval results, and then in the search results retrieved through a second hand way, so you can increase the retrieval accuracy.

References

1. Funkhouser, T., et al.: A search engine for three-dimensional models. ACM Trans. Graph. **22** (1), 83–105 (2003)
2. Min, P.: A 3D model search engine. Citeseer (2004)
3. Funkhouser, T., et al.: A search engine for 3D models. ACM Trans. Graph. **22**(1), 83–105 (2003)
4. Min, P., et al.: Early experiences with a 3D model search engine. ACM (2003)

5. Min, P., Chen, J., Funkhouser, T.: A 2D sketch interface for a 3D model search engine. ACM (2002)
6. Funkhouser, T., et al.: Modeling by example. ACM (2004)
7. Igarashi, T., Matsuoka, S., Tanaka, H.: Teddy: a sketching interface for three-dimensional freeform design. ACM (2007)
8. Chen, D., et al.: On visual similarity based 3D model retrieval. Blackwell Publishing, Inc. (2003)
9. University of Konstanz, G. <http://3d-search.iti.gr/3DSearch>. 24 Mar 2010
10. Princeton. <http://shape.cs.princeton.edu/search.html>. 24 Oct 2016

Design of Civil Aviation Cargo Transportation Information Exchange Platform Based on 5G and Edge Computing

Feilong Chen[✉]

Sanya Aviation and Tourism College, Sanya City 572000,
Hainan Province, China

Abstract. The Internet age is an innovation of science and technology, which has changed our basic necessities of life, food, housing and transportation, and also promoted the development of various industries. In China's transportation, road freight transport still accounts for a large proportion. Therefore, the highway freight transportation also needs a trans era change. This paper mainly analyzes the significance of the construction of the freight information exchange platform in the Internet era, and the development prospect of the freight information exchange platform.

Keywords: Construction system · Multi-objective optimization · Ant colony algorithm

1 Introduction

Under the influence of technology, information has reached instant messaging. It greatly improves the communication of information. The network can easily understand the transportation information of goods, and even locate the location of goods in real time, which can greatly improve the economic growth of goods. Each city is adjusting its own city to optimize.

2 Current Situation of Cargo Transportation

2.1 China's Freight Transport and the Main Mode of Transport

It can be seen from Table 1 that during 2014–2015, the freight volume increased by 85.9 million tons, with a growth rate of 1002613% and a growth rate of 0.2061289%. From 2015 to 2016, the freight volume increased by 2108.77 million tons, the development speed was 1050.4998%, and the growth rate was 504.987%. From 2016 to 2017, the freight transport volume increased by 4180.87 million tons, the development speed was 1095306%, and the growth rate was 9.53% [1]. From 2017 to 2018, the freight volume increased by 3478.82 million tons, the development rate was 1072402%, and the growth rate was 7.2402%.

J. C. Hung et al. (Eds.): FC 2021, LNEE 827, pp. 907–911, 2022.
https://doi.org/10.1007/978-981-16-8052-6_115

Table 1. China's freight transportation and main modes of transportation from 2014 to 2018

	2018	2017	2016	2015	2014
Cargo transportation volume (10000 tons)	5152732	4804850	333186	4175886	4167296
Railway freight volume (10000 tons)	402631	368865	334259	335801	381334
Highway freight transportation volume (10000 tons)	3956971	3686858	3341259	3150019	3113334
Water cargo transport volume (10000 tons)	702684	667846	638238	613567	598283

2.2 Comparison of Highway Freight Transportation Volume and Other Transportation

In 2014, the proportion of road freight transportation was 74.708%, that of railway freight transport was 91.506%, and that of water transport was 14.356%. In 2015, highway freight transportation accounted for 75.433%, railway freight transportation accounted for 8.041%, and water cargo transportation accounted for 14693%. In 2016, the proportion of road freight transport was 76.166%, that of railway freight transport was 75.92%, and that of water cargo transport was 144.549%. In 2017, the proportion of road freight transport was 76.732%, that of railway freight transport was 7.6769%, and that of water cargo transport was 13899%. In 2018, the proportion of road freight transport volume was 76791%, that of railway freight transport was 7.8139%, and that of water cargo transport was 13637%.

Can be very intuitive to see that in today's road freight traffic still occupies a very large proportion and advantages.

2.3 Highway Freight Transportation Volume in Recent Five years in China

It can be seen from Fig. 1 that from 2014 to 2015, the growth of highway freight transportation was 366.85 million tons. The development rate was 101 1783%, and the growth rate was 1.17832%. From 2015 to 2016, the highway freight transportation showed an increase of 191.2 million tons, with a development rate of 106071% and a growth rate of 607107% [2]. From 2016 to 2017, the growth volume of highway freight transportation was 3455.99 million tons, with a development speed of 110.3433% and a growth rate of 10343%. From 2017 to 2018, the growth of highway freight transport volume was 2700.13 million tons, the development rate was 107.3236%, and the growth rate was 7.3236%.

Highway freight transportation volume is increasing year by year, with a total increase of 8435.37 million tons in five years, with a growth rate of 27.0943%.

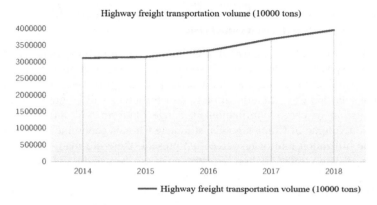

Fig.1. Road freight transport volume in recent five years

2.4   Recommendation Forecast and Evaluation

The average absolute error 3mae is one of the criteria to evaluate the quality of the recommended algorithm. It evaluates the accuracy of the experimental results by calculating the difference between the predicted score and the actual data value. The smaller the value, the higher the accuracy. If $S = \{s_1, s_2, \ldots, s_n\}$ represents the set of predicted values and the corresponding actual score set is $A = \{a_1, a_2, \ldots, a_n\}$, then the specific Mae calculation formula is as follows:

$$MAE = \frac{\sum_{i=1}^{n} |a_i - s_i|}{n} \tag{1}$$

3   Background and Significance of Freight Information Exchange Platform

3.1   Background of Freight Information Exchange Platform

The most effective transportation of goods is a very complex system, which is related to China's economic growth. In today's new century, we need a new combination of cargo transportation and Internet information exchange mode, and connect through the Internet to realize the intellectualization and systematization of cargo transportation information. Through the freight information exchange platform, we can improve the efficiency of cargo transportation and make effective use of resources by using the Internet with strong timeliness. According to the data of the National Bureau of statistics, China's freight transport volume is increasing year by year. In 2018, China's freight transport volume was 515527.32 million tons, and in 2017, it was 48048.5 million tons, In 2016, the freight volume was 43867.63 million tons, in 2015, 41758.86 million tons, and in 2014, 4167.296 million tons.

3.2   Freight Information Exchange Platform can Improve the Deficiencies of Cargo Transportation Supervision

From the data collected, we can see that China's freight transport volume is gradually increasing every year, and the growth is still quite amazing. But now, because most of the trucks are delivering goods to a certain area, there is a great possibility that they will return empty, In this way, the utilization rate of resources will be greatly reduced [3]. The construction of cargo information exchange platform will further enhance the transportation volume of China's goods, stimulate China's economic development and improve the utilization rate of resources. At present, there are many loopholes in the supervision of cargo transportation industry, such as overweight and overrun. It has a great impact on drivers and us. For example, the collapse of the curium free viaduct may not be the last straw that killed the camel. In other words, there will also be more overweight straw on the road. Our freight information exchange platform can strictly supervise the operation of freight cars and standardize the industry rules of the market. What kind of car, how much cargo should be loaded. If drivers or enterprises have violations, we can stop them in time and even pull them into the blacklist of our users. Through the cargo information exchange platform, we can improve the regulatory loopholes of China's current cargo transportation industry.

3.3   Freight Information Exchange Platform can Improve the Income of Drivers

Most of the transportation by large trucks is long distance transportation, and the drivers' fuel and road tolls are relatively high. They may return to their own cities empty cars, which will seriously affect the income of truck drivers. Through our freight information exchange platform, we can choose the appropriate freight order information, return to our own city, or areas close to our city, directly increase the income of drivers.

3.4   Freight Information Exchange Platform can Reduce the Transportation Cost of Enterprises

This freight information exchange platform is mainly aimed at small and medium-sized enterprises, because the difference between small and medium-sized enterprises and large-scale enterprises is that large-scale enterprises will have their own independent transportation teams, while small and medium-sized enterprises may not have enough ability to set up their own transportation teams, and they can only contact the outsourced transportation teams with the help of a third party, which will increase the transportation costs of enterprises [4]. If it is conducive to the freight information exchange platform, it will reduce the cost of their own enterprises, so as to improve the profits of enterprises.

4 Conclusion

From the current form of analysis, China is in great need of building a freight information exchange platform to innovate China's highway transportation. With the continuous development of China's economy, the industries of various cities are being adjusted and the circulation of commodities in each city has been accelerated. To stimulate the further development of China's freight transport, we should use the Internet, a highly time-effective communication mode, to change the cross city freight mode, improve the efficiency of freight transport, and make effective use of resource.

References

1. Guang, S.: Construction of urban freight public information platform. Digit. Technol. Appl. **11**, 221 (2016)
2. Xiaoxue, R.: Analysis on Influencing Factors of highway freight transportation development. Mod. Mark. (Bus. Ed.) **8**, 66 (2018)
3. Li, D., Yu, N., Yin, H.: An internet hotspot mining algorithm in Web 2.0 environment. Acta Electron. Inf., **32**(05), 1141–1145 (2010)
4. Peng, W.: Research and implementation of Internet news hotspot mining system. Harbin Institute of Technology (2010)

E-commerce Big Data Classification and Mining Algorithm Based on Artificial Intelligence

Hongfeng Chen[✉]

Jiangxi Vocational Technical College of Industry and Trade, 699 Jiayan Road, Nanchang city 330038, Jiangxi, China

Abstract. Some languages have been applied to people's daily life and learning, which has a profound impact on and learning, and also profound impact on the development of Chinese language and literature teaching. Therefore, on the basis of expounding the characteristics of network language, this paper analyzes from both negative and positive aspects, and makes a strategic analysis on how to better literature with the help of network language, so as to better promote education.

Keywords: Chinese language and literature · Influence · Significance

1 Introduction

People have made some breakthroughs in the field of scientific research, but in the information age, the use of words is still in a basic functional state. With the development of the Internet age, network language appears in the field of text language application, such as "blue thin mushroom", "little pot friend", etc. the emergence and development of network language not only changes people's language expression, but also has a profound impact on people's ideas.

2 Overview of Network Language

Network language is a language form produced in a certain environment in the development process of modern network. There are great differences between network language and traditional language form in application and development. The most obvious difference is that network language has strong creative characteristics, and the object of using network language is the social masses. use rate of language communication and express the meaning of language more clearly, we need to simplify the language expression itself and express some content more clearly by simplifying the language.

The development characteristics of network language are shown in the following aspects: first, strong simplification. The traditional language chat will make the language communication between people lack of interest, but the use of network language can further enhance the creativity of the overall language expression, change the meaning and form of expression of Chinese characters in the past, and improve the efficiency of communication between people. Network language includes homophonic, homophonic, digital and other types of language. At the same time, some words with similar pronunciation are introduced in the expression of meaning, which has a distinct referential meaning [1]. For example, "microblog" is used by people to express themselves in the network language "Weibo"; second, there is a strong degree of mixing. The rise and development of the Internet is inseparable from the support of the post-90s and post-90s. In particular, the post-90s often use some Martian characters, English words, graphic symbols, etc. in the process of chatting, such as "PFPF" stands for (admire), and some youth conferences use some pictures to express their ideas, which leads to a large number of expression packs. Network language has a simple form and input mode, so it has been widely spread on the Internet since it came into being, showing strong flexibility in the expression of words and meaning.

3 Artificial Intelligence Algorithm

3.1 Introduction of Artificial Intelligence Algorithm

Image inpainting algorithm. It is usually used to repair the incomplete RGB image and supplement the incomplete position, so as to make the repaired image reasonable or close to the original image before damage. In recent years, deep learning technology and computer parallel computing ability, the effect of image inpainting algorithm using deep learning technology has also been greatly improved. which can only get fuzzy results, the latest image inpainting algorithm has been able to get semantic reasonable results, In color, texture and even structure information, it can reach the level that human eyes are not easy to distinguish. With the wide application of deep learning technology in RGB images. Depth image processing algorithm is also developing rapidly. Thanks to the popularity of consumer level depth cameras such as Microsoft Kinect and Intel realsense, more and more researchers begin to focus on various depth image processing algorithms.

$$v_{ver} = [0, 2, z_{(x,y+1)} - z_{(x,y-1)}] \tag{1}$$

$$v_{hor} = [2, 0, z_{(x+1,y)} - z_{(x-1,y)}] \tag{2}$$

$$v_{nor} = v_{ver} \times v_{hor} \tag{3}$$

Threshold convolution image inpainting algorithm.

Among various image inpainting algorithms using deep learning technology, threshold convolution 8 is a relatively simple but effective algorithm. When the damaged depth image is input into the repair network, not every pixel can provide enough effective information for network learning and speculation. For example, pixels in the damaged area at the same time, pixels with different distances from the damaged area also provide different value information for network prediction. Therefore, threshold convolution learns a mask of each layer of features [2]. so as to assign different weights to the channels of each feature. After multiplication, it can highlight the features that are conducive to inferring the defective pixels, and weaken or even ignore the invalid features. In recent years, it has become a common strategy for advanced image inpainting algorithms to guide the image inpainting process through some auxiliary information. For depth image, there is a natural and superior auxiliary information, that is surface normal image. Because the surface normal graph takes the three components of the surface normal as RGB channels, it can express the same color in the plane facing the same direction in the image scene, as shown in Fig. 1, which can provide powerful semantic information. As shown in Fig. 1, if the depth image is damaged, it is preferred to repair the surface normal image with the same color, so as to guide the depth image repair network to repair the plane completely.

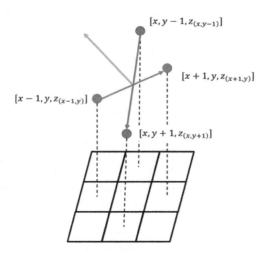

Fig. 1. Calculation diagram of surface normal

4   The Influence of Network Language on the Development of Chinese Language and Literature

4.1   Positive Impact

The development of Chinese language and literature will be further enhanced[3]. Therefore, under the background of network language development, relevant personnel need to be able to deeply understand the current situation and problems.

4.2   Negative Influence

First, guide students to write wrong characters. Network language is a kind of language produced in the process of small group chat, so it is not standardized in the use and creation of language. Especially under the influence of excessive novelty development of network language, some languages are often not understood and recognized after they are formed. At the same time, in the process of using and spreading network language, some wrong words and other words will be used intentionally, which deviates from the application standard and basic significance of Chinese language and literature. Teenagers don't have a systematic understanding of the use of language in the process of development. Under the influence of network language, they are more likely to misinterpret their understanding of language, which is not conducive to their correct use of all kinds of languages in the future. Secondly, it is conducive to the memory of words. The popularization and development of the network makes people more dependent on the use of computer network for the transmission of words. In the long run, there will be the case of forgetting words. In addition, under the generation and influence of network words, people will unconsciously express "I" when they want to express "I".

5   The Optimization of Chinese Language Development Under the Network Language Environment

5.1   Learn from Others' Strong Points and Close the Gap

The development of society has a profound impact on the development of network language, and the emergence of mass media has also promoted the spread of network language. From this point of view, the emergence and development of network language development and dissemination of culture, which can make people express their thoughts and feelings in a more concise language [4]. At the same time, the network language itself is vivid and interesting, which provides an opportunity for the spread and development of. Therefore, in the new era, when we use and develop network language, we need to establish the learning model of quoting.

5.2   Network Language

The extensive use of network language in Chinese language and literature guiding in social development, and then help people better express their feelings and enhance the richness and interest of language use. Therefore, in the development Chinese language and literature.

6   Characteristics of Network Language

6.1   Dependence

The network language has dependence language. many young people will use the network language when they communicate with each other, but they often make mistakes. This is because language depends on the network environment to a certain extent, and in the case of leaving the network environment, people will have some problems when they use the network language again. In addition, when chatting on the Internet, we usually add various expressions, but in the actual chat, the network language will be ambiguous without the help of expressions. Although network language can break the Convention of Chinese language and literature and make Chinese language and literature more novel, it also makes the meaning of many words change greatly and distorts the usage of words, so that students will make mistakes in sentences and words when answering questions, which also brings a lot of inconvenience to educators, It has a negative impact on the language education of teenagers [5]. Because the best period for people to learn is the youth period. In this period, the youth's thoughts are in the most active state, and they have a strong ability to accept new things, but they lack the ability to distinguish right from wrong. In the network environment, young people are more likely to form bad writing habits, improper use of network language will lead to young people can not correctly identify and appreciate the content of Chinese language and literature, which has a great impact on the growth and learning of young people.

6.2   Creativity

Network language is created when people communicate through the network, so the creativity of network language is strong. When people have a complete grasp of the Chinese language, and then communicate through the network, they will actively mobilize their thinking, give full play to their imagination, and then create new sentences and words with special meanings. In the process of creating new sentences and words, people will create a sense of freshness in the process of chatting. The use of network language in communication will make chatting more pleasant and relaxed. In the network chat, the use of traditional Chinese language often makes people feel monotonous because of its more simple and other reasons. With the expansion of the application scope of network language, it has deeply affected people's life. At present, the network language has been fully promoted in the network environment, and by virtue of the advantages of network communication has been very popular. In people's life, the network language can give full play to its practical role, fully reflect people's real psychological feelings and specific expressions, and is loved by most young people.

7   Concluding Remarks

To sum up, in the network environment, the emergence and development of network language has become an inevitable social phenomenon. The rapid development of network language has both advantages and disadvantages. On the one hand, it can simplify people's communication language and promote people's emotional communication, on the other hand, depth neural network can repair the depth image better by surface normal guidance. Because the real surface normal image can not be obtained in the actual damaged image restoration scene, it has better practical significance to repair the surface normal first and then guide the strategy of repairing the depth image. In the following research, we can continue to optimize the repair effect of surface normal repair network and depth image repair network by adding confrontation training. More close to the real surface normal image will bring greater improvement to the depth image repair.

References

1. Yixin, Z.: The influence of network language on the development of Chinese language and literature and the countermeasures. Northern Lit. **17**, 123–124 (2020)
2. He, Y., Zhang, X., Tang, Y., Li, S., Wang, Z.: Analysis of the influence of network language on the development of Chinese language and literature. China New Commun. **22**(10), 149 (2020)
3. Ping, W.: On the influence of network language on the development of Chinese language and literature in the new era. Young Writer **12**, 57 (2020)
4. Wu, Y.: On the influence of network language on the development of Chinese language and literature in the new era. 100 Proses (Theor.), (04), 136 (2020)
5. Huang, M.: Analysis of the impact of network language on Chinese language and literature. Shen Hua (1, 04), 61 (2020)

Traffic Signal Timing Optimization Based on Genetic Algorithm

Xiaomin Hu[✉]

School of Transportation, Xi'An Traffic Engineering Institute, Xi'An 710300, Shaanxi, China

Abstract. With the rapid development of China's society and science and technology, automobile has become an indispensable means of travel for people, which has caused a sharp increase in the number of vehicles on the road, and brought great challenges to the city's transportation system. Traffic congestion has hindered the rapid development of the city. Aiming at the traffic signal timing problem, this paper uses the model to predict the traffic flow. Based on the error estimation model, this paper takes the average vehicle delay as the performance index to optimize the signal intersection timing. Finally, the improved genetic algorithm is used to output the comparison results. With the continuous improvement of computer, software technology and intersection capacity control level, urban traffic signal control has become the research focus of scholars all over the world.

Keywords: Genetic algorithm · Traffic lights · Timing optimization · Intelligent transportation

1   Introduction

In the report of Global Association for mobile communication systems (GSMA) in 2019, it is emphasized that in the 5C era, any road user (including vehicles, non motor vehicles and pedestrians) can exchange information with the surrounding environment in real time and share the real situation of the road.

5G mobile communication network refers to the fifth generation mobile network communication technology, which integrates large-scale antenna array, ultra dense networking, terminal direct connection, cognitive radio and other advanced technologies, providing the possibility to solve the current problems and challenges of the Internet of vehicles. In the high-speed mobile environment, 5G on-board devices of the Internet of vehicles can get better performance. At the same time, 5G does not need to build a separate base station and service infrastructure, which is conducive to the large-scale deployment of the Internet of vehicles environment, and provides a new idea for the construction of the Internet of vehicles environment.

The traditional research object of urban traffic signal timing control is often a single intersection, and it often uses the idea of timing control to design the traffic signal timing scheme. Sometimes it also considers the significant difference of the whole day traffic flow, divides a day into several time periods, and then uses the method of time division to design the timing scheme, or all day timing scheme unchanged. No matter which scheme does not consider the randomness of traffic flow. Such a timing scheme with fixed time and interval of traffic lights is difficult to adapt to the situation of obvious fluctuation of traffic flow, which often causes traffic jams at actual intersections. With the deepening of the research, some big cities begin to coordinate and optimize the traffic lights of multiple intersections [1], so as to form an ideal green wave band between the intersections with high correlation. To a certain extent, the traffic flow is smoother and the time delay of vehicles is reduced. However, this control scheme is fixed and can not fundamentally solve the problem, and it is still easy to cause frequent traffic jams. And when considering the performance index optimization, only a single index is considered, which is obviously unreasonable. Aiming at the disadvantages of the traditional traffic signal timing control strategy, this paper intends to study the dynamic optimization problem of traffic signal timing scheme, that is, traffic lights can adjust the timing scheme intelligently according to traffic factors and real-time traffic flow, and carry out a number of performance index evaluation, so as to enhance the traffic flow passing capacity at the intersection as far as possible, reduce the delay time and parking times, and reduce the platoon Team length.

Using the principle of replication, crossover and mutation of genetic algorithm, we can not only search the solution of complex nonlinear model, but also ignore the internal role and relationship between various factors. Neural network has the characteristics of self-learning ability. It can adjust the traffic control continuously through continuous self-learning in intelligent transportation system to achieve ideal results. Although neural network and genetic algorithm provide excellent methods for complex traffic signal optimization, but at the same time, the shortcomings and shortcomings are also obvious. Neural network is essentially the idea of gradient method, and the results are often obtained in the local region without considering the global situation. The influence of the initial state on the convergence of neural network is critical. The disadvantages of genetic algorithm are also obvious. In addition to the shortage of easy convergence to the local optimal solution, it is not suitable for the new space. Secondly, when a large number of individuals are involved in the calculation, the calculation time is relatively long. It is difficult to deal with and optimize problems with high dimension and nonlinear constraints. And genetic algorithm is a kind of random algorithm, the results of each solution may be different, often need to carry out multiple operations to get more ideal results. As we all know, modern transportation system has uncertain factors which are difficult to complete statistics, and it is difficult to solve this problem with general algorithm theory. Fuzzy control theory can solve a lot of uncertain resource information according to the designed fuzzy control rules. Therefore, the application of fuzzy control theory to traffic signal timing has great advantages.

2 Optimization Analysis of Traffic Signal Timing

By summarizing and analyzing the existing research results, we can see that these schemes consider multiple control performance indexes of intersections, but ignore the real-time change of traffic flow, and consider the real-time change of traffic flow, but ignore the influence of various traffic factors. Obviously, these two independent cases can not fully meet the needs of traffic demand, and can only achieve traffic adaptive control to a certain extent. In other words, it is difficult to adapt to different traffic states only by controlling a single index, and it is also difficult to adapt to the characteristics of randomness and uncertainty of modern traffic flow only by considering timing control. And the coordinated control of multiple intersections is bound to become another research trend. On the basis of the existing research, this paper considers the comprehensive evaluation of capacity, delay time, queue length and parking times of multiple intersections, and introduces pedestrian and non motor vehicle indicators. It intends to use fuzzy control algorithm and multi-objective optimization method to study the dynamic timing scheme of single point intersection, and use numerical method to solve the initial optimal phase difference to study multiple intersections The goal of the scheme is to achieve the optimal control effect.

2.1 Traffic Signal Phase

The first consideration of traffic signal timing optimization is the intersection phase setting. Different types of intersections adapt to different phases. The setting of phase usually depends on the historical data of traffic flow and natural road conditions. Generally, the phase setting of intersections with left turn direction traffic flow is relatively complex. The more phase setting is, the less conducive to traffic flow. If the traffic flow is low saturation, the multi-phase design can facilitate people's travel. If the traffic flow is approaching saturation, the multi-phase design will easily lead to long queue at the entrance of the intersection, resulting in traffic congestion. Then we need to consider is to determine the phase sequence, a reasonable phase sequence can not only reduce the time of pedestrians and vehicles at the intersection, but also improve the capacity of the intersection. Whether the design of phase and phase sequence is reasonable or not is directly related to the effect of traffic signal timing optimization. If the left turn traffic flow in both directions is large, the four phase scheme is better. As shown in Fig. 1, the left turn traffic flow in the east-west direction is relatively large, so the intersection has a dedicated left turn lane on the East-West entrance road. The setting of left turn phase can dredge the left turn traffic flow with large flow in the East and West entrances.

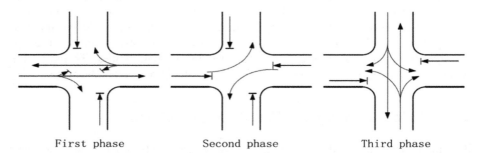

First phase   Second phase   Third phase

Fig. 1. Three phase scheme of left turn phase

2.2   Time Parameter

Intersection signal period is also an important parameter of timing optimization. The optimization effect of traffic signal timing is closely related to the signal cycle [2].

Green signal ratio refers to the ratio of effective green time and cycle of each phase in a traffic signal timing cycle, expressed by λ.

$$\lambda = \frac{g_c}{C} \tag{1}$$

Where: g_c effective green time, unit: s.

Green signal ratio is the most important time parameter in traffic signal timing optimization. According to this parameter, the effective green time of each phase can be obtained when the signal period is known.

Lost time refers to the time that is not fully used or wasted due to no traffic flow in the period when the signal phase can pass, which is generally caused by the driving characteristics and safety factors of vehicles.

$$\sum_{k=1}^{R} \lambda_k = \frac{C - L}{C} \tag{2}$$

Capacity refers to the maximum number of vehicles that a road or a road section can pass in unit time when the vehicle is driving at the permitted speed under the natural conditions and traffic regulations of the existing road.

It is shown that. Numerically, capacity is equal to the product of saturation flow and green signal ratio.

$$Q = S * \lambda = S * \frac{g_c}{C} \tag{3}$$

The capacity of an intersection is determined by the capacity of the entrances in different directions of each signal phase. The capacity of the entrance road is positively proportional to the green signal ratio.

3   Timing Method of 5G Intelligent Traffic Lights

Considering the difficulty of using real-time traffic flow data and the initial stage of 5G Internet of vehicles system, it is difficult to provide detailed and complete traffic volume information. Therefore, this paper uses the grey prediction model to predict the short-term traffic volume, and the prediction value is used to optimize the information allocation of the nodes and the surrounding arterial intersections.

When Gu Huaizhong and Yan Yanxia studied traffic signal timing optimization, they introduced capacity based on delay time and parking times, and refined each index to the average level of each phase. The model is as follows:

$$minf(g_l) = \sum\nolimits_{i=1}^{n} \{a_i D_i + \beta_i H_i - \gamma_i Q_i\} \tag{4}$$

Different weights are determined according to the peak and flat peak characteristics of traffic flow at intersections and the importance of the three indicators, and the exact timing optimization model is obtained. Simulated annealing algorithm and ant algorithm are used to solve the model respectively, and the control effect is better than that of Webster method.

In order to improve the utilization efficiency of limited road resources, it is always an important research direction to reduce the vehicle delay on the entrance road of each phase of the entrance and exit as far as possible. The vehicle delay D on the road is mainly caused by the consistent delay D_ WI, and random delay d_i. It consists of two parts, d_{wi} is the delay with constant arrival rate, d_i is the delay with inconsistent vehicle arrival rate. So the average delay time of phase I is d_{ri} is

$$d_i = d_{wi} + d_{ri} = \sum\nolimits_j \frac{C(1 - \lambda_i)^2}{2(1 - y_{ij})} + \frac{y_{ij}^2}{2\lambda_i q_{ij}(\lambda_i - y_{ij})} \tag{5}$$

Important performance parameters of signalized intersection timing: capacity, delay, queue length, number of stops and saturation, etc. In this paper, the four performance parameters of improving traffic capacity, reducing delay, reducing queue length and reducing the number of stops are taken as the indicators of signal timing. Among them, capacity, delay and parking times are weighted as the parameters of multi-objective joint optimization [3]. The queue length is used as the fuzzy control index, and the effective green time is added. Because of different units, the three indexes of joint optimization are dimensionless. Different weights are used to express the importance of indicators. Therefore, the joint optimization function of signal timing is as follows:

$$minPI = a\frac{D}{D_0} - \beta\frac{Q}{Q_0} + \gamma\frac{H}{H_0} \tag{6}$$

The selection of weighting coefficient should consider the actual situation of traffic flow. When the traffic flow is low, the delay and parking of vehicles at the intersection should be reduced as much as possible. When the traffic flow is approaching saturation and oversaturation, the capacity of the intersection should be improved as much as possible. Each phase has two directions of traffic flow, so the coefficient is multiplied by 2.

$$a_i = 2sy_i(1 - Y) \tag{7}$$

$$\beta_i = 2 \times \frac{YC}{3600} \tag{8}$$

In the genetic algorithm, the fuzzy dynamic optimization takes the queue length of the current phase and the next phase as the fuzzy control variables. According to the theory of double input and single output in the fuzzy control, the queue length of the current phase and the queue length of the next phase are first fuzzed, and the fuzzy synthesis relationship is obtained through the operation of the fuzzy rule table. Then, taking the queue length of the current phase and the next phase of the actual inter-section as the input variables, the fuzzy value of the green light additional time is obtained by the fuzzy control rules, and the green light additional time is obtained by the anti fuzzy processing.

4   Regional Coordinated Control of Intelligent Traffic Lights Based on 5G

At present, the research on regional traffic signal control is actually the research on traffic signal optimization control of a control sub area in a large area. By dividing the urban road network from a large area into several sub areas, we can complete the traffic signal control of the control sub area and complete the traffic signal coordination and optimization of the large area. Therefore, the research object of this paper is mainly the control sub area. Generally speaking, the control sub area describes the urban traffic road area including multiple continuous adjacent intersections, and the number of intersections in the selected sub area is 8 to 10, as shown in Fig. 2.

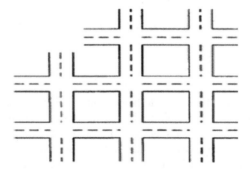

Fig. 2. Example of sub district traffic network

Before studying the traffic signal control of control sub area [4], it is necessary to abstract the road network model of control sub area, and then design the control strategy according to the road network model. Generally, circles are used to replace the intersections of the road network, and the connecting lines can be regarded as the road sections between intersections. The simplified road network diagram of the control subarea is shown in Fig. 3.

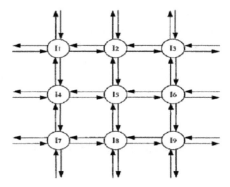

Fig. 3. Road network structure diagram of typical sub areas

The role of urban trunk road in urban road network determines that it is the trunk line carrying the main traffic. In a region, large traffic flow is its most significant feature, and the condition of regional traffic depends on it to a certain extent. In fact, the signal coordination and optimization system of urban main road is a "by wire" system. Compared with the traditional urban trunk road, the urban trunk road can be a broken line or a straight line. The control sub area is divided into three parts: main road, secondary road and branch road. The signal coordination optimization of main road, secondary road and branch road is carried out in the order of first main road and then secondary road. So as to realize the signal coordinated control of the whole region.

5 Conclusion

Based on the development trend of intelligent transportation in 5G era, this paper puts forward the optimization scheme of traffic signal timing, aiming at the problems of time-consuming of traffic volume data transmission and processing, immature technology and insufficient popularity in the initial stage of Internet of vehicles. The research on the coordinated timing control scheme of multiple intersections should be based on the dynamic timing scheme of single intersection. Adjust the system cycle to the intersection cycle, change the green time accordingly, recalculate the phase difference, and give the timing scheme of each intersection. Considering the situation of mixed traffic flow in China, a multi-objective optimization model in mixed traffic environment is established. It is better to divide the sub regions, especially the dynamic division of sub regions, and then use the coordinated control strategy based on trunk road priority for multiple intersections in the sub region. This new idea of dividing the regional signal coordinated control into two parts is conducive to reducing the difficulty of regional traffic coordinated control and improving the effect of the whole regional traffic control.

References

1. Zheng, C., Xiaoyi, L., Dandan, L., et al.: Intelligent traffic signal timing optimization based on video recognition. Autom. Technol. Appl. **37**(11), 139–142 (2018)
2. Chengyu, Z., Qiao Xiaokang, W., Xianjin, et al.: Traffic signal timing optimization model based on traffic flow density. Sci. Technol. Innovation Guide **15**(5), 167–169 (2018)
3. Wang, X.: Research on Data Fusion Trust Model in The Environment of Internet of Vehicles. Chang'an University, Xi'an (2014)
4. Xiaofei, W.: On the development and application prospects of internet of vehicles based on 5g technology. Commun. World **26**(10), 151–152 (2019)

E-commerce Customer Segmentation Based on RFM Model

Jiangwei Ma[(✉)]

Wuhan Huaxia University of Technology, Wuhan 430223, China

Abstract. It is an important link in the development of e-commerce to accurately segment the customers of e-commerce enterprises and adopt corresponding marketing strategies. Based on the traditional RFM model of customer segmentation in retail industry, this paper introduces the attribute of total profit, creates an RFP model, uses data mining k-means algorithm to cluster the customers of an e-commerce enterprise, compares it with RFM model, and analyzes the influence of the correlation of model attributes on the clustering results, and obtains six conclusions and four marketing strategies for model comparison, Be able to put forward relevant marketing strategies for e-commerce industry and other sales industry.

Keywords: E-commerce · Data mining · RFM model · Clustering analysis · Customer segmentation

1 Introduction

In recent years, e-commerce has been booming at an amazing speed in just a few years. Its rapid development has led to the continuous innovation and progress of transaction mode, circulation mode and business mode. On the one hand, it avoids the capital risk of entrepreneurs due to physical store investment; on the other hand, it improves the level of customer service and more opportunities to choose goods. With the rapid increase of e-commerce transaction volume and the fierce competition in the market, e-commerce field needs to subdivide customers like traditional marketing, to meet the growing personalized needs of customers in the past, and to attract customers through differentiated service targeted strategy, to form long-term purchasing behavior, to improve customer loyalty, and to remain invincible in the fierce market competition. This paper introduces the rate of return on investment to establish the rfm-roi model to analyze the customer churn in the securities industry; combines the RFM (recency, freq μ uency, monetary) model with the original collaborative filtering mechanism to formulate a differentiated e-commerce recommendation strategy; proposes a customer behavior clustering method based on empirical mode decomposition (EMD) and k-means, It provides the basis for the business promotion. However, most of the domestic scholars mainly improve the algorithm, without a more realistic description of the model, the marketing strategy still has a superficial stage, and the profit detailed analysis model is very few.

2   Technical Route and Model Background

2.1   Research Ideas and Methods

This paper establishes the RFP (recall, frequency, profit) model, then collects and cleans the sales data of a B2C e-commerce enterprise (such as excellence, Suning e-commerce, etc.), makes K-means clustering analysis on RFM model and RFP model, and finally compares the effects of the two models; at the same time, puts forward corresponding marketing strategies for profit analysis, It can provide customer differentiation basis for large e-commerce enterprises, shopping websites and other sales industries [1].

2.2   Background of RFM Model

RFM model is a quantitative analysis model in the field of customer relationship management. It describes the importance and type of customers through three attribute values, namely, the latest purchase time (R), the number of purchases in a certain period (f), and the total amount of purchases in a certain period (m). It is mainly used in the traditional retail industry and has good representativeness in reflecting customers' purchase preference. It is found that the smaller R (or the larger f or m value), the more likely customers are to enter into new transactions with enterprises. It is often used in data mining customer segmentation.

2.3   Background of Improved RFP Model

Foreign scholars believe that the construction of customer segmentation model directly affects the accuracy of data mining technology. The more accurate the model description is, the better the effect of data mining is. The customers with high purchase amount but low profit are not necessarily big customers, which may not meet the big customer strategy of the enterprise. Although high sales improve the capital turnover rate of the enterprise, the profit is the fundamental benefit to the enterprise, which should not be ignored in the model, especially for the e-commerce enterprises with rich products, large profit difference and high profit. If an enterprise has several customers with similar purchase amount but different profits, the RFM model can not accurately distinguish them. For several customers with large difference in purchase amount, if the profit of customers with large purchase amount is still less than that of customers with small purchase amount, RFM model may even cause wrong clustering results and lose customers by using wrong marketing methods. There are some difficulties in the analysis of profit. Most scholars describe profit as the difference between sales and cost. However, in view of the new mode of e-commerce, most businesses often pre estimate the profit margin of goods, so we can consider that the profit is the product of sales and profit margin [2].

3   RFM Model

In many customer analysis models of customer relationship management, RFM analysis is a popular analysis method and an important evaluation index to measure customer value. RFM model was first put forward by Hughes in 1994 and has been widely used in the field of direct selling. It includes three variables R (recall), f (frequency) and m (monetary). R is the latest purchase time, also called recency. Theoretically, the closer the latest purchase time is, the more likely the user will respond to the provision of real-time goods or services, so the smaller the R, the better; F is the number of times the consumer purchases in a certain period of time, also called frequency. The more frequent the consumer purchases, the higher the customer loyalty, so the larger the F is, the better; M refers to the total amount of customers' purchase in a certain period of time, also known as monetary. The larger the purchase amount, the greater the value it brings to the enterprise. Therefore, the larger the m, the better. That is, the customer's value is inversely proportional to R, and is directly proportional to f and m. Enterprises can use RFM model to measure customer value, and use RFM model indicators to classify customers. The formula of customer value calculated by RFM model is as follows (1).

$$RFM = \omega_R \times R + \omega_F \times F + \omega_M \times M \tag{1}$$

RFM refers to the comprehensive RFM value of the customer.

4   Weight Analysis

In 1994, Hughes proposed that the three indicators should be treated equally and given the same weight. When stone studied and analyzed the credit card related information of customers in 1995, combined with the particularity of the industry, he thought that the consumption frequency in RFM model was the most important, followed by the latest consumption time, and finally the consumption amount. The traditional weight calculation mostly uses the analytic hierarchy process and expert consultation to determine, this method has a strong subjective color, is not accurate, this paper uses the quartile method, make the weight selection more scientific. This paper uses two methods to calculate customer value and segment customers, one is K-means clustering, the other is quartile method to calculate customer value [3].

4.1   K-means Clustering

K-means clustering is the most famous partition clustering algorithm, because of its simplicity and efficiency, it has become the most widely used clustering algorithm. K-means clustering algorithm is an iterative clustering analysis algorithm. Its steps are to randomly select k objects as the initial clustering center, then calculate the distance between each object and each seed clustering center, and assign each object to the nearest clustering center. Cluster centers and the objects assigned to them represent a cluster. Each time a sample is allocated, the cluster center of the cluster will be recalculated according to the existing objects in the cluster, and this process will be repeated until a termination condition is met.

Calculate the distance from each local point to each mean value, and the shortest distance is classified into one category. In this paper, Euclidean method is used to calculate the distance:

$$d = \sqrt{(r_i - \overline{r_j})^2 + (f_i - \overline{f_j})^2 + (m_i - \overline{m_j})^2} \tag{2}$$

$j = 1, 2, \ldots, k, i = 1, 2, \ldots, n$. This paper will use Python software, K-means clustering, and then through the elbow curve, more scientific to determine the K is 3.

4.2 Summary of This Chapter

This chapter starts with the widely used RFM model, analyzes the problems and shortcomings encountered in the use of large retail enterprises, and puts forward two variables, total integral and duration, and forms a new customer value calculation model: RFJ model. The RFJ model is combined with K-means clustering method to aggregate several kinds of customers with different properties. Finally, the paper discusses the function of classification by taking the classification of purchasing behavior as an example [4].

5 Empirical Analysis

This paper analyzes the changes of consumer value. According to the clustering results, the customers are subdivided, and the targeted marketing strategies are extracted for each customer group. This part of the research results can be used for reference in the future research of customer value in the field of retail. At the same time, it will deepen the understanding of enterprises to existing customers, and provide reasonable theoretical basis for enterprises to formulate marketing mix strategy and company reform and innovation.

5.1 Data Integration

The main purpose of this step is to combine different records of the same customer. Due to the long time of data statistics, there is a case that a customer has several membership cards in the data, but only the name is used as the standard to merge the records, which will lead to the confusion of customers with the same name. In this paper, the name and ID number are combined to form the unique identification of the record. After the merger, there are several situations: the ID number is the same, the name is homophonic, and the words are different. We think that this is due to the oral record in the process of handling membership card, and admit that it is the same record; only the name or ID number, according to statistics, there is no consumption record in the record with ID number, and it is deleted; there is name and consumption record at the same time, The new field new ID is used as the unique identification by adding the ID number to the name.

In the traditional RFM model, M represents the consumption amount of customers in a certain period of time. Generally speaking, the higher the consumption amount per unit time, the greater the value of customers to the enterprise. In this study, integral J is used to replace consumption amount m in RFM model. The reason is: when customers buy products in the enterprise, the enterprise database not only stores the consumption amount of customers, but also stores the relevant information of points. According to the importance of the product, the enterprise stipulates the proportion of integral feedback that customers can get when they buy the product. The more important a product is to an enterprise, the higher the proportion of feedback that customers get points when they buy it. For example, product a costs 100 yuan, and product B also costs 100 yuan, but the enterprise stipulates that the importance of product a is higher than that of product B, so that customers can get 100 points of feedback when they buy product a, and the feedback proportion is 100% of the product price. When they buy product B, they can only get 50 points of feedback, and the feedback proportion is 50% of the product price. If two consumers have the same amount of consumption, but have different points, the greater the value of customers with higher points to the enterprise. Therefore, in terms of customer value to the enterprise, points are more representative than consumption amount. The specific operation steps of calculating integral.

5.2   Calculation of Integral (J)

In the traditional RFM model, M represents the consumption amount of customers in a certain period of time. Generally speaking, the higher the consumption amount per unit time, the greater the value of customers to the enterprise. In this study, integral J is used to replace consumption amount m in RFM model. The reason is: when customers buy products in the enterprise, the enterprise database not only stores the consumption amount of customers, but also stores the relevant information of points. According to the importance of the product, the enterprise stipulates the proportion of integral feedback that customers can get when they buy the product. The more important a product is to an enterprise, the higher the proportion of feedback that customers get points when they buy it. For example, product a costs 100 yuan, and product B also costs 100 yuan, but the enterprise stipulates that the importance of product a is higher than that of product B, so that customers can get 100 points of feedback when they buy product a, and the feedback proportion is 100% of the product price. When they buy product B, they can only get 50 points of feedback, and the feedback proportion is 50% of the product price.

6   Conclusion

Based on RFM model, this paper uses K-means clustering and quartile method to segment customers, to help enterprises find high-quality customers and potential customers, and to identify customer value. The identification results are objective and reliable. The k-means method finds the classification quantity K scientifically through the elbow rule, while the new method quartile method proposed in this paper further subdivides the customers. Enterprises can subdivide customers according to their own needs, and the results can be used for the fine management and precision marketing of members. Establishing a stable relationship with high-value members is an effective way for enterprises to develop better.

References

1. Xu, X., Wang, J., Tu, H., et al.: E-commerce customer segmentation based on improved RFM model. Comput. Appl. **32**(5), 1439–1442 (2012)
2. Huiting, L., Zhiwei, N.: Effective clustering of customer behavior. Comput. Eng. Appl. **46**(4), 12–14 (2010)
3. Zhiqiang, B., Yuanyuan, Z., Yan, Z., et al.: Baidu takeout customer segmentation based on RFA model and cluster analysis. Comput. Sci. **45**(S2), 436–438 (2018)
4. Yi, Z., Hao, X.: Research on the relationship between service cost and customer value from the perspective of RFM model. Value Eng. **38**(30), 1–4 (2019)

Cultural and Creative Design Method and Its Application Based on 3D Technology and Virtual Simulation Technology

Pin Li[✉]

Jiangxi Tourism and Commerce Vocational College, NanChang 330100, JiangXi, China

Abstract. In recent years, the design and development of cultural and creative products are becoming more and more single and homogeneous. How to make cultural and creative products with both cultural and modeling characteristics needs to be solved. With the rapid development of social economy and science and technology, products form an effective combination in function display, modeling aesthetics, product life cycle and other aspects through designers, so as to adapt to the current social changes and meet the needs of different consumers and spiritual and cultural goals. In this paper, based on 3D technology combined with virtual simulation technology, this new product innovation development mode is explored. This paper expounds its development and main application fields, and discusses the limitations of traditional design and development mode of cultural and creative product modeling and the advantages of 3D printing technology in the development and application of creative product design by using the method of comparative analysis.

Keywords: Cultural and creative products · 3D technology · Virtual simulation technology · Design and development

1 Introduction

3D technology and virtual simulation technology are important parts of Intelligent Manufacturing in the future. virtual simulation technology, innovative 3D digital comprehensive solutions are continuously launched in the fields of industrial manufacturing, biomedical, teaching, creative design and so on. The emergence of 3D printing technology, especially for the future of the industrial design industry has brought unlimited development opportunities. Although there are still many places to be improved in the process, it has had a profound impact on the product life cycle from design, processing, manufacturing, circulation, consumption and so on. This paper mainly discusses the design strategy and application of 3D technology combined with virtual simulation technology [1].

J. C. Hung et al. (Eds.): FC 2021, LNEE 827, pp. 932–936, 2022.
https://doi.org/10.1007/978-981-16-8052-6_119

2 The Development of the Market of Cultural and Creative Product Design

As the representative of modern product design in the era of functional expressionism and rational expressionism, it pays attention to function and displays its form with functional features. Of course, when the mainstream is in the process of development, the mainstream designers are also thinking that the development of functionalism brings more coldness and singleness to products, and obliterates the characteristics of design, It has lost the expression of traditional and ethnic regional characteristics, and the design style lacks the attention from the aspects of humanistic emotion and personalization. Therefore, cultural creative products were born, whose central idea is a kind of emotional expression, advocating a kind of emotional expression of different cultural and regional characteristics, which needs to express the needs of industrial products with humanistic emotion and personality. Cultural and creative products are mainly products serving consumers. They should pay attention to the cultural characteristics and artistic expression, which is the higher grade demand of buyers after meeting the basic needs of life. of different cultures, we should. At the same time, combined with the needs of the times, establish a new cultural characteristics and brand image needs, express the cultural connotation through the product, make it become a representative symbol. Some people compare cultural and creative products as a kind of cultural business card, which is gorgeous and elegant, and has the value of collection and appreciation; but it is not a simple product, which condenses the cultural characteristics and precipitates the memory of history [2].

3D printing technology has become the mode research and opportunity research in promoting the design and development, providing more vision and inspiration for the design.

3 3D Technology and Virtual Simulation Technology

3.1 3D Technology

3D is the abbreviation of "3 dimensions" in English. In Chinese, it refers to three dimensions, three coordinates, i.e. length, width and height. In other words, it is three-dimensional [3].

In 3D technology, 3D printing is a disruptive innovation technology, which supports the rapid development of products and leads the change of production mode. In the future intelligent manufacturing industry, 3D printing, together with various emerging high technologies, will create an intelligent factory for personalized manufacturing in the future, and provide customers with a distributed manufacturing industry model based on 3D printing ecosystem. Although 3D printing technology can print a variety of products, it is a significant change and beneficial supplement to the traditional production mode. The 3D printing process in Fig. 1.

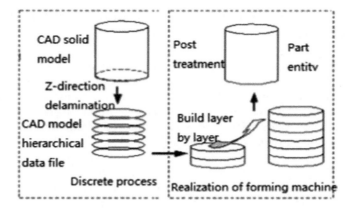

Fig. 1. 3D printing process

3.2   Virtual Simulation Technology

Entities interact in the virtual environment or interact with the virtual environment to show the real characteristics of the objective world. The design process of virtual simulation is shown in Fig. 2. The characteristics of integration, virtualization and networking fully meet the development needs of modern simulation technology.

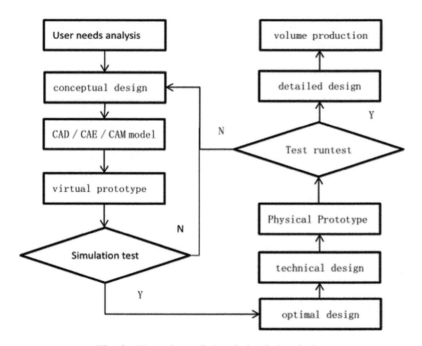

Fig. 2. Flow chart of virtual simulation design

4 Application of 3D Technology and Virtual Simulation Technology in Cultural and Creative Product Design

4.1 Fashionable and Creative Appearance

The emergence of 3D technology and virtual simulation technology has broken the difficulty and design constraints of complex modeling processing in traditional manufacturing process, given designers unlimited design possibilities, and brought strong visual impact to modern design creativity. In particular, parametric design method is widely used in architectural design, but also brings new design perspective and design method to industrial product design. In the process of product design, designers condense design ideas into parametric elements, and establish high-precision virtual simulation models through the powerful surface modeling ability and geometric operation ability of 3D digital technology. 3D printing technology processes complex surfaces with its advantages of easy processing and effective control of manufacturing cost, So as to design a unique style of product form [4].

4.2 Product Design Focuses on Lightweight Structural Design

3D technology has different processing methods. Choosing different processing methods will affect the final presentation quality. The lightweight requirements of product design is also an important part of 3D technology, which runs through the whole design process. The pursuit of lightweight product design, on the one hand, according to the material and structural characteristics of comprehensive mechanical analysis, through the structural design to achieve product lightweight; on the other hand, due to the lightweight structural design, the surface texture of the product appearance can also be redesigned, in the product lightweight at the same time taking into account the beauty of the shape.

4.3 Provide Personalized Design Services

The meet the cultural experience of different users. How to meet the personalized needs of users is one of the future development directions of modern cultural and creative products. Too many homogeneous products and too slow product change restrict and affect the development of cultural and creative products. At present, foreign jewelry design, the introduction of 3D technology customized design service platform customers, which is worthy of cultural and creative product design reference. We can parameterize different user needs through virtual simulation technology and parametric design means, and get many optimized cultural and creative product design schemes through logical calculation.

Integrating information tools: mainly for e-learning and web-based collaborative inquiry activities. With the help of these tools, the information can be analyzed and displayed in an interesting form, so as to promote the cooperation among members, build a network-based community of practice, and integrate virtual enterprises. Communication tools: can realize the transmission of text, image, video, audio or virtual reality based on the network, so as to build a platform for communication between users. It can run on mobile phones, computer desktops, web pages or intelligent learning terminals, which can facilitate marketing, entertainment, mobile learning and timely communication. Event scheduling tools: schedule events and activities, such as meetings, business, travel, etc. Widget contains a wealth of scheduling tools, which can meet the needs of different activities. Its personalized settings also fully reflect the user-centered idea.

5   Conclusion

To development of 3D technology and virtual simulation technology, 3D technology is more and more applied to cultural and creative products, which provides new design ideas for the development and innovation of cultural and creative products. Although 3D printing technology needs to be improved in materials, technology and other aspects at the present stage, with the application of 3D technology combined with virtual technology in the new.

Acknowledgements. 《Research on the Brand Image Construction and Promotion of Hakka Enclosed house in Southern Gannan under the Cultural Background of "Four Colors"》, Social Science Planning Fund project of Jiangxi Province in 2020.

References

1. Kong, Y.: Development and application of 3D printing technology in cultural and creative product design. Design, (10) (2017)
2. Kong, Y.: The development and design of Shu Brocade cultural creative products. Sichuan drama **3**, 107–109 (2014)
3. Zhou, J., Chang, H., Chen, X.: Application of 3D printing ceramics in modern product design. Ceramics. (05) (2018)
4. Jin, X., Zhai, R.R.R.: Research on innovative design strategies of huagudeng art and cultural products. J. Liaoning Univ. Technol. (Soc. Sci. Ed.) (03) (2019)

Green Building Aided Optimization System Based on BIM+GIS and Case-Based Reasoning

Bing Liu[✉], Chunhua Li, and Kaijie Wei

Panjin Vocational and Technical College, Panjin 124000, Liaoning, China

Abstract. The construction industry is one of the industries with high energy consumption in China. People pay more and more attention to the realization of sustainable development in the construction industry. In recent years, China's green building has made great development, there will be more sustainable buildings in the future. In order to ensure the healthy and rapid development of China's green building, China promulgated and implemented the "green building evaluation standard" as an authoritative document to guide and support the sustainable development of the construction industry. There are many problems in the process of traditional green building design, including the lag of green building design guidance, designers' vague understanding of green building concept, and the lack of overall design method and platform to integrate the information interaction process between various design professionals and non design professional project participants, thus affecting the overall sustainable effect of the project. With the continuous updating and improvement of BIM (building information model) technology, the informatization level of the construction industry in the world continues to improve. BM, as a revolutionary technology in the industry, can be applied to the whole life cycle of construction and create value for project participants at all stages. BIM re integrates the architectural design process and pays attention to building life cycle management together with green buildings.

Keywords: Building Information Model (BIM) · Green building design · Evaluation criteria · Information interaction

1 Introduction

BIM originated in the United States, and is widely used and promoted by Europe, America, Japan, Australia, Singapore and other developed countries. Compared with foreign countries, the research on BIM in China started late, but with the improvement of the understanding of Bim in recent years, BIM has developed rapidly in China. BIM model contains a wealth of parameter information, including geometric parameters and non geometric parameters, which help professional designers to complete more and more accurate building performance analysis; green building design is a cross stage and cross professional process, and BIM software has good interaction, which can help multi professional designers to share model information. This can not only reduce the data loss and damage in information interaction, but also provide shared data foundation support for multi professionals to complete collaborative work. At present, the

J. C. Hung et al. (Eds.): FC 2021, LNEE 827, pp. 937–942, 2022.
https://doi.org/10.1007/978-981-16-8052-6_120

auxiliary design software in the field of green building on the market in our country is simply nesting the evaluation standards in our country, and it does not really realize the compliance inspection of the model in the design stage. The designers still can not fully and deeply understand the items, thus affecting the sustainable effect of the project [1].

2   Creation and Application of BIM+GIS Model

2.1   Research on Data Fusion of BIM+GIS Site Model

At present, there are two main types of modeling data acquisition: one is three-dimensional laser scanning, which is different from the traditional single point measurement. Three dimensional laser scanning can directly collect the three-dimensional information of the real building, quickly copy the solid object for three-dimensional modeling, and quickly extract the surface coordinate information of the object, so as to quickly and accurately model. Pulse wave method, phase difference method and trigonometric ranging method are common methods to obtain data. 3D laser scanning modeling generates dense point clouds on the surface of the object through data acquisition by the scanner, then denoises and mends the holes on the point cloud data, leaving effective point cloud data, and then carries out the corresponding point cloud registration through the target control points to determine the position coordinate information of the object. Finally, the point cloud is reconstructed and repaired to generate a 3D model.

2.2   Data Fusion Method of Bim and GIS Platform Model

Geological data are generally obtained through drilling, pit exploration, trenching, geophysical exploration and other data means. Through geological logging, ore grade, lithology, fault distribution and other information correspond to different minerals. The pattern of different mining areas is different, but generally speaking, some basic information is essential, such as the project name, sampling location, analysis grade and lithology in the information record. In order to establish the spatial model of borehole in 3D mine, we must first create the borehole database. After obtaining geological data by data means, the drilling database is created. The database mainly includes four tables: positioning table, inclinometer table, lithology table and coal quality information table [2].

3   Research on Achievement Management of Construction System

According to different contents of each stage of intelligent mine construction system, combined with bm+gis application, the document content and submission format of each stage are formed. 5. It provides the bottom data content for the two subsystems, namely, production and construction process authorization system and whole process construction achievement management system based on bm + gis. The content form of each stage determines the document format, which is very important for the following formation of intelligent mine construction management system.

3.1   The Format and Management of the Results Submission in the Stage of Investment Planning

The main document achievements in the investment stage are determined by the work tasks of this stage, which mainly include: project proposal, environmental impact assessment report, energy saving assessment report, feasibility study, social stability risk assessment report, water and soil conservation scheme, geological disaster risk assessment report, traffic impact assessment report, etc. [3].

The project quality management in the implementation stage is to supervise the Contractor's construction according to the drawings, specifications, procedures and standards according to the entrustment of the investor and the construction contract of the construction project, so as to ensure the orderly construction and installation, and finally form a qualified project with complete use value. The backup of drawings and documents as well as the on-site comparison of the construction drawing level of the Revit model can play a very important role. The progress management in the project implementation stage is mainly to track and check the progress, control the progress, adjust the progress, and ensure that the construction project is completed within the construction period agreed in the contract. The 4D schedule control management can be realized by combining the revit model with the time line. NavisWorks platform links the model with time and cost. It can display the planned progress and actual progress on the platform, and realize dynamic control and timely correction in the construction stage. As shown in Fig. 1.

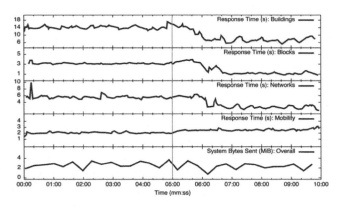

Fig. 1. Format and management of results submission in planning stage

3.2   Submission Format and Management of Project Scrapping Results

The contents of the specific documents to be filed in each stage of the smart mine construction system are summarized, and the formats involved in the BIM+GIS platform simulation are summarized to form the electronic archives in the process of coal mine engineering construction management, and the storage unit and storage period are summarized. The file format of key achievements provides the format support basis for the development of management system platform; the filing classification of each stage provides the document classification basis for the development of management system platform; the preparation, review, confirmation and other contents in the table provide the authorization basis for the development of management system platform. This chapter provides the basis for the development of each subsystem of intelligent mine management system based on bim-gis.

4   Application Analysis of Coal Mine Safety Based on BIM+GIS

Safety in production has long been a basic national policy of our country. It is an important work to protect the safety and health of workers and develop productivity. It is also the basic condition to maintain social stability and unity and promote the stable, sustainable and healthy development of the national economy. The essence of safety production is to prevent all kinds of accidents in the production process. Safety work is the lifeline and happiness line of coal enterprises. Without safety, there will be no production, no benefit, no stable development of the mining area and the happiness of the workers' families. Therefore, only by firmly grasping the safety production work, can we comprehensively improve the safety management level of enterprises, ensure the safety production of enterprises, realize the maximization of enterprise benefits, and better maintain the strong development momentum of coal enterprises. The essence of coal mine safety is human safety.

4.1   Analysis on Application Points of BIM+GIS in Coal Mine Safety

Modern mine production is a complex system. Mine production is an integrated whole composed of different kinds of production operations which are interdependent and mutually restricted. Each production operation contains many elements of equipment, material, personnel and operation environment. The occurrence of a mine casualty accident is often the result of the complex action of many factors. Therefore, it is necessary to comprehensively use various mine safety technologies to effectively reduce the casualty accidents. An important content of mine safety is to forecast and analyze the whole period of mine planning, design, construction, production and scrapping based on the understanding of the mechanism of casualty accidents and the application of the principles and methods of system engineering, so as to strengthen the supervision of coal mining enterprises, evaluate the existing unsafe factors, and comprehensively use various safety technical measures to eliminate and control the dangerous factors, Create a safe production conditions [4].

4.2   Analysis of Application Points Based on Accident Occurrence Theory

In view of the lack of on-site safety knowledge, real-time simulation can be carried out in fuzor. Employees can not only actively operate roaming, but also increase the enthusiasm of learning on-site knowledge, but also simulate escape roaming experience in emergency. When the employees are in real face of disaster, calm the mentality of employees, improve the success rate of escape and reduce casualties. In physiological state such as vision and hearing loss, the scene state of physiological function is simulated by adding harmful sound source in the platform of fuzor, adjusting light illumination to simulate the state of physiological function scene, and can make the escape method or add safety facilities for the special physiological people when they encounter disaster, and for the material stacking in the narrow space, the material is simulated and placed at the end of the model, Make reasonable use of space to reduce the safety risk caused by unreasonable space planning; for the problems of missing on-site equipment and tools and safety protection equipment, the light-weight model is displayed at the mobile equipment end, and the quantity of the model and the actual site facilities is unified and whether there is any defect in the safety protection device on site is checked, Thus, the probability of unsafe state of the material is reduced to improve the safety of the coal mine project site. The staff responsible for on-site inspection shall record the integrity of the equipment on site, such as wear, aging, corrosion, fatigue, etc. the status of the equipment that reduces safety is counted in the model by adding shared parameters, and the model equipment status is updated regularly to brush new details, maintain and replace relevant equipment, and improve the efficiency and level of safety management of site management [5].

5   Conclusion

Based on the analysis of intelligent mine system composition, this paper constructs the intelligent mine construction system based on BIM+GIS. This paper uses WBS and flow chart to sort out and supplement the work flow of smart mine construction system, and analyzes the application points based on BIM+GIS. It realizes the creation and application analysis of macro model and micro model respectively from 3DMine platform and Revit platform, and studies and sorts out the achievements of smart mine construction system, It provides the bottom data for the development of each subsystem of intelligent mine construction management system based on BIM+GIS.

According to the attribute information in ontology and BIM software function, the design identification items in the evaluation standard are evaluated, and the items to realize the compliance inspection potential in BIM environment are determined, and the family component parameters involved in the evaluation standard are summarized, which provides the basis for the system design and development. In this study, using software engineering ideas and a variety of BIM software development tools, we designed and developed the compliance check plug-in system of the Revit side and the information interaction system of the web side.

References

1. Zheng, P., Wang, X., Li, J.: Exploration and practice of curriculum ideological and political construction reform——take information security course as an example. ASP Trans. Comput. **1**(1), 1–5 (2020). https://doi.org/10.52810/TC.2021.100020
2. Luo, Z.: Application and development of electronic computers in aero engine design and manufacture. ASP Trans. Comput. **1**(1), 6–11 (2021). https://doi.org/10.52810/TC.2021.100025
3. Ying, C., Shuyu, Y., Jing, L., Lin, D., Qi, Q.: Errors of machine translation of terminology in the patent text from English into Chinese. ASP Trans. Comput. **1**(1), 12–17 (2021). https://doi.org/10.52810/TC.2021.100022
4. Xiao, J., Dai, Y., Shi, X.: Translation and influence of one two three… infinity in China. ASP Trans. Comput. **1**(1), 18–23 (2021). https://doi.org/10.52810/TC.2021.100021
5. Liang, L., Yin, Q., Shi, C.: Exploring proper names online and its application in English teaching in university. ASP Trans. Comput. **1**(1), 24–29 (2021). https://doi.org/10.52810/TC.2021.100024

English Automatic Translation Platform Based on BP Neural Network for Phrase Translation Combination

Zeng Wenjie[✉]

Nanchang Institute of Science and Technology, Nanchang 330108,
Jiangxi, China

Abstract. In order to further improve the intelligent level of English translation system and enhance the accuracy of English machine translation, this paper designs an intelligent English automatic translation system for phrase translation combination. The system includes two parts: machine algorithm and software design. Through semantic feature analysis and phrase translation combination mode, the automatic translation algorithm is optimized, the translation algorithm program is loaded, and the automatic translation system software is designed based on embedded environment. On this basis, the automatic design of translation system is carried out by means of cross compiling and multi thread phrase translation loading. The system test results show that the system has high translation accuracy and good intelligence.

Keywords: Intelligent translation · Translation combination of phrases · English translation system · Automation

1 Introduction

With the development of information technology, machine translation has become a new research field with the birth of computer technology. Its purpose is to use computer to realize the process of translating one natural language into another. It is an inter-disciplinary subject based on linguistics, mathematics and computer science. With the development of information technology. Machine translation has gradually become a hot topic in information science. It is of great significance to the development of modern information society.

A good translation system should describe the source language and the target language appropriately, which should be independent of each other. The automatic English translation system is transformed into analyzing natural language sentences according to the rules provided by a certain grammar theory, obtaining the internal representation of the syntactic structure of the sentence, and then generating the syntactic structure of the target sentence according to the grammatical rules of the target, Finally, the translated sentences composed of the target language words are obtained, and combined with the automatic analysis method of semantic fuzzy matching phrase, large-scale automatic translation of sentences and words is carried out, and the

J. C. Hung et al. (Eds.): FC 2021, LNEE 827, pp. 943–949, 2022.
https://doi.org/10.1007/978-981-16-8052-6_121

accuracy and reliability of translation are guaranteed through the combination of phrase translation [1].

2   Intelligent Automatic Translation

2.1   Features of Automatic Translation System

(1) It can store a lot of information. With the development of technology, computers can store all kinds of information in large capacity without any limitation, so we can store and concentrate enough translation knowledge.

(2) Computability means that information can be operated by computer program. With the development of computer technology, the processing ability of computer has been able to meet the calculation of a large number of information.

(3) Multidisciplinary cooperation.

(4) Difficult to understand. Because the cognitive process of human language is not clear, it is impossible for computer to reach the level of human language. Therefore, machine translation is known as one of the scientific and technological problems to be solved in the 21st century. The main difficulty is the ambiguity of natural language at all levels, also known as ambiguity or polysemy. The fundamental task of machine translation is to gradually eliminate these ambiguities in the process of processing.

(5) Practicality. Although there are great difficulties in the research of machine translation, people hold high hopes for it. As foreign experts say, machine translation researchers play a dual role of science and commerce, and can make the right compromise between the bottomless hole of language and its users at any time.

2.2   The Current Development of Machine Translation

Throughout the history of machine translation, it has gone through decades of theoretical research, and has moved towards practicality. The United States, Britain, France, Russia, Germany and other countries have carried out the corresponding machine translation of mother tongue and "foreign language". The research and cooperation in this field is very active and has made a lot of achievements. China is the fourth country in the world to carry out machine translation research after the United States, Russia and Britain. However, machine translation is indeed a very difficult topic, including the study of grammar, semantics, pragmatics and other basic issues, which requires the technical achievements of computer science, linguistics, cognitive psychology and other disciplines. So far, machine translation of various languages has not achieved satisfactory results [2].

It must be affirmed that the theory and technology of machine translation have developed very rapidly in recent ten years, and the value of machine translation system in social life has become increasingly apparent. If the effect is satisfied, it can not only be used by government agencies, but also promote scientific and cultural exchanges in the world; On the other hand, because the creation and use of natural language is the manifestation of human's high intelligence, the theoretical part of machine translation helps to uncover the mystery of human intelligence and deepen our understanding of language ability and the nature of thinking. It can be said that the research of machine translation is of great significance in both theory and application.

3 Construction of Intelligent English Automatic Translation Platform Based on Phrase Translation Combination

3.1 System Design Route

Based on the features of English machine translation with semantic information, the syntax is preprocessed to form an English phrase tree; The decoding sentences are constructed by training the features; Test decoding statements and output test results; Words are aligned with syntax; According to the alignment feature, the part of speech feature is labeled, and the output is the node attribute of English phrase tree. The technical route is shown in Fig. 1.

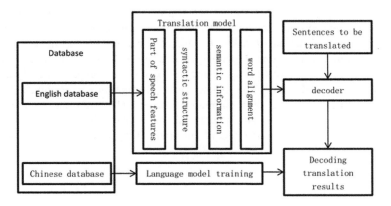

Fig. 1. Technical route of translation system

3.2 Overall Frame Design

The design of English automatic translation system mainly includes two parts: algorithm and software. The system software is designed based on embedded environment, which involves vocabulary acquisition, information processing, vocabulary scheduling, automatic control and other modules [3]. In order to extract the information that can reflect the characteristics of the system association rules, information fusion and intelligent scheduling are used to realize the intelligent management and scheduling of the system. The software of automatic translation system is designed and developed based on embedded ARM environment. TinyOS is used to design the system network component interface, and Linux kernel is used to control the cross compilation of information management system software, so as to improve the intelligent level of system management. The overall structure of English automatic translation system is shown in Fig. 2.

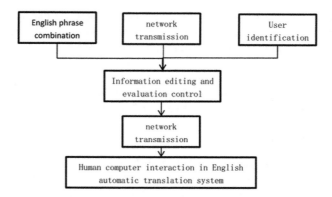

Fig. 2. Overall structure of the system

As far as the overall structure of the system is concerned, the automatic translation system software is developed based on B/S framework system, and the system intelligent management system network structure and database system construction are realized. The main function of MIS of automatic translation system is information collection, fusion, transmission, integrated scheduling and so on. The fuzzy matching and status information characteristics of the translation are comprehensively analyzed through the FIFORAM buffer. The management information of the control system is remotely transmitted by the host computer module. The information transmission adopts the three-tier system mode, namely the basic layer, the middle layer and the application layer. The specific architecture information transmission model is shown in Fig. 3.

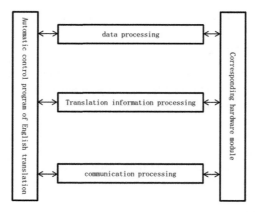

Fig. 3. Three tier architecture information transmission model

Based on TinyOS, the network component interface is designed, and the architecture is optimized through application layer, business layer and data layer. Through the base station transmission design phrase translation combination information, the use of sensor nodes to disseminate information, with high-speed A/D converter as the carrier to convert control digital to analog, and adopt poor compilation and multi-threaded phrase translation loading way, realize the program loading, for this, the automatic translation system intelligent management function software development.

3.3 Design of Machine Dictionary

Machine dictionaries are different from ordinary dictionaries. The entries in ordinary dictionaries can be explained in detail by natural language, but it is difficult for machine dictionaries to use natural language as the explanation of entries at present. Therefore, the information representation of machine dictionaries needs to use certain symbols to represent the specific explanation of each entry, that is, the encoding of dictionaries. Generally speaking, machine dictionaries need to consider the accuracy of knowledge, the granularity and redundancy of knowledge, the scalability of dictionaries and the space-time cost of dictionary utilization. The machine dictionary of this system is designed with full consideration of Chinese grammar information, English grammar information, Chinese English conversion information and semantic information. For any entry w in the dictionary, it can be expressed as:

$$w\{CYF, YF1 - V1, YF2 - V2, \ldots YF_K - V_K\} \tag{1}$$

Among: CYF = {#PD, #PF, #V, #N, #D, #A...}, YFi-VI (i = 1, 2, ... K) denotes the value of the entry in different grammatical formats.

For example: the entry "we", its corresponding machine dictionary code is: w (we) = {#PD, G1- We, G2-us}, the code explanation is as follows: the entry "we", the Chinese grammar is the plural of personal pronouns, translated into we in the nominative case, translated into us in the accusative case.

Due to the lack of tense change in Chinese, tenses are generally expressed by time nouns (such as yesterday, last month, next year, etc.) and temporal auxiliary words (Zhe, Le, Zai, Zheng, Guo, etc.). Therefore, the nouns representing time and the auxiliary words representing tense are separated separately in this machine dictionary. Chinese scholars generally classify time words separately. The singular and plural of nouns, uncountable, the past tense, progressive tense, perfect tense of verbs and other details are reflected in the corresponding YFi-Vi, which will not be elaborated here [4].

3.4 Grammar Rule Design

How to select the specific value of each phrase entry and combine it into a whole English sentence is the last step of the system. Due to the great differences between Chinese and English grammar, such as Chinese emphasis on semantics, English emphasis on structure; Chinese more supplement, English more ellipsis; Chinese more noun, English more pronoun and other different grammatical phenomena, if the target value of phrase entry is directly taken out and connected into a sentence, it obviously can not become a complete English sentence.

According to the differences between Chinese and English grammar, the system designs targeted grammar translation rules, so that it can correctly select the value of phrase entry, and make appropriate combination to become a complete English sentence. The rules designed by the system mainly include personal pronoun subject / object translation selection rules, verb predicate translation selection rules, noun object translation selection rules and time adverbs. The rules of translation selection, auxiliary word deletion and so on can correctly reflect the grammatical differences between Chinese and English, so that the proper combination can be made and translated into the final result on the basis of correctly taking out the entry value.

4 System Test

In order to further test the performance of the English automatic translation system based on the combination of intelligent translation and phrase translation, the simulation experiment is carried out. Based on the visual DSP++ simulation platform, it is clear that the word center frequency of the translated phrase is 12 kHz, the maximum set value of phrase combination length is 2000bit, the English semantic concept set is set to 245 samples, and the vocabulary entity set is based on HowNet, In order to achieve the goal of automatic English translation, the accuracy of translation output and recall rate of English semantic information are selected as the indicators to test and obtain the results. The system test shows that the translation accuracy and recall rate of this system are very high, and the level of intelligence and automation is relatively high.

5 Conclusion

In a word, through the research on the construction of intelligent English automatic translation platform for phrase translation combination, the level of intelligence and automation of English translation can be further improved. The system design mainly includes two parts: software system design and machine translation algorithm design. The automatic translation algorithm is optimized by semantic feature analysis and phrase translation combination, and the software design of automatic translation system is realized based on embedded environment. The system test results show that the translation accuracy and recall rate of this system are very high, and the level of intelligence and automation is relatively high.

References

1. Zhiwei, F.: Fundamentals of Computational Linguistics. Commercial Press, Beijing (2001)
2. Rui, Z.: Design of statistical machine translation system based on phrase similarity. Autom. Instrum. **8**, 66–67 (2017)

3. Lu, Y.: Research and application of machine translation based on statistics. Xi'an University of Technology, Xi'an (2016)
4. Haiyang, Z.: Research on machine translation method based on semantic selection. Autom. Instrum. **8**, 29–32 (2018)

Personalized Melody Recommendation Algorithm Based on Deep Learning

Zeng Lingqiong[✉]

Jiangxi Teachers College, Yingtan 335000, Jiangxi, China

Abstract. The rapid development of Internet technology and electronic information technology provides huge computing power for the whole era. Personalized recommendation system has become the epitome of the product of the times. In this paper, based on the deep learning method, combined with the common core algorithm of recommendation system, the music automatic recommendation algorithm is studied, and an improved algorithm for personalized music is designed. This algorithm uses the user information for deep learning, uses the candidate matrix compression method for recommendation optimization, and uses the accuracy, recall rate and other parameters as the evaluation criteria. Finally, the music recommendation model is completed, and the appropriate music is selected around the model.

Keywords: Deep learning · Recommendation system · Individualization · Music

1 Introduction

With the rapid rise of Internet technology and electronic information technology, the development of big data technology, cloud computing technology, robot technology, artificial intelligence technology and deep learning technology is particularly prominent, which provides a huge computing power for the progress and development of the whole information age. In such a large amount of information, it is more and more important and valuable to find the information quickly and accurately. The recommendation system, which is born from this, has become a bridge between user needs and content, which can not only meet the users' interest in potential content, but also better display the cold door content and explore potential users.

Nowadays, the society has more powerful inclusiveness, and different fields also show unique personalization and diversification. Personalized recommendation system can meet the needs of different users and provide users with better experience accurately, thus creating huge business value and becoming the "cake" for Internet enterprises to compete for. At present, personalized recommendation system has been widely recognized and quietly integrated into our life. Music, as an ancient art form, can bring people pleasure. But it is necessary to find the music that meets the needs of users accurately from a large number of music works. Then, the personalized music recommendation system should select personalized music suitable for users according to user behavior, meet the needs of users in the situation at that time, so as to achieve the goal of "everyone can adjust" [1].

© The Author(s), under exclusive license to Springer Nature Singapore Pte Ltd. 2022
J. C. Hung et al. (Eds.): FC 2021, LNEE 827, pp. 950–955, 2022.
https://doi.org/10.1007/978-981-16-8052-6_122

2   Recommendation System

2.1   Basic Concept of Recommended System

Recommendation system is an intelligent software technology, which can provide personalized recommendation service for users based on user interest preferences and characteristics. There are many definitions of recommendation system, different areas have different definition rules. The informal definition given by Resnick and Varian is as follows: "recommendation system is a technical system which uses e-commerce website to provide commodity information and suggestions to customers, helps users decide what products to buy, and simulates the sales staff to help customers complete the purchase process".

Adomavicius and tuzhilin formally define recommender system as: suppose that C represents user set, and S is the set of all items that may be recommended to users, such as music, movies, books, etc. As a utility function, u is used to evaluate the utility of item s to user c. The calculation process can be expressed as u: $C \times S \rightarrow R$, where R is the sorted item set. For any user $c \in C$, the goal of recommendation algorithm is to recommend the item $s' \in s$ with maximum utility to user c. the formal formula is as follows:

$$\forall c \in C, s'_c = \arg \max_{s \in S} u(c, s) \tag{1}$$

2.2   Traditional Recommendation Algorithm

Among the traditional recommendation algorithms, pop is the simplest one. It mainly recommends hot items, which can solve the problem of cold start, but there is no way to provide personalized recommendation to users. Collaborative filtering is widely used in the recommendation algorithm, which is generally used in the user's scoring system. The score is used to describe the user's preference for items, but it has the shortcomings of sparse data and redundant information, which leads to sparse scoring elements in the matrix [9.20]. Matrix factorization is a widely used method for collaborative filtering, because matrix factorization approximately decomposes the original large scoring matrix into the product of small matrices, which can solve the problem of data sparsity to a certain extent [2].

3   Recommendation System and Core Algorithm Based on Deep Learning

3.1   Deep Learning

Deep learning (DL) is a more important concept in learning contemporary science. In 1976, ference Martin, an American scholar, proposed in "the essential difference of learning: results and processes". The research of deep learning in China started late. It was suggested that deep learning is based on understanding learning. Learners can critically learn new ideas and knowledge, combine existing cognition with them, connect with diversified ideological structure, and transfer existing knowledge to new situations, which is a learning method to solve problems and decision-making problems. In many applications, deep learning has been greatly successful, such as machine translation, computer vision, etc. At present, in the field of recommendation system, deep learning technology is widely used, which can effectively capture the project relationship of non-ordinary and non-linear users, and represent abstract and complex high-level data. In-depth learning can learn more complex user preferences from the original data, and enhance the recommendation effect [3].

3.2 Recommendation System and Core Algorithm

The recommendation system in e-commerce is widely used. With the continuous penetration and development of Internet in various fields, music recommendation system also reflects the eye. According to user preferences, music description information and other content, the recommendation model is constructed to push out the music content that meets the needs of users. At present, the common recommendation methods are mainly divided into three types: content-based recommendation, collaborative filtering recommendation and hybrid recommendation.

(1) Content based recommendation algorithm

The basic idea of this method is to analyze the user preference behavior according to the historical information of the user, and get the user preference set. Then, the information matching between the set and the substitute recommendation content can be realized. The common music recommendation algorithms include the content annotation based recommendation algorithm and the music algorithm based on the music characteristics. Taking the recommended algorithm based on music annotation content as an example, the preference types of music for user a and user B are European and American, electric and Chinese, and ballads, among which European and American, electronic, Chinese and ballads represent the style and type of songs. When new songs C appear, Chinese and folk songs become the marking contents of the song, which belongs to the feature information, The recommendation system will give priority to user B according to these characteristics information, so as to achieve accurate recommendation, as shown in Fig. 1.

(2) Collaborative filtering recommendation algorithm

In view of the universal adaptability of collaborative filtering algorithm, the algorithm is widely used in many fields. The core idea of the algorithm is to use the similarity or similarity of user preferences for content recommendation. Collaborative filtering recommendation algorithm mainly includes three types: user based collaborative filtering recommendation algorithm, article based collaborative filtering recommendation algorithm and model-based collaborative filtering recommendation algorithm.

Taking the collaborative filtering recommendation algorithm based on users as an example, users A and C are more similar in song preferences. As shown in Fig. 2.

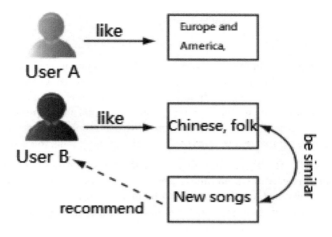

Fig. 1. Content based recommendation map

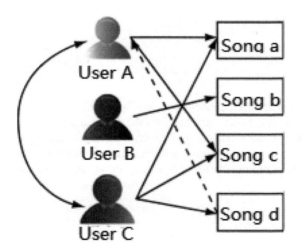

Fig. 2. Content based recommendation map

4 Personalized Music Recommendation Algorithm

The audio features of structured support vector machine (SSVM) are more abundant and diversified than those of ordinary support vector machine, and the relationship between them can be described accurately; The good generalization performance of the learning model is guaranteed by its maximum boundary objective. Therefore, the music recommendation model of this study adopts structured support vector machine [4].

In this study, the fixed value of C is 1, and Eq. (2) is the potential function $\varphi f_i^u, f_i^v$ definition:

$$\varphi f_i^u, f_i^v = \left[s_i \cdot f_i^u \cdot f_i^v \right] \tag{2}$$

s_i in Eq. (2); It refers to the prediction content of the i-segment scene annotation, and the acquisition of training data can be realized on the basis of linear support vector machine. Therefore, the potential function essentially maps the input to a feature vector, which is obtained by concatenating audio features. For a piece of music, its music features are represented by f_i^u, from which we can see that compared with a score of randomly generated music, i.e. $w^\tau \varphi(f_i^u, f_i^v)$, in the training data, the score of matched music is $w^\tau \varphi(\overline{f_i^u}, f_i^v)$ It's bigger than that. After the constraint is imposed, the recommendation model is generated, and the score of the matched music is greater than the score of the mismatched music, so as to obtain the music with the best matching degree. According to formula (1) and formula (2), for the labeled training sample set, we can directly use structured support vector machine to determine the decision parameter w on the basis of learning. On this basis, the candidate audio sets of new clips are sorted to find the best music. In order to enhance the comprehensibility of the model, considering a single video clip, the optimization problem of the learning model can be transformed into Eq. (3) equivalently

$$\min_{w,\xi} \left\{ \frac{1}{2} \|w\|^2 + C \sum_{i=1}^{n} \left[w^\tau \varphi(\overline{f_i^u}, f_i^v) - w^\tau \varphi(f_i^u, f_i^v) \right] \right\} \tag{3}$$

When we get the music model of w, we first need to extract the f^u of the target music; Then, around the relevance between the scene and the model, the scene feature vector s is analyzed comprehensively; Search the music library and combine it with music clips, so that users can score it and determine the score of recitation. The process of calculating the score of alternative music is shown in formula (4)

$$Score(a_j) = w^{*\tau} \left[s, f_j^u, f_j^v \right] = C \sum_{i=1}^{n} \left[w^\tau \varphi(\overline{f_i^u}, f_i^v) - w^\tau \varphi(f_i^u, f_i^v) \right]^\tau \left[s, f_j^u, f_j^v \right] \tag{4}$$

In the formula, a_j is used for the j-th alternative music; Representative, from a; The music features extracted from the model are used in the experiment; Representatives. It can be seen from formula (4) that music recommendation should be based on linear structured support vector machine, that is to say, in the selection process, positive samples and training data are actually consistent, that is to say $\varphi(\overline{f_i^u}, f_i^v) \left[s, f_j^u, f^v \right]$ is larger, but the difference between training data and negative samples is far, that is, the best recommendation is $\varphi(\overline{f_i^u}, f_i^v) \left[s, f_j^u, f^v \right]$ The value of music is small.

5   Conclusion

This paper briefly introduces the core algorithms commonly used in recommendation system. Based on deep learning method, personalized music recommendation algorithm is studied. On the basis of recommendation model, personalized music is established by combining structured support vector machine to ensure the completion of music recommendation model, and appropriate music is selected around this model.

Acknowledgements. Provincial subject of teaching reform in Jiangxi Province in 2020JXJG-20-47-4 Research on College "Music+" Aesthetic Education Teaching.

References

1. Zhu, Y., Sun, J.: Research progress of recommended system. Comput. Sci. Explor. **9**(5), 513–525 (2015)
2. Yaxi, C.: Key technology research of mixed music recommendation system in public environment. Comput. Appl. Res. **29**(11), 4250–4253 (2012)
3. Xu, Z.: The design and research of mobile learning recommendation system from the perspective of deep learning theory. Intell. Comput. Appl. **4**(2), 5758 (2014)
4. Deng, T.: Research on personalized music recommendation system. South China University of technology, Guangzhou (2018)

Design of Deep Learning Algorithm Composition System Based on Score Recognition

Shuyan Wu[✉]

Jiangxi Teachers College, Yingtan 335000, Jiangxi, China

Abstract. After algorithmic composition has become a hot research direction in the field of computer music creation, in recent years, with the rise and application upsurge of artificial intelligence, deep learning and other technologies, the research and exploration of related theories and technologies have been deepened, and the multi-dimensional integration of technology and music has entered a new historical period. Artificial intelligence composition will become the main research branch of algorithmic composition in the future. The purpose of this paper is to realize a deep learning algorithm composition system based on score recognition.

Keywords: Deep learning algorithm · Composition system · Score recognition

1 Introduction

Since the middle of the 20th century, the research on music science and technology has developed rapidly with the progress of computer science. The digital music creation mode has changed the traditional music industry and greatly improved the quality and production efficiency of music works. The derived scientific and technological progress, historical process and cultural progress have made the relevant professional fields begin to explore the use of computer to create automatic music, so the concept of algorithmic composition is born.

In recent years, the related technical aspects of AI are becoming more and more complete, and its broad interdisciplinary development prospects. The research on algorithm composition and artificial intelligence composition has attracted extensive attention in the academic circles and promoted a new round of application upsurge in the industry. At present, music artificial intelligence has developed rapidly in Europe, America, Japan and other regions, and has become an important branch of artificial intelligence. However, the research related to this is just in the initial stage in China. From the perspective of application, artificial intelligence can effectively save human costs and improve the efficiency of music analysis and music creation. In the social environment of "Internet plus" and "industrial manufacturing 4", AI system with deep learning ability has entered the trend of music creation, music education, music research and analysis, music performance and commercial application [1].

© The Author(s), under exclusive license to Springer Nature Singapore Pte Ltd. 2022
J. C. Hung et al. (Eds.): FC 2021, LNEE 827, pp. 956–962, 2022.
https://doi.org/10.1007/978-981-16-8052-6_123

Therefore, the research on the algorithm composition system based on deep learning can not only explore rich theoretical value in the field of music creation, but also help and summarize the process of continuous cognition and innovation in the process of creation. It has positive research significance, extensive research branches and sufficient space for sustainable rise, and also has a wide range of social application prospects.

2 Deep Learning

2.1 Overview of the Principle of Deep Learning

Deep learning aims at establishing deep structured model. Machine learning, generally agreed model contains at least three hidden layers. This kind of multi hidden layer network is difficult to work with the training algorithm of common neural network, such as BP algorithm. It is not only because of the large sample data requirement, slow training process, but also the parameter converges easily in local rather than global best, so it is not practical.

As the basic structure of single layer network, the training of deep network is realized by layered training, and the effectiveness of this method is demonstrated by automatic coding machine, thus the concept and method of deep learning are established. There are three main links in deep learning: first, the system is trained in unsupervised way, that is, it is extracted layer by layer with a large number of unlabeled samples, and the non guidance automatically forms the characteristics. Second, adjust. In this process, some marked samples are used to classify the features, and the system parameters are further adjusted according to the classification results, and the performance of the system in distinguishing different types of information is optimized. Third, test the learning effect of the system with the sample data not recognized by the system, such as the accuracy of sample classification, the relevance between quality assessment and subjective evaluation [2].

2.2 Model and Training of Deep Learning

Restricted Boltzmann machine is an improvement of Boltzmann machine, which is a random network. Because of the interconnection between the inner layer units, as shown in Fig. 1 (a), the process of network training is slow. In 1986, somlensky introduced a restricted Boltzmann machine, which contains an explicit layer and a hidden layer. There is no interconnection relationship between the elements in the layer as shown in Fig. 1 (b). Thus, it is very efficient to use RBM to calculate reasoning.

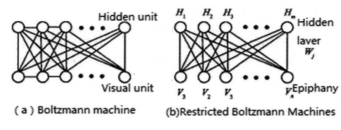

Fig. 1. Boltzmann machine and restricted Boltzmann machine

Suppose an RBM contains n visible units and M hidden units, we can use vectors v and h to represent the states of visible units and hidden units respectively, where v_i represents the state of the ith visible unit and h_j represents the state of the jth hidden unit. Then, for a given set of states v, h, the joint probability distribution of RBM as a system can be expressed by energy function:

$$E(v, h) = -\sum_{i=1}^{m}\sum_{j=1}^{m} W_{ij}v_ih_j - \sum_{i=1}^{m} v_ib_i - \sum_{j=1}^{m} h_jc_i \tag{1}$$

Where w_{ij} and c_i are the parameters of RBM. Where w_{ij} is the connection strength between the visible unit i and the hidden unit j, b_i is the offset of the visible unit i, and c_j is the offset of the hidden unit j. The task of learning RBM is to get the values of these parameters to fit the given training data.

3   The Types of Existing Algorithmic Composition

3.1   Composition with Randomly Generated Parameters

The charm of music often comes from the unpredictability of creative motivation and development mode, and the new auditory experience it brings can also cause people's emotional resonance. From the perspective of composition technology, creative motivation can show the tension of art form through deformation, variation, tone sandhi and form expansion. From the perspective of perceptual thinking, motivation can construct the inner tension of art through extension, diffusion, breaking, reorganization and representation. Stochastic algorithm is a transition between States, which is determined in a random way, and the range of determining the random interval is often limited. Random algorithm is often used in algorithmic composition. In this kind of algorithmic composition system, we can input a random parameter to the system, and then generate a sequence of notes through several limited random processes. As early as the 18th century, Austrian classical music master Wolfgang Amadeus Mozart (1756–1791) created musikalisches wurfelspiel, k516f based on random probability, with the subtitle "musical dice game", as shown in Fig. 2.

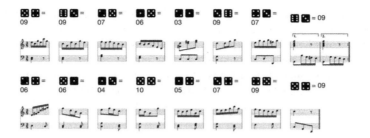

Fig. 2. Simulation results of Mozart music dice game system

However, considering from the perspective of composition technology, the music materials used for random generation often come from a large number of music material libraries, and they are not summarized and classified, so they do not have the above-mentioned Mozart's creation details and overall conception. Moreover, it is inherent in the randomness of mathematics, and lacks the ability to analyze and define music motivation. Of course, it can not ensure the formation of audible music works.

3.2 Algorithm Composition Based on Traditional Model

In the field of algorithmic composition, most of the traditional models used to generate music are statistical models, and the more common one is Markov model (mm) based on Markov chain (MC). Markov chain is a stochastic process with Markov property in probability theory and mathematical statistics, which exists in discrete exponential set and state space. Hidden Markov model (HMM) is a branch of HMM, which is often used in music creation, as shown in Fig. 3.

Markov model has a strong ability to analyze time series problems. Similarly, the music information that needs to be calculated has the characteristics of multi-level and complex information structure, such as notes, rhythm, beat, harmony combination, sentence theme, etc. The Hidden Markov model included in the Markov model can describe not only the transient random process, but also the dynamic (random process transfer) random process, which just reflects and corresponds to the multi-level of music information.

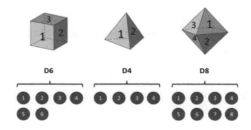

Fig. 3. Probability in hidden Markov model

But an important feature of Markov chain is memoryless (including long-term and short-term memory). All the possibilities that may need to be used to predict the data of the next time point have been limited to the possibility of the current state, and no new probability information can be obtained from the history of the event for backtracking. Therefore, Markov chain does not have the ability to create a large number of music learning, can only complete simple local statistics and probability distribution, and can not obtain the potential laws contained in music data [3].

4 Design of Score Recognition Composing System Based on Deep Learning Algorithm

This paper studies an algorithm composition system with the ability of music data sampling and deep learning. It trains the standardized data of deep neural network input preprocessing. Its purpose is to realize the music generation based on deep learning in digital environment. In order to more effectively realize the function of each module in the system and give consideration to the overall design, this paper analyzes and combs the composition model based on different algorithms and the creation process of computer automatic composition system, summarizes the advantages and disadvantages of various algorithm composition methods, so as to provide more comprehensive theoretical support for the development of the system.

4.1 Overall Function Analysis of the System

The overall framework of composition system based on deep neural network is shown in Fig. 4. The whole system function realization process is mainly divided into seven stages: data acquisition, data set production, data preprocessing, design network, training network, evaluation network and music generation. Since the task of this study is to implement a music deep learning network based on double layer LSTM, we need to obtain the corresponding music score data and MIDI format music data first; After obtaining the music data, we need to make a music data set according to the standard; After making the data set, we need to perform some preprocessing operations on the data input to the network, such as data normalization, extracting notes from MIDI music, etc.; When the required data is ready, we need to use different neural network layers to design the music generation network. The main network layers used in this paper include LSTM, dropout, density, softmax, etc.; After building the corresponding music generation network, we need to use the corresponding deep learning framework to train the network. The specific training scheme is to use the loop operation to continuously optimize iteratively until we get a satisfactory result or reach the set number of iterations. After network training, the training results will be saved as a model file. In order to verify the performance of the network, we need to use the corresponding indicators to evaluate the performance of the network, and use the final model to do a prediction and result visualization work [4].

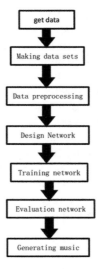

Fig. 4. System flow frame diagram

4.2   Recognition of Music Score

The recognition object of music score recognition is mainly defined as paper music score and digital image music score. At present, the main way is to convert the digital image through various hardware optical scanning equipment to generate binary map. Then, the binary image is further processed, and the paper music score is transformed into digital data information through the deletion of the staff line and the location of the notes. Figure 5 shows the overall block diagram of the music score recognition system.

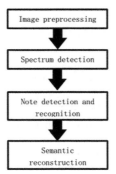

Fig. 5. System workflow framework of music score recognition

The whole algorithm includes four key parts: image preprocessing, score table detection, note detection and classification, and semantic reconstruction. The purpose of image preprocessing is to remove some interference information from the image and facilitate the implementation of subsequent algorithms, including image enhancement, image blur, image denoising and image binarization; The main purpose of score table detection is to accurately detect the specific position of the score table in the music score. Because all the notes are drawn on the score table, it is a basic work to accurately detect the score table in the music score; The main purpose of note detection and classification is to detect and identify the corresponding notes in the spectrum. Detection can determine the specific position of notes, while classification can determine the specific category of notes; The main purpose of semantic reconstruction is to detect the specific relationship between different notes. Because the whole score is formed by different notes in a specific order, only the correct recovery of the relationship between different notes can ensure the quality of electronic score.

5 Conclusion

Nowadays, deep learning technology is not only a scientific research, but also an accelerating penetration into various categories of art creation and art application fields. In the field of algorithmic composition, composition technology and deep learning technology are the hot spot and focus of the industry development, and will inevitably be the focus and direction of this subject. In the future, AI music can help human composers to fill in the thinking gap in creation on the level of computer-aided music creation. Users can choose the most suitable candidate sequence to increase their works or inspire them. They can also re create new music according to the style of specific music performers and composers, And this kind of deep learning can be accumulated.

Acknowledgements. Provincial subject of teaching reform in Jiangxi Province in 2020 JXJG-20-47-4 Research on College "Music+" Aesthetic Education Teaching.

References

1. Yao, Z.: The influence of new technology media on music creation in digital age. J. Nanjing Acad. Art (Music Perform. Ed.) **02**, 51–57 (2008)
2. Zhuhongyuan: Research on automatic composition based on deep learning. Master's degree thesis of China University of Science and Technology (2019)
3. Wu, Y.: Algorithm based spectrum composition technique. Huang Zhong (J. Wuhan Conserv. Music, China), (04), 3–10+203 (2010)
4. Xie, L., Zhuang, Y.: Music creation and phenomenon analysis under the new technology media environment. Huang Zhong (J. Wuhan Conserv. Music, China), (03), 152–158+1 (2012)

Seismic Deformation Monitoring Based on GPS and InSAR

Xiaoguang Ding, Yi Zhang$^{(\boxtimes)}$, Yang He, and Yongqi Zhang

Shaanxi Earthquake Agency, Xian 710068, Shaanxi Province, China

Abstract. The application of GPS HSAR integration technology for surface deformation monitoring is a potential research direction at present. This paper analyzes the characteristics and complementarity of G $and five SAR technologies, and puts forward the data fusion scheme of GPS HSAR integration technology. Through the analysis of research examples at home and abroad, it shows that the application of GPS HSAR integration technology for surface deformation monitoring is feasible and has broad application prospects.

Keywords: Global positioning system · Synthetic aperture radar interferometry · Surface deformation · Integration technology

1 Introduction

Five dimensional synthetic aperture radar (HSAR) is a new technology with great potential for surface deformation monitoring. More and more countries and regions use five dimensional SAR to detect surface deformation phenomena caused by mining, earthquake and volcanic movement. Five SAR technology not only has sub centimeter detection accuracy, but also has the ability of low cost, near continuity and long-range remote sensing detection. However, due to the influence of atmospheric delay (tropospheric delay, ionospheric delay, etc.), satellite orbit error, surface condition and time-varying decorrelation, it is easy to lead to wrong interpretation of blood SAR image, The blood SAR data itself can not solve the above problems, and other earth observation technologies are urgently needed. The application of GPS technology in the field of surface deformation monitoring is becoming more and more mature [1]. It not only has high positioning accuracy, but also can accurately determine the troposphere and ionosphere parameters. It is the most widely used earth observation technology at present.

Therefore, the integration of the two technologies will break through the application limitations of a single technology, give full play to their respective advantages, and greatly improve the spatial and temporal resolution, so as to give full play to the detection potential of high altitude coverage and sub millimeter accuracy in the field of seismic exploration.

2   Technical Characteristics and Complementarity of GPS and InSAR

2.1   Technical Characteristics of GPS

As a kind of earth observation technology, GPS has been widely used. Its positioning methods are divided into absolute positioning and relative positioning. The former can accurately and quickly obtain the three-dimensional coordinates of a point in space, and the latter can accurately measure the relative change of the distance between two points. Deformation monitoring is one of the applications. The steps of monitoring implementation are to select or establish a reference station at an appropriate position far away from the deformation area, set up a number of monitoring points in the deformation area, and install GS receivers on the reference station and monitoring points respectively for continuous observation, And the observation data are automatically sent to the data processing center for analysis and processing through appropriate communication means. Through long-time continuous observation, the millimeter level high-precision deformation monitoring can be realized. Its remarkable feature is that the time resolution is very high, and the sampling rate can reach 20 Hz. However, due to the limitation of the number of GPS receivers and the network array, the network density can not be very high, and the spatial resolution may be tens or hundreds of kilometers, so the continuous positioning results can not be obtained [2]. If we want to comprehensively grasp and analyze the large-scale ground deformation, we must lay a large number of monitoring points, which leads to long field operation cycle and high operation cost.

2.2   Technical Characteristics of InSAR

InSAR is a newly developed space remote sensing technology, which can obtain the radar backscatter intensity information of ground resolution elements and the phase information related to slant distance by actively imaging the earth surface with microwave. Through the joint processing of multiple radar images, the digital elevation model can be established to obtain the deformation of the earth surface of centimeter level or less. It is characterized by active remote sensing, all-weather imaging, high spatial resolution and wide coverage. Its inherent limitations are that it is very sensitive to the changes of atmospheric parameters, satellite orbit parameters and surface coverage, which makes it difficult to interpret InSAR images; the selection of space baseline and time baseline between image pairs is also limited; in deformation monitoring, the selection of space baseline and time baseline is limited, The time resolution of the archived data can not meet the requirements, and although the coverage is large, the accuracy of each point information change is not as good as that of GPS positioning. The interferogram generated by filtering and coherence calculation contains noises generated by various channels. The phase residuals caused by these noises directly affect the accuracy of phase unwrapping. Therefore, it is necessary to filter the interferogram. Similarly, boxcar method is used to calculate the shortest line between the unwrapping phase gradient and the inconsistent points of the unwrapping phase gradient. The calculation formula is as follows:

$$\sum\nolimits_{ij}(\Delta\varphi_{ij}^x + \Delta\varphi_{ij}^x)^0 + \sum\nolimits_{ij}(\Delta\varphi_{ij}^y + \Delta\varphi_{ij}^y)^0 = \min \quad (1)$$

After filtering, we get the result of filtering. In order to measure the accuracy of interferometry, it is necessary to calculate the coherence of the interferogram. Finally, the coherence coefficient can be obtained. The calculation formula is as follows:

$$\gamma = \frac{E[C_1 \cdot C_2]}{\sqrt{E[|C_1|^2]E[|C_2|]}} \quad (2)$$

3 Complementarity of GPS and InSAR Technology

Comparing the technical characteristics of GPS and five InSAR, we can see that the two earth observation technologies are complementary: ① GPS is an ideal point positioning system, especially when the relative positioning mode is adopted, the positioning accuracy is high, but the spatial resolution of GPS is low, and the baseline length of GPS network used to monitor the surface deformation is at least tens to hundreds of kilometers, It is not enough to meet the needs of high-resolution deformation monitoring. HSAR provides continuous information on the whole area, and its spatial resolution can even reach 20 m × 20 mm11: ② GPS obtains high-precision absolute coordinates, while InSAR only provides relative coordinates; ③ due to the relationship of incidence angle, HSAR is particularly sensitive to elevation information, especially using DHAR for deformation monitoring, and the accuracy can reach sub centimeter level, which is just the weakest link of GPS; ④ GPS can provide observation data with high temporal resolution, and the sampling rate is 10 Hz or even 20 Hz [3]. Five SAR is regarded as instantaneous observation, and the repetition period is about 35d. It is difficult to provide enough temporal resolution, and it is easy to be affected by time-varying decorrelation. The fusion of GS and five SAR data can not only correct the error that is difficult to eliminate in the HSAR data itself, but also realize the effective unification of GPS technology with HSAR technology in high temporal resolution and high plane position accuracy. Therefore, the integration technology of GPS and InSAR will play an important role in the field of surface deformation detection.

4 Data fusion Scheme of GPS and HSAR Integration Technology

4.1 Geometric Correction of Interferometric Data Based on GPS Measurement

Because only the orbit parameters provided by spaceborne SAR data are used for positioning, the error may reach several kilometers. Even if the orbit parameters are accurately corrected, the accuracy often can not meet the application requirements. The direct combination of GPS technology in SAR application is accurate geometric positioning or geocoding for the processing results of HSAR data. The usual step is to set the corner reflector in advance and install it properly. When SAR imaging, the electric waves emitted by the corner reflector will be strongly reflected, and obvious feature points will appear in the image, or permanent scatters (PS) similar to the corner reflector will be found, PS is a kind of ground object with strong reflection of electromagnetic wave, most of which are artificial objects. Their geometric shape and physical characteristics will not change obviously for a long time, such as exposed rocks, tall buildings, lighthouses and so on. In this way, the geometric position relationship of these point targets can be accurately determined. As control data, the DEM and deformation distribution map obtained from HSAR data can be accurately corrected to eliminate (weaken) the atmospheric influence, the uncertainty of orbital parameters and other system errors. The method to study the surface deformation information is to use the three orbit method of d-nsar, which requires at least three images, one as the main image, and the other two as the secondary image, combined with the earthquake time point of July 5 in the study area and the orbit parameters of radar satellite. The optimal time series archive information is shown in the following Table 1.

Table 1. Sentinel-1sar data acquisition scheme

Product level	Polarization mode	Orbital mode	Angle of incidence	Resolving power
SLC L1	VV	De orbiting	54	5 m × 20 m
SLC L1	VV	De orbiting	54	5 m × 20 m
SLC L1	VV	De orbiting	54	5 m × 20 m

4.2 InSAR Based on Permanent Scatterers and Corner Reflectors

In areas with abundant vegetation, such as southwest China and South China, the coherence of interferometric imaging is generally poor. In this type of area, the application of InSAR technology is difficult to follow the general way. Permanent scatterers are one of the choices for deformation observation. This avoids the difficulty of unwrapping the whole interferogram when the interference quality is not high [4]. In order to estimate the micro deformation of the earth's surface, the phase change is calculated only at some discrete points. On the one hand, this method needs to accumulate many SAR images in the same imaging area in order to get high-precision estimation; on the other hand, in some areas, there may be lack of exposed artificial objects and rocks, so it is difficult to have scatterers. For those artificial objects that may be lack of exposure, there will be no stable point targets in the InSAR image.

5 Conclusion

Taking advantage of the complementary characteristics of GPS and five SAR, we can not only meet the timeliness of surface deformation monitoring, but also accurately monitor a large range of surface deformation. It is of great scientific and theoretical significance to enrich and improve the SAR data processing theory and improve the accuracy and reliability, It will also play an important role in the urban land subsidence, the surface subsidence caused by the exploitation of groundwater, oil and natural gas, and the slight and continuous surface displacement caused by landslides.

Acknowledgements. The Innovation Fund supported by Shaanxi Earthquake Agency (QC202014), China Scholarship Council & China Earthquake Administration Talent Plan (201904190014), The Earthquake Science and Technology Spark Program supported by China Earthquake Administration (XH21032).

References

1. Ming, H.: InSAR monitoring of coseismic deformation and inversion of fault slip distribution. Chang'an University (2019)
2. Wang, Z., Zhang, R., Wang, X., Liu, G.: InSAR coseismic deformation monitoring and fault inversion of Menyuan earthquake in 2016. Remote Sens. Inf. **33**(06), 103–108 (2018)
3. Yu, J.: Research on InSAR atmospheric correction technology and its application in earthquake deformation monitoring. China University of Petroleum (East China) (2017)
4. Hou, L.: Study on 2-D coseismic deformation field of Yushu earthquake monitored by insar-mai technology. Surv. Mapp. Eng. **24**(11), 64–67+72 (2015)

Seismic Deformation Field Extraction and Fault Slip Rate Inversion Based on D-InSAR Technology

Xiaoguang Ding, Yi Zhang[✉], Fuqiang Shi, Hongguang Zhai, and Zengji Zhen

Shaanxi Earthquake Agency, Xian 710068, Shaanxi, China

Abstract. There are a large number of nearly north-south trending faults in the Qinghai Tibet Plateau, and there are great differences in the research and understanding of these normal faults. Using traditional geodetic data and spatial geodetic data to invert the dynamic parameters of faults is of great significance for understanding the dynamic process of fault sliding and exploring the relationship between crustal movement and earthquakes. This paper mainly analyzes and studies the seismogenic faults of the Dangxiong earthquake in Tibet. Using the C-band radar images of ENVISAT satellite before and after the earthquake, the co seismic deformation is extracted by using the two orbit difference method. Then, based on the Okada elastic half space model, the three-dimensional slip rate of the seismogenic fault is inversed by using the high-precision deformation information.

Keywords: Synthetic aperture radar interferometry · Dangxiong earthquake · Coseismic deformation · Okada · Fault parameters · 3D slip rate

1 Introduction

With the rapid development of modern space geodesy technology, it is widely used in the field of Surveying and mapping. The rapid development of technology makes the acquisition of high spatial resolution earth observation data easier, and enables earth scientists to study the geodynamic phenomena related to seismicity fault layers from a new perspective. The existence and dislocation of seismic fault will cause lithosphere deformation in fault zone. The high-precision and high spatial-temporal resolution surface deformation observation data obtained by InSAR technology, combined with the results of Geology and Geophysics, and the comprehensive study of source parameters will undoubtedly make the earth scientists more fully and accurately understand. In order to determine more accurate source parameters as the core research goal, this paper studies InSAR inversion of source parameters. On the basis of previous studies, this paper summarizes the research results at home and abroad, summarizes the mathematical model of InSAR inversion of source parameters and its error sources, carries out theoretical and simulated inversion analysis on the influence of model error on InSAR inversion estimation of source parameters, and gives the model suitable for

© The Author(s), under exclusive license to Springer Nature Singapore Pte Ltd. 2022
J. C. Hung et al. (Eds.): FC 2021, LNEE 827, pp. 968–972, 2022.
https://doi.org/10.1007/978-981-16-8052-6_125

InSAR inversion of source parameters [1]. Finally, the Dangxiong Mw6.3 earthquake is taken as an example to study.

2   Basic Theory and Data Products of D-InSAR

2.1   InSAR Data Theory

D-InSAR obtains two SAR images of the same area through the receiving antenna mounted on the sensor. There is a difference between the echo signals of the two antennas, and the interferogram is formed after differential processing. According to the geometric position relationship of receiving antenna, general SAR sensor has three imaging modes, namely cross track mode, single antenna repeated track mode and along track mode. Cross track mode: two antennas are carried on one flight platform and observed at the same time. The two antennas are perpendicular to the flight direction with high accuracy and good maneuverability. The main application fields are topographic mapping and monitoring of surface height change, The geometric principle is shown in Fig. 1.

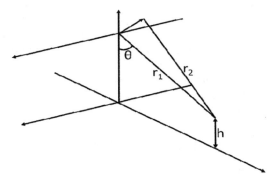

Fig. 1. Cross track mode

2.2   Basic Principle of InSAR

In SAR image data, image pixels contain not only ground backscatter intensity information, but also phase information related to slant distance. The interferometric phase map is obtained by subtracting the phase values of two images covering the same area. The separation of useful information is based on these interference phase signals. This paper mainly introduces that the echo signal of the ground target in the repeated orbit mode includes not only the amplitude value information a, but also the phase value information φ [2]. The backscattering information of each pixel in SAR image can be expressed in the form of complex AE. The phase information includes not only the distance between SAR system and target, but also the scattering characteristics of surface target.

$$\Phi = -\frac{4\pi}{\lambda} R + \Phi_{obj} \tag{1}$$

The expression of interference phase is:

$$\Delta\varphi = -2\frac{Bqs}{R} \tag{2}$$

3   Theory of Okada Elastic Half Space Dislocation Model

3.1   Okada Elastic Half Space Model

There are two ways to develop the research of seismic source. One is to describe the seismic source by the system of acting force at the seismic source, the other is to describe the discontinuity of displacement or strain on both sides of a certain plane at the seismic source. In 1958, Steketee proposed the three-dimensional elastic dislocation theory of seismic source, which unifies the above two description methods. Since then, many seismologists have applied and developed this theory. Okada model assumes that the earth model is a uniform and elastic half space body, and roughly simulates the real earth model. Okada model is mainly based on isotropic, elastic and half space medium fault model, without considering crustal stratification, which will bring model error to the model simulation results, Compared with the previous dislocation model, Okada dislocation model is more suitable for the study of coseismic deformation of point source which is difficult to identify the seismogenic signal and the study of coseismic deformation of limited plane element.

3.2   Extraction of Coseismic Deformation Field

In this paper, the two orbit differential interferometry is used to process the five images before and after the earthquake to obtain the coseismic displacement field of the earthquake. The interference scheme is used to process the differential interferometry to obtain the deformation results before, after and after the earthquake. The details are as follows: take 20080921.sl image which has the shortest baseline before the earthquake and the earthquake as the main image, and register the other four images to 20080921. The registration method uses precise track data, intensity cross-correlation and offset tracking method, and the registration accuracy can reach 1/10 pixel. The phenomenon of decoherence is serious, but on the whole, the coseismic deformation fields of the four images are highly consistent. From the above four images, we can see that the maximum deformation value of the left wall (hanging wall) in the subsidence area is about 30 cm, which is much larger than that of the right wall (footwall), and the maximum deformation value of the uplift area is about 5 cm. The deformation field of the subsidence area almost covers the whole Yangyi basin, and the maximum deformation value of the right wall (footwall) is about 5 cm [3]. The macro epicenter is located in the center of the subsidence area.

4 Simulation of Seismic Deformation Field and Inversion of 3-D Slip Rate of Fault

4.1 Forward Modeling of Seismic Deformation Field

Based on the InSAR deformation observation data, the inversion to determine the source parameters of a seismic event mainly includes the following four steps: establishing the function model and random model, determining the optimization criteria of parameter estimation, determining the source parameters and evaluating the inversion results. Based on InSAR deformation observation data inversion to determine the source parameters, the least square criterion is mostly used as the optimization criterion of source parameter estimation in some studies. Assuming that there is no error or significance in the mathematical model, the source parameter solution estimated based on the least square criterion has good statistical properties such as optimal unbiased, posteriori unbiased unit weight variance and asymptotic optimality. However, due to the complexity of earth structure and fault geometry, there must be some approximation or error in function modeling. In addition, although InSAR technology can obtain massive observation data at the same time, due to its high degree of automation, and it is difficult to remove various environmental factors that affect the observation, there are inevitably some gross errors in the observation data or the corresponding variance covariance matrix is not reasonable. Due to the above factors, there must be some errors in the function model and random model. If it is not considered and compensated scientifically and reasonably, it will have an adverse effect on the estimation of source parameters, and then affect the subsequent geophysical interpretation.

4.2 Downsampling of Coseismic Displacement Field Data

InSAR Data covers a wide range and has a large amount of data. If all the results are used for inversion, the workload is very huge. After data processing, the number of ground observation points is up to $10°$. If all the data are used as constraints, the noise data far from the epicenter will be brought into the calculation process, which will affect the accuracy, There are three common data sampling methods in seismic source research: quadtree sampling, spatial data resolution based sampling and uniform sampling. Considering that the seismogenic fault of this earthquake is single, the surface deformation is relatively regular, and the influence range of surface deformation can be roughly determined, this paper uses the data of epicenter deformation area, $29,1n \sim 30° n$, The basic idea of uniform sampling is: divide the interference image into several equal interval sub intervals, extract the median or average value in each small window for inversion calculation, and select the median coordinate in the window. It is obvious that the size of the divided window will directly affect the number of observations involved in the final inversion [4]. In this paper, we use the scale of 8 to sample [5].

5 Conclusion

In this paper, the coseismic deformation field of the Dangxiong earthquake is extracted from the level 0 data of the en VI isat satellite provided by ESA. The constraint parameters of the fault are obtained by forward modeling using the resampled InSAR deformation observation data. The particle swarm optimization algorithm is used to combine with the dislocation model, The main conclusions of this paper are as follows: the basic principle of InSAR is described in detail, the main errors of InSAR deformation measurement are analyzed, and the noise of interferogram is reduced by filtering, unwrapping and removing baseline errors, The coseismic deformation of the Dangxiong earthquake is obtained by using the two orbit difference method.

Acknowledgements. The Earthquake Science and Technology Spark Program supported by China Earthquake Administration (XH21032), The Talent Plan supported by China Scholarship Council & China Earthquake Administration (201904190014).

References

1. Zhang, L., Wang, X., Dong, X., Sun, L., Cai, W., Ning, X.: Finger vein image enhancement based on guided tri-Gaussian filters. ASP Trans. Pattern Recognit. Intell. Syst. 1(1), 17–23 (2021). https://doi.org/10.52810/TPRIS.2021.100012
2. Cai, W., Wei, Z., Liu, R., Zhuang, Y., Wang, Y., Ning, X.: Remote sensing image recognition based on multi-attention residual fusion networks. ASP Trans. Pattern Recognit. Intell. Syst. 1(1), 1–8 (2021). https://doi.org/10.52810/TPRIS.2021.100005
3. Tong, Y., Lina, Y., Li, S., Liu, J., Qin, H., Li, W.: Polynomial fitting algorithm based on neural network. ASP Trans. Pattern Recognit. Intell. Syst. 1(1), 32–39 (2021). https://doi.org/10.52810/TPRIS.2021.100019
4. Ning, X., Wang, Y., Tian, W., Liu, L., Cai, W.: A biomimetic covering learning method based on principle of homology continuity. ASP Trans. Pattern Recognit. Intell. Syst. 1(1), 9–16 (2021). https://doi.org/10.52810/TPRIS.2021.100009
5. Lina, Y., Tao, S., Gao, W., Limin, Y.: Self-monitoring method for improving health-related quality of life: data acquisition, monitoring, and analysis of vital signs and diet. ASP Trans. Pattern Recognit. Intell. Syst. 1(1), 24–31 (2021). https://doi.org/10.52810/TPRIS.2021.100018

Printed by Printforce, the Netherlands